HANDBOOK OF COMMUNICATION AND AGING RESEARCH

LEA'S COMMUNICATION SERIES

Jennings Bryant/Dolf Zillmann, General Editors

Selected titles in Applied Communication
(Teresa L. Thompson, Advisory Editor) include:

For a complete list of titles in LEA's Communication Series, please contact Lawrence Erlbaum Associates, Publishers, at www.erlbaum.com.

HANDBOOK OF COMMUNICATION AND AGING RESEARCH

Second Edition

Edited by

Jon F. Nussbaum
Pennsylvania State University

Justine Coupland
Cardiff University

LONDON AND NEW YORK

Senior Acquisitions Editor: Linda Bathgate
Assistant Editor: Karin Wittig Bates
Cover Design: Sean Trane Sciarrone
Textbook Production Manager: Paul Smolenski
Full-Service Compositor: TechBooks
Text and Cover Printer: Hamilton Printing Company

This book was typeset in 11/13 pt. Dante, Bold, Italic.
The heads were typeset in Franklin Gothic, Bold, and Bold Italic.

Photo Credit: Original painting "Late Afternoon" by Edward Mattil, Foxdale Village,
State College, PA, U.S.A.

First published 2004 by Lawrence Erlbaum Associates, Inc.

Published 2008 by Routledge
2 Park Square, Milton Park, Abingdon, Oxon OX14 4RN
605 Third Avenue, New York, NY 10017

Routledge is an imprint of the Taylor & Francis Group, an informa business

Copyright © 2004 Taylor & Francis

Library of Congress Cataloging-in-Publication Data

Handbook of communication and aging research / edited by
    Jon F. Nussbaum, Justine Coupland.—2nd ed.
        p.   cm.
    Includes bibliographical references and index.
      ISBN 0-8058-4070-2 (casebound : alk. paper)—ISBN 0-8058-4071-0
(pbk. : alk. paper)
        1. Aged—Communication.   2. Interpersonal communication.
    3. Aging—Psychological aspects.   4. Aging—Abstracts.
    I. Nussbaum, Jon F.   II. Coupland, Justine.
    HQ1061.H3365   2004
    305.26—dc22   2003020517

ISBN 13: 978-0-8058-4071-1 (pbk)

For Bill and Amanda

Contents

VII: SENIOR ADULT EDUCATION

Preface

In the first edition of the *Handbook of Communication and Aging Research*, we wrote that a simultaneous growth in the discipline of communication and in the population of older individuals across the world has produced an explosion of research investigating various aspects of communication and aging. This has taken place against a backdrop of traditional gerontology, which has for the most part been heavily influenced by theories in mainstream biology, psychology, sociology, and medicine, and has focused on the demography, personality, use of social services, uptake of and satisfaction with health care, and intergenerational attitudes. The field of social gerontology is relatively new, innovative, and prolific, but still not well organized. Perhaps because of this, the centrality of communication in the studies of aging is not yet well established. Yet, communication and aging is at the forefront of the new wave of gerontology, giving priority to aging as one of the range of issues currently being scrutinized by the social sciences. Since the publication of the first handbook nearly 10 years ago, the study of aging by social gerontologists has become more focused, more theoretically diverse, and more methodologically interesting. This second edition of the *Handbook of Communication and Aging Research* is our attempt to capture this ever-changing and expanding domain of research. Five new chapters have been added, numerous chapters have received major revision, and two classic chapters appear with only slight modification.

Since it was first recognized that there is more to social aging than demography, gerontology has needed a communication perspective. Much like the first edition of the handbook, the second edition sets out to demonstrate that aging is not only (or not even in the main) an individual but an interactive process. It is the study of communication that can lead us to understand what it means to grow old. It is the social imagery of aging and the way in which aging is culturally constituted and constructed that gives us access to social reality (see Berger & Luckman, 1966; Potter & Wetherell, 1987). We may age physiologically and, of course, chronologically, but our social aging—how we behave, as social actors, toward others, and even how we align ourselves with or come to understand the signs of difference or change as we age—are phenomena achieved primarily through communication experiences. For Shotter and Gergen (1989), "cultural texts furnish their inhabitants with the resources for the formation of selves; they lay out an array of enabling potentials, while simultaneously establishing a set of constraining boundaries beyond which selves cannot easily be made" (p. ix).

The means by which we, as social actors, provide or constrain ways of being old, and in turn can come to understand aging in others as well as in ourselves, is through both verbal and nonverbal communication practices. Communicative processes define, form, maintain, and dissolve all social relationships. More particularly, to give but a few examples of the social contexts examined in this volume, communication is the means by which doctor–patient consultations are negotiated, how we perform parenthood and grandparenthood, and how images of aging are portrayed in the media.

The aim of this second edition is to synthesize the vast amount of research that has been published on communication and aging in numerous international outlets over the last three decades. The book is organized into sections that to some extent reflect the divisional structure of the International Communication Association. Chapters are authored primarily by scholars from North America and Britain, who are active researchers in the perspectives covered in their particular chapter. Authors of chapters review the literature analytically, in a way that reveals not only current theoretical and methodological approaches to communication and aging research, but also sets the future agenda.

The second edition of the handbook is theoretically diverse in that it reflects the perspectives of multiple disciplines and the achievements of a range of different methodologies. What the introduction and the various section prefaces aim to do is frame the following chapters within the broadly constructionist approach to aging we have already briefly outlined. Some of the chapters are indeed written to represent this approach explicitly (Baker, Giles, & Harwood, chap. 6; Bernard & Phillipson, chap. 14; Coupland, chap. 3; Henwood, chap. 8; Hepworth, chap 1; Rawlins, chap. 14), but even those that represent other psychological or communication theorists' research, mainly by using experimental and quantitative methods, are utilizing phenomena highly relevant to constructionism. For example, if we question participants on their relationships, responses to such questions work to reconstruct those relationships. Our access as researchers is only through communicated construals.

There are various metaphors, prevalent in the literature, of aging and its experiences as uncharted waters, or of aging as a battle against uncertainty. It is at least possible to interpret the relatively recent concentration on the elderly as a social group (as opposed to class, ethnicity, and gender) as evidence of an academic silence brought on by the fear of aging (cf. Butler, 1969). Certainly, the responsibility we bear to communication in aging is not only a scholarly but also a personal endeavor. Cameron, Frazer, Harvey, Rampton, and Richardson (1993) spoke of researchers whose commitment to the participants in their research goes beyond the ethical consideration of not harming informants, and who feel more of a positive desire to take up the position of advocate, engaging in research not only on social participants but also for them. In the case of research on aging, there is a special salience for all researchers, where we have at least the potential to approach and align with the research not only as researchers but on behalf of our aging selves. Indeed, as editors, we have found many of the chapters that follow to be not only an academic, but also a personal, challenge. Many of the chapters work to deny earlier images of aging as involving normative decrement, and to provide us with a picture of aging as a process of development involving positive choices and providing new opportunities.

A theme that recurs in many of the chapters is that of the heterogeneity of the group of people who are variously categorized as "older," "aged," "elderly," or over 65. Again and again, the authors here are at pains to remind us that there is nothing intrinsic about being older that predicts difference from those who are younger in any easily categorizable way.

Bernard and Phillipson make the point that future research needs to attend to the ways in which age, gender, race, and class interact to help construct human identity.

Chronological age, then, must be seen as one variable with potential salience to an individual's identity, along with others such as ethnicity, gender, social class, health status, professional capacity, family role, religious affiliation, and so on. This is equally the case for older and for younger individuals. We hope that if this book cannot be seen as one that depicts an optimistic scenario, then at least it predicts one for future conceptions of aging. We must make a general proviso here that the research this volume represents is centered on transatlantic, mostly White cultures and norms. To begin a larger dialogue concerning culture, communication, and aging, we have added a chapter coauthored by Loretta Pecchioni, Hiroshi Ota, and Lisa Sparks. We feel that the communication approach to aging is arguably better suited to cross-cultural comparisons than any other. At the same time, we think it is crucial for all aging research to incorporate a cultural reflexivity, identifying how ideologies of aging are inevitably embedded with specific cultural frameworks of time and place.

REFERENCES

Berger, P. L., & Luckman, T. (1966). *The social construction of reality*. Harmondsworth, UK: Penguin.

Butler, R. N. (1969). Age-ism: Another form of bigotry. *The Gerontologist, 9*, 243–246.

Cameron, D., Frazer, E., Harvey, P., Rampton, B., & Richardson, K. (1993). Ethics, advocacy and empowerment: Issues of method in researching language. *Language and Communication, 13*(2), 81–94.

Potter, J., & Wetherall, M. (1987). *Discourse and social psychology*. London: Sage.

Shotter, J., & Gergen, K. (1989). *Texts of identity*. London: Sage.

List of Contributors

Valerie Barker School of Communication, San Diego State University, San Diego, CA 92182

Anne L. Balazs Division of Business and Communication, Mississippi University for Women, Box W-940, Columbus, MS 39701

Doreen K. Baringer Department of Communication Art and Sciences, Pennsylvania State University, University Park, PA 16802

Analee E. Beisecker formerly of University of Kansas, Lawrence, KS 66045

Mark J. Bergstrom Department of Communication, University of Utah, Salt Lake City, UT 84112

Miriam Bernard School of Social Relations, Keele University, Keele, Staffordshire, England ST55BG

Jaye L. Bonnesen Department of Communication, Georgia State University, Atlanta, GA 30303

R. Suzanne Caverly Department of Communication, California State University, Long Beach, CA 90840

Peter Coleman Department of Psychology, Southampton University, University Road, Highfield, Southampton, England SO17 1BJ

W. Timothy Coombs Department of Communication, Eastern Illinois University, Charleston, IL 61920

Justine Coupland Centre for Language and Communication Research, Cardiff University, Humanities Building, PO Box 94, Colum Drive, Cardiff, Wales CF13XB

Nikolas Coupland Centre for Language and Communication Research, Cardiff University, Humanities Building, PO Box 94, Colum Drive, Cardiff, Wales CF13XB

Mary Anne Fitzpatrick Department of Communication, University of Wisconsin, Madison, WI 53706

Jane Garner Department of Communication, University of Oklahoma, Norman, OK 73019

Teri A. Gartska Communications Studies Department, University of Kansas, Bailey Hall, 1440 Jayhawk Blvd., Room 102, Lawrence, KS 66045

Howard Giles Department of Communication, University of California, Santa Barbara, CA 93016

Frank Glendenning 32 Dartmouth Avenue, Newcastle Under Lyme, Staffordshire, England ST53NY

Karen Grainger School of Cultural Studies, Sheffield Hallam University, Howard Street, Sheffield, England S11WB

Jake Harwood Department of Communication, University of Arizona, Tucson, AZ 85721

Karen L. Henwood School of Health and Public Policy, University of East Anglia, Norwich, England NR 47TJ

Mike Hepworth Department of Sociology, University of Aberdeen, Aberdeen, Scotland, AB24 3QY

Sherry J. Holladay Department of Communication, Eastern Illinois University, Charleston, IL 61920

Michael E. Holmes Department of Communication, Ball State University, Muncie, IN 47306

Mary Lee Hummert Communications Studies Department, University of Kansas, Bailey Hall, 1440 Jayhawk Blvd., Room 102, Lawrence, KS 66045

Lynda Lee Kaid College of Journalism and Communication, University of Florida, 2000 Weimer Hall, Gainsville, FL 32611

Amanda L. Kundrat formerly of Department of Communication Art and Sciences, Pennsylvania State University, University Park, PA 16802

Marie-Louise Mares Department of Communication, University of Wisconsin, Madison, WI 53706

Valerie Cryer McKay Department of Communication, California State University, Long Beach, CA 90840

Jon F. Nussbaum Department of Communication Art and Sciences, Pennsylvania State University, University Park, PA 16802

Ann O'Hanlon Department of Psychology, Southampton University, University Road, Highfield, Southampton, England SO17 1BJ

Hiroshi Ota Faculty of Study in Contemporary Society, Aichi Shukutoko University, Nagakuto-cho, Aichi-gun 480-1197, Japan

Loretta L. Pecchioni Department of Communication, Louisiana State University, Baton Rouge, LA 70803

Chris Phillipson School of Social Relations, Keele University, Keele, Staffordshire, England ST55BG

James L. Query Department of Communication, University of Houston, Houston TX 77204

William Rawlins Department of Communication Studies, Ohio University, Athens, OH 45701

James D. Robinson Department of Communication, University of Dayton, Dayton, OH 45469

Ellen Bouchard Ryan Department of Psychiatry, McMaster University, Hamilton, Ontario, Canada.

Tom Skill Department of Communication, University of Dayton, Dayton, OH 45469

Lisa Sparks Department of Communication, George Mason University, Fairfax, VA 22030

Teresa L. Thompson Department of Communication, University of Dayton, Dayton, OH 45469

Jeanine W. Turner School of Business, G-04 Old North, Georgetown University, Washington, DC 20057

Angie Williams Centre for Language and Communication Research, Cardiff University, Humanities Building, PO Box 94, Colum Drive, Cardiff, Wales CF13XB

Kevin B. Wright Department of Communication, University of Memphis, Memphis, TN 38152

HANDBOOK OF COMMUNICATION AND AGING RESEARCH

I

The Experience of Aging

The dynamic perspective on aging that is one of the principle foci of this volume is well introduced in Hepworth's classic chapter, here in the sense of interaction between the individual and visual representations of elderliness—in particular, the older body. In this argument, learning what it means to grow old, and our own orientation to and understanding of aging, is achieved at least partly through images portrayed in paintings, photographs, or literary descriptions of appearance, which provide us with a cultural repertoire. Hepworth's analysis reveals a range of readings of the look of age, from, for example, negative narrative representations of the older body as "[in] ruins" as a "wreck" or "relic," to positivity, invoking "grandeur," "serenity," and "calm." Hepworth thus displays how images do not simply reflect the facts but project beliefs about aging and the place of old people in society. Hepworth's notion of the geriatric gaze takes us through the prevalent decremental model of aging (Coupland, chap. 3) that predicts decline in sensory, mobile, and cognitive thresholds of ability, with ensuing problems in interpersonal communication resulting in isolation from social interaction and increasing social distance from younger people. A theme of power and powerlessness can be traced through this chapter, written from a sociological–critical theory perspective. In relation to depictions of elderly communicative powerlessness, Hepworth makes a reference to social exchange theory, which captures the belief that the older person becomes costly to interact with (see also Bourdieu, 1991). The perceived costliness of communicating with older people might lead to avoidance and marginalization, or to deindividualization (see also Hummert, Garstka, Ryan, & Bonnesen, chap. 4; Williams & Harwood; Baker, Giles, & Harwood, chap. 6).

The notion of impairment and resulting powerlessness can be related to Fairclough's (1998) constraint on discourse, where older people (particularly those in advanced age) can be seen as being constrained to take up relatively powerless positions, or social roles, relative to younger people. We may conclude that this has more to do with asymmetries in expectations and opportunities than with salient communicative abilities or preferences, at least for older adults.

1

Hepworth points out that to be a young self in an old-looking body is problematic in a society whose dominant images of old age are negative. Postmodern attitudes toward the human body have led us to attempt to align external appearance with inner self-concept. Here the notion of repertoire again becomes relevant, with the self, both in terms of appearance and communicative behavior, being expressed as a kind of social performance. The notion of the variability inherent in social behavior is a theme taken up by numerous other chapters in this second edition of the handbook as well.

How people manage the later part of their lives is a theme taken up by Coleman (chap. 2), who takes an explicit life span perspective. For Coleman, life history must take its place alongside biological and psychological processes, and societal and cultural expectations and ideals are very much part of the experience of aging. Stress in the old has to be seen in the context of the lifelong drama and extended to the experience of dying.

Coleman and O'Hanlon have produced a major revision of Coleman's original chapter, found in the first edition of the handbook. The current chapter focuses on the positive as well as negative attitudes that adults have toward growing old. They postulate that the way we think about the aging process and the attitudes we construct through our experience with the world can have significant effects on health and psychological well-being. Coleman and O'Hanlon review the various problems and challenges that all of us may one day experience as we age. The relationships we engage in, our ability to cognitively function, our ability to cope and adapt to the various age-related changes, and our ability to find enjoyment and pleasure in later life, are viewed as codeterminants in how we manage old age. The extent to which we can construct a positive later life may directly influence the happiness and positive experience we find as we age.

REFERENCES

Bourdieu, P. (1991). *Language and symbolic power.* Cambridge, UK: Polity.
Fairclough, N. (1989). *Language and power.* London: Longman.

Images of Old Age

Mike Hepworth

University of Aberdeen

How do we know old age? How do we know when other people are old? How do we know when we are old ourselves?

One answer to these questions is that there is really no problem: Old age is obvious. As we move chronologically through our lives, old age becomes increasingly and inescapably apparent in the physical changes that take place in our faces and bodies. We believe, we have sure and certain knowledge of the reality of old age through observable changes in the face, including wrinkling and lining of the skin, graying of the hair, and, in the case of men, hair loss. As one of the rapidly expanding number of humorous birthday cards puts it (in this case of representing the image of a balding middle-aged man), "You know when you're getting old when you look into the mirror and see . . . Your Father!" But this answer is in many ways unsatisfactory because it merely leads us to ask another question: What is the "look" *of age*? What do we see when we see old age? A cursory examination of, for example, the family photograph album reveals that not everyone will agree that Aunt Mary looks her age or that Uncle Joe is looking a lot older than he did last year. Controversy concerning the interpretation of photographic evidence of the passage of time on the human face touches the lives of us all.

In his pictorial and documentary biography of the composer and pianist Franz Liszt, Burger (1989) interpreted the photographic portraits of Liszt in old age as revealing "the superior but simple dignity of regal modesty, an irresistible unity of noble bearing and submission to fate" (p. 298), and he recorded the description made by Carl von Lachmund, a pupil of Liszt, of his first meeting with the maestro in April 1882, when Liszt was 71:

> A door opened, and Liszt stood there in its frame; motionless, like a picture, with his long white hair falling almost to his slightly stooping shoulders, a compelling grandeur in his features—and yet his face seemed serene and friendly as his keen eyes observed the

stranger standing before him. During the silence, which lasted no more than a few seconds, I thought a celestial being stood before me . . . My restless imagination had depicted him as the youthful Liszt, the perfect lissome cavalier and conqueror. The man who stood before me now was another Liszt, crowned by his seventy years, gently stooping, and with a sacred calm unclouded by arrogance. (p. 300)

It is interesting to compare this account of the look of Liszt in old age with the comments made several years before in 1867, when Liszt was 56, by a contemporary who described a photograph of the celebrated musician by Joseph Albert as showing "one of Rome's most remarkable ruins . . . : One produced by the ravages of time with all the conscientious accuracy of modern photography. We refer to Abbe Liszt" (Burger, 1989, p. 240). To this less than reverent viewer, Liszt at 56 was already an aged ruin, a relic of his former self. Yet for Carl von Lachmund, the appearance of the composer was that of a manifestation of an ageless spiritual force, although by then he was much older in years. As these examples begin to show, wide variations in interpretations or readings of the look of age are possible (Blaikie & Hepworth, 1997).

In this chapter, I address the controversial question of the look of old age through an examination of selected images of aging in Western culture dating from the 16th century to the present day. Because of the central role of the visual sense in human perception, and because we live increasingly in a visual culture, the images used as evidence in this chapter are taken primarily from representations of aging in paintings, drawings, photographs, and the popular media, with occasional reference to images of aging in the relatively new discipline of literary gerontology (Hepworth, 1993; Hepworth, 2000; Porter & Porter, 1984; Sokoloff, 1987; Woodard, 1991). Research into visual images of aging (Achenbaum, 1982; Blaikie, 1993, Blaikie, 1999; Covey, 1991; Featherstone & Hepworth, 1990; Hepworth & Featherstone, 1982) has revealed the complexity of such images both in terms of their form and content and in terms of everyday life. Within the space available, it is only possible to examine certain specific themes, and for that reason three key images of old age have been chosen: (1) the geriatric body, (2) the physiognomic body, and (3) the mask of old age.

These images have been primarily selected to demonstrate the contribution of visual representations to the social construction of old age. I argue that images of old age are essentially moral categories of the body created within specific historical contexts to enable societies and individuals to make sense of the biological changes that are taking place in and on their bodies as they move through the socially determined life course. The term *moral* is used in this context first in the conventional ethical sense (virtuous versus vicious), and second in the sociological sense to refer to changes that take place in individual personal and social identity as individuals make transitions from one situation to another. Goffman (1968) employed the term *moral career* to indicate "any social strand of any person's course through life" (p. 119), the word *moral* referring specifically to changes in a person's self image. Changes in the social situation of any individual can result in changes in the ways in which he or she experiences the self, and viewed from Goffman's perspective, aging can be conceptualized sociologically as a process of encountering new images of the self as interpreted by others in relationship to observable alterations in the face and body. To clarify the processes that constitute the moral interpretation of old age, and as a preface to a detailed discussion of the three key images, it is necessary to look more closely at the sociological characteristics of images as they have been analyzed in recent years by cultural critics, historians, and students of mass media (Harrison, 2002; Hepworth, 2002).

IMAGES AND IMAGING

The dictionary defines *image* as "an artificial imitation or representation of the external form of any object especially of a person" (*The Oxford Universal Dictionary*, 1965, p. 958). As this definition implies, images are conventionally treated in everyday language as material artifacts—paintings, sculptures, photographs, films, television programs, advertisements—that have an independent physical existence. As historical survivals, they have the capacity to outlive their creators and, moreover, to persist as eloquent testimony to their separate temporal existence from the audience that views them. The contemporary passion for antique objects and for heritage sites and buildings is a case in point.

Because of their material existence, images appear to have a relatively fixed or permanent meaning that is unchanged over time, and this is particularly the case with images of old age. The works of the 17th-century Dutch painter Rembrandt (especially his remarkable series of self-portraits in later life) have proved attractive to students of old age because they are perceived to be a timeless record of the experience of the aging process, which has an urgent relevance for the experience of old age in contemporary society (Breytspraak, 1984). A detail from Rembrandt's *Self Portrait Aged 63*, for example, is reproduced on the front cover of an important manual on group work for professionals working with older people. Presumably, the painting was chosen because it was regarded as a highly appropriate image with which to endorse the title of the book, *Confronting Aging* (Biggs, 1989). As we look at this contemporary reproduction of a 17th-century image it is useful to reflect that Rembrandt's self-portraits do not seem to have attracted the same amount of attention during his own time, being "barely mentioned by any seventeenth-century writer and only very occasionally . . . identified in collections of the period" and "none were described in the inventory of the artist's own possessions" (White, 1984, p. 138). Another example of the interpretation of visual images of aging as essentially timeless— in this case, an image of a woman—can be found in Rodin's 1885 sculpture, *The Old Courtesan* (Hale, 1969). Here is an apparently ageless three-dimensional representation of the aging female body: a bowed and seated figure with sagging breasts and flabby belly. But our response to this image is specifically directed by the description *courtesan*, which invites comparison to a gendered stereotype of sexual desirability as essentially youthful and, of course, beautiful. The label courtesan is not associated in the popular imagination with old age and plain looks. In this image, the viewer is made starkly aware of the facticity of old age by the implicit contrast made by the sculpture between the intrusive reality of the physically aging female body and the idealized youthful body of the female nude of aesthetic convention.

The facticity of the images of old age, therefore, often seems incontestable. This belief is reinforced by the view, particularly with regard to images of the body of old age, that as socially produced artifacts they faithfully reflect a fundamental and inescapable condition of human existence: the inevitability of physical decay (Gullette, 1997). As

Lowenthal (1985) reminded us, the awareness of things past "derives from two distinct but often conjoined traits: antiquity and decay. Antiquity involves cognizance of historical change, decay of biological or material change" (p. 125). The concept of decay, which is closely related to the dominant Western concept of time as an arrow (Gould, 1988) moving relentlessly forward from the past, through the present, to a point in the future, is defined as an ineluctable characteristic of natural life, and this imaging serves to reinforce our belief in the timelessness of Western images of old age. During the 19th-century, when realism came into prominence as an artistic style, efforts were made to "revitalise the centuries-old artistic tradition of accurate, truthful recording of the world and to give this tradition contemporary relevance" (Weisberg, 1980, p. 1). In pursuit of this goal, some artists such as Edououd Manet sought to record in graphic detail the lineaments of old age on the human face and figure, and Ribot attempted to master these signs in his 1869 painting of four bearded old men, which he called *The Philosophers* (Weisberg, 1980). The physical signs of aging were regarded as a challenge to the realist artist's skills of hand and eye: The imaging of aging was seen as yet another test of the ability to record in accurate detail what is really there. Thus, Weisberg commented as follows on Jean-François Millet's *Portrait of Madame Roumy* (c. 1842), depicting the grandmother of Millet's first wife: "Millet sympathetically posed the old woman in a three-quarter position, showing her facial characteristics—her sunken lips, deep-set eyes, bulging forehead, and shrunken features—to underscore her advanced age" (p. 163).

The invention of photography, of course, further advanced the belief in the facticity of visual images. As a technical device, the camera was regarded as a challenge to the role of the artist as a recorder of external reality; a challenge that was especially formidable with regard to making a record of the human body. To the artist, as Scharf (1974) observed, photography offered

> The first comprehensive alternative to forms fixed by antique tradition. Despite the fact that in posing their subjects photographers were as much governed by conventional criteria as were painters, the inevitable vulgarities of real life—the inelegancies, the misproportions, the coarse blemishes—ludicrously asserted themselves on the sensitive plates. The crudities of actuality in photographs of nudes especially did not blend very elegantly with the antique, and photographs of this kind were an effrontery to men and women of good taste. (p. 134)

These comments are significant for the study of images of old age, because they remind us that the aging body was often shunned in Western art precisely because the natural qualities of decay introduced a discordant note into the aesthetics of visual perfection. In this sense, then, the visible changes in the body through time represented an unwelcome intrusion into the prescribed canons of artistic perfection, and the new photographic technique was disturbing precisely because it made possible the apparently endless reproduction of images of the unaesthetic real world, evident, for example, in the displeasing imagery of the aging body (Hepworth, 2003). The gaze of the photographer through the eye of the camera was regarded as clinically realistic, resulting in the production of an image defined as "a slice, at right angles as it were, across the dimension of time, exposing the nerve ends of reality" (Bartram, 1985, p. 11). Although artists of the realist school also set out to produce a natural record (and Rodin's sculpture

The Old Courtesan may in one sense be regarded as an accurate reproduction of the external features of the aging female body), it was the technological apparatus of photography and the procedures it entailed that gave camera images a scientific authority that was distinct from an artistic and interpretative ones.

The photographer therefore challenged an important element of artistic convention, because he or she was now able to create what was believed to be a faithful record of the physical evidence of old age. Commenting on the capacity of the camera to facilitate the relentless scrutiny of the aging face, Maddow (1977) referred to a photograph made by the pioneering French photographer Nadar, of the composer Franz Listz in later life: "In Nadar's photograph of Listz, he does not spare us the famous wens on the face of one who was the romantic fantasy of every middle class woman in Europe, the very image of four-handed love at the piano" (p. 159). Nadar used the technology of the camera to present the viewer with an image of age unmediated by sentimental or romantic artistic convention.

I have argued so far that the definition of images as material artifacts that are inscribed with meanings considered timeless or historical is particularly appropriate to any consideration of images of old age. First, images of old age such as Rembrandt's self-portraits and Rodin's sculptures are perceived to be communicating messages concerning the essential nature of old age, which transcends time, place, and culture. They confirm, so to speak, the eternal facticity of human aging. Second, because the signs of aging are first and foremost perceived as changes that are visible on the human body, they disturb the image, deeply entrenched in Western culture, of an idealized male and female form (Marwick, 1988). According to aesthetic conventions, images of old age are therefore disruptive and only to be sanctioned under specifically prescribed iconographic rules. One example is the image of Father Time, represented since the Renaissance as a partially clothed or almost naked old man usually equipped with a scythe (Panofsky, 1962).

It is the existence of these rules that helps to explain the commonly expressed embarrassment or revulsion at the sight of the naked body of an aging man or woman. In her study of the nude in art, Saunders (1989) reproduced one of Diane Arbus's photographs, *Two Nudists*, dated 1963. This shows two middle-aged naturists, who, we infer from the conventional domestic sitting room in which they are sitting, are a married couple. The man is fat with a protruding belly, and the woman is similarly shaped, with noticeably sagging breasts. In other words, both of these people look their age. Saunders commented:

> The two spreading and aging bodies seem pitiful to us, uncomfortably out-of-context because they do not conform to the usual public image of nudity—young, perfect, almost exclusively female. Significantly, such an image, a sultry and pneumatic fifties starlet, hangs on the wall behind them. Thus we have nudity in its publicly sanctioned form, provocative, passive, sexually desirable, contrasted with nudity as it is, imperfect shaming and shameful, private made shockingly public. (p. 8)

According to this interpretation, the photograph of Arbus deliberately alerts the viewer to the unaesthetic private reality of the naked bodies of the older man and woman by framing them against the public and gendered aesthetic of the eroticized nude.

Broadly speaking, the conventions governing the portrayal of nudity in Western art, those "unwritten rules concerning such matters as pose, gesture and the absence of body hair" (Pointon, 1990, p. 83), have imposed significant restrictions on the imaging of the body of old age and consequently molded the cultural construction of old age. Rules governing the public and private exposure of the body from which the clothing has been removed raise questions concerning the conditions under which the imaging of the body of old age is acceptable, and the prescribed form and content of such imaging. In the introduction to her book on representations of old age in 20th-century Western fiction and in psychoanalysis, Woodward (1991) described the negative reaction of a group of scholars to a photograph of a naked old man in an exhibition of photographs of the body. This photograph of "a thin old man . . . sitting on the side of his bed, his knees wide apart, his body naked except for the shuffling of his feet" (p. 1) evoked a mixed response of anger, aversion, and moral condemnation. Woodward wrote:

> I talked about the photograph with a professor from Tours, a tall, finely featured man with a deeply lined face. His sensitivity and intelligence were striking. He was thoughtful, not abruptly dismissive, either of the photograph or of my reflections. But no, he did not want to look at this picture. No he did not want to contemplate the body of old age. He would, he said, live that time when it came to him . . . We may conclude that his response to the photograph reflected his expectations about old age, the fears of a middle-aged man. (p. 2)

In a similar vein, though moving toward a more positive conclusion, Sarah Kent introduced Cotier's (1991) collection of photographs of naked older men and women, *Nudes in Budapest*, with a reflection on the differences between the naked (i.e., unclothed) bodies of ordinary men and women and the idealized images of naked (i.e., the nude) flesh to which, as I have noted, we are normally accustomed:

> Because, one rarely sees naked strangers rather than idealised images, it is easy to forget that the majority of the population is imperfect . . . Traditionally nudes are dressed in an idealising mantle. The model is posed according to set aesthetic and moral codes that place the image within a cultural context which confers meaning—rather than being denuded, martyrs, mythic heroes, strong men and beauties are metaphorically clothed by their symbolic or erotic status. (p. 3)

As a consequence, Cotier's images have a reassuring quality: they remind us of the normality of physical imperfection and the possibility of imaging the unclothed body of old age in a positive and dignified manner. One source of this reassurance is Cotier's style of photography, in which he does not simply focus on bodies but identifies each one as an individual subject. His photographs are not exposures of the curiosities of the aging body, but have been taken in such a way that the individual personality of each embodied person is expressed in the image. In photographs of persons rather than bodies, age becomes salient insofar as it is indicative of individuality. The brief captions accompanying several of the photographs reinforce this interpretation: Plate 19, for example, a full-frontal photograph of Ibolya Zadorvolgyi, is accompanied by the caption "She is a widow and a kindergarten teacher. Her joys are crosswords." Plate 17, a photograph of Jeno Lazar is captioned "He travels with the operatta theatre and collects cigarette lighters." In the words of Sarah Kent, the facticity of the aging body is compensated by a "rather formal presentation that lends a simple dignity to each sitting" (cited in Cotier, 1991, p. 3).

This discussion of the independent reality of images, both as artifacts and as meanings, clearly suggests that imaging is a much more complex process than the taken-for-granted definition of image with which we began suggests. As art historian Pointon (1990) noted, it is now widely recognized that even photography, as a product of science, "does not offer unmediated truth or special access to history" (p. 86). Photographs, like paintings, sculptures, and other images, should be regarded as constructed products whose form and content reflect the beliefs and attitudes of their makers and therefore their historical culture. Thus, the historian of U.S. culture Banta (1987) added a further dimension to the definition of images when she wrote that imaging is not only "the making of visual and verbal representations (by sculptors, painters, illustrators, writers, advertisers, theatre people, journalists)," but also includes the "responses to these artifacts at every level of society" (p. xxvii). Any definition of images of old age must therefore be informed by an interactive or processual definition of imaging: the image as artifact interpreted in terms of a dynamic relationship with a particular audience.

If we adopt a definition of visual images of old age as dynamic interactive processes, we are then free to regard them as cultural resources we can draw on to make sense of everyday experience. The painted, photographed, and printed images of old age provide form and content to the answers we give to questions concerning aging and old age. As a cultural resource, images do not simply reflect the facts about the aging body but project moral ideas and beliefs about aging and the place of older people in the wider society. Although the ideas projected are those of their specific culture and time, we as human beings have the unique ability to detach them from their historical context and adapt them to our own ends (Hepworth, 2002). The image of St. Jerome as an old man contemplating the skull, a traditional image of death, by van Leyden (1494–1533) in the Ashmolean Museum, Oxford, has clearly been created within the context of the Christian theology of its time, which urged that it was the earthly duty of all human beings to prepare for death and the life hereafter. But this image may also be displayed and exchanged in the contemporary society as simply yet another image of an old man. The fact that the historical context of the original picture is of life as a spiritual journey in the well-established tradition of *The Ages of Man* (Sears, 1986) does not prevent its adaptation and reinterpretation to give meaning to contemporary experience. Similarly Victorian paintings of old ladies in picturesque English cottage gardens (Wood, 1988) are reproduced in contemporary birthday cards specifically designed for older people. In her book *Imaging American Women*, which looks at the role of ideas and ideals in U.S. history, Banta (1987) reproduced the painting by Sarah Eddy, *Susan B. Anthony, on the Occasion of her 80th Birthday*. Susan B. Anthony was a vigorous political activist who is nevertheless portrayed, according to Banta's reading, in the act of "receiving floral tributes from adoring children" (p. 71). Her image is thus constructed in the tradition of the saintly grandmother rather than the feisty political activist. The image of Susan B. Anthony in old age is framed according to the conventions of the period to reinforce the idealized late-19th-century definition of old age as a time of repose and contemplation. The image conveys a particular ideology of old age and especially the prescribed old age of women (Mangum, 1999).

Images have an important part to play in structuring the moral perception of the self in relation to others in the social world. Images are an integral feature of human communication and play an important constructive role in what art historian Hollander (1988) described as "self-image-making," a social process that, she argued, is "the acknowledged activity of us all" (pp. 414–415). In this construction, images of the face and body mediate between the inner subjective (private) world and the other observable (public) world and give structure and meaning to the impressions of the senses. Hollander cited the mirror as significant evidence of this creative process: When we look into a mirror we do not simply see a reflection of selves as we really are. What we see is an image of the face as we imagine it appears to other people. In other words, the looking glass is not a glass we look into to see ourselves but a glass through which we look to gain an impression of ourselves through other people's eyes. The image we see in the mirror is recognizable in terms of the visual imagery of our culture. Without visual images, we do not know what we look like in mirrors: "We may be tall or short or old or young, but what we think we look like must align in some way both with what others look like and with what contemporary pictures tell us is the truth of looks" (Hollander, 1988, pp. 416–417). Looking into a mirror, therefore, is an act of self-comparison and classification. Mirrors are not simply reflecting surfaces; they reflect the cultural repertoire of visual images already embedded in our imagination. Consciousness of aging is a good example of this process. In one of the Sherlock Holmes stories by Sir Arthur Conan Doyle (1923, 1979), Dr. Watson, Holmes' confidant and chronicler, describes meeting a former college friend he has not seen for several years:

> Promptly at ten o'clock next morning Ferguson strode into our room. I had remembered him as a long, slab-sided man with loose limbs and a fine turn of speed, which had carried him round many an opposing back. There is nothing in life more painful than to meet the wreck of a fine athlete whom one has known in his prime. His great frame had fallen in, his flaxen hair was scanty, and his shoulders were bowed. I fear that I roused corresponding emotions in him. (p. 280)

In this fictional example, we see the way in which the perception of the physical appearance of a friend from youth, who has not been seen for several years, is a reflexive process: In the course of reacting to Ferguson's appearance and categorizing him as aging, Watson simultaneously perceives the way he must look through Ferguson's eyes, and this face-to-face encounter is essentially one of mutual recognition. To perceive the aging process in this interpersonal way it is necessary, of course, to have a preexisting mental memory of the past image of an individual. Watson is made acutely aware of Ferguson's aging precisely because he has a memory of the younger man, a memory that Sherlock Holmes, who is also present, does not share.

We have seen that both Banta and Hollander draw attention to the interactive or relational nature of imaging, defined as the interaction between the perceiver and the image perceived. If we follow this definition, we can see that answers to questions concerning the nature of aging and old age are not found on or in the body but in the meanings we give to the observation of biological changes that we are able to make. My central argument, therefore, is that perceptions of old age in everyday life are articulated through culturally prescribed visual and verbal images that are sociologically reflexive and have, therefore, significant implications for the moral career of individual older persons. *Moral career*, as noted earlier, refers to the changes in a person's self image that occur during the life course. The question about how we know old age must be answered in terms of the repertoire of socially constructed imagery, which is essentially metaphorical in nature. In using the term *metaphorical*, I am following Hockey and James (1993), who defined a metaphor "as making one thing, one stage, one condition knowable through reference to another" (p. 39). The problem with old age, to paraphrase Oscar Wilde, is not that we are old but that we are young. Being old does not seem comprehensible in its own terms but must be comprehended in relationship to or comparison with other experiences and the experiences of others. Existentially speaking, as we see later in the discussion of the image of the mask of aging, old age seems to be one of the great unrealizables, often being described in terms of what it is *not*: as not feeling healthy, as not feeling young, or as not feeling oneself. Hockey and James stressed that the use of metaphor cannot simply be dismissed as "a poetic flourish with which we embellish our more mundane thoughts and feelings" (p. 39)—that is, that some kind of clear boundary can be drawn between image as a metaphor and the real world—but is an indispensable aid to thinking about experience and in particular to giving meaning to experiences which, as is typical in the case of old age, often seen almost impossible to express. It is metaphor that allows us to know about aging and to know about it in particular ways, and the history of gerontology may accurately be described as the history of the social construction of meaningful images or metaphors of old age.

We are now in a position to proceed to examine in closer detail our three key images of old age, which represent a prescribed metaphorical structure for everyday negotiations of the meanings of the experience of old age—the cultural context, in other words, of the moral enterprise of aging. As noted at the outset, these images are: (a) the geriatric body, (b) the physiognomic body, and (c) the mask of aging. They are selected because they represent three interpretations of the relationship between the body, the self, and society that have exercised a significant influence over definitions of old age in Western culture. To highlight their distinctive moral characteristics, these images are examined under separate headings, but it becomes apparent that in actual practice they are closely interrelated and that elements of all three can be detected in empirical studies of communication and interaction among older people (Coupland, Coupland & Giles, 1991; Jerrome, 1992; Stephens, 1980; Williams, 1990). In the first key image—the geriatric body—we look more closely at the "geriatric gaze" (Armstrong, 1983, p. 89), at bodily decay. In the second and third images, we look at two variations of the theme of the moral relationship between the body, the self, and society. Taken together, they represent three intersecting strands of Western cultural history of old age that help to account for the ambiguous and elusive nature of old age as it continues to be experienced in early 21st century society.

The Geriatric Body

The word *geriatric* is not neutral; it conjures up immediate images of physical and mental decay. Because of its close association with a medically defined array of problems of the aging body—cerebral impairment, autonomic disorders, sensory impairments, mental confusion, incontinence, and other ailments (Bromley, 1988)—the dominant image of the geriatric body is both a body and a self that are deteriorating irremediably (Katz, 1996). It is not surprising that the word most overwhelmingly rejected by readers of a magazine for the "over-50s" who responded to a questionnaire concerning preferred terms of address for older people was *geriatric* (Ransley, 1993).

During the second half of the 19th century, efforts to establish geriatric medicine as a specialization in its own right ensured that the professionally structured geriatric gaze emerged as a clinical technique for scrutinizing, categorizing, and imaging the aging body (Katz, 1996). As such, it became preoccupied with immobility, instability, incontinences, and intellectual impairment and resulted in a clinically endorsed deficit model of old age that is pervasive in its influence and widely reproduced. Thus, for example, the student textbook *Body and Personality* (Wells, 1983) includes a chapter on aging and dying in which aging is defined as

> Very largely due to the gradual accumulation of defects resulting from errors and failures in the cellular repair system. Its consequences are familiar enough: a loss of strength and elasticity in the muscles, arteries and skin; decline of the sensory and nervous systems, and progressive inability of such organs as lungs, liver and kidneys to function efficiently. Moreover, even the body's defensive mechanisms become impaired so that illnesses, and the damage they create, continuously add to the decay. (p. 146)

The image of the geriatric body is one in which physical decline has rendered normal interpersonal communication difficult or virtually impossible. The geriatric patient is one who has been isolated from full social interaction. When the Duchess of Windsor died in 1988, her former nurse described how, in her final years, the once elegant and fashionably distinctive duchess had joined the ranks of the undistinguished geriatric. She had become a helpless invalid, isolated in the bedroom of her Paris flat and neglected by her former friends and admirers. The moral of this particular image of the geriatric body is that old age is the great leveler: it reduces everyone, rich or poor, to the status of dependent invalid, disqualified from stimulating and self-enhancing social relationships. In the gerontological literature, a series of decremental changes vividly described as the aversive properties of old age. As described by Bromley (1978), these properties include the effects of physical decline on the ability of the body to act as an agent of interpersonal communication. Decreasing mobility of the head, neck, arms, and hands; impairments to vision and hearing; and slower mental reactions impede communicative competence and thus inevitably lead to an increasing social distance between older and younger people who are not similarly handicapped. Older people therefore find it increasingly difficult to express themselves, especially to younger people whose physical powers of self-expression are unimpaired. Younger people are correspondingly embarrassed by the strained interaction that ensues. Strained interaction erodes the perception of the older person as a fully human person, and the result is that the older person is gradually isolated from social interaction with those who are seen as normally competent. This process of social disengagement was described by Elias (1985) as "the loneliness of the dying." In this powerful clinical imagery of medicalized old age, "the crude cost–benefit terms of social exchange theory . . . provides an interesting and useful framework for the analysis of human social relationships, (but) the elderly patient may seem to have little or nothing to give in exchange for any services rendered by a younger person, and he is costly for the younger person to interact with" (Bromley, 1978, p. 26).

The clinical imagery of the geriatric body as physically cared for but socially neglected is frequently reproduced as a dominant image of old age in the popular press. Indeed, this depressing image of old age as physical and mental deterioration is so frequently highlighted that it is a source of persistent concern to all interest groups concerned with the care of older people (Gullette, 1997; Hepworth, 1988). Yet another dramatic example of the pervasive and yet misleading nature of this imagery can be found in Gilbert's (1946/1987) crime novel *The Spinster's Secret*. The spinster of the title, Miss Martin, is examined by a doctor experienced in the treatment of older patients: "It was his experience that a good many elderly people became queer, and the best place for senile cases was the institution, unless there was some convenient relative. From what the Matron had told him, he was prepared to find Miss Martin peculiar and to sign her up, in which case there would be little difficulty in getting his opinion confirmed by another local doctor" (pp. 116–117). The dramatic irony in this story is that, as the reader is only too aware—having privileged access to the spinster's secret of the title—Miss Martin is much more conscious of what is going on around her than those who unintentionally misperceive her actions. In this fictional example, the geriatric gaze is a blinkered gaze, constrained by the stereotypical assumptions about the geriatric body and mind that consequently act as a powerful barrier to communication between Miss Martin and her relatives and associates. This story is another example of the way in which the word *geriatric* has become part of the "vast lexicon of ageist language in everyday usage within which most terms are used to describe or refer to the elderly in a pejorative way, or at least carry negative overtones" (Coupland, Coupland, & Giles, 1991, p. 15).

The history of old age shows that although geriatric medicine began to emerge as a specialization scientific discipline during the early years of the 19th century, the term *geriatrics* did not "enter the medical lexicon until 1909" (Covey, 1988, p. 295; Haber, 1983, p. 47). The clinical label "geriatric," therefore, is the product of several decades of scientific inquiry during which the clinical gaze turned toward the aging body in an attempt to investigate the nature of the biological processes of aging and determine the precise nature of the diseases (if any) that could be regarded as specific to old age. Prior to the 19th century, old age was not seen as an identifiably distinct biological phase of the life cycle, but from the mid-19th century, increasing scientific interest and expertise in probing the internal structures and processes of the human body helped to reinforce the view that aging was either a disease or that aging brought a predisposition to disease. Kirk (1992) located the turning point in the clinical description of the aging body and the diseases of aging in 1839, when the German doctor C. F. Canstatt published a textbook on the clinical medicine of old age, a text that may "justifiably be regarded as a marker of the birth of geriatric medicine because it seeks to identify criteria for the aging process; it seeks to define the relationship between 'involution' and disease; it describes general and specific manifestations of disease in old age; and it emphasizes the need to deal seriously with elderly patients" (p. 487).

The problem is, however, that innovations in the medical construction of the geriatric body did not herald a significant departure in moral terms from the pre-19th-century perspective on the body of old age. The emergence of geriatric medicine did not offer a more hopeful or consoling vision of old age to replace the image of old age as a physical burden than had existed previously. In other words, the imagery of the geriatric body confirmed and reinforced traditional images of old age as physical decline and had little further to offer by way of consolation and compensation. Kirk (1992), the Danish historian of geriatric medicine, concluded that "the scientific images of old age of the nineteenth century were definitely not positive ones" (p. 494). One reason for this gloomy conclusion is evidence of the continuing influence on science of "ancient notions of the aging process" (p. 488). Early scientific images of the body of old age bore the signs of their historical heritage, namely, images of aging based on the Christian iconography of the *Ages of Man*, an iconography that combines images of time as both an arrow and a cycle (Gould, 1988): an arrow insofar as life is considered to be moving as a trajectory through one unrepeatable stage of life to another (cradle to grave) and a cycle insofar as life is believed to be essentially repetitious, with all human beings living out the same cyclical cosmic order as reflected, of course, in the Christian images of the resurrection.

The typical image of old age in the pregeriatric "Ages of Man" model of the lifestyle is one of a final stage of physical decrepitude, symbolized by a stick or a crutch, and a close proximity to death usually portrayed as a skeleton and a tomb (the moral human cycle is gendered in these images as Man, as is Father Time and his close associate, Death). At the end of the arrow's flight is death, but death is only a prelude to rebirth or the soul's release. In her study of medieval interpretations of the *Ages of Man*, Sears (1986) showed how the images of the human life cycle created by the European artists and writers in the Middle Ages always included at the end of their moralistic trajectory the image of decrepitude that, as we have noted, was imaged in the masculine gender. A central feature of the medieval European model of the life cycle is the contrast between the pleasures of youth, the responsibilities of middle age, and the "woes of old age" (Sears, 1986, p. 131). The consistent feature of the model is of a life lasting 70 years (during a period, of course, when the average life expectancy was considerably shorter), which moves in its early stages upward toward the prime of life (middle age) with an ensuing and inevitable downward movement toward death, the proximity of which is inscribed on the body. Not surprisingly, therefore, representations of the outward appearance of old age on the face and body are mirror images of the physical appearance of death. The following description by Karp (1985) of Giovanni Francesco Barbieri's (1591–1666) *Caricature of An Emaciated Old Man* is an apt illustration:

> With great deftness (Barbieri) captured the physical quality of the aging. The old man with sunken cheeks has lost his teeth, causing his chin to protrude and his lower lip to rest in front of the upper. The broad, flat chin with the narrowing of the face at the mouth suggest degradation of the jaw. Reduced visual acuity may be implied by the simplified rendering of the eye sockets without clear notation of the eyes; the result is a furrowed squint, suggesting

that visual difficulties of old age associated with cataracts or severe refractive error. The exaggerated quality of the snoutlike nose and the unusually squared jaw may also place the image within the genre of caricature, although the face is very much rooted in observation from life. (p. 209)

The face of the old man is almost the face of a skull.

Karp's interpretation of Barbieri's *Caricature* was made, of course, over 300 years after Barbieri completed the work, but the correspondences between a medical reading in the 1980s and that of a more religious period are striking. Both interpretations share the view of old age as a period of decline. In the geriatric body of medical science, the process becomes known as involution, but in both cases these physical transformations are regarded as inevitable, and because of their physical consequences, unwelcome. The difference between the two interpretations can be located in the consolation of religion. The model of the *Ages of Man* is a spiritual one: The implication is "that a man should look beyond the worldly life, so transitory, so deceitful, to a spiritual existence which is not bound by time, the vagaries of chance, and the rhythms of growth and decay" (Sears, 1986, p. 145).

This close association between images of old age and images of death establishes a line of continuity between geriatric medicine and the earlier Western moral tradition of imaging old age as the transition stage from life through death to eternal life, epitomized through the 18th century as the *vanitas* theme ("remember you must die"). By way of illustration of the *vanitas* theme of life as a preparation for death and the life thereafter, art historian Hollander (1988) reproduced a 16th-century painting by Laux Furtenagel, *Hans Burgkmair and His Wife*. In this image, the couple, who are middle aged, are portrayed side by side, their faces turned toward the viewer. The wife holds in her hand a looking glass in which we see the reflection of two skulls. In Hollander's words, the artist has constructed his portrait to convey the moral message that "skulls in the mirror are the real truth" (p. 395). The viewer is exhorted to keep his or her eye fixed on the transience of the human body and the immortality of the soul, a message that was also displayed before the eyes of spectators at the Renaissance anatomy theater at the University of Leiden, where "around the dissection area, and on the interior walls of the building, were ranged a collection of human and animal skeletons, some of them bearing appropriately moralising instructions. These mottos (e.g., *Pulvis et Umbris Sumus*—we are dust and shadows) develop the theme of death and dissolution" (Sawday, 1990, pp. 131–132).

The *vanitas* theme is an integral feature of the *Ages of Man* image of the life cycle as a spiritual or moral journey that continues to provide an interpretative framework within which the geriatric body is diagnosed and old age is clinically treated. The geriatric body is in many respects one of the most obvious survivors of what Sears (1986) described as "the moralised life in images" (p. 134). In another analysis of life cycle imagery in the United States from the 17th century to the present day, Kammen (1980) showed how the European tradition of the religious taxonomy of the stages of life influenced representations of the life cycle in the United States: "Colonial views of the life cycle were derived from European perceptions" (p. 36). He further argued that these European images continued to influence U.S. perceptions into the 19th century and through to the present day. The imagery is now familiar to us: From the cradle to the grave, human life passes prescriptively through a fixed series of stages, each of which has its prescribed rules of moral conduct. In the 17th century, for example, Sir Thomas Browne, an English physician, observed that the distinctions of life—youth, adolescence, manhood, and old age—each had their proper virtues: " Let each distinction," he wrote, " have its salutary transition and critically deliver these from the imperfections of the former, so ordering the whole, that prudence and virtue may have the largest section" (Haber, 1983, p. 49). At the end of life's journey we encounter the infirmities of old age, which represent a final test of personal moral fiber. As in the case of the geriatric body, old age is considered to be a burden that should be borne with uncomplaining fortitude; the compensations are intimations of a state of heavenly grace for those who have lived a virtuous life and thus, as we still observe, have aged gracefully. Kammen (1980) quoted from Anne Bradstreet's collection of poems, *The Tenth Muse Lately Sprung Up in America*, written between 1643 and 1644, in which old age "takes pride in the accomplishments of a lifetime and yet recognises the ephemeral folly of moral achievements. Neither knowledge nor honor can forestall death. Hence nostalgia, plus a modicum of wistful wisdom and a sheaf of wheat, are emblems of the elderly" (p. 40). To age gracefully is to accept the pains of the aging body in the sure and certain hope of the soul's ultimate release to a heavenly reward. In contrast, to age viciously is to reject the contemplative mode of old age and to continue to indulge unrepentantly (or to attempt to indulge) in the pleasures of the body normatively associated with an earlier and more physically urgent stage of life. The image of disreputable old age is one to which I return in the next section.

The message that the quiet contemplation of one's mortality is the appropriate accompaniment of physical decline is strikingly similar to the one that emerged from the research of the 19th-century pioneer observers of the body of old age. To make sense of their close clinical scrutiny of the physical changes taking place within the aging body, specialists in the new medical science of aging returned to the ready-made structures of the *Ages of Man* that provided, as we have seen, a vocabulary of resignation and contemplative disengagement in later life. Here was a framing or construction of knowledge that reproduced traditional moral distinctions between graceful and disgraceful aging. If the physical decline of the body was irreversible—a process of inescapable involution—as the pioneering geriatricians concluded, then there were only two courses of action for the patient and his or her carers: a calm acceptance and preparation for death or an unruly and rebellious reaction. In effect, the geriatric profession was reinforcing the distinction between what historian Schama (1988) graphically described in his observations on visual representations of old age in 17th-century Holland as the "wrinkles of vice and wrinkles of virtue" (p. 430).

The Physiognomic Body

We already noted the preoccupation of the Dutch painter Rembrandt with self-portraiture in later life. We also noted that these images are open to a number of readings and that they have been used most recently to endorse late-20th-century gerontological prescriptions for positive aging. In this positive imagery, the face of Rembrandt is often celebrated as the face of virtuous aging. In his study of 17th-century Dutch culture, Schama (1988) argued that it was during this period that artists noticeably began to concentrate on "an entire geography of the ancient face" (p. 430), painstakingly delineating every line and wrinkle. In this reference, Schama described the ways in which artists of the time went to considerable lengths to draw and paint distinctions between two kinds of old women: "the solitary matron saying grace over a humble meal" and the "leering procuress presiding over a sexual transaction" (p. 431). In this imagery, aging is not presented as a homogenous process that reduces each individual to the same level, but a discriminatory one, during the course of which essential differences in character are revealed. The true inner self, so to speak, is ultimately revealed on the visible face of old age.

The belief that the true inner character or self is revealed on the visible face and body is one of the central tenets of the ancient pseudoscience of physiognomy. Defined literally as "the art of judging character and disposition from the features of the face or the form and lineaments of the body generally" (*The Oxford Universal Dictionary*, 1965, p. 1494), physiognomy is deeply embedded, as Schama's comments imply, in Western culture (Magli, 1989), but its greatest impact followed toward the end of the 18th century from the work of the Zwinglian Swiss pastor Johan Caspar Lavatar. As a man of religion, Lavatar believed that each human being was a unitary self and that as a consequence visible external appearance and inner moral character were one. His method, first published in 1775 and subsequently an international bestseller, was to create a visual record of the distinctive characteristics of each individual, a method that was enhanced by developments in the art of engraving. Another technique was the silhouette, for which he had some preference, because the silhouette reproduced in stark outline what were regarded as the essential features of the character of an individual as revealed in his or her face. Moreover, he also favored the dead because they "furnish a new subject for study. Their features acquire a precision and an expression which they had not when either asleep or awake. Death puts an end to the agitations to which the body is a perpetual prey . . . It stops and fixes what was before vague and undecided. Everything rises or sinks to its level; all the features return to their true relation" (Stafford, 1991, p. 106). One of the attractions of Lavatar's work for his contemporaries was his privileging the trained observer who, through the adoption of the principles of physiognomy, became skilled in the interpretation of human character.

The relevance of the physiognomic gaze for the imaging of old age can be located in the interpretation of wrinkles. In the physiognomy of Lavatar and his followers, the lines of old age on the face do not so much provide evidence of closeness to death as a clear indication of the moral nature of the life that was lived. The ability to read the lines on a face conferred on the practitioner of physiognomy the capacity to discriminate between different types of older persons. Founded as it was on the belief that the face and body as expressice instruments revealed the essentials of personal identity, physiognomy continues to be regarded as a significant development in the study of human communication and interaction. In his adherence to principles of observation of external aspects of the face and body, Lavatar's "position was that external behaviour is a direct expression of the interior of man" (Birdwhistell, 1975, p. 51), a position that exercised considerable influence, as Cowling (1989) showed, over artists of the Victorian period who were able to draw on physiognomic principles to create images of essential moral distinctions between older men and women. In these images, aristocratic types of both men and women, for example, exhibited quite different physiognomic characteristics than those of rustic and working men and women. I have already referred to the predilection in Victorian paintings for an English countryside populated with old ladies in cottage gardens. In such images we can find the prototypes of the virtuous "grannie" whose physiognomy displays the comforting qualities of benevolence and physical quietude in the "eventide of life": an image that continues to adorn advertisements for retirement homes. In her analysis of the reaction of art critics to the painting by John Millais, *Moses Supported by Aaron and Hur,* executed in 1876, Cowling (1989) noted that some critics were unhappy with the portrayal of Moses as an apparently enfeebled old man who required the physical support of two sturdier figures. Not only did such critics find fault with the representation of his body as suggestive of "signs of bodily decay and debility which can be explained only as the results of 'an evil and intemperate youth,'" but they also resented the shape of Moses' head, seen as "flat and contracted, the volume of the brain altogether too small" (p. 107) and consequently insufficiently intellectual.

Although the scientific basis for physiognomy has long been discredited, the belief that external appearance represents the inner self continues to influence our conceptualization of the relationship between the body and the self. In a society where, as was noted earlier, visual imagery has become predominant, it becomes increasingly necessary to be able to discriminate between individuals in terms of their outward appearance. Most of us no longer live in small isolated communities where face-to-face interaction is the norm, but in large highly mobile urban environments where the majority of people are strangers whose identities can only inferred from the clues their physical appearances present. This form of social relations puts a premium on the presentation of self and, in particular, in a society where aging and old age are increasingly regarded as a social problem, a high value is placed on the cultivation and preservation of a body and, physiognomically speaking, a self that is aging aesthetically.

In their analysis of the hidden aesthetic of medical practice, Stafford, La Puma, and Schiedermayer (1989) also argued that the physiognomic tradition is alive and well: "Unrecognized presuppositions about patient appearance," they noted, "have become increasingly important in medicine, medical ethics, and medical law" (p. 213). The body of old age, as was noted in the discussion of the geriatric body, is not simply perceived as a biological fact but is interpreted according to interweaving aesthetic and moral perceptual rules that have their roots in the past. Stafford, La Puma, and Schiedermayer (1989) noted a "contempt for the signs of aging" (p. 228) that reflects a physiognomic preoccupation with an outward appearance unmarred, as in the case of Millais' painting of the figure Moses, by signs of weakness or imperfection. The pressures to keep up appearances and not let ourselves go are contemporary variations on the physiognomic theme. If this were not the case, cosmetic surgery would scarcely qualify as a growth industry.

Prior to the emergence of geriatric medicine in the 19th century, the dominant model of the human body was of one supplied with a limited amount of energy. According to vitalism theory, the supply naturally waned as one reached the end of the prescribed 70 years, when the only way to control the aging process was the practice of moderation with the aim of postponing old age as long as possible (Haber, 1983). This model of growing old gracefully through the conservation of energy was endorsed by the new profession of geriatric medicine, in which scientific observations did nothing to discredit the fear of old age and implicitly encouraged the fabrication of images of old age that reflected physiognomic principles. In more recent years, these principles have been realized in the encouragement of programs of positive aging in which new techniques are applied to the traditional ideal of resisting old age as long as possible and, in effect, maintaining a kind of perpetual middle age, (Hepworth & Featherstone, 1982). In her study of European images in *The Ages of Man*, Sears (1986) reproduced the *Steps of Life* by Jorg Breu the Younger, dated 1540. In the center, on the highest point and the widest step, sits a representation of a man who is literally located in middle age. He is completely absorbed in a manuscript and apparently oblivious of the skeletal figure of death who, suitably equipped with a bow and arrows, stands immediately behind him. To his left, the steps descend toward drowsy old age, a coffin, and ultimately resurrection. Over 400 years later, the stated aim of the prescriptions for positive aging in contemporary society is to stay at the midpoint of life—at one's peak, as we say—as long as possible and to close our eyes to the image conjured up by the coffin (Hepworth & Featherstone, 1982). These efforts to maintain middle age can be described as a physiognomic exercise, that is, as a struggle to maintain a close correspondence between an idealized inner self—the self in its prime—and a socially constructed body that proclaims these virtues to the world. In common with earlier times, the body of old age is still regarded as less than desirable, but the difference is that it is no longer regarded as a cruel necessity. We have, we believe, the knowledge and the technology to resist old age until we reach a (hopefully) quick and peaceful end.

Indications of this belief abound in the contemporary images of positive aging. It is a belief that is strongly influenced by what have been described as postmodern attitudes toward the human body (Featherstone & Hepworth, 1990). In his sociological analysis of fitness and the postmodern self, Glassner (1989) defined the increasing concern with the pursuit of physical fitness as a postmodern pursuit, by which he meant the attempt of fitness programs to narrow the gap between the body and the self—to integrate external appearance with one's inner self-concept. In the discussion of the geriatric body, it was noted that the aversive properties of old age include a decline in the physical ability of the body to function as a mechanism of normal human communication and that the quality of social interaction is consequently impaired. The fitness programs of contemporary society reflect the belief that the body and the self can be united in one smooth operation that will maintain the expressive powers of the body in one extended physiognomically, rewarding middle age. The message is not that you are as old as you feel, but that you are as old as you look.

THE MASK OF OLD AGE

"You're as old as you feel" is, of course, a common everyday definition of the experience of aging, which shares with the equally commonplace response, "I don't feel old," a strong element of denial of both the geriatric and the physiognomic body. Implicitly or explicitly, these observations are firm rejections of the belief that old age can be adequately identified in terms of visible changes on the face and body. "I don't feel old," argued Thompson (1992), is a cry of protest against the mythology of old age that is constructed almost entirely from misleading images of outward physical appearance.

The mythology in question is the physiognomic tendency to regard the external appearance of aging as indicating a simultaneously aging inner self. Body and self are drawn together by external observers to construct a kind of geriatric unity and physiognomic union. Coupled with the traditional Western belief in old age as a condition of geriatric disengagement from full active social life, the image of the physiognomic body creates a trap or prison that constitutes an effective barrier of communication with the wider world (Hepworth, 1991). This is the mask of old age: an image that reflects the subjective experience that many older people describe as being constrained by the expectations of others (often younger people) into wearing a mask or disguise of physical aging that, unlike the actor at the close of a theatrical performance of old age, they cannot remove and leave behind in the dressing room (the youthful Olivier as King Lear, for example, or Orson Welles at age 25 in the elaborate makeup for his cinematic role as the aging Citizen Kane). All of the world's a stage, but when it comes to old age, some roles are for real.

The source of the difficulty lies in the existential questions with which I began. Our knowledge of the meaning of old age, especially when we are younger, is constructed from the images or metaphors society makes available to us. It is secondhand knowledge. As we become chronologically older, we discover that a subjective grasp of the meaning of old age is equally elusive. Although our bodies may display signs of the passing years, the subjective self is not experienced as correspondingly old, and in a society where the dominant images of old age are negative, this tension between the inner personal and the outer social identity may cause us considerable distress. Outside the geriatric and *Ages of Man* models of old age, we have few positive models of what an old self should be. To carry the dramatic metaphor a little further, our awareness of old age, either with reference to ourselves or to others, is the product of the range of roles we play throughout the life course and the scripts that are available. But, as we have seen, the models of the life course we have inherited from our cultural past do not offer a great deal of scope to those of advancing years. The predominant images of old age are not derived, as Thompson (1992) argued, from the diversity of subjective experiences of life—that is, from qualitative subjective experience—but from the public imagery of old age as a medical, spiritual, and, increasingly in the contemporary world, a social problem. The result is a mask: a subjective sense of distance between the inner or private self and the outer or physically observable social self, indicated by the geriatric and physiognomic bodies.

Miss Marple, Agatha Christies's formidable spinster detective, perpetually solving murders the (male) police fund impenetrable, is a significant example of the gendered mask of old age. In these popular crime stories, Miss Marple has the appearance of the archetypal old lady of limited means and faded gentility. Hart (1985) wrote:

> Her hair was usually described as white, occasionally grey, her face as pink and crinkled, and her teeth as ladylike. She was tall and thin and had very pretty china-blue eyes, which could look innocent or shrewd depending on one's point of view. Her general expression was usually described as sweet, "with her head a little on one side looking like a amiable cockatoo," but this could change when she was on the trail of someone evil. Inspector Craddock noticed in "A Murder is Announced," "the grimness of her lips and the severe frosty light in those usually gentle eyes." (p. 74)

In appearance, Miss Marple is a stereotype of the "wrinkles of virtue": guileless, harmless, and endlessly preoccupied with her knitting, and blending totally into her genteel surroundings. But, as all readers of Agatha Christie are aware, there is much more to her than meets the eye, for she is wearing what Craig and Cadogan (1981) described as a "grandmotherly disguise" (p. 164). It has been argued that in creating this disguise, Agatha Christie has probably contributed to "the stereotype of geriatric foolishness and risibility in her presentation of a dithering old lady whose hesitations and self-interruptions are intended to mask a cold, detached intellect" (Shaw & Vanacker, 1991, p. 90). But this strategy undoubtedly capitalizes on the space for the misrepresentation of old age that images of geriatric and physiognomic bodies allow.

The image of the mask of old age implies that the physical changes that accompany a life of 70 (plus) years are superficial or surface changes that conceal from the casual observer or the untrained eye essentially unchanging qualities beneath. The mask is, of course, a dualistic concept of aging conceived as a process characterized by a progressive and possibly traumatic sense of disengagement of the inner from the outer social self. Signs of the unchanging element of identity—the inner self—can occasionally be glimpsed when the mask slips or when, as in the case of one of Miss Marple's denouements, it is deliberately removed. At these moments, that which is believed to be unchanging receives recognition and is confirmed; a good example is found in the care with which caregivers for those suffering from Alzheimer's disease will watch for fleeting signs of a remembered loved one occasionally peeping out from behind a mask of mental incoherence (Gubrium, 1986). Indeed, it is increasingly acknowledged that the imaginative employment of communication skills based on the "dialectical interplay between neurological and sociopsychological factors" (Kitwood, 1990, p. 177) can help to recover and sustain elements of the personhood of a dementia sufferer. In this model of psychosocial treatment, dementia is treated precisely as a kind of mask or barrier separating the personal self, or "I", from the social selves or roles. Treatment is directed toward ensuring that sufferers are not disqualified as a consequence of impaired communication skills from playing desirable social roles or expressing their social selves (Sabat & Harré, 1992). The intention here is to sensitize the listener or observer to "the thread of reason" (p. 460) that can be detected behind apparently confused communication, with the wider aim of ensuring "that the only prison in which the sufferer would dwell would be that created by the boundaries of the brain injury and not, in addition, the confinement that is brought about by the innocently misguided positioning and story lines created by others" (p. 461).

Before we move on to explore key social and personal implications of the mask of aging, it is necessary to note that in Western culture, the word *mask* has two meanings. In addition to the commonly accepted meaning of a mask as "something which covers or hides from view" (*The Oxford Universal Dictionary*, 1965, p. 1212), it may also refer back to the Latin source word *persona*, which is derived from the masks worn by actors in early forms of drama to reveal an identity to their audience. According to this usage, the purpose of the mask is not to conceal but to reveal the true nature of the self. In the Western cultural tradition, therefore, masking is a complex psychosocial process of manipulating and reconciling the interdependent elements of concealment and display. It is this characteristic feature of the mask that was described by Goffman (1971) as a role "performance." In his book *The Presentation of Self in Everyday Life*, he drew attention to T. E. Park's observation that "It is probably no mere historical accident that the word *person*, in its first meaning, is a mask. It is rather a recognition of the fact that everyone is always and everywhere, more or less consciously, playing a role . . . It is in these roles that we know each other; it is in these roles that we know ourselves." (p. 30). In other words, the self can only be expressed as a kind of social performance: We obtain our knowledge of the inner personal identities of other people from their observable social conduct, and such conduct can be manipulated to conceal or reveal aspects of the self. Similarly, we express our own inner consciousness of selfhood in terms of our social relations with other people. Such performances depend on the physical ability to communicate, both verbally and nonverbally, and it is for this reason, as noted earlier, that any impairment of the physical ability to communicate is threatening to full social acceptability and, therefore, the self.

The role of performances through which self-awareness is sustained are the product of socially acceptable levels of individual control over the body and the emotions. Cohen-Solal (1987) described, in her biography of Sartre, the sad final days of the philosopher's life when, despite his obvious infirmities, he was still going out in public:

> His presence started, gradually and almost naturally, fading away. His voice remained in our ears, but it also grew feebler, as did his anger. As for his physical strength, that was also slowly flagging. Each of his public appearances left those who loved him feeling uneasy, dismayed, and angry. Where was the Sartre they knew? Was he still somewhere inside that wobbly, puffy, blind little man? (p. 509)

The willingness of others to persist in the quest for those underlying signs of constancy that Gombrich (1982) saw as the evidence of the persistence of individual identity is, in the circumstances described by Cohen-Solal, by no means guaranteed, even when dealing with a world-famous philosopher. Cohen-Solal wrote approvingly of the friends of Sartre, such as Francoise Sagan, who steadfastly provided him with letters and other reminders of her affection. But there were also many who were disconcerted by Sartre's appearance or angered by his physical decay.

The problem of discovering the traces of an enduring inner selfhood in an individual who is obviously declining physically is also highlighted in Jean Renoir's description of his father, the painter Pierre-Auguste Renoir, in old age. Although dreadfully crippled by rheumatism and looking very frail, the artist was still painting:

> His hands were terribly deformed. Rheumatism had cracked the joints, bending the thumb towards the palm, and the other fingers towards the wrist. Visitors who were not used to this mutilation could not take their eyes off it. Their reactions, which they did no dare express, was "It's not possible. With those hands he can't paint these pictures. There's a mystery." (Denvir, 1987, p. 203).

The mystery was, as Jean recorded, Renoir himself. His determination to paint overrode his physical disabilities, and thus the mask of aging Renoir, which threatened to annihilate his public image, simultaneously revealed the continuing presence of the artist.

Renoir still retained control over the performance of his artistic skills and thus kept his public memory alive.

We have already observed that the Western concept of the mask allows for the dual functions of concealment and display. Over the centuries, the early theatrical origins of the mask have been elaborated and extended into everyday life, with the result that a sensitive awareness now exists of the range of masks available, and thus possible variations in the presentation of self, from, for example, artful trickery to sincerity; from playful improvisation to sinister manipulation; and from an existential sense of authenticity to anguished feelings of impermanence. It has also been suggested that in contemporary society, successful performances require minimum levels of physical and emotional control. At the same time, it is important to recognize that the range of performances open to older men and women is not limited to only a single mask of aging; there is a more extensive repertoire that includes several masks of aging. In the discussion that follows, I explore two further variations: the playful mask and the youthful mask. These variations are not meant to be exhaustive, but simply to draw attention to some of the ways in which age-related boundaries around the self may be creatively manipulated by embodied actors.

The Playful Mask

In her study of images of old age in literature, Woodward (1991) argued that there are significant differences in the way in which time and age are inscribed on the outside and inside of the body. To open up what we have called the geriatric body risks exposure to the possibility of seeing the "disturbing disjunctions between the chronological age of the body . . . and its . . . outside" (p. 136). One way of manipulating this tension between the invisible interior of the body and its outer appearance is, Woodward suggested, to engage in a youthful masquerade. And as a significant literary example of the masquerade of youth, she chooses the middle-aged Aschenbach, a central figure in Thomas Mann's short novel, *Death in Venice*.

On a visit to a luxurious hotel in Venice, Aschenbach falls in love with a young boy, and in an anguished attempt to change his aging appearance, makes youthful touches to his clothing, has his hair dyed, and makeup applied to his face. The result is a grotesque mask because, Woodward argued, Aschenbach has fallen victim to "that seductive chestnut: We are all as old as we feel, but no older, and grey hair can misrepresent a man worse then dyed. To a certain extent he is wrong. In addition to being a state of mind, aging is a biological phenomenon and a social construction" (p. 122). In the novel and in the version filmed by Visconti in 1971, the mask worn by Aschenbach is hauntingly tragic because he is attempting pathetically to deny the fact that his body is already "inhabited by death" (p. 121).

As an alternative interpretation of the masquerade of youth, Woodward chose "an act of defiance" or resistance to "dominant social codes" (p. 125). To illustrate this, she cited the older, wealthy widow, Elizabeth Hunter, in Patrick White's novel *The Eye of the Storm*. Imprisoned in a decrepit body, the physically dependent Hunter sardonically bullies her nurse into transforming her appearance with dress, cosmetics, and a wig into a garish travesty of a youthful natural self. The result, said Woodward, is a revelation rather than tragedy: the revealed truth of the mask that "young and old do coexist in the body in their own peculiar way" (p. 125). In the words of Patrick White: "Momentarily at least this fright of an idol became the goddess hidden inside: of life, which you had longed for, but hadn't dared embrace, of beauty such as you imagined, but had so far failed to embrace . . . and finally, of death which hadn't concerned you except as something to be tidied away, until now you are faced with the vision of it" (p. 125).

To this example of the bleakly humorous manipulation of the mask of old age, we can add Penelope Lively's fictional exploration of the inner life of an older hospitalized woman who, unknown to those around her, embarks on an imaginative reconstruction of her past history. Apparently slipping away from the world, Claudia hovers between periods of awareness of her external surroundings and periods of absorption in memories of episodes in her past life. During one of her periods of consciousness of her surroundings, she deliberately pretends to be away:

> Sylvia [Claudia's sister-in-law whom she has always disliked] came to see me last week. Or yesterday. I pretended I wasn't there.
>
> "Oh dear," says the nurse. "I'm afraid it's one of her bad days. You never know with her . . ."
>
> She leans over the bed.
>
> "Here's your sister-in-law, dear, aren't you going to say hello? Wake up dear."

The nurse leaves Sylvia with the "unconscious" Claudia and Sylvia potters nervously around the room: "There is a twitch from the bed. Sylvia drops the card and scuttles back to the chair. Claudia's eyes are still closed, but there comes the sound, unmistakably of a fart. Sylvia, red in the face busies herself with her handbag, hunting for a comb, a hankie." (Lively, 1987, pp. 21–22).

Claudia has no illusions about her age. In common with many of the characters in Penelope Lively's novels, she possesses a sophisticated awareness of the social construction of history and biography. In Lively's view, we do not have direct access to the past as it really was but reconstruct it from the evidence available in the present. This evidence includes personal reminiscences, those residues we draw on to remind ourselves of our past identity and its continuity (or otherwise) with the present and the future. Claudia is ironically aware that only the remotest vestiges of her youthful good looks remain. She is also aware of similar changes in her children and other friends and family, but she manipulates the knowledge with fortitude and uses her reduced physical condition as a last resource to sustain a place in the shifting balance of relationships.

Lively's Claudia does not, at other times in the novel, playfully flaunt her physical decrepitude. Apart from the brief incident when she breaks wind and disconcerts her sister-in-law, the novel does not dwell on the internal physical processes of the geriatric body. There are, however, several additional fictional and nonfictional examples of what we may describe as the "flaunt," or playful manipulation, of the "disgust function." In these instances, the physical decrements of old age are deliberately and often sardonically displayed in an attempt to produce an aversive reaction in others. Datan (1986), in an example of what we may call "disreputable aging," described the humorous flaunting of physical decrepitude on the part of older people themselves as one of the only forms of power remaining to a group whose social position is progressively marginalized. She quoted a poem, "The Senior Citizen Gets It All Together," written in the form of a grotesque parody by Leland Taylor, who was then almost 80 years old. In theses verses, Taylor satirizes the sentimental view of old age as the "best" years of life through a precise recital of the physical indignities of the aging body. Such verbal exposure of the geriatric body is evidence, Datan argued, of the power of marginal identity: The old have at least the advantage of inside knowledge of the aging process. The aging humorist combines "marginal identity with the immunity of the jester" and is thus enabled to "voice the painful truths of the aging through the medium of power—the power of the survivor" (p. 171).

It is not, of course, necessary to be old to flaunt one's bodily functions in a playful fashion. There are many comedians and comic actors who have adopted this ploy on stage in order to get a laugh. In Britain, one famous exploiter of the decrepitude attributed to old age was the comedian Frank Randle (Nuttall, 1978). Among the stage characters for which he was famous during his heyday (he died in 1957) were three very old men notable for their pleasure in alcohol and "lecherous senility" (p. 50). They "all suffered sudden rebellious leaping of the limbs as though gripped by an autistic spasm. They were all bubbling fountains of digestive juices, and they were all drunk" (p. 50). Randle's performances were liberally punctuated with references to the number of belches in a bottle of beer, highly sexed women, and jubilation over the miracle of such expressions issuing from the mouth of a man of "Eighty-two. Dammit!" (p. 50).

The Youthful Mask

The exploitation of the repellent aspects of physical decline to dramatize, challenge, or embarrass stands in stark contrast to the youthful mask of aging. The youthful mask has, of course, a great deal in common with the masquerade defined by Woodward (1991) as the "denial of age, an effort to erase or efface age (or to put on youth)" (p. 148). But the type of mask wearing to which I now refer requires more than the maintenance of an illusion of youth—the adoption of a disguise that simultaneously conceals and reveals—it also requires a faith in the reversibility of the process of physical aging, a faith in the ability of science to prolong a youthful existence (Gruman, 1966). This version of the mask takes us closer to the original theatrical mask of antiquity, the purpose of which is not to conceal but to reveal and project essential character. It is, of course, a physiognomic mask.

In his essay on the wealthy socialites of Palm Beach, journalist Alan Whicker (1983) described the endless preoccupation of many older women with the quest for youth, a quest that extends the struggle against the limitations of aging well beyond the surface transformations of cosmetic surgery and into the realm of blood changes, hormone injections, and other techniques that claim to modify human physiology. For those who have the necessary financial resources, the essence of this contemporary manifestation of an age-old pursuit is the desire to detach chronological from physiological aging, and thus to neutralize conventional, social age gradings: "Hostess Helen Tuchbreiter was quite firm: 'There's nobody old in Palm Beach. I'm 39 and holding. Alternatively, I'm between 21 and death' " (Whicker, 1983, p. 376). On the evidence of outward appearance, the results can, according to Whicker, be remarkable:

> The only reigning social Princess who refused to talk with me before my camera murmured she was afraid her wrinkles might show. Leaving gallantry aside, I could see no wrinkle—not even the scars she was actually worried about. She looked around 32, or maybe 39 in a hard light; then she confided she had just become a great-grandmother. Never underestimate the power of new blood. (p. 377)

As we have seen, the image of the mask has therapeutic significance, as well as alerting us to the pleasures and dangers of unmasking. The ideal aim of the youthful mask is to create more than a temporary and vulnerable presentation of a youthful self; it is to arrest the geriatric processes of normal aging through contemporary techniques of body maintenance and repair (Hepworth & Featherstone, 1982). In this particular physiognomic image of the aging process, age is not merely skin deep. External appearance and inner physiology are all of the piece: the integrity of the one reflects that of the other.

The youthful mask, then, can be described as an idealized image of the positive aging that is only the latest development in response to the challenge of living in an aging world. In addition to the scientific advances that make the new technology of the body of old age possible, there is, first, the influence of the consumer culture, with its high valuation of youth and beauty (Featherstone, 1982). Second comes the metaphor of the body as a machine that, like any other machine, requires proper maintenance to function at optimal levels. Third is the concept of the performative, or protean, self that is, the self that is flexible and variable as distinct from having a single uniform identity, fixed for all time. Fourth, there is the erosion of traditional forms of social inequality and the emergence of conceptions of human equality in a context of postmodern diversity. Fifth, there is the promotion of positive aging that makes a crucial distinction between processes of disease-free normal aging and the final stage of accelerated decline that should ideally be postponed until the eighth decade of life.

As the composite product of these changes, the youthful mask has much in common with its tragic and playful counterparts. Indeed, as we can see from the account by Whicker (1983) of high social life in Palm Springs, youthful mask wearing is distinguishable from other forms in terms of the extensive precautions taken to ensure that it cannot slip, that what you see is what you get. In this respect, the youthful mask is the consumer culture version of positive aging. The conjunction of the time-honored value placed on youth and beauty (Marwick, 1988) with technical developments in processes of body maintenance and repair holds out, in true consumer culture fashion, the promise that aging can be truly defeated and no one need grow old anymore. As such, it is an indication of the anxiety aging continues to create in a society where conceptions of the ideal body as active, youthful and, as a consequence, sexually desirable exercise a considerable influence over images of positive aging. Yet the situation is by no means static. Because images, open as they are to interpretation and reinterpretation, are a malleable human resource, images of old age have considerable potential for change. Cotier's (1991) collection is only one example of the positive images that may be made of the aging body. One of the encouraging signs of our times is the increasing awareness of the contribution the analysis of processes of human communication can make toward the understanding of the social construction of old age. From these sources, the message is that images of aging are flexible, plural, and diverse, and it is through research and practice in the subtleties of communication, verbal and nonverbal, that the key to the enhancement of the quality of later life will be found.

REFERENCES

Achenbaum, W. A. (1982). *Images of old age in America: 1790 to the present*. Ann Arbor: University of Michigan, Wayne State University.

Armstrong, D. (1983). *Political anatomy of the body: Medical knowledge in Britain in the twentieth century*. Cambridge, UK: Cambridge University Press.

Banta, M. (1987). *Imaging American women: Idea and ideals in cultural history*. New York: Columbia University Press.

Bartram, M. (1985). *The Pre-Raphaelite camera: Aspects of Victorian photography*. London: Weidenfeld & Nicolson.

Biggs, S. (1989). *Confronting aging: A groupwork manual for helping professionals*. London: Central Council for Education and Training in Social Work.

Birdwhistell, R. L. (1975). Background considerations to the study of the body as a medium of "expression." In J. Benthall & T. Polhemus (Eds.), *The body as a medium of expression* (pp. 36–58). London: Allen Lane.

Blaikie, A. (1993). Images of age: A reflexive process. *Applied Ergonomics, 24*(1), 51–57.

Blaikie, A., & Hepworth, M. (1997). Representations of old age in painting and photography. In A. Jamieson, S. Harper, & C. Victor (Eds.), *Critical approaches to aging and later life* (pp. 102–117). Buckingham, UK: Open University Press.

Blaikie, A. (1999). *Aging and popular culture*. Cambridge, UK: Cambridge University Press.

Breytspraak, L. M. (1984). *The development of self in later life*. Boston: Little, Brown.

Bromley, D. B. (1978). Approaches to the study of personality change in adult life and old age. In A. D. Isaacs & F. Post (Eds.), *Studies in geriatric psychiatry* (pp. 17–40). Chichester, UK: Wiley.

Bromley, D. B. (1988). *Human aging: An introduction to gerontology* (3rd ed.). Harmondsworth, UK: Penguin.

Burger, E. (1989). *Fransz Liszt: A chronicle of his life in pictures and documents*. Princeton, NJ: Princeton University Press.

Cohen-Solal, A. (1987). *Sartre: A life*. London: Heinemann.

Conan Doyle, A. (1979). The adventure of the Sussex vampire. In *The Sherlock Holmes Omnibus* (pp. 277–287). London: Murrary & Cape. (year pub 1923).

Cotier, J. (1991). *Nudes in Budapest*. London: Aktok.

Coupland, N., Coupland, J., & Giles, H. (1991) *Language, society and the elderly: Discourse, identity and aging*. Oxford, UK: Blackwell.

Covey, H. C. C. (1988). Historical terminology used to represent older people. *The Gerontologist, 28*, 291–297.

Covey, H. C. C. (1991). *Images of older people in Western art and society*. New York: Praeger.

Cowling, M. (1989). *The artist as anthropologist: The representation of type and character in Victorian art*. Cambridge, UK: Cambridge University Press.

Craig, P., & Cadogan, M. (1981). *The lady investigates: Women detectives and spies in fiction*. London: Gollancz.

Datan, N. (1986). The last minority: Humour, old age and marginal identity. In L. Nahemow, K. A. McCluskey-Fawcett, & P. E. McGee (Eds.), *Humour and aging* (pp. 161–171). London: Academic Press.

Denvir, B. (Ed.). (1987). *The impressionists at first hand*. London: Thames & Hudson.

Elias, N. (1985). *The loneliness of the dying*. Oxford, UK: Basil Blackwell.

Featherstone, M. (1982). The body in consumer culture. *Theory, Culture, and Society, 1*(2), 18–33.

Featherstone, M., & Hepworth, M. (1990). Images of aging. In J. Bond & P. Coleman (Eds.), *Aging in society: An introduction to social gerontology* (pp. 250–275). London: Sage.

Gilbert, A. (1987). *The spinster's secret*. London: Pandora. (Original work published 1946)

Glassner, B. (1989). Fitness and the postmodern self. *Journal of Health and Social Behavior, 30,* 180–191.

Goffman, E. (1968). *Asylums: Essays on the social situation of mental patients and other inmates*. Harmondsworth, UK: Penguin.

Goffman, E. (1971). *The presentation of self in everyday life*. Harmondsworth, UK: Penguin.

Gombrich, E. H. (1982). The mask and the face: The presentation of physiognomic likeness in life and art. In *The image and the eye: Further studies in psychology of the pictorial representation* (pp. 105–136). Oxford, UK: Phaidon.

Gould, S. J. (1988). *Time's arrow, time's cycle: Myth and metaphor in the discovery of geological time*. Harmondsworth, UK: Penguin.

Gruman, G. J. (1966). *A history of ideas about the prolongation of life*. Philadelphia: American Philosophical Society.

Gubrium, J. F. (1986). *Oldtimers and Alzheimer's: The descriptive organization of senility*. Greenwich: CT: JAI.

Gullette, M. M. (1997). *Declining to decline: Cultural combat and the politics of midlife*. Charlottesville, VA, and London: University Press of Virginia.

Haber, C. (1983). *Beyond sixty-five: The dilemma of old age in America's past*. London: Cambridge University Press.

Hale, W. H. (1969). *The world of Rodin 1840–1917*. Netherlands: Time-Life International.

Harrison, B. (2002). Seeing health and illness worlds—using visual methodologies in a sociology of health and illness: A methodological review. *Sociology of Health and Illness, 24,* 856–872.

Hart, A. (1985). *The life and times of Miss Jane Marple*. London: Sphere.

Hepworth, M. (1988). *Age conscious? An illustrated look at age prejudice*. Edinburgh, Scotland: Age Concern Scotland.

Hepworth, M. (1991). Positive aging and the mask of age. *Journal of Educational Gerontology, 6*(2), 93–101.

Hepworth, M. (1993). Old age in crime fiction. In J. Johnston & R. Slater (Eds.), *Aging and later life* (pp. 32–37). London: Sage.

Hepworth, M. (2000). *Stories of aging*. Buckingham, UK: Open University Press.

Hepworth, M. (2002). Using "cultural products" in researching images of aging. In A. Jamieson & C. R. Victor (Eds.), *Researching aging and later life* (pp. 80–95). Buckingham, UK: Open University Press.

Hepworth, M. (2003). Framing old age in Victorian painting 1850–1900: A sociological perspective. In J. Hughson & D. Inglis (Eds.), *The sociology of art*. London: Palgrave.

Hepworth, M., & Featherstone, M. (1982). *Surviving middle age*. Oxford, UK: Basil Blackwell.

Hockey, J., & James, A. (1993). *Growing up and growing old: Aging and dependency in the life course*. London: Sage.

Hollander, A. (1988). *Seeing through clothes*. Harmondsworth, UK: Penguin.

Isaac. (1992). *The challenge of geriatric medicine*. Oxford, UK: Oxford Medical.

Jerrome, D. (1992). *Good company: An anthropological study of old people in groups*. Edinburgh, Scotland: Edinburgh University Press.

Kammen, M. (1980). Changing perceptions of the life cycle in American thought and culture. *Proceedings of the Massachusetts Historical Society, 91,* 35–66.

Karp, D. R. (1985). *Ars, medica: Art, medicine and the huma condition*. Philadelphia: Phliadelphia Museum of Art.

Katz, S. (1996). *Disciplining old age: The formation of gerontological knowledge*. Charlottesville and London: University Press of Virginia.

Kirk, H. (1992). Geriatric medicine and the categorization of old age—The historical linkage. *Age and Society, 12,* 483–497.

Kitwood, T. (1990). The dialectics of dementia: With particular reference to Alzheimer's disease. *Aging and Society, 10,* 177–196.

Lively, P. (1987). *Moon tiger*. London: Andre Deutsch.

Lowenthal, D. (1985). *The past is a foreign country*. Cambridge, UK: Cambridge University Press.

Maddow, B. (1977). *Faces: A narrative history of the portrait in photography*. New York: New York Graphic Society and Little, Brown.

Magli, P. (1989). The face and the soul. In M. Fehrer (Ed.), *Fragments for a history of the human body, Part two* (pp. 87–129). New York: Zone.

Mangum, T. (1999). Little women: The aging female character in nineteenth-century British children's literature. In K. Woodward (Ed.), *Figuring age: Women, bodies, generations.* Bloomington and Indianapolis: Indiana University Press.

Marwick, A. (1988). *Beauty in history: Society, politics and personal appearance c. 1500 to the present.* London: Thames & Hudson.

Nuttall, J. (1978). *King Twist: A portrait of Frank Randle.* London: Routledge & Kegan Paul.

Oxford universal dictionary. (1965). Oxford, UK: Clarendon.

Panofsky, E. (1962). *Studies in iconology: Humanistic themes in the art of the renaissance.* New York: Harper Touch.

Pointon, M. (1990). *Naked authority: The body in Western painting 1830–1908.* Cambridge, UK: Cambridge University Press.

Porter, L., & Porter, L. M. (Eds.). (1984). *Aging in literature.* Troy, MI: International Book Publishers.

Ransley, C. (1993, Summer). Sticks and stones . . . *050, 19,* 72–73.

Sabat, S. R., & Harré, R. (1992). The construction and deconstruction of self in Alzheimer's disease. *Aging and Society, 12,* 443–461.

Saunders, G. (1989). *The nude: A new perspective.* London: Herbert.

Sawday, J. (1990). The fate of Marsyas: Dissecting the renaissance body. In L. Gent & N. Llewllyn (Eds.), *Renaissance bodied: The human figure in English culture c. 1540–1660* (pp. 111–135). London: Reacktion.

Schama, S. (1988). *The embarrassment of riches: An interpretation of Dutch culture in the golden age.* London: Fontana.

Scharf, A. (1974). *Art and photography.* Harmondsworth, UK: Penguin.

Sears, E. (1986). *The ages of man: Medieval interpretations of the life cycle.* Princeton, NJ: Princeton University Press.

Shaw, M., & Vanacker, S. (1991). *Reflecting on Miss Marple.* London, Routledge.

Sokoloff, J. (1987). *The margin that remains: A study of aging in literature.* New York: Lang.

Stafford, B. M. (1991). *Body criticism: Imaging the unseen in enlightenment: Art and medicine.* Cambridge, MA: MIT Press.

Stafford, B. M., La Puma, J., & Schiedermayer, D. (1989). One face of beauty, one picture of health: The hidden aesthetic of medical practice. *The Journal of Medicine and Philosophy, 14,* 213–230.

Stephens, J. (1980). *Loners, losers, and lovers: Elderly tenants in a slum hotel.* Seattle: University of Washington Press.

Thompson, P. (1992). "I don't feel old": Subjective aging and the search for meanin in later life. *Aging and Society, 12,* 23–47.

Weisberg, G. P. (1980). *The realist tradition: French painting and drawing 1830–1900.* Bloomington: Cleveland Museum of Art and Indiana University Press.

Wells, B. W. P. (1983). *Body and personality.* London: Longman.

Whicker, A. (1983). *Within Whicker's world.* London: Longman.

White, C. (1984). *Rembrandt.* London: Thames & Hudson.

Williams, R. (1990). *A Protestant legacy: Attitudes to aging and illness among older Aberdonians.* Oxford, UK: Clarendon.

Wood, C. (1988). *Paradise lost: Paintings of English country life and landscape 1850–1914.* London: Barrie & Jenkins.

Woodward, K. (1991). *Aging and its discontents: Freud and other fictions.* Bloomington: Indiana University Press.

Attitudes Towards Aging: Adaptation, Development and Growth Into Later Years

Ann O'Hanlon and Peter Coleman
University of Southampton

INTRODUCTION

Advancements in science, medicine and social services mean that people are living longer than ever before. This century alone has seen an increase of 30 years in the average life expectancy at birth from 47 years in 1900 to 77 years today. Similarly, the number of people in England and Wales aged over 70 years is continuing to rise from just 1 in 36 at the turn of the century, to 1 in 8 in 1991. With these demographic achievements however comes a more urgent need to identify the factors likely to influence health, autonomy and well-being into later years. One such factor is the level of favourability or challenge being associated with the latter part of the life course, that is adults' attitudes to their own aging and future old age.

A better understanding of adults' attitudes to aging can have applied value in adding quality of life and health to increased years. Adults who hold negative attitudes about aging for instance, may be less likely to engage in educational or travel opportunities Adults with more negative age-associated attitudes can also be less motivated to prepare financially, or to engage in more healthy lifestyles including exercise and diet behaviours (O'Hanlon & Coleman, 2002). Adults who are anxious about aging may be less able to attend to relationships with close others; this can be especially serious given the significant and vital role older people can play in the lives of their children (Gutmann, 1987; 1997) and grandchildren (e.g., Lavers & Sonuga-Barke, 1997). Data from longitudinal research also indicates that adults' attitudes towards their own aging and future old age can have a significant effect on later health. Empirical evidence from Coleman, Ivani-Chalian, and Robinson (1998) indicated that the attitudes participants had about their own aging and future old age was the best predictor in the maintenance of self-esteem over a thirteen year period. Similarly, Levy, Slade, Kunkel and Kasl (2002) found that age-associated attitudes were strongly predictive of mortality rates up to 23 years later.

Understanding the ways adults evaluate aging and its associated constraints and challenges can also have theoretical value in improving our understanding of adult development and aging. This is because such evaluations are likely to develop over many decades and become "inherently intertwined with and indistinguishable from what we ordinarily think of development and aging. Therefore, when we examine changes prompted by stress, we may at the same time be observing changes that can also be described as life-course developments" (Pearlin & McKean Skaff, 1996). Theories also allow professionals to explain experiences in logical ways, and develop interventions to solve problems and improve quality of life and health. (See also Bengtson, Rice & Johnson, 1999; Coleman, O'Hanlon, & Ruth, in press; Lynott & Lynott, 1996.) Theories however continually need to be tested and refined, so that we know the contexts and conditions under which they are most valid and predictive of other experiences.

Attitudes have been defined as multi-dimensional constructs reflecting a tendency to evaluate a given entity or experience with a given level of favourability or unfavourability (see Eagly & Chaiken, 1993). Attitudes can be further divided into three distinct but related themes; beliefs, feelings and behaviours. Currently, however, it is not clear whether some attitudinal sub-components are more significant than others, in influencing general attitudes to aging; similarly, further research will need to take place to identify the ways these components manifest themselves, whether these change over time, their possible basis in earlier life experiences, and their consequences on later health and psychosocial functioning.

There is evidence that adults can evaluate their own aging and future old age in negative and anxiety-provoking ways. According to Brandtstädter and colleagues for instance, the prospect of one's own aging and future old age can be a significant and worrying issue for many adults (e.g., see Brandtstädter, Rothermund & Schmitz, 1998; Brandtstädter & Wentura, 1995). Similarly, in a study with a group of nurses, Bernard (1998) found that many of the nurses viewed their own aging with 'trepidation', particularly changes associated with identity, appearance, and choice into their later years. The following quote by the gerontologist Rabbitt (1999) provides further evidence that later life can be evaluated in very negative ways; "early recognition that aging is not at all a benevolent process compelled me to realise that a determination to stress any possible (positive) aspect of this condition . . . is a betrayal of science and of responsibility. . . All of us who are lucky must put up with aging as best we can, but to do anything useful about it we must recognise and understand all of the extraordinary unpleasant things that it does to us and the ways in which these contract the scope of our lives' (pp. 180–181).

Surprisingly, however, despite the potential significance of research examining the subjective experience of aging and old age, a number of researchers have noted the paucity of research in this area (e.g., Biggs, 1993; Coleman 1993a; de Beauvoir 1972; Quirouette & Pushkar, 1999). Currently we know little about the attitudes or knowledge adults hold about the latter part of their lives, the basis for evaluating later life in given ways, or the mechanisms through which attitudes can influence later health and development. This paucity of research is especially acute in the context of British gerontology. For instance according to Treharne (1990) "little research is published in this country about the social and psychological issues faced by the elderly" (780). Similarly, the sociologist Thompson and colleagues (Thompson, Itzin, & Abendstern, 1990) conclude that their book on the experience of aging stands 'strangely alone' (p. 26). Perhaps one of the reasons for this paucity of research might be because researchers and gerontologists themselves view old age in threatening and aversive ways. For instance, according to Bernard (1998) 'if we indeed are so afraid to ask it of ourselves (old age), then perhaps this helps explain why we have so little research on this issue' (p. 635).

The chapter to follow addresses this gap in current literature to ask two main questions; what level of challenge does later life hold for us, and how do adults evaluate and experience those age-associated challenges. In discussing later life a mixed image will emerge. It will be suggested that many challenges can occur in the latter part of the lifecourse, but that this time of life also has much potential for continued development and growth. The attitudes adults have about their own aging have significant effects for the ways later life is experienced and managed.

PROBLEMS AND CHALLENGES

Later life is viewed as occurring from about age 70 to 75 years and upwards; further distinctions are sometimes made between the 'young old' and the 'old-old' i.e., adults in the age period of about 70–84 and adults aged about 85 years and over. Chronological age is an important social and personal marker of identity, behaviour, expectations, experiences, and preferences. Adults in their seventies for instance often score lower than their younger counterparts on measures of negative affect including worry and anxiety (e.g., see Powers, Wisocki, & Whitbourne, 1992); older adults can also score more highly than younger adults on measures of agreeableness and consciousness (see McCrae et al., 1999). Similarly, adults in mid- and later life are more likely to be generative than their younger counterparts; in fact although younger people can be altruistic they cannot be generative as it is only with experience and time that this vital role of teaching, guiding and supporting others can occur (e.g., see Erikson, 1950; McAdams et al., 1998).

Nevertheless, although often a useful starting point, definitions of old age based solely or mainly in terms of chronological years can be problematic given the increasing heterogeneity that can occur between people. For instance, an individual could be aged 85 chronologically, but have the energy, physique, or health of someone aged about 60. Similarly, Thompson et al. (1990) noted that while some respondents were looking forward to visits from the researchers and reluctant to see them depart, others were so busy that it was difficult for the researchers to book time with participants for the interviews to take place. Definitions of old age based solely on chronology may also be less than helpful given the growing recognition by many researchers that there can occur a blurring of the life-course (Biggs, 1993; 2000; Kaufman, 1986). Older adults, for instance, are not necessarily a unique group of adults qualitatively different from other age groups; this is because many of the issues often associated with later life are issues pertinent at any age throughout the life-course, for example, the need to maintain a balance between autonomy and dependence, and the need for secure, warm and accepting relationships with others. Furthermore, although old age is noted as a time for increasing losses and constraints on development, losses and challenges occur throughout the life-course and not just in later life. The section to follow will consider the level of challenged posed by physical, social, and psychological aspects of later years.

Physical Health Into Later Years

Biologically, or physically, later life has been associated with multiple aversive and un-favourable experiences, decline and loss which puts the self under stress. Bromley (1988) for instance, defines aging 'as a complex, cumulative, time-related process of psychobiological deterioration occupying the post-developmental (adult) phase of life' (p. 30). Similarly, images of aging and later life on greeting cards and on books can also portray later life in very negative ways (e.g., de Beauvoir, 1972). There is evidence to support such stereotypes. In explaining why we age, Cristofalo et al. (1999) discuss stochastic theories of aging within which 'insults' from the environment eventually reach a level incompatible with life. As an example, radiation can lead to cell mutation that in turn can mean that somatic cells eventually fail and so the life-course is reduced. In contrast, developmental programmed theories of senescence assume that aging is a continuation of development. Within this framework and towards the latter part of the life-course, problems can occur such as telemetric shortening on chromosomes, or damage to proteins caused by free radicals which in turn can cause vision for instance to deteriorate.

Researchers however have argued for an important distinction to be made between normal, pathological, and optimal or successful aging (Baltes & Baltes, 1990; Rowe & Kahn, 1987). Normal aging refers to the now common human expectation of reaching 70 or 80 years of age without experiencing any significant physical or mental health problems. In contrast, optimal aging refers to the best health possible into later life, in other words, the experience of old age where health, energy, and fitness are better than that found within the normative range. Finally, pathological aging refers to a process of aging where there is clear evidence of physical or mental deterioration. These concepts of normal, pathological, or successful aging offer very different pictures of later life. On one hand, for instance, an optimal view of old age assumes that adults can live healthy, independent, and active lives into advanced old age. The alternate model is that of a pathological old age; within this model adults are highly dependent on others for help in meeting physical, social, and emotional needs.

Researchers working within the context of optimal or successful old age are looking for ways to off-set problems or challenges to the physical self and to find ways to help adults function positively and optimally into advanced old age (Baltes, 1987; Baltes & Baltes, 1990). An optimal outcome for the physical self into later life is one in which health and vigor remains high and ill-health is compressed into a very short period at the end of the life-cycle. Good physical health is one of a number of key components for a successful or optimal old age (Baltes, 1987; Baltes & Baltes, 1990; Rowe & Kahn, 1987). For instance, Valiant (1991) carried out a study, within which good physical health, defined in terms of longevity and biological health, was the outcome measure of successful old age. In contrast to Hall's hill metaphor of aging, this model of aging with a gradual assent to young adulthood and then a plateau to advanced old age without any protracted biological decline or physical health problems is also known as the square wave trajectory model of aging (Eisdorfer, 1983). Within this model, the goal of many health professionals is to compress morbidity into the very end of the life-cycle, achievable for instance through the use of a more healthy and active life-styles. Kirkwood (1999) also argues that the period of ill health or disability prior to the point of death is continually getting much shorter as the life-span increases. This is reason for optimism.

Pathology, when it occurs in later life, can be related to a wide range of factors not intrinsic to the process of aging. The occurrence of ill-health for instance may be related to factors such as treatment availability, geographical location and the availability of funding for healthcare. Diet, social support, and physical activity can also influence health into later life (Morgan, Dallosso, & Bassey et al., 1991). Other factors that affect health and well-being in later life include marital status, gender, knowledge, or attitudes about aging. These factors challenge negative stereotypes of aging and later life which assume that later life is intrinsically a time of decline, loss, and deterioration.

Nevertheless, old age can hold some serious challenges for the physical self that can impact adversely on health and psychosocial well-being. Specifically, although advancements in science and medicine have increased the life-span, this has only extended what is genetically pre-determined, that is, as noted by Fries (1983) few people have lived beyond the age of 85 years irrespective of the quantity of vitamins or exercise taken. Consequently, the potential threat and the sad reality is that physical health will decline in advanced old age, and that ultimately, all human beings will die.

A number of postmodern researchers have sought to challenge ageist stereotypes by arguing that general consumerism within a postmodern world offers new opportunities for a reinvention of the physical self in later life. For instance, Featherstone and Hepworth (1989) argue that the rejection of negative and fatalistic images of later life within postmodernism offers the potential for viewing older adults as consumers of lifestyles more typical of younger adults. Specifically, basic requirements can be purchased enabling many older adults not only to wield power and influence in society, but also to live more active and productive lives for much longer. In addition, the growing use of the world wide web through telephones and televisions means that many more older adults have the opportunity to create identities for themselves in cyberspace or virtual worlds where changes in physical appearance, increasing physical frailty or even immobility may have less relevance; within this environment, the expertise and strengths of older adults can also be recognised and utilised. However, many people neither use nor enjoy technology; also with cohorts of adults becoming increasingly heterogeneous into the latter part of the life course, many people are likely to be without the knowledge or economic resources to benefit from possible consumerist advantages.

To summarise, in considering the level of challenge later life poses for the physical self, evidence is mixed. On one hand, there are many reasons to be optimistic. Although some adults need more support and help into later years, many others function well and have a good quality of life. Similarly, evidence that health is related to a wide range of factors not intrinsic to the process of aging (e.g., diet, social support) means that adults now have increasing control over their own health and experiences into later years. Nevertheless, later life can hold many challenges. Chronological age remains the best predictor of a range of health problems in later life and ultimately all human beings will die. Although technology offers new opportunities for many people, these opportunities are not available to everyone, nor do they alter unwanted changes in appearance or health. Surprisingly however, we have little information about the ways adults evaluate physical changes associated with the latter part of the life cycle; nor is it as yet known what effect such evaluations have on other aspects of health and well-being.

Social Relationships and Role Changes

As social beings, most adults need to be in close relationships with other people, for enjoyment, for meaning and purpose in life, and also as a means of learning more about ourselves and the world within which we live. Warm and close relationships with other people have also been described as being a healthy necessity from 'the cradle to the grave' (Bowlby, 1979: p. 129). Relationships, are very important for our health and well-being particularly reciprocal relationships which also serve a protective function such as relationships with a spouse or equivalent figure (see Ryff, 1991). In addition, Crittenden (1997) notes it is often within close relationships that we offer and receive support, particularly at times when the risk of threat to the actual or representational self is higher, for example when taking on new projects or goals. Relationships with close others can also be a significant source of pleasure, enjoyment, and meaning for adults.

Most older adults need and enjoy relationships with younger generations; so too do communities need relationships with older adults. This point is made strongly by Gutmann (1987) who draws on ethnographic, cross-cultural, and anthropological data to argue that as a consequence of maturation and experience older adults have their own unique strengths and talents that can and should be used in the social context, and particularly in helping, supporting, and teaching the next generation. Gutmann takes this perspective, particularly in light of the 'parental emergency' (p. 7), in other words, the difficulties and problems involved in raising emotionally healthy children without the support of the extended family and friends. In this way, Gutmann (1987) echoes Erikson's notion of generativity (Erikson, Erikson, & Kivnick, 1986), to argue that even when adults are less physically fit, older people still have a vital and necessary role to play in society, particularly in taking on the role of 'emeritus parents' (p. 214). That these generative opportunities are encouraged and provided, is likely to be important not only for the health and well-being of individuals but also Gutmann argues, for the very stability and development of society and all its members.

In later life, however, relationships with other people can be lost, constrained or the quality of relationships impaired. In later life, for instance, adult children may divorce and separate so that older people may lose generative links with younger family members (e.g., Drew & Smith, 1999). Similarly, into the latter half of the life cycle adults are more likely to experience the loss of parents, spouses, and/or the loss of other close attachment relationships through bereavements. Also, personal work roles that the individual had and enjoyed may no longer be salient or relevant such as when children are grown up and leave home. Furthermore, given compulsory retirement, the friends and acquaintances that one had through the work role may have to be surrendered.

Although adults can experience significant losses and constraints in their relationships with others, these can be a consequence of the interactions and expectations of others, rather than being intrinsic to the aging process. Society for instance may not always be supportive of the needs of its older members. For instance, society can impose constraints on relationships into later life in terms of expectations for relinquishing the work role or even about the aptness of certain sexual relationships. This is particularly the case given the potential for ageism within which older adults can be disadvantaged with subsequent loss of health or even loss of life itself, for example, when older adults are denied health treatment or counseling because of the negative views health professionals can have of older people.

Evidence examining the ways in which older adults are viewed by society is mixed. Some researchers have found that many older people are viewed in very negative ways (e.g., Bytheway & Johnson, 1990; Cohen, 1996; Kogan, 1961; Palmore, 1981). In contrast, other researchers have found that the results of studies on attitudes to older adults are influenced by the specific attributes being assessed. Slotterback and Saarnio (1996) for instance asked a group of undergraduates (aged 17–24) to rate their attitudes towards young, mid-life, and older adults across cognitive (intellectual abilities or information processing), personal-expressive (personality or interpersonal attributes), and physical attributes relating to physical health, behaviour, or movement. Results indicated that attitudes towards the physical attributes of older adults were more negative by comparison with young and mid-life adults. In contrast, no significant difference was found for personal-expressive attributes, but there was a main effect for age on cognitive attributes. A post hoc test indicated that the attitudes of participants towards midlife adults were significantly more positive than were those for younger adults (mean -0.48) or for older adults (mean $-.027$). Although these insights are useful, this study was carried out using undergraduate students as participants, thus making generalisability difficult. Researchers in this field should attempt to replicate the above study with other participant groups.

Some social losses and threats that assume crises in mid- and later life (e.g., 'empty nest syndrome') have little empirical evidence to support them (Hunter & Sundel, 1989). A contrasting view is that there are challenges associated with each age period around which adults must adapt. This contrasting view does not remove the reality of social stressors, but instead, recognises that these occur throughout the life course, and not just old age.

Studies have shown that the quality of social networks of older people have many similarities to those of younger age groups, but that the number of contacts within the networks are fewer for the old. Socioemotional selectivity theory (Carstensen, 1991) attempts to describe the functional declines in social contact throughout adulthood. This theory describes the practical aspects of social interactions to include information acquisition, identity maintenance, and emotion regulation. The essential premise of the theory is that the relative important of those goals changes as a function of perceived time, and that these goals influence and explain declines in social contact across adulthood. When time is perceived as being largely open-ended, future-orientated goals such a information acquisition are of paramount importance. However, when time is perceived as being limited, present-orientated goals such as emotional goals, are rated as being most important. Consequently, adults into later years are seen as actively preferring certain types of social contact (e.g., emotionally satisfying contact) over others such as information gathering.

There are substantial differences in adults' preferences for social activities and social contacts, however, and some of these differences can be explained by gender, ethnicity, and physical health. It is well known that women typically have more social contact than do men, especially more intimate friends or confidants. Confidants are typically women, in part given the difference in mean ages between the sexes. Age differences between the sexes in western countries is around five years, but in some countries the difference is larger, such as Finland, where the difference is eight years. This means that older women often live alone, whereas older men are married. It also means that women are more likely to be caregivers by comparison with their male counterparts. Health is another significant factor explaining reductions in social activity in later years. The patterns of not initiating new contacts with non-familial persons is clearly visible in nursing homes and other institutions. There can be a significant risk involved in contact seeking in old age; conversations can become difficult for instance if others have sensory difficulties, particularly hearing loss.

To summarise, relationships in later years can be an intense source of both pleasure and distress. Relationships with others can be a significant source of enjoyment and meaning in life. Relationships with others are also crucial not only for the well-being of individuals but also of whole communities given that society is often dependent on the experience and expertise that older adults develop over many decades of life (e.g., see Bellah et al., 1991). However, in later life as throughout life, the relationships and social roles that adults have with others can be compromised, constrained, or even lost. These losses can arise through bereavement or because society does not always provide its older members with adequate support and resources. Understanding the ways adults evaluate and regulate changes in relationships is likely to be crucial if health professionals are to help facilitate better health and well-being for more people for longer into the latter part of the life cycle. Future research in this area should be carried out to refine and clarify the nature of relationships across adulthood, and the impact that social motives and other factors might play in influencing the quality and nature of relationships. Future research is also needed to explore and examine the positive qualities within close family relationships, particularly between parents and their adult children, and siblings who have shared experiences over many decades of life.

Psychological Functioning Into Later Life

The level of challenge for the psychological self into later life is now considered with particular emphasis on intelligence, memory, and psychosocial functioning. A mixed picture again emerges; although many negative stereotypes need to be revised, there are challenges into later years that necessitate adaptation.

Intelligence

Images of cognitive decline and loss into later life are widespread and may pose a serious challenge to the self. There are two main models of intellectual functioning into later life. The first stability model posits that intellectual functioning remains relatively stable throughout the life course and that within the normative range, any changes in functioning will be slight. In contrast, the change model of intellectual functioning assumes that into old age and particularly into advanced old age, one can expect significant declines in functioning. This latter change model can be further separated into the information processing tradition and the psychometric tradition. In cognitive psychology, the information processing approach suggests that limitations in functioning with age occurs as a consequence of increasing resource limitations such as reduced attentional capacity and working memory capacity. These in turn are hypothesised to impair encoding and retrieval strategies. In contrast, the emphasis within the psychometric tradition, is on examining changes in intelligence with age. The practice of psychometric testing has flourished given the use of more sophisticated statistical tests and the practical applications of such measures across a range of health, occupational, and clinical settings.

However, in contrast to early views of intelligence as being a unitary 'g' factor, recent researchers have made distinctions between fluid and crystallised intelligence, the latter of which is hypothesised to continue to grow and develop into advanced old age. Fluid intelligence is taken to reflect the neurological mechanics or structures of the mind and to include factors such as perceptual speed and reasoning. In contrast, crystallised intelligence refers to culture based pragmatics such as knowledge, comprehension, and verbal fluency. Crystallised intelligence within this perspective is hypothesised to increase into advanced old age and also to lead to the accumulation of expert skills and attributes such as wisdom. Understanding further the nature of fluid and crystallised intelligence was one aspect of research reported by Lindenberger and Baltes (1997). Specifically, as part of the multidisciplinary Berlin Aging Study, they sought to explore the relationship of fluid and crystallised intelligence to socio-biographical (e.g., SES) and biological or sensory-motor variables such as balance-gait, hearing, and vision. Results from this study with a group of adults aged 70–103 (n = 516) indicated a negative age relationship for both fluid and crystallised intelligence although the effect was stronger for the fluid measures of intelligence (r = −.59 to −.49).

Despite the advantages of the fluid/crystallised component model of intellectual functioning, it is possible that this dual model of intellectual functioning is not sufficiently complex or detailed to adequately describe the processes of intellectual functioning into later life. Specifically, people function intellectually within a wide range of areas or domains (e.g., music, art, or bridge), yet experience or expertise within each of these domains follows different neural pathways. Traditionally, there has also been a greater emphasis on problems in intellectual functioning, rather than strengths with age and specifically, with the possibility that adults into later life may be functioning in qualitatively different and superior ways by comparison with younger adults within certain domains or areas. As an example, Gisela Labouvie-Vief (Labouvie-Vief et al., 1995) discussed positive development into later life as a moving away from categorical thinking and towards a much more complex understanding of the world which includes ambiguity and uncertainty.

There are other potential problems and difficulties with the psychometric approach to intellectual functioning. Specifically, there are a range of factors not intrinsic to the process of aging, which can influence the performance of older adults on these tests. For instance, vagueness or inactivity, particularly in advanced old age can be a consequence of boredom or even the use of sedatives. Even at less extreme levels, the expectations people have about the nature of old age and about their own abilities and strengths are likely to influence the ways in which adults function on psychometric tests, in other words, researchers may need to offer more encouragement to some older adults to gain more accurate data on their cognitive functioning. In contrast, younger adults may be very motivated to outperform older adults in ways congruent with their expectations about intellectual performance in young and late adulthood. Physical health problems such as poor hearing or poor eyesight will also influence performance on a range of intelligence tests into later life and these variables need to be controlled.

Nevertheless, some adults in later years do experience declines in cognitive functioning that can necessitate reliance on others. Intellectual declines may be a challenge given their importance and centrality to everyday functioning. For instance, as noted by Teri and colleagues (1997), intellectual functioning is essential given that 'every action of every day is mediated and influenced by our cognitive abilities. From the most sophisticated human activity to the most mundane, everything depends on our capacity to understand and interact with our environment' (p. 269). If there are any declines intellectually, then adults are at greater risk of threats from a range of sources or people in the environment. Furthermore, any significant impairment intellectually, is likely to be a source of inconvenience and threat given the associated high risk of dependence on others for help and even protection.

To summarise, although images of decline in intellectual functioning are widespread, people can and do continue to function well intellectually into later life. Intellectual functioning will also be influenced by a range of other variables including interest in the topic under investigation, perceived self-efficacy, and self-esteem. However, declines in some areas of intellectual functioning can occur and be a significant challenge needing support and adaptation for many people.

Memory Functioning

As noted by Teri et al. (1997) memory is a complex process incorporating abilities to learn new information (recent memory), remember future events (prospective memory), perform familiar activities (procedural memory) and remember past experiences (autobiographical memory). There are a number of theories which suggest that adults can have problems with memory functioning into later life. In reviewing theories of memory impairment in later life for instance, Light (1991) states that theories of memory functioning can range from models which are fairly positive through to models which are fairly negative. In the most positive models of memory functioning with age, slowing of memory is hypothesised to occur as a consequence of inefficient strategies of encoding and retrieval. Although this model still assumes that deficits in memory functioning do occur in later life, this model is more positive in that deficits are assumed to be off-set by mnemonic training and other intervention strategies. In contrast, much more pessimistic models of memory functioning in later life suggest that memory declines with age occur as a consequence of normative but irreversible and unavoidable changes in the mechanics of the mind. Theories by which this might occur include limitations in processing resources such as reductions in attentional capacity or working memory capacity.

Although these models assume that with age comes decline in memory functioning, it is possible that lower levels of performance can be explained in other ways. For instance, the content of the topic or list of items to be memorised may be of little interest to older participants. Consequently, boredom or lack of interest might mean that older adults on these tests are using this time to think about more interesting or salient experiences or events such as activities with other family members or the finite nature of time. In addition, however, it is important to note that younger adults also fail to remember on occasions. Unlike with older adults, however, failure to remember certain events or experiences by younger adults is unlikely to be viewed in pathological ways.

Slowing down the rate of information recall could also reflect qualitative differences or changes in priorities, an increased awareness of the complexities in the material being presented, or both. As already discussed, there is a paucity of research exploring the positive attributes or qualities that can occur into later life. However, in research by Adams et al. (1990) a group of younger (aged 18–22) and older adults (age 55–81) were given either a fable or non-fable story presented in both written and oral formats. When participants were asked to repeat the story, results indicated that older adults exhibited 'a more integrative or interpretative style that did the younger adults, whose style was primarily detailed and text based' (p. 24). Consequently the exciting possibility remains that in the latter part of the life-course, adults are able to think in wider and more complex ways by comparison with their younger counterparts.

Researchers need to recognise the limited generalisability and cohort effects of research findings with current generations of older adults. Specifically, older adults today have been through many unique experiences hopefully unknown to future generations of older adults, such as world wars, extreme poverty, and lower levels of education, nutrition, and health. Similarly, the advent of mass access to information via the world wide web and other mechanisms is also likely to raise the expectations of younger generations towards old age; this may include lower levels of regarding limitations in health, social, or financial resources. Combined, these factors mean that old age for future generations may be a different experience to that of current generations of older adults.

To summarise, there are multiple theories of memory functioning into later life which suggest that with age does come memory loss. There is evidence to support these models. However the consequence of age on the different mechanisms or neural pathways involved in these different memory systems is unclear. In total, 'despite the phenomenological and empirical reality of age-related memory loss and the breadth of attempts to explain it, much work remains be done to understand why it occurs' (Luszcz & Bryan, 1999; p. 3). The fact remains that any actual or potential impairment in memory functioning with age is likely to be a significant source of threat for some adults. Only by carrying out empirical research will it be possible to identify in more detail the nature, duration, intensity, and basis of this and other age-associated challenges.

Psychosocial Functioning

There are many changes and challenges associated with physical and social aspects of later life which impose 'considerable strain on the individual's construction of self and personal continuity' (Brandtstädter & Greve, 1994: p. 52). Potential age-associated physical and social challenges have already been discussed; although there has been little systematic work in this field, additional psychological challenges possible into later life can include anxiety about the finite nature of time, identity changes and / or personal conflicts including reconciliation of past choices and relationships.

Although serious problems and challenges can occur into later life as throughout life, older adults are typically highly competent and creative in managing age-associated problems and constraints (e.g., see Freund & Baltes, 1998; Wrosch et al., 2000). Several of these coping strategies will be discussed in the section to follow; these strategies offer many useful insights into the experience of aging and later life, however it is very important that much more research gets carried out in this area. Possible questions to be addressed include greater detail about each coping strategies, their origins and development over time, the mechanisms underpinning their use, and their consequences on later psychosocial functioning, for example, health and / or relationships. Understanding the skills and strategies older adults use to regulate challenges associated with aging and their own prospective old age is information which can be used to help adults who are adapting less well or to younger generations facing other psychosocial challenges.

It is possible that positive development and growth can arise from successful management and adaptation around age-associated challenges and problems. Into later life, for instance, Erikson and colleagues (Erikson et al., 1986) describe adults struggling 'to accept the inalterability of the past and the unknowability of the future (with the realisation) . . . for the first time . . . that death may come sooner rather than later' (Erikson et al., 1986; p. 56). Addressing these conflicts and threats can lead to further genuine psychological development and growth and particularly towards positive attributes often associated with the latter part of the life cycle such as wisdom, integrity, and authenticity (see later section).

To summarise, adults into later years can face many psychological challenges and problems that can be an intense source of stress: these include cognitive and psychosocial challenges. However, as discussed in the following section, older adults are typically very creative and active in managing constraints and losses when these occur. In addition, the strategies used to adapt to challenge and constraints can also be related to the occurrence of further development and growth into the latter part of the life course.

COPING AND ADAPTATION

The ways adults experience and perceive problems associated with own future old age will be influenced by the strategies being used to manage or regulate those problems.

Selective Optimisation With Compensation

In their model of selection, optimisation, and compensation, Baltes and colleagues outline a general theory for adapting to constraints, losses and potential threats into later life and also for creating more favourable outcomes for the self (e.g., see Baltes, 1987; Baltes & Baltes, 1990; Freund & Baltes, 1998; Marsiske et al., 1995).

According to Baltes and colleagues, selection involves individuals focusing attention on the experiences or activities which are most important to them or which give them the greatest enjoyment and pleasure. Consequently, an individual with a broad range of interests and commitments might select only the most rewarding ones that can be performed without great effort when restrictions occur. This process of selection implies a continual narrowing in the range of alternative options open to the individual. Although this narrowing of options may occur throughout the life-course including childhood, it is likely to be more salient into old age given the pressure of increasing constraints on the self as a consequence of decreases in biology, plasticity, time, and the effectiveness of the culture in supporting older adults (Baltes, 1987). Nevertheless, by reducing activities to high-efficacy domains, preferred activities can be as enjoyable as they were before being reduced in number (Baltes & Carstensen, 1996). The process of selection is initiated by the anticipation of change and restriction in functioning brought by age when limited resources are concentrated onto specific areas viewed of most importance to the individual. As an example, a performer such as a singer or a musician may find it fruitful to select a more limited repertoire, performing only those pieces that were always performed well. Pianist Arthur Rubinstein has described how he actively selected amongst the piano repertoire that he performed in his older days and how he at that time abstained from performing very tricky pieces (Baltes & Baltes, 1990).

In contrast, optimisation is linked to behavioural plasticity and the ability of the individual to modify the environment both to create more desired outcomes for the self and also to meet the continual challenges and changes being experienced in daily life. As noted by Marsiske et al. (1995), examples of optimising outcomes can be understood at an age graded level (e.g., maturation and the accumulation of experience) or at a history graded level (e.g., improvements in health care and education). Examples of optimisation strategies can also be understood at physical, psychological and social levels. An example within the physical sector would be a person who is overweight and whose health therefore is in danger. Optimisation in this case would be adherence to a diet or the engagement in exercise routines. In contrast, an example from the social field would be an aging individual who has considerable difficulties in maintaining functional autonomy in everyday living activities. Optimisation and functional autonomy would mean asking for more help from the spouse, getting home help care from a providing agent, or, if living in an residential or care home, forming an alliance with the care staff and delegating the performance of household activities to them (see M. Baltes, 1996).

When some capacities are reduced and lost in old age the third principle compensation will be used to aid adaptation. The principle of compensation involves using alternate means of reaching a goal making increased use of the 'tricks of the trade' to keep performance at desired levels. The strategy of compensation reflects the need for adults to respond to recognised constraints or losses by taking counter steps so that any potential impairment is lessened. Examples of compensatory mechanisms include the use of hearing aids, glasses, or walking sticks. Similarly, a compensatory mechanism for the aging pianist Rubinstein was to slow down his performance prior to a fast passage, to give the impression the fast passage was being played fast than was actually the case; in this way Rubinstein can continue to perform at top level despite slowness brought on by age.

Heckhausen and Schultz have developed the above theory further with their model of optimisation of primary and secondary control through the lifecourse. Primary control strategies include strategies which enable the individual to overcome obstacles and attain their chosen goals. Examples of primary control include persistence, strategies of selective optimisation with compensation, and tenacious goal pursuit. In contrast, secondary control strategies are hypothesised to include strategies the individual uses to modify their goals or expectations in the face of obstacles and constraints. Secondary control strategies are believed to include positive reappraisals, lowered goals, and strategies of accommodation. (See also Heckhausen, 1997).

Although Marsiske et al. (1995) note that there has been little empirical research with this model, this situation is slowly changing as this model gets tested and used in a range of contexts and with different age groups including young adults (e.g., see Lerner, Freund, De Stefanis, & Habermas, 2001). Researchers in this field need to test this model further and attempt to explain how these strategies emerge and why some people may be more successful than others. For instance, although goals may need to be reduced with age, the mechanisms involved or conditions needed to achieve these are not yet sufficiently well understood, in other words, given the possibility that these strategies are effective in managing threats into later life it is not yet clear why some people might be more successful than others in using these strategies. Scales for these strategies have been developed by Baltes and colleagues which should facilitate further work in this field (see Baltes & Baltes, 1990).

Assimilation and Accommodation

In their largely theoretical work, Brandtstädter et al. (1993; 1998) have attempted to explain the ways in which developmental losses or self-discrepancies with age can be reduced; they have come up with two interrelated processes: assimilation and accommodation.

Assimilative coping refers to strategies where the individual actively attempts to change the environment in ways congruent with their own goals and expectations. Strategies of assimilation can include behavioural changes in eating habits or the level of exercise taken. According to Brandtstädter and colleagues, other sub-patterns of assimilation include the processes of selection, optimisation with compensation (Baltes & Baltes, 1990), and Carstensen's (1993) socio-emotional selectivity theory. Brandtstädter and colleagues argue that these strategies function to enable the individual to engage in their preferred activities at a high level. In addition, these strategies are hypothesised to enable the individual to 'realize, maintain, and stabilize established self-definitions' (Brandtstädter et al., 1997; p. 108).

However, when challenges or losses become too demanding and too difficult to maintain, Brandtstädter and colleagues argue that it may be necessary for the individual to change the strategy towards processes of accommodation. Accommodative coping refers to strategies of readjusting goals or aspirations downwards in the context of constraints and limitations in the environment or the self such as physical ill health or reductions in mobility. Examples of accommodative strategies can include a reappraisal of experiences or the attribution of positive meaning to other goals or experiences, and making self-enhancing comparisons (Brandtstädter & Greve, 1994).

Empirically, research exploring these strategies is still scarce. However, Brandtstädter and colleagues have developed two scales to test the above two processes; the Tenacious Goal Pursuit (TCP) and Flexible Goal Adjustment (FGA) scales. In cross sectional pooled research with nearly 4000 participants, Brandtstädter and Greve report a linear relationship with age for both the TCP (r = .19, p< .001) and FGA (r = −.22, p< .001). Specifically, these results indicate that with age adults are increasingly likely to engage in accommodative processes. In contrast, younger adults are more likely to engage in strategies of assimilation. In addition, both scales were positive correlated with measures of life-satisfaction, optimisation, and the absence of depression (Brandtstädter & Greve, 1994).

The theory just described of assimilation and accommodation is useful and does make important contributions to our understanding of the nature of later life. This is especially the case around threats, constraints, and limitations experienced in later years. The above theory, like that of selective optimisation and compensation, assumes that individuals play an active role in their own development and experiences. Despite its strengths, the above model also has some limitations, not least in explaining the mechanisms involved in the above strategies. Brandtstädter et al. (1998), for instance, argue that one of the key factors in the development of strategies of assimilation and accommodation necessitates flexibility in adjusting goals and expectations in light the context the individual finds him / herself. However, researchers need to be more precise about the mechanisms involved and why some people might be more successful in using these strategies than others.

A further potential problem with both the above models is that they do not consider the role of false or deceptive information, for instance, the self is a complex organism which can distort experiences. Individuals can adapt to adversity even to the point of distorting and denying reality. As such, any understanding of the ways in which adults view their own prospective old age must take place with consideration these kinds of discrepancies and distortions. Given that adults can overtly behave in ways that are not necessarily reflecting the reality underpinning their emotions (e.g., Crittenden & Clausen, 2000), the distinction between appearance and reality is critical. Researchers must look beyond what is said and towards the discourse being used which is much less amenable to conscious control.

Much more research needs to be undertaken examining and exploring this strategy, and within a range of contexts, including with adults from different SES groups. Future research may also wish to consider the relationship in more detail between strategies of assimilation and accommodation and other strategies of adaptation, including selective opitmisaiton with compensation (see Baltes & Baltes, 1990) or primary and secondary control (Heckhausen, 1997; Wrosch et al., 2000). In addition, we need more research on strategies of adaptation with adults in advanced old age. Similarly, there is little research exploring or examining the mechanisms underlying the above strategies, for example, we need to know how adults accommodate when faced with problems or challenges and why this might be somewhat easier for some people and not others.

Adaptation in Relationships

Within the context of relationships, there are a range of strategies adults can use to adapt to age-associated challenges; these can include selective social contact and selective social comparison. By making downward social comparisons with those who appear to be worse off, adults can feel much better about their own experiences and lives. (See also Heckhausen & Krueger, 1993; Heidrich & Ryff, 1993; Taylor & Lobel, 1989). In addition to their practical value, Crittenden (1997) noted that relationships can also aid adaptation when risks to the representational self may be higher as when taking on new projects or goals. Actual or remembered relationships with parents or with adult children can serve similar functions, including into advanced old age.

It has already been noted that adults in later life often prefer the company and support of close family and friends. According to Carstensen (1991), withdrawal from other types of relationships represents an adaptive response to growing awareness or even worry about the finite nature of time. A central principle of Carstensen's selectivity theory is our cultivation of social contacts into the latter part of the life cycle to minimise distress and maximise positive states: 'older people appear to be emotionally conscious, making judicious decisions about activities and giving thoughtful consideration to their functions as affect regulators (Carstensen, 1991). Actively choosing social partners reflects our adaptive efforts to optimise the positive outcome of our social interaction in old age.

Distancing is a related strategy of adaptation which can be used in the context of close relationships. For instance, in exploring age-associated attitudes including adults' fears and hopes for the future, researchers have noted that even adults in advanced old age (e.g., late 80s and above) often distance themselves from other older adults or the experience of being old; later life can often be seen instead as a state more pertinent to other people rather than the self (e.g., see Thompson et al., 1990; Williams, 1990). Researchers have also found that older adults do not typically think of a future that is problematic or distressing; instead they can focus attention on the present, and on maintaining current positive levels of functioning (e.g., see Dittmann-Kohli, 1990). Differences in design and measures means that it is difficult to make comparisons between participant groups. Nevertheless, the point of interest is that these coping strategies assume in different ways that later life has problems and challenges which need to be managed in order to maintain a sense of control, predictability, and safety.

A growing volume of work is being done on adaptation within the context of close relationships. Patricia Crittenden for instance, a past student of Mary Ainsworth, has developed and extended work on close relationships to consider patterns of attachment in terms of their adaptive value (e.g., see Crittenden, 1995; 1997; 2002). Within Crittenden's dynamic maturational model of attachment, patterns of attachment reflect different strategies of adaptation to any situation where the self feels or experiences danger or threat; for some people this may include challenges being associated with their own aging and own future old age. Drawing on the work of Ainsworth and Bowlby, Crittenden has developed a range of measures to assess patterns of attachment from infancy to old age including the Adult Attachment Interview (AAI, see Crittenden, 2002).

Within Crittenden's dynamic maturational framework (e.g., see Crittenden, 1995; 1997; 2002), relationships using the AAI can be classified into one of three 'A', 'B', or 'C' patterns of attachment, with additional subcategories for greater refinement. For instance when people are anxious or distressed and others reliably offer support, such adults learn to respond to challenges by using and trusting both cognitive and affective sources of information (Type B pattern of attachment). However, when the display of negative affect leads to prompt and predictable but unpleasant consequences (e.g., anger, mockery) the

Type A response strategy to these reinforced patterns is to inhibit or dismiss negative feelings and rely more heavily on procedural or semantic sources of information such as verbal generalisations and distorted memory (e.g., idealisation, the stereotypic 'British stiff upper lip'). In contrast, adults classified as Type C adults have had to learn to deal with ambiguity; when under pressure or in need of help such adults find others to be supportive sometimes, in ways that are only sometimes comforting. Given this unpredictability in others' responses, it is not possible for Type C adults to adapt around cognition as do Type A adults; instead the strategy becomes one of adaptation around own their own affective state, particularly feelings of anger, fear, and the desire for comfort. Within the AAI, this strategy can be seen in the use of distorted affect to guide later behaviour and a blurring of people, time, and experiences.

Within Crittenden's developmental framework, all patterns of attachment have adaptive value in specific contexts; avoidant adults for instance can often be very good organisers and managers, while preoccupied C adults can often be very good musicians, actors/actresses, and fire fighters. These information processing strategies allow individuals to maintain a sense of control, predictability, and safety. A strength and limitation of Crittenden's model is its complexity; this complexity is necessary given the diversity of relationships people have and the wide heterogeneity which occurs between groups of adults. However, researchers and practitioners need extensive training to use the Adult Attachment Interview; this interview also takes many hours to administer, transcribe, and analyse. Nevertheless, given the detailed insights available through this model, future researchers should be able to develop more easy to use measures.

To summarise, although ageist stereotypes do need to be challenged, problems can occur into later years that necessitate adaptation. Researchers have identified a number of strategies that can be used to facilitate better health and well-being in later years; these include selective optimisation with compensation (Baltes & Baltes, 1990); assimilation and accommodation (Brandtstädter et al., 1993; 1997; 1998), and a range of strategies in the context of relationships. Much more research in this field is needed.

DEVELOPMENT AND MEANING

Optimal functioning and well-being into later years is not solely about adaptation to challenges but also about one's ability to find enjoyment and pleasure in life. Pleasures adults' associate with later years include warm relationships with children and grandchildren, increased time for leisure pursuits, and continued development and growth (O'Hanlon & Coleman, 2001). There is a surprising paucity of research and theories about positive and optimal development into later years; nevertheless, the seminal work of Erikson (1950), Levinson and colleagues (Levinson et al., 1978; Levinson & Levinson, 1996), and Tornstam (1989; 1994) will be summarised next.

Erik Erikson: Eight Stage Theory of Development

One of the most influential theorists in the field of adult development and aging is that of Erik Erikson (1902–1994). Erikson formulated an eight stage theory of lifespan development in which certain psychological tasks had to be completed at different stages through the life cycle. In early childhood, for instance, the key task is that of developing a sense of trust rather than mistrust in oneself and others, particularly with caregivers. As one progresses through childhood and adolescence, other key tasks or challenges became more salient; these include tasks around each of 'autonomy,' 'initiative,' 'industry,' and 'identity.' In young adulthood, the central issue is that of intimacy versus isolation, while the psychological task in mid-life relates to adults' ability and need to care, guide, and support the next generation, in other words, generativity versus stagnation. The final stage in Erikson's theory of adult development is that of finding integrity over despair. From each of these psychological conflicts can emerge adaptive strengths including care fostered by generativity and wisdom as a result of attaining ego-integrity in old age (Erikson, 1950/1995, Erikson, Erikson, & Kivnick, 1986).

Within Erikson's model, generativity is believed to be the most salient psychological issue for adults in mid-life. Generativity is defined as 'primarily the concern in establishing and guiding the next generation' (Erikson, 1950/1995: p. 240). Essentially, generativity describes the adult's need to assume social, work, and community responsibilities that will be advantageous to others. Adults who are generative may wish to take on projects or goals which will promote or favour the next generation. The ways in which generativity manifests itself is likely to be related to innate and learned priorities, wishes, and motivations of the individual. Kotre, for instance, has divided generativity into biological, parental, technical, and cultural. Each aspect of generativity is likely to manifest itself in different ways through different times of the life cycle. Generative adults, for instance, may take a greater interest in products and institutions that will outlive them. They may wish to transmit knowledge and values to younger generations, as well as engage in political or social events to improve life for the future. These activities are recognised as being necessary for the collective survival and increased development of our society (Kotre 1995; Bellah et al., 1991).

Not everyone, however, is generative. To the contrary, each of Erikson's stages takes place on a continuum where a conflict occurs between two polar opposites. For some people, generativity may not occur, and instead, some adults might find themselves with a persistent sense of stagnation. Within this mode, there is an increased risk for poor mental and physical heath as well as a reduction in the quality of interpersonal relationships. Currently, however, there is little research on generativity, and it is not yet clear whether some aspects of generativity are more important than others for health, or even whether particular kinds of generativity are more salient for adults at different ages through the life-course. Much more research in this area is needed.

The psychological task into later years for Erikson is that of attaining 'ego integrity', an assured sense of meaning and order in one's life and in the universe, as against despair and disgust. This involves 'acceptance of one's one and only life cycle as something that had to be and that, by necessity, permitted of no substitutions' (Erikson, 1950/1995, p. 251). Integrity shows many positive features of earlier developmental stages such as trust, autonomy, identity, intimacy, and generativity. The developmental parts of life are fused into a whole. Integrity involves acceptance of one's life as it has been lived and also feeling grateful for past experiences and relationships.

Despair, the opposite of ego integrity, involves many of the negative features of earlier developmental stages such as mistrust, shame, isolation, and self-absorption. Adults who are despairing in later years can find it difficult to live with past experiences and relationships and often blame others for the mishaps and troubles they have encountered. Anger, regret, and a sense of failure may also be compounded by the urgent awareness that time is running out; choices are constrained and many opportunities are lost for ever. Also present can be disgust with other people, especially the young.

Successfully reaching the last stage of life in search for a balance between integration and despair the final strength of wisdom will develop. Wisdom is defined as a 'detached concern with life itself, in face of death .. and in spite of decline of body and mental function' (Erikson, Erikson, & Kivnick, 1986, pp. 37–38). The emotional and cognitive integration that wisdom is built upon is the result of vital involvement in diverging social spheres in earlier periods of life, such as politics, religion, economy, technology the arts and the sciences (Erikson, 1950/1995). Wisdom is gaining the insight that a single individual is just one member of the human family that shares characteristics in different generations and cultures. In reflecting upon the life lived and the predicament of people, a heightened awareness or an existential identity develops. Wisdom, Erikson says is "truly involved disinvolvement" (Erikson, Erikson and Kivnick, 1986, p. 51). Development then into later life is not only a continuation of the productivity and procreativity of previous stages, it is a new way of being.

Erikson's theory of life-span development has been influential, but also criticised. The developmental periods are quite general and Erikson never put out any clear time limits for when one starts and another begins. Mid-life for instance can span over four decades of the adult life. Further criticisms are that Erikson is not precise in describing each developmental crises or the strengths that result after a successful completion of a developmental stage. Hansen Lemme (1995) has advocated these types of criticism and asks further questions concerning the generativity concept in particular including its relationships with gender, culture, and social class. Several researchers have also highlighted dark generativity, or the reality that people can leave legacies of destruction, doubt, and fear (see Kotre, 1995).

To summarise, Erik Erikson proposed an eight-stage theory of adult development which extends into advanced old age. Each of Erikson's stages involves a psychological conflict from which a positive or adaptive strength can occur. The key conflicts into mid and later life are 'generativity versus stagnation' and 'ego integrity versus despair'; the strengths from these conflicts can be care and wisdom. This theory has many strengths including its ability to challenge negative stereotypes of later life. However, much more research is needed to examine and test aspects of this theory, to encourage more adults to volunteer their skills and experience, and ultimately, to facilitate better health and psychosocial well-being into later years.

Daniel Levinson

Daniel Levinson (like so many others) builds on Erikson's thinking: he also finds developmental periods in his study, but they are larger in number and more delineated than in

Erikson's theory. Levinson and colleagues drew upon intensive interviews with men and women to propose that development into adulthood progresses through a series of stages (see Levinson, Darrow, Klein, Levinson, & McKee, 1978; Levinson & Levinson, 1996). One of the key aspects of Levinson's theory is the life structure, the underlying pattern of an individual's life which includes family and work experiences. Levinson divides developmental phases into two basic entities: large eras and more limited periods, within eras. According to Levinson, there are four eras lasting about 25 years each: childhood and adolescence (aged 0–22), early adulthood (aged 17–45), middle age (aged 40–65) and late adulthood (60–). (In an earlier paper, Levinson (1977) mentions also a fifth era "Late, late adulthood (aged 80+), but he never elaborated upon this). The big transitions are turning points between the eras. Also the developmental periods vary in dynamism; some have to do with finding one's way in a totally new era, some are to do with settling down in an already established era of life. The stable periods ordinarily last about 6–8 years, and the transitional periods 4–5 years.

As previously noted, the most important concept in Levinson's theory of development is the life structure. The life structure consists of three elements: the socio-cultural world (e.g., class, religion, occupation), the self (e.g., personality, wishes, and conflicts, both conscious and unconscious), and participation in the world (which includes both possibilities and obstacles for development). Within the life structure, both the self or the individual and the world or the society are interwoven such that the life structure is seen has having a significant impact on the process of adult development.

In their book *The Seasons of a Man's Life* Levinson and colleagues (1978) presents data from the life-stories and in depth interviews of 40 American men (aged 35–45); participants were a selected group of male workers in industry, academic biologists, business executives, and novelists. The method used in this study was a biographical one, where in depth interviews for a total of 10–12 hours per person were conducted; issues discussed included education, work, marriage, friendships, leisure, political, and religious activities. For each person, a biography was later reconstructed based on these interviews (Levinson, et al., 1978). Like Erikson, Levinson believes that successful adjustment and optimal development involve mastering specific tasks at each stage of life, through adulthood. In early adulthood the men had begun careers and families; after an evaluations of themselves at around age 30, the men settled down and worked towards career advancement. Another transition occurred at about age 40, as men realised some of their ambitions would not be met. During middle adulthood, men worked towards cultivating their skills and assets, whereas in later adulthood, the transition was seen as a time to reflect upon successes and failures and enjoy the rest of the life.

Levinson & Levinson (1996) also interviewed 45 women between the ages of 35 and 45 years. One–third were homemakers, one–third college instructors, and one-third businesswomen. In general, although the life stages of women were more closely tied to relationships and family, Levinson and colleagues found that women go through the same type of cycles as men.

Although there can be much diversity between people, there can also be a striking congruence when comprehensive life histories are developed for each participant and the life-structures compared. The timing and the phases of the developmental tasks seems similar. There also seems to be a typical age for onset for each developmental period, as well as a fixed time for its closure. The periods did not vary for more than 2–3 years between individuals, according to Levinson and colleagues. The developmental periods of this theory, thus are closely linked to chronological time, in contrast to Erikson, whose developmental stages were much more loosely linked to time. The slight variation of some years gives room for some diverging impact of the biological, psychological, and social conditions in adults lives.

The mid-life transition is seen a period of de-illusionment. Adults must acknowledge that some of the earlier life structures were not built upon realistic goals and perhaps that relationships or activities cherished were not that rewarding in the end. At the same time, there is the danger of stagnation or regression in this passage of life. At this time, parents may fall ill or die, adults might contract a fatal illness causing concern and distress. Adults can start to think upon their own mortality as they enter 'the dominant generation' (Levinson et al., 1978). It is not yet too late to create new dreams, to change life's path, to take up new interests, friendships, love relationships. It is a time for termination of dysfunctional involvements and for initiation of new goals and commitments. To reflect on the existential questions in life makes the man to confront and reintegrate four polarities that form the developmental tasks of the mid-life transition: 1) young/old, 2) destruction/creation, 3) masculine/feminine, 4) attachment/separateness (Levinson et al., 1978).

A shift in the masculine/feminine balance in mid-life is an issue brought up by Levinson when both men and women can take on characteristics more typical o the opposite gender. Levinson's thinking is as follows: In earlier years when a young man is in the proces of 'becoming his own man', he was drawing quite heavily on the central cultural stereotype of what a man is. However, in mid-life both the sexual drive and the macho script start loosing is pressing influence. It is now time to give room for nurturing and caring, not just achieving and doing. Also, the body might signal that the machismo days are over and the developmental task now to explore the area of 'feeling', not only the area of 'thinking.' In this phase of life a man might change his relationship to his wife, as well as to his mother. He gets a deeper understanding of their world views and behaviour. He might also gain an insight that the tenderness and nurturance that he earlier considered female features also resides within himself. He might also form his first, non-sexual intimate relationship with a woman, as friend. Mentorship grows out of this, mostly as a mentor for a man, but in some cases, also as a mentor for a woman.

Levinson's theory has been criticised for many things, for the "soft" method used making replication difficult and for not giving exact markers as to how development shifts from one era to another (Hansen Lemme, 1995). While the developmental stages of Erikson have be criticised for being "fuzzy" and not exact enough, Levinson has also been criticised for tying his developmental periods too tightly to chronological time creating the opposite problem, a 'scheduling dilemma' (Hansen Lemme, 1995). The diversity in human development is obscured in this system and the fixed periods might become highly normative. Nevertheless, Daniel Levinson's developmental theory on the eras and periods of life gives an in-depth picture of developmental tasks and the search for meaning that are actualised at different points of life. Levinson's study of men has been an influential one, and since he also completed an in depth study of women, it now remains to be seen what impact this study will have on future developmental research.

To summarise, Levinson's theory of development does make many valuable contributions to the current field; he helps to challenge ageist view of later years to argued that the latter part of the life cycle has much potential for continued development and growth. Nevertheless, much more research is needed, to examine these stages further and to consider their universality for adults in different countries and different cohorts. It is the universality claim that seems to be the most unwarranted in both Erikson's and Levinson's theories. Such claims are still plaguing some otherwise quite creative gerontological theories of to day, as we will see is the section to follow.

Lars Tornstam: Theory of Gerotranscendence

An additional theory of development and growth into later years comes from Lars Tornstam. In his theory of gerotranscendence (Tornstam, 1989; 1994), the gerotranscendent individual is said to experience a redefinition of time, space, life and death, and the self. To use the words of Tornstam, 'gerotranscendence is a shift in meta perspective, from a materialistic and rational vision to a more cosmic and transcendent one, accompanied by an increase in life-satisfaction' (Tornstam, 1996).

The signs of gerotranscendence include an increased feeling of communion with the sprit of the universe and a redefinition or blurring of time, space, and objects such that an affinity with earlier times and other cultures develops in the aging person. The borders between the self and others also become diffuse which may lead to a decreased self-centeredness and to the development of a more cosmic self. Additional signs of geo-transcendence include a redefinition of life and death and a decreased fear of death as it is not the individual life but the general flow of human life that becomes important. Gero-transcendent adults also develop an increased affinity with past and coming generations. This notion, as well as the forgoing on, closely resembles Erikson's conceptualisations. As part of the increased ego-integrity, Erikson postulates both an increased feeling of affinity with past and coming generations and a feeling of death being quite a normal end-stage of the life-span development for human beings.

This theory draws upon many earlier theoretical formulations such as Erikson's (1950/1995) ego-integrity, Cumming and Henry's (1961) disengagement theory, as well as Gutmann's (1978) passive and magical mastery, but it is a creative reformulation of them. According to Tornstam, the debate on the limitations of disengagement theory might have been led astray by mistakenly projecting mid-life patterns and values into old age, such as advocating continuous activity as the ideas normal for old age. Tornstam further draws upon the life-philosophy of the Zen Buddhists whereby the developmental tasks of later life would be withdrawal from the mundane world, a withdrawal into spiritual meditation reaching a stages of cosmic transcendence over time where past, present, and future becomes one. Tornstam further refers to Jung's concept of the collective unconscious where the individual reaches affinity with universal psychological structure that have characterised man in different cultures.

Tornstam states that the development of gerotranscendence is neither a defense mechanism nor a coping process. The reason for this assumption he states, is that defense mechanisms and coping processes develop 'in the same meta-theoretical frame, while a radical shift in development is needed for gerotranscendence' (p. 78). These assumptions can, however, be debated. It may be that there is not such a clear difference between coping or adapting and the process of gerotranscendence. In coping with serious life distress, a re-organisation of values and behaviour can often be seen after the stress period is over. This can constitute a qualitative shift or a change in meta-theoretical framework.

The process of gero-transcendence can be slowed down by gerontologists or caretakers that are constantly striving for an 'activation' of an old person says Tornstam. The integration of personality, typical of old age according to Erikson's (1950/1995) theory, is thus obstructed. Pondering on life, meditation is a pre-requisite for reaching integrity and thereby wisdom. Also in connection with this issue, Tornstam would like to make demarcation line between Erikson theory and his own by stating that gero-transcendence presupposes the growth of a new meta-theoretical framework. Nevertheless, according to both theories, a qualitative shift in thinking will be needed for the development of wisdom.

To summarise, although Tornstam's theory of gerotranscendence makes a useful contribution to the current field much more research is needed. As has been noted, traditionally there has been an overemphasis in gerontological research on problems and setbacks into later years, and few researchers have examined or tested theories about positive development and growth. Future questions to be addressed in this field include the relationship between gero-transcendence and other aspects of health, adaptation, and psychosocial functioning. It would also be interesting and useful to know the factors that influence or predict gerotranscendence; possible factors include demographic variables, attitudes to aging, religiosity, personality traits, or experiences in close relationships.

BELIEFS AND ATTITUDES TOWARD AGING

Although problems and challenges can occur into later years as throughout life, the argument in this chapter is that many positive attributes and experiences can also occur into later years, including continued development and growth. In the section to follow, of interest are the ways adults themselves view the latter part of their own lives and particularly whether the latter part of the life cycle is being evaluated in unfavourable ways.

Beliefs and Expectations About Own Future Old Age

There has been little research exploring or examining the beliefs and expectations adults have with regard to the latter part of their own lives, particularly attitudes towards possible positive attributes and experiences. Evidence from studies to date, however, indicates that many adults hold negative expectations for their own future old age. Downward expectations and beliefs cross many areas of functioning include personality change (e.g., Heckhausen & Baltes, 1991; Krueger & Heckhausen, 1993), physical health (e.g., Ardelt, 1997; Keller, Leventhal & Larson, 1989; Rabbitt, 1999), social relationships (e.g., Quam & Whitford, 1992), and enjoyed physical activities (e.g., Stuart-Hamilton, 1998). For instance, Keller et al. carried out a study with 32 community based adults aged 50 to 80 years. Although participants recognised some positive attributes into later life, the changes they expected were 'uniformly negative' (p. 67). Similarly, in a study involving 37 Finns aged 73 to 83 years, Ruth & Öberg (1996) found that some respondents viewed the latter part of their lives as being a time of problems, powerlessness, and lost control.

Additional evidence that adults have negative beliefs about later life comes from the work of Heckhausen et al. (1989). In this study, participants were recruited via newspapers and grouped into three cohorts; young adults (aged 20–36), mid-life adults (aged 40–55) and older adults (aged 60–85 years). In two separate group sessions, participants were asked to rate a number of adjectives (e.g., intelligent, absent-minded, patient, wise) in terms of their desirably (very undesirable–very desirable), the adult chronological age at which the attribute was first likely to occur, and the age at which the attribute was likely to finish. Participants were also asked to rate the list of attributes in terms of their controllability and in terms of their significance. The researchers found that participants in each of the age groups generally reported similar expectations in the distributions of ratings. Specifically, results indicated that more losses and fewer gains were associated with the latter part of the life cycle although some positive attributes were also expected into later life, namely wisdom and dignified with age onset at 55 years. Into advanced old age (over 80 years), losses were perceived to outnumber gains.

Adults' beliefs and expectations for their own aging and future old age (rather than aging for people generally) were also negative. For instance, in a study examining adults' expectations for future health and well-being, Ryff (1991) found that whereas young and mid-life adults expected improvements in all dimensions of functioning, older adults expected declines on these same dimensions. Although older adults scored significantly higher on some aspects of present functioning (particularly self-acceptance and environmental mastery), results indicated that the older adults, unlike the younger generations, expected decline in most aspects of future functioning including self-acceptance, environmental mastery, and purpose in life. Further, some attributes, like personal growth, were rated highest in the past suggesting declines were seen to have occurred from the past to the present. This study draws on a number of theoretical frameworks including possible selves (e.g., Markus & Nurius, 1986) and is one of the first systematic attempts to examine beliefs and expectations towards psychological functioning into the latter part of one's own lifecycle.

Future research urgently needs to consider the implications and consequences of negative age-associated evaluations on health and psychosocial functioning. If significant relationships are found between age-associated attitudes and psychosocial functioning, further research is warranted examining in more detail the nature of adults' beliefs and expectations, their possible origins in earlier life experiences, and the contexts and conditions under which they carry a dysfunctional effect.

Age-Associated Attitudes: Fears and Worries

There is evidence that adults can evaluate the experience of their own future old age with some degree of worry and anxiety. In a study with a group of nurses, for instance, Bernard, 1998) found that many were highly anxious about their own aging and future old age and viewed their own aging with 'trepidation' (p. 637). Although there is little research exploring the ways in which individuals or society more collectively views later life, the quote to follow by De Beauvoir (1972) provides further evidence from within the social context that later life can be a significant source of threat:

> Acknowledging that I was on the threshold of old age was tantamount to saying that old age was *lying there in wait* for every woman, and that it had already laid hold upon many of them. Great numbers of people, particularly old people, told me kindly or angrily but always at great length and again and again, that old age simply did not exist! There were some who were less young than others and that was all it amounted to. Society looks upon old age as a kind of shameful secret that it is unseemly to mention. (de Beauvoir 1972: 7–8)

Although the above studies provide evidence that adults have anxieties and fears about their own aging and future old age, more research is urgently needed given methodological constraints, particularly in relation to measures being used. For instance, in a study with 188 Bostonions aged 14 to 83 years, Montepare and Lachman (1989) used the Attitudes to Aging scale of the Aging Opinion Survey (Kafer et al., 1980) to examine fears that participants had for their own aging and future old age; internal reliability scores for this scale however were just .52 and .60 for men and women respectively. Similarly, many researchers (e.g., Mosher-Ashley & Ball, 1999; Watkins, Coates, & Ferroni, 1998) used single item measures with no attempt to examine their validity. In addition, there is very little research on age-associated attitudes on community based adults of all ages; instead, researchers in this field have tended to concentrate on specific populations including groups of older adults (e.g., Heikkinen, 1993; Watkins et al., 1998), students (e.g., Lasher & Faulkender, 1993; Mosher-Ashley & Ball, 1999) and/or groups of nursing professionals (e.g., Bernard, 1998).

Although people do have worries and concerns about their future old age, it should not be assumed that age-associated losses and constraints will always be viewed in threatening ways, by everyone, all the time. Not only do people cope well with problems and challenges (e.g., see the earlier section on adaptation) but there is evidence from other disciplines including history and anthropology (e.g., Garliski, 1975; De Vries, 1995) which give testament to the enormous creativity and resilience of human beings, not only in overcoming adversity or threat but also in creating more favourable outcomes for the self and others, including progeny. Nevertheless, as discussed next, it is surprising that adults do not experience the level of anxiety or depression that one might have expected.

The Aging-Health Paradox

Contrary to expectations, there is evidence from a range of detailed, rigorous, and sometimes multi-disciplinary studies that self-esteem and well-being generally remains high into later life (e.g., Baltes & Baltes, 1990; Carstensen, 1991; 1993; Dietz, 1990; Dittmann-Kohli, 1990: Herzog & Rogers, 1981). For instance, in the Southampton Aging Study, participates were followed up longitudinally 10 and 13 years later. Results from this study indicated 'remarkable stability' of self-esteem and well-being into later life, in other words, rather than becoming more anxious or depressed, participants in this study generally retained high self-esteem and autonomy (Coleman et al., 1993). Similarly, in a study carried out by Thompson and colleagues (Thompson, 1992; Thompson, Itzin, & Abend-stern, 1990), results indicated that participants generally had good health and a positive sense of identity, until/unless they were physically in poor health.

In evaluating the above aging–health paradox, one must consider warnings given by a number of researchers about the dangers in relying solely on the use of self-report measures (e.g., see Baltes & Baltes, 1990; Coleman, 1996; Ryff & Essex, 1992). For instance, participants may not wish to communicate their true thoughts or feelings, perhaps because they are concerned about being thought foolish by others. This issue is particularly crucial given the argument by Biggs (2000) that adults may use masquerades or personas to respond to questions in ways more favourable than participants actually feel. However, this explanation is unlikely. For instance, in carrying out interviews with older adults Coleman (1986) reports taking time to build a close one-to-one relationship with participants to encourage them to speak about 'matters that interested or concerned them' rather than having a preset agenda (p. 18). Given these realities, it seems unlikely that the aging–health paradox can be explained by methodological limitations around the expression of negative affect.

An additional methodological issue to be considered in interpreting the aging–health paradox are possible biases with sample groups, particularly in qualitative studies. Specifically, although the use of interviews can access a detailed and a rich set of data, the generalisabilty of these interviews can sometimes be more limited, especially given that the numbers of participants for such research are generally small and may be biased towards adults more altruistic, more healthy, or with more positive age-associated attitudes. To address the latter possibility, researchers in this field should consider using some kind of incentives or rewards to widen participation in research, particularly amongst adults less altruistic or generative.

It is surprising that the aging–health paradox exists; future research in addressing this discrepancy may suggest that ideas about the nature of later years need to be revised. Equally, it may be that gerontologists need to use more sophisticated assessment procedures to assess health into later years; the latter is especially important given increasing suicide rates for older people (e.g., see Gallagher-Thompson & Osgood, 1997; De Leo & Ormskerk, 1991; Pampel & Williamson, 2001).

To summarise, although there is very limited research in this field, there is evidence that people can experience anxiety and fears about their own aging and future old age. Surprisingly, however, there is little evidence to support high levels of anxiety or depression into later years; although several methodological and theoretical explanations for the aging health paradox were considered, a satisfactory explanation will necessitate further research and study.

Attitudes to Aging Measures

Given the potential for age-associated attitudes to impact adversely on health, particularly when negative, along with the reality that many adults do have worries and fears about their own future old age, it is vital that further research takes place in this field; we urgently need to understand the origins, nature, and consequences of age-associated attitudes. In an attempt to facilitate this work, the section to follow will consider briefly some of the measures available to researchers. For additional measures, see Gething, 1994; Kogan, 1961; Pratt, Wilson, Benthin, & Schmanll, 1992; Ryff, 1991; Tuckman & Lorge, 1953.

Facts on Aging Quiz (FAQ, Palmore 1977).   Palmore's Facts on Aging Quiz is used mainly to measure the level of knowledge adults have about aging and later life. Palmore argues that quiz items cover three areas of knowledge (physical, mental, and social). A strength of Palmore's measure is his strategy of systematically documenting each scale item in empirical research. However, Norris, Tindale, and Matthews (1987) question the content validity of this scale, given that Palmore does not discuss the way in which the scale items were derived, how the above categories emerged, or the theoretical rationale behind these questions. Other researchers have found that the FAQ does not correlate with other knowledge-based measures (e.g., see O'Hanlon, Camp, & Osofsky, 1993; Norris et al., 1987); in examining the validity of a given scale, one would expect it to correlate with other scales measuring similar experiences. In response to these psychometric criticisms, however, Palmore argues that the use of modern statistical methods are inappropriate with his scale given that the FAQ was 'designed to examine performance in terms of current knowledge or skill rather than in relation to others within the group'.

Ontario Opinions About People Scale (1971).   This scale consists of 5 items, all of which are graded on a 5-point Likert-type scale from strongly disagree (1) to strongly agree (5). The items for this measure include: 'I always dreaded the day I would look in the mirror and find a gray hair', and 'the older I become the more anxious I am about the future'. A strength of this scale is its shortness; short scales can be particularly useful given space constraints often typical within questionnaire packs. However, in a study investigating aging anxiety in a national survey of caregivers, several researchers report low internal reliability for this scale (Cronbach's alpha .52–.62, see Montepare & Lachman, 1989; Wullschleger et al., 1996).

The General Attitudes to Aging Scale (O'Hanlon & Coleman, submitted).   The GAAS is a 5-item measure of the general attitudes adults have towards their own aging and future old age. Participants are asked to rate all items on a 5-point Likert-type scale from strongly agree to strongly disagree in the direction of negative attitudes to own old age. Across a wide range of sample groups, this short measure was found to have good psychometric properties (see O'Hanlon & Coleman, submitted).

The Anxiety About Aging Scale (AAS, Lasher & Faulkender (1993).   This 20-item scale measures the anxiety people feel about their own aging. Although this scale is long, it does have good reliability and validity, with four interpretable factors (fear of older adults, psychological concerns, physical appearance, and fear of loss) which are summed to gain a mean score of anxiety. Internal reliability for this scale is also good (Cronbach's alpha score for each of the subscales ranged from .69–.78). There are some conceptual difficulties, however, in documenting the ways anxiety about aging might manifest itself. For instance, some older adults might be reluctant to express negative age-associated fears, given fears about being thought foolish. Similarly, when anxiety does occur, it may manifest itself as somatic problems which can be functional in allowing older adults to gain needed support and help. These potential difficulties however simply need further study and this Anxiety about Aging Scale may offer many researchers a way to do just that.

To summarise, there are few available measures for researchers interested in age-associated attitudes; with the exception of the GAAS, shorter measures do not tend to have such good psychometric properties as the longer scales. One of the challenges for researchers in the field of adult development and aging is to develop new measures that have good psychometric properties, that are short, easy to use, and simply to interpret, and which can facilitate research exploring and examining adults' attitudes towards the experience of their own aging and future old age.

CONCLUSION

There is an urgent need to understand the ways adults experience and evaluate the latter part of their own lives. Although there is a paucity of research in this field, there is data to indicate that adults' attitudes to aging can be a significant factor in influencing later health and mortality rates. In examining the nature of later life, different pictures emerge. While some people are plagued by ill health and other problems, many others are in exceptionally good health, with very active social and personal lives. When challenges occur, adults are often very skilled in managing these; theories of adaptation considered included selective optimisation with compensation and assimilation and accommodation. Later life can also offer opportunities for continued development and growth, towards generativity, wisdom, and gero-transcendence. Attitudes to aging will influence the ways adults manage and experience the latter part of their lives; further work on attitudes has the potential to facilitate optimal well-being for more older people.

REFERENCES

Adams, C., Labouvie-Vief, G. Hobart, C. J., & Dorosz, M. (1990). Adult age group differences in story recall style. *Journal of Gerontology: Psychological Science, 45,* 17–27.

Ardelt, M. (1997). Wisdom and life satisfaction in old age. *Journal of Gerontology, 52b*(1), 15–27.

Baltes, M., & Carstensen, L. I. (1996). The process of successful ageing. *Ageing and Society, 16,* 397–422.

Baltes, M. M. (1996). *The many faces of dependency in old age.* Cambridge, UK: Cambridge University Press.

Baltes, P. B., & Baltes, M. M. (1990). *Successful Aging: Perspectives from the Behavioral Sciences.* Cambridge, UK: Cambridge University Press.

Baltes, P. B. (1987). Theoretical propositions of life-span developmental psychology: On the dynamics between growth and decline. *Developmental Psychology, 23,* 611–626.

Baltes, P. B. (1991). The many faces of human aging: Toward a psychological culture of old age. *Psychological Medicine, 21,* 837–854.

Baltes, P. B. (1996). On the incomplete architecture of human ontogeny; Selection, Optimization and compensation as foundation of developmental theory. *American Psychologist, 52*(4), 366–380.

Bellah, R. N., Madsen, R., Sullivan, W. M., Swidler, A., & Tipton, S. M. (1991). *The good society.* New York: Knopf.

Bengtson, V. L., Rice, C. J., & Johnson, M. L. (1999). Are theories of aging important? Models and explanations in gerontology at the turn of the century. In V. L. Bengtson & K. W. Schaie (Eds.), *Handbook of Theories of Aging* (pp. 3–21). New York: Springer.

Bernard, M. (1998). Backs to the future? Reflections on women, ageing and nursing. *Journal of Advanced Nursing, 27,* 633–640.

Biggs, S. (1993). *Understanding ageing-images, attitudes and professional practice.* Berkshire, UK: Open University Press.

Biggs, S. (2000). The blurring of the life course: Narrative, memory and the question of authenticity. *Journal of Aging and Identity, 4*(4).

Bowlby, J. (1979). *The making and breaking of affectional bonds.* London: Routledge.

Brandtstädter, J., & Greve, W. (1994). The aging self: Stabilizing and protective processes. *Developmental Review, 14,* 52–80.

Brandtstädter, J., & Wentura, D. (1995). Adjustment to shift possibility frontiers in later life: Complementary adaptive modes. In R. A. Dixon & L. Bäckman (Eds.), *Compensating for Psychological Deficits and Declines: Managing Losses and Promoting Gains* (pp. 83–106). Lawrence Erlbaum Associates.

Brandtstädter, J., Rothermund, K., & Schmitz, U. (1997). Coping resources in later life. *Revue Européenne de Psychologie Appliquée 47,* 107–113.

Brandtstädter, J., Rothermund, K., & Schmitz, U. (1998). Maintaining self-integrity and efficacy through adulthood and later life: The adaptive functions of assimilative persistence and accommodative flexibility. In Heckhausen, J. & Dweck, C. S. (Eds.), *Motivation and self-regulation across the life-span* (pp. 365–421). Cambridge, UK: Cambridge University Press.

Brandtstädter, J., Wentura, D., & Greve, W. (1993). Adaptive resources of the aging self: Outlines of an emergent perspective. *International Journal of Behavioral Development, 16,* 323–349.

Bromley, D. B. (1988). *Human ageing: an introduction to gerontology.* 3rd Edition. London: Penguin.

Bytheway, B., & Johnson, B. (1990). On defining ageism. *Critical Social Policy, 29, 10*(2), 27–39.

Carstensen, L. L. (1991). A Life-span approach to social motivation. In J. Heckhausen, J., & Dweck, C. S. (Eds.), *Motivation and Self-Regulation Across the Life-Span,* (pp. 341–364). Cambridge, UK: Cambridge University Press.

Carstensen, L. L. (1993). Social and emotional patterns in adulthood: Support for Socioemotional Selectivity Theory. *Psychology And Aging, 7,* 331–338.

Cohen, G. (1996). The special case of mental health in later life. *The American Journal of Geriatric Psychiatry, 4*(1), 17–23.

Coleman, P. G. (1985). *Ageing and reminiscence processes: Social and clinical implications.* Chichester: Wiley.

Coleman, P. G., Ivani-Chalian, C., & Robinson, M. (1998). The story continues: Persistence of life themes in old age. *Ageing and Society, 18,* 389–419.

Coleman, P. G. (1993). Adjustment in later life. In J. Bond, P. Coleman, & Peace, S. (Eds.), *Ageing in Society: An introduction to social gerontology* (pp. 97–132). Second Edition. Beverly Hills, CA: Sage.

Coleman, P. G. (1993). Psychological ageing. In J. Bond, P. Coleman, & Peace, S. (Eds.), *Ageing in Society: An introduction to social gerontology* (pp. 68–96). Second Edition. Beverly Hills, CA: Sage.

Coleman, P. G. (1996). Identity management in later life. In B. Wood (Ed.), *Handbook Of The Clinical Psychology Of Ageing* (pp. 93–113). New York: Wiley.

Coleman, P. G., Aubin, A., Robinson, M., Ivani-Chalian, C., & Briggs, R. (1993). Predictors of depressive symptoms and low self-esteem in a follow-up study of elderly people over 10 years. *International Journal of Geriatric Psychiatry, 8,* 343–349.

Coleman, P. G., & O'Hanlon, A. (in press). *Ageing and Development: Theory and Research.* Santa Cruz, CA: Arnold Publishing.

Cristofalo, V. J., Tresini, M., Francis, M. K., & Volker, C. (1999). Biological theories of senescence. In V. L. Bengtson & K. W. Schaie (Eds.), *Handbook of Theories of Aging* (pp. 98–112). New York: Springer.

Crittenden, P. M. (1995). Attachment and the risk for psychopathology: The early years. *Developmental and Behavioural Pediatrics, 16*(3), S12–S16.

Crittenden, P. M. (2002). Adult Attachment Interview Coding Manual. Unpublished manual available from the author.

Crittenden, P. M., & Claussen, A. H. (Eds.) (2000). *The organization of attachment relationships: Maturation, context, and culture.* New York: Cambridge University Press.

Crittenden, P. M. (1997). The effect of early relationship experiences on relationships in adulthood. In S. Duck (Ed.), *Handbook of Personal Relationships,* (pp. 100–119). New York: Wiley.

Cumming, E. & Henry, W. E. (1961). *Growing Old: The Process of Disengagement*. New York: Basic Books.

De Beauvoir, S. (1972). *Old Age*. London: Penguin Books.

De Leo, D. D., & Ormskerk, S. C. R. (1991). Suicide in the elderly: General characteristics. *Crisis*, 3–17.

De Vries, S. (1995). *Strength of spirit: Pioneering women of achievement from first fleet to federation*. Monterey, CA: Millenium Books.

Dietz, B. (1990). The relationship of aging to self-esteem: The reflective effects of maturation and role accumulation. *International Journal of Aging and Human Development*, 43(3), 249–266.

Dittmann-Kohli, F. (1990). The construction of meaning in old age: Possibilities and constraints. *Ageing and Society*, 10, 279–294.

Drew, L., & Smith, P. K. (1999). The impact of parental separation/divorce on grandparent-grandchild relationships. *International Journal of Aging and Human Development*, 48, 191–216.

Eagly, A. H., & Chaiken, S. (1993). *The psychology of attitudes*. Fort Worth, TX: Harcourt Brace Jovanovich College Publishers.

Eisdorfer, C. (1983). Conceptual models of ageing: The challenge of a new frontier. *American Psychologist*, February, 197–202.

Erikson, E. H. (1950/1995). *Childhood and Society*. (2nd ed.). New York: Norton. Reprinted by Vintage.

Erikson, E. H., Erikson, J. M., & Kivnick. H. (1986). *Vital involvement in old age. The experience of old age in our time*, London: Norton.

Featherstone, M., & Hepworth, M. (1989). Ageing and old age: Reflections on the postmodern lifecourse. In B. Bytheway, T. Kiel, P. Allat, & A Bryman (Ed's), *Becoming and Being Old*. London: Sage

Freund, A. M., & Baltes, P. B. (1998). Selection, optimization and compensation as strategies of life management: Correlations with subjective indicatory of successful aging. *Psychology and Aging*, 13(4), 531–543.

Fries, J. F. (1983). The compression of morbidity. *Milbank Memorial Quarterly/Health and Society*, 61(3), 397–419.

Gallagher-Thompson, D., & Osgood, N. J. (1997). Suicide in later life. *Behavior Therapy*, 28, 23–41.

Garliski, J. (1975). Fighting Auschwitz: The resistance movement in the concentration camp. London: Orbis Books.

Gething, L. (1994). Health professional attitudes towards ageing and older people: Preliminary report of the Reactions To Ageing Questionnaire. *Australian Journal of Ageing*, 13, 77–81.

Gutmann, D. (1987). *Reclaimed powers; toward a new psychology of men and women in later life*. New York: Basic Books.

Gutmann, D. (1997). The human elder in nature, culture and society. Boulder, CO: Westview Press.

Hansen Lemme, B. (1995). *Development in Adulthood*. Boston: Allyn & Bacon.

Heckhausen, J., & Baltes, P. B. (1991). Perceived controllability of expected psychological change across adulthood and old age. *Journal of Gerontology: Psychological Sciences*, 46, P165–P173.

Heckhausen, J., & Krueger, J. (1993). Developmental expectations for the self and most other people. Age grading in three functions of social comparison. *Developmental Psychology*, 29, 539–548.

Heckhausen, J., & Schulz, R. (1993). Optimisation by selection and compensation: Balancing primary and secondary control in life span development. *International Journal of Behavioral Development*, 16(2), 287–303.

Heckhausen, J. (1997). Developmental regulation across adulthood: Primary and secondary control of age-related challenges. *Developmental Psychology*, 33, 176–187.

Heckhausen, J., Dixon, R. A., & Baltes, P. B. (1989). Gains and losses in development throughout adulthood as perceived by different adult age groups. *Developmental Psychology*, 25, 109–121.

Heidrich, S. M., & Ryff, C. D. (1993). The role of social comparisons. processes in the psychological adaptation of elderly adults. *Journal of Gerontology: Psychological Sciences*, 48, 127–136.

Heikkinen, R. (1993). Patterns of experienced ageing with a Finnish cohort. *International Journal of Ageing and Human Development*, 36(4), 269–277.

Herzog, A. R., & Rogers, W. L. (1981). Age and satisfaction: Data from several large surveys. *Research on Aging*, 3, 142–165.

Hunter, S., & Sundel, M. (Eds.) (1989). *Midlife Myths: Issues, Findings and Practice Implications*. Beverly Hills, CA: Sage.

Kafer, R. A., Rakowski, W., Lachman, M., & Hickey, T. (1980). Aging Opinion Survey: A report on instrument development. *International Journal of Aging And Human Development*, 11, 319–333.

Kaufman, S. R. (1986). *The ageless self: Sources of meaning in late life*. Madison: The University of Wisconsin Press.

Keller, M. L., Leventhal, E. A., & Larson, B. (1989). Aging: The lived experience. *International Journal of Aging and Human Development, 29*(1), 67–82.

Kirkwood, T. (1999). *Time of our lives; the science of human ageing*. London: Weidenfeld & Nicolson.

Kogan, N. (1961). Attitudes toward old people: The development of a scale and an examination of correlates. *Journal of Abnormal and Social Psychology, 62*, 616–622.

Kotre, J. (1995). Generative outcome. *Journal of Aging Studies, 9*(1), 33–41.

Krueger, J., & Heckhausen, J. (1993). Personality development across the adult life span: Subjective conceptions vs. cross sectional contrasts. *Journal of Gerontology: Psychological Sciences, 48*, P100–P108.

Labouvie-Vief, G., Chiodo, L. M., Goguen, L. A., Diehl, M., & Orwoll, L.(1995). Representations of self across the life span. *Psychology and Aging, 10*, 404–415.

Lasher, K. P., & Faulkender, P. J. (1993). Measurement of aging anxiety: Development of the anxiety about aging scale. *International Journal of Aging and Human Development, 37*(4), 247–259.

Lavers, C., & Sonuga-Barke, E. J. S. (1997). Annotation: The grandmother's role in her grandchildren's adjustment. *Journal of Child Psychology and Psychiatry, 38*, 747–754.

Lerner, R. M., Freund, A. M., De Stefanis, I., & Habermas, T. (2001). Understanding developmental regulation in adolescence: The use of the selection, optimization, and compensation model. *Human Development, 44*, 29–50.

Levinson, D. J., & Levinson, J. D. (1996). *The season's of a woman's life*. New York: Knopf.

Levinson, D., Darrow, C., Klein, E., Levinson, M., & McKee, B. (1978). *The Seasons of a Man's Life*. New York: Knopf.

Levy, B. R., Slade, M. D., Kunkel, S., & Kasl, S. V. (2002). Longevity increased by positive self-perceptions of aging. *Journal of Personality and Social Psychology, 83*(2), 261–270.

Light, L. L. (1991). Memory and aging: Four hypotheses in search of data. *Annual Review of Psychology, 42*, 333–376.

Lindenberger, U., & Baltes, P. B. (1997). Intellectual functioning in old age very old age: Cross sectional results from the Berlin Aging Study. *Psychology and Aging, 12*, 410–432.

Luszcz, M., & Bryan, J. (1999). Toward understanding age-related memory loss in late adulthood. *Gerontology, 45*, 2–9.

Lynott, R. J., & Lynott, P. P. (1996). Tracing the course of theoretical framework in the sociology of ageing. *The Gerontologist, 35*(6), 749–760.

Markus, H., & Nurius, P. (1986). Possible selves. *American Psychologist, 41*, 954–969.

Marsiske, M., Lang, F. R., Baltes, P. B., & Baltes, M. M. (1995). Selective optimization with compensation: Life-span perspectives on successful human development. In R. A. Dixon & L. Bäckman (Eds.), *Compensating for Psychological Deficits and Declines: Managing Losses and Promoting Gains* (pp. 35–79). Lawrence Erlbaum Associates.

McAdams, D. P., Diamond, A., De St. Aubin, E., & Mansfield, E. (1997). Stories of commitment: The psychosocial construction of generative lives. *Journal of Personality and Social Psychology, 72*(3), 678–694.

McCrae, R., Costa, P., de Lima, et al. (1999). Age differences in personality across the adult life span: Parallels in five cultures. *Developmental Psychology, 35*(2), 466–477.

Montepare, J. M., & Lachman, M. E. (1989). "You're only as old as you feel": Self-perceptions of age, fears of aging, and life satisfaction from adolescence to old age. *Psychology and Aging, 4*(1), 73–78.

Morgan, K., Dallosso, H., Bassey, E. J., Ebrahim, S., Fentem, P. H., & Arie, T. H. D. (1991). Customary physical activity, psychological well-being and successful ageing. *Ageing and Society, 11*, 399–415.

Mosher-Ashley, P. M., & Ball, P. (1999). Attitudes of college students toward elderly persons and their perceptions of themselves at age 75. *Educational Gerontology, 25*, 89–102.

Norris, J. E., Tindale, J. A., & Matthews, A. M. (1987). The factor structure of the facts on aging quiz. *Gerontology, 27*(5), 673–676.

O'Hanlon, A., & Coleman, P. G. (2001). Exploring the joys and pleasures of later life; developing and testing a new measure. In S. Tester, Archibald, C., Rowlings, C., & Turnerm S. (Eds), *Quality in Later Life: Rights, Rhetoric and Reality; Proceedings from the 30th Annual Conference of the British Society of Gerontology* (pp. 131–134). Stirling, Scotland, UK: University of Stirling.

O'Hanlon, A., & Coleman, P. G. (2002). Facilitating a more active old age; examining the influence of attitudes to ageing on health behaviours. *Active ageing–Myth or reality? Proceedings from the British Society of Gerontology Annual Conference* (pp. 71–73). University of Birmingham.

O'Hanlon, A., & Coleman, P. G. (submitted). The General Attitudes to Ageing Scale. *The Gerontologist.* Manuscript available from the authors.

O'Hanlon, A. M., Camp, C. J., & Osofsky, H. J. (1993). Knowledge of and attitudes toward aging in young, middle-aged and older college students: A comparison of two measures of knowledge of aging. *Educational Gerontology, 19,* 753–766.

Opinions about people [OAP] (1971). Ontario Welfare Council. In D. J. Mangen & W. A. Peterson (1982). *Research instruments in social gerontology* (pp. 556–561, 598–599). Minneapolis, MN: University of Minnesota.

Palmore, E. (1977). Facts on Aging: A short quiz. *The Gerontologist, 17*(4), 315–320.

Palmore, E. (1980). The facts on aging quiz: A review of findings. *The Gerontologist, 20*(6), 669–672.

Palmore, E. (1981). The facts on aging quiz: Part two. *The Gerontologist, 21*(4), 431–437.

Pampel, F. C., & Williamson, J. B. (2001). Age patterns of suicide and homicide mortality rates in high-income nations. *Social Forces, 80,* 251–282.

Pearlin, L., & McKean Skaff, M. (1996). Stress and the life course: A paradigmatic alliance. *The Gerontologist, 36*(2), 239–247.

Powers, C. B., Wisocki, P. A., & Whitbourne, S. K. (1992). Age differences and correlates of worrying in young and elderly adults. *The Gerontologist, 32,* 82–88.

Pratt, C. C., Wilson, W., Benthin A., & Schmall, V. (1992). Alcohol problems and depression in later life: Development of Two knowledge quizzes. *The Gerontologist, 32*(2), 175–183.

Quam, J. K., & Whitford, G. S. (1992). Adaptation and age-related expectations of older gay and lesbian adults. *The Gerontologist, 32*(3), 367–374.

Quirouette, C., & Pushkar, D. (1999). Views of future aging among middle-aged, university educated women. *Canadian Journal on Aging, 18*(2), 236–258.

Rabbitt, P. (1999). Why I study cognitive gerontology. *The Psychologist, 12*(4), 180–181.

Rowe, J. W., & Kahn, R. L. (1987). Human aging: Usual and successful. *Science, 237,* 143–149.

Ruth, J.-E., & Öberg, P. (1996). Ways of life: Old age in a life history perspective. In Birren et al. (Eds.), *Ageing and Biography: Explorations in Adult Development* (pp. 167–186). New York: Springer.

Ryff, C. (1991). Possible selves in adulthood and old age: A tale of shifting horizons. *Psychology and Aging, 6,* 286–295.

Ryff, C. D., & Essex, M. J. (1992). The interpretation of life experience and well being: The sample case of relocation. *Psychology and Aging, 7,* 507–517.

Slotterback, C. S., & Saarnio, D. A. (1996). Attitudes toward older adults reported by young adults: Variation based on attitudinal task and attribute categories. *Psychology and Aging, 11*(4), 563–571.

Stuart-Hamilton, I. (1998). Women's attitudes to ageing: Some factors of relevance to educational gerontology. *Education and Ageing, 13*(1), 67–87.

Taylor, S., & Lobel, M. (1989). Social comparison activity under threat: Downward evaluation and upward contacts. *Psychological Bulletin, 96,* 569–575.

Teri, L., McCurry, S. M., & Logsdon, R. G. (1997). Memory, thinking and ageing–what we know about what we know. *Western Journal Of Medicine, 167*(4), 269–275.

Thompson, P. (1992). "I don't feel old": Subjective ageing and the search for meaning in later life. *Ageing and Society, 12,* 23–48.

Thompson, P., Itzin, C., & Abendstern, M. (1990). *I don't feel old: The experience of later life.* Oxford: Oxford University Press.

Tornstam, L. (1994). *Gerotranscendence–A Theoretical and Empirical Exploration.* In Thomas, L. E., Eisenhandler, S. A. (Eds), Aging and the Religious Dimension, Westport, CT: Greenwood Publishing Group.

Tornstam, L. (1996). Gerotranscendence–a theory about maturing into old age. *Journal of Aging and Identity,* 1:37–50.

Tornstam, L., 1989, Gero-transcendence; A Meta-theoretical Reformulation of the Disengagement Theory, Aging: Clinical and Experimental Research, Vol 1, Nr. 1: 55–63; Milano.

Treharne, G. (1990). Attitudes towards the care of elderly people: Are they getting better? *Journal of Advanced Nursing, 15,* 777–781.

Tuckman, J., & Lorge, I. (1953). Attitudes of the aged toward the older work: For institutionalised and non-institutionalised adults. *Journal of Gerontology, 7,* 559–564.

Vaillant, G. E. (1991). The association of ancestral longevity with successful aging. *Journal of Gerontology: Psychological Sciences, 46*(6), P292–298.

Watkins, R. E., Coates, R., & Ferroni, P. (1998). Measurement of aging anxiety in an elderly Australian population. *International Journal Of Aging And Human Development, 46*(4), 319–332.

Williams, R. (1990). *A Protestant Legacy: Attitudes to illness and death among older Aberdonians.* Oxford: Clarendon Press.

Wrosch, C., Heckhausen, J., & Lachman, M. E. (2000). Primary and secondary controls strategies for managing health and financial stress across adulthood. *Psychology and Aging, 15*(3), 387–399.

Wullschleger, K. S., Lund, D. A., Caserta, M. S., & Wright, S. D. (1996). Anxiety about aging: A neglected dimension of caregivers' experiences. *Journal of Gerontological Social Work, 26*(3/4), 3–18.

II

Language, Culture and Social Aging

This section in various ways fills out the critical linguistic dictum that "language is not a clear window, but a refracting, structuring medium" (Fowler, 1991, p. 10). Coupland (chap. 3) presents aging research from a sociolinguistic perspective. Sociolinguistic research on age and aging is rare and concentrated, and when done conceptualizes the process of aging mainly in relation to changes in early life. Coupland sets forth an impressive agenda for sociolinguists to clarify the radical uncertainty of what social aging means and how age is negotiated in language and discourse. Sociolinguists, as well as all gerontologists interested in studying the social processes of aging, should recognize that older adults are a minoritized and "other" social group, but that old age is far more than a progressing disability. An understanding of the aging process involves an interplay between human developmental processes, historical cohort influences, and the shifting contexts of social structure and culture, all subject to creative interpretation. Coupland places the study of age and aging at the forefront of the sociolinguistic research agenda.

Hummert, Garstka, Ryan, and Bonnesen (chap. 4) suggest four research questions for future research in their chapter that appeared in the first edition of this handbbook:

1. What factors in the communication situation lead to positive or negative stereotyping of older adults?
2. What individual differences between communicators are related to positive or negative stereotyping of older adults?
3. What communication behaviors are associated with positive or negative stereotyping of older adults?
4. How do the communication behaviors of the conversants affect stereotyping and self-perception processes?

The revision of this chapter found in this handbook considers the recent research on age stereotypes and communication that attempt to answer these questions. The most recent research on implicit age stereotyping and self-stereotyping receives special attention from the authors. Two models of age stereotyping in communication, the communication predicament of aging model and the age stereotypes in interactions model, are described and help to explain the communication process of stereotyping as we age. Hummert et al. conclude their chapter by suggesting that older adults can interrupt the possibility of negative cycles of stereotypes in several important ways. This notion that individuals of all ages can have a positive impact on the direction of their lives is a common theme throughout the handbook.

It seems to us that the issues Hummert propose as crucial to the future research on stereotypes of the elderly and communication might be aided by using the communication and social cognition approach in conjunction with a sociolinguistic approach. Hummert's own emphasis on the importance of providing a contextualized account of cognitions, beliefs, and behaviors as part of a self-renewing process would be supported by, for example, Ervin-Tripps (1964) sociolinguistic framework, which studies verbal behavior "in terms of the relations between the setting, the participants, the topic, the functions of an interaction, the form, and the values held by the participants of each of these" (p. 192). For a recent treatment of the interrelations between language and context, see Duranti and Goodwin (1992).

Williams and Harwood (chap. 5) consider whether age groups are meaningful "groups" in the social identity sense of the term in this new chapter to the handbook. Intergroup and social identity theory provide a solid foundation for the content of this chapter that considers the possible self-alignment that all of us engage in as we segregate into age categories. The authors consider how soft these age categories may be and the possible effects age segregation has on our ability to communicate across the generations. In addition, Williams and Harwood discuss the possibility that intergenerational communication within a family context provides a rather unique set of issues and outcomes when compared to interactions between strangers of differing ages.

Flowing quite nicely from the previous three chapters, Baker, Giles, and Harwood (chap. 6) trace the history of intergroup and communication theory with respect to intergenerational interaction. Readers of this handbook should now be familiar with the great influence that social identity theory and communication accommodation theory has had on the study of communication and aging. These authors suggest a new model to explain and help us to understand intergenerational communication. The integrated model of inter- and intragenerational communication presents a holistic description of intergenerational communication, recognizes facets of social structure and cultural variability, and assumes that age distinctions within intergenerational communication may be advantageous to competent interactive behavior.

The final chapter in this section, authored by Pecchioni, Ota, and Sparks (chap. 7), considers the role that culture plays in communication and aging. The authors organize their chapter around four primary issues: (1) Chronological age is a poor predictor of aging-related behaviors, (2) definitions of and meanings assigned to the aging process vary culturally, (3) age may be treated as a coculture in its own right; and (4) the increasing percentage of older individuals within the population is likely to impact culture itself. The authors argue strongly and persuasively that any discussion of successful aging must include the dynamic and often elusive construct of culture.

REFERENCES

Duranti, A., & Goodwin, C. (1992). *Rethinking context: Language as an interactice phenomenon*. Cambridge, UK: Cambridge University Press.

Ervin-Tripp, S. (1964). An anlysis of the interaction of language, topic, and listener. *American Anthropologist, 6*(2), 86–102.

Fowler, R. (1991). *Language in the news: Discourse and ideology in the press*. London: Routledge.

Age in Social and Sociolinguistic Theory

Nikolas Coupland

Cardiff University

Age is sociolinguistics' underdeveloped social dimension.[1] Sociolinguistics has, for example, made an outstanding contribution to gender research, providing one of the cornerstones of modern feminist scholarship. Ethnic and race-related cultural patterns and differences and politics are, similarly, focal concerns in sociolinguistics with diverse and compelling literatures. Theoretical and empirical research centering on social class was, especially in the United Kingdom, one of the discipline's earliest, and ultimately most controversial, passions. By comparison, sociolinguistic research on age and ageing is, with only a few specific exceptions, rare. The most important exceptions are paradigms that have dealt with early parts of the life course—language development, variation and use in childhood and in adolescence, early bilingualism, and language use in educational settings. As this suggests, sociolinguistics has conceptualized the process of ageing mainly in relation to changes in early life maturation and the acquisition and deployment of communicative competences.

When sociolinguistics has taken its data from adult populations, age as a social issue has generally been neutralized. That is, adulthood hasn't been construed as a life stage, but rather as an unmarked demographic condition whose marked alternative is youth. It has usually been assumed that adulthood (implying young and midadulthood) is the empty stage on which the social dramas of gender, class, and ethnicity are played out in their various contexts. As Eckert says, in sociolinguistics, "only the middle-aged are seen as engaging in mature use, as 'doing' language rather than learning or losing it" (1997, p. 157). There has been little interest in the social experiencing of adulthood, its

---

[1] An earlier version of this chapter appeared in Coupland, Sarangi, and Candlin. (Eds.). (2001). *Sociolinguistics and social theory*. London: Longman.

relational demands and opportunities, or its identificational possibilities. But language is undoubtedly central to the lived experience of age and aging, and age is as potent a dimension of social identity as gender, class, or race. The politics of age and ageism, often mediated by language, impinge sharply on social life, and show signs of becoming both more prominent (in an "ageing society") and more diverse.

When we consider old age, the sociolinguistic slate is almost entirely clean. An important exception is the "apparent time" methodology (cf. Labov 1972, 1994) for studying language change in progress. This is the methodological device that captures language and especially dialect change by sampling the speech of different age cohorts at one point in time. For the apparent time device to work, sampling older informants has of course been an essential part of empirical designs, in documenting how specific patterns of linguistic usage have changed either quantitatively or qualitatively across successive generations. Beyond the language change paradigm, however, it is difficult to identify any substantial body of research driven by the wish to explore old age as a sociolinguistically significant life stage. The apparent time device uses old people as informants heuristically rather than focally.

There are many reasons for the absence of sustained, focal interest in the sociolinguistics of aging. One reason, as colleagues and I have previously suggested (Coupland, Coupland, & Giles, 1991), is the climate of veiled antipathy that surrounds the very concepts of ageing and late life in contemporary Western society. In this regard, unfortunately, sociolinguistics is more of a barometer of societal ageism than a force for understanding social ageing. The relevant theoretical concept is gerontophobia. Because we all age and fear its implications, repression is a predictable response (Becker, 1973), individually but perhaps also institutionally. This explanation imputes a critical naivety to sociolinguists, perhaps unfairly. But it is quite striking how even the most committed and libertarian theorists of gender, race, and class can sometimes be blind to the social politics of age—a point also made by Woodward (1995, p. 88).

A second, related reason is the sequestering of research issues to do with ageing into social policy and social services/welfare research literatures. Social gerontology, which has recently contributed important new critical and theoretical treatments, is a rapidly growing discipline (see the range of sources drawn on later). But it has had to resist the assumption that social ageing is an inherently "applied" concern—in that narrow and limiting sense of "applied" that excludes theory. That is, it is has been too readily assumed in the past that research on late life must address "the social problems of old age" and their remediation. It would be foolish to deny that certain health and social problems are probabilistically associated with old age. But to preset an agenda for ageing research in terms of such problems, and these alone, is to contribute to social disenfranchisement, when processes of disenfranchisement are part of the proper agenda for critical research.

A third reason may be the highly diverse social environments of old age in Western societies. We might legitimately ask whether the old are a powerful or a powerless social group. Is late life a time of fulfillment in retirement or one of penury and social exclusion? Because we all age and generally fear our own ageing, isn't old age a condition we must accept, the bad along with the good? So where is the moral imperative? It is possibly true that sociolinguistics has been more easily drawn to social themes in which it can construe large-scale, near universal, within-category social inequalities and establish its agenda relative to these boundaries. This would be a blinkered orientation, and there are many reasons to challenge the assumption that any demographic group is essentially defined by its disempowerment, and that this is the required motivating criterion for researching it. There is certainly no shortage of social practices and arrangements that fall within the remit of the term *ageism*, but the fact that not all old people are equally and unremittingly prey to societal ageism should not deflect research interest. In any case, social inequality is not the only motivating issue for sociolinguistics, which should be interested in documenting and modelling the changing social and ideological configurations of late life, as mediated by language.

The further reason for sociolinguistics' neglect of old age in many ways subsumes those mentioned above. It is that sociolinguistics has operated with an extremely impoverished social theory of ageing, and has not been sufficiently alert to how its own simple, implicit theory has limited its endeavors so far. Age itself may have seemed an uninteresting, almost mechanistically linear demographic dimension, which would be unrewarding except in possibly being able to index graduated linguistic change. On the contrary, modern social theory offers rich insights into ageing as a multidimensional and historically shifting human, social, and cultural process. Appreciating the theoretical richness of social ageing should expose the importance of a sociolinguistic contribution to the social science of age, and the importance of age to sociolinguistics. Before expanding on this, I shall comment on the social theory that implicitly drives the "apparent time" method, which defines the sociolinguistic status quo on researching age in mid- and late adulthood.

AGE IN THE APPARENT TIME DESIGN

Chambers discusses age-related research in sociolinguistics in his volume, *Sociolinguistic Theory* (Chambers, 1995, chap. 4). "Age," Chambers writes, "exerts an irrepressible influence on our social being. Our age is an immutable social fact" (1995, p. 146). Unlike social class, where mobility is possible and indeed common, and unlike gender, where gender roles are nowadays "less confining," Chambers argues that "our ages remain fixed," and that age "plays an almost autocratic role in our social lives" (Chambers, 1995, chap. 4). Already, in Chambers' introductory comments, we see what appears to be a principled rationale for treating age differently from other social dimensions in sociolinguistics, because of its immutability. In fact, Chambers implied that age is not a truly social dimension at all, if by this we mean one that is socially negotiable. He wrote that "By and large, the physical indicators of age are shared by all people in all cultures" (p. 147). If our age is physically imprinted on us, age can indeed appear immutable and its influence "autocratic."

Against this backdrop, Chambers then identifies "the three sociolinguistically crucial ages": the time of first exposure to social pressures from parents and peers; adolescent networks; and early adulthood. It is worth quoting his summary:

> It is clear, in the first place, that vernacular variables and style-shifting develop along with phonology and syntax from the very beginning of acquisition. From that point, there appear to be three formative periods in the acquisition of sociolects by normal individuals. First, in childhood the vernacular develops under the influence of family and friends . . . Second, in adolescence vernacular norms tend to accelerate beyond the norms established by the

previous generation, under the influence of dense networking . . . Third, in young adulthood standardization tends to increase, at least for the sub-set of speakers involved in language-sensitive occupations in the broadest sense of the term . . . After that, from middle-age onward, speakers normally have fixed their sociolects beyond any large-scale or regular changes. (Chambers, 1995, pp. 158–159)

In this view, the assumption that mid- and late life are not "sociolinguistically crucial" is concretized, although this is a theoretical position reached by a selective route. It depends on a prior definition of sociolinguistics as exclusively concerned with dialectal / sociolectal (and within this, mainly phonological) dimensions of language, and on a model of dialects becoming "fixed for life" by early adulthood. Limits on space preclude me from discussing the limitations of a dialect-only perspective here, also the limitations of a language-centred rather than a user- and use-centred perspective (see Coupland, 2000a). Also, we can note in passing that focusing on acquisition prioritizes within-culture rather than multicultural or intercultural processes (although Chambers does discuss data from several multilingual settings) and underplays geographical and sociocultural mobility during adulthood. It also ignores all manner of social-contextual motivations for dialect variation, including variation across speech (including performance) repertoires. In fact, details of so-called stylistic variation receive little attention in Chambers' review generally. (See Coupland, 2000a, for discussion of "style" in sociolinguistics.)

I am more concerned here with the general, basal theory of ageing, which has prevailed in sociolinguistics, and Chambers is rather explicit about it. Chambers distinguishes two kinds of age-related changes that relate to language. One is *sound change in progress*. Small deviations from the cultural norm (for example, in pronunciation) that are sometimes introduced during and after socialization mean that a community's language tends to change with succeeding generations. Younger people's speech tends to drift in specific respects away from that of their older family members, whose speech is assumed to remain largely unchanged through their lives. Sound change in progress can be established in "real time" or in "apparent time." Real-time studies sample the speech of the same individuals over time, longitudinally. Apparent time studies sample different but matched samples of people at different ages, creating an apparent time-depth effect. Chambers holds that demonstrating sound change in progress, as variationist sociolinguistics has done (Labov, 1994, is the standard exposition) is "perhaps the most striking accomplishment of contemporary linguistics" (Chambers, 1995, p. 147).

The other sort of age-related change, in Chambers' and many other sociolinguists' terms, is *age grading*, which he says is "relatively rare" (p. 147).[2] This is "maturational

---

[2]On the other hand, Chambers characterizes the physical and cultural indicators of ageing as overwhelmingly age graded:

> . . . advancing age brings wrinkling skin, weight re-apportionments in chest and abdomen, greying hair and, for men, receding hairlines . . . only old women dress in solid black to go shopping, only young men wear team sweaters, only little girls wear buckled sandals, only teenagers wear ripped jeans, only old men wear felt fedoras to sit in the sun in the park . . . (p. 148).

These are what Chambers called "infallible indicators" of old age, although this account now seems remarkably traditionalised and stereotyped.

change repeated in successive generations" (Chambers, 1995)—when people, of more or less whatever birth cohort, and their behavior generally change as they age. Age grading is a developmental process that, we might say, allows individuals to escape their historical and cultural rootedness. It implies that a chronological age designation, like "being 47," may not reflect membership of a specific culture-bound cohort born 47 years ago and which has maintained its socialized speech characteristics. Rather, it reflects the changes that individuals have undergone as they age, individually or collectively, and the behavioral gap between what one does at age 27 and 47.

Chambers offers two examples of the apparently rare sociolinguistic phenomenon of age grading (pp. 188–193). One is that children regularly use the form "zee" (for the last letter of the alphabet) in southern Ontario, Canada, then replace it with "zed" before adulthood. (Use of the "zee" form is speculated to be triggered by *Sesame Street* on TV, accounting for its use by the very young, who then drop out of using it as they age.) The other example is the use of glottal stops in Glasgow, Scotland, and elsewhere, based on evidence in Macaulay's (1977) study. Middle-class 10-year-olds use stigmatized glottal stops but lose this feature by the age of 15. Vocal creak and pitch are age-graded prosodic and paralinguistic features and certainly mark age, and in an "age-grading" sort of way. But Chambers (I think wrongly) says that creak and pitch "do not carry any special social significance" (p. 151), feeling that they cannot be stigmatized because they affect the whole population. Chambers considers that the scarce evidence of age grading may be misleading; it might just be that few instances of age grading have been noticed by sociolinguists and reported. Age grading, as a conceptual model of change, undermines the apparent time perspective, which needs to assume that speech changes during adulthood are minimal.

Other sociolinguists are much less confident about the preponderance of cohort effects—the persistence of socialized patterns of language use throughout the life span—over age-grading effects. For example, Eckert (1997) argues that the extent of social contextual changes over an adult's life course will necessarily impact on speech characteristics: "Progress through the life course involves changes in family status, gender relations, employment status, social networks, place of residence, community partici-pation, institutional participation, engagement in the marketplace—all of which have implications for patterns of [sociolinguistic] variation" (Eckert 1997, p. 152).

Eckert reviews several sociolinguistic variation studies that, by revisiting earlier re-search communities and studying the same or closely comparable populations, have found evidence of both historical changes in the speech of the community (differences between successive cohorts a the same age), and also evidence of age grading (differences within cohorts between the two times). Cedegren's study of Panama is discussed as a clear instance by both Eckert (1997) and in detail by Labov (1994, pp. 94–97); I will not summarize the data and findings here.

Labov's own overview also entertains complex possibilities about linguistic changes in real time and apparent time, and about how our understanding of change in gen-eral is hampered by methodological and practical problems. Labov suggests, "Many well-established sociolinguistic variables exhibit . . . age-grading, where adolescents and young adults use stigmatized variants more freely than middle-aged speakers, especially when they are being observed" (1994, p. 73). He gives the examples of the stereotyped (*dh*) variable (using stops in place of fricatives in the words *these*, *them*, *those*, etc.) in Philadelphia and New York City. But his general conclusion is that there is evidence of considerable stability across the adult life course, particularly in the use of phonological features below the level of speakers' conscious awareness. He writes that sociolinguistic differences generally "indicate that generational change rather than communal change is the basic model of sound change" (p. 112). That is, linguistic change is achieved more by new generations adopting qualitative and quantitative speech features that are different from those used by older generations, rather than by whole communities (including all age groups) simultaneously adopting new features.

DAMAGING EXTRAPOLATIONS: "AGED ALIENS"

In one sense, the success of the apparent time device is its own justification. It has clearly been a key methodological resource for investigating sound change that, as I mentioned, has been the dominant concern of variationist sociolinguistics. With the caveats expressed, the evidence does indeed suggest that, in relation to dialect, older adults do largely maintain their socialized speech characteristics into later life. But, as Eckert's quoted comments imply, what we might call the *cohort continuity model* is highly inappropriate as a general model of adult sociolinguistic development, because it edits out all manner of sociorelational changes in which language use is predictably implicated. It is, in fact, a model of nondevelopment in adulthood—of stasis. In Chambers' words, "those inferences [about language change] depend on the validity of a particular hypothesis, namely that the linguistic usage of a certain group will remain essentially the same for that group as they grow older" (pp. 193–194). Whatever its factual status for dialect change research, the model is potentially damaging as a general theoretical orientation, because it accords with a partial truth (dialect stasis does not imply sociolinguistic stasis) and with a virulently pervasive social stereotype, certainly of late ageing (older people being "set in their ways" and "inflexible").

Compare, for example, one nonlinguist sociologist's characterization of the cohort continuity perspective:

> The aged may be seen as travellers in time . . . They have come down to our own time from what we, the younger generations, think of as the distant past, long before we ourselves were born. Some even come from the last century, quite a few from before World War I. In a sense the generational gap is the equivalent to cultural difference. (Westin, 1994, p. 134)

Westin's characterization captures the social alienation that a strong version of the cohort continuity model imputes, but that Chambers and many sociolinguists (through the filter of a dominant concern with dialect change) take to be the normal case. Westin talks about "the aged," but if early adulthood defines the end of what is "sociolinguistically crucial" about age, it follows that, for most of our adult lives, we are relegated to the status of sociolinguistic aliens who seem to "travel in time." In this view (but using a slightly more appropriate metaphor), it is "time" or "being in time" that "travels" past us. The assumption that maturation and enculturation "fix" our sociolinguistic and social selves implies that, as ageing adults, we drift further and further away from the presumed mainstream. Culture, including language usage norms, moves on beyond us. By implication, the older we become, the less culturally attuned to the present we seemingly become. We, presumably, travel increasingly in other people's time. As we will see, this is a dated and damaging model of social ageing.

The "out-of-time-ness" of older people is, however, a recurrent element of age mythology and of ageist stereotyping in the West. The fund of age-prejudicial referring expressions ("old fogeys," "old farts," and "old biddies," used in the UK—see Nuessel, 1982, 1984) distances them from the mainstream. Some older people will themselves recycle tropes such as "I've seen better days" and "in our day," where the days invoked are noncurrent, "long gone." In a darkly humorous key, older people may refer to their earlier life stages as "when I was alive." A sense of disjunction from present sociocultural concerns is of course a potential effect of a long-lived life. Social arrangements and lifestyles may conspire to dislocate individuals from what is held to be the mainstream. Some older people might in fact value having left the mainstream, in certain respects. But adaptation and sociocultural "age grading," including sociolinguistic adaptation, are equally feasible even if we assume that the cultural core is occupied by youth or young adults. In fact, there are many reasons to suppose that older people occupy a more mainstream cultural position nowadays than previously, also that the core–periphery structure is generally looser than we have traditionally assumed. I will come back to these issues below.

In any event, the important sociopolitical point about the out-of-time and out-of-culture status of old age is that, however accurate or inaccurate a claim it is, social dislocation has been institutionally promoted in the West, tending to naturalize the notion of social as well as occupational "retirement." As Phillipson (1987) notes, "The emergence of retirement has been one of the most significant social trends of the past fifty years ... The *expectation* of retirement at a fixed age (or earlier) now affects the thoughts and plans of millions of men and women [in the UK]" (Phillipson 1987, p. 156, with original italics). In terms of personal responses, retirement can be either a welcome release from constraint or exclusion from financial security, status, opportunity, and a social network. But either way, in the economic framing of human worth that developed within capitalism, it has been occupational activity and "productivity" that has come to define a boundary between the core and periphery of society.

The pattern of enforced retirement from work at age 60 or 65 (see later discussion), especially as it developed at the height of the midmodern industrial period in the middle years of the 20th century, produced what Parsons (1942, p. 616) called, in an American context, a "structural isolation" from work and the community. Correspondingly, the social category of "elderly people," criterially defined as those who are no longer "economically productive," was in large part created by the new requirement to retire from working at a set age. As Parsons also commented, this structural development also engendered the view of late life as a "problem" and triggered the first evidence of resistance against societal ageism.

But before any form of anti-ageist program or even the notion of ageism itself (Butler, 1969; Bytheway & Johnson, 1990; Williams & Giles, 1998) could consolidate, developmental social science coined wholesale theories of late life based around the notions of "rolelessness" and "disengagement" (e.g., Rosow, 1974; Cumming & Henry, 1961;

Phillipson, 1987, p. 178ff.). Cumming and Henry (1961) argued that disengagement from demanding social roles was actually an avenue to improved morale in late life. It was portrayed as a natural, desirable, and culturally universal condition. Even Erikson's (1980) famous taxonomy of developmental stages of "identity and the life cycle" posited that each stage of life held out "appropriate" tasks that needed to be completed before a person can move on to the next stage (see also Blaikie, 1999, p. 7). Later studies, including Phillipson's, stressed the disruptive effects of occupational retirement, and even its propensity to engender "shock" and "crisis," and it is evident that retirement is, for many, an undesirable or at least ambivalent social transition. Current sociological theory stresses how the emergence of a retirement norm was part of a wider process of institutionalizing the life course (Kohli 1991, p. 277ff.). The work economy came to define people as either "productive" (in their young and midadulthood), or as preparing for this status (through education and work apprenticeship), or as having left this status (retired). These processes imbued the notions of life stages and chronological age with rather rigid meanings and reified them at the level of social structure and practice. As Kohli suggests, this construction was also conducive to seeing social groups, theoretically, in terms of temporal cohorts, moving in succession through the economy. As J. Coupland also points out (in a personal note), it is also a predominantly male-focused construction, because the work economy was, until relatively recently, one from which women were allowed limited participation.

Here we seem to have moved some considerable distance away from the apparent time method and variationist sociolinguistic interests in language change. But it is important to appreciate how the emphases of this sort of sociolinguistics in many ways perpetuate a "modernist" (as opposed to "late-modern") model of ageing and the life course—one based on chronological age categories and on tracking the characteristics of successions of age cohorts. The sociolinguistic model expressed through studies of dialect change presumes a strongly institutionalized and regimented life course, where age-in-years predicts social characteristics. More than that, variationist sociolinguistics has inherited the "time-traveling" perspective, according to which older people are expected to be (and are then shown to be, in certain very specific respects) progressively distanced from contemporary sociocultural norms. But whereas the modernist perspective constructed retirement from work as the most distinctive alienating moment, variationism has established early adulthood as the onset of cultural alienation, in terms of dialect use. It is important to repeat that variationism has needed to stand by this model, because it is consistent with its most common pattern of results. At the same time, sociolinguistics needs to consider, and contribute, other theoretical models of social ageing, and especially those more attuned to contemporary, late-modern social arrangements.

THE NEW OLD AGE: SOCIAL AGEING IN LATE MODERNITY

Theories of late modernity (e.g., Giddens, 1990, 1991) are based on observation of rapidly shifting social patterns and priorities, involving critical reappraisal of traditional social orders. In some discussions, they admittedly also include pockets of millennial ideology and hyperbole. It would be wrong to assume that the social conditions of the early 21st century have as yet been comprehensively or convincingly theorized, although some trends are well established.

It has been clear for some time that Western societies have moved out of an epoch of full-blooded industrialization and developed much more active service economies. Linked to this key shift, traditional, rigid patterns of social organization, most obviously class and gender configurations linked to occupational roles, have loosened. As a result, late modernity can generally be characterized as a more flexible social order, although by no means necessarily a more liberal or equal one. The growing service sector of the economy requires and gives increasing capital value to the management of social relationships, perhaps to informality and intimacy, and certainly to communication and language (Bourdieu, 1991). Cameron (1995), on the other hand, shows how economic priorities can lead to repressive workplace practices, including the scripting of repetitive communication tasks for low-paid workers, for example in fast-food outlets or in telephone sales and other sorts of call-center work. Developments in communication technology and cheaper travel have broken many of the constraints on how people associate and communicate. Globalizing modes of communication are likely to impact on the directions and rates of language change, but may also produce new styles, genres, and functions of language. Like all social behavior, language is also increasingly reflexive (Beck, Giddens, & Lash, 1994; Giddens, 1994; Jaworski, Coupland, & Galasinski, in press; Lucy, 1993), in the sense that we see social life relayed back to us, reformatted, mimicked, or analyzed by photographs, radio, video, television, and the new media.

This brief and informal overview is perhaps enough to suggest that the social organization of the life course is also beginning to take new forms. Giddens (1991), for example, made the case that social identities in late modernity are best viewed as "projects." We reflexively manage our social identities, including our identities in age terms, updating them progressively and particularly at salient life-transitional points:

> The reflexive project of the self, which consists in the sustaining of coherent, yet continuously revised, biographical narratives, takes place in the context of multiple choice as filtered through abstract systems. In modern social life, the notion of lifestyle takes on a particular significance. The more tradition loses its hold, and the more daily life is reconstituted in terms of the dialectical interplay of the local and the global, the more individuals are forced to negotiate lifestyle choices among a diversity of options. (Giddens, 1991, p. 5)

This implies a process of deinstitutionalization of ageing and of old age, and at least potential access to greater self-definition and social opportunities:

> Once institutionalised through retirement, later life is now being deinstitutionalised. This fracturing suggests that rather than focus on the social construction of the life-cycle, as a fixed set of stages occupied by people of particular age bands, we should analyse the ways in which it is being deconstructed by individual elders, or groups of older people, negotiating their own life courses. (Blaikie 1999, p. 59)

Blaikie's suggestion is a resonant one for language and communication research, because it shows that there is work to be done on the discursive negotiation of life course identities. It is work that contributes directly to the social theorizing of age, rather than being conditioned by preexisting social theories, implicit in its designs. Communication research can explore the extent to which, and the ways in which, age is (re)negotiated in different settings, under what constraints, and with what consequences. This dynamic perspective on identity and age identity in particular has an established pedigree. It can be found very explicitly, for example, in Mead's (1932) process model of identity, which assumed that the self emerges through interactional experience. A thread of social psychological theory has for some time stressed the on-the-ground, negotiative nature of age identity (Ainley & Redfoot, 1982; see also Taylor, 1989). I refer below to some of my own work with colleagues on age identity in talk.

But late life is not radically freed from all its social constraints by the shifting forms of industrial capitalism. As Blaikie and others have pointed out, the late-modern social order is still heavily structured and unequal, albeit rather less deterministically so. Social differentiation and inequalities persist, but they are visible more through patterns of consumption and other aspects of lifestyle rather than in the social fabric of class heredity and occupational status. Social differences are often referred to as lifestyle choices, although which groups have the economic resources and the will to choose specific lifestyles and consumer goods remains a crucial question. As fixed patterns of work in adulthood begin to break down, and as people retire earlier, and as leisure becomes more common, more diverse, and itself more institutionalized (Coupland & Coupland, 1997), retirement certainly loses much of its inherently sequestered nature. Older people—or at least those with adequate resources—can merge with the late-midlife early retirees or those in part-time or flexible work who define themselves through possessions, club membership, travel, sports, socializing, and other forms of recreation. In refreshing contradiction of Chambers' essentialist reading of the styles of old age (see footnote 2), Featherstone and Hepworth (1993) comment that dress styles may be becoming "uni-age," just as they have at various times been "unisex." But, as previously mentioned, it would be naïve to claim that late life as a whole has thrown off its economic and ideological constraints, and opened up an older person's playground for leisure and self-definition. It is important to keep track of the economics and demography of old age, and to observe the discursive working through of ideologies in that context. Let us consider economics and demography first.

Irwin (1999) summarizs recent demographic data for the United Kingdom, where, among men aged 60 to 64 in 1975, 1985, and 1995, respectively, economic participation rates (as they are conventionally called) declined from 84% to 53% to 50% (see also Arber & Evandrou, 1997). That is, according to this trend, less than half of U.K. males aged 60 to 64 are currently in paid work, compared with well over 80% in 1975. These data graphically illustrate the extent of the social changes we have been discussing, but they also foreground the question of social inequality. So-called early retirement is occasioned by both affluence and poverty, for different subgroups. Much-discussed data showing the significant increase in the number of "pensionable" people as a proportion of total populations have clearly established that the post-retirement segment of society is growing rapidly—for example, 2.7 to 9.7 million in the United Kingdom between 1901 and 1981. But other data (which Irwin discusses) also show that working-age groups, in the United Kingdom and in Europe generally, have become significantly less equal socioeconomically over the last 20 years. This, in turn, implies that inequalities among post-retirement populations have increased and will continue to do so.

This is an important generalization. As we have already seen, social theory has tended to orient to ageing in age-categorial terms, attempting to generalize about age cohorts and their social environments. Irwin's perspective reminds us that any age-stratificational model of ageing, and any generalizations about "the elderly," under-represent important social divisions within this chronologically widespread and itself economically highly stratified population. This is also therefore an important qualification of Laslett's well-known distinction between the "third age" of postwork fulfillment and the "fourth age," which Laslett called "an era of final dependence, decrepitude and death" (Laslett, 1989, p. 4). Neither of these "ages" will be experienced homogenously by age cohorts. For the poor, the "third age" will not exist. Sociolinguistics and communication science need to operationalize age in a correspondingly complex way.

Similarly, generalizations about the new economic power and influence of older people in the marketplace need to be qualified. In 1991, in her chapter titled "Gold in Gray," Minkler wrote that "the elderly population [in the United States] . . . is considered the newest exploitable growth market in the private sector" (p. 81). It is undoubtedly true that older people, overall, are more affluent, not least in the United Kingdom as a result of increased rates of home ownership through the period of extreme house-price inflation in the late 1980s. Minkler noted that, in the United States, the incomes of families headed by people aged 65 and above rose by 54% between 1970 and 1986 (Minkler, 1991, p. 82). But she also confirmed that "There is tremendous income variation within the elderly cohort . . . and deep poverty pockets continue to exist" (p. 83). Many older people occupy the socioeconomic mainstream nowadays and can therefore resource the lifestyle choices they make. But, for example in the United Kingdom, the steady decline in the (non-means-tested) state pension, at a time when many people approaching retirement age are not members of occupational or private pension schemes, leads one to expect increasing poverty, as well as increasing affluence in this sector.

IDEOLOGIES OF OLD AGE

The changing sociostructural arrangements in which age is embedded should influence social scientific research agenda, but should also be part of them. If we take sociolinguistics to be the study of social life through language, as much as the social study of language change, a wide range of age-related issues presents itself for sociolinguists. Part of this broader agenda for the sociolinguistics of ageing lies in the analysis of ideology—to consider how the changing social meanings and values of old age are conveyed in language and related symbolic practices. Several studies of ageing and later life are first responses to this challenge (e.g., Cole & Gadow 1986; Green, 1993; Featherstone & Wernick, 1995, Jamieson, Harper, & Victor, 1997), although they are not sociolinguistic projects, as conventionally defined.

The Cole and Gadow collection, for example, starts from the assumption that the fundamental meanings of ageing and of old age are unclear and changing. Our understandings have been obscured, partly, by biomedical concerns that have swamped gerontology (see also Estes & Binney, 1991, on the "biomedicalization of ageing"; also Green, 1993). As sociostructural binds have loosened, asking (in Cole's & Gadow's title) "What does it mean to grow old?" has a very contemporary salience, and we would expect different clusters of responses. At the same time, structures of meaning and value ("discourses" in Foucault's abstract sense—see, e.g., Foucault, 1980) can often be more tightly woven than the social conditions they represent and influence. They can act as a repressive force, traditionalizing the experience of a social condition even when social circumstances and structures have moved on. The meanings of old age are therefore to be found not only in the economic and lifestyle characteristics of different groups of older people, but also in the recurrent discourses of age that we regularly imbibe and perhaps recycle.

Ageism, in Butler's (1969) original definition of the concept, was the complex of material disadvantages and prejudicial attitudes and predispositions of younger people toward older people (see also Butler, 1975). But ageism is probably better viewed as a complex ideological formation—a structured, historically formed set of myths or discourses that endorse the subordinate or marginal positions and qualities of the old. Even more affluent and less structurally constrained older people will be prey to the effects of an ageist ideology, and they may confirm elements of it in their own talk and practices. If we assume that ideologies will occasionally surface as linguistic and visual representations, as part of discourse in the less abstract sense (Jaworski & Coupland, 1999), however mitigated or recontextualised, then analyzing the discourse of ageism is another key project for language-based research. Several potentially productive subthemes for a critical, ideological analysis of ageing and ageism have emerged. My own work with colleagues has addressed some of them, and I also reference some other recent studies. But what follows is more an agenda for future research than a list of completed studies. Because of limits on space, I can only list some key issues here, without extended commentary.

Gerontophobia and the "Othering" of the Old

I referred earlier to a generalized societal gerontophobia—fear of or revulsion toward old age, which is the clearest manifestation of ageism. Its psychoanalytic origins are in the repression of death, to which "deep old age" is a precursor. In an illuminating critical study of responses to artistic and literary images of old age, Woodward (1991) developed an account of the "unwatchability" of the elderly nude as a classic gerontophobic impulse. Physical, including physiognomic, characteristics of old age have been recognized as powerful cultural markers (Blaikie, 1999, chaps. 4 & 6), and probably underlie the most negative stereotyped reactions to old age. Sociolinguistics has not systematically examined linguistic–textual parallels, which may be more mitigated and subtle than visual representations. The "othering" of older people is an urgent agenda for sociolinguistics (for a review of issues, including gerontological ones, and a suggested taxonomy of "othering" strategies, see Coupland, 2000b).

We should not expect language practices to be transparent carriers of ageism, however. Alienation through discourse can be achieved by silence or underrepresentation (see, for example, Rodwell, Davis, Dennison, Goldsmith, & Whitehead, 1992), and even by apparently supportive styles and stances (see references below on patronizing talk). Featherstone and Hepworth (1995) wrote that "The increasing preoccupation in social gerontology with positive ageing . . . arises out of the critical belief that we live in an ageist society, one in which the predominant attitude towards older people is coloured by a negative mixture of pity, fear, disgust, condescension and neglect" (p. 30). We need to read the ideological contexts in which age politics are transacted with some sophistication. Even so, it is also true that even the most transparent forms of ageist reference to old people have yet to be studied systematically (the studies I mentioned by Nuessel are a useful first contribution).

As this work develops, cross-cultural perspectives will be important to offset the tendency for Western experiences and shifts to dominate theory. The rather static and traditional conceptualizations of age that I commented on above have remained unchallenged for want of considering alternatives. The research themes I am setting out here all need to be contexualized by research in different cultural contexts. Ethnographic approaches developed within sociolinguistics (Hymes, 1974) are well suited to taking on this task.

Age-appropriateness and the (II)legitimacy of Old Age

As the demography of late life changes and as lifestyle options increase, at least for wealthier old people, a conservative discourse continues to structure old age by imposing norms of age-appropriateness. As I argued, the traditional cultural order, mainly through its hierarchical occupational structure, dictated fixed ways of being old, and ways of speaking are no doubt included in them. Retirement has traditionally been constructed as a form of withdrawal—from status, fulfillment, and from some sorts of social relationships. Although no sociolinguistic research has taken up this theme explicitly, it seems true that people in retirement have been expected to be less vocal, for example, in the political arena. But even in contemporary life, discourses of age-appropriateness are commonplace. There are many discursive ways, not yet systematically examined, in which the potential freedoms of late-modern old age are undermined, for example, in media texts or in humor. Many rock and pop music icons of the 1960s and 1970s are regularly pilloried as "dinosaurs of rock," "repaying the debts of yesteryear," and so on. Age boundaries continue to be policed in the popular press, for example, through the ubiquitous practice of what we might call "age tagging." People featured in news stories are very commonly introduced by name plus an appositional age-in-years tag ("Mary McGregor, 36"). The discourse function of the age tag is often to indicate some element of surprise or even indignation at an action linked to an age category, such as a teenager giving birth or an old person apprehending a burglar. The same textually manufactured disjunction generates headings, like, "Geriatric Mum" or "Nintendo Granny." Physical attractiveness and certainly sexuality are generally portrayed as illegitimate in later life (Gibson, 1993). The physical "youthfulness" of Joan Collins, Jane Fonda, and others (cf. Blaikie, 1999, p. 107) is not without its detractors, who might argue that theirs is an "unnatural" or "undignified" version of later life.

Although older people can be penalized for acting out of age, it is unclear what is taken to be the legitimate condition of old age. This is part of the definitional ambiguity of old age. Featherstone (1995, p. 25; see also Featherstone & Hepworth, 1990, 1993, 1995) argues that contemporary Western societies endorse two contrasting sets of images. In the first set are the "heroes of ageing" who are "positive" about ageing, in the tautological sense of maintaining "youthful" habits and demeanor. In the second are those whose physical selves have degenerated to the extent of being incompatible with and disguising their inner selves, the so-called mask of ageing. Neither condition implies that there is a social space that old people can legitimately occupy as old people (see later discussion).

We have previously shown how presumed norms for health and well-being in later life strongly color older people's representations of themselves, for example, in responses to "How are you?" questions in varying settings. "Not bad for 75" and the general strategy of relativizing health to presumed, normative decrement with age are conventional responses (Coupland, Coupland, & Robinson, 1992). Elaborated accounts structured on similar lines are commonly also found in medical situations (Coupland, Robinson, & Coupland, 1994). We have also traced the ideological bases of notions of "premature ageing" and "anti-aging protection" as they surface in the texts of skin-care product advertising (J. Coupland & N. Coupland, 2000; Coupland, 2003), and Coupland and Williams (2002) have explored the ideological underpinnings of texts on the menopause and their ageist constructions of gendered life span development.

The "Inverted U" Stereotype of the Life Course

The visual image of two bent, huddled old people crossing a road is strongly iconic of old age in the United Kingdom, where it still features as a road sign warning of an "old people's home." Physical smallness is a stereotyped (and, no doubt, quantitatively, an actual) characteristic of both the old and the young. These groups are often represented as being similar in other ways, for example, in terms of incompetence, dependency, and societal marginality. We have referred to this association as the "inverted U" model of the life course (Coupland, Coupland, & Giles, 1991; Hockey & James, 1993, 1995), and old people are commonly said to be in their "second childhood." On the other hand, it may be that children are presumed to be legitimately dependent, and old people to be illegitimately dependent (Hockey & James, 1995, p. 140). Correspondingly, a very active research paradigm has developed accounts of the sociolinguistic infantilization of old people under the rubrics of "secondary baby-talking" (Caporael, Lucaszewski, & Culbertson, 1983), "overaccommodation" (Coupland, Coupland, Giles, & Henwood, 1988), and "patronizing talk" (e.g., Hummert, 1994). This last approach, mainly developed through experimental research in the social psychology of language, takes an explicitly evaluative and political line on "intergenerational miscommunication."

Ageism and Anti-ageism

All the themes that I am summarizing here help to fill out our understanding of societal ageism as an ideology and a related set of sociolinguistic practices. But as I suggested, sociolinguistics needs to be cautious in its appeal to the notion of ageism and needs to develop theory in this area too. For example, ageism is not simply or uniformly "something done to older people by younger people," as a generational conflict theory would assume, because many of the conventional tropes of intergenerational talk involve older people in a form of self-disenfranchisement ("What can I expect at my age?" etc).

Again, because ageism is an ideological formation and not just a set of prejudicial personal beliefs, it can be conveyed even by well-meaning and would-be liberal people. We have to recognize forms of ageism conveyed, for example, in oversimplistic and romantic images of old age (Achenbaum, 1995). There is a recurring semantic that associates late life with warmth and comfort ("golden agers," "the Golden Girls," "the sunset years") and another that imputes dignity and wisdom ("senior citizens," "seniors"). "Positive" though these qualities are, they may well have been coined in the knowledge that the circumstances of old age are *less than* rosy and statusful. In a reflexive sociolinguistic society, these historico-political machinations of reference can often be understood. It is therefore difficult to disentangle "compensatory anti-ageism" in address or reference from ageism itself (cf. Binstock, 1983, p. 140, on "compassionate ageism").

In analyses of first-acquaintance conversations between women in their 30s and 40s, and others in their 70s and older, we found a propensity for younger women to presume speaking rights and conversational access to their older partners, nonreciprocally (Coupland, Coupland, & Giles, 1991, chap. 4). The data suggested that the younger women's conversational styles were intrusive or negative face threatening, and yet some older people approved of them. They read their younger partners' actions as signaling interest in their personal lives, which they welcomed. Age telling by older people ("I'm seventy-four next month, actually") and age complimenting by younger partners ("Gosh, you don't look it!") is another ritualized intergenerational discourse practice (Coupland, Coupland, & Giles, 1991, chap. 6). These conventionalized responses are denials of ageing, which may well be locally motivated by anti-ageist intents, and again they may be well received. But denial of ageing returns us to the issue, already mentioned, of ageism working subtly and often in multiply embedded ways to deny a legitimate social space for old age, with its distinctive advantages and disadvantages.

OVERVIEW

In the previous, necessarily eclectic, review, I hope to have shown the yawning gap between a highly circumscribed mainstream sociolinguistic paradigm concerned with age (what Chambers, 1995, refers to as the study of "accents in time") and a complex theoretical conception of the sociolinguistics of ageing. Ageing is, with little doubt, a future priority for sociolinguistics. This is partly because of the much discussed, massively shifting demography of the life course in the West. As many writers have suggested, social science cannot ignore a post-retirement segment of society that outnumbers working adults. But the apocalyptic tenor of many such reports ("We must attend to the looming elderly health disaster and the onward march of the ageing hoards") should not color research priorities. The general imperative is for sociolinguistics to take on the task of clarifying the radical uncertainty of what social ageing means and how it is negotiated in language and discourse. In Western societies, the social context will be detraditionalization and the new politics of age. Other societies will have radically different histories of intergenerational relations and other problems and possibilities.

Sociolinguistic research into ageing needs a sociopolitical dimension. As I sketched it in the previous section, sociolinguistics needs to address a "disenfranchisement agenda" of old age, much as it has done in relation to social class, gender, and ethnic relations. The patterns and modes of disenfranchisement will, however, be distinctive to the ageing domain and will vary—from one cultural context to another, and from one social group to another within particular societies. The language change paradigm has presented a thoroughly *de*policitized perspective on ageing, and this is how Chambers and others can claim that mid- and late life are not "sociolinguistically crucial" periods. Age is undeniably the basis of social inequalities, for the young as well as the old, although I have not considered the former case in this review. Sociolinguistics has demonstrated that power, control, and authority are routinely transacted through language. Many older people are socially disadvantaged, and poverty in late life still has a structural social basis—for example, because the United Kingdom still allows the working-age poor to enter retirement without an adequate financial safety net. However, as I suggested above in commenting on socioeconomic trends in the United Kingdom and the United States, it will be important not to overgeneralize about "the disadvantaged elderly." We need to be clear to distinguish the specific older populations for whom disadvantage is socially structured and the larger number for whom disadvantage is ideologically ascribed. Language, discourse and communication will be implicated in both cases.

Beyond issues of disenfranchisement, the "age-identity agenda" is tantalizingly poised for sociolinguistics to explore. In my own work with Justine Coupland, Jon Nussbaum, and other colleagues, I have suggested that age identity is often nearer to the surface of talk and text than other dimensions of social identification. There are very many social encounters where age is made immediately and obviously salient, and where it becomes a thematic resource for talk. In most public encounters between nonacquaintances, gender, race, and class will be covertly negotiated, and their salience and consequentiality will be difficult to read at the surface of talk. But age-work, especially if we mean in the context of young adult–old adult interaction, is often overt, for example, in the actual telling of age in years (mentioned earlier) or in talk about social or personal change. Age can be told because it plausibly claims a dimension of status—experience achieved through longevity, or the presumption of earned respect. But it can also serve to account for incompetence, incapacity, or frailty. Another age-related genre we have examined is troubles telling, or painful self-disclosure in intergenerational discourse (Coupland & Coupland, 2002; Coupland, Coupland, & Giles, 1991). But these are very specific foci, textually and socioculturally, and a much wider view is needed. A general finding in the sociolinguistic analysis of age-identity may prove to be that age values in discourse are radically out of step with the new social conditions of late life in the West.

When sociolinguists write about identity, they often interpret the term in a rather anodyne way, as if "having an identity" or "negotiating an identity" were selecting and displaying options from a repertoire of equally plausible alternatives. Antaki and Widdicombe (1998), for example, write about "using identities" as the unmarked case. Although sociolinguists have always known that identity "matters," contemporary theorizing of identity stresses contingency and how a multiplicity of identificational symbols are both projected and ascribed. There are important and credible claims in these literatures (Duranti & Goodwin, 1992, provide an excellent overview). But in the context of ageing, identity work assumes a more profound personal importance than this model proposes. The negotiation of one's identity as an older person subsumes the self-appraisal of one's own worth as a "time-traveling" person. We readily reach almost ineffable questions, such as "What is it all about?" and "What sort of a life has it been anyway?" The word *essentialist* is commonly used in disparagement of theoretical orientations that fix social identities too rigidly and are unresponsive to social contextual processes (see, e.g., Rampton, 1995), and again, this is an important insight.

But age-identities are, in another sense, "essential." They are the products of the evaluative component of our own life narratives (Gubrium, 1976, 1993; Linde, 1993), the cumulative assessment of where we stand, developmentally—as individuals and in relation to our social environments. This isn't to say that people do not play identificational games with and around their age-identities or that talk exposes our essential understandings of our ageing selves. But, as we have repeatedly found in the geriatric medical discourses we have studied (e.g., Coupland & Coupland, 1998), identity in ageing ultimately connects to morale and well-being. There is an intensity of personal consequence when old people perceive their identities to be spoiled or their narratives (to use Linde's term) to be "incoherent."

For three decades, sociolinguistics has assumed that age is mainly of interest as a resource for the study of language change, and then that only some life stages are of interest. We need to discredit the assumption that ageing is only of interest at the margins. In Eckert's words, "a balanced view of sociolinguistic ageing must merge a developmental perspective with a mature-use perspective for all age groups . . . The developmental perspective recognizes that development is lifelong" (1997, p. 157). Sociolinguistics, I would certainly agree, needs a theory of lifelong ageing rather than a theory of socialization. But this chapter's review also suggests that it needs to consider a theoretically much *richer* "change agenda," as well as a *longer* one. As we have seen, the main conceptual distinction for age-related sociolinguistics at present is between "real time" and "apparent time." Then there is the further distinction between researching age cohorts through either panel studies (sampling data from the same informants over time) or trend studies (sampling from different but matched individuals across time periods, Eckert, 1997, p. 153; Labov, 1994, p. 77). This metalanguage shows how resolutely sociolinguistics has defined its orientation to ageing in terms of cohorts and the analysis of group norms.

In social gerontology, on the other hand, it is acknowledged that the cohort perspective for ageing research has been overplayed. Riley and Riley (1999, p. 125), in fact, refer to the "fallacy of cohort centrism," which they define as "erroneously assuming that members of all cohorts age in the same fashion as members of the cohort under scrutiny." Sociolinguistics has successfully navigated around another fallacy—Riley's and Riley's (1999) "life-course fallacy"—the assumption that cross-sectional age differences must indicate ("internal") processes of human ageing. But it remains "cohort centric," asserting that there are universal chronological principles guiding the acquisition and consolidation of cultural norms and practices, such as dialect. As Riley and Riley (1999) go on to say for social gerontology generally, social explanation ultimately lies beyond cross-sectional and cohort studies, which entail a neglect of local social structure and its linkage to individual experience:

> The major challenge for future research on age, as we see it, is to examine, not lives or structures *alone*, but the dynamic *interplay* between them as each influences the other. In the continuing dialectic between lives and structures, it is not only lives that change: structures

also change. And full understanding requires understanding the linkages—both method-ological and substantive—between ageing and the process of change in the surrounding structures. (pp. 126–127, with original italics)

As this implies, the experience of ageing lies in how changing social conditions (e.g., a looser structure for post-retirement lives or structured poverty in retirement) interact with historical cohort experiences (e.g., being a babyboomer or having built a career under a right-wing political regime) and a host of other social factors impinging on individual identities (gender, class, network, ethnicity, sexuality, and so on).[3] The forum for all these interactions is discourse.

Sociolinguistics currently proposes that in late adulthood we are, at least in terms of speech style, pretty much what we were socialized to be. This is a depressing and repres-sive assumption, even if dialect is only a small part of discursive practice. I previously argued that ageism is best seen as a complex ideology, tending to naturalize conservative, repressive models of ageing. But people will accommodate or resist these discourses differently and in different ways, as their resources and personal creativity allow. Resis-tance is being mobilized in increasingly organized ways (Tulle-Winston, 1999; see also Wallace, Williamson, Lung, & Powell, 1991, on "senior power" and the Gray Panther movement in the United States). For those who successfully resist, age itself starts to seem like a metaphorical and ideological ascription. Other than in having our age in years thrust before us (decade birthday cards, enforced retirement ages, the marketing of holidays for "over-50s," age-stereotyped humor, etc.), and assuming we can avoid the health and social problems that have been overstated in the popular account of old age, "being old" can seem a refreshingly empty category. Healthy and socially advantaged people in later life can establish their own nontraditional identity spaces, if they can ward off ageist discourses that interpret age nonconformity as the illegitimate aping of youth. Alternatively, old age can be constituted as a life stage distinct from mid-adulthood, with unique competences and opportunities. That is, the new old age does not have to be "youth with wrinkles." Disadvantaged older people will find it far harder to resist living out the stereotyped attributes of an ageist and gerontophobic society.

The same challenge, more or less, confronts those who research ageing. Researchers need to reset the tolerances on their own conceptualizations of later life. Green (1993), for example, noted how even the term *life cycle* suggests "externally set mechanisms" regulating human development and change (cf. also the term *menopause*; Coupland & Williams, 2002), whereas metaphors of the life course suggests a passage or journey or convoy, with individuals more actively responsible for their development and social events as the "landscape" (p. 131 ff.; see also Hagestad & Neugarten, 1985, p. 35). How

---

[3]Westin (1994, quoted earlier) similarly identified three sets of explanatory factors that relate to human and social time (p. 135), and cohort effects again constitute only one of the three. He distinguished *ageing effects* (changes related to individual ageing and maturation processes, including health—cf. age grading) from *period effects* (changes related to significant events in society, or in individual life, during the period between measurements, including social policy changes), and *cohort effects* (the unique historical experiences of people born at a certain time within a cultural group). This taxonomy originally appears in Bengston, Cutler, Mangen and Marshall (1985); see Coupland, Coupland, and Nussbaum (1993) for a related discussion of inherent, developmental and environmental factors in social ageing.

the object of inquiry is framed in language tends to determine the questions addressed and therefore what we learn from research. In the ageing context, and especially at this early stage of sociolinguistic research into ageing, there are many reasons to avoid all rigid (pre)conceptualizations. We should recognize that old people often are a minoritized and "othered" social group, but also that there is far more to social ageing than disadvantage. We should refine our understanding of societal ageism and explain its linguistic–discursive characteristics, but without oversimplifying its moral complexites. We should continue to model ageing in terms of change, but we should recognize that change involves an interplay between human developmental processes, historical cohort influences, and the shifting contexts of social structure and culture, all subject to creative interpretation by people in their other demographic categories. As in the case of language change, these social configurations will certainly leave their mark on speech and language, and there are still large gaps in our knowledge about the sociolinguistic differentiation of age groups. But a more ambitious agenda would see sociolinguistics operating in the middle of social theoretic concerns, examining the voices that express the conflicting meanings of ageing and that will determine our individual experiences of age in the future.

REFERENCES

Achenbaum, A. (1995). Images of old age in America, 1790–1970: A vision and a re-vision. In M. Featherstone & A. Wernick (Eds.), *Images of ageing: Representations of later life*. (pp. 19–28) London: Routledge.

Ainley, S. C., & Redfoot, D. L. (1982). Ageing and identity-in-the-world: A phenomenological analysis. *International Journal of Ageing and Human Development 15*, 1–15.

Antaki, C., & Widdicombe, S. (Eds.). (1998). *Identities in talk*. London: Sage.

Arber, S., & Evandrou, M. (1997). Mapping the territory: Ageing, independence and the life course. In S. Arber, & M. Evandrou (Eds.), *Ageing, independence and the life course* (pp. 9–26). London & Bristol, PA: Kingsley.

Beck, U., Giddens, A., & Lash, S. (1994). *Reflexive modernization: Politics, tradition and aesthetics in the modern social order*. Cambridge, UK: Polity.

Becker, E. (1973). *The denial of death*. New York: Free Press.

Bengston, V. L., Cutler, N., Mangen, D., & Marshall, V. (1985). Generations, cohorts and relations between age groups. In R. H. Binstock & E. Shanas (Eds.), *Handbook of aging and the social sciences*. (pp. 304–338). New York: Van Nostrand Reinhold.

Binstock, R. H. (1983). The aged as scapegoat. *The Gerontologist, 23*, 136–143.

Blaikie, A. (1999). *Ageing and popular culture*. Cambridge, UK: Cambridge University Press.

Bourdieu, P. (1991). *Language and symbolic power*. London: Polity.

Butler, R. N. (1969). Age-ism: Another form of bigotry. *The Gerontologist, 9*, 243–246.

Butler, R. N. (1975). *Why survive? Being old in America*. New York: Harper & Row.

Bytheway, B., & Johnstone, J. (1990). On defining ageism. *Critical Social Policy, 29*, pp. 27–39.

Cameron, D. (1990). Demythologizing sociolinguistics: Why language does not reflect society. In J. E. Joseph & T. J. Taylor (Eds.), *Ideologies of language* (pp. 79–93). London: Routledge.

Cameron, D. (1995). *Verbal hygiene*. London: Routledge.

Caporael, L., Lucaszewski, M. P., & Culbertson, G. H. (1983). Secondary babytalk: Judgements by institutionalized elderly and their caregivers. *Journal of Personality and Social Psychology, 44*, 746–754.

Chambers, J. K. (1995). *Sociolinguistic theory: Linguistic variation and its social significance*. Oxford, UK: Blackwell.

Cohen, S., & Wills, T. A. (1985). Stress, social support, and the buffering hypothesis. *Psychological Bulletin, 98*, 310–357.

Cole, T. R., & Gadow, S. (Eds.). (1986). *What does it mean to grow old: Reflections from the humanities*. Durham, NC: Duke University Press.

Coleman, P. (1993). Adjustment in later life. In J. Bond, P. Coleman, & S. Pearce (Eds.), *Ageing in society* (pp. 89–122). London: Sage.

Coulmas, F. (1997). Introduction. In F. Coulmas (Ed.), *The handbook of sociolinguistics* (pp. 1–11). Oxford, UK: Blackwell.

Coupland, J. (2003). Ageist ideology and discourses of control in skin care product advertising. In J. Coupland & R. Gwyn (Eds.), *Discourse, identity and the body* (pp. 127–150). London: Palgrave.

Coupland, J., & Coupland, N. (2000). Selling control: Ideological dilemmas of sun, tanning, risk and leisure. In S. Allan, B. Adam, & C. Carter (Eds.), *Environmental risks and the media.* (pp. 145–159). London: Routledge.

Coupland, J., Coupland, N., & Robinson, J. (1992). "How are you?" Negotiating phatic communion. *Language in Society, 21,* 201–230.

Coupland, J., Robinson, J., & Coupland, N. (1994). Frame negotiation in doctor–elderly patient consultations. *Discourse and Society, 5,* 89–124.

Coupland, J., & Williams, A. (2002). Conflicting discourses, shifting ideologies: Phramaceutical, "alternative" and feminist emancipatory texts on the menopause. *Discourse and Society, 13,* 419–445.

Coupland, N. (1997). Language, Ageing and ageism: A project for applied linguistics? *International Journal of Applied Linguistics, 7,* 26–48.

Coupland, N. (2000a). Language, context and the relational self: Re-theorising dialect style in sociolinguistics. In P. Eckert & J. Rickford (Eds.), *Style and sociolinguistics.* (pp. 185–210). Cambridge, UK: Cambridge University Press.

Coupland, N. (2000b). "Other" representation. In J. Verschueren, J.-O. Ostman, J. Blommaert, & C. Bulcaen (Eds.), *Handbook of Pragmatics, Instalment 2000* (pp. 21–26). Amsterdam & Philadelphia: Benjamins.

Coupland, N., & Coupland, J. (1997). Bodies, beaches and burntimes: "Environmentalism" and its discursive competitors. *Discourse and Society, 8*(1), 7–25.

Coupland, N., & Coupland, J. (1998). Reshaping lives: Constitutive identity work in geriatric medical consultations. *Text, 18*(2), 159–189.

Coupland, N., & Coupland, J. (1999). Ageing, ageism and anti-ageism: Moral stance in geriatric medical discourse. In H. Hamilton (Ed.), *Language and communication in old age: Multidisciplinary perspectives* (pp. 177–208). New York & London: Garland.

Coupland, N., & Coupland, J. (2000). Relational frames and pronominal address / reference: The discourse of geriatric medical triads. In S. Sarangi & M. Coulthard (Eds.), *Discourse and social life* (pp. 207–229). London: Longman.

Coupland, N., & Coupland, J. (2002). Language, ageing and ageism: A social action perspective. In P. Robinson & H., Giles (Eds.), *Handbook of language and social psychology* (2nd ed., pp. 465–486). London: Wiley.

Coupland, N., Coupland, J., & Giles, H. (1991). *Language, society and the elderly.* Oxford, UK, & Cambridge, MA: Basil Blackwell.

Coupland, N., Coupland, J., Giles, H., & Henwood, K. (1988). Accommodating the elderly: Invoking and extending a theory. *Language in Society, 17*(1), 1–42.

Coupland, N., Coupland, J., & Nussbaum, J. (1993). Epilogue: Future prospects in lifespan sociolinguistics. In N. Coupland & J. Nussbaum (Eds.), *Discourse and lifespan identity* (pp. 284–293). Newbury Park, CA: Sage.

Coupland, N., & Nussbaum, J. (Eds.). (1993). *Discourse and lifespan identity.* Newbury Park, CA: Sage.

Cumming, E., & Henry, W. (1961). *Growing old.* New York: Basic Books.

Duranti, A., & Goodwin, C. (Eds.), (1992). *Rethinking context.* Cambridge, UK: Cambridge University Press.

Eckert, P. (1997). Age as a sociolinguistic variable. In F. Coulmas (Ed.), *The handbook of sociolinguistics* (pp. 151–167). Oxford, UK, & Cambridge, MA: Blackwell.

Eckert, P. (2000). *Linguistic variation as social practice.* Maldon, MA, & Oxford, UK: Blackwell.

Eckert, Penelope and John Rickford (Eds.), (2001). *Style and Sociolinguistic Variation.* Cambridge: Cambridge University Press.

Erikson, E. H. (1980). *Identity and the life cycle* (a reissue). New York: Norton.

Estes, C. L., & Binney, E. A. (1991). The biomedicalization of aging: Dangers and dilemmas. In M. Minkler & C. Estes (Eds.), *Critical perspectives on aging: The political and moral economy of growing old* (pp. 117–134). Amityville, NY: Baywood.

Featherstone, M. (1995). *Undoing culture*. London: Sage.

Featherstone, M., & Hepworth, M. (1990). Ageing and old age: Reflections on the postmodern life course. In B. Bytheway, T. Keil, P. Allatt, & A. Bryman (Eds.), *Becoming and being old: Sociological approaches to later life* (pp. 133–157). London: Sage.

Featherstone, M., & Hepworth, M. (1993). Images of ageing. In J. Bond & P. Coleman (Eds.), *Ageing in society: An introduction to social gerontology* (pp. 250–275). London: Sage.

Featherstone, M., & Hepworth, M. (1995). Images of positive ageing: A case study of retirement *Choice* magazine. In M. Featherstone & A. Wernick (Eds.), *Images of ageing: Representations of later life* (pp. 29–47). London: Routledge.

Featherstone, M., & Wernick, A. (Eds.). (1995). *Images of ageing: Representations of later life*. London: Routledge.

Foucault, M. (1980). *Power/knowledge: Selected interviews and other writings 1972–1977* (C. Gordon, Ed.) Brighton, UK: Harvester.

Gibson, H. B. (1993). Emotional and sexual adjustment in later life. In S. Arber & M. Evandrou (Eds.), *Ageing, independence and the life course* (pp. 104–118). London & Bristol, PA: Kingsley.

Giddens, A. (1990). *The consequences of modernity*. Cambridge, UK: Polity (in association with Basil Blackwell).

Giddens, A. (1991). *Modernity and self-identity: Self and society in the late modern age*. Cambridge, UK: Polity (in association with Basil Blackwell).

Giddens, A. (1994). Living in a post-traditional society. In U. Beck, A. Giddens, & S. Lash (Eds.), *Reflexive modernization: Politics, tradition and aesthetics in the modern social order* (pp. 56–109). Cambridge, UK: Polity.

Green, B. S. (1993). *Gerontology and the construction of old age: A study in discourse analysis*. New York: DeGruyter.

Gubrium, J. F. (Ed.). (1976). *Time, roles and self in old age*. New York: Human Sciences Press.

Gubrium, J. F. (1993). *Speaking of life: Horizons of meaning for nursing home residents*. Hawthorne, NY: DeGruyter.

Hagestad, G. O., & Neugarten, B. L. (1985). Age and the life course. In R. H. Binstock & E. Shanas (Eds.), *Handbook of aging and the social sciences*. New York: Van Nostrand Reinhold.

Hockey, J., & James, A. (1995). Back to our futures: Imaging second childhood. In M. Featherstone & A. Wernick (Eds.), *Images of ageing: Representations of later life* (pp. 135–148). London: Routledge.

Hockey, J., & James, A. (1993). *Growing up and growing Old: Ageing and dependency in the life course*. London: Sage.

Hummert, M. L. (1994). Stereotypes of the elderly and patronising speech. In M. L. Hummert, J. M. Wiemann, & J. F. Nussbaum (Eds.), *Interpersonal communication in older adulthood: Interdisciplinary theory and research* (pp. 162–184). Thousand Oaks, CA: Sage.

Hymes, D. (1974). *Foundations in sociolinguistics: An ethnographic approach*. Philadelphia: University of Pennsylvania Press.

Irwin, S. (1999). Later life, inequality and sociological theory. *Ageing and Society, 19*, 691–716.

Jamieson, A., Harper, S., & Victor, C. (Eds.), (1997). *Critical approaches to aging and later life*. Buckingham, UK, & Philadelphia: Open University Press.

Jaworski, A., & Coupland, N. (Eds.). (1999). *The discourse reader*. London: Routledge.

Jaworski, A., Coupland, N., & Galasinski, D. (Eds.). (in press). *The sociolinguistics of metalanguage*. The Hague, the Netherlands: Mouton.

Kohli, M. (1991). Retirement and the moral economy: An historical interpretation of the German case. In M. Minkler & C. Estes (Eds.), *Critical perspectives on aging: The political and moral economy of growing old* (pp. 273–292). Amityville, NY: Baywood.

Labov, W. (1972). *Sociolinguistic patterns*. Philadelphia: University of Pennsylvania Press.

Labov, W. (1994). *Principles of Linguistic Change: Vol. 1. Internal Factors*. Cambridge, MA, & Oxford, UK: Blackwell.

Laslett, P. (1989). *A fresh map of life: The emergence of the third age*. London: Weidenfeld & Nicolson.

Linde, C. (1993). *Life stories: The creation of coherence*. New York & Oxford, UK: Oxford University Press.

Lucy, J. A. (1993). Reflexive language and the human disciplines. In J. A. Lucy (Ed.), *Reflexive language: Reported speech and metapragmatics* (pp. 9–32). Cambridge, UK: Cambridge University Press.

Macaulay, R. K. S. (1977). *Language, social class and education: A glasgow study*. Edinburgh, Scotland: The University Press.

Mead, G. H. (1932). *Philosophy of the present*. LaSalle, II: Open Court.

Minkler, M. (1991). Gold in gray: Reflections on business' discovery of the elderly market. In M. Minkler & C. Estes (Eds.), *Critical perspectives on aging: The political and moral economy of growing old* (pp. 81–93). Amityville, NY: Baywood.

Nuessel, F. (1982). The language of ageism. *The Gerontologist, 22,* 273–276.

Nuessel, F. (1984). Ageist language. *Maledicta, 8,* 17–28.

Parsons, J. T. (1942). Age and sex in the social structure of the United States. *American Sociological Review, 7,* 604–616.

Phillipson, C. (1987). The transition to retirement. In G. Cohen (Ed.), *Social change and the life course* (pp. 156–183). London: Tavistock.

Rampton, B. (1995). *Crossing: Language and ethnicity among adolescents.* London: Longman.

Riley, M. W., & Riley, J. W. (1999) Sociological research on age: Legacy and challenge. *Ageing and Society, 19,* 123–132.

Rodwell, G., Davis, S., Dennison, T., Goldsmith, C., & Whitehead, L. (1992). Images of old age on British television. *Generations Review, 2,* 6–8.

Rosow, I. (1974). *Socialization to old age.* Berkeley: University of California Press.

Taylor, C. (1989). *Sources of the self.* Cambridge, MA: Harvard University Press.

Tulle-Winston, E. (1999). Growing old and resistance: Towards a new cultural economy of old age? *Ageing and Society, 19,* 281–300.

Wallace, S. P., Williamson, J. B., Lung, R. G., & Powell, L. A. (1991). A lamb in wolf's clothing? The reality of senior power and social policy. In M. Minkler & C. Estes (Eds.), *Critical perspectives on aging: The political and moral economy of growing old* (pp. 95–114). Amityville, NY: Baywood.

Westin, C. (1994). Discussion: Some reflections on age and identity occasioned by reading Nikolas Coupland and Justine Coupland's paper. In *Health care encounters and culture* (pp. 129–137). Proceedings of the 1992 Summer University of Stockholm Seminar, Botkyrka.

Williams, A., & Giles, H. (1998). Communication and agism. In M. Hecht (Ed.), *Communicating prejudice* (pp. 136–160). Thousand Oaks, CA: Sage.

Woodward, K. (1991). *Aging and its discontents: Freud and other fictions.* Bloomington: Indiana University Press.

Woodward, K. (1995). Tribute to the older woman: Psychoanalysis, feminism and ageism. In M. Featherstone & A. Wernick (Eds.), *Images of ageing: Representations of later life* (pp. 79–96). London: Routledge.

The Role of Age Stereotypes in Interpersonal Communication

Mary Lee Hummert and Teri A. Garstka
University of Kansas

Ellen Bouchard Ryan
McMaster University

Jaye L. Bonnesen
Georgia State University

In the first edition of this handbook, we noted that much of the literature on communication and aging pointed to negative age stereotypes as a major influence on speech accommodations to older adults (Coupland & Coupland, 1989; Coupland, Coupland, Giles, & Henwood, 1988; Harwood, Giles, & Ryan, 1995; Hummert, 1994a; Ryan, Giles, Bartolucci, & Henwood, 1986). Accordingly, our chapter reviewed the literature on age stereotypes, age attitudes, beliefs about aging, and communication (see Hummert, Shaner, & Garstka, 1995). That review highlighted the complex nature of age stereotypes, showing that they include positive as well as negative views of older individuals. It also revealed that the relationship between age stereotypes and communication was more complex than initially conceptualized. As a result, we suggested four research questions for future research:

(1) What factors in the communication situation lead to positive or negative stereotyping of older adults? (2) What individual differences between communicators are related to positive or negative stereotyping of older adults? (3) What communication behaviors are associated with positive or negative stereotyping of older adults? and (4) How do the communication behaviors of the conversants affect stereotyping and self-perception processes? (Hummert, Shaner, et al., 1995, p. 125).

In this chapter, we consider recent research on age stereotypes and communication relevant to these questions. The chapter begins with an overview of the literature on age stereotypes (for comprehensive reviews, see Hummert, Shaner, et al., 1995; Hummert, 1999). Two recent advances in research on age stereotyping receive special attention: implicit age stereotyping and self-stereotyping. The chapter then outlines two models of how age stereotypes are involved in the communication process, illustrating how recent research speaks to these models and our four guiding questions about age stereotypes and communication. We conclude with a consideration of directions for future research that, we believe, will help us build the insights necessary to reduce the negative effects of age stereotyping in communication.

THE NATURE OF AGE STEREOTYPES

The cognitive perspective (Ashmore & Del Boca, 1981; Hamilton & Trolier, 1986) constitutes the predominant approach to the study of age stereotypes today (Hummert, Shaner, et al., 1995; Hummert, 1999). From this perspective, age stereotypes are viewed as person perception schemas that use age as the primary categorization principle, with "older adult" operating as a superordinate category that subsumes several subordinate categories or subtypes of older adults.

Researchers have used trait generation, trait sorting, and photograph sorting tasks to investigate the structure and content of age stereotype schemas (Brewer, Dull, & Lui, 1981; Brewer & Lui, 1984; Hummert, 1990; Hummert, Garstka, Shaner, & Strahm, 1994; Schmidt & Boland, 1986). It became clear through this research that age schemas encompass both positive and negative subcategories, or stereotypes, of older adults. Further, the similarities and differences in the stereotype schemas of young, middle-aged, and older persons suggest that these schemas become increasingly complex over the course of a lifetime. The findings of the most recent of these studies (Hummert et al., 1994) are illustrative.

Hummert et al. (1994) first asked one group of young, middle-aged, and older individuals to generate traits that they associated with the category *older adult*. Next, a separate group of young, middle-aged, and older adults sorted the resulting 97 traits (adjectives describing general physical health, and cognitive and personality characteristics) into groupings that described individual older persons. Cluster analysis of the trait groupings showed that participants of all ages had both positive and negative age stereotypes, and that the complexity of the stereotype schemas increased across the three age groups. That is, older participants had more stereotypes (12) than did the middle-aged and young participants (10 and 8, respectively), and the additional stereotypes included in the sets of the middle-aged and older participants were subcategories of broader stereotypes held by the young.

Despite these differences in complexity, similarities across the age groups were apparent. As shown in Table 4.1, seven stereotypes emerged in the stereotype sets of all three age groups: 3 positive (Golden Ager, Perfect Grandparent, John Wayne Conservative) and 4 negative (Severely Impaired, Despondent, Recluse, Shrew/Curmudgeon). As Hummert et al. (1994) stated, these stereotypes appear to represent cultural archetypes of aging. That is, adults of widely varying ages have stereotypes corresponding to these categories even though their individual cognitive schemas for the stereotypes may include slightly different configurations of traits. Further, only a few of these stereotypes describe older adults with cognitive, emotional, and physical problems. Other stereotype research from the cognitive perspective has shown that attitudes toward the stereotypes vary

TABLE 4.1
Stereotypes of Older Persons Held by Young, Middle-Aged,
and Older Adults*

Negative Stereotypes and Traits	*Positive Stereotypes and Traits*
Severely Impaired: Slow thinking, incompetent, feeble, incoherent, inarticulate, senile	**Golden Ager:** Lively, adventurous, alert, active, sociable, witty, independent, well-informed, skilled, productive, successful, capable, volunteer, well-traveled, future-oriented, fun-loving, happy, curious, healthy, sexual, self-accepting, health-conscious, courageous
Despondent: Depressed, sad, hopeless, afraid, neglected, lonely	**Perfect Grandparent:** Interesting, kind, loving, family-oriented, generous, grateful, supportive, trustworthy, intelligent, wise, knowledgeable
Shrew/Curmudgeon: Complaining, ill-tempered, prejudiced, demanding, inflexible, selfish, jealous, stubborn, nosy	**John Wayne Conservative:** Patriotic, religious, nostalgic, reminiscent, retired, conservative, emotional, mellow, determined, proud
Recluse: Quiet, timid, naive	

Note: Trait sets include traits grouped with the stereotype by those in all three age groups plus additional traits grouped with the stereotype by those in two age groups.

*As reported in Hummert et al. (1994).

consistently with the valence of the traits, and that both positive and negative stereotypes are viewed as equally representative of the general older adult population (Hummert, 1990; Hummert, Garstka, Shaner, & Strahm, 1995; Schmidt & Boland, 1986).

Implicit Age Stereotypes

Lately, attention has turned to the ways in which age stereotypes may function automatically (i.e., outside conscious awareness) in perception to affect judgments and behaviors. Greenwald and Banaji (1995) defined implicit stereotypes as "the introspectively unidentified (or inaccurately identified) traces of past experience that mediate attributions of qualities to members of a social category" (p. 15). Perdue and Gurtman (1990) found evidence for implicit age stereotypes in two studies. In the first, participants primed with *old* remembered more negative trait words in a subsequent recall task, whereas those primed with *young* remembered more positive traits. In the second, participants made semantic judgments about negative traits more quickly when primed with *old*, but they made judgments about positive traits more quickly when primed with *young*.

Recent studies have attempted to measure implicit age attitudes and stereotypes, focusing on the strength of implicit cognitions and their content. This interest developed as research on other stereotypes suggested that automatic or implicit attitudes may differ from those participants offer on standard questionnaire measures, primarily because responses on questionnaires (and similar direct, self-report measures) are sensitive to self-presentational effects (Greenwald & Banaji, 1995). For example, in a study of racial prejudice, Devine (1989) found that indirect measures of attitudes revealed implicit racial prejudice among participants who reported low prejudice on a standard prejudice questionnaire. The studies of implicit age attitudes and stereotypes have used the Implicit Association Test (IAT; Greenwald, McGhee, & Schwartz, 1998), a computer-based task that was developed specifically to measure the strength of automatic or unconscious social perceptions. We provide a brief overview of the IAT here. (For a full discussion of the IAT and its development see Greenwald et al., 1998.)

Participants completing the IAT are asked to assign target items (e.g., photos of young and old persons) to category poles (e.g., young or old) as quickly as possible. The computer records the time it takes to make each judgment. Making such single category assignments is quite easy. However, the critical IAT tasks require that participants consider target items from two bipolar categories within the same set of trials. As an example, the categories for an age attitude IAT might be age (old or young) and pleasantness (pleasant or unpleasant). Target items for the age category would be photos of old and young persons, whereas target items for the pleasantness category would be pictures of pleasant and unpleasant objects (e.g., teddy bear, roach, etc.). The poles of the categories are assigned to response keys in pairs so that the pairings are either consistent with presumed implicit associations (e.g., young/pleasant and old/unpleasant) or inconsistent with them (e.g., young/unpleasant and old/pleasant). The IAT is based on the assumption that paired judgments consistent with implicit associations are easier to make, and thus made more quickly, than judgments that are inconsistent. Therefore the difference in the time it takes to make consistent and inconsistent judgments on the IAT (the IAT effect) provides a measure of the strength of implicit associations. A growing body of research attests to the validity of the IAT as a measure of implicit social cognitions (Dasgupta, McGhee, Greenwald, & Banaji, 2000; Greenwald & Farnham, 2000; Greenwald et al., 1998; Rudman, Greenwald, & McGhee, 2001; see Greenwald & Nosek, 2001, for a review based on more than 30 published or in press articles using the IAT).

Studies of implicit age attitudes (Dasgupta & Greenwald, 2001; Hummert, Garstka, O'Brien, Greenwald, & Mellott, 2002; Karpinski & Hilton, 2001; Mellott, Greenwald, Hummert, & O'Brien, 2001) using the IAT have found a negative bias toward older people in comparison to young people that is consistent with that reported in prior research on general attitudes toward older people (Crockett & Hummert, 1987; Kite & Johnson, 1988; Kite & Wagner, 2001). However, these implicit attitudes had only low and nonsignificant correlations with standard questionnaire measures of age attitudes and, in some cases, indicated differing attitudes from the questionnaire measures. For instance, young adults' IAT responses showed more favorable attitudes toward young people than old people, but their responses on attitude questionnaires were more favorable toward old people than young people (Dasgupta & Greenwald, 2001; Hummert et al., 2002). As Hummert et al. concluded, the latter finding confirms the value of the IAT in assessing attitudes that participants may hesitate to express on questionnaires.

The IAT has also been used to assess the strength of implicit positive and negative age stereotypes (Hummert, Garstka, O'Brien, Savundranayagam, Zhang, & Geiger, 2003), providing information helpful in interpreting conflicting evidence from prior stereotype, attitude, and communication research. For instance, although the evidence for multiple positive and negative age stereotypes is substantial, reviews of the age attitude literature have shown that negative attitudes seem to dominate in people's perceptions about older adults (Crockett & Hummert, 1987; Kite & Johnson, 1988; Kite & Wagner, 2001). Likewise, much of the research on intergenerational communication has suggested that negative stereotypes have more influence on communication behavior than do positive stereotypes (e.g., leading to patronizing communication; Caporael, 1981; Hummert & Shaner, 1994; Hummert, Shaner, Garstka, & Henry, 1998; Kemper, Ferrell, Harden, Finter-Urzcyk, & Billington, 1998). The initial studies of implicit age stereotypes indicate that differences in the strength of positive and negative stereotypes may explain this paradox.

For example, Hummert and colleagues (2003) asked young (ages 18–25), middle-aged (ages 30–59), young-old (ages 60–74), and old-old (ages 75 and older) participants to complete age-stereotype IATs: old-positive (old as wise, young as foolish) and young-positive (young as open, old as closed). Results revealed that those in all four age groups had equivalent implicit stereotypes of young people as open (and older people as closed), but that only those in the three older age groups had an implicit positive stereotype of older people as wise (and young people as foolish). Even so, their implicit stereotype of older people as wise was significantly weaker than their stereotype of young people as open. This research confirmed the post hoc explanation for the seeming conflicts between the multiple stereotype, age attitude, and communication and aging research: Positive stereotypes of older people are less accessible than those of young people, particularly for young individuals. As we will discuss later, this finding has clear implications for the role of age stereotypes in interpersonal communication.

Implicit Age Stereotypes and Self-Stereotyping

The relationship of implicit age stereotypes to self-stereotyping in older adults is another line of research that has developed since the publication of the first edition of this handbook. Levy (1996) introduced this topic with a study that measured memory performance of older and younger participants before and after exposure to subliminal priming of positive (e.g., wise) or negative (e.g., senile) age stereotypes. Results showed that older participants primed with negative age stereotypes showed poorer pre- to posttest memory performance, whereas those primed with positive age stereotypes (e.g., wise) showed improved memory. The memory performance of young participants was unaffected by the priming manipulation. Levy and her colleagues have also reported similar effects of implicit stereotyping on older participants' handwriting (Levy, 2000), their cardiovascular responses to stress (Levy, Hausdorff, Hencke, & Wei, 2000), and their responses to hypothetical medical situations (Levy, Ashman, & Dror, 2000). In each of these studies, the implicit positive age stereotypes had beneficial effects on participants' behavior (i.e., more controlled handwriting, lower levels of cardiovascular response to stress, greater acceptance of interventions to prolong life), but the implicit negative stereotypes had detrimental effects (i.e., shakier handwriting, higher levels of cardiovascular response, and greater refusal of interventions to prolong life). The implications of the self-stereotyping research for communication will be addressed later in this chapter.

Summary

By considering age stereotypes as knowledge structures or schemas possessed and employed at some level by all perceivers, the cognitive perspective on age stereotypes offers several advantages to scholars of communication and aging (Hummert, 1999). First, it emphasizes that stereotyping is a process that is instantiated in interactions. Second, it invites study of the various aspects of that process: how age stereotypes become salient during interactions, when they affect perceptions and behaviors, and how they develop over time. Third, by identifying stereotyping as a normal part of the perceptual process, the cognitive perspective can help to remove the stigma that may interfere with individuals' acknowledging—and perhaps changing—their own age stereotypes. Finally, attention to the role of implicit stereotypes and self-stereotyping may help us to clarify the ways in which age stereotypes at times function outside conscious awareness, becoming incorporated into the self-perceptions of older individuals with potentially detrimental effects. We turn now to two models of communication and aging that have adopted the social cognitive perspective on age stereotypes as the key to identifying how age stereotyping processes are implicated in interpersonal interactions.

MODELS OF AGE STEREOTYPES IN COMMUNICATION

Communication Accommodation Theory (CAT; Giles, Coupland, & Coupland, 1991) focuses on "the social cognitive processes mediating individuals' perceptions of the environment and their communicative behaviors" (Giles, Mulac, Bradac, & Johnson, 1987, p. 14). Appropriately, CAT provides the theoretical framework for much of the research on communication and aging, including the two models on which we focus here: the Communication Predicament of Aging Model (Ryan et al., 1986) and the Age Stereotypes in Interactions Model (Hummert, 1994a). Both models provide accounts of the role of age stereotypes in communication that have received considerable empirical support. In recent years, the models have been extended to other cultures (e.g., Giles, Harwood, Pierson, Clément, & Fox, 1998; Lin & Harwood, 2003; Noels, Giles, Gallois, & Ng, 2001; Zhang & Hummert, 2001), as outlined by Pecchioni, Ota, and Sparks (Chapter 7, this volume). Our review in this chapter highlights studies conducted in the Western cultures where the models originated.

Communication Predicament of Aging Model

The Communication Predicament of Aging Model (CPA; Ryan et al., 1986) outlines how intergenerational encounters may unfold when negative age stereotyping occurs, evolving into a negative feedback cycle for the participants (Rodin & Langer, 1980). As shown in Fig. 4.1, the recognition of age cues (physical features, voice, context, etc.) in an encounter with an older individual may make negative age stereotypes salient to younger persons. Such age stereotypes are likely to carry with them beliefs about deficits in communication skills (e.g., poor hearing) and cognitive abilities (e.g., inability to recall names) that often accompany aging. Consistent with the tenets of Communication

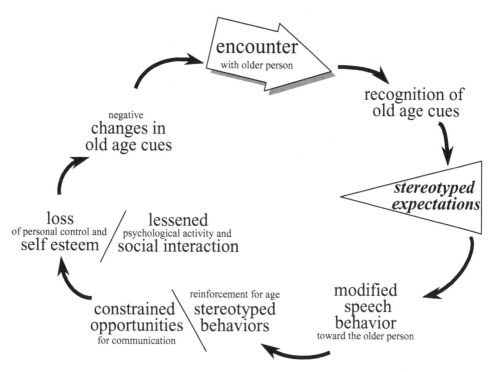

FIG. 4.1. Communication Predicament of Aging Model (Ryan, Meredith, MacLean, & Orange, 1995, p. 91).

Accommodation Theory (Giles et al., 1987; Giles et al., 1991), younger individuals may adapt their communication to the older individual to meet these presumed deficits. In other words, they may overaccommodate to negative age stereotypes in their talk, perhaps by speaking more loudly than normal, rather than adapt their communication to the actual competencies of older communicators. Such overaccommodations contribute to a negative feedback cycle by reinforcing negative age stereotypes for both participants. For older individuals, such reinforcement may lead to constrained opportunities for communication, lower self-esteem, and functional declines in physical and cognitive abilities consistent with those stereotypes.

The CPA model has provided a useful heuristic for research on communication and aging, with over 90 citations in refereed journal articles in the last 10 years. Empirical support for the relationships represented in the model is substantial.

Support for the CPA Model. Evidence that the recognition of age cues leads to perceptions consistent with negative age stereotypes comes from research on age-related characteristics of voice, facial structure, and nonverbal behavior. Vocal cues to age can suggest negative age stereotypes to listeners. A number of studies have demonstrated that listeners evaluate older speakers more negatively, in comparison to younger speakers, on personality and attitude scales, particularly on competence dimensions (Giles, Henwood, Coupland, Harriman, & Coupland, 1992; Ryan & Capadano, 1978; Ryan & Johnston, 1987; Ryan & Laurie, 1990; Stewart & Ryan, 1982).

Age-related changes in facial structure can also suggest negative age stereotypes. Physical changes in facial features with age include the size and placement of the eyes, nose, and mouth; the proportional relationships of the chin, cheeks, forehead, and skull; and the characteristics of the skin (e.g., wrinkling) and hair (e.g., graying; Berry & McArthur, 1986). A number of studies have demonstrated that people with young facial features are associated with more positive traits and behaviors by college students and children than are persons with older facial features (Berry & McArthur, 1985, 1986, 1988; McArthur, 1982). For example, faces that were perceived younger (or more "babyfaced") with large, round eyes, high eyebrows, and a small chin were positively correlated with perceptions of the stimulus person's naivete, honesty, kindness, and warmth (Berry & McArthur, 1985). In addition, Montepare and Zebrowitz-McArthur (1988, 1993) found that nonverbal behavior like walking can influence perceivers' evaluations. When participants observed 5- to 70-year-old walkers depicted in point-light displays, they perceived the younger walkers as more powerful and happier than older walkers.

Other research shows that declines in communication competence are believed to accompany old age (Giles, Coupland, & Wiemann, 1992; Ryan & Cole, 1990; Ryan, Kwong See, Meneer, & Trovato, 1992). Ryan and colleagues (1992) developed the Language in Adulthood questionnaire (LIA) to assess beliefs about the communication competence of older adults. They asked young and older adults to provide assessments of their own language abilities on the LIA, as well as those of either a typical 25-year-old or a typical 75-year-old. Consistent with the CPA model, those in both age groups rated typical 75-year-old targets as experiencing more problems in receptive and expressive communication than typical 25-year-old targets. Receptive problems included difficulty in understanding others in noisy situations, losing track of the topic in conversation, losing track of who said what in conversation, and so on. Expressive problems included declining use of difficult words, dominating the conversation, finding it hard to speak when pressed for time. The older targets were rated as more skilled than younger targets on only two items: sincerity in speaking and story telling ability, both expressive communication skills.

As Ryan and colleagues (1992) pointed out, beliefs about the communication skills of older adults may suggest appropriate communication strategies to use with them. That is, if one believes that older adults have difficulty hearing, remembering words, and processing language, one might believe that it is appropriate to speak loudly, slowly, and in simple sentences to them. These adaptations are associated with the overaccommodations to age stereotypes in the CPA model (Ryan et al., 1986). These overaccommodations have been variously termed elderspeak (Cohen & Faulkner, 1986; Kemper, 1994, 2001; Kemper & Harden, 1999), dependency-supporting communication (Baltes, Neumann & Zank, 1994; Baltes & Wahl, 1996), and patronizing talk (Hummert & Ryan, 1996; Ryan, Hummert, & Boich, 1995; see Hummert & Ryan, 2001, for a review). Secondary baby talk (Caporael, 1981) or infantalizing speech (Whitbourne, Culgin, & Cassidy, 1996) is an extreme form of patronizing talk that might be used with highly impaired older adults.

A variety of research demonstrates that these overaccommodations contribute to the negative feedback cycle that is the essence of the CPA model (Ryan et al., 1986). First, experimental studies that have manipulated the type of communication addressed to older targets show that patronizing talk reinforces negative age stereotypes in both observers and older recipients. When asked to evaluate older recipients of patronizing communication in comparison to recipients of nonpatronizing communication, observers rate the recipients of patronizing communication as less satisfied, more dependent and less competent (Harwood, Ryan, Giles, & Tysoski, 1997; La Tourette & Meeks, 2000; Ryan, Meredith, & Shantz, 1994) and select traits of negative age stereotypes to describe them (Hummert & Mazloff, 2001). The work of Kemper and colleagues (Kemper & Harden, 1999; Kemper, Othick, Warren, Gubarchuk, & Gerhing, 1996) shows that the negative stereotyping extends to self-perceptions of older listeners: Older participants who were addressed with the modifications of patronizing communication (elderspeak) within the context of a referential communication task gave lower assessments of their own communication abilities on the Language in Adulthood Questionnaire (Ryan et al., 1992) than participants who did not experience the patronizing talk.

Second, O'Connor and Rigby (1996) provide suggestive evidence supporting the CPA model's hypothesized link between age-modified communication and lower self-esteem in older individuals. In that study, community-dwelling seniors and nursing home residents completed measures of perceptions of secondary baby talk (patronizing talk) and neutral talk scenarios, frequency of exposure to secondary baby talk, and self-esteem. Results showed that for those with negative perceptions of baby talk, more exposure to this communication style was related to lower self-esteem. However, this pattern did not hold true for those with positive perceptions of baby talk, although this does not preclude other negative effects of baby talk (e.g., self-stereotyping, Levy, 1996) on these individuals.

Finally, other research shows that modifications based on an age stereotype of dependency contribute to functional decline in older individuals, as suggested by the CPA model. For example, Baltes and colleagues (1994) used behavioral analysis to examine interaction patterns between older adults, their caregivers, and their family members. Baltes and Wahl (1996) summarized the results of a series of such studies that found an increase in dependent behaviors of nursing home residents when caregivers used a dependency-supporting script. Within families as well, care-recipients who perceived their family caregivers' communication as patronizing reported reduced well-being (Edwards & Noller, 1998).

The Age Stereotypes in Interactions Model

Despite its power, the CPA model applies only to those situations in which negative age stereotyping occurs. Hummert (1994a) proposed a model of the role of age-related stereotypes in interactions that extends the CPA model in two ways. First, it specifies characteristics of the communicators and the context that contribute to negative age stereotyping in interactions. Second, it accounts for the influence of positive as well as negative age stereotyping in the communication process. As a result, this model allows not only for the negative feedback cycle of the CPA model, but also for a positive feedback cycle and for disruption of the negative feedback cycle. In Fig. 4.2 we present the essential elements of the Age Stereotypes in Interactions Model (ASI), adapted from the complete model as presented in Hummert (1994a). The ASI model highlights the perspective of an individual communicating with an older person. For the sake of clarity,

Age Stereotypes in Interactions Model

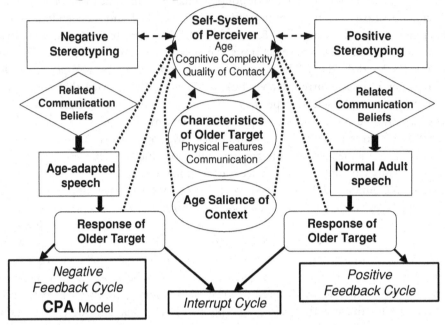

FIG. 4.2. Age Stereotypes in Interactions Model (Hummert, 1994a; modified to highlight perspective of perceiver and connection to the Communication Predicament of Aging Model).

that person is termed the Perceiver and the other person is designated the Older Target, although we realize these designations do not acknowledge the transactional nature of the communication process. As the ASI model shows, the self-system of the perceiver, the characteristics of the older target, and the context play significant roles in whether the perceiver categorizes the older target positively or negatively.

First, the ASI model posits that three aspects of the perceiver's self-system are centrally involved in the stereotyping process: age, cognitive complexity, and quality of prior contact with older adults. These characteristics have implications for the nature of the perceiver's age stereotype schemas and for the tendency to make category-based judgments of others. For example, older individuals have more complex age schemas (Heckhausen, Dixon, & Baltes, 1989; Hummert et al., 1994) and more accessible positive age stereotypes (Hummert et al., 2003) than do younger ones; and those higher in cognitive complexity are more likely to use person-centered communication strategies than are those lower in cognitive complexity (Burleson, 1984; Delia & Clark, 1977; Denton, Burleson, & Sprenkle, 1995). Research has shown that the quality of contact with older adults, rather than the frequency, is positively related to attitudes toward and perceptions of older adults (Hale, 2000; Knox, Gekoski, & Johnson, 1986; Rose-Colley & Eddy, 1988). Therefore, as perceiver age, cognitive complexity, and the quality of contact with older people increase, negative age stereotyping of the older target should be reduced and positive age stereotyping should be enhanced.

Second, the model shows how the characteristics of the older target may influence the stereotyping process, with the target's physical features and communication behaviors suggesting either positive or negative age stereotypes. Based upon research on beliefs about age-related changes over the life-span (Heckhausen et al., 1989; Ryan, 1992; Ryan & Kwong See, 1993), the ASI model predicts that the negative stereotyping of the CPA model should occur primarily when the age cues of the older target suggest advanced age (i.e., the old-old age category, 75 and over; Neugarten, 1974). Likewise, negative stereotyping may be based on other physical features (e.g., frowning facial expression, stooped posture, etc.) and communication behaviors (e.g., painful self-disclosures; Bonnesen & Hummert, 2002; Coupland, Coupland, Giles, Henwood, & Wiemann, 1988) of older individuals that are consistent with the traits of negative stereotypes. Conversely, positive stereotyping is more likely when the individual's physical features and communication behaviors (e.g., young-old age of 60–74, smiling expression, interesting anecdotes) are consistent with the traits of positive age stereotypes (Bonnesen & Hummert, 2002; Hummert, Garstka, & Shaner, 1997).

Third, the ASI model includes the influence of context on the stereotyping process. Contexts may vary in the extent to which they (a) make age stereotypes salient and (b) feature the positive or negative aspects of those stereotypes. As an example, a grocery store is an age-neutral context in the sense that the setting is not linked (in any obvious way) with age stereotype schemas. A nursing home, however, not only makes age salient but also calls forth the negative aspects of age stereotypes, as suggested by the incidence of secondary baby talk in nursing home settings (e.g., Caporael, 1981; Sachweh, 1998; see Ryan et al., 1995). In contrast, a 50th anniversary celebration may favor positive age stereotyping (e.g., the Perfect Grandparent).

According to the ASI model, these three elements (perceiver's self-system, characteristics of the older target, context) are jointly implicated in the stereotyping process. The perceiver's self-system is central to the process, however, because it provides the cognitive framework that selects, weighs, and interprets the other two elements. It is only when the stereotyping process leads to negative categorization of the older target that the negative feedback cycle of the CPA model is initiated. Positive stereotyping, in contrast, is followed by beliefs about communication that are consistent with the use of normal adult speech. In this way, the ASI model shows how a positive feedback cycle may be initiated. Finally, by including the response of the older target in the model, the ASI model acknowledges that the communication behaviors of the individuals may serve to reinforce or alter initial perceptions. As the two individuals converse, their communication behavior itself may suggest positive or negative age stereotypes. Through communication, the older target may change the perceiver's original categorization to a positive one, thereby interrupting the negative feedback cycle of the CPA model. The alternative—interruption of a positive feedback cycle—is also possible, of course.

Research has provided support for the ASI model (e.g., Bonnesen & Hummert, 2002; Bieman-Copland & Ryan, 2001; Mulac & Giles, 1996; Harwood & Williams, 1996; Hummert et al., 1997; Ryan, Kennaley, Pratt, & Shumovich, 2000; Thimm, Rademacher, & Kruse, 1998). Here we discuss that research as it relates to the following aspects of the model: characteristics of older individuals and the activation of age stereotypes, age stereotypes as antecedents of communication beliefs and behaviors, perceiver age, and the role of older individuals' responses in interrupting the negative feedback cycle of the CPA.

Characteristics of Older Individuals and the Activation of Age Stereotypes. Like the CPA model, the ASI model views the characteristics of the older individual as an important influence on the activation of age stereotypes in a perceiver, but it specifies how those characteristics may lead to positive or negative stereotyping. Support for the model comes from studies of physiognomic cues to age (Hummert, 1994b; Hummert et al., 1997), vocal cues to age (Hummert, Mazloff, & Henry, 1999; Mulac & Giles, 1996), and perceptions of age-related communication behaviors (Bieman-Copland & Ryan, 2001; Bonnesen & Hummert, 2002; Ruscher & Hurley, 2000; Ryan, Bieman-Copland, Kwong See, Ellis, & Anas, 2002).

Hummert et al. (1997) gave young, middle-aged and older participants sets of 18 photographs of older men and women (with either smiling or neutral facial expressions) and asked them to pair the photographs with trait sets of positive and negative age stereotypes from Hummert and colleagues (1994). Results supported the ASI model in that photographs of those who appeared to be in their 80s and 90s were paired most often with traits of negative stereotypes. In contrast, photographs of those who appeared to be in their sixties were paired most often with positive stereotypes. Facial expression moderated the strength of the association between perceived age and negative stereotyping, with a smile reducing the negative stereotyping of those who looked the oldest, particularly if they were female, and a neutral facial expression increasing the negative stereotyping of those who looked the youngest.

Results of studies that have examined the relationship between the perceived age of an older speaker and evaluations of the speaker also show support for the ASI model. Mulac and Giles (1996) collected young listeners' social perceptions of older speakers ranging in age from 59 to 92 and found that vocal variables (unclear, strained, vowel elongation, and lack of coarse voice) predicted the perceived age of the speaker, and that perceived age was correlated with negative stereotype traits (frail, illnatured, subdued, incompetent, and dependent). Similarly, Hummert et al. (1999) asked young listeners to make stereotype judgments of 30 older speakers (aged 61–89; 15 men and 15 women). As predicted within the ASI model, participants selected more negative stereotypes for the voices of old-old speakers and more positive stereotypes for the voices of young-old speakers, although this effect was significant only for judgments of female speakers.

Evidence that conversational behaviors may serve as age cues and trigger stereotyped expectations of an older adult's competence comes from recent studies of painful self-disclosures (Bonnesen & Hummert, 2002), age excuses for memory failures (Ryan et al., 2002), repetitious verbal behavior (Bieman-Copland & Ryan, 2001), and off-target verbosity (Ruscher & Hurley, 2000). In general, perceivers associated each of these communication behaviors with negative age stereotypes. To illustrate, we consider the research on painful self-disclosures and age excuses.

Painful self-disclosures (PSDs) are negative, intimate revelations that some researchers have labeled an older adult phenomenon (Coupland, Coupland, Giles, Henwood, et al., 1988). Through discourse analysis of peer and intergenerational conversations between young (in their 30s) and older women, Coupland et al. found that the older women revealed significantly more information about poor health, bereavement, loneliness, financial troubles, and similar problems than did the young women. In subsequent discussions with the researchers, the young women reported that they found these disclosures underaccommodative, and viewed them as typical older adult behavior (Coupland, Henwood, Coupland, & Giles, 1990).

Bonnesen and Hummert (2002) manipulated disclosure type (PSDs vs. Non-Painful Self-Disclosures) in videotaped scenarios to investigate the link between PSDs and negative age stereotypes. Participants included both young and old respondents. Those in both age groups found the PSDs more negative, more intimate, and less appropriate than the nonpainful disclosures (NPSDs), and rated those who revealed PSDs higher on negative stereotype traits and lower on the positive stereotype traits than those who revealed NPSDs.

With regard to age excuses, Ryan and colleagues (2002) asked young and older participants to evaluate forgetful older targets in their 70s who used their age, lack of ability, lack of effort, or the situation to explain forgetting. Consistent with the ASI model, participants viewed the targets who used an age excuse as older than their peers. In addition, young participants were particularly sensitive to the negative consequences of using such an excuse. They rated the user of an age excuse as more likely to forget in the future than the target using no excuse, and as more likely to elicit frustration from others than the target using an ability, effort, or situation excuse. The authors pointed out that these findings emerged even though participants viewed the age excuses as more believable and socially adept than the other excuse types.

Age Stereotypes as Antecedents of Communication Beliefs and Behaviors.   Once a positive or negative age stereotype has been activated in a perceiver (e.g., based on the older individual's characteristics or behaviors, the context, etc.), the ASI model posits that the stereotype will lead to beliefs about the target's communication competencies that are consistent with the stereotype's valence. To test this hypothesis, Hummert, Garstka, and Shaner (1995) used the LIA (Ryan et al., 1992) to assess beliefs of young, middle-aged and older adults about the communication skills of four older targets representing two positive (Golden Ager, John Wayne Conservative) and two negative (Despondent, Shrew/Curmudgeon) stereotypes of older adults (see Table 4.1). As expected under the ASI model, judgments of the communication abilities of the targets differed according to the nature of the stereotype they represented: The two positive targets were viewed as experiencing significantly fewer communication problems and having better communication skills than the negative ones. Similarly, Harwood and Williams (1996) found that young respondents believed that interactions with a Despondent older target would be less satisfactory and produce more anxiety than interactions with a Perfect Grandparent target.

A key feature of the ASI model is its link of negative and positive age stereotypes not only to different beliefs about the communication competence of older persons, but also to age-adapted or normal adult communication styles. Studies that have examined the link between age stereotypes and communication behavior have found general support for the model, but also have shown that positive stereotypes may sometimes engender age-adapted talk (Hummert & Shaner, 1994; Hummert et al., 1998; Thimm et al., 1998). For example, Thimm et al. reported that young participants' oral instructions about using a clock radio contained fewer patronizing features when directed to a competent older target than to a less competent one, or to an older person described simply by age (82 years old). However, the instructions to the competent older target contained more age-adapted characteristics than those to a 32-year-old target.

Hummert and colleagues (1998) found that context affects the extent to which positive and negative age stereotypes are associated with age-adapted or normal adult communication. Participants (young, middle-aged, and older) were placed in hypothetical situations and asked to give oral persuasive messages to two older targets. One target fit the Despondent stereotype, and the other the Golden Ager stereotype (see Table 4.1). For half the participants, the targets were presented in a context consistent with the stereotypes, so that the Despondent target was in a hospital setting and the Golden Ager target in a community setting. For the remaining participants, the context served to undermine the stereotypes, with the Despondent target presented in a community setting and the Golden Ager target in a hospital. Analysis of the resulting messages revealed three message types: affirming, a style analogous to normal adult-to-adult talk; overly nurturing, a patronizing style including some of the features of secondary baby talk; and directive, a patronizing style that was cold and controlling.

Results supported the predictions of the ASI model regarding the effects of initial categorizations on the incidence of age-adapted speech. Specifically, participants gave more patronizing messages (primarily overly nurturing) to the Despondent target than to the Golden Ager. The hospital context appeared to reinforce the Despondent stereotype, with the Despondent target receiving more patronizing messages in that setting (53%) than in the community setting (42%). However, messages to the Golden Ager target varied to an even greater extent across the two settings, with the percentage of patronizing messages increasing from only 30% in the community setting to 47% in the hospital setting. Further, the patronizing messages to the Golden Ager target were primarily of the directive style, which may be more dissatisfying at an interpersonal level than an overly nurturing style. As Hummert and colleagues (1998) concluded, this study showed not only that context and stereotypes may interact to affect expectations about appropriate communication with older individuals, but also suggested that positive categorizations of older persons may be less stable than negative categorizations. The apparent instability of positive categorizations may be associated with the differences in the strength of implicit positive and negative age stereotypes discussed earlier (Hummert et al., 2003).

Perceiver Age and the ASI Model. Several of these studies have included participants varying in age from young to old, providing information about the ASI model's prediction that older participants should be less likely to engage in negative age stereotyping and to use age-adapted communication than younger ones. This prediction was based on evidence that older individuals have more complex age schemas than do younger persons (Brewer & Lui, 1984; Heckhausen et al., 1989; Hummert et al., 1994). The age differences observed provide mixed support for the model. For instance, contrary to the predictions of the model, older participants were more likely than young and middle-aged participants to associate the physical features of advanced old age with negative age stereotypes (Hummert et al., 1997). On the other hand, several studies show that age differences in communication beliefs and behaviors are consistent with the ASI model (Bonnesen & Hummert, 2002; Hummert et al., 1998; Kemper et al., 1998; Ryan et al., 2002). Bonnesen and Hummert (2002) found that the older respondents rated the PSD disclosers less negatively than did the young respondents, and Ryan and colleagues (2002) found that older participants were less likely to engage in negative age stereotyping based on age excuses than were the young participants. In terms of patronizing talk, Hummert et al. (1998) reported that only 25% of older participants' messages to the Despondent target were in the patronizing categories, whereas 60% of middle-aged messages and 58% of young-adult messages were patronizing. Together, these results suggest that the differences between age groups lie not in their acceptance of negative age stereotypes, but in their beliefs about the implications of negative age stereotypes for communication.

Responses of Older Targets: Interrupting the Negative Feedback Cycle of the CPA.
As illustrated in the ASI model, the way in which the older target responds to the other individual's communication can either reinforce the existing feedback cycle or interrupt it. Ideally, the response should reinforce a positive cycle initiated by the other's use of a normal adult communication style, but disrupt a negative cycle begun by the other's use of age-adapted or patronizing talk. Several studies have examined how older individuals can accomplish the latter goal (Harwood & Giles, 1996; Harwood, Giles, Fox, Ryan, & Williams, 1993; Harwood et al., 1997; Hummert & Flora, 1999; Hummert & Mazloff, 2001; Ryan et al., 2000).

In their review of the literature on patronizing communication, Hummert and Ryan (2001) outlined six different ways of responding to patronizing communication that have been investigated: passive, appreciative, assertive, humorous, condescending, and ignoring. To illustrate, we consider these response types in relation to a patronizing message from a nurse to an older patient who has asked about side effects of a new medication: "Now, dearie, why don't we just let the doctor worry about that." Passive responses involve unquestioning meek acceptance of patronizing (e.g., "Okay."), and the ignoring response offers an unrelated comment as if the older person has not heard the patronizing statement (e.g., "Oh, what did you say my blood pressure was?"). The other four response types challenge the message of incompetence inherent in patronizing talk, but vary in the directness and politeness of the challenge (Hummert & Mazloff, 2001; Ryan et al., 2000). Assertive responses state the challenge directly without embellishment: "I need to know that information myself." Appreciative responses also state the challenge directly, but consider the feelings of the patronizer: "I know that you're only trying to protect me, but I would also like to know that information." Humorous responses are less direct, but as a result may be more polite than assertive responses: "Oh, I think the doctor has enough to worry about! I'd better do my part as well." Finally, condescending responses counter the original patronizing remark with an attack on the patronizer: "Well that's just the type of insulting comment I've come to expect from you. Please remember to treat me like an adult in the future."

These response styles may vary in their ability to interrupt the negative feedback cycle of the CPA, however. Hummert and Ryan (2001) pointed out that the extremes of passive acceptance and confrontation carry obvious risks of confirming negative stereotypes, and therefore these extreme styles may reinforce the negative cycle rather than disrupt it. At the same time, assertive responses may also be perceived negatively and thus not be effective in establishing a more positive feedback cycle. For instance, in nursing home contexts where passivity is the norm, assertive responders have been perceived more negatively than passive responders (Hummert & Flora, 1999; Ryan et al., 2000). This finding extends to community contexts as well. Assertive responders have been evaluated as less polite and respectful than passive responders in community contexts, even though the assertive responders were seen as more competent (Harwood & Giles, 1996; Harwood et al., 1993; Harwood et al., 1997). According to Hummert and Ryan (2001), the appreciative and humorous response styles seem to have the most potential for reversing the negative feedback loop in the CPA model because they project a competent image while avoiding potential face threat to the patronizer.

CONCLUSIONS AND DIRECTIONS FOR FUTURE RESEARCH

The research investigating the role of age stereotypes in interpersonal communication has provided useful insights into the four issues we raised at the conclusion of our chapter in 1995: the factors that lead to positive or negative stereotyping, individual differences between communicators that are related to age stereotyping in interaction, what communication behaviors toward older adults are based on positive and negative stereotyping, and the relationship of communication behaviors of both young and older persons to stereotyping and self-perception. At the same time, this research has neglected some aspects of the communication and aging process as represented in the CPA (Ryan et al., 1986) and ASI models (Hummert, 1994a), and has demonstrated the difficulty of disrupting the negative feedback cycle of the CPA model.

Future research should continue to test and refine the CPA (Ryan et al., 1986) and ASI models (Hummert, 1994a). For instance, attention to the role of individual differences in cognitive complexity and interpersonal contact with older persons would advance our understanding of the ASI model. Recent research suggests that the grandparent-grandchild relationship may provide useful insights on how these individual differences affect age stereotyping and intergenerational communication (Harwood, 2000; Soliz & Harwood, in press; see Williams & Harwood, this volume). The roles of situational context and culture in the age-stereotyping and communication process also require additional study. Context has been manipulated primarily through verbal labels, which may not provide an ecologically valid test of its contribution to the process, and recent research on culture has revealed cross-cultural variations that require integration into the models (see Baker, Giles, & Harwood, this volume; Pecchioni et al., this volume).

Implicit stereotyping offers particular promise as a focus for future studies. In most discussions of the CPA and ASI models, scholars stress that speakers' overaccommodations to age stereotypes are often based on good intentions (Hummert & Ryan, 1996, 2001; Kemper, 2001; Ryan et al., 1995) as they try to adapt their communication appropriately to their partners. However, this does not mean that speakers consciously draw on these stereotypes in all instances. Instead, as Hummert (1999) argued, the adaptations may reflect the implicit (automatic, unconscious) operation of age stereotypes. With the development of the Implicit Association Test (Greenwald et al., 1998) and its use to measure age differences in implicit stereotypes, attitudes, and age identity (Dasgupta & Greenwald, 2000; Hummert et al., 2002; Hummert et al., 2003; Mellott et al., 2001), future research may be able to address the contribution of implicit social cognitions to the communication process. Such research also has the potential to clarify the factors that underlie the observed age differences in perceptions of and communication with older individuals (e.g., Hummert et al., 1995, 1998; Ryan et al., 2002).

Research on implicit stereotyping should pay special attention to implicit self-stereotyping in the communication process. A major advance in research on communication and aging has been the move to consider how the communication of older adults (e.g., painful self-disclosures, Bonnesen & Hummert, 2002; repetitious verbal behavior, Bieman-Copland & Ryan, 2001; age excuses, Ryan et al., 2002; off-target verbosity, Ruscher & Hurley, 2000) may elicit age stereotypes in others. The ways in which these and other communication behaviors may contribute to or reflect implicit self-stereotyping have received less attention. For example, Ryan and colleagues (2002) noted that although age excuses are believable, they carry with them a message of incompetence. The danger is that the individual who uses the age excuse may start to believe it, and act accordingly. This danger derives from the recursive relationship between communication and individual cognitions as outlined in Giddens' (1979) Structuration Theory and illustrated by the negative feedback cycle of the CPA model (Ryan et al., 1986). According to Giddens, cognitions (e.g., stereotype schemas) are called forth and instantiated in interaction, with the result that the cognitions not only affect interaction but also are modified and reinforced in that process. Thus, by self-stereotyping, older individuals perpetuate negative age stereotypes not only in listeners, but also in themselves, contributing to the negative feedback cycle of the CPA model.

A recent study by Bonnesen and Burgess (in press) of the meanings and functions of the phrase *senior moment* reinforces this notion. *Webster's College Dictionary* defines *senior moment* as "a brief lapse in memory or a moment of confusion, especially in an older person" (p. 1198). Bonnesen and Burgess' analysis revealed more nuances in meanings of the phrase than this definition might suggest, but showed that the majority of meanings and uses stressed stereotypes about age-related memory loss. Although there were a few examples in which *senior moment* referred to a positive development in an older person's life (e.g., winning a senior golf tournament), it was used most often as an age excuse for a negative event that occurred to the speaker (e.g., putting the newspaper in the refrigerator). From a social identity perspective, by using *senior moment* to describe their own behavior, these individuals emphasized their shared membership in the category *older adult* and the negative traits associated with that membership rather than their individuating traits. In Hogg and Turner's (1987) terms, they depersonalized the self and engaged in self-stereotyping.

As Levy's research suggests (Levy, 1996; Levy, Ashman, et al., 2000; Levy, Hausdorf, et al., 2000), negative self-stereotyping in communication is potentially harmful to older persons, but positive self-stereotyping may entail benefits for older individuals. For example, older adults may emphasize their age group's experience and maturity—and by extension, their own—in a context in which those attributes matter (e.g., political positions). Such self-stereotyping may increase their personal self-esteem and self-confidence, rather than induce the feelings of decline that may accompany painful self-disclosures or age excuses. Both forms of self-stereotyping communication behaviors should be addressed in future research.

In addition, to these issues, continued emphasis on ways of interrupting and reversing the CPA cycle is warranted. Baltes and colleagues (Baltes et al., 1994; Baltes & Wahl, 1996) have used a behavioral intervention to turn the dependency-supporting script into an independence-supporting script, with resulting gains in the functional health of nursing home residents. Interventions of this type may be usefully employed with professional caregivers, and, perhaps, with family caregivers. Perspective taking is another intervention that may reduce negative stereotyping. In two experiments, Galinsky and Moskowitz (2000) demonstrated that participants who were encouraged to take the perspective of an older individual evaluated the target more positively and less stereotypically in written descriptions, and saw the target as more similar to themselves than did participants who were encouraged to suppress the stereotype. However, additional attention to how older individuals themselves may intervene to change the nature of the interaction is needed. The studies of responses to patronizing communication begun in the 1990s must be continued, with a move from the assessment of written scripts to the assessment of oral responses.

Here, too, older persons may either help to interrupt a negative feedback cycle for their peers or contribute to it. For example, in a study reported by Duval, Ruscher, Welsh, and Catanese (2000), young listeners who heard an older woman describe an older target echoed that description in a subsequent evaluation of a new older target. This held true regardless of whether the earlier description was stereotypically positive (e.g., mature, dignified), stereotypically negative (e.g., slow, stubborn), counterstereotypically positive (e.g., healthy, never forgets anything), or counterstereotypically negative (e.g., inexperienced, immature). However, hearing the same descriptions from a young speaker did not affect evaluations of the new target. These results suggest that as in-group members, older individuals may be influential in perpetuating or changing age stereotypes as they are applied to their peers. This possibility requires further study, but it illustrates that older persons need not be seen as victims of age stereotyping in communication. Instead, they may productively challenge those stereotypes through and in their communication.

Of course, the ultimate goal would be to avoid the CPA cycle entirely. Ryan, Meredith, MacLean, and Orange (1995) conceptualized how this could be accomplished in the Communication Enhancement Model (See Fig. 4.3). In this model, Ryan and colleagues envision the communication encounter with an older person as a positive feedback loop. This positive cycle would be achieved through emphasizing a person-centered, as opposed to category-based, approach to communication with older individuals. This approach requires not only a consideration of the individual characteristics of an older conversational partner at the beginning of an interaction, but also a constant reassessment of the interaction as it progresses. If the partners engage in appropriate adaptations, the enhancement model sees positive outcomes for both parties in terms of empowerment, increased competence, satisfaction, health, and effective communication. Clarifying how

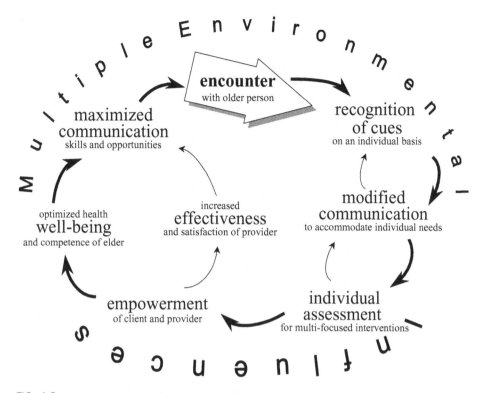

FIG. 4.3. Communication Enhancement Model (Ryan, Meredith, MacLean, & Orange, 1995, p. 96).

communication partners of all ages can enact this model must remain our goal. It is our hope that by moving research on age stereotyping forward in the ways we have outlined, we will gain insights that will help to meet this goal.

ACKNOWLEDGMENTS

Preparation of this manuscript was supported by Grant AG016352 from the National Institute on Aging/National Institutes of Health to the first author.

REFERENCES

Ashmore, R. D., & Del Boca, F. K. (1981). Conceptual approaches to stereotypes and stereotyping. In D. L. Hamilton (Ed.), *Cognitive processes in stereotyping and intergroup behavior* (pp. 1–35). Hillsdale, NJ: Lawrence Erlbaum Associates.

Baker, V., Giles, H., & Harwood, J. (in press). Inter- and intragroup perspectives in intergenerational communication. In J. F. Nussbaum & J. Coupland (Eds.), *Handbook of communication and ageing research* (2nd ed.). Mahwah, NJ: Lawrence Erlbaum Associates.

Baltes, M. M., Neumann, E. M., & Zank, S. (1994). Maintenance and rehabilitation of independence in old age: An intervention program for staff. *Psychology and Aging, 9,* 179–188.

Baltes, M. M., & Wahl, H. W. (1996). Patterns of communication in old age: The dependence-support and independence-ignore script. *Health Communication, 8,* 217–231.

Berry, D. S., & McArthur, L. A. (1985). Some components and consequences of a babyface. *Journal of Personality and Social Psychology, 48,* 312–323.

Berry, D. S., & McArthur, L. A. (1986). Perceiving character in faces: The impact of age-related craniofacial changes on social perception. *Psychological Bulletin, 100,* 3–18.

Berry, D. S., & McArthur, L. A. (1988). What's in a face? The impact of facial maturity and defendant intent on the attribution of legal responsibility. *Personality and Social Psychology Bulletin, 14,* 23–33.

Bieman-Copland, S., & Ryan, E. B. (2001). Social perceptions of failures in memory monitoring. *Psychology and Aging, 16,* 357–361.

Bonnesen, J. L., & Burgess, E. O. (in press). "Senior moments": An analysis of an ageist attribution. *Journal of Aging Studies.*

Bonnesen, J. L., & Hummert, M. L. (2002). Painful self-disclosures of older adults in relation to aging stereotypes and perceived motivations. *Journal of Language and Social Psychology, 21,* 275–301.

Brewer, M., Dull, V., & Lui, L. (1981). Perceptions of the elderly: Stereotypes as prototypes. *Journal of Personality and Social Psychology, 41,* 656–670.

Brewer, M., & Lui, L. (1984). Categorization of the elderly by the elderly. *Personality and Social Psychology Bulletin, 10,* 585–595.

Burleson, B. R. (1984). Age, social-cognitive development, and the use of comforting strategies. *Communication Monographs, 51,* 140–153.

Caporael, L. R. (1981). The paralanguage of caregiving: Baby talk to the institutionalized aged. *Journal of Personality and Social Psychology, 40,* 876–884.

Cohen, G., & Faulkner, D. (1986). Memory for proper names: Age differences in retrieval. *British Journal of Developmental Psychology, 4,* 187–197.

Coupland, N., & Coupland, J. (1989). Language and later life: The diachrony and decrement predicament. In H. Giles & P. Robinson (Eds.), *Handbook of language and social psychology* (pp. 451–468). London: Wiley.

Coupland, N., Coupland, J., Giles, H., & Henwood, K. (1988). Accommodating the elderly: Invoking and extending a theory. *Language and Society, 17,* 1–41.

Coupland, N., Coupland, J., Giles, H., Henwood, K., & Wiemann, J. (1988). Elderly self-disclosure: Interactional and intergroup issues. *Language and Communication, 8,* 109–133.

Coupland, N., Henwood, K., Coupland, J., & Giles, H. (1990). Accommodating troubles-talk: The young's management of elderly self-disclosure. In G. M. McGregor & R. White (Eds.), *Reception and response: Hearer creativity and the analysis of spoken and written texts* (pp. 112–133). London: Croom-Helm.

Crockett, W. H., & Hummert, M. L. (1987). Perceptions of aging and the elderly. *Annual Review of Gerontology and Geriatrics, 7,* 217–241.

Dasgupta, N., & Greenwald, A. G. (2001). On the malleability of automatic attitudes: Combating automatic prejudice with images of admired and disliked individuals. *Journal of Personality and Social Psychology, 81,* 800–814.

Dasgupta, N., McGhee, D. E., Greenwald, A. G., & Banaji, M. R. (2000). Automatic preference for White Americans: Eliminating the familiarity explanation. *Journal of Experimental Social Psychology, 36,* 316–328.

Delia, J. G., & Clark, R. A. (1977). Cognitive complexity, social perception, and the development of listener-adapted communication in six-, eight-, ten-, and twelve-year old boys. *Communication Monographs, 44,* 326–345.

Denton, W. H., Burleson, B. R., & Sprenkle, D. R. (1995). Association of interpersonal cognitive complexity with communication skill in marriage: Moderating the effects of marital distress. *Family Process, 34,* 101–111.

Devine, P. G. (1989). Stereotypes and prejudice: Their automatic and controlled components. *Journal of Personality and Social Psychology, 56,* 5–18.

Duval, L. L., Ruscher, J. B., Welsh, K., & Catanese, S. P. (2000). Bolstering and undercutting use of the elderly stereotype through communication of exemplars: The role of speaker age and exemplar stereotypicality. *Basic and Applied Social Psychology, 22,* 137–146.

Edwards, H., & Noller, P. (1998). Factors influencing caregiver-carereceiver communication and its impact on the well-being of older carereceivers. *Health Communication, 10,* 317–341.

Galinsky, A. D., & Moskowitz, G. B. (2000). Perspective-taking: Decreasing stereotype expression, stereotype accessibility, and in-group favoritism. *Journal of Personality and Social Psychology, 78*, 708–724.

Giddens, A. (1979). *Central problems in social theory: Action, structure and contradiction in social analysis.* Berkeley: University of California Press.

Giles, H., Coupland, N., & Coupland, J. (1991). Accommodation theory: Communication, context, and consequence. In H. Giles, J. Coupland, & N. Coupland (Eds.), *Contexts of accommodation: Developments in applied linguistics* (pp. 1–68). Cambridge, UK: Cambridge University Press.

Giles, H., Coupland, N., & Wiemann, J. M. (1992). "Talk is cheap" but "my word is my bond": Beliefs about talk. In K. Bolton & H. Kwock (Eds.), *Sociolinguistics today: Eastern and western perspectives* (pp. 218–243). London: Routledge.

Giles, H., Harwood, J., Pierson, H. D., Clément, R., & Fox, S. (1998). Stereotypes of older persons and evaluations of patronizing speech: A cross-cultural foray. In R. K. Agnihotri, A. L. Khana, & I. Sachdev (Eds.), *Social psychological perspectives on second language learning* (pp. 151–186). New Delhi: Sage.

Giles, H., Henwood, K., Coupland, N., Harriman, J., & Coupland, J. (1992). Language attitudes and cognitive mediation. *Human Communication Research, 18*, 500–527.

Giles, H., Mulac, A., Bradac, J. J., & Johnson, P. (1987). Speech accommodation theory: The last decade and beyond. In M. L. McLaughlin (Ed.), *Communication yearbook 10* (pp. 13–48). Beverly Hills, CA: Sage.

Greenwald, A. G., & Banaji, M. R. (1995). Implicit social cognition: Attitudes, self-esteem, and stereotypes. *Psychological Review, 102*, 4–27.

Greenwald, A. G., & Farnham, S. D. (2000). Using the Implicit Association Test to measure self-esteem and self-concept. *Journal of Personality and Social Psychology, 79*, 1022–1038.

Greenwald, A. G., McGhee, D. E., & Schwartz, J. L. K. (1998). Measuring individual differences in implicit cognition: The implicit association test. *Journal of Personality and Social Psychology, 74*, 1464–1480.

Greenwald, A. G., & Nosek, B. A. (2001). Health of the Implicit Association Test at age 3. *Zeitschrift für Experimentelle Psychologie, 48*, 85–93.

Hale, N. M. (2000). Effects of age and interpersonal contact on stereotyping of the elderly. *Current Psychology: Developmental, Learning, Personality, Social, 17*, 28–47.

Hamilton, D. L., & Trolier, T. K. (1986). Stereotypes and stereotyping: An overview of the cognitive approach. In J. F. Dovidio & S. L. Gaertner (Eds.), *Prejudice, discrimination, and racism* (pp. 127–163). Orlando, FL: Academic Press.

Harwood, J. (2000). Communicative predictors of solidarity in the grandparent-grandchild relationship. *Journal of Social and Personal Relationships, 17*, 743–766.

Harwood, J., & Giles, H. (1996). Reactions to older people being patronized: The roles of response strategies and attributed thoughts. *Journal of Language and Social Psychology, 15*, 395–421.

Harwood, J., Giles, H., Fox, S., Ryan, E. B., & Williams, A. (1993). Patronizing young and elderly adults: Response strategies in a community setting. *Journal of Applied Communication Research, 21*, 211–226.

Harwood, J., Giles, H., & Ryan, E. B. (1995). Aging, communication, and intergroup theory: Social identity and intergenerational communication. In J. L Nussbaum & J. Coupland (Eds.), *Handbook of communication and aging research* (pp. 133–159). Hillsdale, NJ: Lawrence Erlbaum Associates.

Harwood, J., Ryan, E. B., Giles, H., & Tysoski, S. (1997). Evaluations of patronizing speech and three response styles in a non-service-providing context. *Journal of Applied Communication Research, 25*, 170–195.

Harwood, J., & Williams, A. (1996). Expectations for communication with positive and negative subtypes of older adults. *International Journal of Aging and Human Development, 47*, 11–33.

Heckhausen, J., Dixon, R. A., & Baltes, P. B. (1989). Gains and losses in development throughout adulthood as perceived by different adult age groups. *Developmental Psychology, 25*, 109–121.

Hogg, M. A., & Turner, J. C. (1987). Intergroup behavior, self-stereotypes, and the salience of the intergroup context. *British Journal of Social Psychology, 4*, 325–340.

Hummert, M. L. (1990). Multiple stereotypes of elderly and young adults: A comparison of structure and evaluations. *Psychology and Aging, 5*, 182–193.

Hummert, M. L. (1994a). Stereotypes of the elderly and patronizing speech style. In M. L. Hummert, J. M. Wiemann, & J. F. Nussbaum (Eds.), *Interpersonal communication in older adulthood: Interdisciplinary theory and research* (pp. 162–185). Newbury Park, CA: Sage.

Hummert, M. L. (1994b). Physiognomic cues to age and the activation of stereotypes of the elderly in interaction. *International Journal of Aging and Human Development, 39,* 5–20.

Hummert, M. L. (1999). A social cognitive perspective on age stereotypes. In T. M. Hess & F. Blanchard-Fields (Eds.), *Social cognition and aging* (pp. 175–195). New York: Academic Press.

Hummert, M. L., & Flora, J. (1999, July). *Responses to patronizing talk: Perceptions of college students and nursing home staff.* Paper presented at the Fourth International Conference on Communication, Aging, and Health, Brisbane, Australia.

Hummert, M. L., Garstka, T. A., O'Brien, L. T., Greenwald, A. G., & Mellott, D. S. (2002). Using the Implicit Association Test to measure age differences in implicit social cognitions. *Psychology and Aging, 17,* 482–495.

Hummert, M. L., Garstka, T. A., O'Brien, L. T., Savundranayagam, M., Zhang, Y. B., & Geiger, W. (2003). *Strength of positive and negative implicit age stereotypes across age groups.* Unpublished manuscript, University of Kansas, Lawrence, KS.

Hummert, M. L., Garstka, T. A., & Shaner, J. L. (1995). Beliefs about language performance: Adults' perceptions about self and elderly targets. *Journal of Language and Social Psychology, 14,* 235–259.

Hummert, M. L., Garstka, T. A., & Shaner, J. L. (1997). Stereotyping of older adults: The role of target facial cues and perceiver characteristics. *Psychology and Aging, 12,* 107–114.

Hummert, M. L., Garstka, T. A., Shaner, J. L., & Strahm, S. (1994). Stereotypes of the elderly held by young, middle-aged, and elderly adults. *Journal of Gerontology: Psychological Sciences, 49,* P240–P249.

Hummert, M. L., Garstka, T. A., Shaner, J. L., & Strahm, S. (1995). Judgments about stereotypes of the elderly: Attitudes, age associations, and typicality ratings of young, middle-aged, and elderly adults. *Research on Aging, 17,* 168–189.

Hummert, M. L., & Mazloff, D. C. (2001). Older adults' responses to patronizing advice: Balancing politeness and identity in context. *Journal of Language and Social Psychology, 20,* 167–195.

Hummert, M. L., Mazloff, D. C., & Henry, C. (1999). Vocal characteristics of older adults and stereotyping. *Journal of Nonverbal Behavior, 23,* 111–132.

Hummert, M. L., & Ryan, E. B. (1996). Toward understanding variations in patronizing talk addressed to older adults: Psycholinguistic features of care and control. *International Journal of Psycholinguistics, 12,* 149–169.

Hummert, M. L., & Ryan, E. B. (2001). Patronizing. In W.P. Robinson & H. Giles (Eds.), *The new handbook of language and social psychology* (2nd ed.) (pp. 253–269). Chichester, UK: Wiley.

Hummert, M. L., & Shaner, J. L. (1994). Patronizing speech to the elderly as a function of stereotyping. *Communication Studies, 45,* 145–158.

Hummert, M. L., Shaner, J. L, & Garstka, T. A. (1995). Cognitive processes affecting communication with older adults: The case for stereotypes, attitudes, and beliefs about communication. In J. F. Nussbaum & J. Coupland (Eds.), *Handbook of communication and aging research* (pp. 105–131). Mahwah, NJ: Lawrence Erlbaum Associates.

Hummert, M. L., Shaner, J. L., Garstka, T. A., & Henry, C. (1998). Communication with older adults: The influence of age stereotypes, context, and communicator age. *Human Communication Research, 25,* 124–151.

Karpinski, A., & Hilton, J. L. (2001). Attitudes and the implicit association test. *Journal of Personality and Social Psychology, 81,* 774–788.

Kemper, S. (1994). "Elderspeak": Speech accommodations to older adults. *Aging and Cognition, 1,* 17–28.

Kemper, S. (2001). Over-accommodations and under-accommodation to aging. In N. Charness, D. C. Parks, & B. Sabel (Eds.), *Communication, techonology, and aging: Opportunities and challenges for the future* (pp. 30–46). New York: Springer.

Kemper, S., Ferrell, P., Harden, T., Finter-Urczyk, A., & Billington, C. (1998). The use of elderspeak by young and older adults to impaired and normal older adults. *Aging, Neuropsychology, and Cognition, 5,* 43–55.

Kemper, S., & Harden, T. (1999). Experimentally disentangling what's beneficial about elderspeak from what's not. *Psychology and Aging, 14,* 656–670.

Kemper, S., Othick, M., Warren, J., Gubarchuk, J., & Gerhing, H. (1996). Facilitating older adults' performance on a referential communication task through speech accommodations. *Aging and Cognition, 3,* 37–55.

Kite, M. E., & Johnson, B. T. (1988). Attitudes toward older and younger adults: A meta-analysis. *Psychology and Aging, 3,* 233–244.

Kite, M. E., & Wagner, L. S. (2001). Attitudes toward older adults. In T. D. Nelson (Ed.), *Stereotyping and prejudice against older persons* (pp. 129–161). Cambridge, MA: MIT Press.

Knox, V. J., Gekoski, W. L., & Johnson, E. A. (1986). Contact with and perceptions of the elderly. *The Gerontologist, 26,* 309–313.

La Tourette, T. R., & Meeks, S. (2000). Perceptions of patronizing speech by older women in nursing homes and in the community. *Journal of Language and Social Psychology, 19,* 463–473.

Levy, B. (1996). Improving memory in old age through implicit self-stereotyping. *Journal of Personality and Social Psychology, 71,* 1092–1107.

Levy, B. (2000). Handwriting as a reflection of aging self-stereotypes. *Journal of Geriatric Psychiatry, 33,* 81–94.

Levy, B., Ashman, O., & Dror, I. (2000). To be or not to be: The effects of aging stereotypes on the will to live. *Omega: Journal of Death and Dying, 40,* 409–420.

Levy, B., Hausdorf, J., Hencke, R., & Wei, J. Y. (2000). Reducing cardiovascular stress with positive self-stereotypes of aging. *Journals of Gerontology: Psychological Science and Social Sciences, 55B,* P205–P213.

Lin, M.-C., & Harwood, J. (2003). Predictors of grandparent-grandchild relational solidarity in Taiwan. *Journal of Social and Personal Relationships, 20,* 537–563.

McArthur, L. A. (1982). Judging a book by its cover: A cognitive analysis of the relationship between physical appearance and stereotyping. In A. Hastorf & A. Isen (Eds.), *Cognitive social psychology* (pp. 149–211). New York: Elsevier/North Holland.

Mellott, D. S., Greenwald, A. G., Hummert, M. L., & O'Brien, L. T. (2001). *But I don't feel old! Implicit balance among ageism, youthful identity, and self-esteem in the elderly.* Unpublished manuscript, University of Washington, Seattle, WA.

Montepare, J. M., & Zebrowitz-McArthur, L. A. (1988). Impressions of people created by age-related qualities of their gaits. *Journal of Personality and Social Psychology, 55,* 547–556.

Montepare, J. M., & Zebrowitz, L. A. (1993). A cross-cultural comparison of impressions created by age-related variations in gait. *Journal of Nonverbal Behavior, 17,* 55–68.

Mulac, A., & Giles, H. (1996). "You're only as old as you sound": Perceived vocal age and social meanings. *Health Communication, 8,* 199–215.

Neugarten, B. L. (1974). Age groups in American society and the rise of the young-old. *The Annals of the American Academy of Political and Social Science, 415,* 187–198.

Noels, K., Giles, H., Gallois, C., & Ng, S. H. (2001). Intergenerational communication and psychological adjustment: A cross-cultural examination of Hong Kong and Australian Adults. In M. L. Hummert & J. F. Nussbaum (Eds.), *Aging, communication, and health: Linking research and practice for successful aging* (pp. 249–278). Mahwah, NJ: Lawrence Erlbaum Associates.

O'Connor, B. P., & Rigby, H. (1996). Perceptions of baby talk, frequency of receiving baby talk, and self-esteem among community and nursing home residents. *Psychology and Aging, 11,* 147–154.

Pecchioni, L., Ota, H., & Sparks, L. (in press). Cultural issues in communication and aging. In J. F. Nussbaum & J. Coupland (Eds.), *Handbook of communication and aging research* (2nd ed.). Mahwah, NJ: Lawrence Erlbaum Associates.

Perdue, C. W., & Gurtman, M. B. (1990). Evidence for the automaticity of ageism. *Journal of Experimental Social Psychology, 26,* 199–216.

Rose-Colley, M., & Eddy, J. M. (1988). Interactions of university students with elderly individuals: An investigation into the correlates. *Educational Gerontology, 14,* 33–43.

Rodin, J., & Langer, E. J. (1980). Aging labels: The decline of control and the fall of self-esteem. *Journal of Social Issues, 36,* 12–29.

Rudman, L. A., Greenwald, A. G., & McGhee, D. E. (2001). Self-esteem and gender identity are manifest in implicit gender stereotypes. *Personality and Social Psychology Bulletin, 27,* 1164–1178.

Ruscher, J., & Hurley, M. (2000). Off-target verbosity evokes negative stereotypes of older adults. *Journal of Language and Social Psychology, 19,* 141–149.

Ryan, E. B. (1992). Beliefs about memory changes across the adult life span. *Journal of Gerontology: Psychological Sciences, 47,* P41–P46.

Ryan, E. B., Bieman-Copland, S., Kwong See, S. T, Ellis, C. H., & Anas, A. P. (2002). Age excuses: Conversational management of memory failures in older adults. *Journal of Gerontology: Psychological Sciences and Social Sciences, 57B,* P256–P267.

Ryan, E. B., & Capadano, H. L. (1978). Age perceptions and evaluative reactions toward adult speakers. *Journal of Gerontology, 33*, 98–102.

Ryan, E. B., & Cole, R. L. (1990). Evaluative perceptions of interpersonal communication with elders. In H. Giles, N. Coupland, & J. Wiemann (Eds.), *Communication, health, and the elderly*, Fulbright Series #8 (pp. 172–190). Manchester, UK: Manchester University Press.

Ryan, E. B., Giles, H., Bartolucci, G., & Henwood, K. (1986). Psycholinguistic and social psychological components of communication by and with the elderly. *Language and Communication, 6*, 1–24.

Ryan, E. B., Hummert, M. L., & Boich, L. (1995). Communication predicaments of aging: Patronizing behavior toward older adults. *Journal of Language and Social Psychology, 14*, 144–166.

Ryan, E. B., & Johnston, D. G. (1987). The influence of communication effectiveness on evaluations of younger and older adult speakers. *Journal of Gerontology, 42*, 163–164.

Ryan, E. B., Kennaley, D. E., Pratt, M. W., & Shumovich, M. A. (2000). Evaluations by staff, residents, and community seniors of patronizing speech in the nursing home: Impact of passive, assertive or humorous responses. *Psychology and Aging, 15*, 272–285.

Ryan, E. B., & Kwong See, S. (1993). Age-based beliefs about memory change for self and others across adulthood. *Journal of Gerontology: Psychological Sciences, 48*, P199–P201.

Ryan, E. B., Kwong See, S., Meneer, W. B., & Trovato, D. (1992). Age-based perceptions of language performance among young and older adults. *Communication Research, 19*, 423–443.

Ryan, E. B., & Laurie, S. (1990). Evaluations of older and younger adult speakers: Influence of communication effectiveness and noise. *Psychology and Aging, 5*, 514–519.

Ryan, E. B., Meredith, S. D., & Shantz, G. B. (1994). Evaluative perceptions of patronizing speech addressed to institutionalized elders in contrasting conversational contexts. *Canadian Journal on Aging, 13*, 236–248.

Ryan, E. B., Meredith, S. D., MacLean, M. J., & Orange, J. B. (1995). Changing the way we talk with elders: Promoting health using the Communication Enhancement Model. *International Journal of Aging and Human Development, 41*, 87–105.

Sachweh, S. (1998). Granny darling's nappies: Secondary babytalk in German nursing homes for the aged. *Journal of Applied Communication Research, 26*, 52–65.

Schmidt, D. F., & Boland, S. M. (1986). The structure of impressions of older adults: Evidence for multiple stereotypes. *Psychology and Aging, 1*, 255–260.

Soliz, J., & Harwood, J. (in press). Communication in a close family relationship and the reduction of prejudice. *Journal of Applied Communication Research.*

Stewart, M. A., & Ryan, E. B. (1982). Attitudes toward younger and older adult speakers: Effects of varying speech rates. *Journal of Language and Social Psychology, 1*, 91–109.

Thimm, C., Rademacher, U., & Kruse, L. (1998). Age stereotypes and patronizing messages: Features of age-adapted speech in technical instructions to the elderly. *Journal of Applied Communication Research, 26*, 66–82.

Whitbourne, S. K., Culgin, S., & Cassidy, E. (1996). Evaluations of infantilizing intonation and content of speech directed at the aged. *International Journal of Aging and Human Development, 41*, 109–116.

Williams, A., & Harwood, J. (in press). An intergroup perspective on intergenerational relationships in the family. In J. F. Nussbaum & J. Coupland (Eds.), *Handbook of communication and aging research* (2nd ed.). Mahwah, NJ: Lawrence Erlbaum Associates.

Zhang, Y. B., & Hummert, M. L. (2001). Harmonies and tensions in Chinese intergenerational relationships: Young and older adults' accounts. *Journal of Asian and Pacific Communication, 11*, 199–226.

Intergenerational Communication: Intergroup, Accommodation, and Family Perspectives

Angie Williams
Cardiff University

Jake Harwood
University of Arizona

As described by Harwood, Giles, and Ryan (1995) in the previous edition of this handbook, age groups offer some interesting challenges to traditional intergroup theories, most notably social identity theory (SIT: Tajfel & Turner, 1979, 1986). Social identity theory contends that an individual's self concept comprises two parts: personal identity and social identity. Personal identity is that part of self that includes personal characteristics, likes and dislikes, and idiosyncracies. Social identity, however, is our identity as members of particular social groups. These groups result from the psychologically inevitable process of categorization (McGarty, 1999; Taylor, 1981). Tajfel and Turner argued that we compare our own group's position in society with that of other groups and seek to obtain positive distinctiveness for our in-group. As an example of this, if we are men, we might seek information or hold beliefs that suggest that, on certain important dimensions, we are superior to women. These positive comparisons provide us with positive social identities that, in turn, bolster our self-esteem. Although a number of critiques of the theory exist (Billig, 1985; Rabbie, Schot, & Visser, 1989), there is considerable evidence of a tendency to seek positive distinctiveness for one's in-group—no matter how mundane the group may be (e.g., a group to which you have been arbitrarily assigned: Tajfel, Billig, Bundy, & Flament, 1971). Evidence also exists that we obtain some level of self-esteem from the assignment of positive features or rewards to the in-group and the denial of these to the out-group (see Branscombe & Wann, 1994; Finchilescu, 1986; Hogg & Abrams, 1990; Hogg & Turner, 1985; Lemyre & Smith, 1985; Oakes & Turner, 1980; Sachdev & Bourhis, 1987, for debate on this point).

This chapter examines the ways in which age groups constitute meaningful social categories in communication. We review the evidence for considering age groups as social categories, and some of the challenges they provide for intergroup theory, such as SIT, as well as talking about the ways in which age becomes a salient categorization dimension in interaction. We then talk about the ways in which individuals categorize themselves and others into age categories, and the ways in which people respond when their self-categorizations according to age are personally threatening. Next, we consider the ways in which these categorizations and identifications influence the intergenerational communication process between younger and older people who are strangers and those who are family members, with a particular focus on communication accommodation theory. We close with some reflections on the ways in which areas of "intergroup communication" and "family communication" might benefit from some consolidation and theoretical integration to understand communication between younger and older people.

First, it is important to consider whether age groups are meaningfully "groups" in the social identity sense of the term. We would argue that they are. Extensive research exists demonstrating the pervasiveness of broad age categorization labels ("young," "middle-aged," and "old") (Coupland & Coupland, 1990; Giles, 1999), as well as more sophisticated delineations in certain areas (infant, toddler, adolescent, e.g., Williams & Garrett, 2002a). It is also clear that age categorizations have distinct affective implications (Taylor, 1992), and hence it is likely that considerable energy may be expended in managing one's own age categorization (see Williams & Garrett, 2002a). Age has also been shown to be a fundamental element in our categorizations and judgments of others (Coupland, chap. 3, this volume; Hummert, Garstka, Ryan, & Bonnesen, chap. 4, this volume; Kite, Deaux, & Miele, 1991).

That said, although age group categorizations are ubiquitous, they also have an inherent ambiguity. Of course, a certain "fuzziness" may exist between groups in other intergroup situations (for instance, with bicultural identities), but the blurring of boundaries between age groups seems especially significant. Age categories are more obviously abstracted from a continuous variable (chronological age), and hence are perhaps more contextually and communicatively negotiable than most other social group memberships (Coupland, 1999). As a result, "identity," as a psychological feature of group membership, may be more crucial in the realm of age, as opposed to, for instance, gender differences where the physical characteristics underlying categorization are often fairly clear. Although men and women may be concerned about being viewed as "real" men or "real" women, it is unlikely that in their daily lives they will be mistaken for a member of the other sex. However, it is considerably less unusual for individuals to be categorized into age groups other than those with which they identify. Boundaries between some social groups and categories are relatively inflexible (e.g., race, gender), and individuals crossing those boundaries in any sustained way are relatively rare. However, the definitions for what counts as a particular age group are more socially fluid and negotiable in daily interaction (Coupland & Nussbaum, 1993).

Considerable work has examined the ways in which age may become salient in interaction. Coupland, Coupland, Giles, and Henwood (1991) described some ways in which older adults mark their age and have shown such marking to be frequent in their conversations. The clearest example of such marking is disclosure of chronological age (Coupland, Coupland, & Giles, 1989), but it can also be found in older adults' use of age-related role references for themselves (e.g., "pensioner"), their references to the past, or their descriptions of age-related health complaints, for instance (Coupland et al., 1991; see also Harwood, 2002; Lin, Hummert, & Harwood, 2002). Harwood and Giles (1992) examined similar age-marking phenomena in the media, looking at ways in which age was marked in the television show *The Golden Girls*. Their analysis suggested that age marking (and particularly counterstereotypical age marking) was often associated with humor in the shows, hence reducing the potential for the show to have a positive influence on viewers' attitudes. Less systematic work is available on the early stages of the life span, but the obsession with age among younger children is self-evident. Likewise, teens are often engaged with issues of age when it comes to doing certain activities (seeing movies, buying beer, etc.). The ways in which age is marked at that point in the life span, and the ways in which it is dropped from "normal" interaction as we enter adulthood, are worthy of further examination, as indeed are the ways in which it enters adult discourse surreptitiously (e.g., through ageist birthday cards, see Giles, 1999).

Thus, we conceive of the boundaries between age groups as somewhat "soft" and "open" (cf. Banton, 1983); however, this permeability is not unconditional. That is, there are certain "qualifications" for passing through the boundaries: One is unlikely to be classified as "middle-aged" as an unmarried male student with a full head of (not gray) hair. Similarly, the boundaries are only truly open in one direction. Once you have moved from one age group to the "next" in self- or other-perception, it will be difficult to move back again. That said, age identities do offer the unique possibility of presenting very different identities in different contexts (e.g., being "old and dependent" for certain discursive purposes, being "young and vibrant" when that is more appropriate; see Coupland, Coupland, & Grainger, 1991). Therefore, the option is open to attempt to move back to earlier stages of the life span through adopting different styles of self-presentation (e.g., makeup, clothing) or peer affiliations (e.g., seeking out new and younger friends).

PERCEPTIONS OF CHRONOLOGICAL BOUNDARIES ASSOCIATED WITH AGE LABELS

An investigation of the chronological ages associated with certain age labels (Harwood, Giles, Clément, Pierson, & Fox, 1994) asked young people from the United States to estimate age in years associated with "young," "middle aged," and "elderly." Perceptions of young adulthood were found to stretch from the midteens to around age 30. Middle age was perceived as ranging from 30 to 50, with "old" adulthood starting in the early 50s(!). Previously, Harwood and Giles (1993) found 37 to 53 years to be the boundaries of "middle age" as estimated by young California students.

A recent cross-cultural study by Giles asked young participants from 11 nations (Canada, Australia, New Zealand, Taiwan, the People's Republic of China, Japan, South Korea, the Philippines, Singapore, and India) to estimate the ages between which adults could be classified as "young," "middle-aged," and "older." Generally, Western respondents across these nations thought that young adulthood began at 17 years and ended at 28 years, middle age began at 29 years and ended at 48 years, and old age began at about 51 years. For Eastern respondents, young adulthood began at 17 and ended at 29 years, middle age began at 31 and ended at 51 years, and old age began at 52. Participants were all young university students under 30 years old, with an average age of 20.

Williams and Garrett (2002a) asked adults 20 to 60 years of age to provide the upper and lower age boundaries associated with the age labels "adolescence," "young adulthood," "middle age," and "old age." Adults of all ages perceived adolescence to be between 12 and 17 years of age and "young adulthood" to be between age 18 and 26. However, there was no such consensus on "middle age" and "old age." Young adult respondents (aged 20–29) suggested that middle age begins at 37 years old and ends at 54 years old. Older respondents' (aged 40–60) estimates were significantly different from those of the young cohorts (aged 20–29 and 39–39). The older cohorts were inclined to raise the boundaries of "middle age," moving the lower boundary to 40 or 41 and the upper boundary to 57. In other words, as you get older and approach "middle age," you might be inclined to raise its lower limit to ensure that you could not be included in that category. With increasing age, one then might be tempted to raise the upper limit of middle age to ensure your continued membership of that category, because moving out of middle age into old age is seen as an even more negative self-categorization.

None of the respondents subjectively belonged to the group labelled "old age," although the lower boundary of old age also varied according to respondent age. For the youngest respondents (aged 20–29), old age began at 62, but for the oldest group of respondents (who were aged between 50 and 59 and were therefore in imminent danger of belonging to this category), "old age" began at 66. Apart from the details, these findings support the view that perceptions of chronological age associated with age group labels vary depending on how old you are. Moreover, if you are "in danger" of being classified as belonging to a less positively perceived age group, you may have a tendency to adjust the perceptual boundaries to avoid such classification. As we might expect from popular notions about the double jeopardy of being older and a woman, women in this study were inclined to adjust the onset of middle and old age in more dramatic fashion than men (Williams & Garrett, 2002a). These findings align well with intergroup and social identity theory, which postulates that people would be disinclined to self-categorize in terms of negatively perceived groups (Garrett, Giles, & Coupland, 1989; Tajfel & Turner, 1986; Williams & Giles, 1998).

Intergroup theory provides an explanation for the prevalence of negative out-group stereotypes in numerous domains of social life, and particularly in the current context, age. Specifically, younger adults have been shown to hold negative stereotypes of older adults using numerous methodologies and across cultures (Harwood et al., 1996; Kite & Johnson, 1988; Perdue & Gurtman, 1990). From a communication perspective, we should note that these negative stereotypes extend to beliefs about older adults' communication abilities and intergenerational conversation (e.g., Hummert, Gartska, & Shaner, 1995; Ryan, Kwong See, Meneer, & Trovato, 1992). Under some circumstances, similar negative attitudes toward aging have been found in older adults' beliefs (Harwood et al., 2001; Ryan et al., 1992). Similarly, negative attitudes toward younger adults are also common, as evidenced by portrayals of Generation X in the media (Williams, Coupland, Folwell, & Sparks, 1997); adults' comments about adolescents' communication styles (Drury & Dennison, 1999, 2000); and young peoples' own reports about the way adults patronize them (Giles & Williams, 1994).

Given this emphasis on obtaining positive comparisons between one's own group and the outgroup, SIT addresses the problems faced by members of subordinate groups who are not able to obtain positive distinctiveness. Three strategies are suggested: social mobility, social creativity, and social competition (Tajfel & Turner, 1979). Social mobility is an individualistic strategy in which a person "leaves"—psychologically, physically, or communicatively—a group that is not providing positive social comparisons and moves to a group that can provide more in the way of social rewards. This transition is, of course, easier in some situations than others, with the relative softness and openness of boundaries between the groups playing an important role (see Giles, 1979; Giles & Johnson, 1981). In the context of aging, examples of social mobility are common (Williams & Giles, 1998). Younger individuals, who often perceive their group to be devalued in terms of status or respect, will often attempt to self-present as older than they actually are. This occurs in many ways, from young girls wanting to wear makeup or jewellery, to young men lying about their age to join the armed services or buy beer. The use of an "adult" speech style can often contribute to these attempts. Similarly, as we enter middle age and older adulthood, attempts to appear younger become common (e.g., plastic surgery, makeup, hair dyes, lying about one's age). It is notable that the use of makeup operates at both ends of the age continuum for social mobility purposes, suggesting that issues surrounding cosmetics and other beauty products may be a productive avenue for examining the aging process (Coupland, in press).

Social creativity is a strategy in which individuals attempt to realign their intergroup comparisons so as to provide more positive group-based self-esteem. In other words, social creativity strategies allow the individual to gain positive social identity but tend to leave the underlying power structure relatively unchanged. For older adults, these strategies might be manifest in the substitution of ingroup comparisons for outgroup comparisons. For instance, younger (or more healthy) older adults might choose to compare their status and abilities with those of older or less healthy older adults, and hence emerge with a positive impression of their in-group (Paoletti, 1998; Williams & Guendouzi, 2000). For both younger and older individuals, these comparisons might include carefully selecting the dimensions along which intergroup comparisons are going to occur. For instance, older individuals might characterize other age groups as "reckless" or foolish, or "in too much of a hurry," and gain a positive group identity as being individuals who are wise, sensible, and take appropriate time over tasks. Likewise, younger individuals may concede an older person's superior societal status but claim to be in tune with the modern world, fashionable, or good-looking, and derogate the older out-group as ugly, asexual, and out of touch (see also Carver & de la Garza, 1984; Williams & Giles, 1998).

Finally, social competition strategies emphasize a group member's desire to radically change societal organization to give a low status group more overall power. Groups engaging in challenges for civil rights are engaging in social competition. *Social competition* strategies are seen as most likely when the underlying system is seen by the subordinate group as both unstable and illegitimate (see Turner & Brown, 1978). These conditions lead to an awareness of alternatives to the current intergroup situation, which will foster group mobilization and social competition in the subordinate group (see Gurin, Miller, & Gurin, 1980). Older adults have shown remarkable strength on these fronts, with a number of pressure groups now actively lobbying national governments for increased political representation and legislative attention to matters affecting older adults. In the United States, the American Association of Retired Persons, the Grey Panthers, 65-plus, and the National Council of Senior Citizens are some of the more active and important groups, although there are many others (see also Williams & Giles, 1998). In the United Kingdom, parallels exist with Age Concern, Saga, and the Association of Retired Persons (ARP), although arguably these organizations have less visibility than some of those in the United States. Youth advocacy groups also exist, although they appear to have less sustainability, perhaps because "youth" as a period of life is relatively short. Hence, many such groups are not run by "youths" (e.g., the American Youth Policy Forum in the United States) and hence would not count as activism in the way described by Tajfel and Turner (1986).

The next section examines some of the ways in which these theoretical bases have been applied in the communication literature, and specifically to intergenerational communication issues. As we have become more aware of the societal salience of age and aging issues, and the importance of age identities to individuals, the way in which intergenerational communication dynamics can be explained by age identity and stereotyping has become a focal research area.

COMMUNICATION ACCOMMODATION THEORY

In the communication literature, the chief theoretical perspective to emerge from the intergroup approach has been communication accommodation theory (CAT; Gallois, Giles, Jones, Cargile, & Ota, 1995; Giles, Coupland, & Coupland, 1991; Shepard, Giles, & LePoire, 2001). CAT explores the ways in which individuals attune their speech to others based on a variety of interpersonal and intergroup factors. Early in its development, CAT was particularly focused on convergence and divergence in speech styles—particularly as these were influenced by local intergroup (generally ethnocultural) relations (e.g., Bourhis & Giles, 1977). At the broadest level, the theory argues that social group identifications are negotiated and expressed in intergroup conversations and that the local organization of such encounters may reflect the larger intergroup dynamics and individuals' perceptions of their own position within those dynamics.

Recent work has developed the theory to explain multiple elements of intergenerational relations, and in particular the ways in which intergenerational conversational dynamics reflect individuals' attempts to negotiate their own age identities in the context of societal age relations (Coupland, Coupland, Giles, & Henwood, 1991; Giles et al., 1991; Williams & Guendouzi, 2000). Details of this work are provided in subsequent sections examining the ways in which intergenerational communication phenomena both within and outside of the family can be explained by intergroup theory in general, and CAT specifically.

CAT maintains that when we wish to signal ingroup solidarity, express personal affiliation, and cooperate, we attempt to accommodate our speech and communication to others, but when we wish to indicate out-group membership, distance ourselves personally, or employ either of these behaviors, we would be less inclined to accommodate. Sometimes individuals wish to emphasize the difference between themselves and their partner, and this would be characteristic of many intergroup encounters in which identity is salient and behavior is negatively attributed and evaluated by recipients.

According to CAT (see Giles et al., 1991), communication behavior with people from other groups reflects the stereotypes we hold about those groups. Thus, corresponding with the negative stereotypes of old age, research indicates that younger people may "overaccommodate" to older adults in their effort to communicate with them (Coupland, Coupland, Giles, & Wiemann, 1988). Typically, overaccommodation occurs when we attempt to converge to what we perceive as a partner's communicative needs but we "overdo it," making more adjustments than the person actually needs. For example, young people may be excessively concerned with ensuring that their message to elders is clear and simple, as characterized through louder and slower speech, simplified grammar and vocabulary, and repetition (Giles & Coupland, 1991; Ryan, Hummert, & Boich, 1995). This behavior is particularly, but not only, evident when negative substereotypes are activated (Hummert, 1994; Harwood & Williams, 1998). Overaccommodation often occurs regardless of the older person's functional autonomy and is neither perceived as appropriate, nor valued by socially and cognitively active older people (for review, see Hummert & Ryan, 2001; Ryan, Hummert, & Boich, 1995). Likewise, if younger people perceive older people in terms of cognitively and communicatively challenged stereotypes, they would expect their older partner's communication to be overly under- or nonaccommodative (i.e., unable to meet the younger person's communicative needs; Giles & Coupland, 1991). Satisfaction with intergroup communication should be a consequence of healthy intergroup relations and acceptance of social diversity. But for young people at least, patterns of overaccommodation and evaluations of elders as stereotypically underaccommodative might be associated with unsatisfying intergenerational relationships, intolerance, and even avoidance of older people.

How widespread is the notion that older people are underaccommodative to young peoples' conversational needs? Do young people see this as a problematic aspect of intergenerational talk? What other problems do young people experience when conversing with older people? What are the positive aspects of intergenerational talk? These were the questions behind studies by Williams (1992) and Williams and Giles (1996).

INTERGENERATIONAL COMMUNICATION SATISFACTION AND DISSATISFACTION

An initial questionnaire study of young people's perceptions of intergenerational communication satisfaction (Williams, 1992) found that young people (18–20) rated intergenerational interactions (with adults aged 65–75) as less satisfying overall than interactions with same-aged peers. Four factors emerged that differentiated satisfying intergenerational encounters from less satisfying ones. "Old underaccommodative negativity" was noted for situations in which the older person talked excessively and exclusively about his or her own problems. "Mutuality" was when the age gap was diminished, and "elder individuation" occurred when the older person was treated as an individual rather than an "old person." Finally, "young individuation" was when the young person felt individuated—not treated in a stereotypical fashion. These dimensions were related to reported emotions in predictable directions. For example, "old underaccommodative negativity" was associated with increased anger, frustration, and decreased relaxation, but was not related to happiness.

The findings of this study relate in interesting ways to Hewstone's and Brown's (1986) intergroup contact theory. In line with Hewstone and Brown, intergroup intergenerational contact was defined as occuring when respondents perceived age as salient, and themselves and the older person as typical of their respective age groups. As might be expected on the basis of intergroup theory, dissatisfying conversations were rated higher than satisfying conversations on these intergroup dimensions. In addition, young people who reported high levels of intergroup salience were less inclined to perceive satisfying or dissatisfying conversations as "mutual," were less inclined to agree that the older person was individuated, and were more inclined to agree that the older person was negative and underaccommodative in dissatisfying conversations than those who reported relatively low levels of intergroup salience. Thus, high intergroup salience may lead to more negative ratings of intergenerational conversations (Williams, 1992: see also Harwood, Hewstone, Paolini, & Hurd, 2002). These results are consistent with intergroup theory and indicate that intergroup salience may be an important factor in evaluations of intergenerational communication (see also Insko & Schopler, 1987, for similar findings in other intergroup contexts). Of course, these data leave open the possibility that negative encounters are retrospectively evaluated (and attributed) in intergroup terms—the causal order here is yet to be disentangled.

A further study investigating perceptions of intergenerational communication from an accommodation theory perspective (Williams & Giles, 1996) asked college students to recall and recount, in open-ended written format, recent satisfying and dissatisfying conversations with nonfamilial elders (65–75 years old). Results from content analysis supported and expanded previous research revealing that satisfying conversations were those where older people were reportedly accommodative to the needs of the young person. In doing so, they were supportive, listening, and attentive to the younger person, giving compliments and telling interesting stories. Satisfying conversations were also those where a mutual understanding was achieved and both the old and the young person expressed positive emotions. It is interesting that age differences in these conversations were frequently discounted by the young—the older people in satisfying encounters were either perceived to violate ageist expectations, or else age was completely discounted and thought to have no bearing on the conversation.

Reports of dissatisfying conversations included frequent characterizations of older people as being underaccommodative (i.e., inattentive, nonlistening, closed-minded, out of touch, and forcing unwanted attention on young interlocutors). Older people were also portrayed as complaining either in an angry accusing fashion (e.g., in conflict situations) or despondently about their ill health and problematic life circumstances (see also Coupland et. al., 1988). Moreover, a number of young participants felt that older people negatively stereotyped the young as irresponsible, or naïve or both. Young people tended to describe themselves as "reluctantly accommodating" to older dissatisfying partners—they had to restrain themselves by "biting their tongue" and felt under obligation to show respect for age. In other words, they often did not "talk back" even when they felt they had a right to, and this led to considerable resentment.

Some very distinct profiles for intergenerational communication emerged from these studies, not least of all that older people were not universally viewed negatively. Some older people were perceived in extremely positive ways, coinciding with stereotypes of older people as benevolent and wise. Particularly salient in the satisfying conversations in the Williams and Giles study were those where the younger respondent described their older conversational partner as "like a typical grandparent." This aligns very well with research by Hummert and colleagues (Hummert, 1990, 1994; Hummert et al., chap. 4, this volume) that shows that adults have multiple positive and negative stereotypes of older people.

Considering this, Harwood and Williams (1998) investigated young people's expectations for conversations with two different hypothetical elder (71 years old) substereotypes—the Despondent and the Perfect Grandparent (see Hummert, 1994; Hummert et al., chap. 4, this volume). In addition to perceptions of intergenerational communication, communication anxiety (Booth-Butterfield & Gould, 1986), communication satisfaction (Hecht, 1978), and attitudes toward aging (Braithwaite, Lynd-Stevenson & Pigram, 1993) were measured. The stereotypes were manipulated through the use of photographs and descriptions, including stereotype-associated traits. Although the nature of the older target subtype played a significant role in evaluations of older people, general attitudes to them were also very important predictors of perceptions and expectations in this study. For example, participants who were presented with the Despondent subtype described themselves as more anxious, and described the older adult as less accommodative and more likely to complain, as compared to those participants who were presented with the Perfect Grandparent subtype. Despite such effects for the substereotypes, more positive general attitudes toward contact with older adults were related to more positive expectations for these hypothetical interactions independent of the specific nature of the older adult target. It is interesting that younger adults described very little variation in judgments of their own behavior across the target types—variation in perceptions of the older adults' behaviors were more important in predicting perceptions of communication satisfaction.

Therefore, and not surprisingly, we can conclude that young people have more negative conversational expectations for negative older targets, and they expect them to complain more and to be less competent than positively characterized older people, who are thought to be more sociable and more accommodative. However, expectations such as these, which varied dramatically across targets, were not necessarily those that predicted overall satisfaction. This could be interpreted as meaning that younger people will tolerate an older person's complaints providing he or she meets the young persons' communicative needs. In addition, for some conversational expectations, general stereotypes of older people may be stronger predictors than substereotypes. This suggests that in some circumstances, specific individuating information about elders may be overlooked in favor of general positive or negative attitudes toward age (Caporael, 1981).

Harwood, McKee, and Lin (2000) extended Hummert's research by examining multiple communication schemas for intergenerational communication. From open-ended interviews and an extensive coding process, these authors derived a series of descriptions of "types" of intergenerational conversations, which ranged from positive to negative for both young and old respondents. For instance, a "helping" schema emerged. This involved an abstract representation of very positive interactions featuring mutual warmth and caring. The older person was described in very positive terms, and the younger person mentioned that the older person enjoyed the conversation and was helped by it. As a result of this, the younger people often reported "feeling good about themselves." An example from the older respondents involved a "no connection" schema. In these conversations, the older person was highly aware of a generation gap between the younger and older individuals. They described an inability to relate to the younger person and generally described the younger person in largely negative terms. These schemas incorporate traits along the same lines as Hummert's stereotype work, but also incorporate communicative behaviors, feelings, and the like (see also Lin, 2002).

Overall, it seems that older adults are a marked communication outgroup as characterized by young people's descriptions and accounts of intergenerational communication. Notably, intergenerational communication can be satisfying (although within certain limits), as well as dissatisfying. The literature from the older adults' perspective is less extensive, but work such as that by Harwood et al. (2000) indicated similar issues. That said, communication with peers is also bound to show features that are satisfying and dissatisfying, so the question remains: Do people evaluate age peers any differently from the way they evaluate age outgroup members? Until this is answered, we can not really say for sure that intergenerational communication is very different to intragenerational communication.

The answer can be gleaned from further research on intergenerational communication perceptions that has been conducted around the Pacific Rim (e.g., Hong Kong, Taiwan, Australia, South Korea, Japan, the United States, New Zealand, Canada, the Philippines). A questionnaire developed from the themes identified in the Williams and Giles study previously described (e.g., Giles et al., 2002; Williams et al., 1997) was used in an ongoing cross-national research program. This program was initially designed to test assumptions that older people in Eastern societies are venerated and respected much more than those in the West (see Pecchioni, Ota, & Sparks this volume, for a review of this work). For the purposes of this chapter, we can summarize the results of these studies as follows. In Western (non-European) contexts, nonfamily elders are perceived as more nonaccommodative (i.e., closed-minded, out-of-touch, complaining) and less accommodative (i.e., supportive, attentive, etc.) than same-aged peers (although see McCann, Giles, Ota, & Caraker, 2002), for a recent U.S. anomaly to this pattern). Younger people feel more respect and obligation to nonfamily older people than to peers, and a stronger desire to avoid or end conversations with nonfamily elders than they do with peers (for a review, see Giles, McCann, Ota, & Noels, 2002).

In addition, a vignette study comparing an elder versus a peer who was portrayed as nonaccommodative indicated that older people are more likely to be forgiven for their nonaccommodations than peers, because elder nonaccommodation is stereotype consistent and much easier to explain away (Williams, 1996). When older people are noticeably not nonaccommodative, it is likely that positive stereotypes are activated, and younger people may then have a tendency to view older people in overly positive and accommodative ways—as kindly and benevolent. In this way, negative and positive stereotypes of older people are kept in play as "ready made" attributions for dissatisfying or satisfying conversations (Williams, 1996). Older people seeking individuation may find themselves unable to step outside this no-win situation. This is a special case of in-group–out-group attributions—usually bad out-group behavior is attributed negatively as a fixed and stable characteristic of the group, and good out-group behavior is explained away as due to situational and unstable factors (the "ultimate attribution error": Hewstone, 1990; Pettigrew, 1979).

It should be noted that the younger people in the studies described above have, for the most part, been young college students (for an exception, see Cai, Giles, & Noels, 1998). Adults of different ages and life stages may well view intergenerational interactions differently according to their life stage or cohort group. Indeed, in the interests of facilitating a life span approach to language and communication generally, Williams and Garrett (2002b) recently investigated perceptions of nonfamily elders (aged 65–85 years old) and teenagers (aged 13–16 years old) among a (nonstudent) group of British adults aged from 20 to 60 years old.

The nature and functions of intergenerational communication both within and outside the family would be expected to change dramatically across the life span as peer reference groups, social networks, roles, and responsibilities change and develop. In very simple terms, whereas younger people may negatively stereotype older people (the out-group) and favor their peers (the in-group) as we have demonstrated, we might expect this pattern to reverse as people get older and thus closer to the age out-group. Increasing age might trigger a shift to favor older people (as they become the in-group) and negatively stereotype young people (see also Noels, Giles, Cai, & Turay, 1999). A recent survey with British adult participants aged 20 to 60 years old by Williams and Garrett (2002b) investigated this possibility. This study supported and extended previous findings that older people (over 65) are perceived as more nonaccommodative than same-age peers or teenagers. However, the finding that adults between 30 and 50 years old rated elders as more accommodative than peers, and adults aged 50 to 60 years old rated elders equally as accommodative as peers, may reflect life span changes in relationships with older adults. In addition, ratings of elder nonaccommodation decreased with increasing cohort age—the older adults viewed older people as less nonaccommodative than did the younger adults. Older people were more likely to be afforded respect and obligation than either peers or teenagers, and this only declined very slightly with increasing cohort age, whereas respect and obligation afforded to peers increased linearly across the cohorts. Again, in support of previous findings, these adults reported making the most communication adjustments to elders.

These results can be interpreted in the context of frequency of contact data that show that young adults have the most contact with peers, and intergenerational contact becomes more varied across the life span (Williams & Garrett, 2002b). Young adults in their 20s might be more inclined to orient to their own age group, seeking support, advice, and so forth primarily from within their age group. As people have families, build careers, and interact in multigenerational settings, they may be more likely to rely on older adults as mentors, for advice, support, and so forth. Evaluations of elders reflect this changing picture across the cohorts studied by Williams and Garrett (2002b).

Evaluations of teenagers was a new venture in this program of intergenerational communication research. Although stereotypes of youth and intergenerational communication with teenagers have not been systematically researched by language and communication scholars, there is evidence that adults do indeed stereotype the young (Drury & Dennison, 1999; Giles & Williams, 1994). In the Williams and Garrett (2002b) study, stereotypes of teenagers (aged 13–16 years old) as sullen, tongue-tied, or arrogant and self-centered were invoked by the finding that they were rated as less accommodative, more noncommunicative (i.e., dried up in conversation, gave short answers, and were uncommunicative), and more self-promotional (e.g., overconfident, trying to impress, talking about their own life and acting superior) than either peers or elders. More adjustments were made to teenagers than to peers, but older cohorts were less likely than the younger adults to agree that they made adjustments to teenagers (again, a linear pattern across cohorts was observed). Although findings indicate that as people get older they rate elders more favorably, the ratings for teenagers showed less change across the age groups.

One of the many questions raised by the results of this study is: Did the older participants affiliate with older targets? As described earlier, there is some emerging evidence that people who believe they are on the cusp of "middle" and "old age" might be motivated to negotiate the upper and lower chronological boundaries of these categories so that they could be included in the younger age bracket (Williams & Garrett, 2002b). So affiliation with a category labeled "older adults aged 65 to 85" may be a somewhat contentious and sensitive issue for the oldest people in this study (aged 60). Respondents from the target group (i.e., over 65) were not included in the sample.

Some indications can be gained from other studies that have accessed elders' (70-80-year-olds') views of their intra- and intergenerational communication experiences (Cai et al., 1998; Noels, Giles, Cai, & Turay, 1999; Ota, Giles, & Gallois, 2002). It seems that the communication gap is reciprocally felt to the extent that both age groups (20- and 60-year-olds) perceived their own age peers as more accommodating to them than the other age group (see Harwood et al., 2000). But a number of older respondents reported communication problems with people of their same age group, because they perceived elders to be more nonaccommodative than younger people. Thus, for some elders in some circumstances, intragenerational communication can be dissatisfying (see also Williams & Guendouzi, 2000). As we have suggested, whether all older people consider others of their supposedly same age bracket as being "age peers" is an important consideration (see Paoletti, 1998.)

The results for accommodation in these studies do support the ingroup notion that people favor their own age group. In this case, both younger and older people feel that their own age peers are more supportive, attentive, complimenting, and so forth. However, the finding that older respondents perceived their own age group to be nonaccommodating is interesting in that it could be interpreted in several ways. Either older people are aware of other elders' communication problems and report them as nonaccommodations, or they are endorsing a commonly held negative stereotype of older people to distance themselves from a perceptual out-group—perhaps a form of social mobility (discussed previously).

Williams' and Guendouzi's (2000) recent interviews with older people (aged 78–90) in a sheltered retirement community may shed additional light on this point. The interviewed residents, who were all in good health, all discussed difficulties and dissatisfaction with peer communication and relationships. In contrast with communication with younger people (which respondents claimed was highly satisfactory), peers were characterized as nonaccommodative along various dimensions. However, the interviewees were careful to construct positive identities for themselves as physically active, cognitively alert, and able. Apart from demonstrating the very poignant difficulties for able older people living in sheltered residential communities, this study illustrates that these people wished to distance themselves from a negatively stereotyped group—institutionalized elders.

Of course, the quality of intergenerational contact is not merely important in and of itself, it also may have consequences in terms of more general attitudes toward aging and older adults. The "contact hypothesis" suggests that contact with outgroup members has the potential to change attitudes (Allport, 1954; Brewer & Miller, 1988). Support for this hypothesis has emerged in the intergenerational sphere. For instance, among student participants, those who report more frequent contact with older people tend to disagree that they are nonaccommodative and are more likely to agree that older people are accommodative. They are also more likely to indicate respect for elders but less likely to feel obligated and less likely to endorse the idea that large age differences matter in intergenerational interaction (Giles et al., in press; Williams et al., 1997). Other investigations have revealed similar effects (e.g., Knox, Gekoski, & Johnson, 1986; Silverstein & Parrott, 1997; Soliz & Harwood, 2002), although the literature also features some mixed results (for a review, see Fox & Giles, 1993; Williams & Garrett, 2002b).

One problem in examining contact theory issues in the context of intergenerational relations is that intergenerational contact appears to be relatively rare. Sociologists have commented that modern postindustrial societies are increasingly age segregated (Chudacoff, 1989). They argued that contact between people of vastly different ages may be in decline partly as a result of increased geographic mobility, which means that families are geographically scattered, and more generally because of age-segregated living arrangements (e.g., university accommodation, retirement communities, child-free apartment living). In our studies, college students report spending as little as 5% of their time communicating with people over 65 (Williams, 1992), although recent work suggests that community living adults have more diverse intergenerational contact than has been previously found for young college students (Williams & Garrett, 2002b).

All of the research reviewed so far is relevant to interactions involving elders who are not family, and chiefly that between young people and nonfamily elders. It is probably much easier to apply generalized stereotypes to strangers and acquaintances who are not well-known and not individuated than it is to family members who may be more individuated. Family elders may be viewed in different ways, perhaps because they are more intimately known. In addition, the family appears to be one of the few places in which intergenerational contact is relatively frequent (Harwood, 2000a).

Intergenerational Communication in the Family

Some recent intergenerational studies have afforded comparisons of communication variables as experienced with and by family elders with those of same age peers and nonfamily elders (over 65 years old, Giles et al., 2002). These studies have been conducted around the Pacific Rim, but only results for Western nations are relevant for this discussion (for cross-cultural comparisons, readers should see Pecchioni, Ota, & Sparks, chap. 7, this volume). Overall, the findings endorse the view that conversations with family elders would be perceived more positively by young people than those with nonfamily elders. Reported conversations with family elders tend to be judged as more accommodative (in Canada and the United States) and equally as positive and nonaccommodative as those with peers (see also Ng, Liu, Weatherall, & Loong, 1997). Westerners also indicate high respect for family elders. These findings are not surprising given that Western younger people tend to enjoy warm and intimate relationships with family elders (Somary & Stricker, 1998), particularly grandparents (Harwood & Lin, 2000).

According to social theorists concerned with family relations (e.g., Bengtson, Marti, & Roberts, 1991), families are defined by a number of independent, complex, and stable features of solidarity (defined as close feelings and cohesion). In fact, the family itself is an in-group with an associated familial identity, albeit overlayed and interwoven with other group allegiances that emerge across the life span (Harwood, 2002). Just as solidarity is important to the emergence and coherence of any social group or category, so it is essential and functional for families. In the familial context, solidarity has a number of components—associational, affectional, consensual, functional, and normative (Bengtson et al., 1991).

Associational solidarity refers to the frequency and pattern of interactions. Older people attempt to portray their familial interactions as frequent (subjectively defined) and harmonious—although they are not entirely resistant to discussing tensions in interactions (see Williams & Guendouzi, 2000). Affectional solidarity refers to the type and degree of positive sentiment held about family members and the degree of reciprocity. Consensual solidarity is the degree of agreement on values, attitudes, and beliefs among family members—again the subject of much management by elders. Functional solidarity is the degree to which family members exchange services. Normative solidarity is the perception and enactment of norms of the family (like rituals at Christmas), and intergenerational family structure refers to the number type and geographic proximity of family members.

Theory and research on intergenerational solidarity (Bengtson, Olander, & Haddad, 1976) and life span attachment (e.g., Cicirelli, 1991) support the suggestion that relationships with elder family members may be viewed more positively than those with elder strangers (Williams & Nussbaum, 2001). But solidarity must be achieved by continual maintenance and management by family members. Knowing when not to express hostile thoughts and reaching a consensus on values, beliefs, and opinions directly implies familial accommodation processes. In some way, then, families manage to achieve an accommodative working consensus that works for them most of the time. We might also suggest that, in many cases, both adult children and elderly parents exercise a form of accommodative censorship that protects the solidarity of the relationship. Each party knows what topics not to discuss in front of the other and in this way a protective veneer of consensus is created and sustained (see Noller & Bagi, 1985, for an example of this between late adolescents and parents).

Williams' and Guendouzi's (2000) interviews with elderly residents of a sheltered retirement community examine the ways that the older interviewees discursively portray their interactions with family members. Although difficulties with family members were not uncommon, interviewees were very keen to present their own family as relatively harmonious and themselves as relatively privileged in relation to their peers. Interviewees typically presented themselves as operating a "norm of noninterference" in their relationships with family members, especially when there was underlying conflict or disapproval. This noninterference appears to be designed to achieve several goals: to avoid conflict, to conform to socially accepted norms for how families should interact, and to respect other familial relationships, such as parenting hierarchies. One of the most revealing aspects of these interviews was the suggestion that these older people struggled not to interfere in conflict and disapproval situations. In other words, they struggled to accommodate conversationally to their family members, but there was evidence that disapproval was "leaked" verbally and nonverbally. Among other things, this leakage may well be one means of transmitting values to younger family members and may serve as one way that families try to control potentially risky behaviors without engaging in direct conflict.

Harwood (2000a) took a different approach to this same issue. Using a questionnaire measure of accommodative behaviors, he found that both grandparents' and grandchildren's levels of satisfaction in the relationship were influenced by the other's perceived involvement and accommodation in interactions (see Harwood & Williams, 1998, for similar results). Some of the accommodation behaviors previously identified as crucial in intergenerational relations (e.g., older underaccommodation) were not found to be significant predictors of satisfaction or closeness. Similar results emerged in a recent replication conducted in Taiwan (Lin & Harwood, 2003). However, in the Taiwanese context it was found that perceptions of one's own accommodation were the most influential in predicting satisfaction and closeness. Lin and Harwood explain the anomaly by suggesting that the importance of collectivist beliefs in Asian cultures, and the family as the primary collective, result in a belief in personal responsibility for maintaining harmony and stability in the family. Hence, perceptions of closeness and satisfaction are tied more closely to one's own behaviors.

Families increasingly consist of several layers of multigenerational interaction. Simple dyadic relationships are rare in such settings. For example, middle-aged people may be managing their own relationships with their children and their own aging parents while managing the relationship between the aging grandparents and their grandchildren (e.g., by encouraging contact, negotiating misunderstandings, or even translating). Similarly, older people frequently manage their relationships with grandchildren in the context of multiple relationships with other grandchildren, as well as their own adult children (e.g., seeking to avoid any perceptions of favoritism while treating each person as an individual). Each relationship influences the other to a greater or lesser degree in a multiplex network, and communication reflects this. One place in which dyadic communication is frequent is over the telephone, and it is interesting that Harwood (2000b) found that telephone communication between grandparents and grandchildren was a more significant (positive) predictor of satisfaction than communication in other media (including face-to-face communication). That is, increased use of the telephone was associated with more positive evaluations of the relationship to a greater degree than increased use of other media.

From our earlier discussion of the nature of intergroup communication (i.e., age is salient and the older person is seen as typical), it is clear that intergroup interaction can occur in the context of family relationships. The question of whether intergroup theory is valid for family interaction becomes less interesting than other questions, for example, when and under what circumstances do family members engage an intergroup stance, what local and more global purposes does it serve, and what consequences does it have. In this regard, it may be best to view group salience in terms of an intergroup–interindividual dialectic. Thus, relationships with an elderly family member, for example, can be characterized as swinging between two opposing dialectical poles, and family members can be conceived as struggling with this dialectic at certain crucial moments in time (e.g., extreme ill health and incapacity of an older person). In some contexts, the older family member is highly individuated and age does not matter, whereas in other circumstances age does matter and the older individual is seen as a rather typical elder. In fact, Coupland and Coupland (2000) provided an excellent example of this in the discourse between a daughter and her mother's doctor during a medical examination of the mother. In the presence of the mother, the doctor and the daughter discuss the mother as a typical representative of old people, generalizing her individual behavior as typical of old people as a group.

Although family members may be concerned to pursue the best interests of their loved ones (e.g., an older parent or a grandparent), they may also have some fairly stable and stereotypical expectations about "old age." For example, they may expect their parent or grandparent to behave in age-appropriate ways and be shocked and ashamed if he or she does not. Or they may expect a certain level of functioning associated with a particular age and be unable to accept that their older person does not conform to their expectations.

Various theoretical perspectives come into play here. First, Gaertner's and colleagues' common ingroup identity model describes the ways in which focusing on those group memberships that are shared rather than those that are not shared can result in more positive interactions (Gaertner et al., 2000). Gaertner's model provides one account of why intergenerational contact within the family may tend to have more positive outcomes than those outside of the family. The family provides a context within which establishing a common ingroup identity is relatively easy (Banker & Gaertner, 1998). Focusing on the shared family issues will undoubtedly provide some sense of solidarity and may deemphasize the extent to which age operates as a distancing mechanism. Family has been shown to operate as an important topic resource in grandparent–grandchild interactions (Lin, Harwood, & Bonneson, 2002), which supports the notion that this is a solidarity-creating shared identity that is drawn on in interaction.

One criticism of models such as Gaertner's, however, is that they suggest positive consequences for ignoring or deemphasizing important group identities. From the perspective of social identity theory, denying group memberships may lead to an identity-based backlash. In addition, maintaining the salience of group identities has been shown to be important in linking intergroup contact to attitude change. Perspectives grounded in social identity theory (e.g., Brown, Vivian, & Hewstone, 1999; Hewstone & Brown, 1986) and social cognition (Rothbart & John, 1985) agree that generalization from a specific intergroup contact situation to more general intergroup attitudes is only likely if group memberships maintain some level of salience in the contact situation. In other words, a grandparent–grandchild interaction in which the shared family identity is paramount and age is not salient may be a positive interaction. However, it is unlikely to result in positive consequences in terms of the grandchild's general attitudes toward older adults. Support for this perspective has been provided by Harwood, Hewstone, Paolini, and Hurd (2002), who found a positive relationship between the general quality of grandparent–grandchild contact and attitudes toward older adults only among grandchildren who find their grandparents' age to be a salient part of the interaction. Ironically, and as noted earlier, contact in which group memberships are salient is not always the most positive. In the same data, Harwood et al. (2002) found that age-salient interactions tend, in general, to be less positive despite their potential for a positive impact on attitudes (see also Williams & Giles, 1996). Clearly, the family context requires more careful theorizing in these terms, but for now there does appear to be a clear tension between the need for quality interactions in the family and the need to maintain awareness and salience of age such that these positive interactions might have positive consequences for more general attitudes toward aging.

A final area in which the work on intergenerational communication within the family might make links to the intergroup literature is in terms of the ways in which the various strategies outlined by social identity theory for dealing with low status might play out in the family context. In particular, various recent research has pointed to the grandparent role as one that may provide positive connotations associated with age, in contrast to the generally negative associations with age (Harwood, 2002; Harwood & Lin, 2000). Indeed, it is notable that one of Hummert's Garstka's, Shaner's, and Strahm's (1994) positive stereotypes of older adults was labelled as the "perfect grandparent" stereotype, even though these were examined outside of the family context. We would hence suggest that behaviors within the grandparent role might serve to provide rewards to older adults experiencing identity threat (see Harwood, 2002). In some circumstances, these might resemble social mobility strategies—grandchildren may offer an opportunity to engage in activities or behaviors characteristic of younger age groups, either by engaging in childhood activities or parenting activities. A grandparent doing parenting might be seen as having "left" the group of older adults and joined the middle aged. This, of course, might be particularly true for grandparents raising their grandchildren, who would have substantial contact with significantly younger parents of their children's peers. At other times, grandparenting may offer social creativity options to grandparents. For instance, being a "good grandparent" might be a way to achieve positive identity and leave the status quo unchallenged. Similarly, the family may offer a context in which older adults are able to claim status, seek respect, and even control decision making—roles that are often denied outside of the family context. Finally, grandparents may sometimes engage in actions best characterized as social competition. For instance, recent Supreme Court decisions concerning grandparents' custody rights have affirmed the ability of grandparents to challenge the status quo and achieve victories for their group (*Troxell v. Granville*, 2000; see Holladay & Coombs, 2001). A challenge for the future is clearly to disentangle the extent to which being a grandparent can be usefully considered as a social identity (i.e., are grandparents joining together, addressing issues of mutual collective concern, sharing in other ways that indicate a sense of group identity with others in the same role, or performing all of these functions?). Further, if we identify a meaningful sense of social identity among grandparents, then what are the ways in which that social identity might interact with the more general identification as an older adult (given that many, but not all, grandparents are older adults; Harwood, 2002).

CONCLUSIONS

Drawing on and extending the work of Harwood et al. (1995), the primary goal of this chapter has been to present an intergroup perspective on intergenerational communication. In addition, we have presented some ways in which that perspective might inform our examinations of intergenerational interaction in family contexts. Clearly, there is theoretical work to be done. As described, the extent to which age identities interact with other social identities to exacerbate intergroup tensions is a complex issue. Likewise, more work is needed to explore further the ways in which age identities are salient in family interaction, and the impact of that salience on the experience and effects of family communication. For now, we would simply draw attention to the fact that family communication often involves meaningful intergenerational issues and that these need to be integrated with a more conventional "relational" approach to such phenomena. A number of areas seem ripe for future research in this area, and we would particularly encourage research on the following.

First, the previous research on age marking in discourse would benefit from an extension into family contexts. The ways in which grandparents "claim" age and use age marking to their discursive benefit is worthy of examination. Likewise, this work would probably benefit from a broader life span focus. In other words, it is likely that older adults are not the only individuals in families that use age as a "token." Middle-aged adults may well use age to justify authority to their children and to claim status with their own parents ("I'm an adult now!"). Children may also use it creatively to claim dependence or independence, depending on the context. Recent work by Harwood (2002) may be suggestive here. He examined Web pages created by older adults about their grandchildren, or the grandparenting experience. Explicit references to age were rare in the Web pages, except in situations where age was in some way unusual or "off schedule" (e.g., by people who became grandparents particularly young). This is in accordance with Coupland et al.'s (1991) suggestion that the role reference of being a grandparent carries with it automatic age associations.

Second, the role of age-based stereotypes in the family is in need of further examination. The ways in which such stereotypes are invoked in family interactions is important, particularly as they are used in interactions concerning decisions about older adults. Frequently, younger adults in families make decisions about institutionalization, driving abilities, and other aspects of the lives of their elders. Age stereotypes might well influence such decisions, resulting in the premature curtailing of older adults' independence. Stereotypes may be invoked in the context of family relationships in ways that constrain the older adults' options to reject them. For instance, Ann Landers (a popular U.S. advice columnist) recently presented a list of "guidelines for older parents" in her column. The list, apparently submitted by a frustrated adult child, featured such elements as "When my children tell me I should no longer drive, I will believe them and quit, because I know they love me" ("Guidelines for Elderly Parents," 2001). Invoking "love" in the context of the message curtails responses that might reject the somewhat more ominous messages here. The message clearly draws on stereotypes of inevitable decline with age ("When my children . . ." not "If . . . ") and assumptions about children assuming authority over dependent older adults ("tell me" as opposed to more negotiative options).

In a similar vein, there may be scope for examining the ways in which individuals self-stereotype and act in age-typical ways in the family context. For instance, adolescents acting in a rebellious fashion or older adults telling lengthy stories about family history might be engaging in role- and age-related self-stereotyping. Such patterns would lead to reinforcement of stereotypical beliefs in the family, and probably an increased reliance on stereotype-driven patterns of interaction.

In other words, and at a more general level, our hope for the future is an increasing emphasis on the ways in which group memberships, identities, and stereotypes might influence the minutiae of family interactions, and more study of the ways in which family dynamics might feed back and influence identities and stereotypes. The family is an inherently intergenerational setting, and a dynamic one in which to examine some of the important issues of age relations facing us.

REFERENCES

Allport, G. W. (1954). *The nature of prejudice*. Reading, MA: Addison-Wesley.

Banker, B. S., & Gaertner, S. L. (1998). Achieving stepfamily harmony: An intergroup-relations approach. *Journal of Family Psychology, 12*, 310–325.

Banton, M. (1983). *Racial and ethnic competition*. Cambridge, UK: Cambridge University Press.

Bengtson, V. L., Marti, G., & Roberts, H. E. L. (1991). Age-group relationships: Generational equity and inequity. In K. Pillemer & K. McCartney (Eds.), *Parent–child relations throughout life* (pp. 253–278). Hillsdale, NJ: Lawrence Erlbaum Associates.

Bengtson, V. L., Olander, E. B., & Haddad, A. A. (1976). The generation gap and aging family members: Towards a conceptual model. In J. E. Gubrium (Ed.), *Time, roles and the self in old age* (pp. 237–263). New York: Human Sciences Press.

Billig, M. (1985). Prejudice, categorization and particularization. *European Journal of Social Psychology, 15*, 79–103.

Booth-Butterfield, S., & Gould, M. (1986). The communication anxiety inventory: Validation of state- and context-communication apprehension. *Communication Quarterly, 34*, 194–205.

Bourhis, R. Y., & Giles, H. (1977). The language of intergroup distinctiveness. In H. Giles (Ed.), *Language, ethnicity, and intergroup relations* (pp. 119–135). London: Academic Press.

Braithwaite, V., Lynd-Stevenson, R., & Pigram, D. (1993). An empirical study of ageism: From polemics to scientific utility. *Australian Psychologist, 28*, 9–15.

Branscombe, N., & Wann, D. L. (1994). Collective self-esteem consequences of outgroup derogation when a valued social identity is on trial. *European Journal of Social Psychology, 24*, 641–657.

Brewer, M. B., & Miller, N. (1988). Contact and cooperation: When do they work? In P. A. Katz & D. A. Taylor (Eds.), *Eliminating racism* (pp. 315–326). New York: Plenum.

Brown, R., Vivian, J., & Hewstone, M. (1999). Changing attitudes through intergroup contact: The effects of group membership salience. *European Journal of Social Psychology, 29*, 741–764.

Cai, D., Giles, H., & Noels, K. (1998). Elderly perceptions of communication with older and younger adults in China: Implications for mental health. *Journal of Applied Communication Research, 26*, 32–51.

Caporael, L. R. (1981). The paralanguage of caregiving: Baby talk to the institutionalized elderly. *Journal of Personality and Social Psychology, 40*, 876–884.

Carver, C. S., & de la Garza, N. H. (1984). Schema-guided information search in stereotyping of the elderly. *Journal of Applied Social Psychology, 14,* 69–81.

Chudacoff, H. P. (1989). *How old are you? Age consciousness in American culture.* Princeton, NJ: Princeton University Press.

Cicirelli, C. (1991). Attachment theory in old age: Protection of the attached figure. In K. Pillemar & K. McCartney (Eds.), *Parent–child relationships throughout life* (pp. 2–42). Hillsdale, NJ: Lawrence Erlbaum Associates.

Coupland, J. (2003). Ageist ideology and Discourses of control in skin care marketing. In J. Coupland & R. Gwyn (Eds.), *Discourse, the body, and identity* (pp. 127–150). London: Macmillan/Palgrave?

Coupland, J., Coupland, N., Giles, H., & Henwood, K. (1991). Formulating age: Dimensions of age identity in elderly talk. *Discourse Processes, 14,* 87–106.

Coupland, J., Coupland, N., Giles, H., & Wiemann, J. M. (1988). My life in your hands: Processes of self-disclosure in intergenerational talk. In N. Coupland (Ed.), *Styles of discourse* (pp. 201–253). London: Croom Helm.

Coupland, J., Coupland, N., & Grainger, K. (1991). Intergenerational discourse: Contextual versions of aging and elderliness. *Aging and Society, 11,* 189–208.

Coupland, N. (1999). "Other" representation. In J. Verschueren, J.-O. Ostman, J. Blommaert, & C. Bulcaen (Eds.), *Handbook of pragmatics* (pp. 1–24). Amsterdam: Benjamins.

Coupland, N., & Coupland, J. (1990). Language and later life: The diachrony and decrement predicament. In H. Giles & W. P. Robinson (Eds.). *The handbook of language and social psychology* (pp. 451–468). Chichester, UK: Wiley.

Coupland, N., & Coupland, J. (2000). Relational frames and pronominal address/reference: The discourse of geriatric medical triads. In S. Sarangi & M. Coulthard (Eds.), *Discourse and social life* (pp. 207–229). London: Longman.

Coupland, N., Coupland, J., & Giles, H. (1989). Telling age in later life: Identity and face implications. *Text, 9,* 129–151.

Coupland, N., Coupland, J., Giles, H., & Henwood, K. (1988). Accommodating the elderly: Invoking and extending a theory. *Language in Society, 17,* 1–41.

Coupland, N., & Nussbaum, J. F. (Eds.). (1993). *Discourse and life-span identity.* Thousand Oaks, CA: Sage.

Drury, J., & Dennison, C. (1999). Individual responsibility versus social category problems. Benefit officers' perceptions of communication with teenagers. *Journal of Youth Studies, 2,* 171–192.

Drury, J., & Dennison, C. (2000). Representations of teenagers among police officers: Some implications for their communication with young people. *Youth & Policy, 66,* 62–87.

Finchilescu, G. (1986). Effect of incompatability between internal and external group membership criteria on intergroup behavior. *European Journal of Social Psychology, 16,* 83–87.

Fox, S., & Giles, H. (1993). Accommodating intergenerational contact: A critique and theoretical model. *Journal of Aging Studies, 7,* 423–451.

Gaertner, S. L., Dovidio, J. F., Nier, J. A., Banker, B. S., Ward, C. M., Houlette, M., & Loux, S. (2000). The common ingroup identity model for reducing intergroup bias: Progress and challenges. In D. Capozza & R. Brown (Eds.), *Social identity processes: Trends in theory and research* (pp. 133–148). London: Sage.

Gallois, C., Giles, H., Jones, E., Cargile, A. C., & Ota, H. (1995). Accommodating intercultural encounters: Elaborations and extensions. In R. L. Wiseman (Ed.), *Intercultural communication theory* (pp. 115–147). Thousand Oaks, CA: Sage.

Garrett, P., Giles, H., & Coupland, N. (1989). The contexts of language learning. In S. Ting-Toomey & F. Korzenny (Eds.), *Language, communication and culture* (pp. 201–221). London: Sage.

Giles, H. (1979). Ethnicity markers in speech. In K. R. Scherer & H. Giles (Eds.), *Social markers in speech* (pp. 251–290). Cambridge, UK: Cambridge University Press.

Giles, H. (1999). Managing dilemmas in the "silent revolution": A call to arms! *Journal of Communication, 49,* 170–182.

Giles, H., Coupland, J., & Coupland, N. (Eds.). (1991). *Contexts of accommodation: Developments in applied sociolinguistics.* Cambridge, UK: Cambridge University Press.

Giles, H., & Coupland, N. (1991). *Language: Contexts and consequences.* Pacific Grove, CA: Brooks/Cole.

Giles, H., & Johnson, P. (1981). The role of language in ethnic group relations. In J. C. Turner & H. Giles (Eds.), *Intergroup behaviour* (pp. 199–243). Oxford, UK: Basil Blackwell.

Giles, H., McCann, R., Ota, H., & Noels, K. (2002). Challenging intergenerational stereotypes across Eastern & Western cultures. In M. S. Kaplan, N. Z. Henkin, & A. T. Kusano (Eds.), *Linking Lifetimes: A global view of intergenerational exchange* (pp. 13–28). Honolulu: University Press of America.

Giles, H., Noels, K., Williams, A., Ota, H., Lim, T.-S., Ng, S. H., Ryan, E. B., & Somera, L. B. (2003). *Intergenerational communication across cultures: Young peoples' perception of conversations with family elders, non-family elders & same-age peers. Journal of Cross-cultural Gerontology, 18,* 1–30.

Giles, H., & Williams, A. (1994). Patronizing the young: Forms and evaluations. *International Journal of Aging and Human Development, 39,* 33–53.

Guidelines for elderly parents. (2001, August 4). *The Kansas City Star,* p. E6.

Gurin, P., Miller, H., & Gurin, G. (1980). Stratum identification and consciousness. *Social Psychology Quarterly, 43,* 30–47.

Harwood, J. (2000a). Communication media use in the grandparent–grandchild relationship. *Journal of Communication, 50*(4), 56–78.

Harwood, J. (2000b). Communicative predictors of solidarity in the grandparent–grandchild relationship. *Journal of Social and Personal Relationships, 17,* 743–766.

Harwood, J. (2002). *Relational, role, and social identity as expressed in grandparents' personal Web sites.* Unpublished manuscript, University of Arizona.

Harwood, J., & Giles, H. (1992). "Don't make me laugh": Age representations in a humorous context. *Discourse and Society, 3,* 403–436.

Harwood, J., & Giles, H. (1993). Creating intergenerational distance: Language, communication and middle-age. *Language Sciences, 15,* 1–24.

Harwood, J., Giles, H., Clément, R., Pierson, H., & Fox, S. (1994). Vitality of age groups across cultures. *Journal of Multilingual and Multicultural Development, 15,* 311–318.

Harwood, J., Giles, H., McCann, R. M., Cai, D., Somera, L. P., Ng, S. H., Gallois, C., & Noels, K. (2001). Older adults' trait ratings of three age-groups around the Pacific Rim. *Journal of Cross-Cultural Gerontology, 16,* 157–171.

Harwood, J., Giles, H., Ota, H., Pierson, H. D., Gallois, C., Ng, S. H., Lim, T. S., & Somera, L. (1996). College students' trait ratings of three age groups around the Pacific Rim. *Journal of Cross-Cultural Gerontology, 11,* 307–317.

Harwood, J., Giles, H., & Ryan, E. B. (1995). Aging, communication, and intergroup theory: Social identity and intergenerational communication. In J. F. Nussbaum & J. Coupland (Eds.), *Handbook of communication and aging research* (pp. 133–159). Hillsdale, NJ: Lawrence Erlbaum Associates.

Harwood, J., Hewstone, M., Paolini, S., & Hurd, R. (2002). *Intergroup contact theory, the grandparent–grandchild relationship, and attitudes towards older adults.* Unpublished manuscript, University of Arizona.

Harwood, J., & Lin, M-C. (2000). Affiliation, pride, exchange and distance in grandparents' accounts of relationships with their college-age grandchildren. *Journal of Communication, 50*(3), 31–47.

Harwood, J., McKee, J., & Lin, M.-C. (2000). Younger and older adults' schematic representations of inter-generational conversations. *Communication Monographs, 67,* 20–41.

Harwood, J., & Williams, A. (1998). Expectations for communication with positive and negative subtypes of older adults. *International Journal of Aging and Human Development, 47,* 11–33.

Hecht, M. L. (1978). The conceptualisation and measurement of interpersonal communication satisfaction. *Human Communication Research, 4,* 253–264.

Hewstone, M. (1990). The "ultimate attribution error"? A review of the literature on intergroup causal attribution. *European Journal of Social Psychology, 20,* 311–335.

Hewstone, M., & Brown, R. J. (1986). "Contact is not enough": An intergroup perspective on the contact hypothesis. In M. Hewstone & R. Brown (Eds.), *Contact and conflict in intergroup relations* (pp. 1–44). Oxford, UK: Basil Blackwell.

Hogg, M. A., & Abrams, D. (1990). Social motivation, self-esteem, and social identity. In D. Abrams & M. A. Hogg (Eds.), *Social identity theory: Constructive and critical advances* (pp. 28–47). London: Harvester Wheatsheaf.

Hogg, M. A., & Turner, J. C. (1985). Interpersonal attraction, social identification and psychological group formation. *European Journal of Social Psychology, 15,* 51–66.

Holladay, S. J., & Coombs, W. T. (2001, November). *Media portrayals of the intergenerational battle over "grandparents' rights": An examination of the Troxel v. Granville case.* Paper presented at the annual meeting of the National Communication Association, Atlanta, GA.

Hummert, M. L. (1990). Multiple stereotypes of elderly and young adults: A comparison of structure and evaluations. *Psychology and Aging, 5,* 182–193.

Hummert, M. L. (1994). Stereotypes of the elderly and patronizing speech. In M. L. Hummert, J. M. Wiemann, & J. F. Nussbaum (Eds.), *Interpersonal communication in older adulthood: Interdisciplinary research* (pp. 162–184). Thousand Oaks, CA: Sage.

Hummert, M. L., Garstka, T. A., & Shaner, J. L. (1995). Beliefs about language performance: Adult's perceptions about self and elderly targets. *Journal of Language and Social Psychology, 14,* 235–259.

Hummert, M. L., Garstka, T. A., Shaner, J. L., & Strahm, S. (1994). Stereotypes of the elderly held by young, middle-aged and elderly adults. *Journal of Gerontology: Psychological Sciences, 49,* 240–249.

Hummert, M. L., & Ryan, E. B. (2001). Patronizing. In W. P. Robinson & H. Giles (Eds.), *The new handbook of language and social psychology* (pp. 253–270). Chichester, UK: Wiley.

Insko, C. A., & Schopler, J. (1987). Categorization, competition, and collectivity. In C. Hendrick (Ed.), *Group processes. Review of personality and social psychology* (Vol. 8, pp. 213–251). Beverly Hills, CA: Sage.

Kite, M. E., Deaux, K., & Miele, M. (1991). Stereotypes of young and old: Does age outweigh gender? *Psychology and Aging, 6,* 19–27.

Kite, M. E., & Johnson, B. T. (1988). Attitudes toward older and younger adults: A meta-analysis. *Psychology and Aging, 3,* 233–244.

Knox, V. J., Gekoski, W. L., & Johnson, E. A. (1986). Contact with and perceptions of the elderly. *The Gerontologist, 26,* 309–313.

Lemyre, L., & Smith, P. M. (1985). Intergroup discrimination and self-esteem in the minimal group paradigm. *Journal of Personality and Social Psychology, 49,* 660–670.

Lin, M.-C. (2002). *Systematic examinination of intergenerational communication schemas in Taiwan and the United States.* Unpublished manuscript, University of Kansas.

Lin, M.-C., & Harwood, J. (2003). Predictors of grandparent–grandchild relational solidarity in Taiwan. *Journal of Social and Personal Relationships, 20,* 537–563.

Lin, M.-C., Harwood, J., & Bonneson, J. L. (2002). Topics of conversation in the grandparent–grandchild relationship. *Journal of Language and Social Psychology, 21,* 302–323.

Lin, M.-C., Hummert, M. L., & Harwood, J. (2002). *Discursive representations of age identity in an on-line discussion group.* Unpublished manuscript, University of Kansas.

McCann, R. M., Giles, H., Ota, H., & Caraker, R. (2003). *Perceptions of intra- and intergenerational communication among young adults in Thailand, Japan & the USA. Communication Reports, 16,* 1–23.

McGarty, C. (1999). *Categorization in social psychology.* Thousand Oaks, CA: Sage.

Ng, S. H., Liu, J. H., Weatherall, A., & Loong, C. S. F. (1997). Younger adults' communication experiences and contact with elders and peers. *Human Communication Research, 24,* 82–108.

Noels, K., Giles, H., Cai, D., & Turay, L. (1999). Intergenerational communication and health in the United States and the People's Republic of China. *South Pacific Journal of Psychology, 10,* 120–134.

Noller, P., & Bagi, S. (1985). Parent–adolescent communication. *Journal of Adolescence, 8,* 125–144.

Oakes, P. J., & Turner, J. C. (1980). Social categorization and intergroup behaviour: Does minimal intergroup discrimination make social identity more positive? *European Journal of Social Psychology, 10,* 295–301.

Ota, H., Giles, H., & Gallois, C. (2002). *Perceptions of younger, middle aged and older adults in Australia and Japan: Stereotypes and age group vitality.* Unpublished manuscript, University of California, Santa Barbara.

Paoletti, I. (1998). *Being an older woman: A case study in the social production of identity.* Mahwah, NJ: Lawrence Erlbaum Associates.

Perdue, C. W., & Gurtman, M. B. (1990). Evidence for the automaticity of ageism. *Journal of Experimental Social Psychology, 26,* 199–216.

Pettigrew, T. (1979). The ultimate attribution error: Extending Allport's cognitive analysis of prejudice. *Personality and Social Psychology Bulletin, 5,* 461–476.

Rabbie, J. M. Schot, J. C., & Visser, L. (1989). Social identity theory: A conceptual and empirical critique from the perspective of a behavioral interaction model. *European Journal of Social Psychology, 19,* 171–202.

Rothbart, M., & John, O. P. (1985). Social categorization and behavioral episodes: A cognitive analysis of the effects of intergroup contract. *J. of Social Issues, 41,* 81–104.

Ryan, E. B., Hummert, M. L., & Boich, L. H. (1995). Communication predicaments of aging: Patronizing behavior toward older adults. *Journal of Language & Social Psychology, 14,* 144–166.

Ryan, E. B., Kwong See, S., Meneer, W. B., & Trovato, D. (1992). Age-based perceptions of language performance among younger and older adults. *Communication Research, 19,* 311–331.

Sachdev, I., & Bourhis, R. Y. (1987). Status differentials and intergroup behavior: Does minimal intergroup discrimination make social identity more positive? *European Journal of Social Psychology, 17,* 277–293.

Shepard, C., Giles, H., & LePoire, B. A. (2001). Communication accommodation theory. In W. P. Robinson & H. Giles (Eds.), *The new handbook of language and social psychology* (pp. 33–56). Chichester, UK: Wiley.

Silverstein, M., & Parrott, T. M. (1997). Attitudes toward public support of the elderly: Does early involvement with grandparents moderate generational tensions? *Research on Aging, 19,* 108–132.

Soliz, J., & Harwood, J. (2002). *Communication in a close family relationship and the reduction of prejudice.* Unpublished manuscript, University of Kansas.

Somary, K., & Stricker, G. (1998). Becoming a grandparent: A longitudinal study of expectations and early experiences as a function of sex and lineage. *The Gerontologist, 38,* 53–61.

Tajfel, H., Billig, M. G., Bundy, R. P., & Flament, C. (1971). Social categorization and intergroup behaviour. *European Journal of Social Psychology, 1,* 149–178.

Tajfel, H., & Turner, J. C. (1979). An integrative theory of intergroup conflict. In W. C. Austin & S. Worchel (Eds.), *The social psychology of intergroup relations* (pp. 33–53). Monterey, CA: Brooks/Cole.

Tajfel, H., & Turner, J. C. (1986). The social identity theory of intergroup behavior. In S. Worschel & W. G. Austin (Eds.), *The social psychology of intergroup relations* (2nd ed., pp. 7–24). Chicago: Nelson-Hall.

Taylor, B. C. (1992). Elderly identity in conversation: Producing frailty. *Communication Research, 19,* 493–515.

Taylor, S. E. (1981). A categorization approach to stereotyping. In D. L. Hamilton (Ed.), *Cognitive processes in stereotyping and intergroup behavior* (pp. 83–114). Hillsdale, NJ: Lawrence Erlbaum Associates.

Troxell v. Granville, 99–138 S. Ct. (2000).

Turner, J. C., & Brown, R. J. (1978). Social status, cognitive alternative, and intergroup relations. In H. Tajfel (Ed.), *Differentiation between social groups* (pp. 201–234). London: Academic Press.

Williams, A. (1992). *Intergenerational communication satisfaction: An intergroup analysis.* Unpublished masters thesis, University of California, Santa Barbara.

Williams, A. (1996). Young people's evaluations of intergenerational versus peer underaccommodation: Sometimes older is better? *Journal of Language and Social Psychology, 15,* 291–311.

Williams, A., Coupland, J., Folwell, A., & Sparks, L. (1997). Talking about Generation X: Defining them as they define themselves. *Journal of Language and Social Psychology, 16,* 251–227.

Williams, A., & Garrett, P. (2002a). *Moving the goalposts: Adults' estimates of chronological age corresponding with age labels.* Unpublished manuscript, Cardiff University.

Williams, A., & Garrett, P. (2002b). Communication evaluations across the lifespan: From adolescent storm and stress to elder aches and pains. *Journal of Language and Social Psychology, 21,* 101–126.

Williams, A., & Giles, H. (1996). Intergenerational conversations: Young adults' retrospective accounts. *Human Communication Research, 23,* 220–250.

Williams, A., & Giles, H. (1998). Communication of ageism. In M. Hecht (Ed.), *Communicating prejudice* (pp. 136–160). Thousand Oaks: Sage.

Williams, A., & Guendouzi, J. (2000). Adjusting to "the home" dialectical dilemmas and personal relationships in a retirement community. *Journal of Communication, 50,* 65–82.

Williams, A., & Nussbaum, J. F. (2001). *Intergenerational communication across the lifespan.* Mahwah, NJ: Lawrence Erlbaum Associates.

Williams, A., Ota, H., Giles, H., Pierson, H. D., Gallois, C., Ng, S. H., Lim, T.-S., Ryan, E. B., Somera, L., Maher, J., & Harwood, J. (1997). Young people's beliefs about intergenerational communication: An initial cross-cultural comparison. *Communication Research, 24,* 370–393.

Inter- and Intragroup Perspectives on Intergenerational Communication

Valerie Barker
San Diego State University

Howard Giles
University of California, Santa Barbara

Jake Harwood
University of Arizona

In recent times, growing scholarly interest in the health, socio-economic, psychological, and communicative aspects of aging has occurred, in part from increasing longevity across societies (Giles, 1999). In the second half of the 20th century, 20 years were added to the average life span, and within 30 years a *third* of the population in the more developed countries will be over age 60. Moreover, it is estimated that, globally, the number of elderly persons will reach that proportion by 2150 (Annan, 1999). Under these circumstances, the concerns and conditions of older individuals come into sharp relief. Indeed, Giles, Williams, and Coupland (1990) have argued that compared to younger generations "not only do the elderly emanate from different historical periods with their own cohorts, values and predispositions (communicative and non-communicative), but they also have different problems (some existential) to which to adjust, somatically and life-historically" (p. 10).

Much of the discussion about improved intergenerational communication in *applied* contexts has been a-theoretical (Abrams & Giles, 1999). However, because of the above-mentioned differences between younger and older generations, some researchers have begun to view intergenerational communication as analogous to intergroup or inter-cultural communication (Giles et al., 1990; Harwood, Giles, & Ryan, 1995; Williams & Nussbaum, 2001). In adopting an intergroup perspective, the (often) problematic nature of intergenerational communication with regard to negative outcomes for older persons is a pressing pragmatic concern as well as a theoretical and research imperative (e.g., Harwood, Giles, Fox, Ryan, & Williams, 1993; Ryan & Cole, 1990).

Several existing intergroup/intercultural theories have been employed to describe and *explain* intergenerational communication (Harwood et al., 1995). More recently communication scholars have provided theoretical models analyzing both the processes by which negative intergenerational communicative outcomes occur (Hummert, 1994; Ryan, Giles, Bartolucci, & Henwood, 1986) and how they may be avoided (Ryan, Meredith, Maclean, & Orange, 1995). None of these models provides a holistic explanation of intergenerational communication that includes elucidation of negative processes *as well* as potentially positive processes and outcomes (although for intercultural competence—see Hajek & Giles, 2003). Moreover, they do not attend to cultural variability nor acknowledge the importance of intragenerational interaction.

To that end, this chapter comprises two parts. Later, a model is introduced that integrates aspects of existing intergroup and communication theory with other useful constructs. Specifically, the model will be used to analyze how communication between- and within-generations may be understood and facilitated or hindered by communicative processes born out of categorization and stereotyping. But first, to understand the origins and rationale of this model, it is necessary to trace the history of intergroup and communication theory with respect to intergenerational interaction.

THE GENEALOGY OF INTERGROUP THEORY AND INTERGENERATIONAL COMMUNICATION

Aging presents an interesting case with regard to intergroup theory (Harwood et al., 1995; Hogg & Abrams, 1990). That is, barring illness or tragedy, most of us will grow old. In contrast to other forms of group categorization such as sex or race, we make a transition over the life span from groups labeled "young" to others labeled "old." This transition is arguably one of the most pivotal in terms of social judgment because it often serves as a basis for categorization and behavioral expectations. Indeed, some scholars have proposed that age explains more of the variance in people's attributions about others than characteristics such as gender or ethnicity (Montepare & Zebrowitz, 1998). That is, movements, vocal qualities, and facial and bodily appearance draw attention to age-related physical qualities yielding social judgments related to those qualities. For example, for young adults who listened to interviews with older men and women, Mulac and Giles (1996) found that how old a person *sounded* predicted negative psychological judgments.

Attributions about age and aging are socially constructed (Coupland & Nussbaum, 1993). Over the course of the life span and from a variety of sources (e.g., family, culture, mass media), we learn what it is to be "old" (Gerbner, Gross, Signorielli, & Morgan, 1980; Harwood & Anderson, 2002; Harwood & Roy, 1999; Levy & Langer, 1996). Definitions of age, appropriate age-related roles, and the outcomes associated with aging vary within and between cultures (Giles et al., 2000; Levy & Langer, 1994) and this is inevitable given differential life expectancies.

Whatever the form of attribution or definition of "being a grand age," an intergroup perspective provides *one* framework for understanding the origins and communication of ageism, negative and positive intergenerational stereotypes about aging, and how these may be supported or undermined (for additional approaches see, for example, Montepare & Zebrowitz, 2002; Zebrowitz & Montepare, 2000).

Hajek and Giles (2003) define intergroup communication as "any communicative behavior exhibited by one or more individuals toward one or more others that is based on the individuals' identification of themselves and others as belonging to different social categories". Relatedly, Williams and Coupland (1998) aptly characterize aging as a *social* as opposed to a biological process by describing aging as "the accumulation of local and historical experiences which conspire to define a person in late life distinctively as a member of a social group" (p. 144). Therefore, younger people and older people may seem to live alongside each other in society, but they actually operate in different, barely interacting cultural spaces (Harwood et al., 1995; Williams & Nussbaum, 2001).

Intergroup theories speak to the concept of identity and personal worth we obtain from association with certain social groups and the comparisons we make with others. One such theory (social identity theory: Tajfel, 1978; Tajfel & Turner, 1986) has been applied to intergenerational communication with, arguably, some success (e.g., Fox, 1999; Fox & Giles, 1993; Hajek & Giles, 2002; Williams & Giles, 1996). Social identity theory (SIT) posits that positive social identity is achieved through psychological distinctiveness with regard to members of other groups (Giles & Coupland, 1992). Some of the basic tenets of social identity theory (Tajfel, 1978) are at the heart of communication accommodation theory (CAT) (e.g., Coupland & Giles, 1988; Giles, Coupland, & Coupland, 1991; Gallois, Giles, Jones, Cargile, & Ota, 1995). CAT explains how social identity is made manifest in communicative behavior using the linguistic strategies of accommodation; that is, we gain positive social identity by linguistically converging toward, or diverging from members of other social groups. Theoretical extensions of CAT relating specifically to intergenerational interactions include the stereotype activation model (Hummert, 1994), the communication predicament model (Ryan et al., 1986), and the communication enhancement model (Ryan et al., 1995). These and other relevant theoretical developments are discussed in more detail below.

Social Identity Theory

Social identity theory (Tajfel, 1978) describes and explains intergroup (including intercultural) interaction, rivalry, and discrimination. The theory states that individuals inevitably categorize themselves and others as members of groups, and by comparing their group position with that of others attempt to achieve a sense of positive identity. Attempts to achieve positive identity often result in discrimination, as favoritism towards the ingroup results in a negative orientation to the outgroup. This implies the presence of conflict in all intergroup relationships, a view not shared by other intergroup theorists (e.g., Hogg & Turner, 1985; Hogg & Abrams, 1993). According to Brewer (1999a), members of a group may feel a sense of positive identity when compared to other groups, but while this may mean ingroup favoritism and positive discrimination in favor of ingroup members, it does not necessarily lead to negative discrimination or hostility toward other groups. This, Brewer contends, is triggered by factors (among others) such as perceived threat and changes in political power or status. Similarly, social dominance theory (Sidanius & Pratto, 1999) suggests that intergroup conflict is related to the presence or absence of economic surplus.

In the event of such socio-structural triggers, SIT describes certain change strategies utilized by social groups whose members perceive their social identity to be less than positive (Tajfel & Turner, 1986). These strategies, called "social mobility", include the process of distancing or denial where individuals classify themselves as "different" from the rest of their ingroup. For example, Paoletti (1998) describes how some young-old Italian women distanced themselves from aging and aging attributions by defining their interactions with other older adults as inter- rather than intragenerational. The others they classified as "old" were the frail or institutionalized elderly (see also Coupland, Coupland, & Giles, 1991). In addition, some members of the currently aging baby boomer generation may become "socially mobile" by the engaging in healthy behaviors incorporating exercise, low fat diets, or age-defying medical procedures such as plastic surgery (Williams & Nussbaum, 2001). By taking "you but not me" or "staying as young as possible" approaches, younger-old people may be seen to conspire in, and to perpetuate, existing intergroup stereotypes about age decrement.

Alternatively, members of negatively-stereotyped groups may engage in "social creativity" strategies by improving group-based self-esteem and investing in, rather than dissociating from, their age group. This could be manifest in emphasizing wisdom, experience, having lived "in the good old days", surviving despite immense struggles (social and health ones), and so forth. Such techniques reinforce more positive attitudes towards the ingroup, but may perhaps have little effect on the status quo.

Finally, when members of social groups perceive the system to be unstable and illegitimate, they may mobilize to facilitate change. We see evidence of this in organizations like the Gray Panthers and the American Association of Retired People who represent a formidable political lobby. These social competition strategies are aimed at challenging the status quo and realizing material improvements in the ingroup's social position.

It is important, however, to avoid the simplistic notion that individuals develop only *one* social identity. Older people, for example, are as heterogeneous as any other demographic group—young old, old-old, older women, older men, persons of color, healthy or sick, rich or poor. Brewer (1999b) argues that individuals typically belong to several social groups and develop social identities related to them (see also, Giles & Johnson, 1981). Although one particular social identity may generally predominate, it may not be salient in *all* social and communicative contexts.

Work stemming from SIT has paid particular attention to the ways in which identities are managed in conversational interaction, and particularly the ways in which intergroup attitudes influence intergroup communication. Recently, communication scholars have been at pains to point out two processes. First, language and communication styles are often core dimensions of ingroup identity such that they are the very features differentiated in intergroup exchanges (e.g., Cargile, Giles, & Clément, 1996). Second, social categorization and social identity, among other cognitive constructs, are not immutable but, rather, are triggered, constructed, sustained, and ultimately modified through communicative processes (Abrams, O'Connor, & Giles, 2002; Hecht, 1993). A central theoretical statement in this tradition is outlined next (Giles et al., 1991).

Communication Accommodation Theory

Communication accommodation theory (CAT) posits that the situation, our similarity to others, our attraction to them, and our motivation to communicate with them will influence our tendency to accommodate their communicative style. Accommodation refers to the way in which we adapt our speech style in response to others in an interaction. To illuminate the role of language and communication in the process of categorization and stereotyping, CAT brought various other new concepts into play (Shepard, Giles, & Le Poire, 2001). We may converge (adopt a similar speech style), diverge (adopt a dissimilar style of speech), or maintain one form of speech in the face of another (a type of divergence). In addition, we may adjust our speech, nonverbals, vocals, and discourse in ways that relate (or fail to relate) to our interlocutor's needs in the encounter. Two such forms of accommodation associated with intergenerational communication are over- and underaccommodation (see Williams & Nussbaum, 2001).

Overaccommodation has often been examined in the context of secondary baby talk (baby talk directed toward an older person, see Caporael, 1981) or patronizing speech, a communication style "characterized by slower speech, exaggerated intonation, higher pitch, repetition, vocabulary simplification, and reduced grammatical complexity" (Fox & Giles, 1993, p. 443; Hummert & Ryan, 2001; for a typology of its verbal and nonverbal constituents, see Ryan, Hummert, and Boich, 1995). Within the context of a study examining reactions to patronizing speech, Giles, Fox, and Smith (1993) found that 35% of American elderly informants claimed to have been the personal recipient of it but claimed that, for older people in general, the figure was closer to 65%. While an array of studies report that such overaccommodators are viewed negatively as inconsiderate or incompetent (Hummert & Ryan, 2001), the previously-cited investigation found that older observers of patronizing talk find it particularly problematic. Elderly people were far more prone than younger people to evaluate the "victim" of overaccommodation as incompetent and weak. Nonetheless, not all elderly people find this form of talk demeaning, especially those who are socially inactive and frail (Ryan & Cole, 1990). Among other things, this led Ytsma and Giles (1997) to underscore that patronizing talk does not objectively exist as a communicative style *per se*, rather, it is a situated subjective attribution; the same communicative patterns can be variably labeled accommodations or overaccommodations (or even underaccommodations). Hence, when overaccommodation is inferred to be motivated by caring and nurturing rather than controlling and dominating intentions, its negative effects can be ameliorated (Harwood & Giles, 1996; Hummert & Ryan, 1996).

Interestingly, recipients of overaccommodation can be positively evaluated when they respond assertively to the patronizer and call into question their assumptions (e.g., Harwood et al., 1993). However, data also show that this threatening of the latter's face can cause its own communicative management problems and even further reinforce agist stereotypes in labeling the assertor as grouchy and abrasive. As one antidote, Ryan, Kennaley, Pratt, and Shumovich (2000) have argued that reactance via humor may sometimes help re-orient the partnership into a more socially adjusted relationship. Of course, overaccommodations are not the sole prerogative of younger people. Giles and Williams (1994) reported that young students claim to be widespread recipients of it too (e.g., over-protectiveness) and in ways that frustrate them.

Underaccommodation is defined as failing to be sensitive to the conversational needs of others (e.g., deflecting, ignoring). This form of speech is identified as a common response of older individuals to younger speakers (Harwood et al., 1995; Hummert & Ryan, 1996; Ryan et al., 1986). Research has documented a tendency for elderly persons to disclose personally painful information (Coupland, Coupland, Giles, Henwood, & Wiemann, 1988; Grainger, Atkinson, & Coupland, 1990). Such patterns of disclosure are generally regarded as inappropriate in initial interactions (Altman, Vinsel, & Brown, 1981; Berger & Bradac, 1982; Gilbert, 1977). Indeed, younger people are said to find (especially "out-of-the-blue) painful disclosures like "communicative grenades" and hence unnerving and awkward (Henwood, Giles, Coupland, & Coupland, 1993; Nussbaum, Hummert, Williams, & Harwood, 1996). This uncertainty often results in minimal responses on the part of the younger person such as 'mmm,' 'good heavens' or 'oh dear' (Coupland, Coupland, & Giles, 1991).

Underaccommodation is also present in communication from young to old. For instance, research suggests that health professionals practice underaccommodation (as well as overaccommodation) when interacting with elderly individuals (e.g., Street, 1991). For instance, in studies investigating medical encounters, Greene and colleagues found that doctors' speech to patients over sixty-five was rated as less egalitarian, patient, responsive, and respectful than speech directed to patients under forty-five (e.g., Greene, Adelman, Charon, & Hoffman, 1986; Greene, Hoffman, Charon, & Adelman, 1987).

Grainger, Atkinson, and Coupland (1990) investigated responses to elderly troubles-talk in caring situations including home help contexts and long-stay geriatric wards. The research examined how troubles-talk management contributes to the qualitative experience of health care for elderly people. The authors found many examples of caregiver deflection of elders' troubles talk in their data. Deflection, arguably a form of under-accommodation, was defined as occurring where caregivers "avoid fully and directly accepting the impact of their elderly charges' troubles-talk and its implication for future action" (p. 198). Based on a qualitative analysis of discourse, the findings indicated that communication deflection strategies epitomized the routine practices of care-delivery. This emphasis on deflection was seen to impair the quality of interactional experiences for older people. Grainger and colleagues (1990) contend that routine deflection by care-givers "drives a wedge between carees and carers, denying each group the potential fulfillment and health- and identity bolstering of supportive discourse" (p. 210).

Studies show that both overaccommodation *and* underaccommodation can result in negative outcomes for elderly persons such as low self-esteem or lessened psychological and physical health (Levy & Langer, 1996; Ryan et al., 1986). Such consequences have been emphasized in the theoretical perspective outlined next.

The Communication Predicament Model

The communication predicament model (CPM) of aging (Ryan et al., 1986) is derived from CAT as well as the social breakdown syndrome (Rodin & Langer, 1980). The CPM illustrates the way negative age stereotypes constrain communicative behaviors in intergenerational interactions. According to the model, when a younger individual encounters an older person certain old age cues (physiological, communicative, contextual) activate age-related stereotypes (see Fig. 6.1). Therefore, older persons whose appearance, voice, behavior, or social roles imply disability will be vulnerable in intergenerational encounters because of negative assumptions about old age on the part of the younger person (Ryan & Butler, 1996).

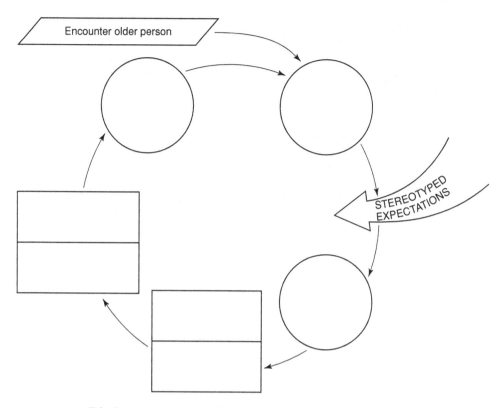

FIG. 6.1. Communication Predicament Model (from Ryan et al., 1986).

The outcome may be a form of negatively modified speech (e.g., overaccommodation or underaccommodation) which leads to constrained opportunities for communication, and reinforcement of age stereotyped behaviors. Where negative assumptions about age are made salient, older people may begin to self-stereotypically enact behaviors (see Turner, 1987) considered symptomatic of old-age (e.g., form and tone of speech, type of movement, demeanor: Avorn & Langer, 1990; Giles, Fox, Harwood, & Williams, 1994; Whitbourne & Wills, 1993). In turn, this leads to negative outcomes such as loss of personal control, low self esteem, lessened psychological activity, and social interaction with accompanying decrements in psychological and physiological health of the older person (Avorn & Langer, 1982; Langer & Rodin, 1977). In sum, the process of self-stereotyping can be an important communicative ingredient of the social construction of aging, worsening as it can memory performance and self-efficacy (Levy, 1996). This is a phenomenon we have termed "instant aging" elsewhere (Giles & Condor, 1988) and may accelerate psychological aging, the willingness to live (Levy, Ashman, & Dror, 2000), and physical demise.

While not denying the existence of some positive attributions about aging, the emphasis on negative stereotypes in this model reflects research findings that confirm the prevalence of negative stereotypes about elderly people (Williams & Nussbaum, 2001). Williams and Ylänne-McEwen (2000) state that "Our research has been populated by elders who are patronized, treated benignly as sweet old dears, or more negatively as helpless and decrepit, or as struggling heroically with the rigors of ill health and decrement" (p. 5). Such stereotypes are so embedded as to be below awareness for many people (Hummert & Ryan, this volume; Levy & Langer, 1996). That said, the role of *positive* sub-stereotypes of aging is important, and has been addressed in models such as that described next.

Stereotype Activation Model

Hummert and colleagues (Hummert, Garstka, Bonneson, & Strahm, 1994) found evidence of positive and negative stereotypes and associated traits of elderly people through research using trait sorting tasks and experimental manipulation. Their findings identified four predominantly negative old-age stereotypes (severely impaired, despondent, shrew/curmudgeon, and recluse) and three positive old-age stereotypes (perfect grandparent, John Wayne conservative, and golden ager: see Harwood, McKee, & Lin, 2000, for similar work on intergenerational communication schemas). With respect to communicative outcomes related to these old-age stereotypes, Hummert's (1994) stereotype activation model addresses how initial activation of stereotypes about an elderly person in an intergenerational interaction may (or may not) lead to problematic communication (see Fig. 6.2).

The model, grounded in CAT, posits that individuals develop cognitive schemas about aging. The perceiver's quality of contact with older adults and level of cognitive complexity affect the valence of these schemas. Individuals who exhibit greater cognitive complexity are able to form sophisticated and varied schemas about aging, and thus can communicatively accommodate older interlocutors more sensitively than individuals who lack cognitive complexity. Physical cues in the older person (e.g., presence or absence of gray hair) and context (e.g., nursing home or university) also act as catalysts for activation of stereotypes. Thus, for example, younger individuals who experience low quality contact, and who exhibit low levels of cognitive complexity are more likely to negatively stereotype older persons, especially those who appear handicapped in service contexts. According to Hummert (1994), negative stereotyping results in age-adapted communication styles (e.g., patronizing speech and secondary baby talk) on the part of the younger person in the interaction. By contrast, if the perceiver has experienced high quality contact with older persons, is cognitively complex, the older person taking part in the interaction is healthy, and the context is not age-salient then positive stereotypes may be activated and normal adult speech characterizes the intergenerational encounter.

Some scholars question whether the positive old-age stereotypes and responses identified in research associated with intergenerational communication (particularly health communication) are truly "positive" in nature (for a related point, see Sontag, 1978). Such conceptions of elderly people may be superficially benign, but they nevertheless pigeonhole older adults, and may lead to inappropriate and patronizing communication in intergenerational encounters (Harwood, 1998; Williams & Giles, 1998). Avorn and Langer (1982, 1990) suggest that older people are typically seen as in need of help. They state that "... although well meant, such attention may communicate belief in inability and inadequacy of the recipient. If the person faces no difficulty, if there are no challenges,

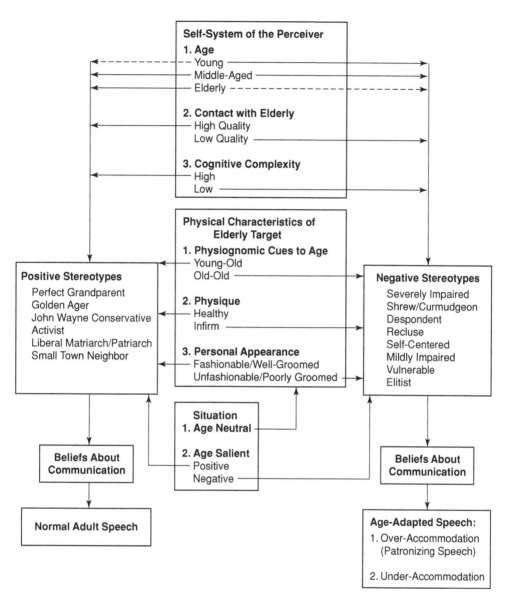

FIG. 6.2.   Stereotype Activation Model (from Hummert, 1994).

large or small, feelings of mastery are precluded and consequences such as involution, depression, and morbidity are real possibilities" (Langer & Avorn, 1985, p. 466).

In other words, how do we know that all elderly women who look like "sweet old ladies" (or perfect grandparents) want to be treated like sweet old ladies? An older person's communicative requirements and perceptions about him/herself might well be at odds with those of younger persons in an intergenerational encounter, and they may resent the imposition of a stereotypical image on themselves by younger people (cf. Louw-Potgieter & Giles, 1987). The theoretical model discussed next addresses how individuals participating in an intergenerational encounter may avoid imposing social identities on others based on the damaging stereotypes discussed so far.

The Communication Enhancement Model

The communication enhancement model (Ryan et al., 1995) was developed as a theoretical and pragmatic solution to the communicative predicament of aging (see Fig. 6.3). The solution revolved around facilitating intergenerational communication on the basis of personal identity rather than social identity (i.e., reducing the importance of young and old age stereotypes). Modified communication should take place as it relates to *individual* need assessment on behalf of both young and old (e.g., health care provider and client). Sensitivity to the other person's communicative needs through active listening is emphasized as opposed to attention to ill-informed, preconceived notions. Participants in an intergenerational encounter attune to each other's communication style, giving a sense of reciprocity or *optimal* convergence. This may involve amending communication style and *content*, but it is clear that convergence along too many dimensions simultaneously may result in evaluations of talk as patronizing or overaccommodative (Giles & Smith, 1979).

Optimal communication during an intergenerational encounter is said to lead to a sense of support for the older person and a willingness to engage in future interaction. Giles and colleagues (1990) contend that "feeling supported may be a function, to a greater of lesser extent, of the degree of attuning one receives, and so those who are known or perceived to possess high attuning skills may be preferentially sought out as supporters"

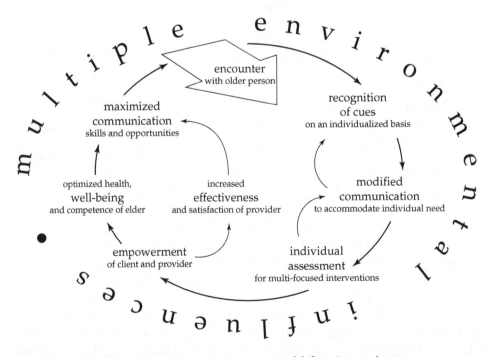

FIG. 6.3. Communication Enhancement Model (from Ryan et al., 1995).

(p. 13). The outcome in such a situation is posited to be positive for *both* parties—resulting in maximized communication, empowerment, increased well-being of the elderly person, increased effectiveness/satisfaction of the health provider or younger person.

A NEW MODEL OF INTER- AND INTRAGENERATIONAL COMMUNICATION

The above review of some important intergroup and communication theories reveals an abundance of useful concepts and perspectives with which to understand and explain intergenerational interactions and their outcomes. However, a close examination reveals lacunae, and some faulty assumptions about the nature of intergenerational communication as an intergroup phenomenon. Recognition of such inadequacies is not new—previous work has attempted to integrate intergroup theory and communication models as a way to facilitate intergenerational interaction and outcomes (e.g., Fox & Giles, 1993) or improve upon existing models (e.g., Harwood et al., 1993). However, a *parsimonious* model that both describes and explains intergenerational communication and that has pragmatic implications for interaction and outcomes remains elusive. The communication predicament model provides a valued explanation of the often-negative outcomes of intergenerational communication, while the communication enhancement model offers a guide to ameliorating such negative outcomes. One obvious (rhetorical) question is whether a single integrated model could serve these functions better than the variety of models described above? The following heuristic attempts to integrate elements of the previous models, as well as introduce some key themes that have been under-attended to in the literature.

In part as a response to a call for the emerging of aspects of the communication predicament and enhancement models into a single integrated framework (Giles, 2000), Barker and Giles (2003) formulated a new integrative framework (that they invoked to analyze the dire situation of elderly Native American health care). In this, a number of other important theoretical concepts (e.g., mindfulness and group vitality) were brought together to provide a holistic explanation of intergenerational communication. Two main processes are addressed: those involved in the recognition of cues and those of modifying communication (Edwards & Giles, 1998). In what follows, we introduce the Barker and Giles model, yet modestly revise and extend it in socially significant directions. This is pursued so as to address the plea by Noels, Giles, Gallois, and Ng (2001) that we go beyond current models given they "... need to be infused not only with culturally mediating mechanisms (e.g., filial piety and the like), but also require attention to *intragenerational concerns*" (our italics, p. 268).

The latter concern arises from work that was designed to determine whether nonaccommodating intergenerational communication climates would negatively influence subjective health outcomes for older people. Research has shown that being the recipient of accommodative behavior from young people is predictive of higher self-esteem, greater life satisfaction and lower depression scores among older people in different cultures (Cai, Giles, & Noels, 1998; Noels et al., 2001). However, *intra*generational communication climates were also predictive of these outcomes and especially among Asian older people where intergenerational experiences were actually non-predictive (Cai et al., 1998; Noels, Giles, Cai, & Turay, 1999; Noels et al., 2001; Ota, 2001).

We are not, of course, alone in documenting the importance of same-aged peer interactions for older adults (e.g., Auslander & Litwin, 1991; Ryff & Seltzer, 1996). As Rawlins (1995, p. 252) commented, ". . . despite their limitations, friends usually play vital roles in sustaining older persons' feelings of well-being and life satisfaction. Friends are uniquely valued to talk, reminisce, and judge with, and to keep confidence. They. . . .help with incidental needs, connect individuals to larger communities, and foster their ongoing enjoyment of life." Furthermore, there are tendencies for: some old people not to socially identify (or even be identified) with others their own age anyway (as documented above); friends often to be in short supply for this sector of the population (Patterson, Bettini, & Nussbaum, 1992); and otherwise-bolstering strong family ties to be commensurate with over-protection and a loss of autonomy (Nussbaum, 1985). Hence, we see a need to at least begin acknowledging the importance of *intra*generational communication in the development of future theoretical frames.

This new model (see Fig. 6.4) incorporates the notion that communication proceeds at, and is influenced by, multiple levels of social interaction (cf. the integrative model of levels of analysis of "miscommunication"—Coupland, Wiemann, & Giles, 1991). Therefore, interpersonal differences (e.g., in the capacity to be mindful) affect how well individuals communicate. Intergroup and cultural differences function to enhance or hinder communication. At the societal level, prevalent *stereotypes* about aging often reify forms of communicative overaccommodation or underaccommodation when the socially

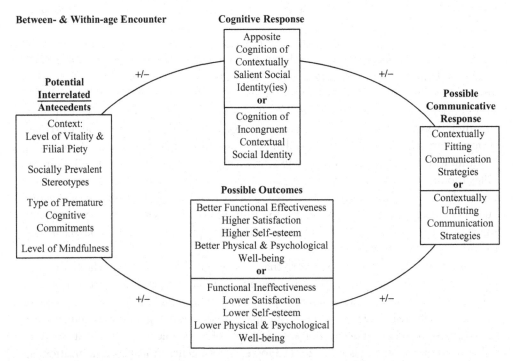

FIG. 6.4. An integrated model of inter- and intragenerational communication (revised after Barker & Giles, 2003).

dominant communicate with the disenfranchised or marginalized (e.g., for gendered communication, see Boggs & Giles, 1999). Also implicit in the model is the view that for positive outcomes in inter- and intra generational interaction, optimal forms of accommodation are required. That is, individuals should accommodate inter- and intragroup differences to a point of communication effectiveness, not to the point of assuming the other's social identity.

Antecedents

Taking these factors into account, the model posits that among the many precursors to an encounters involving older people, there are (at least) four *interrelated* sets of potentially powerful antecedents that, positively or negatively, influence the way the interaction proceeds.

The **first** antecedent is the context in which the encounter takes place (e.g., a doctor's office with elder as patient or a professor's office with elder as full professor). Certainly how communicators' subjective construals of a situation (as well as its goals) has been recognized for quite some time (see Giles & Hewstone, 1982). What has not received sufficient empirical attention is how interactants of different ages may conceptualize the supposedly same context in very different ways. Nevertheless, as both CAT and the stereotype activation model suggest, *where* the interaction takes place is influential in its process and outcome (Giles et al., 1991; Hummert, 1994) and, in particular, in the recognition of contextually-salient age, as well as health, identities (see Giles and Harwood [1997] for a typology of contexts which could make age group membership salient).

While afforded status as an independent antecedent in Barker and Giles (2003), we prefer to discuss *age vitality* as but one dimension of context. Not unrelatedly, all our antecedents are inter-related with communicative outcomes and climates helping to fashion them. The notion of "group vitality" was first introduced by Giles, Bourhis, and Taylor (1977) as a means of conceptualizing some of the sociostructural factors which could be important to contextualize interethnic relations, and therefore influence any communicative exchange between groups. Vitality theorists became particularly involved in how communicators socially construct their respective ingroup and outgroup vitalities rather than merely paying attention to factors that could be objectively measured (Harwood, Giles, & Bourhis, 1994). Since then there has been evidence to support the notion that the more vitality you construe your group as having—that is, the more sociostructural factors you believe to be in your group's favor—the more likely you are to invest psychologically in your group's identity and act collectively for it (for recent developments, see Giles, 2001). Conversely, lower vitality perceptions may lead to a loss of ingroup communication patterns (Giles & Byrne, 1982).

Vitality theory has been applied to a variety of intergroup contexts beyond the interethnic, including intergenerational relations (Harwood, Giles, Clément, Pierson, & Fox, 1994) where middle aged people are reported as having considerable perceived vitality (that is, societal status and institutional support). In a study spanning four Western and seven Eastern sites, this finding was universally supported but with more vitality accorded middle-aged and elderly targets in the former than the latter sites (Giles et al., 2000). Interestingly, the vitality profiles of the different-aged targets varied across cultural contexts and so did the vitality accorded elderly people. Fig. 6.5 is a selection of three

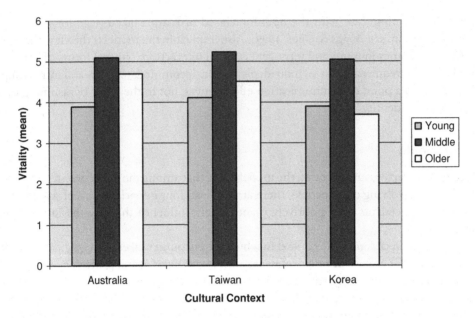

FIG. 6.5.   Vitality as a function of cultural context and target group.

contexts demonstrating decreasing vitality—at least as rated by young adult informants—in Australia, Taiwan, and South Korea respectively. We would argue that the more vitality is seen to drop from middle-age to elderliness—and especially in contexts where it dives below that of young adults (see Korea in Figure 5 and Hong Kong in Harwood et al., 1994), the more intergenerational communication problems that will emerge for older people given their decreasing societal recognition.

Another important feature of cultural variability we wish to subsume under the rubric of context (and not highlighted explicitly in Barker and Giles, 2003) is that of *filial piety* (see Figure 4). Contexts that ideologically support this notion are those that demand veneration of, and support for, older people, particularly (albeit not restricted to) older family members. Filial piety, as instantiated in Confucian doctrine, describes formal obligations that protect the power of parents relative to their children throughout their lives, and which carry concomitant obligations on parents to play significant social roles within the family (Chow, 1996). Although most work hitherto has been biased to Anglophone contexts, there is every reason to believe that intergenerational relations may not be the same in filially pious contexts, extending beyond Confucian societies to others (such as in The Philippines and Thailand) that have their own labels for this philosophy. Given that many individuals raised in East and South Asian countries are exposed to this value of appreciation for older age, it might be expected that there would be more positive relations, along with more accommodative communication, between the generations here than in western contexts. Of course, these notions raise the important issue of considering differences in treatment of, and attitudes towards, older adults who are family members as compared to those who are not (Harwood & Lin, 2000; Ng, Liu, Weatherall, & Loong, 1997).

However, this vision of honored elders in this part of the world has been viewed as some (e.g., Tobin, 1987) as an *American* idealization. Moreover, several authors argue that a positive conception of elderly people may be a myth at worst or a diluted feature of the past at best (e.g., Koyano, 1989; Tien-Hyatt, 1987). Regarding the latter, scholars have pointed to the erosion of filial piety given sudden changes in family structure to the nuclear model, relative increases in educational and income differentiations between younger and older people, and modernization (Chow, 1999; Ingersoll-Dayton & Saengtienchai, 1999). Interestingly, Gallois et al. (1999) in a study of filial piety around Pacific Rim contexts found that while East Asian young people reported greater willingness to support older people instrumentally and tangibly, they were less likely than Westerners to provide communicative support. In addition, this pattern is complemented by a growing number of studies showing older people are viewed less favorably (e.g., Harwood et al., 1996; Harwood et al., 2001) and as more nonaccommodating in East Asian than Western settings (e.g., Noels et al., 1999, 2001). The directionality of this variability is dismaying for those scholars seeking to locate societies where more positive models of aging exist than the one currently being portrayed by Western empirical research.

At this juncture, it is important to highlight that the cross-cultural differences in age vitality and filial piety discussed above are apparent and contribute to the mosaic of ethnic heterogeneity among older people within any society. In this regard, the U.S. Bureau of the Census (1993, p. 1) pointed out that "each age, gender, race, and ethnic group has distinctive characteristics and the experience of aging is different among the demographic groups".

Second, prevalent *stereotypes* about aging (Hummert, 1994) are posited to influence the likelihood that interactants will show the ability to recognize salient identities - and these could well be fashioned by communicated forces of perceived vitality and filial piety. For example, negative sub-stereotypes about older persons invoked will increase the probability that participants in intergenerational interactions will ignore evidence of salient identities not consistent with such stereotypes. Greene, Adelman, Charon, and Friedmann (1989) cite evidence that medical providers view older patients as hypochondriacs with hard-to-treat conditions, as asexual, preoccupied with death, and more likely to be difficult in a medical interaction. Exactly how such expectations may ultimately affect outcomes for elderly persons is illustrated by Levy and Langer (1994), who show that negative self-stereotyping is associated with poor memory performance.

In other words, and issues of ethnic heterogeneity notwithstanding (e.g., McNamara, 1999), more and more people are resisting any imposition of stereotypic labels such as "elderly" or "older worker" and, not inconsequentially, having problems relating with those they deem "old". Undeniably, this can cause stressful intragenerational exchanges as some older same-aged peers disavow (while others do not) membership in a stigmatized group (cf. Zebrowitz & Montepare, 2000). Consider also the following:

> I accept the challenges of growing old and I adjust. I move forward, not back. It's like I always say: Old people are all different. Some people resent getting old and many people handle it fine. But one thing, for sure, is that we aren't all bald and grumpy, sitting in a wheel chair (80 year-old senior activist quoted by Quintanilla, 1995, p. E5).

Third and relatedly, yet purportedly distinct from stereotypes, are interlocutors' *premature cognitive commitments* about aging (Langer, Perlmuter, Chanowitz, & Rubin, 1988).

These refer to beliefs about the inevitability of decrement associated with aging that can influence the cognitive responses of interactants. For example, interactants (younger and older) who have learned that aging is a generally painful experience—perhaps from recognizing their low vitality, the prevalence of negative stereotypes, and the eroding of filial piety—are more likely to communicate based on that view of elder identity. That is, a potentially incongruent identity is imposed on the older person regardless of whether the individual accepts it. Such attitudes about aging and aging stereotypes begin in early life (Langer et al., 1990) and develop over the life span (Giles et al., 1990). These premature cognitive commitments can often result in intergenerational miscommunication and negative outcomes for many (albeit not at all) older persons (Whitbourne & Wills, 1993).

Fourth, as suggested by Langer and colleagues (e.g., Langer, Bashner, & Chanowitz, 1985), interlocutors' levels of *mindfulness* (active distinction making) also influence the way such interactions procede. Participants who are highly mindful of distinctions are posited to be more likely to be able to communicate in a contextually fitting manner because they are aware of currently salient social identities. The inclusion of mindfulness in the current model is a departure in some ways from the stereotype activation model's inclusion of the concept of cognitive complexity (as the ability to form sophisticated and varied schemas). Although cognitive complexity is intuitively applicable to intergenerational communication (and indeed all forms of communication), it begs certain questions. For example, are cognitively complex persons cognitively complex all the time, and if so, how did they get to be that way? In other words, is this really the traditional trait we have been led to believe it is, or is, rather, actually a state? With regard to processing intergroup *differences*, Ryan, Hewstone, and Giles (1984) reason that "the developing intergroup belief structures and strategies attending group memberships will inspire progressively more differentiated and hierarchically organized evaluative dimensions in order to function so as to preserve or enhance positive social identities" (p. 149). However, Ryan et al. also believe that when identity is really threatened individuals tend to revert to simply categorizing on the basis of outgroup characteristics learned at an earlier stage.

In relation to categorizing differences, the theory of mindfulness (Langer, 1989), as a theory of cognitive development across the life span, involves the ability to make distinctions, form new categories, and novel solutions to problems (Demick, 2000). Mindfulness offers an alternative to the constructivist concept of cognitive complexity as an aid to understanding personal differences as they relate to intergenerational communication. Mindfulness appears superficially similar to cognitive complexity in positing that people develop cognitive commitments about the way things are based on early experiences determining to an extent whether the response to aging (their own and others) will be novel or over-learned. However, the capacity to be mindful (or mindless) can be learned at any point across the life span. That is, people at various stages of life span development have the capacity to learn to be mindful in certain situations while operating mindlessly in others (e.g., Langer et al., 1985).

Mindlessness, in contrast, occurs when "individuals consider available information and alternatives incompletely, rigidly, reflexively and thoughtlessly" (Burgoon & Langer, 1994, p. 105). Burgoon, Berger, and Waldron (2000) believe that "communication that is planful, effortful, creative, strategic, flexible, and/or reason-based (as opposed to emotion-based) would seem to qualify as mindful, while communication that is reactive, superficially processed, routine, rigid and emotional would fall toward the mindless end of the continuum" (p. 112). Langer argues that individuals often communicate mindlessly when they think in absolute (rather than conditional) terms, that is people respond based on long established categories or stereotypes—premature cognitive commitments (Levy & Langer, 1994). In general, people "typically speak from a single perspective, ignoring the richness of our language which makes available ways of saying the same thing" (Langer, 1992, p. 324). Competent communicators use creative planning, consider the potential effect of their communication practices, develop alternative communication strategies, and are responsive to context (Burgoon et al., 2000). Relatedly, in the arena of intercultural communication, anxiety uncertainty management theory (AUM) (Gudykunst, 1995) suggests that to be mindful when communicating with persons from another culture "we need knowledge of strangers' cultures and group memberships . . . knowledge of their language, dialect, slang, and/or jargon facilitates management of our anxiety and uncertainty" (p. 39). Additionally, competent intercultural communicators are able to adapt, empathize, manage anxiety in new situations, tolerate ambiguity, and make accurate predictions and explanations (see also Hajek & Giles, 2003).

Our approach here appears to partially buttress Hummert's (1990, 1994) view that those who are cognitively complex tend to be less reliant on stereotypes in intergenerational communication. In other words, individuals who are cognitively complex are *also* mindful–literally making distinctions. Yet this is markedly different from operating on the basis of personal identity, because it presupposes conscious consideration of one's own and outgroup other's social identity in a communicative context.

Consequently, with regard to intergenerational communication, when interlocutors communicate mindfully they are sensitive and accommodating to differences (and do not rely on negatively-valenced "premature cognitive commitments"). They develop the ability to *read identity salience* and know how and when to employ or alter certain communication strategies in order to accommodate identity. Any model of intergenerational communication must take into account levels of mindfulness and flexibility (or a lack of it) in the strategies used by interlocutors (young and old) and how they feed into outcomes relating to future interactions and those affecting elder psychological and physiological well being.

Cognitive Response

Based on these antecedent factors, intergenerational interactants will either understand which social identities are contextually salient (e.g., wheel chair occupant or highly regarded professional person, or both), or participants will simply be cognizant of one contextually *incongruent* social identity (e.g., an elderly, handicapped person).

Some intergroup scholars posit that one way to avoid discriminatory outcomes in social interaction is for group members to interact with outgroup member on the basis of individual or personal identity where group attributes are not salient (i.e., decategorization: Brewer & Miller, 1994; Gaertner et al., 2000). Indeed, at times, SIT seems to suggest that there is a plane on which all individuals may communicate outside of group identity. This involves seeing the individual as a person rather than a member of an outgroup (see also in this regard, the process model of communication competence, Hajek & Giles, 2003). From our perspective, we would argue that all communicative encounters—even the *intra*generational where health, aging, self-stereotyping and the like can make age salient (and potentially negatively so)—involve some awareness of group memberships (social identities); indeed, group identities are central to an appropriate interpersonal understanding in a given situation. Ignoring group memberships or underattending to them may be as problematic as overattending to them. Therefore, effective communication seems most likely to occur when interlocutors are able to accommodate group attributes and identities and demonstrate flexibility in their communicative strategies. Recognizing that difference makes a difference in intergenerational interaction, Langer and colleagues (1990) have argued:

> . . . different concerns may make different forms of behavior rational in the same setting. Similarly, identical behaviors do not necessarily arise from common bases, nor do they necessarily convey the same information. What may be desirable to younger persons may be undesirable or uninteresting to an elderly person. (p. 118)

What appears then to be important in effective communication, and also in comprehending miscommunication, is the ability (or inability) to recognize and understand differences, and to accommodate to them *optimally*. This does *not* mean that one interactant should "mimic" another. Research has shown this to be counterproductive (see Giles & Smith, 1979). It does require that interactants to be sensitive to differences and open to them. Such sensitivity may also have positive consequences in terms of allowing generalization from specific contact experiences to more general attitudes (Harwood, Hewstone, & Paolini, 2001).

Communicative Response and Outcomes

As a consequence of the cognitive process described above, individuals' communicative responses in an encounter will incorporate fitting or unfitting strategies, which will have particular consequences within and beyond the encounter. Fitting strategies can be defined as those that are most satisfying, beneficial, and effective for both (all) intergenerational participants. For example, apart from discussing symptoms, a *fitting* communicative strategy on the part of a health care professional might include asking the elder (who is also active in his or her profession) how work is progressing. Street (1991) suggests that during a medical encounter, physician and older patient interaction should engage in communication accommodation featuring relative similarity in expressive behavior. For example, "in medical consultations in which the patient is highly anxious about the medical condition, both doctor and patient may display high levels of involvement through directness of body orientation, forward body leans, gaze toward partners, reciprocal topic development, and facial expressiveness" (p. 138). Additionally, Greene, Adelman, Friedmann, and Charon (1994) found that older patients prefer and feel much more satisfied with medical encounters in which there is physican supportiveness and shared laughter and they are given opportunities to provide information of their own volition. Over time, this type of communicative approach is likely to increase functional effectiveness for both parties, in addition to increasing satisfaction with such interactions, and maintaining self-esteem and general well being.

As a further example, researchers found that a group of institutionalized elderly persons who were encouraged to make decisions for themselves, given decisions to make, and given responsibility for a task became more active and felt happier (Langer & Rodin, 1976). Moreover, when a sub-group of these participants were re-evaluated 18 months later, they showed better health and higher activity patterns and mortality rates lower than residents who had not received the responsibility-induced condition (Rodin & Langer, 1977). Contextually-fitting strategies *intra*generationally might also be attuned to creative compensation for age-related decrements where they really exist (Baltes & Wahl, 1996) and a focus on the conversational benefits of lifespan advancement and not on socially constructed limitations.

By contrast, contextually *unfitting* communication strategies are defined as those that lead to dissatisfaction, lack of understanding, or even the avoidance of future interaction. For example, intergenerational patronizing speech *can* be a particularly inappropriate and unfitting form of communication in intergenerational interaction (e.g., Sachweh, 1998; Thimm, Rademacher, & Kruse, 1998). Ryan and Cole (1990) suggest that "even though an increase in volume while talking with a hearing impaired individual can be helpful, the most frequent response is to shout in a manner which raises the volume too much and which automatically also raises the pitch" (p. 187). They note that the higher pitch of these messages actually makes them more difficult for older adults to comprehend due to high frequency hearing loss.

This type of speech may be used with good intentions, but often it is employed as a way of controlling the elderly person (Lanceley, 1985). Ryan et al. (1986) have argued that "demeaning talk may not only induce momentary feelings of worthlessness in elderly people but may also lead to reduced life satisfaction and mental and physical decline in the long run" (p. 14). Indeed, even when well meant, overaccommodation can have adverse effects. Avorn and Langer (1982) found that disability may be induced by "helping" behaviors. Participants who received intrusive helping in the completion of a psychomotor task (versus verbal encouragement) showed reduced ability and motivation to perform the task. The researchers concluded that excessive infantilization of nursing home residents and overt helping beyond requirements can lead to learned helplessness and further disability.

We contend that contextually unfitting communication can also emerge through self-stereotyping processes with other older people too. One of the current authors' experience of being the recipient of agist comments is not so much from much younger people, but those within his same age bracket! If we accept such sentiments unquestioningly— and we have been socialized to just this—then we are absorbing these verbalized constructions of our age position in a way that can harmful outcomes. In an email sent to same-aged peers recently, one colleague reminded us, "In case you didn't know, Jack is 50 today! Let's *torture* him!" Such inclinations, like the sending of agist birthday cards, allow us to collude in "torturing" each other mercilessly with the consent of the community and without comeback from "victims"; Jack (not atypically) acted graciously, both verbally and nonverbally. Put another way, intragenerational "havens" are not immune from age discrimination and can render potentially harmful communication climates on occasion.

Relatedly, one of us has been collecting e-mails that communicate self-stereotypical accounts of memory problems. The following are a modest sample from extraordinarily competent, very busy, 40–50 year old perpetrators:

"I absolutely spaced out on getting back to you . . . must be old age . . . Anyway, . . .
"Dang! Obviously old age is setting in! I completely forgot about . . .
"I must apologise for not getting this notice about what are doing out sooner. It slipped through my ever widening senior cracks . . .
"Please just chalk this up to the feebleness of old age. Of course I am coming . . .

Of course, expectations within and between cultures vary regarding what is due to the elder population in terms of care and respect. Hence, an understanding of what is a contextually fitting strategy will depend upon the nature of the cultural dynamics as inherent in our model.

Contextually fitting communication responses are likely to lead to increased functional effectiveness for all parties, increased communication satisfaction, and potentially bolstered self-esteem. In the long run, improved or maintained psychological or physical health are potential outcomes (Noels et al., 1999, 2001; Ota, 2001). Contextually unfitting communication strategies may, in contrast, result in a lessening of functional effectiveness for either party (physically, socially or both), as well as decreased communication satisfaction, self-esteem, and physical or psychological health and well being. Moreover, such outcomes are likely to confirm or reinforce antecedent factors experienced by intergenerational interlocutors. That is, participants will enter the next intergenerational encounter with expectations related to this type of context, perhaps operating mindlessly on the basis of this prior experience, with perceptions about each other's group vitality, and with negative social stereotypes confirmed in their minds.

CONCLUSION

The foregoing suggests that the new model possesses the following advantages over other intergroup models of intergenerational communication.

- It presents a holistic description of the processes and outcomes (positive or negative) of intergenerational communication as an intergroup phenomenon.
- It recognizes facets of the social structure and cultural variability as well as the need to monitor subjective changes in them.
- It assumes that making and maintaining distinctions (as opposed to ignoring them) can be advantageous to communication between- as well as within-generations.
- It recognizes that communicative outcomes fashion the very antecedent factors with which we entered into the transactive cycle of the model.

It is anticipated that this model may set an interdisciplinary agenda for revising, elaborating, and testing among scholars interested in investigating and exploring inter- as well as intragenerational communication from an intergroup perspective, and thereby responding to the challenge of illuminating appropriate communication strategies that lead to *positive* health outcomes (Barker & Giles, 2003; Williams & Coupland, 1998). However, the purpose of integrating aspects of the theoretical models and concepts discussed in this chapter is not just to refine ideas and constructs, or to provide a rationale for research into intergenerational communication. As in all cases, the role of theory is to inform practice. It is hoped that medical practitioners, clinicians, gerontologists, and students of aging will be able improve, develop, and make more effective their everyday communication with *all* older persons, especially those belonging to marginalized or minority groups.

REFERENCES

Abrams, J., & Giles, H. (1999). Epilogue: Intergenerational contact as intergroup communication. *Child and Youth Services, 20,* 203–217.

Abrams, J., O'Connor, J., & Giles, H. (2002). Identity and intergroup communication. In W. B. Gudykunst & B. Mody (Eds.), *Handbook of international and intercultural communication* (2nd ed., pp. 225–240). Thousand Oaks: Sage.

Altman, I., Vinsel, A., & Brown, B. (1981). Dialectic conceptions in social psychology: An application to social penetration and privacy regulation. *Advances in Experimental Social Psychology, 14,* 108–159.

Annan, K. (1999). Address at the ceremony launching the international year of older persons (1999). *Journal of Gerontology, 54,* P5–P6.

Auslander, G. K., & Litwin, H. (1991). Social networks, social support, and self-ratings of health among older persons. *Journal of Aging and Health, 3,* 493–510.

Avorn, J., & Langer, E. B. (1982). Induced disability in nursing home patients: A controlled trial. *Journal of American Geriatrics Society, 30,* 397–400.

Avorn, J., & Langer, E. B. (1990). Impact of the psychosocial environment of the elderly on behavioral and health outcomes. In B. B. Hess & E. W. Markson (Eds.), *Growing old in America* (pp. 462–473). New Brunswick, NJ: Transaction Books.

Baltes, M. M., & Wahl, H. W. (1996). Patterns of communication in old age: The dependence-support and independence ignore script. *Health Communication, 8,* 217–231.

Barker, V., & Giles, H. (2003). Integrating the communicative predicament and enhancement of aging models: The case of Native Americans. *Health Communication, 15,* 255–276.

Berger, C. R., & Bradac, J. S. (1982). *Language and social knowledge.* London: Edward Arnold.

Boggs, C., & Giles, H. (1999). "The canary in the coalmine": The nonaccommodation cycle in the gendered workplace. *International Journal of Applied Linguistics, 9,* 223–245.

Brewer, M. B. (1999a). The psychology of prejudice: Ingroup love or outgroup hate? *Journal of Social Issues, 55,* 429–444.

Brewer, M. B. (1999b). Multiple identities and identity transition: Implications for Hong Kong. *International Journal of Intercultural Relations, 2,* 187–197.

Brewer, M., & Miller, N. (1994). Beyond the contact hypothesis: Theoretical perspectives on desegregation. In N. Miller & M. B. Brewer (Eds.), *Groups in contact: The psychology of desegregation* (pp. 281–302). Orlando: Academic Press.

Burgoon, J. K. (1995). Cross-cultural and intercultural applications of expectancy violations theory. In R. L. Wiseman (Ed.), *Intercultural communication theory* (pp. 194–214). Thousand Oaks, CA: Sage.

Burgoon, J. K., Berger, C. R., & Waldron, V. R. (2000). Mindfulness and interpersonal communication. *Journal of Social Issues, 56,* 105–128.

Burgoon, J. K., & Langer, E. J. (1994). Language fallacies, and mindlessness-mindfulness in social interaction. In B. R. Burleson (Ed.), *Communication Yearbook 18* (pp. 105–132). Newbury Park, CA: Sage.

Cai, D., Giles, H., & Noels, K. A. (1998). Elderly perceptions of communication with older and younger adults in China: Implications for mental health. *Journal of Applied Communication Research, 26,* 32–51.

Caporael, L. R. (1981). The paralanguage of caregiving: Baby talk to the institutionalized aged. *Journal of Psychology and Social Psychology, 40,* 876–884.

Cargile, A, Giles, H., & Clément, R. (1996). The role of language in ethnic conflict. In J. Gittler (Ed.), *Racial and ethnic conflict: Perspectives from the social disciplines* (pp. 189–208). Greenwich, CT: PAI Press.

Chow, N. (1996). Filial piety in Asian Chinese communities. *Hong Kong Journal of Gerontology, 10* (Supplement), 115–117.

Chow, N. (1999). Diminishing filial piety and the changing role and status of the elders in Hong Kong. *Hallym International Journal of Aging, 1,* 67–77.

Coupland, N. (1997). Language, aging and ageism: A project for applied linguistics. *International Journal of Applied Linguistics, 7,* 26–48.

Coupland, N., Coupland, J., & Giles, H. (1991). *Language, society, and the elderly: Discourse, identity and aging.* Oxford: Blackwell.

Coupland, N., Coupland, J., Giles, H., Henwood, K., & Wiemann, J. M. (1988). Elderly self-disclosure: Interactional and intergroup issues. *Language & Communication, 8,* 109–133.

Coupland, N., & Giles, H. (1988). Introduction: The communicative contexts of accommodation. *Language & Communication, 8,* 175–182.

Coupland, N., & Nussbaum, J. F. (Eds.) (1993). *Discourse and lifespan identity.* Newbury Park, CA: Sage.

Coupland, N., Wiemann, J. M., & Giles, H. (1991). Talk as "problem" and communication as "miscommunication": An integrative analysis. In N. Coupland, H. Giles, & J. M. Wiemann (Eds.), *Miscommunication and problematic talk* (pp. 1–17). Newbury Park, CA: Sage Publications, Inc.

Demick, J. (2000). Toward a mindful psychological science: Theory and application. *Journal of Social Issues, 56,* 141–160.

Edwards, H., & Giles, H. (1998). Prologue on two dimensions: The risk and management of intergenerational miscommunication. *Journal of Applied Communication Research, 26,* 1–12.

Fox. S. A. (1999). Communication in families with an aging parent: A review of the literature and agenda for future research. In M. E. Roloff (Ed.), *Communication Yearbook 22* (pp. 377–429). Thousand Oaks: Sage.

Fox, S. A., & Giles, H. (1993). Accommodating intergenerational contact: A critique and theoretical model. *Journal of Aging Studies, 7,* 423–451.

Gaertner, S. L., Dovidio, J. F., Nier, J. A., Banker, B. S., Ward, C. M., Houlette, M., & Loux, S. (2000). The common ingroup identity model for reducing intergroup bias: Progress and challenges. In D. Capozza & R. Brown (Eds.), *Social identity processes: Trends in theory and research* (pp. 133–148). London: Sage.

Gallois, C., Giles, H., Jones, E., Cargile, A. C., & Ota, H. (1995). Accommodating intercultural encounters: Elaborations and extensions. In R. L. Wiseman (Ed.), *Intercultural communication theory* (pp. 115–147). Thousand Oaks, CA: Sage.

Gallois, C., Giles, H., Ota, H., Pierson, H. D., Ng, S. H., Lim, Tae-Seop, Maher, J., Somera, L., Ryan, E. B., & Harwood, J. (1999). Intergenerational communication across the Pacific Rim: The impact of filial piety. In J-C. Lasry, J. G. Adair, & K. L. Dion (Eds.), *Latest contributions to cross-cultural psychology* (pp. 192–211). Amsterdam: Swets & Zeitlinger.

Gerbner, G., Gross, L., Signorielli, N., & Morgan, M. (1980). Aging with television: Images on television drama and conceptions of social reality. *Journal of Communication, 30,* 37–48.

Gilbert, S. (1977). Effects of unanticipated self-disclosure on recipients of varying levels of self-esteem: A research note. *Human Communication Research, 3,* 368–371.

Giles, H. (1999). Managing dilemmas in the "silent revolution": A call to arms! *Journal of Communication, 49,* 170–182.

Giles, H. (2000). Review of C. Gallois and V. Callan (Eds.), "Communication and culture: A guide for practice." Chichester: Wiley, 1997. *Journal of Occupational and Organizational Psychology, 73,* 261–263.

Giles, H. (2001). Ethnolinguistic vitality. In R. Mesthrie (Ed.), *Concise Encyclopaedia of Sociolinguistics* (pp. 472–473). Oxford, UK: Elsevier.

Giles, H., Bourhis, R. Y., & Taylor, D. M. (1977). Towards a theory of language in ethnic group relations. In H. Giles (Ed.), *Language ethnicity and intergroup relations* (pp. 307–348). London: Academic Press.

Giles, H., & Byrne, J. (1982). An intergroup model of second language acquisition. *Journal of Multilingual and Multicultural Development, 3,* 17–40.

Giles, H., & Condor, S. (1988). Ageing, technology, and society: An introduction and future priorities. *Social Behaviour: An International Journal of Applied Social Psychology, 3*, 59–70.

Giles, H., & Coupland, N. (1992). *Language: Contexts and consequences.* Pacific Grove, CA: Brooks/Cole.

Giles, H., Coupland, N., & Coupland, J. (1991). Accommodation theory: Communication, context, and consequence. In H. Giles, N. Coupland, & J. Coupland (Eds.), *Contexts of accommodation: Developments in applied sociolinguistics* (pp. 1–68). Cambridge, UK: Cambridge University Press.

Giles, H., Fox, S. A., Harwood, J., & Williams, A. (1994). Talking age and aging talk: Communicating through the lifespan. In M. L. Hummert, J. M. Wiemann, & J. F. Nussbaum (Eds.), *Interpersonal communication in older adulthood: Interdisciplinary theory and research* (pp. 130–161). Newbury Park, CA: Sage.

Giles, H., Fox, S. A., & Smith, E. (1993). Patronizing the elderly: Intergenerational evaluations. *Research in Language & Social Interaction, 26*, 129–149.

Giles, H., & Harwood, J. (1997). Managing intergroup communication: Lifespan issues and consequences. In S. Eliasson & E. Jahr (Eds.), *Language and its ecology: Essays in memory of Einar Haugen* (pp. 105–130). Mouton de Gruyter: Berlin.

Giles, H., & Hewstone, M. (1982). Cognitive structures, speech and social situations: Two integrative models. *Language Sciences, 4*, 187–219.

Giles, H., & Johnson, P. (1981). The role of language in ethnic group relations. In J. C. Turner & H. Giles (Eds.), *Intergroup behavior* (pp. 199–243). Oxford, UK: Blackwell.

Giles, H., Noels, K., Ota, H., Ng, S. H., Gallois, C., Ryan, E. B., Williams, A., Lim, T.-S., Somera, L., Tao, H., & Sachdev, I. (2000). Age vitality across eleven nations. *Journal of Multilingual & Multicultural Development, 21*, 308–323.

Giles, H., & Smith, P. M. (1979). Accommodation theory: Optimum levels of convergence. In H. Giles & R. N. St. Clair (Eds.), *Language and social psychology* (pp. 45–65). Oxford, England: Basil Blackwell.

Giles, H., & Williams, A. (1994). Patronizing the young: Forms and evaluations. *International Journal of Aging and Human Development, 39*, 33–53.

Giles, H., Williams, A., & Coupland, N. (1990). Communication, health, and the elderly: Frameworks, agenda and a model. In H. Giles, N. Coupland, J. M. Wiemann (Eds.), *Communication, health and the elderly* (pp. 1–28). Manchester, England: Manchester University Press.

Grainger, K., Atkinson, K., & Coupland, N. (1990). Responding to the elderly: troubles-talk in the caring context. In H. Giles, N. Coupland & J. M. Wiemann (Eds.), *Communication, health and the elderly* (pp. 192–212). Manchester: Manchester University Press.

Greene, M. G., Adelman, R., Charon, R., & Friedmann, E. (1989). Concordance between physicians and their older and younger patients in the primary care medical encounter. *The Gerontologist, 29*, 808–813.

Greene, M. G., Adelman, R., Charon, R., & Hoffman, S. (1986). Ageism in the medical encounter: An exploratory study of the doctor elderly patient relationship. *Language & Communication, 6*, 113–124.

Greene, M. G., Adelman, R. D., Friedmann, E., & Charon, R. (1994). Older patient satisfaction with communication during a medical encounter. *Social Science & Medicine, 9*, 1279–1288.

Greene, M. G., Hoffman, S., Charon, R., & Adelman, R. (1987). Psychosocial concerns in the medical encounter: A comparison of the interactions of doctors with their old and young patients. *Gerontologist, 27*, 164–168.

Gudykunst, W. B. (1995). Anxiety/uncertainty management (AUM) Theory: Current status. In R. L. Wiseman (Ed.), *Intercultural communication theory* (pp. 8–58). Thousand Oaks, CA: Sage.

Hajek, C., & Giles, H. (2002). The old man out: An intergroup analysis of intergenerational communication among gay men. *Journal of Communication, 52*, 698–714.

Hajek, C., & Giles, H. (2003). New directions in communication competence: The process model. In J. O. Greene & B. R. Burleson (Eds.), *Handbook of communication and social interaction skills* (pp. 935–957). Mahwah, NJ: Lawrence Erlbaum Associates.

Harwood, J. (1998). Young adults' cognitive representations of intergenerational conversations. *Journal of Applied Communication Research, 26*, 13–31.

Harwood, J., & Anderson, K. (2002). The presence and portrayal of social groups on prime-time television. *Communication Reports, 15*, 81–98.

Harwood, J., & Giles, H. (1996). Reactions to older people being patronized: The roles of response strategies and attributed thoughts. *Journal of Language & Social Psychology, 15*, 395–422.

Harwood, J., Giles, H., & Bourhis, R. Y. (1994). The genesis of vitality theory: Historical patterns and discoursal dimensions. *International Journal of the Sociology of Language, 108*, 168–206.

Harwood, J., Giles, H., Clément, R., Pierson, H. D., & Fox, S. A. (1994). Perceived vitality of age categories across California and Hong Kong. *Journal of Multilingual and Multicultural Development, 15*, 311–318.

Harwood, J., Giles, H., Fox, S. A., Ryan, E. B., & Williams, A. (1993). Patronizing young and elderly adults: Response strategies in a community setting. *Journal of Applied Communication Research, 3*, 211–226.

Harwood, J., Giles, H., McCann, R. M., Cai, D., Somera, L. P., Ng, S. H., Gallois, C., & Noels, K. A. (2001). Older adults' trait ratings of three age-groups around the Pacific rim. *Journal of Cross-Cultural Gerontology, 16*, 157–171.

Harwood, J., Giles, H., Ota, H., Pierson, H. D., Gallois, C., Ng, S. H., Lim, T.-S., & Somera, L. (1996). College students' trait ratings of three age groups around the Pacific Rim. *Cross-Cultural Gerontology, 11*, 307–317.

Harwood, J., Giles, H., & Ryan, E. B. (1995). Aging, communication and intergroup theory: Social identity and intergenerational communication. In J. F. Nussbaum & J. Coupland (Eds.), *Handbook of communication and aging research* (pp. 133–159). Hillsdale, NJ: Lawrence Erlbaum.

Harwood, J., Hewstone, M., & Paolini, S. (2001). *Intergroup contact theory, the grandparent-grandchild relationship, and attitudes towards older adults*. Unpublished manuscript, University of Kansas.

Harwood, J., & Lin, M.-C. (2000). Affiliation, pride, exchange and distance in grandparents' accounts of relationships with their college-age grandchildren. *Journal of Communication, 50*, 31–47.

Harwood, J., McKee, J., & Lin, M.-C. (2000). Younger and older adults' schematic representations of intergenerational communication. *Communication Monographs, 67*, 20–41.

Harwood, J., & Roy, A. (1999). Portrayals of older adults in magazine advertisements in India and the United States. *Howard Journal of Communications, 10*, 269–280.

Hecht, M. L. (1993). 2002-a research odyssey: Toward the development of a communication theory of identity. *Communication Monographs, 60*, 76–82.

Henwood, K., Giles, H., Coupland, N., & Coupland, J. (1993). Stereotyping and affect in discourse: Interpreting the meaning of elderly painful self-disclosure. In D. M. Mackie and D. L. Hamilton (Eds.), *Affect, cognition, and stereotyping: Interactive processes in group perception* (pp. 269–296). San Diego: Academic Press.

Hogg, M. A., & Abrams, D. (1990). Social motivation, self-esteem, and social identity. In D. Abrams & M. A. Hogg (Eds.), *Social identity theory: Constructive and critical advances* (pp. 28–47). New York: Harvester Wheatsheaf.

Hogg, M. A., & Abrams, D. (1993). Towards a single-process uncertainty reduction model of social motivation in groups. In M. A. Hogg & D. Abrams (Eds.), *Group motivation: Social psychological perspectives* (pp. 173–190). New York: Harvester Wheatsheaf.

Hogg, M. A., & Turner, J. C. (1985). Interpersonal attraction, social identification and psychological group formation. *European Journal of Social Psychology, 15*, 51–66.

Hummert, M. L. (1990). Multiple stereotypes of the elderly and young adults: A comparison of structure and evaluations. *Psychology and Aging, 5*, 182–193.

Hummert, M. L. (1994). Stereotypes of the elderly and patronizing speech. In M. L. Hummert, J. M. Wiemann, & J. F. Nussbaum (Eds.), *Interpersonal communication in older adulthood* (pp. 162–184). Thousand Oaks, CA: Sage.

Hummert, M. L., Garstka, T. A., Bonneson, J. L., & Strahm, S. (1994). Stereotypes of the elderly held by young, middle-aged, and elderly adults. *Journal of Gerontology: Psychological Sciences, 49*, 240–249.

Hummert, M. L., & Ryan, E. B. (1996). Toward understanding variations in patronizing talk addressed to older adults: Psycholinguistic features of care and control. *International Journal of Psycholinguistics, 12*, 149–170.

Hummert, M. L., & Ryan, E. B. (2001). Patronizing. In W. P. Robinson & H. Giles (Eds.), *The new handbook of language and social psychology* (pp. 253–270). Chichester: John Wiley.

Ingersoll-Dayton, B., & Saengtienchai, C. (1999). Respect for older persons in Asia: Stability and change. *International Journal of Aging and Human Development, 48*, 113–131.

Koyano, W. (1989). Japanese attitudes toward older persons: A review of research findings. *Journal of Cross-Cultural Gerontology, 4*, 335–345.

Lanceley, A. (1985). Use of controlling language in the rehabilitation of the elderly. *Journal of Advanced Nursing, 10,* 125–135.

Langer, E. J. (1989). *Mindfulness.* Reading, MA: Addison-Wesley.

Langer, E. J. (1992). Interpersonal mindfulness and language. *Communication Monographs, 59,* 324–327.

Langer, E. J., & Avorn, J. (1985). Impact of the psychosocial environment on the elderly on behavioral and health outcomes. In B. B. Hess & E. W. Markson (Eds.), *Growing old in America* (pp. 462–473). New Brunswick, NJ: Transaction Books.

Langer, E. J., Bashner, R. S., & Chanowitz, B. (1985). Decreasing prejudice by increasing discrimination. *Journal of Personality and Social Psychology, 49,* 113–120.

Langer, E. J., Chanowitz, B., Palmerino, M., Jacobs, S., Rhodes, M., & Thayer, P. (1990). Nonsequential development and aging. In C. N. Alexander & E. J. Langer (Eds.), *Higher stages of human development: Perspectives in adult growth* (pp. 114–136). New York: Oxford University Press.

Langer, E. J., Perlmuter, L., Chanowitz, B., &, Rubin, R. (1988). Two new applications of mindfulness theory: Alcoholism and aging. *Journal of Aging Studies, 2,* 289–299.

Langer, E. J., & Rodin, J. (1976). The effects of choice and enhanced personal responsibility for the aged: A field experiment in an institutional setting. *Journal of Personality and Social Psychology, 34,* 191–198.

Levy, B. (1996). Improving memory without awareness: Implicit self-stereotyping in old age. *Journal of Personality and Social Psychology, 71,* 1092–1107.

Levy, B., Ashman, O., & Dror, I. (2000). To be or not to be: The effects of aging stereotypes on the will to live. *Omega - Journal of Death & Dying, 40,* 409–420.

Levy, B., & Langer, E. J. (1994). Aging free from negative stereotypes: Successful memory in China and among the American deaf. *Journal of Personality & Social Psychology, 66,* 989–997.

Levy, B., & Langer, E. J. (1996). Reversing disability in old age. In P. M. Kato & T. Mann (Eds.), *Handbook of diversity issues in health psychology* (pp. 141–159). New York: Plenum Press.

Louw-Potgieter, J., & Giles, H. (1988). Imposed identity and linguistic strategies. *Journal of Language and Social Psychology, 7,* 261–286.

McNamara, M. (1999, Novemember 2). For Latinos, retirement years not always so golden. *Los Angeles Times,* pp. E1, E5.

Montepare, J. M., & Zebrowitz, L. A. (1998). Person perception comes of age: The salience and significance of age in social judgments. *Advances in Experimental Social Psychology, 30,* 93–161.

Montepare, J. M., & Zebrowitz, L. A. (2002). A social-developmental view of ageism. In T. D. Nelson, (Ed.) *Ageism: Stereotyping and prejudice against older persons* (pp. 77–125). Cambridge, MA: The MIT Press.

Mulac, A., & Giles, H. (1996). "You're only as old as you sound": Parameters of elderly age attributions. *Health Communication, 8,* 199–215.

Ng, S. H., Liu, J. H., Weatherall, A., & Loong, C. S. F. (1997). Younger adults' communication experiences and contact with elders and peers. *Human Communication Research, 24,* 82–108.

Noels, K. A., Giles, H., Cai, D., & Turay, L. (1999). Perceptions of inter- and intra-generational communication in the United States of America and the People's Republic of China: Implications for self-esteem and life satisfaction. *South Pacific Journal of Psychology, 10,* 120–135.

Noels, K. A., Giles, H., Gallois, C., & Ng, S. H. (2001). Intergenerational communication and psychological adjustment: A cross-cultural examination of Hong Kong and Australian adults. In M. L. Hummert & J. F. Nussbaum (Eds.), *Communication, aging, and health: Multidisciplinary perspectives* (pp. 249–278). Mahwah, NJ: Erlbaum.

Nussbaum, J. F. (1985). Successful aging: A communication model. *Communication Quarterly, 33,* 262–269.

Nussbaum, J. F., Hummert, M. L., Williams, A., & Harwood, J. (1996). Communication and older adults. In B. R. Burleson (Ed.), *Communication Yearbook 19* (pp. 1–47). Newbury Park, CA: Sage.

Ota, H. (2001). *Intergenerational communication in Japan and the United States: Debunking the myth of respect for older adults in contemporary Japan.* University of California, Santa Barbara: Ph.D. dissertation.

Paoletti, I. (1998). *Being an older woman: A study in the social production of identity.* Mahwah, NJ: Lawrence Erlbaum Associates.

Patterson, B. R., Bettini, L. A., & Nussbaum, J. F. (1992). The meaning of friendship across the lifespan: Two studies. *Communication Quarterly, 41,* 145–160.

Quintanilla, M. (1995, April 9). The human tornado. *Los Angeles Times*, pp. E. 1, E5.

Rawlins, W. K. (1995). Friendship in later life. In J. F. Nussbaum & J. Coupland (Eds.), *Handbook of communication and aging research* (pp. 227–258). Mahwah, NJ: Erlbaum.

Rodin, J., & Langer, E. J. (1977). Long-term effects of a control-relevant intervention with the institutionalized aged. *Journal of Personality & Social Psychology, 35*, 897–902.

Rodin, J., & Langer, E. J. (1980). Aging labels: The decline of control and the fall of self-esteem. *Journal of Social Issues, 36*, 12–29.

Ryan, E. B., & Butler, R. N. (1996). Communication, aging, and health: Toward understanding health provider relationships with older clients. *Health Communication, 8*, 191–198.

Ryan, E. B., & Cole, R. L. (1990). Evaluative perceptions of interpersonal communication with elders. In H. Giles, N. Coupland, & J. M. Wiemann (Eds.), *Communication, health and the elderly* (pp. 172–191). Manchester, England: Manchester University Press.

Ryan, E. B., Giles, H., Bartolucci, G., & Henwood, K. (1986). Psycholinguistic and social psychological components of communication by and with the elderly. *Language and Communication, 6*, 1–24.

Ryan, E. B, Hewstone, M., & Giles, H. (1984). Language and intergroup attitudes. In J. R. Eiser (Ed.), *Attitudinal Judgment* (pp. 135–160). New York: Springer Verlag.

Ryan, E. B., Hummert, M. L., & Boich, L., (1995). Communication predicaments of aging: Patronizing behavior toward older adults. *Journal of Language and Social Psychology, 13*, 144–166.

Ryan, E. B., Kennaley, D. E., Pratt, M. W., & Shumovich, M. A. (2000). Evaluations by staff, residents, and community seniors of patronizing speech in the nursing home: Impact of passive, assertive, or humorous responses. *Psychology and Aging, 15*, 272–285.

Ryan, E. B., Meredith, S. D., MacLean, M. J., & Orange, J. B. (1995). Changing the way we talk with elders: Promoting health using the communication enhancement model. *International Journal of Aging & Human Development, 41*, 89–107.

Ryff, C. D., & Seltzer, M. M. (1996). *The parental experience of midlife.* Chicago: University of Chicago Press.

Sachweh, S. (1998). Granny darling's napies: Secondary babytalk in German nursing homes. *Journal of Applied Communication Research, 26*, 52–65.

Shepard, C., Giles, H., & LePoire, B. (2001). Accommodation theory 25 years on. In W. P. Robinson & H. Giles (Eds.), *The new handbook of language and social psychology* (pp. 33–56). Chichester: Wiley.

Sidanius, J., & Pratto, F. (1999). *Social dominance : An intergroup theory of social hierarchy and oppression.* Cambridge, UK: Cambridge University Press.

Sontag, S. (1978). The double standard in ageing. In V. Carver & P. Liddiard (Eds.), *An aging population* (pp. 72–80). Milton Keynes, England: Open University Press.

Street, R. L., Jr. (1991). Accommodation in medical consultations. In H. Giles, N. Coupland &, J. Coupland (Eds.), *Contexts of accommodation: Developments in applied sociolinguistics* (pp. 131–156). Cambridge, UK: Cambridge University Press.

Tajfel, H. (Ed.) (1978). *Differentiation between social groups.* London: Academic Press.

Tajfel, H., & Turner, J. C. (1986). The social identity of intergroup behavior. In S. Worchel & W. G. Austin (Eds.), *Psychology of intergroup relations* (pp. 7–24). Chicago: Nelson.

Thimm, C., Rademacher, U., & Kruse, L. (1998). Age stereotypes and patronizing messages: Features of age-adapted speech in technical instructions to the elderly. *Journal of Applied Communication Research, 26*, 66–82.

Tien-Hyatt, J. L. (1987). Self-perceptions of aging across cultures: Myth or reality? *International Journal of Aging and Human Development, 24*, 129–148.

Tobin, J. J. (1987). The American idealization of old age in Japan. *The Gerontologist, 27*, 53–58.

Turner, J. C. (1987). *Rediscovering the social group: A self-categorization theory.* Oxford: Blackwell.

U. S. Bureau of the Census. (1993). *We the American elderly.* U.S. Department of Commerce: Economics and Statistics Administration.

Walker, B. L., Osgood, N. J., Richardson, J. P., & Ephross, P. H. (1998). Staff and elderly knowledge and attitudes toward elderly sexuality. *Educational Gerontology, 24*, 471–489.

Whitbourne, S., & Wills, K. J. (1993). Psychological issues in institutional care of the aged. In S. Goldsmith (Ed.), *Long term care administration handbook* (pp. 19–32). Gaithersburg, MD: Aspen Press.

Williams, A., & Coupland, N. (1998). Epilogue: The socio-political framing of communication and aging research. *Journal of Applied Communication Research, 26,* 139–154.

Williams, A., & Giles, H. (1996). Intergenerational conversations: Young adults' retrospective accounts. *Human Communication Research, 23,* 220–250.

Williams, A., & Giles, H. (1998). Communication of ageism. In M. L. Hecht (Ed.), *Communicating prejudice* (pp. 136–160). Thousand Oaks: Sage.

Williams, A., & Nussbaum, J. F. (2001). *Intergenerational communication across the lifespan.* Mahwah, NJ: Lawrence Erlbaum.

Williams, A., & Ylänne-McEwen, V. (2000). Elderly lifestyles in the 21st century: "Doris and Sid's excellent adventure". *Journal of Communication, 50,* 4–8.

Ytsma, J., & Giles, H. (1997). Reactions to patronizing talk: Some Dutch data. *Journal of Sociolinguistics, 1,* 259–268.

Zebrowitz, L. A., & Montepare, J. M. (2000). "Too old, too young": Stigmatizing adolescents and elders. In T. Heatherton, R. Kleck, & J. H. Hull (Eds.), *The social psychology of stigma* (pp. 334–373). New York: Guildford.

Cultural Issues in Communication and Aging

Loretta L. Pecchioni
Louisiana State University

Hiroshi Ota
Aichi Shukutoku University

Lisa Sparks
George Mason University

Culture is so pervasive that we are generally unaware of the role it plays in defining our basic assumptions about the nature of human nature, the nature of reality, and our relationships to one another and the environment (Gudykunst, 1998; Hofstede, 1984; Schein, 1985). Language is defined by, reflects, and transmits culture. Our personal and social identities (Tajfel & Turner, 1986; Turner et al., 1987) are formed through interaction with others (Coupland & Nussbaum, 1993; Gergen, 1991; Harré, 1983; Shotter, 1993). These interactions most commonly occur with other individuals who are members of our culture, but sometimes they occur with those who are from outside of our culture. This chapter begins with a discussion of the role that culture plays in communication and aging, from its influences on language to our definitions of self through our social identities as they are acquired across the life span. Next, critical, overarching issues that place the research findings into context are addressed. Then, connections are drawn to the other chapters and sections of this book through a comprehensive review of the pertinent intercultural and cross-cultural literature on communication and aging. The chapter concludes with a discussion of the theoretical, research, and pragmatic concerns that arise during the examination of culture, communication, and aging.

THE ROLE OF CULTURE IN SOCIAL LIFE

Culture can be defined as "a set of fundamental ideas, practices, and experiences of a group of people that are symbolically transmitted generation to generation through a learning process" (Chen & Starosta, 1998, p. 25). Therefore, culture is the collective way in which a group of people share and interpret their experiences of the world. Culture is reflected in the values, beliefs, norms, rules, communicative behaviors, social institutions, and media channels of a group. These underlying values drive people's thinking, reacting, and behaving (Ting-Toomey, 1999). The pervasiveness of culture makes it difficult to answer the question, What is culture? Examining the functions of culture, however, makes it easier to "see" culture. Ting-Toomey (1999) identified five of these functions: (1) identity meaning, (2) group inclusion, (3) intergroup boundary regulation, (4) ecological adaptation, and (5) cultural communication. Each of these functions plays an important role in understanding the social processes of aging and the consequences for communication.

Our discussion will begin with the last of these functions, cultural communication—which is the coordination of communication and culture—and is the mechanism through which the other functions are achieved. Culture and communication are inextricably intertwined, each influencing the other (Chen & Starosta, 1998; Gudykunst, 1998; Markus & Kitayama, 1994). The symbols and messages we exchange and the meanings assigned to them reflect our culture, but they also change our culture (Giddens, 1984; Gudykunst, 1998; Williams & Nussbaum, 2001). Cultural norms are transmitted from one generation to the next through communication. Based on the beliefs and values of our culture, we learn not only what are appropriate interaction scripts within our culture, but also the meanings that should be assigned to these interactions (Ting-Toomey, 1999). For example, we learn how to greet strangers, how older individuals are (dis)valued, and our own role within society. As the next generation learns the expectations of their culture, however, changing environmental factors may lead to changes within the culture—the ecological adaptation function of culture.

Therefore, as cultures undergo transformation, definitions of "old age," for example, change throughout historical times, particularly as life expectancies have increased dramatically over the past 100 years. In addition, cultural boundaries may become blurred as technology leads to greater interaction between previously isolated cultures, leading to changing beliefs and values within a culture. The "Westernization" or industrialization of many cultures through the demands of the global economy is a prime example of the blurring of cultural boundaries. Ecological adaptation applies, however, not only at the level of the larger society, but at the individual level as well. Individual changes may occur based on normative life transitions, for example, taking on the role of "married person," "parent," or "widow." Cultural definitions of these roles help us to know what is acceptable behavior within these roles.

These individual changes are connected to the first function of culture, previously listed, that of identity meaning, helping us to answer the question, Who am I? Our culture defines the roles (see Stryker, 1987; Thois, 1991, for roles and identities) we can expect to take up in our lives and how to competently perform those roles and the meanings attached to them (Ting-Toomey, 1999). As a result, there is social pressure to conform to expectations about being a member of a certain group (such as being an older person). Although we may have the ability to disobey these expectations, there are consequences for such violations (Burgoon, 1995; De Ridder & Tripathi, 1992). Some roles carry automatic age identifiers, such as "child," "adolescent," "adult," "midlifer," and "elder." Other roles may be age-related but carry broader age connotations. For example, fulfilling the role of "parent" suggests a minimum age at which fertility and engaging in certain sexual activity makes becoming a parent possible, but the term *parent* may apply to someone 14 or 84 years old.

The function of group inclusion allows us to identify to which groups we belong, that is, to which others we are similar, satisfying our need for affiliation and belonging (Ting-Toomey, 1999). Identifying with people who are similar to us may help us to reinforce our self-esteem by seeing ourselves as normal as defined by those around us who belong to our group. Interacting with others from our same group is comfortable and safe because we know what to expect and do not have to explain or justify our behaviors (e.g., Gudykunst, 1995). Interacting with individuals from dissimilar groups may provoke anxiety or uncertainty as we become aware that we not only speak different languages but have different interpretations of behaviors and feel the need to explain ourselves (Gudykunst, 1995; Ting-Toomey, 1999). These groups may be at the level of national identity—"I'm American" or "I'm Japanese"—or on a range of smaller levels—"I'm a college graduate," "I'm a teacher," or "I'm a grandparent." Age may serve as one of these group categorizations (Coupland, Coupland, & Giles, 1991; Coupland, Coupland, Giles, & Henwood, 1988; Gudykunst, 1998; Harwood, Giles, & Ryan, 1995). An important point about age as a group category is that people move through a series of group memberships, a process similar to intercultural adaptation (Giles, Fox, Harwood, & Williams, 1994), as opposed to most group categorizations in which one is or is not, fairly permanently, a member—being male or female, being a native of a particular country, and so forth. An individual, however, belongs to any number of groups, and interactions with others may be influenced by not only which group memberships they hold in common, but which groups they do not share. For example, a physician practicing medicine in the United States may see a variety of patients who are also U.S. citizens but who may come from different ethnic backgrounds, have different education levels, or be from a different age group. Which of these similarities and which of these differences will become salient during the interaction may affect the satisfaction each participant has with that interaction.

Membership in a number of groups leads to the function of managing group boundaries (Ting-Toomey, 1999). When we interact with an other, we identify—usually subconsciously—our similarities and differences (see also Tajfel & Turner, 1986). Our culture and group memberships shape our attitudes toward people of dissimilar groups (Allport, 1954; Ting-Toomey, 1999). Because we share a common language and interpretation of meanings, in-group interactions are easy, comfortable, and predictable. Interacting with people from other groups, however, may lead to strong emotional reactions as our expectations, based on cultural norms, are violated (Burgoon, 1995). Individuals tend to think that their own group behaviors are superior and that out-group behaviors are inferior (e.g., ingroup bias, Brewer, 1979), resulting in a reaction of "It's just not right!"

As a consequence of these assumptions about group members, when individuals from different cultural groups communicate, they tend to operate on differing expectations for the interaction. Interactions between such individuals are labeled *intercultural communication* (Ting-Toomey, 1999). Because all cultures do not have the same underlying beliefs and values, we can compare the expectations of cultures—labeled *cross-cultural communication* (Ting-Toomey, 1999). An example will help to differentiate intercultural from cross-cultural communication. A physician from a large urban area in the United

States may be seeing a patient who has immigrated to the United States from a rural area in Colombia. Intercultural aspects of their interaction will include language and status differences and may include age, gender, and socioeconomic differences as well. Cross-cultural aspects of their interaction may be revealed as they discover that each participant has different concepts about the causes and nature of illness and different definitions of and values for aging.

In addition to identifying membership in a culture, it should be pointed out that cultures are not homogeneous (Gudykunst, 1998). For example, the United States is said to have an "American" culture that focuses on individualism, direct communication, and to assume a low power distance between individuals (Hofstede, 1991; Triandis, 1995). Not everyone living in the United States, however, adheres to this dominant style. Subcultures are groups of individuals within a culture whose members share many values with the dominant culture, but also have some values that differ from the larger culture (Gudykunst, 1998). The use of the term *subculture* implies a hierarchy, with some cultures being considered more powerful or at least having more members. The term *coculture* is used to give equal weight to all cultures regardless of the size or power of each group (Al-Deen, 1997; Orbe, 1998). With this basic discussion of the role of culture in social life and definition of terms used in this chapter, we turn to overarching issues before reviewing the literature on communication, culture, and aging.

ISSUES

We understand the pervasive role of culture in our communicative lives. The five key functional features of culture previously discussed draw our attention to, among other things, meanings of age and their impact on self-concept, people's interactional stance and behavior in communication, and their relationship to an overall culture. Based on these assumptions, four primary issues help to place the literature on communication, aging, and culture into context: (a) chronological age is a poor predictor of aging-related behaviors; (b) definitions of and meanings assigned to the aging process vary culturally; (c) age may be treated as a coculture in its own right; and, (d) the increasing percentage of older individuals within the population is likely to impact culture itself. Each of these issues will be discussed in turn.

Chronological Age

When we discuss aging, it is important to point out that chronological age is a poor predictor of behavior. Communication and aging research, however, is commonly char-acterized via chronological age with little regard for distinctions between the biological and social processes involved in social group membership (Williams & Coupland, 1998). In Western contexts, three broad categories of age segregation and appropriate life span normative roles are defined as follows: youth—the period of education; middle age—the period of employment and related productivity; and elderly—the years of retirement (Cole, 1989; Peterson, 2000). As a result of these rather broad age categories, the mean-ings assigned to each stage of life vary. Robertson (1996) argued that the early stages of life focus on the acquisition of knowledge and goods, the middle stages are concerned with making sense of our social experience (in effect, bridging the younger and older generations through deeper understanding of action and interaction), and the later stages with the formal definition of social life and the transference of valuable knowledge and goods to the following generations. These meanings are socially constructed and only partially driven by biology as the individual ages (Robertson, 1996). In addition, the adult years may be further segmented in young adulthood, middle age, and old age; however, definitions of what chronological ages fall into each of these categories vary greatly and change as individuals themselves move through these categories (Goldman & Goldman, 1981; Williams & Garrett, 2001). In spite of the ambiguities that exist around the meaning of age categories, age-based categories, such as retirement ages, social security benefits, and birthdays permeate social life, yet the defining criteria for entering assisted living facilities and geriatric medical clinics is dependency and health status rather than age, and age and health are barely correlated (Giles, Williams, & Coupland, 1990).

Baltes and Baltes (1990) provided convincing evidence for the great variability in aging physically, cognitively, and socially across the life span. Nussbaum and Baringer (2000) concurred that considerable variability and differences exist among older adults, solidifying the notion that we do not age exactly alike, chronologically or otherwise. As a consequence of this variability, other ways of conceptualizing age other than chronological years are needed. Three age-related constructs have been used in the research: age by experience, contextual age, and age identity. Life span developmental psychologists argue that wisdom is developed as individuals have the opportunity to acquire useful experiences as they age and that some individuals may be exposed to more such experiences that facilitate their development of knowledge and skills. Therefore, "age by experience" is a more appropriate measure than is chronological age (Baltes, 1993; Ericsson & Smith, 1991; Salthouse, 1991).

Taking the argument seriously that chronological age is not reflective of behavior, Rubin and Rubin (1981, 1982a, 1982b, 1986) developed the concept of "contextual age." As operationalized by the Rubins (1981, 1982a, 1982b, 1986), contextual age reflects six dimensions of aging that affect the use of mass-mediated and interpersonal communication: interpersonal interaction, social activity, mobility, life satisfaction, health, and economic security. Although these dimensions have proven to be reliable and useful in communication research, these dimensions were developed based primarily on European American participants in the United States. Whether these same dimensions apply to other ethnic groups within the United States or to other cultures requires further research.

The third concept often used in research is age identity (Garstka, Branscombe, & Hummert, 1997; Ota, Harwood, Williams, & Takai, 2000). Age identity is one type of social identity (Tajfel & Turner, 1986) and stems from one's subjective definition of the self as a member of an age group. This identity is constructed based on one's social interaction with others and society (Ryan, Giles, Bartolucci, & Henwood, 1986; Ylänne-McEwen, 1999). One's view of the self is highly relevant to his or her interpretation of a communicative message (Caporael, Lukaszewski, & Culbertson, 1983; Ryan & Cole, 1990). Moreover, age identity mediates one's intergenerational communication experience (Harwood & Williams, 1998). Given these findings, assessment of one's sense of age identification is deemed a viable alternative for the use of chronological age in intergenerational communication research.

Cultural Variation in the Aging Process

Not only does chronological age serve as a poor predictor of aging-related behaviors, but different cultures may define the aging process quite differently. Torres (1999) argued that any construct developed within one cultural setting cannot be assumed to apply to other cultures and inherently reflects assumptions of the culture in which it was developed (see also Berry, Poortinga, Segall, & Dasen, 1992; Long, 2000). Two such concepts that are used frequently in the gerontological literature are "successful aging" and "quality of life."

Research on successful aging has revealed deep cultural differences in the definition of "success," yet the notion of success is a very American one (Torres, 1999). In the United States, success is defined as reaching one's potential, being productive, achieving individual accomplishment, and, for the elderly, exhibiting behaviors that resemble those of young people. According to Torres (1999), this particular definition reflects a master orientation of Kluckhohn's (1950; Kluckhohn & Strodtbeck, 1961) man–nature value. However, older individuals' understanding of successful aging is contingent on their cultural origins. As the Project AGE studies (Keith, Fry, & Ikels, 1990) revealed, older individuals in the United States define successful aging according to standards of self-sufficiency and the ability to live alone; however, older individuals in Hong Kong asked why anyone would want to be self-sufficient and measured successful aging based on their families' willingness to meet their needs. As these studies revealed, successful aging is often equated with functionality, and definitions of functionality vary greatly across cultures (C. L. Fry, 2000). Climbing stairs is not a critical component in cultures where the homes do not have stairs; lifting water is more important when tending to livestock in areas that do not have pumps to help move water; and managing money is only important in cultures that have accumulated wealth in cash terms (C. L. Fry, 2000). In addition, the same activity may have very different meaning in different cultures. Individuals in Africa, Asia, Europe, and North America all indicated walking was an important element of functionality. However, in rural Africa this entailed being able to walk to the water hole or several kilometers to visit with family, whereas walking in more industrialized areas reflected an ability to maintain independence (C. L. Fry, 2000).

Quality of life (QOL) is usually conceptualized as consisting of a number of dimensions, including general life satisfaction, health-related evaluations (either subjective, objective, or both) of physical and cognitive abilities, and social phenomena, such as social support, religion, and interpersonal relationships (P. S. Fry, 2000). P. S. Fry (2000) argued, however, that because not all measures of QOL include all of these dimensions, some measures reveal little about aging among different cultural groups. For example, Adams and Jackson (2000) found that among African Americans, social contact and family support along with hope are the most important factors in determining their evaluations of personal efficacy. Asakawa, Koyano, Ando, and Shibata (2000) found that Japanese elders with mild functional declines are more likely than elders in the United States to select coresidence with family as an acceptable solution, and those individuals with reduced social contact were more likely to be depressed. Berdes and Zych (2000) compared QOL among Poles living in Poland, Polish immigrants, and Polish Americans. They found that Polish Americans had the highest QOL, followed by Polish immigrants, with Poles in Poland reporting the lowest QOL. These authors suggested that this finding reflects the U.S. construct of "vital aging" to which Polish Americans have been exposed all their lives, Polish immigrants for some shorter period of time, and Poles not at all. These findings, however, might also be explained by the measures reflecting a particularly American version of QOL that is not appropriate within Poland. P. S. Fry (2000) argued that measures of QOL should not only include the widest range of dimensions identified to date, but should also ask participants to indicate the intensity and relative importance of each dimension in their own definitions in determining the quality of their lives.

Age as a Coculture

Age may be viewed as a separate culture—Generation X vs. the elderly—or as a process that occurs within a culture (Coupland & Nussbaum, 1993; Giles, Fox, Harwood, & Williams, 1994). An age cohort may be seen as having a culture itself that has distinctive experiences, values, norms, and use of language. Although none of us age and join successive age group cultures in exactly the same manner, individuals often assess others to determine if the other is an age peer or not and indicate, verbally or nonverbally, their age group membership (Coupland, Coupland, & Giles, 1989; Coupland et al., 1991; Giles & Coupland, 1991). This communicative process is akin to one that is often observed at an initial encounter between people from different countries (Gudykunst, 1995). Therefore, we should ask to what extent a person's age influences how she or he responds to or sends messages. Williams and Nussbaum (2001) argued that communicative exchanges across the life span and into older adulthood become increasingly complex as older individuals have a greater range of experiences to draw on while also needing to compensate for declining abilities. Intergenerational conversations, thus, are likely to require unique communicative approaches that account for distinctions between age segments of the population. However, even when age serves as a membership category, individuals are operating within a system of overlapping cultural systems (Hill, Long, & Cupach, 1997). The experience of being an older individual in the United States may be quite different for an urban European American male compared to someone who is a rural First Nations female. The interpenetration and salience of an individual's cultures has implications for the aging process (Hill, Long, & Cupach, 1997).

Older Population's Impact on Culture

The increase in life expectancies and the growing percentage of older individuals in society may influence the nature of cultures as much as culture influences the individuals experiencing the aging process. In the past 125 years, life expectancies have doubled from 40 to 80 years in industrialized countries, changing the definitions of old age (Gibbons, 1990). In fact, the number of older persons is growing so rapidly in these countries that old age itself is being segmented into the young-old (65–75 years) and the old-old (85 years and older). Although our potential life span has probably remained relatively stable, the development of complex social systems, particularly those leading to medical advances, has increased the number of people who can expect to live to their potential oldest age (Robertson, 1996). Thus, the aging of the population is a reality, especially in the industrialized world (Restrepo & Rozental, 1994). Restrepo and Rozental (1994) argued that this new reality will have an impact on social policy as programs are developed to meet the needs of a growing number of older people, especially taking into consideration the number of older women who will be living in urban centers and, most likely, in poverty. The ability of less industrialized countries, such as those in the Caribbean, Latin America, and Asia, to meet these demands may be particularly taxed as the dependency ratio increases and programs must be developed that attempt to agree with the social values and relationship dynamics within each society (Martin, 1988; Phillips, 2000; Restrepo & Rozental, 1994).

These demographic shifts are likely to impact the larger culture as well (Peterson, 2000). Historically, all cultures have developed with a larger percentage of the population being young. Peterson (2000) wondered if an aging society will become less innovative, will avoid risk, and hold to more conservative values, much as older individuals shift in their values as they age. Robertson (1996), attempting to reconcile culture and biology, contemplated the nature of a population with increased life expectancies, particularly surviving past the point of reproductive capability. Darwin's theory of survival of the fittest does not account for survival past the point when one's genes can be passed on to the next generation. Robertson (1996) argued that such a population would be less concerned with sexuality and reproduction and more concerned with somatic survival and consumption. Such a shift might also lead to a shift in what is valued within a culture as well. Young men may no longer battle for supremacy (and the related rights to reproduction), but individuals of all ages may be more concerned with alliance building that benefits all ages.

Along with the changing configuration of the population, examining the strength and status of age groups in society, expressed in age group vitality, helps us to understand the impact of age groups on culture (Giles et al., 2000; Harwood, Giles, & Bourhis, 1994; Harwood, Giles, Pierson, Clément, & Fox, 1994; Ota, Giles, & Gallois, 2002). Age group vitality identifies the salience, strength, and status of young, middle-aged, and older adults in various facets of life, including politics, economics, and education, and thus is considered a barometer of the relative impact that a particular age group has on an overall culture. For instance, middle-aged adults are perceived to be the backbone of society, although older persons' roles in society seem to be changing in a number of countries (Giles et al., 2000). As younger adults gain more vitality, they dominate older people in their cultural influence. Older people's dwindling status and role in society has been pointed out in many Asian countries, and this may be reflected in the political discourse and voices that cry for the need for social policies that aid older persons.

Culture is socially constructed and transmitted; meanings are fluid and in constant flux. The fact that a larger percentage of society is older cannot be denied and how that fact will impact cultural values and their transmission is yet to be seen, but may be indicated in many facets of our social lives, as we see in the subsequent sections of this chapter. Critical analyses of the current political discourse indicate that older people make sense of their aging experiences by examining normalizing processes (Tulle-Winton, 1999).

With a larger number of older people, more models of aging may be available for such comparison.

REVIEW OF LITERATURE

In this section, the review of relevant literature on communication, aging, and culture is organized to reflect the sections of this text. Treatment of the sections is unequal, as some areas have received more attention than others. Within each section, the research on general issues are addressed first, followed by cross-cultural comparisons, and then the consequences of intercultural interactions are discussed.

Cultural Influences on Experiences of Aging

As discussed in the introduction to this chapter, culture plays an important role in defining the self (the identity function of culture) and the meaning of and attitudes toward aging, as well as the roles that are considered to be appropriate (the group inclusion and intergroup-boundary regulation functions). To provide an adequate background for the subsequent discussion of national culture and intergenerational communication, it is important to review how cultural differences have been treated, particularly, in communication and related fields. Indeed, culture has garnered scholarly attention as a factor that mediates people's psychological, sociological, and communicative experiences (Gudykunst, 1983a; Kim & Gudykunst, 1988; Kim & Yamaguchi, 1995; Smith & Bond, 1999; Wiseman, 1995), which includes, of course, the aging process. Among other elements researchers have paid special attention to cultural differences manifested in philosophical and moral orientations (Kim & Yamaguchi, 1995), values (Hofstede, 1991; Kluckhohn & Strodtbeck, 1961; Schwartz & Bilsky, 1990), and views of self (Kanagawa, Cross, & Markus, 2001; Markus & Kitayama, 1991, 1994; see also Gudykunst & Ting-Toomey, 1988; Triandis, 1995). Within this section, we present a brief overview of those cultural differences along these dimensions from an East–West split perspective so as to bring cultural variabilities in communication and aging into sharp relief. This perspective is adopted herein to highlight the differences in philosophical orientations across cultures and their impact on people's intergenerational communication.

Eastern Cultures. Many Eastern cultures (e.g., the People's Republic of China, Japan, Thailand) have been built around Confucianism, Daoism, and Buddhism as moral, ethical, and philosophical principles of life (Chang, 1997; Ho, 1994; Kim & Yamaguchi, 1995; Slote & De Vos, 1998; Yum, 1988). By way of formal and informal education, people are socialized to act toward others with human-heartedness and other-orientation to achieve group, or collective, goals in harmony with others in the same group (Kim & Yamaguchi, 1995; Markus & Kitayama, 1994; Triandis, 1995). Endorsement of group-focused collectivistic values, with family as the center (Kim, 1994; Triandis, 1995), is cultivated. People's self-view is predominantly interdependent rather than independent (Markus & Kitayama, 1991, but see Matsumoto 1999; Takano & Osaka, 1999; for variation within Asia, see Matsudo & Takata, 2000). People develop a strong self-awareness

(Kanagawa, Cross, & Markus, 2001) focusing on a sensitivity to relational expectations (i.e., situational norms; Triandis, 1995). Moreover, reciprocity is strongly expected in many domains of life over a long period of time (Kim & Yamaguchi, 1995; Morisaki & Gudykunst, 1994). Power inequality based on status and age differences is granted as legitimate (Hofstede, 1991), and people pay attention to others' status characteristics in social interaction (Gudykunst, Gao, Nishida, Nadamitsu, & Sokai, 1992). Fulfilling duties associated with one's role and status, illustrated in younger people's obedience to older people and respect for them is deemed highly important (Ho, 1994; Kim & Yamaguchi, 1995; Phillips, 2000; Sung, 1995).

Against this background, people in Eastern nations tend to use high context and indirect communication styles (Hall, 1976; Holtgraves, 1997). Explicit verbal expressions of one's meanings are not considered desirable, because of a high concern for the other's "face" (Kim, 1995; Ting-Toomey, 1988; Ting-Toomey & Kurogi, 1998) and, relatedly, the importance of harmony with others (Hofstede, 1991). However, people are expected to grasp the meanings others express in communication, even without explicit verbal expressions (Gudykunst, Matsumoto, Ting-Toomey, Nishida, & Heyman, 1996; Holtgraves, 1997). As observed in their group orientation, people tend to rely on, and therefore ask for, information about others' group membership to reduce uncertainty in communication (Gudykunst, 1983b). In addition, a strong in-group orientation (Triandis, 1995) makes people distinguish communicatively between their in-group and out-groups (Gudykunst, Yoon, & Nishida, 1987). Use of the distinctive polite register underscores communication with people of out-groups, especially those in different age and status groups (Barnlund & Yoshioka, 1990; Gao & Ting-Toomey, 1998; Holtgraves, 1997; Ide, 1987; Matsumoto, 1988).

Western Cultures.   By contrast, a major philosophical principle strongly held in Western countries (e.g., the United States, Canada, Australia, Germany) is liberalism (Kim & Yamaguchi, 1995). The hallmark of liberalism is an encouragement toward the development of an autonomous individual who is capable of making rational choices of his or her own (Kim & Yamaguchi, 1995; Markus & Kitayama, 1991). Individuals are expected to defend their own rights but respect those of others. They are less likely to be influenced by the situation in defining themselves than are their counterparts in Eastern cultures (Kanagawa, Cross, & Markus, 2001). Personal rights, rather than role-based duties (Kim & Yamaguchi, 1995; Moghaddam, Slocum, Norman, Mor, & Harre, 2000), and personal intentions, rather than social norms (Kashima, Siegal, Tanaka, & Kashima, 1992), tend to be driving forces of individuals' social behaviors. To prove that people can actually do something is of higher importance in Western cultures than is who they are (Kluckhohn & Strodtbeck, 1961). Distribution of power based simply on social categories, such as age and role, is less likely to be endorsed in Western than in Eastern cultures (Hofstede, 1991; Kim & Yamaguchi, 1995).

People in Western cultures, relative to their Eastern counterparts, are likely to be more direct in communication (Hall, 1976; Kim, 1993; Yum, 1988). They tend to explicate their meanings in verbal utterances for the sake of message clarity (Hall, 1976; Kim, 1993) and achievement of tasks (Cocroft & Ting-Toomey, 1994; Kim, 1995). In-group versus out-group boundaries are relatively soft in comparison to those in Asian cultures, and thus people do not communicatively differentiate between those who belong to their in-group as opposed to those who do not (Gudykunst et al., 1987). Status and age differences are less likely to require the use of distinctive linguistic markers in communication than in Eastern cultures (Gudykunst & Ting-Toomey, 1988), although various "face work" is used instead to manage interpersonal relationships (Brown & Levinson, 1987; Cocroft & Ting-Toomey, 1994).

Consequences of Culture for the Experience of Aging.   Culture, then, obviously plays a role in defining the aging experience and providing meaning for that experience. As pointed out in the discussion of overarching issues, individuals in different cultures may define basic concepts of aging quite differently. For example, individuals in Eastern cultures are more concerned with in-group relationships, whereas individuals in Western cultures are more concerned with self-sufficiency (Keith, Fry, & IKels, 1990). As a consequence, definitions of successful aging and quality of life vary across cultures (C. L. Fry, 2000; P. S. Fry, 2000 Torres, 1999). Individuals with similar life experiences and ages, but living in different cultures, may report their satisfaction with their circumstances quite differently. A common phenomenon across cultures is the telling of one's life story to make sense of one's experiences; however, different cultures provide different templates for those narratives (Bruner, 1999). Although most cultures have narrative templates that denigrate aging and narratives that enable or honor aging, which genre is selected by a particular individual at a given time depends on that individual's social position and life experiences (Bruner, 1999). Therefore, how an individual makes sense of his or her own aging is influenced both by his or her own culture and personal experiences. Narratives serve as a method of impression management, used not only to present a selected identity to an external audience, but to present that identity internally to the individual as well (Biggs, 1997). Biggs (1997) argued that in the postmodern world, individuals must manage a difficult mind–body split in which the individual does not feel old but sees an old face in the mirror (see Hepworth, chap. 1 this volume for further discussion of this idea). Thompson (1992) identified this as the "I don't feel old" phenomenon, in which the aging individual must make sense of increasingly disparate views of the self and the body. This phenomenon was reflected in interviews with 80-year-old Finns who did not define themselves as old and did not see themselves as living an old-age existence (Heikkinen, 1993). The role of this postmodern angst in non-Western cultures is unclear but deserves attention with the onset of globalization.

Culture, Language and Social Aging

The discourse of aging is culturally bound (Coupland, Coupland, & Giles 1991; Gudykunst, 1998). Whether our particular age-based identity is accepted depends on the values underlying our cultural expectations (identity function of culture). Our attitudes, beliefs, and stereotypes of aging are also culturally bound and reinforced (the group inclusion and intergroup boundary-management functions of culture, Gudykunst, 1998; Ting-Toomey, 1999). Statements regarding what is appropriate "for someone my age" depend on cultural assumptions of role appropriateness. In addition, culture may indicate how salient age is as a factor contributing to group identification. Cultures that are more aware of the number of years lived may result in age being more salient in interactions, whereas a culture in which age is less important may not have as clear definitions of age-based group memberships (Triandis, 1994). With age, we have increased experiences and abilities to draw on as we face interactional challenges. Being able to bring a wider range of skills and knowledge to interactions may lead to more successfully achieving multiple goals in those interactions. In addition, individuals face differing life events and levels of health, which may in turn further affect their ability to understand and to be understood. Moreover, many miscommunication and sociolinguistic problems have been found to stem from group or cultural differences (Coupland, Wiemann, & Giles, 1991). One area that has received considerable research attention is the nature of intergenerational communication.

Age and Intergenerational Communication.   As mentioned in the previous section, age is an important social marker in all countries. However, it is important in different ways in different countries, carrying different attitudes and connotations. Being old is often associated with stigma in Western cultures (Goffman, 1963), and thus is seen in a rather negative light. Indeed, empirical evidence shows that younger adults make rather negative evaluations of older adults, especially along the physical dimension (Kite & Johnson, 1988; but see Brewer, Dull, & Lui, 1981; and Hummert, Shaner, & Garstka, 1995, for positive attitudes toward older adults). Such negative evaluations can mediate the social judgments people make about incidents that involve older adults (Giles, Henwood, Coupland, Harriman, & Coupland, 1992) and lead to the use of disrespectful and demeaning talk to older adults, referred to as patronizing talk (Caporael, 1981; Giles, Fox, & Smith, 1993; Harwood, Ryan, Giles, & Tysoski, 1997; Hummert & Ryan, 1996) and "elderspeak" (Kemper, 1994). Moreover, younger adults report negative experiences of communicating with older adults and believe intergenerational communication to be difficult, replete with stereotypes and "miscommunication" (Coupland, Wiemann, & Giles, 1991; Giles, Coupland, & Wiemann, 1992; Williams & Giles, 1996; Williams & Nussbaum, 2001). Relatedly, older people are likely to be discriminated against in job situations, because of their age (Braithwaite, Lynd-Stevenson, & Pigram, 1993; Finkelstein & Burke, 1998; McCann & Giles, 2002). To older adults, feelings of alienation (Russell & Schofield, 1999) and lowered sense of self-esteem (Ryan et al., 1986) are plausible outcomes of such negative treatment.

These attitudes toward older adults begin to be developed in childhood. Seefeldt and Ahn (1990) suggested that culture plays a role in children's attitudes. More specifically, they found that children of Korean heritage in the United States rated older adults most positively overall; children in Korea and Korean American children rated older adults as right, good, and friendly compared to European American children; and European American and Korean American children reported older adults as healthier and cleaner than did children in Korea (Seefeldt & Ahn, 1990). Goldman and Goldman (1981) reported that children (5–15 years old) in Western nations tend to view older adults more negatively than positively. In contrast, Nakano (1991) found that children (10–12 years old) had overall positive images of older adults in Japan. Zandi, Mirle, and Jarvis (1990) found that children in the United States of European and Asian Indian descent had overall positive attitudes toward older adults, but that children with Indian heritage focused on their specific behaviors, whereas children with European heritage focused on their feelings towards the group; reflecting cultural values of expressing feelings through actions and words, respectively.

In spite of traditional values that carry socially positive meanings about aging (see Ho, 1994; Keifer, 1990; Sung, 1995), negative attitudes occur in Eastern nations as well as Western ones, and thus put the traditional notion of "respect for the aged" into a precarious status. The indications of stigmatization and peripherization of older adults and their dissociation from social activities have been documented in studies conducted in Eastern cultures (Chow, 1983; Giles et al., 2000; Harwood et al., 1994, 1996; Nakazato, 1990, but see Knodel, Chayovan, Graisurapong, & Suraratdecha, 2000). Older adults' social status and power used to be perceived as the highest among the three age groups (e.g., Keifer, 1990), but they are now at a precarious stage (Giles et al., 2000). Because of that perception, older individuals are more likely to receive public attention as being in need of care and "welfare" provisions from the government, as well as the family in many countries (Bengtson & Putney, 2000; Koyano, 1999; Martin, 1988; Phillips, 2000). True, family continues to be the main source of care and social support across many Asian nations (Phillips, 2000), but the dwindling size of family (e.g., one-child policy in the People's Republic of China) and migration (Hugo, 2000) have made continuation of the traditional family care of older adults difficult. Part of care responsibilities has been transferred to the formal service system (Koyano, 1999; Phillips, 2000; Sung, 1995), which invites some public concern against the adoption of such a Western model (e.g., Martin, 1988). The perception that older adults are a "burden" (Hugo, 2000) and are stigmatized (Goffman, 1963) may be quite valid, even in Eastern cultures. The current cohort of older people is sometimes referred to as "an interim generation" (Phillips, 2000, p. 13) that is "caught between a variety of forces in society, economy and environment" (Phillips, 2000, p. 13). Uncertainty, rather than certainty, may permeate their lives, coming from minimal state provisions, insufficient family care, and unfriendly local environments.

A large-scale cross-cultural investigation was conducted in a number of countries around the Pacific Rim nations with the communication predicament of aging model (CPA; Ryan et al., 1986) and communication accommodation theory (CAT; Coupland, Coupland, Giles, & Henwood, 1988; Gallois, Giles, Jones, Cargile, & Ota, 1995) as an overall framework (see also Giles, McCann, Ota, & Noels, 2002). Younger adults from Western (the United States, Canada, Australia, New Zealand) and Eastern (Japan, Hong Kong, the People's Republic of China, Korea, the Philippines) nations contributed data to this research. Although younger adults do not see becoming older solely in a negative light across cultures (e.g., Giles, Fortman, Honeycutt, & Ota, 2003), advancing age can be a fear- and anxiety-provoking experience (Lynch, 2000; Shimonaka & Nakazato, 1980). First, younger adults in Eastern (Korea, Hong Kong, the Philippines) countries reported less positive stereotypes of older adults than did their Western counterparts, especially along personality characteristics named benevolence (generosity, kindness, wisdom; Giles, Harwood, Pierson, Clément, & Fox, 1998; Harwood et al., 1996). This tendency for less positive evaluation of older adults in Eastern than in Western countries was also observed in a number of other studies conducted in Japan (Ota et al., 2002), Korea (Kim, 2000), and Thailand (Sharps, Price-Sharps, & Hanson, 1998).

Older adults in Eastern countries sense that the traditional notion of respect (e.g., younger people's obedience) is changing these days (Ingersoll-Dayton & Saengtienchai, 1999; Mehta, 1994; see Kauh, 1997, for Korean immigrants to the United States): Nowadays, politeness and courteous behavior to older adults is a more salient feature of filial piety than is obedience. Respect is something that older adults have to earn, and not what they are accorded automatically. In contrast, Eastern younger adults appear to have rather complicated views about the provision of care and respect for older adults, which may illustrate their difficult position in this changing milieu (e.g., Ota et al., 2002). Consistent with the literature (Sung, 1995), Asian younger adults hold more generalized notions of filial piety and are ready to provide care to and respect for both family and nonfamily older adults, unlike Western younger adults, who have a more specialized view of filial piety (i.e., to family older adults, Gallois et al., 1999). Western younger adults' stance regarding filial piety is based on personal choice rather than obligation, whereas their Eastern counterparts tend to understand it as mainly an obligation (Harris & Long, 2000; Ota et al., 1996), as literature on norms and duties suggests (e.g., Miller, 1992; Moghaddam et al., 2000). The consistency with the literature notwithstanding, younger adults in Eastern nations tend to distinguish clearly what they would do for older adults and what older adults expect them to do, whereas Western younger adults saw a congruency between the two components (Gallois et al., 1999). Communication was regarded as a highly important aspect of filial piety in Western nations, but this attitude was attenuated in Eastern nations. Provisions of practical support were more likely than communication to be accorded importance in Eastern nations.

Assessment of communication between younger and older adults added another brush stroke for the less positive intergenerational relationships in the East. Indeed, past studies reported that young adults in Western nations find communication with older adults challenging and difficult. To illustrate the point once again, although using patronizing speech to older adults constrains their communicative choice (Hummert & Ryan, 1996; Ryan, Giles, Bartolucci, & Henwood, 1986), younger adults experience an accommodation dilemma, exemplified by conversations involving older adults' painful self-disclosures (Coupland, Coupland, & Giles, 1991) and older adults' nonaccommodation and overaccommodation (e.g., patronizing speech) to the young (Giles & Williams, 1994; Williams & Giles, 1996), and thus evaluate communication with older adults quite negatively (e.g., Giles, Henwood, Coupland, Harrimon, & Coupland, 1992; Williams & Giles, 1996). Nevertheless, younger adults in Eastern nations such as Japan, Korea, Hong Kong, the People's Republic of China, and, to a lesser extent, the Philippines, follow suit (see also Zhang & Hummert, 2001), but in a more accentuated way. Although great variability was identified within the block of Eastern nations (which will be discussed shortly), communication with nonfamily older adults was judged, in general, as more effortful and problematic than in Western countries, because of the lack of accommodation and higher age saliency in communication (Williams et al., 1997). Further, communication with nonfamily older adults was evaluated less favorably than that with peer age group members and family older adults in both Eastern and Western nations (but see Yue & Ng, 2000, for potential difficulty in communication with family older adults); this tendency was more prominent in Eastern nations than in Western ones (Giles, Liang, Noels, & McCann, 2001; Giles et al., submitted; Ota, 2001; McCann, Giles, Ota, & Caraker, in press;

Ng, Liu, Weatherall, & Loong, 1997; Noels, Giles, Gallois, & Ng, 2001). Although no major cross-cultural differences were reported with regard to intragenerational communication, Eastern younger adults made less favorable judgments of communication with family older adults (especially in terms of accommodation, age saliency, and general positivity) than did younger adults from Western nations (Giles et al., 2001).

In addition, older adults themselves find communication with other older adults to be highly problematic. This tendency was more likely, again, in Eastern nations than in Western nations. In the People's Republic of China (Cai, Giles, & Noels, 1998), despite outstanding accommodation, older adults reported greater nonaccommodation from and, at the same time, greater felt-need to be respectful to other older adults than to younger people. They were more likely to avoid communication with their peers than with younger family members. Similar tendencies were reported with regard to their affective status in communication. Japanese older adults also felt that other older adults were more nonaccommodating and felt more obligation to respect them (Ota, 2001). Similarly, older Hong Kong adults felt greater nonaccommodation and respect, or obligation, to older adults than to younger adults (Noels et al., 2001).

Health consequences of inter- and intragenerational communication were also investigated across cultures. Communication with younger adults had some impact on older adults' psychological health, as measured by self-esteem (Luhtanen & Crocker, 1991; Rosenberg, 1979) and sense of coherence (Antonovsky, 1987). Supporting the CPA, communication with younger adults had a significant impact on older adults' psychological health in the United States and Australia, but its effect was quite limited in the People's Republic of China (Cai, Giles, & Noels, 1998), Hong Kong (Noels et al., 2001) and Japan (Ota, 2001). In Eastern nations, it was mainly communication with other older adults that was important to older adults' psychological health.

As this review indicates, surprisingly limited differences occur when cross-cultural comparisons are made regarding younger adults' attitudes and the consequences that those attitudes have on their communication with older adults. Age as a group category is likely to be problematic in communication in most cultural settings. Despite this universal tendency, younger adults in Eastern cultures are more likely than their Western counterparts to find intergenerational communication with older adults difficult. We may conclude that being older is more problematic in the East than in the West. Nevertheless, as McCann et al. (in press) succinctly stated, young people in Eastern nations may still respect the older tradition and therefore feel obligated to be polite to, and avoid,—that is, distance themselves from—older people to maintain appropriate group boundaries (see Sung, 2001, for the most recent discussion of respect for older adults in Asia).

Another important point is variability within the Eastern and Western block of nations. As briefly mentioned, younger adults' evaluations of communication with older adults varied across cultures, especially in Eastern nations (see Williams et al., 1997; Giles et al., 2003, for details). For example, Filipino respondents showed a relatively positive evaluation pattern regarding intergenerational communication and stereotypes similar to that of Western respondents. The results indicate the importance of treating each culture as independent, as well as part of a larger cultural group (e.g., East or West). In other words, particularization and categorization (Billig, 1987) are both important perspectives to adopt for intergenerational communication research, especially now that influences from different cultures are common through a number of channels at this time of relative borderlessness.

Cultural Influences on the Communicative Construction of Relationships in Later Life

Our culture specifies appropriate definitions of behavior within roles and the dynamics of interpersonal relationships (the identity, group inclusion, and intergroup boundary regulation functions of culture). Even such matters as basic as the definition of what is a family and what is a friend are culturally bound. Do we recognize an individual as an immediate family member or an extended family member (e.g., Kagitcibasi, 1996)? Are cross-sex friendships allowed or taboo? Our interconnectedness with others, as exemplified by the size, composition, and density of our social networks are culturally bound (Triandis, 1995). Cultural definitions of age categories and attitudes toward members of these age groups defines which intergenerational interactions are considered to be intergroup interactions and which are interindividual.

When age is treated as a culture, interactions between different age groups can be considered to be intercultural interactions. Limited research, beyond the type reviewed in the previous section on intergenerational communication, has been conducted. One exception is research assessing age differences in friendships, which revealed that intergenerational friendships in the United States are less intense and close in terms of companionship, satisfaction, intimacy, nurturance, and reliable alliances, yet have a significantly higher admiration quality (Holladay & Kerns, 1999). Thus, friendship appears to serve a different function for age-peer relationships than for age-discrepant relationships, as the nature of the friend relationship contains different qualities across the life span. These findings may in part reflect the tendency of older individuals, at least in the United States, to be more selective in how much energy they invest in relationships (Carstensen, 1991, 1992, 1995). As we develop more skill at developing and evaluating relationships, we identify those relationships that we feel will be "profitable" and focus our energies on them and allow more peripheral relationships to wane (Carstensen, 1992, 1995; Carstensen et al., 2000). Whether these relational dynamics hold true for older individuals in other cultures is unknown, although some research (Barnlund, 1989) suggested there is cultural variation in the tendency of companionship between people of greater age discrepancy. Further, the review of intergenerational communication provided earlier may suggest that communication across age groups is less difficult in Western than in Eastern cultures. This may be interpreted to mean that friendship between younger and older adults is more likely in the former than the latter nations. Culture, again, should not be disregarded.

Family relationships, those between spouses, parents and children, grandparents and grandchildren, and other family members are based on cultural expectations. Although considerable research on families in the United States includes participants of European descent as one group, differences within this group are usually overlooked or explained by social class or immigration status (Merrill & Dill, 1990). Merrill and Dill (1990) found that older Americans of French Canadian and Italian descent had different patterns of coresidence with their children, based on their definitions of what family members do for each other.

Another approach to examining what cultures define as appropriate familial interactions is to study immigrants to the United States and differing generational expectations. Individuals who immigrate to the United States at an older age often do so to follow their children and therefore have more difficulty adapting, because of language barriers and inadequate understanding of the cultural values of the United States (Gelfand, 1989; Kim, Kim, & Hurh, 1991). These difficulties in adapting may lead to greater dependence of the parents on the children, as the parents need the children to translate and explain the new country (Gelfand, 1989; Kim et al., 1991). Gelfand (1989) studied Salvadoran immigrants who were primarily illegal aliens in the United States. In Hispanic cultures, older people are treated with respect and have status and authority within the family. Salvadoran immigrants, however, particularly because of their illegal status, are more dependent on their children because they are not eligible for government assistance, which is available to other older individuals. Kim et al. (1991) studied Korean immigrants to the United States. The expectation of filial piety falls particularly to the eldest son and his wife in caring for his parents. This expectation can lead to intergenerational conflict as the son and his wife, adopting more American ways of family, focus their attention on their own relationship and on their children (see also Kauh, 1997). Hilker (1991) argued that issues of ethnic identity often lead to intergenerational conflict not only when the younger generation adopts the values of a new country, which contrast with the older generation's values from the old country, but also when members of the older generation, in an attempt to adapt to the new country and make the passage easier for their children, suppresses their ethnic identity and do not share the uniqueness of their cultural background with their offspring (see Liu, Ng, Weatherall, & Loong, 2000, for identity and intergenerational communication).

The in-law relationship is one that is often fraught with tension, as two families must coordinate their relational cultures and expectations. In cultures with high expectations of filial piety, such as Korean, intergenerational conflict may arise when the daughter-in-law must care for the mother-in-law (Kim et al., 1991). This conflict may be increased when the family has migrated and the daughter-in-law has additional demands on her time, including work. Some cultures may have developed a system for managing these potential conflicts. For example, in traditional Moroccan Muslim society, the mother-in-law is given considerable status because, on marrying into a family, young brides are tutored and protected by mothers-in-law, for whom in return the young brides perform daily rituals (Mernissi, 1975). In Western society, however, it is quite common for mothers-in-law to have little control over their sons' wives, as the son is expected to shift his primary allegiance from his mother to his wife (Kim et al., 1991).

Research on cultural sharing of traditions, beliefs, and customs between grandparents and adult grandchildren suggests that minority and female participants in the United States are more likely to engage in intergenerational cultural sharing and report more positive statements about this sharing (Wiscott & Kopera-Frye, 2000). The important role of African American grandmothers in family life has been well documented not only in the social science literature, but in literature as well (Hilker, 1991). African American grandparents have more contact with their grandchildren and are more likely to provide care and assistance and to teach important life lessons to their grandchildren, even when living geographically distant in comparison with European American grandparents (Hilker, 1991; Strom, Collinsworth, Strom, & Griswold, 1993). Not only do African American grandparents play a more significant role in the lives of their grandchildren, but the social network plays an important role in the lives of older African Americans. A recent study of middle-income African American women aged 75 years and older indicated that satisfaction with quality of life is associated with belonging to a family and a social support network that includes church, as well as maintaining physical health, independence, and personal safety (Martin-Combs & Bayne-Smith, 2000). These findings concur with those of Adams and Jackson (2000), who found that family support and social contact influence older African Americans' evaluations of their quality of life to a greater extent than they do for older European Americans.

Culture in Organizational Communication and Aging

Our culture defines the role of older individuals within organizations (identity, group inclusion, and intergroup boundary regulation functions of culture). Are older workers viewed as a benefit to the organization because of their years of experience, or a drain on resources because they are not up-to-date? Is an older worker seen as wise or slow? Returning to the ecological adaptation function of culture, notions of retirement and leisure are relatively new concepts, appearing in the last 100 years or so, and are not equally distributed across cultures (the ecological adaptation function of culture). What are the attitudes within a culture about these life stages, and how do these attitudes compare across cultures?

With the increasing percentage of older individuals and longer years of good health, older workers are likely to have a significant impact on the workplace and organizational culture, just as they are expected to have an impact on the larger culture. An additional consequence of changing demographic patterns is that fewer young workers will be entering the workplace; therefore, older workers will be needed to fulfill projected labor shortages (Graham, 1996). Organizations will need to adapt training to the needs of older workers, including continual updating and providing greater opportunity to practice new skills (Beatty & Burroughs, 1999; Graham, 1996). Definitions of jobs may need to be modified as individuals over 65 prefer part-time over full-time work. In addition, organizations will need to modify job duties and make accommodations for workers who have less physical strength, poorer eyesight or hearing, or other health restrictions (Graham, 1996). Organizations can continue to benefit from the experience of older workers by having them serve as mentors or trainers for younger, less experienced workers (Graham, 1996).

The increasing percentage of older workers is an international phenomenon, particularly in industrialized countries ("Boom in Older Workers," 1996; European Community, 1998). The oldest workforces are in Finland, Sweden, Switzerland, Luxembourg, and Japan. These countries are already adapting to the large percentage of older workers. The European Union (consisting of Austria, Belgium, Denmark, Finland, France, Germany, Greece, Ireland, Italy, Luxembourg, Netherlands, Norway, Portugal, Spain, Sweden, and the United Kingdom) has established a variety of programs to address the needs of older workers (European Community, 1998). Older workers have suffered from age discrimination, such as retraining programs being offered to younger workers only, because many of these countries had high unemployment among the young in the 1980s and 1990s, so official policy encouraged early retirement (European Community, 1998). These countries now are changing these policies, because of concerns for the expense related to pension costs (European Community, 1998). Although various programs have been developed, each with strengths and weaknesses, best practices have been identified: targeting services with tailored solutions that benefit both the employer and the employee; motivating older workers to receive more education; tracking skills to match employer needs; developing relationships with employers to facilitate the placement of older workers; and raising awareness about the problems facing older workers (European Community, 1998). For example, Finland has established a program called FinnAge: Respect for Aging, which seeks to improve the work environment, organizational culture, and health promotion activities for older workers ("Boom in Older Workers," 1996). An important aspect of this program is that it focuses not only on sedentary jobs but active ones, such as police and fire personnel, and has had great success in reducing early retirement and retaining older workers.

Older workers have been on the political agenda longer in Japan than in the United States, with a focus on employment versus retirement, in the respective countries (Rix, 1996). One reason for this increased attention is that older workers have been a larger part of the Japanese workforce, and retirement age, until the 1990s, was set at 55 or 60, much younger than the average of 65 in the United States, which reflects the time when individuals may receive retirement benefits. The Japanese government subsidizes companies that reemploy their retirees; however, new contracts are often at reduced wages and benefits (Rix, 1996). Reduced wages are important to the organization, as the Japanese system is based on seniority and employees and employers are likely to have a lifelong contract, so older workers have significantly higher wages than younger workers (Rix, 1996). Additional subsidies exist to help accommodate older workers and to provide training. The government benefits from older people remaining in the workforce, because this delays the onset of benefits. Older individuals in Japan benefit from continued work not only because of continuing wage production, but because of favorable attitudes toward work and relatively negative attitudes toward leisure (Rix, 1996).

Midlife and older displaced workers in the United States and Europe face many of the same problems on their reintegration into the workforce (Bowman, 1997). Workers with lower levels of education, shorter periods of job tenure, and more menial occupations have more trouble gaining reemployment. Women often have less education than men, although the number of individuals with an education that leads to certification is higher in Europe than in the United States (Bowman, 1997). Length of time on the job is often lower for women who delay entry to or leave the workforce to tend to family needs. Older individuals without highly sought after specialized skills also find the job market more competitive, with jobs going to younger workers first (Bowman, 1997).

When considering age as a culture, individuals in different age groups may find their interactions complicated by stereotypes of each other. In the United States, older adults tend to be perceived highly stereotypically (i.e., negatively) and sometimes receive disparaging remarks from people in different age groups (McCann & Giles, 2002). Younger workers in the United Kingdom perceive older workers to be more effective, but less adaptable to change compared with younger workers in Hong Kong (Chiu, Chan, Snape, & Redman, 2001; but also see Finkelstein & Burke, 1998, for older adults' unfavorable evaluations of older workers). For younger workers in both places, stereotypes of older workers affected their attitudes toward training, promotion, and retention of older workers, their willingness to work with older workers, and their support for positive discrimination (Chiu et al., 2001). In addition, a worker's own age predicted positive stereotypes of older workers. On a positive note, younger workers who worked for organizations with antiage discrimination policies were more likely to have more positive attitudes about the adaptability of older workers (Chiu et al., 2001).

Within an organization, age and role may interact to produce problems or solutions to management problems. For example, channels of communication vary according to rank within an organization. Hospital administrators and executives, nurses in middle-management positions, and nursing staff receive messages via different routes (du Pre, 2000). However, the assumption that senior-level individuals in an organization are older and junior-level individuals are younger is quickly becoming a blurred line. As younger individuals take on senior management roles and more mature individuals return to the workforce in lower-level jobs, the structure of management is changing. This shift is likely to be particularly evident in high-demand sectors, such as technology and health care organizations, where younger people are more frequently playing a central role in delivering products and services. McKenzie and Cannon (1998) found that young talent from North American, Latin American, European, and Asian markets tend to show re-markably similar patterns in how they manage their careers. For instance, a 28-year-old Australian was found to have more career-related behavior in common with a Brazilian or American of the same age than with most 45-year-old Australians (McKenzie & Cannon, 1998). Moreover, older and middle-aged members report feeling more of an affinity or al-legiance to their organization and its success (McKenzie & Cannon, 1998). Consequently, managers' structuring of messages may very well impact younger employees differently than older employees.

Culture in Political and Mass Communication and Aging

Culture inevitably becomes entangled in the political structure (the identity, group inclu-sion, intergroup boundary regulation, and ecological adaptation functions of culture). Cultures that place different value on the right to one having a political voice have dis-similar styles of governing. In addition, messages in the media may reinforce or establish cultural expectations about an individual's performance of roles. With the growing num-bers of older individuals, we should expect increasing attention to issues of concern to them in both the political and commercial arenas.

In the political arena, the majority of focus regarding older individuals has been on work-related and retirement issues (see the previous discussion in the section on orga-nizational communication). Another aspect of this issue is how the political economy affects aging (Estes, 1979; C. L. Fry, 2000; Minkler & Estes, 1999). Political economies occur when states are formed and political activity is centralized. When the political economy is associated with industrial capitalism, a market economy results (C. L. Fry,

2000). The transfer of income based on age, such as that associated with Social Security and Medicare in the United States, establishes a political economy of aging (Estes, 1979; Minkler & Estes, 1999). Cultures without such wealth transfer, such as those in Africa included in the Project AGE studies, do not have a political economy of aging (C. L. Fry, 2000). As a result, no part of the economy is focused on the needs of older people, unlike in more industrialized countries with modern medical facilities, retirement and assisted living communities, and nursing homes (C. L. Fry, 2000). Segregation into age-homogenous enclaves does not occur when a market economy does not benefit from the creation of such facilities.

In the commercial arena of the mass media, age is often treated as a coculture as target audiences become narrowly segmented, focusing advertisements and entertainment to very specific niche groups (Sawchuk, 1995). Harwood (1997) found that older adults in the United States, as well as people in other age groups, prefer viewing characters of their own age. In Japan, Katori (2000) found that older adults are likely to watch for information and education purposes besides entertainment, but that they do not necessarily like programs made specifically for older adults. In contrast, Mares and Cantor (1992) found that lonely older adults in the United States prefer to watch negative portrayals of older people. In addition, images in the mass media may help to reinforce stereotypes about aging. First, older people tend to be underrepresented in TV programs (Robinson & Skill, 1995) and TV commercials (Roy & Harwood, 1997). When those images are primarily negative, which is often the case (Miller, Miller, McKibbin, & Pettys, 1999; Vasil & Wass, 1993; but for positive ones, see Harwood, 2000; Harwood & Roy, 1999; Roy & Harwood, 1997), the consequences can be a reinforcement of negative attitudes held by younger people and reduced ability exhibited by older individuals (see discussion in the section on health communication for studies that examine the influence of stereotypes on the cognitive processing of older adults).

When age is treated as a culture, older age groups are treated rather negatively in the mass media, as briefly mentioned in the previous paragraph. Although older adults have established a memorable presence in American television commercials, their portrayal may not be an accurate one (Gerbner, 1993). Gerbner (1993) effectively assessed the powerful impact of television on cultures with that technology:

> Mass media are the most ubiquitous wholesalers of contemporary culture. They present a steady, repetitive, and compelling system of images and messages. For the first time in human history, most of the stories are told to most of the children not by their parents, their school, or their church but by a group of distant corporations that have something to sell. This unprecedented condition has a profound effect on the way we are socialized into our roles, including age as a social role. . . . The world of aging (and nearly everything else) is constructed to the specifications of marketing strategies. (p. 207)

These negative portrayals are not limited to television. The portrayal of individuals aged 50 and older in general audience advertising has traditionally not been positive (Miller et al., 1999; Thomas & Wolfe, 1995). In a recent study of advertisements in *Modern Maturity*, which is targeted to older individuals, Roberts and Zhou (1997) found older adult characters to be portrayed as capable, important, healthy, and socially active;

however, few ethnic minorities were shown in these advertisements. The overall lack of positive images of older individuals, especially in some Eastern nations (e.g., Japan; Holtzman & Akiyama, 1985) in the mass media, reinforces negative stereotypes of aging.

The lack of appropriate coverage of issues affecting older adults may also be influencing the public's knowledge and views of such issues as long-term care, caregiving, pharmaceuticals, and medicines, etc. Moreover, when there is coverage, it tends to be negative. Mebane (2001) found that a small percentage of media coverage concerns long-term care and that most of this coverage deals with nursing homes. A large majority (86%) of Americans aged 45 and older have expressed a desire to avoid an institutional setting when they need help taking care of themselves (American Association of Retired People, 2000). Moreover, a large majority (76%) of adults aged 70 and older who believe they may need long-term care in the coming year do not have the personal resources to pay for care (Mebane, 2001). Of course, older individuals living in cultures without specialized facilities must find other sources of support to meet their needs.

Culture in Health Communication and Aging

Definitions of health and illness and the causes of disease are culturally bound (Witte & Morrison, 1995). Appropriate roles as physician and patient are defined culturally (identity, group inclusion, and intergroup boundary regulation functions of culture). Through communication, we learn what expectations are held for both formal and informal caregivers. Who steps in to provide us with assistance as we age? How do the formal caregiving network and the informal caregiving network interact? Does the culture have institutions for providing care to frail older people or are these individuals cared for informally? This section reviews the rather extensive literature on health and aging within a cultural framework, from issues of biology, physiology, and psychology to general health issues, such as the patient–provider relationship, elder care, health promotion campaigns, and end-of-life decision making.

Although biological imperatives would appear to be the same everywhere, culture interacts with biology to affect the aging process. Although normal aging appears to be tied to the finite life span of cells (Gibbons, 1990), the aging body occurs within the context of its experiences and locality (Kontos, 1999). Although some theorists argued that meanings to biological processes are assigned through culture, others argued that biology and self are inextricably intertwined (Kontos, 1999). Therefore, human behavior is not a result of biology or of culture, but of the interplay between the two (Hinde, 1991). Still others argued that because these two components are so intertwined, culture affects biology as much as biology affects self (Kontos, 1999). These arguments derive from critical gerontology theorists, who examine the meaning of body and the connection of body to place (Kontos, 1999).

One specific area that has received considerable attention by researchers is that of how biology and culture interplay in the development of cognition. Perhaps most important, an individual's cognitive development occurs within his or her social context (Robertson, 1996). Knowledge is not solely a matter of cognitive mechanics (i.e., the neurophysiological architecture of the mind, or fluid intelligence) and, therefore, is not simply determined by biology (Baltes, 1993). Knowledge also consists of cognitive pragmatics (crystallized intelligence), which reflect culture-based growth—knowing what the culture defines as factual and procedural knowledge about the world (Baltes, 1993). Although most researchers concur that culture influences cognitive pragmatics, Park, Nisbett, and Hedden (1999) argued that culture also plays a role in how individuals learn to process information, the supposedly hard-wired, biological component of knowledge (see also Nisbett, Peng, Choi, & Norenzayan, 2001). Eastern cultures encourage the processing of information in a holistic, contextual fashion, whereas Western cultures encourage more analytic, feature-based processing (Park et al., 1999). Measures of memory, therefore, may tap into culturally invariant or culturally saturated aspects of information processing. Measures in which contextual cues are supportive would enhance the ability of individuals from Eastern cultures, whereas measures in which contextual cues are distracting would interfere with the ability of such individuals (Park et al., 1999). However, as individual's cognitive processing ability severely declines, processing styles may become more similar as the strategies that are used become more limited (Park et al., 1999). Culture can affect cognitive performance in other ways as well. Older individuals who were exposed to positive stereotypes of aging performed better on memory tests and assessed their own performance more positively than did older individuals exposed to negative stereotypes of aging (Levy, 1996; Yoon, Hasher, Feinberg, Rahhal, & Winocur, 2000); however, exposure to the same conditions did not effect the performance of younger adults (Levy, 1996). Levy (1996) argued that cultures with pervasive negative stereotypes of older people are handicapping older people's cognitive performance in unnecessary ways (see also Levy & Langer, 1994).

Schulz and Heckhausen (1999) argued that striving for control is a universal human behavior for which humans (and all animals) are hardwired based on evolutionary drives. Although all humans desire control over their environment, how the desire for control is expressed varies by culture (Weisz, Rothbaum, & Blackburn, 1984; see also Chang, Chua, & Toh, 1997) and across an individual's life span as goals shift during different developmental stages. For older individuals, issues of control become important to manage and cope with physical decline (Schulz & Heckhausen, 1999) and negative emotions (Chang, Chua, & Toh, 1997). As C. L. Fry (2000) pointed out, when decline requires assistance, a disability may be frustrating or life threatening, depending on the consequences of that disability within the individual's culture. For older individuals in nontechnological cultures, physical decline may lead to an inability to hunt, collect food, gather firewood and water, build kraals, or take livestock to pasturage. For older individuals in technological cultures, the availability of labor-saving devices and availability of wealth generate a different barrier to control. For these individuals, the critical issue is whether they possess the financial security to purchase needed goods and services.

When adaptive technology is available, however, it can be used to help make life easier and extend the abilities of older individuals. For example, technology has helped to maintain older individuals' ability to drive (Sheldrick, 1992), not to mention to satisfy their social needs (Ryan, Anas, Hummert, & Laver-Ingram, 1998) and psychological positivity (e.g., Cody, Dunn, Hoppin, & Wendt, 1999). Another physical decline which is common for many older individuals—who benefit from technological intervention—is loss of hearing, or presbycusis (Scott, 1993). Studies reveal that in the United States, cocultural differences may impact physiological aging, diet, stress, disease processes, and genetics. Evidence suggests that differences exist in auditory sensitivity as a function of age between male and female populations and between African American and European American populations. Women tend to hear better than men, and African Americans tend to hear better than European Americans (Scott, 1993).

Cultural variability is reflected in general health issues as well. As discussed in the section on overarching issues, different groups may define the process of aging and the elements of successful aging and quality of life quite differently. For all age groups, poverty is health's worst enemy (Kalache, 1995). For older individuals across all cultures, the accumulated years of exposure to poverty can negatively impact health. Wieck (2000) stated that many older people, particularly ethnic minorities living in the inner city in the United States, are trapped, because of limited fixed incomes at or below poverty level, living among crime-filled streets and drug dealers, and loss of family support. Older African American women who lived alone had the highest poverty rates (49.3%, U.S. Bureau of the Census, 1994). In the United States, many ethnic minorities have suffered because of social inequality (Ferraro & Farmer, 1996; Stoller & Gibson, 2000; Wykle & Ford, 1999). In addition, attitudes toward the health care industry may lead to a reluctance on the part of some individuals to use these services (Jitapunkul et al., 1999; Ushikubo, 1998; Wykle & Ford, 1999).

Gender may also be treated as culture (Maltz & Borker, 1982), as different sets of expectations exist for males and females in many aspects of life, including in regard to caregiving for older adults. For instance, females are more likely than males to be expected to provide care to older people in many, especially Eastern, countries (Kending, Koyano, Asakawa, & Ando, 1999; Lee & Sung, 1997, but see Harris & Long, 2000, for male caregivers). Further, gender as a culture becomes a prominent issue in health concerns, with one of the most important age-related differences occurring in women as they enter menopause. Women of the Yanomano culture, who live in the forest area near the border between Brazil and Venezuela, report they are eager to reach menopause, which is the marker of older age, increased status, and decision-making power within the family (Mercer, 1999). Similarly, among the Qemant, a pagan peasant population of Ethiopia, women who have reached menopause and beyond can walk on the ground of a sacred site and partake in ritual food and drink, which is not allowed before this time, because menstruating women are considered unclean (Garnst, 1969). By contrast, in the West, there is no elevated status for menopausal and older women. Older women in Western societies are often referred to as 'old maids' or other derogatory descriptions. Western women are typically only able to achieve status via socioeconomic means, but not merely for having arrived at this time in their lives. Similarly, menopause in Japan is considered a natural life-cycle transition and is not considered to be of great importance (Lock, 1986). Variability occurs within the United States, however. Women of Native American tribes find new meaning in later years through ceremonies that acknowledge their transition from their childbearing years (Tijerina-Jim, 1993). These life transitions may extend beyond menopause to other midlife transitions, such as acquiring the role of "grandmother" (Mantecon, 1993). In the United States, the focus on chronological age as a marker, rather than physiological changes such as menopause, often results in reassessments of self and relationships to others (Niemala & Lento, 1993).

Older individuals are more likely to have medical problems that require visits to health care professionals. Different cultures define patient–provider roles in different ways. Smith and Smith (1999) found that individuals in mainland China rated their doctors' communication, asking for information, explaining, encouraging patient talk, and sharing decisions higher than individuals in either Hong Kong or the United States. The authors also found that doctors and patients in China and the United States need more communication about prescriptions and appropriate use of medicine than those in Hong Kong. Further communication with doctors may become more problematic when doctors use patronizing speech to older adults. Although data for this argument wait to be provided, extrapolating findings about nurse–older patient communication (Masataka, 2000) suggest the presence of problematic issues that need to be tackled across cultures.

In the United States, ethnic minorities have differing attitudes toward health care services and their role in these encounters (Wykle & Ford, 1999). For example, older Hispanics frequently have limited English skills, limited education, and poor socioeconomic status, which impacts their ability to interact in a proactive manner with health care professionals. These status differences, especially when combined with age, lead to encounters of a strong intercultural nature between the patient and the provider. These issues of language frequently apply to older individuals who immigrated late in life. An example of these problems is exhibited by Jewish Russian émigrés, who in many cases moved to the United States with their children and grandchildren (Brod & Heurtin-Roberts, 1992). The Communist system ensured a minimum personal income with basic health services provided in an authoritarian manner. These émigrés find it difficult to get jobs in the United States and do not understand why they do not have greater access to services. In addition, they are not used to preventive care or the bureaucratic system of U.S. health care. Their physicians report feeling frustrated that they cannot meet all of their patients' needs and are expected to provide mental health and social services that are outside their area of expertise (Brod & Heurtin-Roberts, 1992). The issues of language difference is especially important to help older immigrants maintain satisfying communication and psychological health in hospitals and institutions (Ryan, Meredith, MacLean, & Orange, 1995).

Once an older individual needs assistance, how care is provided is influenced by cultural preferences. For example, in the United States, older individuals of European descent are more likely to hire formal caregivers to provide services, whereas members of ethnic and religious minorities (e.g., Jews, African Americans, Hispanics) are more likely to receive considerable instrumental support from an extended family (Cantor & Brennon, 2000). Reliance on family members (especially those coresiding) for informal care is also common in Eastern nations (Chen & Silverstein, 2000; Kending et al., 1999; Koyano, 1996; Phillips, 2000). Because ethnic minorities in the United States are more likely to live in or near poverty, these additional burdens can have a tremendous impact on the economic well-being of the entire family. Even though these families experience considerable stress, they deal with it in ways that are not common among non-Latino White caregivers (Delgado, 1998). For example, Puerto Rican Catholics trust their faith to help them cope with caregiving burden (Delgado, 1998).

For older individuals who enter long-term care facilities, the amount of care being provided continues to increase. Residents and staff often come from different groups, which may lead to instances of intercultural communication. Managing health care environments means taking responsibility for the control of things that are increasingly uncontrollable and difficult to manage (Aroian, Meservey, & Crockett, 1996). Negotiating this responsibility requires considerable effort on the part of the staff and the resident and, sometimes, his or her family. Directors of nursing have to balance the needs of all concerned, using knowledge about clinical practice, as well as nursing and management skills, particularly in long-term care facilities (Aroian, Patsdaughter, & Wyszynski, 2000). Staff nurses and family caregivers of hospitalized older adult patients play crucial roles in the health care and recovery of patients. Nurses are typically involved and clinically informed about patients during inpatient care; however, the majority of care following hospital discharge falls on family caregivers, Organizational factors in large tertiary care hospitals may serve as a deterrent to contact and communication between health care providers, family caregivers, and patients during hospitalization (Rose, Bowman, & Kresevic, 2000), leading to problems in communication and successful management of the older individual's health needs after release from the hospital.

Health promotion campaigns should be tailored to the needs of cultural groups. Culturally competent care takes into consideration the needs of the patient along with his or her family and community (Wykle & Ford, 1999). In the United States, ethnic minorities have been the focus of such campaigns. For example, older native Hawaiians were more likely to attend and more satisfied with health education programs when outreach workers had established meaningful one-on-one relationships with them (Koseki, 1996). Even such activities as providing descriptive health status data should be done with a recognition of how health, culture, and economics intertwine (Mercer, 1994).

Some of the most critical and difficult health-related decisions are those surrounding end-of-life care, such as use of life-extending technology. Culture helps us to make coherent sense of our mortality, suffering, and death (Greenberg, Solomon, & Pyszczynski, 1997; Lynn, 2000). These issues are influenced not just by ethnicity but also by religious beliefs and personal experiences (Lynn, 2000). Ethnic minorities in the United States exhibit a range of preferences for varying reasons. African Americans are more likely to prefer life-sustaining treatments but are reluctant to share their treatment preferences with health care professionals (Mouton, 2000). On the other hand, African Americans are less likely to use hospice care or agree to organ donation (Mouton, 2000). Hispanics are also less likely to agree to organ donation or autopsies (Talamantes, Gomez, & Braun, 2000). For both groups, the main reason for not selecting organ donation is the desire to present a whole body to God (Mouton, 2000; Talamantes et al., 2000). Hispanics, like African Americans, are likely to request life-prolonging treatments (Talamantes, et al., 2000). Both groups express a desire to allow time for God to work miracles. Among Asian Americans and Pacific Islanders, considerable variability reflects the diversity of their countries of origin, their immigration histories, rural versus urban backgrounds, and education levels (Yeo & Hikoyeda, 2000). As reflects their more collectivistic approach to life, however, they are more collectivistic regarding issues surrounding death and prefer for family to be involved in all decision making (Yeo & Hikoyeda, 2000). In general, Asian Americans and Pacific Islanders are more likely to want life support, because longevity is more important than quality of life. To provide a whole body to their ancestors, they avoid autopsy and organ donation (Yeo & Hikoyeda, 2000). American Indians and Alaska

Natives also represent a diverse group, with over 300 federally registered tribes (Van Winkle, 2000). In general, among this group, death is seen as a natural part of the circular pattern of life, so it tends to have more positive attitudes toward dying than most other groups (Van Winkle, 2000).

Culture in Educational Gerontology

The nature and role of learning across the life span is culturally bound (the identity and group inclusion functions of culture). Are individuals encouraged to continue learning, or is learning confined to a formal process during a small number of years early in life? When the life span is divided into life stages and the life stage dedicated to learning is childhood and adolescence (Cole, 1989; Peterson, 2000; Robertson, 1996), then the term *adult learner* takes on a non-normative quality, reinforcing the concept that learning should be limited to the early years. This sense of non-normativity is inherent in Shea's (1985) argument that lifelong learning is seen as a challenge that involves the risk of failure and is, therefore, discouraged in some cultures. These attitudes toward age-appropriate stages of learning lead to an age culture within educational realms.

The tremendous increase in the older adult population will likely lead to an increase in the number of older adults who participate in educational activities. Research reveals that a predictor of older adult participation in education is prior participation (Truluck & Courtenay, 1999). Because educational levels are increasing, we can expect to see an increase in the number of older adults participating in formal educational activities. In addition, educational attainment has been linked to successful aging. Meeks and Murrell (2001) found that education and negative affect both were directly related to health and life satisfaction for older adults, with negative affect mediating the relationship between education and successful aging indicators. Thus, education appears to bring about a lifelong advantage for healthy aging. Higher educational attainment is related to lower levels of trait negative affect, and lower negative affect results in better health and life satisfaction (Meeks & Murrell, 2001).

Although older adults may be entering formal educational opportunities at a greater rate, teaching styles will need to be adapted, because many older adults have learning styles different from those of younger adults. Truluck and Courtenay (1999) found that not all older adult learners are active, hands-on learners as adult education literature suggests, but rather with age there is a tendency to become more reflective and observational in the learning environment. Thus, it is vital that instructors pay attention to and gain a stronger understanding of the varied learning styles and needs in the ever changing learning environments that comprise multiage populations of learners.

Whitehouse, Bendezu, Fallcreek, and Whitehouse (2000) suggested that multiage communities of learners can represent a conceptual and organizational response to the rapid changes in human societies resulting from population growth, technology, and environmental issues, which create a need for new ways of learning and sharing across generations. In an effort to achieve community harmony, the authors provided an educational design of an intergenerational magnet school that focused on literacy, artistic studies, technology, environment, and responsibility for personal wellness. Intergenerational service learning courses that involve planned, ongoing interactions between younger and older adults have been increasing in recent years. Several scholars have noted the value of including intergenerational programs in the curriculum in an effort to increase knowledge of the aging process and to dispel negative stereotypes and myths related to the aging process (see, e.g., Knapp & Stubblefield, 2000).

Much of the discussion about training older adults has occurred within an organizational setting (see the discussion in the section on organizational communication). Beatty and Burroughs (1999) suggested that issues regarding aging be incorporated across the curriculum, but particularly in human resource development programs. Issues regarding older workers in the workplace that need to be addressed in such programs include understanding their role and abilities, understanding changes in motivation that occur with age, identifying unique strategies for training and supervising them, countering negative stereotypes, and reviewing business competencies, such as the legal aspects of discrimination and accommodation (Beatty & Burroughs, 1999).

Research indicates that individuals over the age of 65 can improve their cognitive abilities and memories, and the use of computers has shown promise in helping to maintain cognitive function (Bond, Wolf-Wilets, Fiedler, & Burr, 2000). One pervasive stereotype of older individuals is their reluctance to adopt new technologies (Ryan, Szechtman, & Bodkin, 1992). Adult learners' use of the Internet, however, is determined by access and appropriate training in the new technologies more than age (Cody, Dunn, Hoppin, & Wendt, 1999). In the United States, for example, the problem arises that individuals who have access to computers tend to be young to middle-aged and predominantly Caucasian from middle- to upper-income backgrounds (Anderson, Bikson, Law, & Mitchell, 1995), although this is gradually changing. Members of minority populations, the poor, and older adults tend to be Internet "have-nots," who do not have easy access to online materials and databases (Anderson et al., 1995). Given access to this technology, however, older individuals learn to use it effectively. Cody et al. (1999) found that when training older adults in the use of new technologies, a paced format geared toward different profiles of adult learners that integrated programs to reduce anxiety and build confidence was quite successful. Hollis-Sawyer and Sterns (1999) provided further evidence that a goal-oriented approach for training older computer novices offers opportunities for repeated experiences of mastering skills during the knowledge and skill acquisition process, as well as provides for active participation leading to increased task-proficiency outcomes. Having older individuals in the classroom may lead to opportunities for cultural sharing or barriers based on miscommunication and misunderstanding. Classrooms provide a unique place for overcoming such barriers, which leads to formal intergenerational contact programs.

Intergenerational contact programs provide formal opportunities for children to interact with individuals who are of their grandparents' and great-grandparents' ages (Abrams & Giles, 1999; Fox & Giles, 1993). Younger adults' judgment of communication difficulties may stem from the lack of communicative contact with older adults, high levels of negativity when they do engage in intergenerational interactions, or both of these factors. As a result, children and young adults may develop negative stereotypes of older adults. Integration of younger and older adults by creating opportunities for interaction may be highly beneficial not only in changing negative stereotypes (e.g., Fox & Giles, 1993; Hewstone & Brown, 1986), but also in promoting cultural sharing and mutual understanding (Yamazaki, 1994). Opportunities for positive intergenerational contact and direct communication is also important for older adults. Being able to provide social support to younger people through communication may boost older adults' self-esteem. As this review shows (see also Strom et al., 1995), older adults are evaluated negatively in communication with younger adults. Intergenerational contact programs may provide positive opportunities that help older adults to raise awareness of their own communicative behaviors and also to talk directly with younger people regarding these behaviors and the values that may be the cause of negative evaluation of their competency. Further, using such programs to point out the expectations each group has of the other may potentially lead to the creation of smoother and more comfortable intergenerational interaction (Strom et al., 1999).

SUMMARY

Culture is pervasive. As the functions of culture reveal, cultural values and beliefs define appropriate definitions of self, of roles within the family, workplace and community, and even how to age. Thus, theoretical considerations of aging should place the individual within the context of his or her cultural place and times. Even issues of biology and cognitive function should be framed by these contextual factors. As we have seen, chronological age has relatively little meaning in determining the aging process. Of greater consequence to that process are the meanings that individuals assign to their own aging. As social gerontologists attempt to understand aging, they should listen to the voices of those who are undergoing the experience and incorporate their voices in their theories, acknowledging that these meanings vary across and within cultural groups. As a consequence, complex theoretical stances are needed to develop adequate theories of aging—theoretical stances that can incorporate culture, time in history, place, individual characteristics, and experiences. The purpose of science is not only to predict, but also to explain, that is, to lay bare the fundamental mechanisms of nature (Waldrop, 1992). Human experience is built on a series of small chance events that become magnified through positive feedback (Waldrop, 1992). Although scientists often have difficulty looking at an individual's life and identifying the factors that would predict a given outcome, individuals looking back over their own lives are quite adept at making sense out of what happened. Social gerontologists can advance knowledge about the aging process by identifying the patterns that exist in the narratives of those individuals. Such a theoretical undertaking leads to an ambitious research agenda to identify the key components and patterns that are available at any given time. Of course, with the passage of time, each cohort of older people will bring new experiences and considerations that will need to be examined and understood.

As individuals age, they tend to develop more complex and more positive views about aging. As cultures come to include a larger percentage of older individuals, cultures may come to have more complex and more positive views about aging as well. With more examples of "successful agers" available, more people should be able to successfully age. However, more options is not always a guarantee to finding the right path, and this could lead to greater problems as individuals become confused amidst this array of potentialities. Pragmatically, awareness of the issues that help define the experience of aging may help us to manage the process more successfully, both for ourselves and for those around us.

REFERENCES

Abrams, J., & Giles, H. (1999). Epilogue: Intergenerational contact as intergroup communication. *Child and Youth Services, 20,* 203–217.

Adams, V. H., III, & Jackson, J. S. (2000). The contribution of hope to the quality of life among aging African Americans: 1980–1992. *International Journal of Aging and Human Development, 50,* 279–295.

Al-Deen, H. S. N. (1997). Preface. In H. S. N. Al-Deen (Ed.), *Cross-cultural communication and aging in the United States* (pp. xi–xiii). Mahwah, NJ: Lawrence Erlbaum Associates.

Allport, G. W. (1954). *The nature of prejudice.* Reading, MA: Addison-Wesley.

American Association of Retired People. (2000). *Fixing to stay: A national survey on housing and home modification issues* [online]. Available: http://www.research.aarp.org/il/home_mod.html. Accessed May 16, 2001.

Anderson, R. H., Bikson, T. K., Law, S. A., & Mitchell, B. M. (1995). *Universal access to e-mail: Feasibility and societal implications.* Santa Monica, CA: Rand.

Antonovsky, A. (1987). *Unraveling the mystery of health: How people manage stress and stay well.* San Francisco: Jossey-Bass.

Aroian, J. F., Meservey, P. M., & Crockett, J. (1996). Developing nurse leaders for today and tomorrow: Part 1, Foundations of leadership in practice. *Journal of Nursing Administration, 26,* 18–26.

Aroian, J. F., Patsdaughter, C. A., & Wyszynski, M. E. (2000). DONs in long-term care facilities: Contemporary roles, current credentials, and educational needs. *Nursing Economics, 18,* 149–156.

Asakawa, T., Koyano, W., Ando, T., & Shibata, H. (2000). Effects of functional decline on quality of life among the Japanese elderly. *International Journal of Aging and Human Development, 50,* 319–328.

Baltes, P. (1993). The aging mind: Potential and limits. *The Gerontologist, 33,* 580–594.

Baltes, P. B., & Baltes, M. M. (1990). Psychological perspectives on successful aging: The model of selective optimization with compensation. In P. B. Baltes & M. M. Baltes (Eds.), *Successful aging: Perspectives from the behavioral sciences* (pp. 1–34). New York: Cambridge University Press.

Barnlund, D. C. (1989). *Communication styles of Japanese and Americans.* Belmont, CA: Wadsworth.

Barnlund, D. C., & Yoshioka, M. (1990). Apologies: Japanese and American styles. *International Journal of Intercultural Relations, 14,* 193–206.

Beatty, P. T., & Burroughs, L. (1999). Preparing for an aging workforce: The role of higher education. *Educational Gerontology, 25,* 595–611.

Bengtson, V. L., & Putney, N. M. (2000). Who will care for tomorrow's elderly? Consequences of population aging East and West. In V. L. Bengtson, K.-D. Kim, G. C. Myers, & K.-S. Eun (Eds.), *Aging in East and West: Families, states, and the elderly* (pp. 263–286). New York: Springer.

Berdes, C., & Zych, A. A. (2000). Subjective quality of life of Polish, Polish-immigrant, and Polish-American elderly. *International Journal of Aging and Human Development, 50,* 385–395.

Berry, J. W., Poortinga, Y. H., Segall, M. H., & Dasen, P. R. (1992). *Cross-cultural psychology: Research and applications.* Cambridge, UK: Cambridge University Press.

Biggs, S. (1997). Choosing not to be old? Masks, bodies and identity management in later life. *Ageing and Society, 17,* 553–570.

Billig, M. (1987). *Arguing and thinking.* Cambridge, UK: Cambridge University Press.

Bond, G. E., Wolf-Wilets, V., Fiedler, F. E., & Burr, R. L. (2000). Computer-aided cognitive training of the aged: A pilot study. *Clinical Gerontologist, 22*(2), 19–42.

Boom in older workers goes international. (1996). *Safety and Health, 153,* 42.

Bowman, J. B. (1997). An international perspective: Factors that affect female midlife and older displaced workers' re-employment earnings. *Journal of Consumer Studies and Home Economics, 21,* 105–115.

Braithwaite, V., Lynd-Stevenson, R., & Pigram, D. (1993). An empirical study of ageism: From polemics to scientific utility. *Australian Psychologist, 28,* 9–15.

Brewer, M. (1979). In-group bias in the minimal intergroup situation: A cognitive-motivational analysis. *Psychological Bulletin, 86,* 307–334.

Brewer, M. B., Dull, V., & Lui, L. (1981). Perception of the elderly: Stereotypes as prototypes. *Journal of Personality and Social Psychology, 41,* 656–670.

Brod, M., & Heurtin-Roberts, S. (1992). Older Russian émigrés and medical care. *The Western Journal of Medicine, 157,* 333–336.

Brown, P., & Levinson, S. (1987). *Politeness: Some universals in language usage.* Cambridge, UK: Cambridge University Press.

Bruner, J. (1999). Narratives of aging. *Journal of Aging Studies, 13,* 7.

Burgoon, J. K. (1995). Cross-cultural and intercultural applications of expectancy violations theory. In R. L. Wiseman (Ed.), *Intercultural communication theory* (pp. 194–214). Thousand Oaks, CA: Sage.

Cai, D., Giles, H., & Noels, K. (1998). Elderly perceptions of communication with older and younger adults in China: Implications for mental health. *Journal of Applied Communication Research, 26,* 32–51.

Cantor, M., & Brennan, M. (2000). *Social care of the elderly: The effects of ethnicity, class, and culture.* New York: Springer.

Caporael, L. (1981). The paralanguage of caregiving: Baby talk to the institutionalized aged. *Journal of Personality and Social Psychology, 40,* 876–884.

Caporael, L., Lukaszewski, M. P., & Culbertson, G. H. (1983). Secondary baby talk: Judgments by institutionalized elderly and their caregivers. *Journal of Personality and Social Psychology, 44,* 746–754.

Carstensen, L. L. (1991). Socioemotional activity theory: Social activity in life-span context. *Annual Review of Gerontology and Geriatrics, 11,* 195–217.

Carstensen, L. L. (1992). Social and emotional patterns in adulthood: Support for socioemotional selectivity theory. *Psychology and Aging, 3,* 331–338.

Carstensen, L. L. (1995). Evidence for a life-span theory of socioemotional selectivity. *Current Directions in Psychological Science, 4,* 151–156.

Chang, H. C. (1997). Language and words: Communication and the Analects of Confucius. *Journal of Language and Social Psychology, 16,* 107–131.

Chang, W. C., Chua, W. L., & Toh, Y. (1997). The concept of psychological control in the Asian context. In K. Leung, U. Kim, S. Yamaguchi, & Y. Kashima (Eds.), *Progress in Asian social psychology* (Vol. 1, pp. 139–150). Singapore: Wiley.

Chen, G.-M., & Starosta, W. J. (1998). *Foundations of intercultural communication.* Boston: Allyn & Bacon.

Chen, X., & Silverstein, M. (2000). Intergenerational social support and the psychological well-being of older parents in China. *Research on Aging, 22,* 43–65.

Chiu, W. C. K., Chan, A. W., Snape, E., & Redman, T. (2001). Age stereotypes and discriminatory attitudes towards older workers: An East–West comparison. *Human Relations, 54,* 629–661.

Choi, S.-J. (2000). Ageing in Korea: Issues and policies. In D. R. Phillips (Eds.), *Ageing in the Asia–Pacific Region: Issues, policies, and future trend* (pp. 223–242). London: Routledge.

Chow, N. W. (1983). The Chinese family and support of the elderly in Hong Kong. *The Gerontologist, 23,* 584–588.

Cocroft, B. A., & Ting-Toomey, S. (1994). Facework in Japan and the United States. *International Journal of Intercultural Relations, 18,* 469–506.

Cody, M. J., Dunn, D., Hoppin, S., & Wendt, P. (1999). Silver surfers: Training and evaluating Internet use among older adult learners. *Communication Education, 48,* 269–284.

Cole, T. R. (1989). Generational equity in America: A cultural historian's perspective. *Social Science and Medicine, 29,* 377–383.

Coupland, N., Coupland, J., & Giles, H. (1989). Telling age in later life: Identity and face implications, *Text, 9,* 129–151.

Coupland, N., Coupland, J., & Giles, H. (1991). *Language, society, and the elderly: Discourse, identity and ageing.* Oxford, UK: Basil Blackwell.

Coupland, N., Coupland, J., Giles, H., & Henwood, K. (1988). Accommodating the elderly: Invoking and extending a theory. *Language and Society, 17,* 1–41.

Coupland, N., & Nussbaum, J. F. (1993). *Discourse and lifespan identity*. Newbury Park, CA: Sage.

Coupland, N., Wiemann, J. M., & Giles, H. (1991). Talk as "problem" and communication as "miscommunication": An integrative analysis. In N. Coupland, H. Giles, & J. M. Wiemann (Eds.), *Miscommunication and problematic talk* (pp. 1–17). Newbury Park, CA: Sage.

De Ridder, R., & Tripathi, R. C. (1992). *Norm violation and intergroup relations*. Oxford, UK: Clarendon.

Delgado, M. (1998). *Latino elders and the twenty-first century: Issues and challenges for culturally competent research and practice*. Binghamton, NY: Haworth.

Du Pre, A. (2000). *Communicating about health: Current issues and perspectives*. Mountain View, CA: Mayfield.

Ericsson, K. A., & Smith, J. (Eds.). (1991). *Towards a general theory of expertise: Prospects and limits*. New York: Cambridge University Press.

Estes, C. L. (1979). *The aging enterprise*. San Francisco: Jossey-Bass.

European Community. (1998). *Projects assisting older workers in European countries: A review of the findings of Eurowork age*. Luxembourg: Office of Official Publications of the European Community.

Ferraro, K. F., & Farmer, M. M. (1996). Double jeopardy, aging as leveler, or persistent inequality? A longitudinal study of White and Black Americans. *The Journals of Gerontology: Social Sciences, 51B*, S319–S328.

Finkelstein, L. M., & Burke, M. J. (1998). Age stereotyping at work: The role of rater and contextual factors on evaluations of job applicants. *The Journal of General Psychology, 125*, 317–345.

Fox, S., & Giles, H. (1993). Accommodating intergenerational contact: A critique and theoretical model. *Journal of Aging Studies, 7*, 423–451.

Fry, C. L. (2000). Culture, age, and subjective well-being: Health, functionality, and the infrastructure of eldercare in comparative perspective. *Journal of Family Issues, 21*, 751–776.

Fry, P. S. (2000). Guest editorial, Special issue: Aging and quality of life (QOL)—The continuing search for quality of life indicators. *International Journal of Aging and Human Development, 50*, 245–261.

Gallois, C., Giles, H., Jones, E., Cargile, A., & Ota, H. (1995). Accommodating intercultural encounters: Elaborations and extensions. In R. Wiseman (Ed.), *Intercultural communication theory* (pp. 115–147). Thousand Oaks, CA: Sage.

Gallois, C., Giles, H., Ota, H., Pierson, H. D., Ng, S.-H., Lim, T.-S., Maher, J., Somera, L., Ryan, E. B., & Harwood, J. (1999). Intergenerational communication across the Pacific Rim: The impact of filial piety. In J.-C. Lasry, J. Adair, & K. Dion (Eds.), *Latest contributions to cross-cultural psychology* (pp. 192–211). Lisse, the Netherlands: Swets & Zeitlinger.

Gao, G., & Ting-Toomey, S. (1998). *Communicating effectively with Chinese*. Thousand Oaks, CA: Sage.

Garnst, F. (1969). *The Qemant: A pagan peasantry of Ethiopia*. New York: Rinehart Winston.

Garstka, T. A., Branscombe, N. R., & Hummert, M. L. (1997). *Age group identification in young, middle aged, and elderly adults*. Unpublished manuscript, University of Kansas, Department of Psychology.

Gelfand, D. E. (1989). Immigration, aging and intergenerational relationships. *The Gerontologist, 29*, 366–372.

Gerbner, G. (1993). Learning productive aging as a social role: The lessons of television. In S. A. Bass, F. G. Caro, & Y. P. Chen (Eds.), *Achieving a productive aging society*. (pp. 207–219). Westport, CT: Auburn House.

Gergen, K. J. (1991). *The saturated self: Dilemmas of identity in contemporary life*. New York: Basic Books.

Gibbons, A. (1990). Gerontology research comes of age. *Science, 250*, 622–625.

Giddens, A. (1984). *The constitution of society: Outline of a theory of structuration*. Cambridge, UK: Polity.

Giles H., & Coupland, N. (1991). *Language: Contexts and consequences*. Pacific Grove, CA: Brooks/Cole.

Giles, H., Coupland, N., & Wiemann, J. M. (1992). "Talk is cheap" but "My word is my bond": Beliefs about talk. In K. Bolton & H. Kwok (Eds.), *Sociolinguistics today: International perspectives* (pp. 218–243). London: Routledge & Kegan Paul.

Giles, H., Fortman, J., Honeycutt, J., & Ota, H. (2003). Future selves and others: A life span and cross-cultural perspective. *Communication Report, 16*, 1–22.

Giles, H., Fox, S., Harwood, J., & Williams, A. (1994). Talking age and aging talk: Communicating through the life span. In M. L. Hummert, J. M. Wiemann, & J. F. Nussbaum (Eds.), *Interpersonal communication in older adulthood: Interdisciplinary theory and research* (pp. 130–161). Thousand Oaks, CA: Sage.

Giles, H., Fox, S., & Smith, E. (1993). Patronizing the elderly: Intergenerational evaluations. *Language and Social Interaction, 26*, 126–149.

Giles, H., Harwood, J., Pierson, H. D., Clément, R., & Fox, S. (1998). Stereotypes of the elderly and evaluations of patronizing speech: A cross-cultural foray. In R. K. Agnihotri, A. L. Khanna, & I. Sachdev (Eds.), *Social psychological study of language* (pp. 151–186). New Delhi, India: Sage.

Giles, H., Henwood, K., Coupland, N., Harriman, J., & Coupland, J. (1992). Language attitudes and cognitive mediation. *Human Communication Research, 18,* 500–527.

Giles, H., Liang, B., Noels, K. A., & McCann, R. M. (2001). Communicating across and within generations: Taiwanese, Chinese-Americans, and Euro-Americans' perceptions of communication. *Journal of Asian Pacific Communication, 11,* 161–176.

Giles, H., McCann, R. M., Ota, H., & Noels, K. (2002). Challenging intergenerational stereotypes across Eastern and Western cultures. In M. S. Kaplan, N. Z. Henkin, & A. T. Kusano (Eds.), *Intergenerational program strategies from a global perspective.* (pp. 13–28) Honolulu: University Press of America.

Giles, H., Noels, K. A., Ota, H., Ng., S. H., Gallois, C., Ryan, E. B., Williams, A., Lim, T.-S., Somera, L., Tao, H., Lim, T.-S., & Sachdev, I. (2000). Age vitality across 11 nations. *Journal of Multilingual and Multicultural Development, 21,* 308–323.

Giles, H., & Williams, A. (1994). Patronizing the young: Forms and evaluations. *International Journal of Aging and Human Development, 39,* 33–53.

Giles, H., Williams, A., & Coupland, N. (1990). Communication, health, and ageing: Approaches and theoretical frameworks. In H. Giles, N. Coupland, & J. Wiemann (Eds.), *Communication, health, and the elderly: Fulbright international colloquium* (Vol. 8, pp. 1–28). Manchester, UK: Manchester University Press.

Goffman, E. (1963). *Stigma: Management of spoiled identity.* New York: Simon & Schuster.

Goldman, R. J., & Goldman, J. D. (1981). How children view old people and aging: A developmental study of children in four countries. *Australian Journal of Psychology, 33,* 405–418.

Graham, S. (1996). Debunk the myths about older workers. *Safety and Health, 153,* 38–41.

Greenberg, J., Solomon, S., & Pyszczynski, T. (1997). Terror management theory of self-esteem and cultural worldviews: Empirical assessments and conceptual refinements. In M. Zanna (Ed.), *Advances in experimental social psychology* (Vol. 29, pp. 61–139). San Diego, CA: Academic Press.

Gudykunst, W. B. (1983a). *Intercultural communication theory.* Beverly Hills, CA: Sage.

Gudykunst, W. B. (1983b). Uncertainty reduction and predictabilities of behavior in low- and high-context cultures. *Communication Quarterly, 33,* 49–55.

Gudykunst, W. B. (1995). Anxiety/uncertainty management (AUM) theory: Current status. In R. L. Wiseman (Ed.), *Intercultural communication theory* (pp. 8–59). Thousand Oaks, CA: Sage.

Gudykunst, W. B. (1998). *Bridging differences: Effective intergroup communication* (3rd ed.). Thousand Oaks, CA: Sage.

Gudykunst, W. B., Gao, G., Nishida, T., Nadamitsu, Y., & Sakai, J. (1992). Self-monitoring in Japan and the United States. In S. Iwawaki, Y. Kashima, & K. Leung (Eds.), *Innovations in cross-cultural psychology* (pp. 185–194). Amsterdam: Swets & Zeitlinger.

Gudykunst, W. B., Matsumoto, Y., Ting-Toomey, S., Nishida, T., & Heyman, S. (1996). The influence of cultural individualism-collectivism, self-construals, and individual values on communication styles across cultures. *Human Communication Research, 22,* 510–543.

Gudykunst, W. B., & Ting-Toomey, S. (1988). *Culture and interpersonal communication.* Newbury Park, CA: Sage.

Gudykunst, W. B., Yoon, Y., & Nishida, T. (1987). The influence of individualism–collectivism on perceptions of communication in ingroup and outgroup relationships. *Communication Monographs, 54,* 295–306.

Hall, E. T. (1976). *Beyond culture.* Garden City, NY: Doubleday.

Harre, R. (1983). *Personal being: A theory for individual psychology.* Oxford: Basil Blackwell.

Harris, P. B. & Long, S. O. (2000). Recognizing the need for gender-responsive family caregiving policy: Lessons from male caregivers. In S. O. Long (Ed.), *Caring for the elderly in Japan and the U.S.: Practices and policies* (pp. 248–272). London: Routledge.

Harwood, J. (1997). Viewing age: Life span identity and television viewing choices. *Journal of Broadcasting & Electronic Media, 41,* 203–213.

Harwood, J. (2000). "Sharp!" Lurking incoherence in a television portrayal of an older adult. *Journal of Language and Social Psychology, 19,* 110–140.

Harwood, J., Giles, H., & Bourhis, R. Y. (1994). The genesis of vitality theory: Historical patterns and discoursal dimensions. *International Journal of the Sociology of Language, 108,* 167–206.

Harwood, J., Giles, H., Ota, H., Pierson, H. D., Gallois, C., Ng, S.-H. et al. (1996). College students' trait rating of three age groups around the Pacific Rim. *Journal of Cross-Cultural Gerontology, 11,* 307–317.

Harwood, J., Giles, H., Pierson, H. D., Clément, R., & Fox, S. (1994). Vitality perceptions of age categories in California and Hong Kong. *Journal of Multilingual and Multicultural Development, 15,* 311–318.

Harwood, J., Giles, H., & Ryan, E. B. (1995). Aging, communication and intergroup theory: Social identity and intergenerational communication. In J. F. Nussbaum & J. Coupland (Eds.), *Handbook of communication and aging research* (pp. 133–159). Hillsdale, NJ: Lawrence Erlbaum Associates.

Harwood, J., & Roy, A. (1999). The portrayal of older adults in Indian and U.S. magazine advertisements. *The Howard Journal of Communication, 10,* 269–280.

Harwood, J., Ryan, E. B., Giles, H., & Tysoski, S. (1997). Evaluations of patronizing speech and three response styles in a non-service-providing context. *Journal of Applied Communication Research, 25,* 170–195.

Harwood, J., & Williams, A. (1998). Expectations for communication with positive and negative subtypes of older adults. *International Journal of Aging and Human Development, 47,* 11–33.

Heikkinen, R.-L. (1993). Patterns of experienced aging with a Finnish cohort. *International Journal of Aging and Human Development, 36,* 269–277.

Hewstone, M., & Brown, R. J. (1986). "Contact is not enough": An intergroup perspective on the contact hypothesis. In M. Hewstone, & R. Brown (Eds.), *Contact and conflict in intergroup relations* (pp. 1–44). Oxford, UK: Basil Blackwell.

Hickson, L., Worrall, L., Yiu, E., & Barnett, H. (1996). Planning a communication education program for older people. *Educational Gerontology, 22,* 257–269.

Hilker, M. A. (1991). Generational viewpoints in culturally diverse literature: Commentary. *International Journal of Aging and Human Development, 33,* 211–215.

Hill, L. B., Long. L. W., & Cupach, W. R. (1997). Aging and the elders from a cross-cultural perspective. In H. S. N. Al-Deen (Ed.), *Cross-cultural communication and aging in the United States* (pp. 5–22). Mahwah, NJ: Lawrence Erlbaum Associates.

Hinde, R. (1991). A biologist looks at anthropology. *Man* (N. S.), *26,* 583–608.

Ho, D. Y.-F. (1982). Asian concepts in behavioral science. *Pscyhologica, 25,* 228–235.

Ho, D. Y.-F. (1994). Filial piety, authoritarian moralism, and cognitive conservativism in Chinese societies. *Genetic, Social and General Psychology Monographs, 120,* 349–365.

Hofstede, G. (1984). *Culture's consequences.* Beverly Hills, CA: Sage.

Hofstede, G. (1991). *Culture and organizations: Software of the mind.* New York: McGraw-Hill.

Holladay, S. J., & Kerns, K. S. (1999). Do age differences matter in close and casual friendships? A comparison of age discrepant and age peer friendships. *Communication Reports, 12,* 101–114.

Hollis-Sawyer, L. A., & Sterns, H. L. (1999). A novel goal-oriented approach for training older adult computer novices: Beyond the effects of individual-difference factors. *Educational Gerontology, 25,* 661–684.

Holtgraves, T. (1997). Styles of language use: Individual and cultural variability in conversational indirectness. *Journal of Personality and Social Psychology, 73,* 624–636.

Holtzman, J. M., & Akiyama, H. (1985). What children see: The aged on television in Japan and the United States. *The Gerontologist, 25,* 62–68.

Hugo, G. (2000). Lasnia-elderly people in Indonesia at the turn of the century. In D. R. Phillips (Ed.), *Ageing in the Asia-Pacific region: Issues, policies, and future trends* (pp. 299–321). London: Routledge.

Hummert, M. L., & Ryan, E. B. (1996). Toward understanding variations in patronizing talk addressed to older adults: Psycholinguistic features of care and control. *International Journal of Psycholinguistics, 12,* 149–169.

Hummert, M. L., Shaner, J. L., & Garstka, T. A. (1995). Cognitive processes affecting communication with older adults: The case for stereotypes, attitudes, and beliefs about communication. In J. F. Nussbaum & J. Coupland (Eds.), *Handbook of communication and aging research* (pp. 105–132). Mahwah, NJ: Lawrence Erlbaum Associates.

Ide, S. (1987). Gendaino keigoriron: Nihonto oubeino noukatsue [Contemporary Theories of politeness and honorific expressions: Toward the integration of Japanese and Western theories]. *Gengo, 16*(8), 26–31.

Ingersoll-Dayton, B. & Saengtienchai, C. (1999). Respect for the elderly in Asia: Stability and change. *International Journal of Aging and Human Development, 48*, 113–130.

Jitapunkul, S., Songkhla, M. N., Chayovan, N., Chirawatkul, A., Choprapawon, C., Kachondahm, Y., & Buasai, S. (1999). A national survey of health service use in Thai elders. *Age and Aging, 28*, 67–71.

Kagitcibasi, C. (1996). *Family and human development across cultures: A view from the other side.* Mahwah, NJ: Lawrence Erlbaum Associates.

Kalache, A. (1995). Aging well! *World Health, 48*, 21.

Kanagawa, C., Cross, S. E., & Markus, H. R. (2001). "Who am I?": The cultural psychology of the conceptual self. *Personality and Social Psychology Bulletin, 27*, 90–103.

Kashima, Y., Siegal, M., Tanaka, K., & Kashima, E. S. (1992). Do people believe behaviors are consistent with attitudes? Towards a cultural psychology of attribution processes. *British Journal of Social Psychology, 31*, 111–124.

Katori, A. (2000). *Oi to media* [Aging and media]. Tokyo: Hokuju shuppan.

Kauh, T.-O. (1997). Intergenerational relations: Older Korean Americans experiences. *Journal of Cross-Cultural Gerontology, 12*, 245–271.

Keifer, C. W. (1990). The elderly in modern Japan: Elite, victims or plural player. In J. Sokolovsky (Ed.), *The cultural context of aging: Worldwide perspectives* (pp.181–196). New York: Bergin & Garvey.

Keith, J., Fry, C. L., & Ikels, C. (1990). Community as context for successful aging. In J. Sokolovsky (Ed.), *The cultural context of aging* (pp. 245–261). New York: Bergin & Garvey.

Kemper, S. (1994). "Elderspeak": Speech accommodation to older adults. *Aging and Cognition, 1*, 17–38.

Kending, H., Koyano, W., Asakawa, T., & Ando, T. (1999). Social support of older people in Australia and Japan. *Ageing and Society, 19*, 185–207.

Kim, K. C., Kim, S., & Hurh, W. M. (1991). Filial piety and intergenerational relationships in Korean immigrant families. *International Journal of Aging and Human Development, 33*, 233–245.

Kim, K.-D. (2000). Cultural stereotypes of old age. In V. L. Bengston, K.-D. Kim, G. C. Myers, & K.-S. Eun (Eds.), *Aging in East and West: Families, states and the elderly* (pp. 227–242). New York: Springer.

Kim, M.-S. (1993). Cross-cultural comparisons of the perceived importance of the conversational constraints. *Human Communication Research, 21*, 128–151.

Kim, M.-S. (1995). Toward a theory of conversational constraints: Focusing on individual-level dimensions of culture. In R. L. Wiseman (Ed.), *Intercultural communication theory* (pp. 148–169). Thousand Oaks, CA: Sage.

Kim, U. (1994). Individualism and collectivism: Conceptual clarification and elaboration. In U. Kim, H. C. Triandis, C. Kagitcibasi, S.-C. Choi, & G. Yoon (Eds.), *Individualism and collectivism: Theory, method, and applications.* (pp. 19–40). Newbury Park, CA: Sage.

Kim, U., & Yamaguchi, S. (1995). Cross-cultural research methodology and approach: Implications for the advancement of Japanese social psychology. *Research in Social Psychology, 10*, 168–179.

Kim, Y. Y., & Gudykunst, W. B. (1988). *Theories in intercultural communication.* Newbury Park, CA: Sage.

Kite, M. E., & Johnson, B. T. (1988). Attitudes toward older and younger adults: A meta-analysis. *Psychology and Aging, 3*, 233–244.

Kluckhohn, F. R. (1950). Dominant and substitute profiles of cultural orientations: Their significance for the analysis of social stratification. *Social Forces, 23*, 376–393.

Kluckhohn, F., & Strodtbeck, F. (1961). *Variations in value orientations.* Evanston, IL: Row, Peterson.

Knapp, J. L., & Stubblefield, P. (2000). Changing students' perceptions of aging: The impact of an intergenerational service learning course. *Educational Gerontology, 26*, 611–621.

Knodel, J., Chayovan, N., Graisurapong, S., & Suraratdecha, C. (2000). Ageing in Thailand: An overview of formal and informal support. In D. R. Phillips (Eds.), *Ageing in the Asia–Pacific region: Issues, policies, and future trends* (pp. 243–266). London: Routledge.

Kontos, P. C. (1999). Local biology: Bodies of difference in ageing studies. *Ageing and Society, 19*, 677–689.

Koseki, L. K. (1996). A study of utilization and satisfaction: Implications for cultural concepts and design of aging services. *Journal of Aging and Social Policy, 8*, 59–75.

Koyano, W. (1989). Japanese attitudes toward the elderly: A review of research findings. *Journal of Cross-Cultural Gerontology, 4*, 335–345.

Koyano, W. (1996). Filial piety and intergenerational solidarity in Japan. *Australian Journal on Ageing, 15*, 51–56.

Koyano, W. (1999). Population aging, changes in living arrangement, and the new long-term care system in Japan. *Journal of Sociology and Social Welfare, 26*, 155–167.

Lee, Y.-R. & Sung, K.-T. (1997). Cultural differences in caregiving motivations for demented parents: Korean caregivers versus American caregivers. *International Journal of Aging and Human Development, 44*, 115–127.

Levy, B., (1996). Improving memory in old age through implicit self-stereotyping. *Journal of Personality and Social Psychology, 71*, 1092–1107.

Levy, B., & Langer, E. (1994). Aging free from negative stereotypes: Successful memory in China and among the American deaf. *Journal of Personality and Social Psychology, 66*, 989–997.

Liu, J. H., Ng, S.-H., Weatherall, A., & Loong C. (2000). Filial piety, acculturation, and intergenerational communication among New Zealand Chinese. *Basic and Applied Social Psychology, 22*, 213–223.

Lock, M. (1986). Ambiguities of aging: Japanese menopause. *Culture, Medicine, and Psychiatry, 10*, 23–46.

Long, S. O. (2000). Introduction. In S. O. Long (Ed.). *Caring for the elderly in Japan and the U.S.: Practices and policies* (pp. 1–16). London: Routledge.

Luhtanen, R., & Crocker, J. (1992). A collective self-esteem scale. *Personality and Social Psychology Bulletin, 18*, 302–318.

Lynch, S. M. (2000). Measurement and predictions of aging anxiety. *Research on Aging, 22*, 533–558.

Lynn, J. (2000). Preface. In K. L. Braun, J. H. Pietsch, & P. L. Blanchette (Eds.), *Cultural issues in end-of-life decision making* (pp. ix–xi). Thousand Oaks, CA: Sage.

Maltz, D. N., & Borker, R. A. (1982). A cultural approach to male–female miscommunication. In J. J. Gumperz (Ed.), *Language and social identity* (pp. 195–216). London: Wiley.

Mantecon, V. H. (1993). Where are the archetypes? Searching for symbols of women's midlife passage. *Women and Therapy, 14*, 77–88.

Mares, M.-L., & Cantor, J. (1992). Elderly viewers' responses to televised portrayals of old age: Empathy and mood management versus social comparison. *Communication Research, 19*, 459–478.

Markus, H. R., & Kitayama, S. (1991). Culture and self: Implications for cognition, emotion and motivation. *Psychological Review, 98*, 223–253.

Markus, H. R., & Kitayama, S. (1994). A collective fear of the collective: Implications for selves and theories of selves. *Personality and Social Psychological Bulletin, 20*, 568–579.

Martin, G. (1988). Aging in Asia. *Journal of Gerontology: Social Sciences, 43*, S99–S113.

Martin-Combs, C. P., & Bayne-Smith, M. (2000). Quality of life-satisfaction among Black women 75 years and older. *Journal of Gerontological Social Work, 34*(1), 63–80.

Masataka, N. (2000). *Oiwakoushite tsukurareru: Kokorotokaradano kareihenka* [*People age in this way: Psychological and physical ageing*]. Tokyo: Chukoshinsho.

Matsudo, T., & Takata, T. (2000). *Henbosuru Asia no syakaishinrigaku* [*Changing Asian Social Psychology*]. Tokyo: Nakanishya.

Matsumoto, D. (1999). Culture and self: An empirical assessment of Markus and Kitayama's theory of independent and interdependent self-construal. *Asian Journal of Social Psychology, 2*, 289–310.

Matsumoto, Y. (1988) Reexamination of the universality of face: Politeness phenomena in Japanese. *Journal of Pragmatics, 12*, 403–26.

McCann, R., Giles, H. (2002). Ageism in the workplace: A communication perspective. In T. D. Nelson, (Ed), *Ageism: Stereotyping and prejudice against older persons*. Cambridge, MA: MIT Press. 163–199.

McCann, R., Giles, H., Ota, H., & Caraker, R. (in press). Perceptions of intra- and intergenerational communication among young adults in Thailand, Japan, and the U.S.A. *Communication Report*.

McKenzie, A., & Cannon, D. (1998). Across the ages: Generational communication matters. *Communication World, 15*, 21–24.

Mebane, F. (2001). Want to understand how Americans viewed long-term care in 1998? Start with media coverage. *The Gerontologist, 41*, 24–33.

Meeks, S., & Murrell, S. A. (2001). Contribution of education to health and life satisfaction in older adults mediated by negative affect. *Journal of Aging and Health, 13*(1), 92–119.

Mehta, K. (1997). Respect redefined: Focus group insights from Singapore. *International Journal of Aging and Human Development, 44,* 205–219.

Mercer, C. (1999). Cross-cultural attitudes to the menopause and the ageing female. *Age and Ageing, 28,* 12–17.

Mercer, S. O. (1994). Navajo elders in a reservation nursing home: Health status profile. *Journal of Geronotological Social Work, 23,* 3–29.

Mernissi, F. (1975). *Beyond the veil: Male–female dynamics in a modern Muslim society.* Cambridge, MA: Schenkman.

Merrill, D., & Dill, A. (1990). Ethnic differences in older mother–daughter co-residence. *Ethnic Groups, 8,* 201–214.

Miller, J. (1992). A cultural perspective on the morality of beneficence and interpersonal responsibility. In S. Ting-Toomey (Ed.), *Cross-cultural interpersonal communication* (pp. 11–27). Newbury Park, CA: Sage.

Miller, P. N., Miller, D. W., McKibbin, E. M., & Pettys, G. L. (1999). Stereotypes of the elderly in magazine advertisements 1956–1996. *International Journal of Aging and Human Development, 49,* 319–337.

Minkler, M., & Estes, C. L. (Eds.). (1999). *Critical gerontology: Perspectives from a political and moral economy.* Amityville, NY: Baywood.

Moghaddam, F. M., Slocum, N. R., Norman, F., Mor, T., & Harre, R. (2000). Toward a cultural theory of duties. *Culture & Psychology, 6,* 275–302.

Morisaki, S., & Gudykunst, W. B. (1994). Face in Japan and the United States. In S. Ting-Toomey (Ed.), *The challenge of facework: Cross-cultural and interpersonal issues* (pp. 47–96). New York: State University of New York Press.

Mouton, C. P. (2000). Cultural and religious issues for African Americans. In K. L. Braun, J. H. Pietsch, & P. L. Blanchette (Eds.), *Cultural issues in end-of-life decision making* (pp. 71–82). Thousand Oaks, CA: Sage.

Nakano, I. (1991). Jido no rojin image [Children's image of older adults]. *Shakaironengaku, 34,* 23–36.

Nakazato, K. (1990). Shakaikarano danzetuniyoru komyunikeeshon shougai: Roujinno shitennkara [Alienation from society and communication problems: Older adults perspective]. *Kyouikuto Igaku, 38,* 581–587.

Ng., S.-H., Liu, J. H., Weatherall, A., & Loong, C. F. (1997). Younger adults' communication experiences and contact with elders and peers. *Human Communication Research, 24,* 64–82.

Niemala, P., & Lento, R. (1993). The significance of the 50th birthday for women's individuation. *Women and Therapy, 14,* 117–127.

Nisbett, R. E., Peng, K., Choi, I., & Norenzayan, A. (2001). Culture and systems of thought: Holistic versus analytic cognition. *Psychological Review, 108,* 291–310.

Noels, K. A., Giles, H., Gallois, C., & Ng, S. H. (2001). Intergenerational communication and psychological adjustment: A cross-cultural examination in Hong Kong and Australian adults. In M. L. Hummert & J. F. Nussbaum (Eds.), *Aging, communication, and health: Linking research and practice for successful aging* (pp. 249–278). Mahwah, NJ: Lawrence Erlbaum Assoicates.

Nussbaum, J. F., & Baringer, D. K. (2000). Message production across the life span: Communication and aging. *Communication Theory, 10,* 200–209.

Orbe, M. P. (1998). From the standpoint(s) of traditionally muted groups: Explicating a co-cultural communication theoretical model. *Communication Theory, 8,* 1–26.

Ota, H. (2001). *Intergenerational communication in Japan and the United States: Debunking the myth of respect for older adults in contemporary Japan.* Unpublished doctoral dissertation, University of California, Santa Barbara, Department of Communication.

Ota, H., Giles, H., & Gallois, C. (2002). Perceptions of younger, middle-aged, and older adults in Australia and Japan: Stereotypes and age group vitality. *Journal of Intercultural Studies. 23,* 253–266.

Ota, H., Giles, H., Harwood, J., Pierson, H. D., Gallois, & C. Ng et al. (1996). *A neglected dimension of communication and aging: Filial piety across eight nations.* San Diego: Speech Communication Association.

Ota, H., Harwood, J., Williams, A., & Takai, J. (2000). Age identity in the U.S. and Japan. *Journal of Multicultural and Multilingual Development, 21,* 33–41.

Palmore, E. B. (1999). *Ageism: Negative and positive* (2nd ed.). New York: Springer.

Park, D. C., Nisbett, R., & Hedden, T. (1999). Aging, culture, and cognition. *The Journals of Gerontology, Series B, 54,* P75–P84.

Peng, D., & Zhi-gang, G. (2000). Population aging in China. In D. R. Phillips (Ed.), *Ageing in the Asia-Pacific Region: Issues, policies and future trends* (pp. 194–209). London: Routledge.

Peterson, P. G. (2000). Does an aging society mean an aging culture? *The Futurist, 34,* 20.

Phillips, D. R. (2000). *Ageing in the Asia–Pacific region: Issues, policies and future trends.* London: Routledge.

Restrepo, H. E., & Rozental, M. (1994). The social impact of aging populations: Some major issues. *Social Science and Medicine, 39,* 1323–1338.

Rix, S. E. (1996). The challenge of an aging work force: Keeping older workers employed and employable. *Journal of Aging and Social Policy, 8,* 79–96.

Roberts, S. D., & Zhou, N. (1997). The 50 and older characters in the advertisements of *Modern Maturity:* Growing older, getting better? *The Journal of Applied Gerontology, 16*(2), 208–220.

Robertson, A. F. (1996). The development of meaning: Ontogeny and culture. *Journal of the Royal Anthropological Institute, 2,* 591–610.

Robinson, J. D., & Skill, T. (1995). Media usage patterns and portrayals of the elderly. In J. F. Nussbaum & J. Coupland (Eds.), *Handbook of communication and aging research* (pp. 359–392). Mahwah, NJ: Lawrence Erlbaum Associates.

Rose, J. H., Bowman, K., & Kresevic, D. (2000). Nurse versus family caregiver perspectives on hospitalized older patients: An exploratory study of agreement at admission and discharge. *Health Communication, 12,* 63–80.

Rosenberg, M. (1979). *Conceiving the self.* New York: Basic Books.

Roy, A. Y., & Harwood, J. (1997). Underrepresented, positively portrayed: Older adults in television. *Journal of Applied Communication Research, 25,* 39–56.

Rubin, A. M., & Rubin, R. B. (1981). Age, context, and television use. *Journal of Broadcasting, 25,* 1–13.

Rubin, A. M., & Rubin, R. B. (1982a). Contextual age and television use. *Human Communication Research, 8,* 228–244.

Rubin, R. B., & Rubin, A. M. (1982b). Contextual age and television use: Re-examining a life-position indicator. *Communication yearbook 6,* 583–604.

Rubin, A. M., & Rubin, R. B. (1986). Contextual age as a life-position index. *International Journal of Aging and Human Development, 23,* 27–45.

Russell, C., & Schofield, T. (1999). Social isolation in old age: A qualitative exploration of service providers' perceptions. *Ageing and Society, 19,* 69–91.

Ryan, E. B., Anas, A. P. Hummert, M. L., & Laver-Ingram, A. (1998). Young and older adults' views of telephone talk: Conversation problems and social uses. *Journal of Applied Communication Research, 26,* 83–98.

Ryan, E. B., & Cole, R. (1990). Evaluative perceptions of interpersonal communication with elders. In H. Giles, N. Coupland, & J. M. Wiemann (Eds.), *Communication, health and the elderly* (pp. 172–190). Manchester, UK: Manchester University Press.

Ryan, E. B., Giles, H., Bartolucci, G., & Henwood, K. (1986). Psycholinguistic and social psychological components of communication by and with the elderly. *Language and Communication, 6*(1/2), 1–24.

Ryan, E. B., Meredith, S. D., MacLean, M. J., & Orange, J. B. (1995). Changing the way we talk with elders: Promoting health using the communication enhancement model. *International Journal of Aging and Human Development, 41,* 89–107.

Ryan, E. B., Szechtman, B., & Bodkin, J. (1992). Attitudes toward younger and older adults learning to use computers. *Journal of Gerontology: Psychological Sciences, 47,* 96–101.

Salthouse, T. A. (1991). *Theoretical perspectives on cognitive aging.* Hillsdale, NJ: Lawrence Erlbaum Associates.

Sawchuk, K. (1995). From gloom to boom: Age, identity and target marketing. In M. Featherstone & A. Wernick (Eds.), *Images of ageing.* (pp. 173–187). London: Routledge.

Schein, E. H. (1985). *Organizational culture and leadership.* San Francisco: Jossey-Bass.

Schulz, R., & Heckhausen, J. (1999). Aging, culture and control: Setting a new research agenda. *The Journals of Gerontology, Series B, 54,* P139–P145.

Schwartz, S. H., & Bilsky, W. (1990). Toward a theory of the universal content and structure of values: Extensions and cross-cultural replications. *Journal of Personality and Social Psychology, 58,* 878–891.

Scott, D. M. (1993) Aging and hearing loss: Race and gender differences in African Americans and Euro-Americans. *The Howard Journal of Communications, 4,* 369–379.

Seefeldt, C., & Ahn, U. R. (1990). Children's attitudes toward the elderly in Korea and the United States. *International Journal of Comparative Sociology, 31,* 248–256.

Sharps, M. J., Price-Sharps, J. L., & Hanson, J. (1998). Attitudes of young adults toward older adults: Evidence from the United States and Thailand. *Educational Gerontology, 24,* 655–660.

Shea, P. (1985). The later years of lifelong learning. In F. Glendenning (Ed.), *Educational gerontology: International perspectives* (pp. 58–80). London: Croom Helm.

Sheldrick, M. G. (1992). Technology for the elderly. *Electronic News, 38,* 22.

Shimonaka, Y., & Nakazato, K. (1980). Psychological characteristics of Japanese aged: A comparison of sentence completion test responses of older and younger adults. *Journal of Gerontology, 35,* 891–898.

Shotter, J. (1991). Becoming someone: Identity and belonging. In N. Coupland & J. F. Nussbaum (Eds.), *Discourse and lifspan identity* (pp. 5–27). Newbury Park: Sage.

Slote, W. H., & DeVos, G. A. (1998). *Confucianism and the family.* New York: State University of New York Press.

Smith, D. H., & Smith, S. J. (1999). Chinese elders' communication about medicine. *Health Communication, 11,* 237–248.

Smith, P. B., & Bond, M. H. (1999). *Social psychology across cultures* (2nd ed.) Boston: Allyn & Bacon.

Stoller, E. P., & Gibson, R. C. (2000). *Worlds of difference: Inequality in the aging experience.* Thousand Oaks, CA: Pine Forge.

Strom, R., Collinsworth, P., Strom, S., & Griswold, D. (1993). Strengths and needs of Black grandparents. *International Journal of Aging and Human Development, 36,* 255–268.

Strom, R. D., Strom, S., Collinsworth, P., Sato, S., Makino, K., Sasaki, Y., Sasaki, H., & Nishio, N. (1995). Grandparents in Japan: A three-generational study. *International Journal of Aging and Human Development, 40,* 209–226.

Strom, R. D., Strom S. K., Wang, C.-M., Shen, Y.-L., Griswold, D., Chan, H.-S., & Yang, C.-Y. (1999). Grandparents in the United States and the Republic of China: A comparison of generations and cultures. *International Aging and Human Development, 49,* 279–317.

Stryker, S. (1987). Identity theory: Developments and extensions. In K. Hardley & T. Honess (Eds.), *Self and identity* (pp. 89–104). New York: Wiley.

Sung, K. T. (1995). Measures and dimensions of filial piety. *The Gerontologist, 35,* 240–247.

Sung, K.-T. (2001). Respect for elders: Myths and realities in East Asia. *Journal of Aging and Identities, 4,* 2000.

Tajfel, H., & Turner, J. C. (1986). The social psychology of intergroup behavior. In S. Worchel & W. G. Austin (Eds.), *The psychology of intergroup relations* (pp. 7–24). Chicago: Nelson-Hall.

Takano, Y., & Osaka, E. (1999). An unsupported common view: Comparing Japan and the U.S. on individualism/collectivism. *Asian Journal of Social Psychology, 2,* 311–343.

Talamantes, M. A., Gomez, C., & Braun, K. L. (2000). Advance directives and end-of-life care: The Hispanic perspective. In K. L. Braun, J. H. Pietsch, & P. L. Blanchette (Eds.), *Cultural issues in end-of-life decision making* (pp. 83–100). Thousand Oaks, CA: Sage.

Tan, P.-C., & Ng, S.-T. (2000). Ageing in Malaysia. In D. R. Phillips (Eds.), *Ageing in the Asia-Pacific region: Issues, policies, and future trend* (pp. 284–291). London: Routledge.

Thois, P. A. (1991). On merging identity theory and stress research. *Social Psychological Quarterly, 54,* 101–112.

Thomas, V., & Wolfe, D. B. (1995, May). Why won't television grow up? *American Demographics,* 24–29.

Thompson, P. (1992). I don't feel old: Subjective ageing and the search for meaning in later life. *Ageing and Society, 12,* 23–48.

Tijerina-Jim, A. (1993). Three native American women speak about the significance of ceremony. *Women and Therapy, 14,* 33–39.

Ting-Toomey, S. (1988). Intercultural conflict styles: A face-negotiation theory. In Y. Y. Kim & W. B. Gudykunst (Eds.). *Theories in intercultural communication* (pp. 213–238). Newbury Park, CA: Sage.

Ting-Toomey, S. (1999). *Communicating across cultures.* New York: Guilford.

Ting-Toomey, S., & Kurogi, A. (1998). Facework competence in intercultural conflict: An updated face-negotiation theory. *International Journal of Intercultural Relations, 22,* 197–225.

Torres, S. (1999). A culturally-relevant theoretical framework for the study of successful aging. *Ageing and Society, 19,* 33–51.

Triandis, H. C. (1994). *Culture and social behavior.* New York: McGraw-Hill.

Triandis, H. C. (1995). *Individualism and collectivism.* Boulder, CO: Westview.

Truluck, J. E., & Courtenay, B. C. (1999). Learning style preferences among older adults. *Educational Gerontology, 25*, 221–236.

Tulle-Winton, E. (1999). Growing old and resistance: Towards a new cultural economy of old age? *Ageing and Society, 19*, 281–299.

Turner, J. C., with Hogg, M., Oakes, P., Reicher, S., & Wetherell, M. (1987). *Rediscovering the social group: A self-categorization theory.* Oxford, UK: Basil Blackwell.

U.S. Administration on Aging. (1999). *Profile on older Americans—1999* [on line]. Washington DC: Author Available: http://www.aoa.dhhs.gov/aos/stats/profile/default.htm. Retrieved November 29, 1999; last accessed May 21, 2001.

U.S. Bureau of the Census. (1994). *Marital status and living arrangements: March 1993* (Current Population Rep. P20–478). Washington, DC: U.S. Government Printing Office.

Ushikubo, M. (1998). A study of factors facilitating and inhibiting the willingness of the institutionalized disabled elderly for rehabilitation: A United States–Japanese comparison, *Journal of Cross-Cultural Gerontology, 13*, 127–157.

Van Winkle, N. W. (2000). End-of-life decision making in American Indian and Alaska Native cultures. In K. L. Braun, J. H. Pietsch, & P. L. Blanchette (Eds.), *Cultural issues in end-of-life decision making* (pp. 127–144). Thousand Oaks, CA: Sage.

Vasil, L., & Wass, H. (1993). Portrayal of the elderly in the media: Literature review and implications for educational gerontologists. *Educational Gerontology, 19*, 71–85.

Waldrop, M. M. (1992). *Complexity: The emerging science at the edge of order and chaos.* New York: Simon & Schuster.

Weisz, J. R., Rothbaum, F. M., & Blackburn, T. C. (1984). Standing out and standing in: The psychology of control in America and Japan. *American Psychologist, 39*, 955–969.

Whitehouse, P. J., Bendezu, E., Fallcreek, S., & Whitehouse, C. (2000). Intergenerational community schools: A new practice for a new time. *Educational Gerontology, 26*, 761–770.

Wieck, K. L. (2000). Health promotion for inner-city minority elders. *Journal of Community Health Nursing, 17*, 131–139.

Williams, A., & Coupland, N. (1998). The socio-political framing of aging and communication research. *Journal of Applied Communication Research, 26*, 139–154.

Williams, A., & Garrett, P. (2001, May). *Communication evaluations across the lifespan: From adolescent storm and stress to elder aches and pains.* Paper presented at the annual meeting of the International Communication Association, Washington, D.C.

Williams, A., & Giles, H. (1996). Intergenerational conversations: Young adults' retrospective accounts. *Human Communication Research, 23*, 220–250.

Williams, A., & Nussbaum, J. F. (2001). *Intergenerational communication across the life span.* Mahwah, NJ: Lawrence Erlbaum Associates.

Williams, A., Ota, H., Giles, H., Pierson, H. D., Gallois, C., Ng., S. H. et al. (1997). Young people's beliefs about intergenerational communication: An initial cross-cultural analysis. *Communication Research, 24*, 370–393.

Wiscott, R., & Kopera-Frye, K. (2000). Sharing of culture: Adult grandchildren's perceptions of intergenerational relations. *International Journal of Aging and Human Development, 51*, 199–215.

Wiseman, R. L. (1995). *Intercultural communication theory.* Thousand Oaks, CA: Sage.

Witte, K., & Morrison, K. (1995). Intercultural and cross-cultural health communication: Understanding people and motivating healthy behaviors. In R. L. Wiseman (Ed.), *Intercultural communication theory* (pp. 216–246). Thousand Oaks, CA: Sage.

Wykle, M. L., & Ford, A. B. (1999). *Serving minority elders in the 21st century.* New York: Springer.

Yamazaki, T. (1994). Intergenerational interaction outside the family. *Educational Gerontology, 20*, 453–462.

Yeo, G., & Hikoyeda, N. (2000). Cultural issues in end-of-life decision making among Asians and Pacific Islanders in the United States. In K. L. Braun, J. H. Pietsch, & P. L. Blanchette (Eds.), *Cultural issues in end-of-life decision making* (pp. 101–125). Thousand Oaks, CA: Sage.

Ylänne-McEwen, V. (1999). "Young at heart": Discourses of age identity in travel agency interaction. *Ageing and Society, 19*, 417–440.

Yoon, C., Hasher, L., Feinberg, F., Rahhal, T. A., & Winocur, G. (2000). Cross-cultural differences in memory: The role of culture-based stereotypes about aging. *Psychology and Aging, 15,* 684–704.

Yue, X. D., & Ng, S. H. (2000). Effects of age and relation of intergenerational communication: A survey study in Beijing. *Psychologia, 43,* 102–115.

Yum, J. O. (1988). The impact of Confucianism on interpersonal relationships and communication patters in East Asia. *Communication Monographs, 55,* 374–388.

Zandi, T., Mirle, J., & Jarvis, P. (1990). Children's attitudes toward elderly individuals: A comparison of two ethnic groups. *International Journal of Aging and Human Development, 30,* 161–174.

Zhang, Y. B. & Hummert, M. L. (2001). Harmonies and tensions in Chinese intergenerational communication: Younger and older adults' accounts. *Journal of Asia Pacific Communication, 11,* 203–230.

III

The Communicative Construction of Relationships in Later Life

Each of the chapters within this section of the second edition of the handbook are updates of the chapters that appeared within the first edition. Henwood (chap. 8) reviews the literature on adult parent–child relationships. Historically, as she points out, such research has concerned itself with the changing social, economic, political, and cultural contexts of such relationships, and their connections with family organization and functioning. Set against this, the perspective here is that of a discursive social psychologist interested in the social construction of relationships. Henwood rehearses the ideas of social-psychological research in the "social identity categorization" tradition (cf. Tajfel, 1974). Such a research proposes a self-comprising a range of possible personal and social identifications (see the earlier section of this handbook) that is sensitive to context as influencing behavior and has positively evolved into a individual or dyadic focus that has the advantage of being firmly grounded in the examination of broader social factors.

The motivations, or relational variables, underlying parent–child solidarity are examined in the wake of such puzzling findings as a lack of correlation among contact, association, and affect. Henwood seeks to problematize existing explanatory constructs that have created, as she put it, an overly coherent, unified, and harmonious view of mother–daughter relationships. Henwood's review turns to specifically feminist research on mother–daughter relationships that draws extensively on psychoanalytical theory. A discussion on the notions of mothering, gender and relational identity, autonomy, and the reproduction of mothering leads to an account of her own research. This has investigated the discursive construction by daughters and mothers of their own mutual relationships, interpreted in their social, cultural, and historical contexts. Social constructions of feminine normalcy and discourses of femininity are seen as representing threats to face (Brown & Levinson, 1987) specifically, for example, to older women who may be categorized negatively if they do not fit idealized images. The constraints on potential subject positions consequently open to older recipients of such categorization can make links with the work of Foucault (1971) and Fairclough (1989). Later, she refers to wider, (or public, as Fairclough would call them) discourses with social and regulatory power.

These ideas can make clear theoretical links to Hummert, Garstka, Ryan, and Bonnesen, (chap. 4) in its discussion of negative stereotypes, as the production of such stereotypes can be seen as part of such a public discourse on older people. The kind of regulatory power involved once again revolves around the notion of an elderly identity, in offering to older people the range of personal and social identifications previously mentioned, through which they can communicate their elderliness.

Perhaps Henwood's main aim here is to argue for a particular methodological perspective for future research. She sees the communication and social integration tradition that analyzed variables, searching for causal links with outcomes, often in search of effective communication, as too deterministic. She looks to a more interpretive paradigm of inquiry, equally applicable, as she points out, to aging mothers and their sons, or to aging fathers and their children. In her own research, Henwood (Henwood, & Coughlan, 1993) argued that psychoanalysis and social constructionism enable us to explore the meaning of experience, which includes making sense of our own aging. In contrast to Henwood, Mares and Fitzpatrick (chap. 9) place their work firmly in the tradition of experimental, quantitative social science; most of the investigations they report are based on self-report and self-assessment. Nevertheless, they use some qualitative data to advantage in discussion and explanation. They are able to present a review of work from the last few decades on the aging couple that reflects largely positive findings, but they add the caveat that these findings must be read as relevant to the current cohort of elderly adults, who seem to be quite atypical in their marital constancy at the very least.

It seems to us that many of the effects here stem from changing social constructions. Salient points are that the definition of a happy marriage looks likely to have changed over time, and that divorce has apparently come to be seen as more socially acceptable (this in addition to changes in the incidence of divorce). As a means to explore the effect of changing cultural mores, like these on the aging marriage, Mares and Fitzpatrick recommend the methodologically sound but cumbersome course of longitudinal research. Indeed, there are many areas where longitudinal work could reap rewards, such as the effects on the aging identity and relationships of grandparenthood, retirement, bereavement, or the onset of chronic ill health and institutionalization.

Mares and Fitzpatrick give us a comprehensive review of quantitative research on changes in marital satisfaction across the life span, which has yielded equivocal findings. A striking statistic relating to gender differences concerns survival ages. Whereas most men over 65 spend the majority of their time in old age interacting with their wives, most women of the same age spend their time alone (men marry younger women, and women die later than men). At the upper end of the life span, of adults aged 75 and over, 70% of men and only 22% of women were married, which suggests potentially very different communication experiences across gender in old age.

Some of the findings on decline in sexual activity in old age are attributed to cohort rather than age-related change. The effects of societal beliefs and stereotypes on sexual decline are related to differences between Western societies and preindustrial or traditional cultures, where reports of sexuality frequently indicate far less age-related decline for both males and females. The role of communication in such transitions in later life has not as yet been explored. Mares and Fitzpatrick point out, for example, that how the decline in sexual activity is negotiated has not yet been studied.

Studies that have explicitly set out to examine communication between older couples have discovered historical difference in the amount of conversational contact that couples engage in (dissatisfaction with levels of communication being reported especially by females). Beyond this, research has examined issues such as self-disclosure, expression of feelings, and conflict resolution (all by self-report), and has also coded conversations between young, middle-aged, and older couples. These appeared to show patterns of low engagement among (nonconfrontational) older couples and high engagement among young couples. Mares and Fitzpatrick report some more recent research that seems to support the suggestion that older cohorts may have been less engaged (expressed in terms of reporting less time spent together, fewer feelings being expressed, and fewer problems confided), even in the early stages of their relationship.

It seems to us that what is being indicated here is quite a remarkable shift in patterns of communication across different cohort groups, with more recent marriages apparently putting greater stress on the importance of negotiation through conversation. The authors here suggest a complementary interpretation: that the observed style that has been characterized as disengagement might be because of change over time caused by a history of shared experience. Garfinkel (1967) would see this as likely leading to more implicit rather than explicit reference, with shared knowledge enabling sophisticated readings of what is on the surface to talk, not obvious to those outside the relationship. Future research into shared assumptions and their effects on marital conversations would also need to make reference to pragmatic theory (Levinson, 1983).

McKay and Caverly (chap. 10) provide a review of the research on relationships between grandparents, grandchildren, and older adults siblings. The chapter, like Henwood's, proceeds within a feminist framework that expresses concern about the reductionist perspective of tradition scientific models of social research. It moves toward a more heuristic approach, integrating aspects of gender, culture, and age into the interpretative work. The focus here is on changing norms, citing the shift away from the traditional nuclear family, the convergence of gender roles in the worlds of paid employment and the family, and multicultural variation in worldviews and lifestyles. The authors take us through research on the nature of the grandparenting role that has shown that factors such as proximity, age, family roles, emotional closeness, and parent's marital status enhance or inhibit the development of close relations between grandparents and grandchildren.

Given that 80% of older people living in the West today have living siblings and that most people remain in contact with siblings throughout their lives, it is somewhat surprising that the sibling relationship in later life has only recently been the focus of research attention, and such work is still rather limited. But as McKay and Caverly point out, the number of later life sibling relationships is steadily increasing. The authors describe McKay's work with Black grandparents and their grandchildren that serves to deny earlier myths of the dysfunctional nature of Black family life, raised as a result of comparative research on Blacks and Whites. This type of study leads them to propose a future approach incorporating the practice of feminist, life span, and multicultural ideologies. They suggest that understanding will be enriched through investigating both grandparent–grandchild and sibling relationships as "characterized by life experiences shared through stories and interpersonal discourse."

Rawlins begins his chapter (chap. 11) with the point that friendships of older adults show as much diversity in life circumstances and outlooks as any other developmental period. The positivity communicated through emphasizing development seems fruitful here, with its concomitant notion of change through choices made, as well as through constraints encountered. Rawlings uses a culturally grounded dialectical perspective, seeing friendships as complex interactive achievements and speaking of how friends manage the dialectic tension of the freedom to be independent and the freedom to be dependent. At the level of interaction, a different theoretical framework would have quite a contribution to make to our understanding of these friendship issues; that of Goffman (1967) on face work, and again Brown and Levinson (1987) on politeness. The complex of discourse strategies associated with the simultaneous protection of one's own and the interlocutor's negative face (the wish not to be imposed on) and positive face (the wish to be liked and included) could also inform Rawlings' dialectic of expressiveness and protectiveness, relevant to self-disclosure.

In reviewing gender-linked friendship patterns and the effects of retirement on friendship, Rawlins discusses some interesting differences in male and female patternings of combining commitments to paid employment, family, and friend relationships. Females often develop "interactional habits and types of friendship that foster ongoing intimacy as well as sociability." Rawlings also reports Chappell's findings that "wives . . . serve as their husbands' principle link with neighbors, relatives, and other friends". These views have much in common with those expressed by Bernard and Phillipson (this volume), who propose conversation as one of the main activities of female friendship throughout the life span. This is in comparison to male friendships, which seem to function through organized activities (e.g., sports) and shared experiences. The premium that women place on talk as a shared activity may be related to their being consistently stereotypically linked with so-called small talk, chat, or gossip, which have been pejorated, for example, in the sociolinguistic literature (for a review, see Coupland, Coupland, & Robinson, 1992), although anthropology and feminist literatures have long emphasized the positive functions of such socially cohesive talk.

It would be interesting, then, to see these different communication styles evidenced in close textual study of friendship encounters between male and female dyads across the life span. In all, Rawlins' chapter is a testament to the importance of social interaction in the maintenance of engagement, connectedness, and enjoyment in the later years of life. The functions of talk in friendship are shown in this chapter to include giving pleasure, support, and confirmation of self-worth; providing counseling; and even the redefinition of the self.

REFERENCES

Brown, P., & Levinson, S. (1987). *Politeness: Some universals in language uses.* Cambridge, UK: Cambridge University Press.

Coupland, J., Coupland, N., & Robinson, J. (1992). "How are you": Negotiating phatic Communication. *Language in Society, 21*(20), 207–230.

Fairclough, N. (1989). *Language and power.* London: Longman.

Foucault, M. (1971). Orders of discourse. *Social Science Information, 10,* 7–30.

Garfinkel, H. (1967). *Studies in ethnomethodology.* Englewood Cliffs, NJ: Prentice Hall.

Goffman, E. (1967). *Interaction ritual: Essays on face-to-face behavior.* Garden City, NY: Doubleday.

Henwood, K., & Coughlan, G. (1993). The construction of "closeness" in mother–daughter relationships across the lifespan. In N. Couplan & J. F. Nussbaum (Eds.), *Discourse and lifespan identity* (pp. 191–214). Newbury Park, CA: Sage.

Levinson, S. (1983). *Pragmatics.* Cambridge, UK: Cambridge University Press.

Tajfel, H. (1974). Social identity and intergroup behaviour. *Social Science Information, 13,* 65–93.

Adult Parent–Child Relationships: A View From Feminist and Discursive Social Psychology

Karen L. Henwood
University of East Anglia

INTRODUCTION

One consequence of the aging of Western industrialized societies, together with the tendency for parents to complete their childrearing fairly early on in the life span, is the prolonged length of time many parents and children can share as adult, and even older, individuals. [1] This situation led Shanas (1980) to describe recent cohorts of older children and their parents as "the new pioneers of our era" who have "ventured into uncharted areas of human relationships" (p. 16). A succession of substantial reviews of adult parent–child relationships (see, e.g., Hagestad, 1987; Lye, 1996; Mancini, 1989; Mancini & Bliesner, 1989) testifies to the central place occupied by these relationships within human and social sciences research. However, none of these reviews has been conducted from the perspective that follows—that of a specific interest in the social and discursive construction of relational identities in adulthood and later life.

In a previous publication, I have distinguished between, on the one hand, the tradition of intergenerational family relationships research and, on the other, analyses of adult mother–daughter relationships originating in psychoanalytic and feminist theories of women's psychological development (Henwood & Coughlan, 1993). I have also reported some original analyses of the construction of adult mother–daughter "closeness," drawing on a specifically feminist and discourse analytic approach to the practice and theory of social psychological research (Henwood, 1993). This chapter draws out some aspects of research and theorizing from the aforementioned research traditions that can shed light on the social construction of experiences, identities, and relationships

---

[1] This prolonged sharing of lifetime by adult children and their parents is sometimes called their *cobiography* (Bengston & Dannefer, 1987).

of adult parents and children. I argue that when the affective dimension of adult parent–child ties is being emphasized, this must take account of their psychological complexity, ambivalence, and division, and locate them within the social and discursive relations that continue to structure our experiences and identities later in life. Later in the chapter, I also consider some of the diverse assumptions and practices that can inform communicatively oriented family and aging research.

THE SOCIAL DIMENSION IN RESEARCH ON INTERGENERATIONAL RELATIONSHIPS IN OLDER FAMILIES

As suggested by the general rubric, investigations into adult parent–child relationships within later-life family research tend to be informed by wider analyses of kinship context and family structure. Looking back over the history of such research, three phases have been identified concerning: (a) the implications of industrialization and urbanization, and the emergence of a nuclear family structure, (b) the apparent breakdown of consensus in values between different generations during the 1960s, and (c) the challenge of increased numbers of frail older people, in particular for the provision and quality of care (Hagestad, 1987). An additional area of concern is the influence of the various life events (e.g., divorce, remarriage, widowhood, childlessness) that distinguish and shape the character of older families (see, e.g., Aldous, 1985).

Although identified as phases, all share a common concern for wider social, economic, political, and cultural processes and events, as they connect with family organization and functioning. Research concerned with the processes of industrialization investigated the possibility that it resulted in the social isolation of older people. Lineage and cross-sectional comparisons of family values prompted a reappraisal of the unidirectional model of parent-to-child socialization. An awareness among social scientists of changes in age demography, together with the widespread political and social move to "care in the community," led to research on practical and relational support—often by adult children—of older family members (see, e.g., Cicerelli, 1981, 1989; Noelker & Townsend, 1987; Townsend & Noelker, 1987), followed by investigations into ways of coping with the stresses and strains of family caregiving (Gallagher-Thompson, Coon, Rivera, Powers, & Zeiss, 1998). Interest in later-life family events has emerged partly as a consequence of the take-up of life span theory (see, e.g., Abeles, 1987), which holds that families and individuals continue to develop and change, in response to life span–specific occurrences and tasks, beyond childhood and throughout the whole of life. In addition to this, researchers are now simply more aware of the intellectual vacuum created when the diversity of people's experiences within the family—which may be structured, for example, along the social dimensions of "race," ethnicity, class, and gender, as well as age—is neglected or ignored (Bengston, Rosenthal, & Burton, 1990; see also Thompson & Walker, 1989).

The emphasis on wider social forces and influences has considerable strength when compared with those strands of research in social psychology that apparently deem the social structuring of human bonds to be irrelevant to understanding interpersonal relationships. Within such research, the character of a relationship (e.g., whether it is hostile or friendly) is reduced to matters of geographical proximity, frequency of contact, perceived attractiveness, reward schedules, belief similarity, and need compatibility. However, this is an individualistic approach that fails to take account of the transformations that can occur in human relationships when people interact together not as individuals but in terms of their membership of social categories (Tajfel, 1978; Tajfel & Turner, 1979). Research in the "social identity / categorization" tradition (see, e.g., Hogg & Abrams, 2001) adopts a theoretical model of the self as comprising a range of possible personal and social identifications—identifications that contribute to the determination of behavior according to their relative social or situational salience. This tradition of research stands in contrast to approaches that abstract thought and behavior from their social context, and that understand relational dynamics in purely personal or interpersonal terms.

The research on interpersonal relationships that was the object of Tajfel's and Turner's (1979) critique has now evolved to consider relationship initiation, growth, and dissolution over the life span, as the result of continuity and change in personal and social biography (see, e.g., Duck, 1997, 1998). In this form, it shares many common features with intergenerational family relationships research. Matters such as geographical proximity, association or contact, reward, and exchange retain a central place—but as properties of family systems, and therefore integrated with a concern for family roles, values, social expectations, and norms (see, e.g., Mangen, Bengston, & Landry, 1988). The connection between individual behavior and social functioning is accordingly preserved. The result is research where the primary objective may be, for example, to investigate the quality of dyadic relationships and their implications for psychological distress or health (see, e.g., Umberson, 1993). But this dyadic and individual focus does not displace a concern for broader social factors (in Umberson's research these are family structure, socioeconomic status, and sociocultural influences) that are built into the theoretical model and guide empirical research.[2]

INTERGENERATIONAL SOLIDARITY, AFFECT, AND THE PSYCHOLOGICAL DYNAMICS OF ADULT PARENT–CHILD RELATIONSHIPS

One of the questions with which research on adult parent–child relationships is most concerned is why the relationship endures—what holds the parties in the relationship together. A number of suggestions have been made: the irrevocability of the relative life span positions of parent and child;[3] the functional interdependence of family members for giving and receiving aid and assistance;[4] powerful social norms of parental and filial

---

[2]Umberson (1993) described her research as adopting a "structure and personality" approach to social integration, looking specifically at the intergenerational family tie. Her empirical findings are that aspects of social structure are important in predicting relationship quality, and that only the strains (negative features) of adult parent–child relationships are associated with (poor) psychological well-being.

[3]The life span positions of parent and child are, of course, fixed in relation to one another within one family. Patterns of dependency and codependency are unlikely to remain static, however, as both members of a dyad change their life span position within the trajectory of their cobiography.

[4]It has been observed that the parent–child relationship is unique as a unit of giving and receiving aid throughout life (Lopata, 1978) and that each person's position as mainly a donor or recipient may fluctuate as role definitions, needs, and circumstances change.

obligations that guide action, evoke sanctions for nonconformity, and that may be associated with a personal sense of moral obligation or guilt; and feelings of love and affection. The uniqueness and continuity of the adult child–parent bond has also been attributed to a shared sense of family history (Gubrium, 1988; cited in Brubaker, 1990).

Mangen et al. (1988) incorporated many of these dimensions, plus some additional ones, to give a multidimensional index of intergenerational solidarity, and it is not uncommon for researchers to commend this as one recognizable theoretical development in family aging research (see, e.g., Mancini & Bliesner, 1989). However, Atkinson (1989) maintained that the concept offers little to attempt to understand family relationships, on the grounds that it privileges mechanical solidarity (deriving from similarity and a shared worldview) over organic solidarity (mutual dependence based on performance of different roles or a division of labor); in her view, the latter is far more relevant. She also maintained that the intergenerational solidarity model draws inappropriately on theories of voluntary and temporary relationships. Many researchers have puzzled over the repeated empirical finding that there is a lack of correlation between contact, association, and affect. This may occur where contact and assistance persist in the absence of positive regard, or where an adult child has withdrawn concrete assistance but is still committed to offering companionship to a loved parent. Each of these scenarios can be taken to undermine any assumption that intergenerational solidarity can be assessed simply as a compound product of multiple component dimensions.

Atkinson (1989) concluded from this that the motivation for the continuing involvement of adult parents and children is affective. This view emerged partly because of her conviction that researchers should seek to use concepts that link together dyadic relations in young and older families (see also Hagestad, 1987). The notion of affective involvement is equally relevant at each end of the family life span, and research on infant–parent bonding tends to be considered within the framework of attachment theory. This theory has at least one use when applied to the adult child–parent bond, because it enables the researcher to distinguish between love for a novel (or "interesting"; Troll & Smith, 1976[5]) but replaceable object and a strong feeling of affection or emotional bonding (viz., attachment) for a more familiar but irreplaceable object (Atkinson, 1989). Objects of the latter kind are important in that they can provide a sense of security in infancy and throughout life (Skolnick, 1986).

Atkinson (1989) suggested that two further overlapping concepts may be useful in understanding the connection between parents and children in later life—"crescive bonds" and "situated identities." The former develop into an irreplaceable individual via processes of identification (viz., assimilation of the qualities of a desirable other into the self) and enhancement of self-esteem. They come to exist as a stable aspect of the self-concept, and at this stage no longer depend for their continued existence on further reinforcement of the person's self-esteem. Two important characteristics of these bonds are that the people in the relationship share a sense of responsibility toward one another, together with a view of the future as involving joint activities and accomplishments. The latter are again aspects of the self-concept, which are derived from socially defined categories (e.g., the roles of parent and child), and which provide a basis for continuing self (in-role)

---

[5]Troll and Smith (1976) also described attachment as the opposite of attraction.

knowledge, affirmation, and value in social interaction and throughout life. Both concepts are useful in that they foreground the social roles and definitions of persons and the way they can function at a psychological level. They refer to mechanisms that make it possible to explain the achievement of smooth social transitions (e.g., continuation of a person's identity as their parents' "child" may underpin a transition to caring for them later in life), together with a person's sense of continuity and stability across the life span.

Nonetheless, one reservation must be that the notion of internalized social identification can create an overly coherent, unified, and harmonious view of the formation and functioning of aspects of the social and developmental self (Wetherell, 1996, 1998). Increasingly, researchers are commenting on the ambiguity, complexity, and sometimes conflicted[6] nature of the role of older parents. Older people may, for example, express an emotional commitment to their children that is not reciprocated to the same degree. This makes them vulnerable, although not necessarily less contented (Knipscheer, 1984). Older parents may expect to provide assistance, but simultaneously wish to avoid interfering (Bliesner & Mancini, 1987). Other research has rebutted the frequently held assumption that the morale of older people suffers most when they do not receive expected aid, with the finding that most negative affect tends to be experienced when they are unable to reciprocate in relationships typically characterized by exchange (Stoller, 1985).

Such studies have identified no clearly defined or consistent elderly familial role, as might be suggested by a regular, developmental analysis of a "cycle of dependency" (Lewis, 1990) between parent and child. The situation would seem, if only partially, to be better described by Lee (1988), who pointed to the potentially invidious nature of older persons' dependency, because it contravenes the cultural value that is widely attributed to independence and self-reliance (see also Cohler, 1983). Accordingly, one potentially fruitful strategy in research is to explore the suggestion that struggles over power and autonomy are not necessarily resolved when parents and children are both adults later in life (Hess & Waring, 1978). The notion that exploring the ambivalence of adult parent–child relationships and family ties (see, e.g., Connidis, 2001) offers one means of addressing limited concepts and perspectives in intergenerational family research has gained increased recognition since this chapter was first published in 1995 (see, e.g., Connidis & McMullin, 2002; Lüscher & Pillemer, 1998; Pillemer & Suitor, 2002; Smelser, 1998). The treatment in this chapter, though, draws on a rather different research tradition.

FEMINIST RESEARCH ON ADULT MOTHER–DAUGHTER RELATIONSHIPS: THE SOCIAL AND DISCURSIVE CONSTRUCTION OF RELATIONAL IDENTITIES

The idea that adult parent–child relationships may be informed by analyses of struggles for autonomy and independence has been developed extensively within feminist literature on women's psychosexual development, motherhood, and mother–daughter relationships— literature that draws extensively on psychoanalytic thought (Henwood, 1995). For the

---

[6]The term *conflicted* is used here to allude to the often unconscious psychological and emotional conflicts elderly parents may experience: hence, I draw on the term's specific usage in psychoanalytic thought.

current purposes, one limitation of this research is that has been interested solely in the fate of women family members. However, it must be noted that the symbolic representation of the phallus (viz., masculinity) does, in some accounts, play a vital role and that, characteristically, a special emphasis is given to the interdependence of women and men's social position and conflict at the level of their unconscious psychological formations (Frosh, 1997; Hollway & Featherstone, 1997; Maguire, 1995). For this reason, it also offers a useful starting point for further research into adult mother–son, father–daughter, and father–son relationships.

As is probably widely known, feminism itself has had a stormy relationship with psychoanalysis. The idea that girls develop heterosexual attachments to their fathers through recognition of sexual difference resulting in penis envy has been a major barrier to the take-up of Freudian ideas within feminism. The beginning of attempts to challenge the paternalism of analytic thought came with Melanie Klein's suggestion that psychosexual development is underpinned in important ways by infantile fantasies of the good and bad breast. Since Klein there have been many further developments in "mothering psychoanalysis" (Sayers, 1992) as represented in the work of Margaret Mahler, Daniel Stern, and Donald Winnicott, who together contribute to the "object relations" school of psychoanalytic thought. Herein, Freud's concern for thwarted sexual desires is displaced from center stage by accounts of the gradual unfolding of the child's separate sense of an autonomous self or ego. According to object relations theory, such psychological separation and autonomy can occur only in the context of a satisfactory relationship with the primary caregiver—usually the mother—who, ideally, should provide a secure base from which the child can explore and discover boundaries between itself and other objects in the external world. A child's primary experiences of bonding with, and separation from, the mother are then also deemed to lay down the psychological foundations for his or her capacity to give and receive love, and for the position he or she will take up in the relational dynamics of power (involving both domination–subordination and autonomy–dependency) later in life.

One widely read and acclaimed feminist account of women's relational development as mothers and daughters is to be found in Nancy Chodorow's *The Reproduction of Mothering* (1978). She linked object relations theory to an analysis of patriarchal social structures and relations—where "patriarchal" means ruled by and in the interests of men.[7] Her analysis of patriarchy mainly concerns production relations in the household, and specifically the way that it is mainly women and not men who mother. Given her assumption that young children exist in a state of symbiosis with their primary carer, her observations about the asymmetrical patterns of parenting, and the fact that girls and their mothers have a shared gender identity, she developed the following thesis: that girls will find it more difficult to recognize themselves as separate from their mothers than boys, and accordingly will tend to experience a more prolonged period of psychological symbiosis or mergence with her. At least two interrelated consequences are said to then follow from this. First, girls and women will tend to develop a feminine personality structure, defining themselves through relationships with others. Second, in adulthood and later life, daughters will

---

[7] It is important to note that, in referring to social structures and the practices that instantiate them, the concept of patriarchy does not mean that every man is always in a dominant position and every woman in a subordinate one. Highly sophisticated theories of patriarchy exist today (Walby, 1990).

often continue to experience highly emotionally charged, ambivalent relationships with their mothers, because of unresolved conflicts over a need for autonomy and separation. Some support for these ideas has been forthcoming from psychoanalytic case studies (see, e.g., Eichenbaum & Orbach, 1985), where women have recalled early childhood experiences of having felt pressure to mother their own mothers, and also expressed their fear of loss of maternal love when seeking to act independently. In analytic terms, these experiences are taken as evidence of inadequately defined psychological boundaries and of girls' and women's struggle to achieve psychological separation, respectively.

For many years, nonpsychoanalytic accounts of adult mother–daughter relationships have tended to speak of the unique closeness of the bond, which is also viewed as the lynchpin of family life (see, e.g., Young & Wilmott, 1957). However, the psychodynamic view that such relationships are potentially far more complex, divided, and contradictory does, in the view of many researchers and women themselves, accord more nearly with women's experiences (see, e.g., Burck & Speed, 1995; O'Connor, 1990). Another notable strength of Chodorow's account is that it provides a theoretical framework for linking the social, institutional, and experiential aspects of motherhood and daughterhood,[8] by considering the psychology of a history of object choices in relation to the patriarchal relations of parenting.

Chodorow's original thesis does not offer the definitive account of adult mother–daughter relations, however, because it continues to draw the parallel between the social and the psychological too tightly (Benjamin, 1990; Elliott & Frosh, 1995; Segal, 1987, 1999). As with socialization accounts of role-specific identities, Chodorow did appear to view gender identity as a unitary and coherent feature of women's psychology throughout life. Mothers and daughters may be ambivalent in their feelings for one another, but still her account sees these relationships as propelling each generation of women into the cycle of the reproduction of mothering. Yet many women with bitter memories of childhood in the 1950s postponed motherhood, a small but increasing proportion of women are deciding against motherhood entirely, and for others a distinctive maternal psychology only appears to develop after the birth of a child (Segal, 1987). Bearing these issues in mind, there would seem to be no one universally applicable maternal psychology that is passed down between generations of women family members.[9]

A further criticism of the popular individuation–separation account of women's relational development is that it amounts to mother blaming and does not allow for the possibility that daughters' expectations of their mothers may be simply unreasonable (Hollway, 1993). The charge that it infantilizes women as suffering from stunted emotional development has also had to be met (Bart, 1983). The latter point may be responsible in part for interpretations of the importance adult mothers and daughters often attach to their relationship, not as a sign of poor psychological separation, but as a way of challenging the cultural devaluation of their shared position as women (see, e.g., Apter, 1990; Brown & Gilligan, 1993; Hancock, 1990).

---

[8]The argument for understanding this link is long-standing within feminism (see, e.g., Phoenix, Woollett, & Lloyd, 1991; Rich, 1976).

[9]Part of the reason for this may be that there is an element of self-defining agency in all human subjectivity (Mahony & Yngvesson, 1992), such that we all have the capacity to identify with objects of desire that are not members of our own sex (Segal, 1987).

Accordingly, Chodorow (1989) later argued against any "single cause" theory of women's relational development and identity, based around the social psychological dynamics of mothering in the domestic sphere. However, this involved no retreat on her view that mother–daughter relationships must be understood in terms of the history and emotional potency of women's experiences of these relationships at the level of the psychological unconscious (see also Chodorow, 2002). Rather, the way forward is to consider their construction and reconstruction, in the light of the full variety and specificity of social, cultural, and historical relationships, in the present as well as the past. This point was argued particularly forcefully by Enders-Dragaesser (1988) when she disputed that "the deciding phase of a woman's feminine socialisation is closed off, consigned to the unconscious long before she ever reaches adulthood" (p. 583). As she went on to explain: "This understanding implies a concomitant if unadmitted devaluation of girls' and women's later experience. We cannot earnestly assume that experiences and interactions become less meaningful with increasing age, as if early experiences alone are supposed to be vivid, weighty and durable. The effects of joy, love, fear, mourning, illness, wounds, and trauma are not tied to a particular age, and the specific experiences of girls' and women's reality as they live their lives do not remain without an impact on their development which in turn influences their identity" (p. 583).

In my own research, I have investigated the construction that mothers and adult daughters themselves put on their relationships with their mother and daughters, and sought to interpret this in the light of the wider social, historical, and cultural relations that structure our understandings, conduct, and lives (see also Croghan & Miell, 1995; for the general methodological approach, see Nicolson, 1998; Stoppard, 1997; Ussher, 1991; Woollett & Boyle, 2000). Enders-Dragaesser's argument is that women's identities and relationships are constituted within a "paradoxical lived reality." By this she meant that we are simultaneously "strangers" within a (purportedly gender neutral) world that is defined through masculinist values and "alien," because we are locked within a fictitious social construction of feminine normalcy. There is some evidence to support elements of such a view in my work—particularly for older women. For some older women, the ideal of mother–daughter closeness, which functions as part of a wider discourse of femininity, at times represented an interactional threat to face. That is, it threatened their rights in interaction to be well thought of, and to avoid violations of their personal space (see, e.g., Brown & Levinson, 1987). Where older mothers did not fit an idealized grandmaternal image, it seemed to me that they could struggle to resist a potentially face-threatening categorization as evil maternal figures. Stoical acceptance represented one of the few subject positions open to such women, which could be taken up by them discursively to avoid further interactional threats to their reputation and identity (Henwood, 1993).

CONCEPTUAL AND METHODOLOGICAL ISSUES IN COMMUNICATION AND AGING RESEARCH

At least two influential, underlying currents may be detected in research on later-life family relationships, and specifically on relationships between parents and their adult children. The one I have dealt with most overtly so far in this chapter is a desire on the part of researchers to consider the affective, emotional, or psychological dimension to family relationships, as opposed to their demographic and behavioral features. The other, which is central to the chapter but which has as yet been left far more implicit, is the view that too much attention has been paid to the outcomes of relationships, and too little to the interactive and communicative processes by which they are produced (Hagestad, 1987). Some examples exist where researchers draw these two strands together. For example, Thompson and Nussbaum (1988) were concerned with relational closeness and intimacy, seeing this as not merely reflected by, but as built and created in, communicative interaction. In making this point, these authors gave expression to the more general view that family gerontology and later-life studies stand to benefit from a more communicatively or linguistically informed mode of analysis (see, e.g., Coupland & Coupland, 1990; Tamir, 1979). However, communication and aging research contains within it a diversity of theoretical approaches, assumptions, concepts, and methodological practices. At least some of these must be considered if we are to understand more fully what stands to be gained by approaching adult parent–child relationships from the perspective of discursive social psychology, which guides my own research and hence the perspective adopted in this chapter.

In a good deal of communication and aging research, communication is construed as a vehicle for making interpersonal contact, and for maintaining social cohesion and integration. Accordingly, it draws on a commonsense, relational view of communication, embedded within a model of the interpersonal contributions to a normatively functioning society. The importance of communication is stressed for older and elderly people in particular, given an analysis of their social position as characterized by role loss (Rosow, 1974), social differentiation (Riley, 1971), and isolation, or disengagement (see, e.g., Cumming & Henry, 1961). An associated conviction is that such communication should or will flow mainly between already connected family members, on the assumption that a major function of families is to act as a buffer for individual members against social strains and stress. In a review of such research, Shroeder (1988) pointed to the communicative exchanges that can exacerbate the problem of dependency and autonomy relations in adult parent–child relationships. However, this issue is still approached very much within an overarching framework that assumes that improved communication is both the problem of, and the solution to, the segregation and disengagement of elderly people.

Research within what could be called such a "communication and social integration" tradition tends to involve analyzing causal linkages between fixed (and hence measurable) variables, thereby reducing the symbolic, practical activity of communication to overt behavioral sequences of social exchange (for further comments, see later discussion). This is just one manifestation of the many well-documented blind spots of structural-functionalism, a theoretical perspective that "assumes that social systems are relatively determinate, boundary-maintaining systems in which the parts are interdependent in certain ways to preserve one another and the character of the system as a whole" (Barber, 1956, p. 129), and which imbues the theories of age relations mentioned earlier. As Marshall (1986) argued, support does not tend to be found for the assumption that systems naturally tend toward equilibrium or that social relations are based on no more than societal survival functions. Accordingly, it may be necessary to take a critical step back from taken-for-granted notions of productive or effective communication, exploring instead how mere participation in various social worlds can be a meaningful basis for social identity across the life span (Marshall, 1986).

When communication and aging researchers are concerned with "the symbolic, transactional process of shared meanings" (Galvin & Brommel, 1986; cited in Shroeder, 1988), the work is representative of a more interpretive inquiry strategy. This strategy emerged in the human and social sciences much earlier in the 20th century in the writings of phenomenologists and symbolic interactionists such as Schutz and Blumer, in opposition to the idea that social research could simply imitate concepts and methods of the natural sciences. Criticism of deterministic, rule-following modes of explanation and sense making in sociology was also part of the discussions at this time. Debate has continued regarding the advantages and limitations of the so-called positivistic and interpretative ways of approaching social inquiry, together with attempts to reconcile the two, along epistemological, conceptual, theoretical, and methodological lines (see, e.g., Bryman, 2001; Silverman, 1985). There is also a growing recognition that ignorance of the principles of interpretative work has had serious consequences for disciplines, such as psychology, which trains researchers almost exclusively to use positivist and experimental methods (Camic, Rhodes, & Yardley, 2003; Henwood & Pidgeon, 1992). These consequences are particularly serious when researchers seek to study communication, symbolism, and meaning (Potter & Wetherell, 1987; see also later comments). Shifting epistemologically and methodologically to do more interpretive work has been a distinctive feature of feminist social science. Fortunately, therefore, the relevance of interpretive and feminist methods is now beginning to be appreciated in the full range of areas of family research (see, e.g., Opie, 1992; Thompson, 1992; Henwood and Procter, in press).

Nevertheless, even when communication and aging researchers present their subject matter as negotiated symbols or meanings, these are still often considered as a transparent layer, the analysis of which will reveal the underlying structure and dynamics of family systems. This may be seen, for example, in the continuation of the previously stated quotation from Galvin and Brommel, in which shared meanings are said to "undergird(s) and illuminate(s) the structure of family relationships." According to Gubrium (see, e.g., 1987, 1988), the argument that the social order of the family and the life span can be discerned from careful attention to its component structures and dynamics is overworn, leading to a neglect of its variable contextualization and practical realization. The neglect is challenged in his own (social constructionist) research with Holstein (Gubrium & Holstein, 1999). Together these two authors look at biographical work and life span malleability (Gubrium & Holstein, 1995) and the ways in which, despite variations in the way family relations are experienced, an outward appearance of orderliness is taken for granted as "objective" and "real" (Gubrium & Holstein, 1990).

A second form of communication and aging research is therefore identifiable, which adopts a thoroughgoing view of the social constitution of family relationships as achieved in the interactional domain. Nevertheless, there are various strands of interactionally focussed research where the status attributed to the "objects" of study may at times seem to be somewhat closer to the realist, and further from the constructionist, end of any epistemological dimension. Research originating in collaboration between applied discourse analysts and social psychologists (Coupland, Coupland, Giles, & Henwood, 1991) has focused on the contextual patterning and sequential ordering of the interactional and discursive means of producing social life (see, e.g., Coupland, 2003; Coupland, & Coupland, 1998). Such research leaves space for the interpretation that discursive acts of formulation or enactment can result in functionally autonomous relationships, stereotypes, embodiment, and identities.

Increasingly, concepts and methods in social sciences research are becoming influenced by the cultural and intellectual traditions of postmodernism and poststructuralism. This development adds further to work conducted from the perspective of discourse analysis, by drawing attention to the fictional status of socially organized "grand meta-narratives" (e.g., science) and to their power to regulate or control the thoughts, wishes, and conduct of human participents. Excluding feminist studies of maternal representation (see, e.g., Kaplan, 1992), these ideas are innovations in late-life family studies and represent a potential for a third front of communicatively oriented research in the field. My own research combines a concern for discourse as wider regulatory social narrative with the ethnomethodologically charged and interactionally grounded principle of performative linguistic action (Wetherell, 1998; Wetherell & Potter, 1992; Wetherell, Taylor & Yeates 2001). These ideas are compatible because they have in common the idea that discourse is meaningful, but that it can also function as reality or truth, because of its practical effects.

This dual strategy has enabled me to identify the structuring effects of gendered societal discourses on the subjectivities of family members. It has allowed me to take account of the constitutive effects of the grand meta-narrative of psychoanalysis itself, in providing a powerful dichotomous scheme for interpreting women's psychology within a good–bad mother framework (Henwood, 1993). Latterly awareness of the operation of gendered cultural dichotomies has also pointed to tensions and dilemmas in paternal subjectivity (Henwood and Procter, in press). This is a nonreductionist approach that can integrate a concern for the social, experiential, and psychological dimensions of adult parent–child relationships. Accordingly, it can be recommended as an expression of theoretical and methodological developments within contemporary feminist and discursive social and cultural psychology (see, e.g., Burman, 1992; Henriques, Hollway, Urwin, Venn, Walkerdine, 1998; Squire, 2000; Walkerdine & Lucey, 1989).

CONCLUSION

This chapter examines a number of themes in research on adult parent–child relationships from different theoretical and methodological perspectives. The role played by communicative processes in the formation and functioning of adult parent–child relationships is considered, and three rather different strands in such research are identified. These are: communication as a means of interpersonal support and social integration, pragmatically and ethnomethodologically inspired analyses of the discursive construction of family order and relational identities, and analyses of wider discourses or narratives with social and regulatory power.

Research is burgeoning within late-life family studies, and much of this has the strength of maintaining a clear perspective on the importance of social context in supporting and transforming family ties. However, although a good deal of attention has been paid to demographic and behavioral indexes and conditions of adult–parent relationships, the continuing involvement of parents and adult children is not unreasonably deemed to reflect the affective and emotional dimensions of such ties.

These kinds of issues are not easy to study, but rich insights can be found from within feminist studies of mother–daughter relationships and from research on the development of gendered subjectivities or identities into, and throughout, adulthood. The general approach here is to relate social relations of parenting to the development of complex, partly unconscious, and potentially conflicted psychological bonds. Such research also contains points of relevance for understanding adult mother–son, father–daughter, and father–son relationships.

REFERENCES

Abeles, R. P. (1987). *Life-span perspectives and social psychology.* Hillsdale, NJ: Lawrence Erlbaum Associates.

Aldous, J. (1985). Parent–adult relations as affected by the grandparent status. In V. L. Bengston & J. F. Robertson (Eds.), *Grandparenthood* (pp. 117–132). London: Sage.

Apter, T. (1990). *Altered loves: Mothers and daughters during adolescence.* London: Harvester Wheatsheaf.

Atkinson, M. (1989). Conceptualisations of the parent–child relationship: Solidarity, attachment, crescive bonds, and identity salience. In J. A. Mancini (Ed.), *Ageing parents and adult children* (pp. 81–98). Lexington, MA: Heath.

Barber, B. (1956). Structural-functional analysis: Some problems and misunderstandings. *American Sociological Review, 21*(2) 129–135.

Bart, P. (1983). [Review of the book *The reproduction of mothering*, by N. Chodorow]. In J. Treblicott (Ed.), *Feminist views on mothering* (pp. 147–152). Lanham, MD: Rowan & Allanheld.

Bengston, V. L., & Dannefer, D. (1987). Families, work and ageing: Implications of disordered cohort flow for the 21st century. In A. Ward & S. S. Tobin (Eds.), *Health in ageing: Sociological issues and policy directions* (pp. 256–289). New York: Springer.

Bengston, V. L., Rosenthal, C., & Burton, L. (1990). Families and ageing. In R. H. Binstock & L. K. George (Eds.), *Handbook of ageing and the social sciences* (pp. 269–281). New York: Academic Press.

Benjamin, J. (1990). *The bonds of love.* London: Virago.

Bliesner, R., & Mancini, J. (1987). Enduring ties: Older adults' parental role and responsibilities. *Family Relations, 36,* 176–180.

Brown, L. M., & Gilligan, C. (1993). Meeting at the crossroads: Women's psychology and girls' development. *Feminism and Psychology, 3*(1), 11–35.

Brown, P., & Levinson, S. (1987). *Politeness: Some universals in language use.* Cambridge, UK: Cambridge University Press.

Brubaker, T. H. (1990). Families in later life: A burgeoning research area. *Journal of Marriage and the Family, 52,* 959–981.

Bryman, A. (2001). *Social Research Methods,* Oxford: Oxford University Press.

Burck, C., & Speed, B. (Eds.). (1995). *Gender, power and relationships.* London: Routledge.

Burman, E. (1992). Feminism and discourse in developmental psychology: Power, subjectivity and interpretation. *Feminism and Psychology, 2*(1), 45–60.

Camic, P., Rhodes, J., & Yardley, L. (Eds.). (2003). *Qualitative research in psychology: Expanding perspectives in methodology and design.* Washington, DC: American Psychological Association.

Chodorow, N. (1978). *The reproduction of mothering.* Berkeley: University of California Press.

Chodorow, N. (1989). *Feminism and psychoanalytic theory.* New Haven, CT: Yale University Press.

Chodorow, N. (2002). Response and afterword. In C. Heenan (Ed.), *The reproduction of mothering: Psychoanalysis and the sociology of gender: A reappraisal* [Special issue]. *Feminism and Psychology, 12*(1), 49–54.

Cicirelli, V. G. (1981). *Helping elderly parents: Role of adult children.* Boston: Auburn House.

Cicirelli, V. G. (1989). Helping relationships in later life: A reexamination. In J. Mancini (Ed.), *Ageing parents and adult children* (pp. 167–180). Lexington, MA: Heath.

Cohler, B. J. (1983). Autonomy and interdependence in the family of adulthood: A psychological perspective. *The Gerontologist, 23*(1), 33–39.

Connidis, I. A. (2001). *Family ties and aging.* Thousand Oaks, CA: Sage.

Connidis, I. A., & McMullin, J. A. (2002). Sociological ambivalence and family ties: A critical perspective. *Journal of Marriage and the Family, 64,* 558–568.

Coupland, J. (2003). *Discourse, the body and identity.* Basingstoke, UK: Palgrave, Macmillan.

Coupland, J., Coupland, N., Giles, H., & Henwood, K. (1991). Formulating age: Dimensions of age identity in intergenerational talk. *Discourse Processes, 14*(1), 87–106.

Coupland, N., & Coupland, J. (1990). Language and later life. In H. Giles & P. Robinson (Eds.), *Handbook of language and social psychology* (pp. 451–468). New York: Wiley.

Coupland, N., & Coupland, J. (1998). Reshaping lives: Constitutive identity work in geriatric consultations. *Text, 18,* 159–189.

Croghan, R., & Miell, D. (1995). Blaming our mothers, blaming ourselves: Themes in the relationships of mothers and their incestuously abused daughters. *Feminism and Psychology, 5*(1), 31–46.

Cumming, E., & Henry, W. (1961). *Growing old: The process of disengagement.* New York: Basic Books.

Duck, S. (1997). (Ed.). *Handbook of personal relationships.* Chichester, UK: Wiley.

Duck, S. (1998). *Human relationships* (3rd ed.). London: Sage.

Eichenbaum, L., & Orbach, S. (1985). *Understanding women.* Harmondsworth, UK: Penguin.

Elliott, A., & Frosh, S. (Eds.). (1995). *Psychoanalysis in contexts.* London: Routledge.

Enders-Dragaesser, U. (1988). Women's identity and development within a paradoxical reality. *Women's Studies International Forum, 11*(6), 583–590.

Frosh, S. (1997). Fathers' ambivalence (too)." In W. Hollway & B. Featherstone (Eds.), *Mothering and ambivalence* (pp. 37–53). London: Routledge.

Gallagher-Thompson, D., Coon, D. W., Rivera, P., Powers, D., & Zeiss, A. M. (1998). Family caregiving: Stress, coping and intervention. In M. Herson, & V. M. Hassalt (Eds.), *Handbook of clinical geropsychology* (pp. 469–493). New York: Plenum.

Gubrium, J. (1987). Organisational embeddedness and family life. In T. H. Brubaker (Ed.), *Ageing, health and family* (pp. 23–41). London: Sage.

Gubrium, J. (1988). "Family responsibility and caregiving in the qualitative analysis of the Alzheimer's disease Experience," *Journal of marriage and the family, 50,* 197–207.

Gubrium, J., & Holstein, J. (1990). *What is family?* Mountain View, CA: Mayfield.

Gubrium, J., & Holstein, J. (1995). Life course malleability: Biographical work and deprivatization of experience. *Qualitative Inquiry, 1,* 207–223.

Gubrium, J., & Holstein, J. (1999). Constructionist perspectives on aging. In V. L. Bengston & K. Warner Schaie (Eds.), *Handbook of theories of aging* (pp. 287–305). New York: Springer.

Hagestad, G. O. (1987). Parent–child relations in later life: Trends and gaps in past research. In J. B. Lancaster, J. Altmann, A. S. Rossi, & L. R. Sherrod (Eds.), *Parenting across the life-span: Biosocial dimensions* (pp. 405–433). New York: Aldine.

Hancock, S. (1990). *The girl within: A radical new approach to female identity.* London: Pandora.

Henriques J., Hollway, W., Urwin, C., Venn, C., & Walkerdine, V. (1998). *Changing the subject: Psychology, social regulation and subjectivity* (2nd ed.). London: Routledge. (Original work published 1984).

Henwood, K. L. (1993). Women and later life: The discursive construction of identities within family relationships. *Journal of Ageing Studies, 7,* 303–319.

Henwood, K. L. (1995). Adult mother–daughter relationships: Subjectivity, power and critical psychology. *Theory and Psychology, 5,* 483–510.

Henwood, K. L., & Coughlan, G. (1993) The construction of "closeness" in mother–daughter relationships across the lifespan. In N. Coupland & J. Nussbaum (Eds.), *Discourse and lifespan identity* (pp. 191–214). London: Sage.

Henwood, K. L., & Pidgeon, N. F. (1992). Qualitative research and psychological theorising. *British Journal of Psychology, 83,* 97–111.

Henwood, K. L., & Procter, J. (in press). The 'good father': Reading men's accounts of paternal involvement during the transition to first time fatherhood. *British Journal of Social Psychology.*

Hess, B. B., & Waring, J. M. (1978). Parent and child in later life: Rethinking the relationship. In R. M. Lerner & G. B. Spanier (Eds.), *Child influences on marital and family interaction* (pp. 241–268). New York: Academic Press.

Hogg, M., & Abrams, D. (2001). *Intergroup Relations.* London: Routledge.

Hollway, W. (1993). [Review of the book *Feminism and psychoanalytic theory*, by N. Chodorow]. *Feminism and Psychology, 3,* 259–261.

Hollway, W., & Featherstone, B. (Eds.). (1997). *Mothering and ambivalence.* London: Routledge.

Kaplan, E. A. (1992). *Motherhood and representation.* London: Routledge.

Knipscheer, C. (1984). The quality of the relationship between elderly people and their adult children. In V. Garms-Homolova, E. M. Hoerning, & D. Shaeffer (Eds.), *Intergenerational relationships* (pp. 90–101). Lewinson, NJ: Hogrefe.

Lee, G. (1988). Ageing and intergenerational relations. *Journal of Family Issues, 8,* 448–450.

Lewis, R. A. (1990). The adult child and older parents. In T. H. Brubaker (Ed.), *Family relationships in later life* (2nd ed.), (pp. 68–85). London: Sage.

Lopata, H. Z. (1978). Contributions of extended families to the support systems of metropolitan area widows: Limitations of the modified widow network. *Journal of Marriage and the Family, 40,* 355–364.

Lüsher, K., & Pillemer, K. (1998). Intergenerational ambivalence: A new approach to the study of parent–child relationships in later life. *Journal of Marriage and the Family, 64,* 585–593.

Lye, D. N. (1996). Adult child–parent relationships. *Annual Review of Sociology, 22,* 79–102.

Maguire, M. (1995). *Men, women passion and power: Gender issues in psychotherapy.* London: Routledge.

Mahony, M. A., & Yngvesson, B. (1992). The construction of subjectivity and the paradox of resistance: Reintegrating feminist anthropology and psychology. *Signs, 18*(1), 44–73.

Mancini, J. A. (1989). Family gerontology and the study of parent–child relationships. In J. A. Mancini (Eds.), *Ageing parents and adult children* (pp. 3–12). Lexington, MA: Heath.

Mancini, J., & Bliesner, R. (1989). Ageing parents and adult children: Research themes in intergenerational relations. *Journal of Marriage and the Family, 51,* 275–290.

Mangen, D. J., Bengston, V. L., & Landry, P. H. (1988). *Measurement of intergenerational relations.* Beverly Hills, CA: Sage.

Marshall, V. W. (1986). Dominant and emerging paradigms in the social psychology of ageing. In V. W. Marshall (Ed.), *Later life: The social psychology of ageing* (pp. 9–31). London: Sage.

Nicolson, P. (1998). *Post-natal depression.* London: Routledge.

Noelker, L. S., & Townsend, A. L. (1987). Perceived caregiving effectiveness: The impact of parental impairment, community resources and caregiver characteristics. In T. H. Brubaker (Ed.), *Ageing, health and family* (pp. 58–79). New York: Sage.

Phoenix, A., Woollett, A., & Lloyd, E. (1991). *Motherhood: Meanings, practices and ideologies* London: Sage.

Pillemer, K., & Suitor, J. T. (2002). Explaining mothers' ambivalence toward their adult children. *Journal of Marriage and the Family, 64*(3), 602–613.

O'Connor, P. (1990). The adult mother–daughter relationship: A uniquely and universally close relationship? *Sociological Review, 38,* 293–323.

Opie, A. (1992). Qualitative research, appropriation of the "other" and empowerment. *Feminist Review, 40,* 52–69.

Potter, J., & Wetherell, M. (1987). *Discourse and social psychology.* London: Sage.

Rich, A. (1976). *Of woman born.* London: Virago.

Riley, M. W. (1971). Social gerontology and the age stratification of society. *The Gerontologist, 11,* 79–87.

Rosow, I. (1974). *Socialisation to old age.* Berkeley: University of California Press.

Sayers, J. (1992). *Mothering psychoanalysis.* Harmondsworth, UK: Penguin.

Segal, L. (1987). *Is the future female: Troubled thoughts on contemporary feminism.* London: Virago.

Segal, L. (1999). *Why feminism? Gender, psychology and politics.* Cambridge, UK: Polity.

Shanas, E. (1980). Older people and their families: The new pioneers. *Journal of Marriage and the Family, 42*(9), 9–15.

Shroeder, A. B. (1988). Communication across the generations. In C. W. Botan & R. Hawkins (Eds.), *Human communication and the ageing process* (pp. 129–139). Prospect Heights, IL: Waveland.

Silverman, D. (1985). *Qualitative methodology and sociology.* Aldershot, UK: Gower.

Skolnick, A. (1986). Early attachment and personal relationships across the lifecourse. *Lifespan Development and Behaviour, 7,* 173–205.

Smelser, N. E. (1998). The rational and the ambivalent in the social sciences. *American Sociological Review, 63*(1), 1–16.

Squire, C. (2000). (Ed.). *Culture in psychology.* London: Sage.

Stoller, E. P. (1985). Exchange patterns in the informal support of the elderly: The impact of reciprocity on morale. *Journal of Marriage and the Family, 47,* 335–342.

Stoppard, J. (1997). Women's bodies, women's lives and depression: Towards a reconciliation of material and discursive accounts. In J. Ussher (Ed.), *Body talk: The material and discursive regulation of sexuality, madness & reproduction* (pp. 10–32). London: Routledge.

Tajfel, H. (1978). Intergroup behaviour: Individualistic perspectives. In H. Tajfel & C. Fraser (Eds.), *Introducing social psychology* (pp. 401–422). Harmondsworth, UK: Penguin.

Tajfel, H., & Turner, J. T. (1979). An integrative theory of intergroup conflict. In W. G. Austin & S. Worchel (Eds.), *The social psychology of intergroup relations* (pp. 33–47). Monterey, CA: Brooks/Cole.

Tamir, L. (1979). *Communication and the ageing process.* New York: Pergamon.

Thompson, L. (1992). Feminist methodology for family studies. *Journal of Marriage and the Family, 54,* 3–18.

Thompson, L., & Walker, A. L. (1989). Gender in families: Women and men in marriage, work and parenthood. *Journal of Marriage and the Family, 51,* 845–871.

Thompson, L., & Nussbaum, J. (1988). Interpersonal communication: Intimate relationships and ageing. In C. W. Carmichael, C. H. Boton, & R. Hawkins (Eds.), *Human communication and the ageing process* (pp. 95–109). Prospect Heights, IL: Waveland.

Townsend, A. L., & Noelker, L. S. (1987). The impact of family relationships on perceived caregiving effectiveness. In T. H. Brubaker (Ed.), *Ageing, health and family* (pp. 80–101). New York: Sage.

Troll, L. E., & Smith, J. (1976). Attachment through the lifespan: Some questions about dyadic bonds among adults. *Human Development, 19*(2), 156–170.

Umberson, D. (1993). Relationships between adult children and their parents: Psychological consequences for both generations. *Journal of Marriage and the Family, 54,* 664–674.

Ussher, J. (1991). *Women's madness: Misogyny or mental illness.* London: Routledge.

Walby, S. (1990). *Theorizing patriarchy.* Oxford, UK: Blackwell.

Walkerdine, V., & Lucey, H. (1989). *Democracy in the kitchen: Regulating mothers and socialising daughters.* London: Virago.

Wetherell, M., & Potter, J. (1992). *Mapping the language of racism: Discourse and the legitimation of exploitation.* London: Harverster Wheatsheaf.

Wetherell, M. (1996). Life histories/social histories. In *Groups, identities and social issues* (pp. 299–342). London: Sage.

Wetherell, M. (1998). Positioning and interpretive repertoires: Conversation analysis and post-structuralism in dialogue. *Discourse and Society, 9,*(3) 387–412.

Wetherell, M., Taylor, S., & Yates, S. J. (Eds.). (2001). *Discourse as data: A guide for analysis.* London: Sage.

Wetherell, M. (1999). The psychodynamic approach to family life. In J. Muncie, M. Wetherell, M. Langan, R. Dallos, & A. Cochrane, (Eds.), *Understanding the family* (2nd ed.,). London: Sage.

Woollett, A., & Boyle, M. (Eds.). (2000). Reproduction, women's lives and subjectivities [Special issue]. *Feminism and Psychology, 10,* 307–380.

Young, P., & Wilmott, M. (1957). *Family and kinship in East London.* New York: Penguin.

Communication in Close Relationships of Older People

Marie-Louise Mares and Mary Anne Fitzpatrick
University of Wisconsin–Madison

Since the 1950s, the amount of research on the family situation of older people has increased substantially. In reaction to earlier depictions of older people as socially isolated and undervalued, most of the discussion has focused on evidence that older adults tend to have strong family ties, report considerable satisfaction with their marriages, and remain sexually active well into old age. In this chapter, we review these and other issues relevant to the aging marriage, and the picture typically looks positive.

Unfortunately, we must begin with a cautionary note. Put simply, it is unlikely that future cohorts of older adults will have the same marital experience as the cohorts described in the research reviewed here. Goldscheider (1990) argued that the current older and near-older cohorts are largely atypical. Individuals in these cohorts have tended to marry and have children with whom they remain in close contact; they have tended not to get divorced and not to have extramarital affairs (Carlson, 1979). More recent cohorts are much less likely to follow this path.

These changes obviously present a problem for a chapter about marriage and older people. We review the available research and indicate each point where the findings are likely to become dated. We outline the ways in which future older adults are expected to differ in their marital status from current older adults. Finally, we make suggestions for future investigations of the aging couple.

It is worth noting at the beginning of this chapter, that both authors are social scientists with a commitment to systematic empirical investigations yielding quantifiable results. At times, we refer to qualitative data to illustrate a point and to provide a richer sense of the material. We also use it sometimes as the basis for informed speculation. Overall, the majority of research reported here is quantitative. There are a number of other characteristics of this research that should also be discussed.

First, most of the research was done in the United States, and much of it is on White couples. Clearly, there are other cultures with different relational styles that have not been investigated.

Second, most of the research involves self-assessment: Couples are asked to report on a variety of aspects of their relationship. Self-reports are subject to certain problems, including social desirability, memory distortions, and cognitive biases. There is some evidence that increasing age is associated with a tendency to give conventional or socially desirable responses to questions (Spanier & Cole, 1975).

In addition to these problems, reliance on self-report data makes it difficult to compare cohorts, because the meaning of their responses may differ. For example, cohorts may not (in fact, probably do not) share the same standards for marriage. Because of this, the meaning of high relational-satisfaction scores probably varies with each cohort. In other words, a happy marriage 50 years ago might have been defined by the participants as one in which the husband had a full-time job, brought home his paychecks, and never hit his wife.

Third, the findings of cross-sectional and longitudinal studies are difficult to interpret, because such studies confound the effects of chronological age of individuals studied and the passage of time. For example, with the passage of time, unhappy couples may divorce. If the final pool of older husbands and wives report high levels of satisfaction with their marriages, it is unclear whether this is because unhappy couples have separated or is, in fact, attributable to other age-specific factors such as the length of the marriage. Baltes (1968) and a number of writers since then (e.g., Schaie & Hertzog, 1985) suggested that the solution is to carry out more longitudinal and cross-lagged investigations (essentially, a series of longitudinal studies), but such research is difficult and consequently very rare.

Despite these limitations, the research can and does yield interesting, often counterintuitive, findings. We have attempted to describe the research as it stands, combined with a number of cautionary remarks about the interpretation of results.

Implications of Marital Status

Research suggests that being married has a number of important, unobvious effects on older adults' lives. Put simply, marriage appears to provide a lifestyle and a set of resources that protect older adults from declines in activity and health.

Cutler (1995) used data from a longitudinal study of 1,192 high-functioning adults aged 70 to 79. Individuals were interviewed between 1988 and 1989, and then reinterviewed 3 years later about the amount of time they spent on housework, yard work, child care, paid work, and volunteer work, as well as their health and psychosocial status. Compared to unmarried seniors, married seniors showed only half the decline in time spent on productive activities, even after controls for gender, age, education, and race.

Pienta, Hayward, and Jenkins (2000) used data from the 1992 wave of the Health and Retirement Study, which provided a sample of 9,333 adults between the ages of 51 and 61 who were not in a same-gender union. Almost without exception, married people had the lowest rates of morbidity for fatal and nonfatal diseases, functioning problems, and disabilities, even after controlling for age, gender, and race and ethnicity. Exiting marriage, either through divorce or widowhood, had more profound negative health consequences than never having married at all, particularly for African Americans and Latinos.

Other studies report that married persons experienced lower mortality (i.e., die later) than unmarried persons. Being widowed or divorced, with subsequent depression and loss of lifestyle, appeared to elevate mortality risks, particularly among men (e.g., Goldman, Koreman, & Weinstein, 1995.

Two reasons invoked to explained association between health and marital status are (a) the selection of more healthy individuals into marriage and (b) the protective nature of marital unions. Goldman et al. (1995) observed that, compared with the unmarried, married individuals typically have a greater number of family members (including children) in their networks, are more likely to have an intimate confidant (i.e., the spouse) who provides emotional and instrumental support, and are more likely to have relatives who impose constrains on their behavior, particularly practices that may endanger their health, such as poor eating habits, drug use, drinking, or smoking (see also Sherbourne & Hays, 1990). Research by Ross (1995) suggested that, at least for psychological benefits, actual marital status is less important than having a happy relationship with a partner living in the house. Moreover, having an unhappy relationship was, she observed, worse for psychological well-being than no relationship.

Aside from protective benefits, what characterizes marriage in old age? How do people feel about each other, and how do they behave toward each other after living together for most of their lives? How does the marriage adapt to stressors such as children, retirement, and ill health? How do marriages formed late in life differ from those in which the couple have been married for many years?

MARITAL QUALITY IN OLD AGE

It remains a controversial and important question how best to evaluate the quality of a marriage. Glenn (1998) argued that there are two basic approaches. The first is based on the view that marital quality consists of the way married persons feel about their marriages. Adherents to this view use individual spouses' reports of marital happiness or satisfaction or similar reports of feelings to assess marital quality. The second approach is based on the view that marital quality is a characteristic of the relationship between spouses (the "marital adjustment" school). Adherents to this approach use measures of communication, conflict, and similar relationship qualities. Some multidimensional scales of marital adjustment include both relationship characteristics and reports of the feelings of individuals. More recently, however, there has been a move toward direct observation of marital interactions and analysis of behavioral patterns (see Gottman & Notarius, 2000, for a review).

Unfortunately, research on marital quality in long-term marriages continues to lag behind. Virtually all the observational research is on young couples. As Gottman and Notarius (2000) pointed out, research on the aging marriage tends to be self-report. Moreover, numerous studies are based on one or two simple questions about happiness (e.g., Glenn, 1998) or perceived severity of marital problems. We report the findings that are available, but we continue to mourn the lack of rich, detailed observations of older marriages that are increasingly common in studies of younger populations.

When we first wrote this chapter in 1995, we noted three things. First, cross-sectional studies conducted in the 1950s and 1960s generally showed a monotonic decline in marital satisfaction as couples grew older (e.g., Blood & Woolfe, 1960; Luckey, 1966; Tuckman & Lorge, 1954; see Hicks & Platt, 1970, for a review). Second, cross-sectional studies conducted during the 1970s and 1980s typically showed a U-shaped relationship between age and marital satisfaction: a midlife downturn associated with childrearing responsibilities and career development followed by a later-life upturn (Anderson, Russell & Schumm, 1983; Burr, 1970; Roberts, 1979; Sporakowski & Hughston, 1978; Stinnett, Carter, & Montgomery, 1972; Stinnett, Collins, & Montgomery, 1970). Third, we suggested that the seeming shift in findings were probably because of cohort differences. Ethnographic research by Caplow, Bahr, Chadwick, Hill, and Williamson (1982) suggested that marriages formed in the early part of the 20th century were more distant and less satisfying than marriages in the latter half of the century. Our final conclusion was that marital satisfaction is generally high among older couples.

Since then, more research has been published, and two findings have emerged. The first is that current cohorts of elderly couples continue to show higher levels of marital satisfaction and more positive interaction styles than younger couples. The second is that these are probably cohort differences rather than maturational effects and that current young and middle-aged couples may continue to be relatively dissatisfied even as they grow older together. There may be no late-life upturn for these more recent cohorts.

Cross-sectional research published during the 1990s, like research from the 1970s and 1980s, found that older couples report relatively low levels of conflict and greater positive affect than younger couples. Levenson, Carstensen, and Gottman (1993) studied 156 long-term marriages varying in spouses' age (40–50 years old or 60–70 years old) and relative satisfaction (satisfied and dissatisfied). Compared to younger couples, older couples showed greater pleasure in several areas of their marriage (including children), and showed fewer gender differences in sources of pleasure. They also rated potential sources of conflict as less problematic than younger couples. In another study (Levenson, Carstensen, & Gottman, 1994), younger and older couples were shown videotapes of their interactions. They were asked to rate how they were feeling from moment to moment during the interaction. Older couples indicated feeling more emotionally positive than middle-aged couples.

These findings of relatively low conflict and greater positive affect in old age are consistent with research by Heckhausen and Brim (1997), in which 1,688 adults (Whites, aged 18 and older) rated the seriousness of problems in 12 domains, including their marriage, money, health, friends, and not having enough sex. Marital evaluations remained stable across age groups until age 55, after which ratings of severity of marital problems declined.

Lachman and Weaver (1998) reported on surveys of 3,032 adults aged between 25 and 75 years of age from the MacArthur Studies of Midlife. Respondents were asked to rate the amount of control they had in seven domains, including marriage. Perceived control in domains such as frequency of sex or interactions with offspring decreased with age. However, perceived control over the marriage increased significantly after middle age, particularly among men.

Are these cohort effects? There are relatively few longitudinal studies of marital sat-isfaction, generally with small, unrepresentative sample sizes (e.g., Weishaus' & Field's 1998 study of 17 couples). An alternative to longitudinal research is to do trend analysis, using cross-sectional research conducted at a variety of periods to look for the effects of maturation versus cohort. Recent research using this technique provides a more disturb-ing image of marital satisfaction.

Glenn (1991) analyzed marital satisfaction data from the General Social Survey (GSS) for the 15 years from 1973 to 1988. During this time period, the divorce rate increased dramatically, leveling off at a historic high during the 1980s (Rogers & Amato, 1997). Increased access to divorce should mean that more dissatisfied individuals leave the marital pool, raising the average level of satisfaction remaining in the pool. In fact, Glenn found that the percentage of people describing their marriage as "very happy" as opposed to "pretty" or "not too" happy declined gradually throughout the 15 years of his study.

In a second study, Glenn (1998) revisited the GSS data set to analyze marital success in five 10-year marriage cohorts from 1973 to 1994. That is, he studied respondents who were first married between 1933 and 1942, 1943 and 1952, 1953 and 1962, 1963 and 1972, and 1973 and 1982. Obviously, couples in later cohorts were studied at earlier stages in their marriages than those from earlier cohorts. Glenn reported that tracing the five 10-year cohorts through different, but slightly overlapping, phases of their marriages provided no evidence for an upturn in marital satisfaction and strong support for cohort differences. Older cohorts were more likely to report being "very happy" than younger cohorts. However, the trend in marital happiness was downward in all five cohorts. Glenn argued that across cohorts, marital happiness declined steeply for the first decade of marriage, then declined moderately for two additional decades, and then continued to show a slight decline through the fifth decade.

Other researchers also find cohort differences in marital quality. Rogers and Amato (1997) reported on two waves of nationally representative data: 1,119 individuals aged 20 to 35 interviewed in 1980 (i.e., married between 1969 and 1980), and 154 individuals aged 20 to 35 interviewed in 1992 (i.e., married between 1981 and 1992). Compared with the older group, both men and women in the younger group reported significantly lower levels of marital interaction and significantly higher levels of marital conflict and problems. In a similar study, Rogers and Amato (2000) again used national data to compare a cohort married between 1964 and 1980 and interviewed in 1980, with a cohort married between 1981 and 1997 and interviewed in 1997. Members of the more recent cohort reported significantly more marital discord, although their reported levels of overall happiness were the same as the earlier cohort.

If marital quality is declining, what explains this pattern, and what differentiates the long-term successful marriage from unsuccessful marriages? Several studies have focused on what might loosely be called compatability. Lauer, Lauer, and Kerr (1990) surveyed 100 couples who had been married for 45 years or more (all respondents aged over 65) about what factors they considered critical to a stable and satisfying marriage. Husbands and wives produced similar lists: liking the spouse "as a person," a sense of humor; consensus on goals in life, friends, and decision making; and commitment to the spouse and to marriage. Goodman (1999) examined marital satisfaction and perceptions of reciprocity of social support in 80 long-term marriages (average duration of 40 years). Spouses who perceived reciprocal contributions of love, respect, information, goods, services, and money gave more positive descriptions of their marriages, though reciprocity appeared more important to wives than to husbands. Camp and Ganong (1997) examined the effect of similarity in locus of control (internal vs. external) among 137 couples who had been married for an average of 26 years. Overall, having an internal locus of control was more critical to marital satisfaction than similarity with partner, though couples in which both partners were "internals" were the happiest. The authors suggested that spouses with an internal locus of control felt more responsible for marital problems and perceived the possibility of change.

Rogers and Amato hypothesized that a cluster of social changes related to shifts in gender roles might explain some of the cohort differences in marital satisfaction they observed in adults in their 20s and 30s. In their 1997 paper, the most substantial explanation of declining marital satisfaction, accounting for 38% of the cohort difference, was the increase in work and family demands. Marital interaction and satisfaction were particularly low when the mother was employed and preschool children were in the household, and this situation was more common in the recent cohort. In their 2000 paper, Rogers and Amato again found that increases in work and family demands based on wives' level of employment and the number of preschool-aged children in the household explained some of the gap in marital discord between early and recent marriage cohorts. Although men in the more recent cohort were doing more housework than those in early cohorts, this did not appear to be enough to offset the stressful effects of maternal employment.

Because Rogers and Amato only looked at young adults, we gain no insight from their data sets about what might be occurring among older age groups. Maternal employment would not explain Glenn's findings that even older adults experience a decline in marital quality with each successive year. A relatively old study by Gilford (1984) suggested that predicting marital satisfaction in old age is a complex matter. In her study of 318 married individuals aged 55 to 90, she found that the middle cluster, aged 63 to 69, were the most satisfied, which she took to reflect a "honeymoon" postretirement period. More important, she looked at the relationship between satisfaction and a wide range of variables (including gender role orientations, retirement status, finances, happiness, health, education, and religiosity) within each group. With each successive age group, fewer variables were significant predictors, and less variance overall was explained. The finding that traditional predictors of attitudes and behavior are less explanatory as age increases and individual experiences accumulate has been reported in other areas as well (e.g., predicting the amount of television viewing among older adults; Chayko, 1988).

Where does this leave us? If nothing else, the above studies highlight the dangers of doing cross-sectional research without paying attention to possible cohort effects. For all that we noted the possibility of cohort effects in our previous edition of this chapter, we could not foretell the direction or extent of such effects without the sort of research that has only recently started to be done. For now, our best summary is that there is evidence of a trend toward declining marital satisfaction, but that the current cohorts of older adults typically report high levels of marital satisfaction, declining severity of marital problems, and high levels of control over the marital situation.

TRANSITIONS IN LATER LIFE

We discuss three important transitions in later life: first, the effects of retirement on the relationship; second, the effect of children moving away from the home (and the effects of childlessness in old age); and third, prolonged caretaking of an ailing spouse.

Retirement

Does retirement increase marital satisfaction? Gilford (1984) suggested that older couples may experience a "honeymoon" period after retirement, with renewed levels of affection and intimacy as a result of increased time together. In fact, research suggests that the impact of retirement on marital quality is highly contingent on a variety of factors and that retirement itself is a relatively complex process rather than a single event.

Moen, Kim, and Hofmeister (2001) argued that most researchers have tended to treat retirement as a dichotomous variable and ignore the fact that (a) there is often a long period of transition before and after the actual termination of employment and (b) that increasing numbers of both men and women return to some form of employment after leaving their primary career. In their study of 534 married men and women, they found that more than one fourth of their relatively middle-class sample were reemployed after retiring, though the effects of reemployment were not examined. In addition, Moen et al. (2001) highlighted the need for longitudinal study of the transition process. Using two waves of interview data collected between 1994 and 1997, they compared employed respondents with those who retired within the 4 years of the study and those who had retired before the first wave of interviews. One of their chief observations was that recent retirement of both men and women was a strong predictor of a drop in marital satisfaction and an increase in marital conflict, perceived by both spouses. However, couples for whom retirement had occurred before the data collection period were more satisfied. The authors took this as an indicator that retirement is a stressful process taking even years to complete, but that couples eventually settle back into more stable and satisfying roles.

Other studies are generally less subtle in their treatment of retirement as a variable. Nonetheless, they provide a list of variables that alter the effect that retirement has on marital quality. Myers and Booth (1996) used data from a 12-year longitudinal study of a national sample of married persons interviewed four times between 1980 and 1992. They found that marriages that were initially unhappy, conflict ridden, or unstable became even more negative after retirement. "High-quality" marriages improved still further after retirement. However, because Myers and Booth removed respondents from the data set after the first postretirement interviews, their study provides no data about the extent to which these effects are short-term or endured over time.

The effect of retirement also depends on synchrony between spouses in timing of retirement. Moen et al. (2001) found that both men and women reported increases in conflict when they remained working after their spouse retired. Other studies reported that marital quality is particularly likely to decline when the wife continues to work after the husband retires, especially if the couples are gender-role traditional (Lee & Shehan, 1989; Myers & Booth, 1996; Szinovacz, 1996). Szinovacz and Schaffer (2000) suggested that this is because the situation may not only pose a threat to the husband's role as provider, but can also contribute to disagreements over the division of household labor. As Lee and Shehan pointed out, women who are still working but are expected to do more household work than their retired husbands may become dissatisfied with the relationship. In contrast, Szinovacz (1996) found that traditional-oriented husbands (but not wives) reported higher marital happiness if they themselves continued employment after the wife's retirement, suggesting that the wives' return to a more traditional housewife role contributes to marital quality among gender-role traditional husbands.

Studies conducted during the 1970s and 1980s reported that retirement is seldom accompanied by more egalitarian sharing of household tasks, even though couples typically report expecting greater sharing following retirement (Ballweg, 1967; Brubaker & Hennon, 1982; Keating & Cole, 1980; Rexroat & Shehan, 1987; Suitor, 1991; Szinovacz, 1980). Brubaker (1990) concluded that most couples continue to follow patterns established earlier in their relationship—with disparity in time spent on housework and gender-stereotyped division of tasks. Szinovacz (2000) reported a different route to traditional gender roles. When husbands retired, they spent more time in housework, both in their own and in their wife's domain. However, when wives retired, husbands then spent less time with female chores and "the wife seems again to take charge of her domain" (p. 89).

How do couples spend their time during retirement? It is important to remember that gender differences exist in marital status during retirement. As Altergott (1988) reported, most older men (over 65) spend the majority of time in their old age interacting with their spouses, but most older women (over 65) spend a large part of their old age in solitude. This is because women are more likely to be widows because of shorter life expectancies for men, and also because of the tendency for men to marry younger women.

Altergott (1988) used data from a national survey of time use by U.S. adults done in 1976. She concluded that there are many similarities across age groups in the use of time, but that older adults (those over 65) spent more time sleeping, watching TV, and engaging in religious activities. They spent less time in entertainment that required leaving the house, education, and paid work than did middle-aged adults. Not surprisingly, they spent more time with their spouses (if they were married) and less time with their colleagues than their younger counterparts. Largely because of differences in marital status, older men spent more time per day in marital interaction than older women, whereas women spent more time with friends, relatives, and neighbors than older men.

Children Leave Home

There is a substantial amount of evidence indicating that having children is associated with lower marital satisfaction. Ryder (1973) and Houseknecht (1979) noted that wives who choose not to have children report being more satisfied with their marriage and report being more likely to engage in outside interests with their spouses, exchange stimulating ideas with their partners, and work on projects together.

What about the effects of children on satisfaction later in life? Do children prove to be a blessing in one's old age? Public perception is that children play a crucial role in providing fulfillment and companionship in old age. Blake (1979) reported that over 70% of a middle-aged sample said they thought that the childless were more likely to be lonely in old age. Fifty percent said they considered childless people (of all ages) to have empty lives or to be unfulfilled.

In fact, the evidence about the effects of children on one's satisfaction in later life is mixed. Glenn and McLanahan (1981) analyzed interview data from the 1973 through 1978 General Social Surveys. Their sample consisted of 1,500 adults aged 50 or over. Their results indicated that individuals without children were as happy and showed as much satisfaction on a variety of dimensions as those with children. This did not differ significantly according to marital status; that is, those without a spouse did not show greater detriments from not having a child. In fact, having a child was negatively related to global happiness for men over 50, specifically for Black men and highly educated White men. The authors suggested that rather than reaping rewards to compensate for earlier costs of parenthood, older fathers seemed to be incurring additional costs (though these were unspecified).

Keith (1983) surveyed older adults (aged 72 and older) in small rural towns and found that an equal percentage (35%) of childless elderly adults and parents reported feeling lonely. Similarly, Kivett and Lerner (1980) examined couples in rural towns and found that although childless couples had fewer social contacts, they did not report feeling any more lonely or being less likely to have a confidant than older parents.

Rempel (1985) analyzed data from the 1979 Social Change in Canada Survey (a national probability sample) and found that although there were some costs and benefits attached to parenthood in old age (average of elderly sample was 72.3 years), parental status made relatively little difference in most aspects of life. Older adult parents had a larger network of friends, were more integrated into the neighborhood, and were more likely to own their own homes. However, childless older adults were in better health, and there were no significant differences in reported life satisfaction (measured by asking about satisfaction on seven components of life), happiness, loneliness, availability of confidants, satisfaction with housing, or satisfaction with financial status.

In contrast, Bachrach (1980) found that childlessness was associated with increased isolation in a national sample of adults aged 65 or older. However, this relationship was modified by other variables. Childlessness was associated with a relatively high probability of social isolation among those in poor health and who had working-class backgrounds (i.e., had been employed in manual jobs). This effect was not found for those with few health problems or nonmanual backgrounds. Brubaker (1990) concluded: "Research in the 1980s does not necessarily support the pronatal assertion that children benefit parents during the last half of life.... While [the childless] may be more socially isolated, their financial and health situations are better and they are not more lonely than elderly parents" (p. 969).

Caretaking

What happens when a spouse falls sick? Eighty-six percent of all those aged 65 suffer from at least one chronic condition (Kovar, 1977). How do families cope, and what are the effects on the marital relationship?

Stone, Cafferata, and Sangl (1987) analyzed data from the 1982 National Long-Term Care Study and reported that 36% of the primary informal caregivers of older adults (i.e., over 65) were spouses. These spouses were typically over 65 themselves. This was particularly true for husbands—the mean age of caregiving husbands was 73 years. Husbands and wives were equally likely to quit jobs to take care of an ailing spouse. Caregiving husbands reported spending more time doing household tasks, whereas wives reported spending more time on financial matters than before.

There is some evidence that women are more likely than men to find the task of caretaking stressful and damaging to relational satisfaction. Fitting, Rabins, Lucas, and Eastham (1986) interviewed spouses who were caregivers of a husband or wife with dementia. They found that caregiving wives reported more depression than caregiving husbands. There were also significant gender differences in perceptions of the relationship. Seven of the 28 men (35%) reported that their relationship with their wives had deteriorated since the onset of the illness, compared to 16 of 26 wives (63%). Seven of the men (25%) reported that their relationships with their wives had actually improved since the onset of the illness in comparison with two of the women (8%). This is despite the fact that husbands were more likely than wives to be taking care of spouses with advanced dementia.

Zarit, Todd, and Zarit (1986) also interviewed husbands and wives who were taking care of spouses with dementia. Caregivers were asked to indicate the amount of burden they experienced by reporting the extent to which they felt their emotional health, physical health, social life, and financial status suffered as a result of caretaking. In addition, caregivers reported on the frequency of various memory and behavior problems related to dementia, and the extent to which they felt they could tolerate these problems. Husbands who were caregivers reported less burden than wives and reported greater tolerance of memory and behavior problems. The authors suggested that husbands were more likely to adopt an instrumental approach to daily problems, whereas wives had difficulty maintaining alternative strategies for managing problems. However, when these caregivers were reinterviewed 2 years later, these gender differences were no longer present. Women's perceptions of burden showed a significant decrease. Both men and women reported significantly more tolerance for memory and behavior problems even though the disease had progressed. The authors suggested that the women caregivers had come to adopt a more instrumental approach over time and concluded that both men and women seemed to have established a daily routine of caregiving that was not excessively demanding or burdensome.

Johnson (1985) interviewed 76 husbands and wives whose spouses were recuperating from a hospital stay (all participants were 65 or older). She found that although wives as caregivers were significantly more likely to report strain, both husbands and wives reported considerable continued satisfaction with their marital relationship despite their caregiving role. Moreover, the degree of impairment of the spouse requiring care had little influence on reports of levels of conflict, emotional interdependence, or shared activities. The degree of impairment did affect the perceived power balance between husbands and wives, as the caretakers of severely impaired patients had to make more decisions on their own. In addition, those who took care of patients who were more seriously disabled reported significantly more strain. When interviewed 8 months later, those who were still caretakers reported less strain than before.

In summary, husbands and wives provide a considerable amount of care for ailing spouses and, over time, they seem to adjust to the caretaking role very successfully.

SEXUAL ACTIVITY AND SATISFACTION IN OLD AGE

Most studies indicate a decrease in sexual activity and interest with increasing age (Keller, Eakes, Hinkle, & Hughston, 1978; Palmore, 1985; Pfeiffer & Davis, 1972; see Garza & Dressel, 1983, for a review). Verwoedt, Pfeiffer, and Wang (1969) used longitudinal data collected at three points over a 6-year period as part of the Duke Longitudinal Studies and found that the sharpest declines in sexual activity were noted for men in their mid-70s. However, there were some exceptions to the general rule of declining sexual activity—about one fifth of men aged 80 or over reported having sexual intercourse once a month or less. Women in these samples (aged 60 or over) consistently reported engaging in less sexual activity than reported by men.

Pfeiffer and Davis (1972) found a number of factors contributed independently to lowered sexual activity for men, including age, poor health, use of antihypertension medication, and low socioeconomic status. For women, age and the absence of a socially sanctioned sexual partner were the main predictors of lowered sexual activity (see also Malatesta, Chambless, Pollack, & Cantor, 1988).

There is evidence that decreases in frequency of sexual activity are more attributable to husbands' loss of interest than to a loss of interest among wives. George and Weiler (1981) also reported on longitudinal data from the Duke Longitudinal Studies and found that both men and women overwhelmingly attributed cessation of sexual relations to the attitudes or physical condition of the male (see also Murphy, Hudson, & Cheung, 1980; Roberts, 1979).

None of these studies have examined how couples negotiate decreases in sexual activity. However, Schiavi, Mandeli, and Schreiner-Engel (1994) found that although older couples (husbands aged over 64) reported having sex less frequently than younger couples (those under 65), levels of satisfaction with frequency of sex and enjoyment of sex were the same for both groups. Levels of marital satisfaction were the same in both groups. Libman (1989) had earlier reported the same patterns.

There is some evidence that couples compensate for decreased sexual activity. Roberts (1979) found that couples who had ceased sexual activity reported increased emphasis on emotional intimacy, sitting and lying close to each other, touching, and holding hands.

There is also some evidence that age-related declines in sexual activity are at least partly dependent on cohort and culture. George and Weiler (1981) found that although older members of their study did have lower levels of sexual activity than younger age groups (e.g., those aged 72–77 had lower levels than those aged 66–71, who had lower levels than those aged 60–65), those levels remained stable over the course of the study. As individuals aged, the majority did not report declining interest or activity. The authors concluded that "part of the differences between age groups reflect the effects of cohort-related rather than age-related change" (pp. 17–18).

It also seems probable that cohort differences are responsible for Keller's, Eakes', Hinkle's, and Hughston's (1978) finding that older married women (60 or older) reported more guilt about sex than younger married women (50 or younger). The idea that guilt about sex even within the socially sanctioned confines of marriage would accumulate over the years seems less plausible than the argument that earlier generations held more restrictive attitudes toward sex.

Similarly, it seems possible that Libman's (1989) finding that the sexual repertoire of couples over 65 focused primarily on intercourse and showed typical gender role stereotypes (males initiating sex more frequently, females agreeing to participate in relatively few sexual activities) is related more to cohort than to age or length of marriage.

Finally, it is important to acknowledge that social norms probably play a large role in shaping the course of sexual relationships across the life span. Numerous studies indicate that in Western societies, older adults are regarded as less sexually competent, both by younger adults and by older adults themselves (Cameron, 1970; Damrosch, 1982; La Torre & Kear, 1977; Libman, 1989). In contrast, Winn and Newton (1982) reported that sexuality among older adults of preindustrial and traditional cultures frequently indicated no (or very minimal) age-related decline for both men and women.

COMMUNICATION AND THE AGING COUPLE

In the previous edition of this chapter, we mourned the dearth of studies of communication among aging couples. This still remains true, though some new studies have emerged. The research on communication styles among older married couples reflects the major finding about marital satisfaction—that older adults engage in relatively little conflict. What is more controversial is whether this simply reflects a disengaged, unemotional style of interaction or whether it really indicates fewer problems and more positive affect among older couples.

Zietlow and Sillars (1988) examined differences in marital conflict resolution among individuals ranging in age from 23 to 83 years. As is usual in the research, couples first ranked a series of potential problems in the relationship. Irritability and poor communication were ranked as the most significant problems for all age groups. The reported importance of marital problems in general decreased linearly with age, so older couples, even when they acknowledged that there were problems, felt they were unimportant.

Couples in this study were also asked to discuss together these problematic issues in their marriages. Their conversations were then coded for different conflict resolution strategies. The conflict resolution codes included denial and equivocation, topic management (e.g., shifting the topic of conversation away from the problem), analytic remarks (e.g., nonevaluative remarks describing the problems), and confrontative remarks (e.g., personal criticism, rejection, hostile imperatives, and hostile jokes). Young couples showed fairly high levels of engagement (e.g., openly and directly dealing with problematic issues), regardless of whether they rated the issue as a problem for them or not. When discussing important issues, middle-aged adults moved toward an analytic style of discussion (problem-solving and solution-orientation remarks), with low levels of confrontative remarks (e.g., blaming and disparaging).

Older adults showed different patterns. Much of the time, their conversations were typified by a noncommittal style—they tended to use abstract and hypothetical remarks, statements that are irrelevant to any conflict issue, unfocused questions, and procedural remarks. Zietlow and Sillars characterized this style as a low-risk, low-disclosure style of communication that is much like casual conversation or small talk. Only a few unhappy older adults engaged in high levels of confrontation (see Sillars & Wilmot, 1989, for further discussion).

Carstensen, Gottman, and Levenson (1995) studied the communication styles of four groups of married couples: middle-aged satisfied, middle-aged dissatisfied, older adult satisfied, and older adult dissatisfied. Satisfaction was assessed by averaging scores on the Locke–Williamson and the Locke–Wallace measures of marital satisfaction. Middle-aged couples had been married for a minimum of 15 years; older couples were married for at least 35 years. Couples were observed discussing the events of the day, a problem area of continuing disagreement in their marriage, and a mutually agreed on pleasant topic.

Overall, middle-aged and older adult couples displayed the same patterns as observed in prior research on younger adults: Even after many years of marriage, women were more emotionally expressive than men, and satisfied couples were more positive in their interaction styles than dissatisfied couples. Nonetheless, there were significant age differences in emotional styles. Put simply, older adults were more positive in their interactions. Older speakers expressed significantly more affection toward their listeners than middle-aged speakers, arguing against the claim that older couples are simply less emotionally involved. Older dissatisfied couples engaged in fewer negative start-ups (i.e., sequences in which one spouse's neutral affect was followed by the other spouse's negative affect) than any of the other three experimental groups. The authors argued that older dissatisfied couples may have learned to "leave well enough alone" by staying in affectively neutral interactive sequences and avoiding escalation to negative affect. In contrast, middle-aged couples expressed more emotions overall, including more negative emotions. They were coded as displaying significantly more interest, humor, disgust, belligerence, and whining than older couples, even after controlling for couples' ratings of the severity of marital problems.

Szinovacz and Schaffer (2000) studied the effects of retirement on couples' conflict tactics. They used a sample of 559 continuously married couples in which at least one spouse was employed and aged 50 to 70 at the start of the study. Each couple was interviewed twice between 1987 and 1994. Respondents were asked how often, during serious marital disagreements, they used the following tactics: keeping opinions to themselves, calm discussion of disagreement, and shouting or heated arguments. Szinovacz and Schaffer expected that retirement would reduce work- and household-related stresses, leading to a decline in heated arguments and an increase in calm discussion. In fact, their results were more complex, to a large extent mirroring the finding by Myer and Booth (1996), that retirement reinforced existing marital qualities. Calm discussions increased following retirement if spouses were strongly attached to the relationship and decreased if spouses were low in attachment. Similarly, husbands and wives tended to keep controversial opinions to themselves following retirement unless they felt relatively powerful (i.e., spouse more dependent on the relationship), in which case they apparently spoke their minds. Heated arguments remained relatively stable, except that husbands whose wives retired at the same time as them perceived a subsequent drop in heated arguments. As the authors acknowledged, the intricate interactions found in this study were post hoc and deserve further investigation. Overall, though, the question remains about the extent to which the results of Szinovacz and Schaffer (2000) and Carstensen et al. (1995) are cohort or developmental effects.

Three studies provide rather sketchy evidence of generational changes in expressiveness among couples. Rands and Levinger (1979) had older adults fill out questionnaires describing their close relationships in early and midlife. The descriptions contained fewer accounts of expression of feelings and confiding about problems than young adults reporting on current relationships. Of course, this study involved reconstructive memories of communication in the early years of the older adults' marriage. Van Lear (1992) compared young couples (average age late 20s) and their parents (average age mid-50s) using Fitzpatrick's (1988) Relational Dimensions Inventory (RDI). Van Lear found that young husbands and wives reported spending more time together than did either of their parents, even after controlling for social desirability and (more relevantly) number of years married. Finally, ethnographic research by Caplow et al. (1982) found evidence of substantial changes both in style and amount of marital interaction from the 1920s to the 1970s. Couples in the 1920s reported spending little time in conversation and having numerous taboo topics, including sexual desires and birth control. By contrast, couples in the 1970s talked to each other much more often about a wider range of topics.

Apart from cohort effects, there probably are effects of the length of the marriage, but we have no real tests of them. Carstensen et al. (1995) argued that their findings reflected maturational changes such that emotion regulation assumes increasing importance with age and that older adults actively avoid negative emotional experiences. Without longitudinal research following the move from middle to old age, we cannot really evaluate such a claim. Similarly, Sillars and Wilmot (1989) discussed the probability that over time, couples develop increasingly implicit and idiomatic ways of speaking. Bell, Buerkel-Rothfuss, and Gore (1987) studied couples in their early 20s and found a move toward developing private idioms for sexual references as the relationship developed. More work is clearly needed to explore the effects of years of shared experience on the ways in which couples communicate with each other.

CONCLUSION

On returning to this chapter after 5 years, we were happy to find some evidence that the field had moved forward. In particular, the increasing number of panel, trend, and longitudinal studies is a substantial improvement, and we look forward to further waves of data from these studies. Probably, the major theme in this chapter is the importance of untangling maturational and cohort effects, and we can only begin to do so with these new data sets.

Another encouraging sign is the cluster of studies by Carstensen, Gottman, and Levenson (in various combinations and permutations) that involve detailed observational work of sizable samples of elderly adults. As of 5 or 6 years ago, there was almost no such work. When detailed observations were taken, it was typically of very small, unrepresentative samples. Only now, and still only with these authors, are we beginning to see some of the rigorous theory and techniques often used with younger samples being applied to older populations as well (see Gottman & Notarius, 2000, for a review).

Overall, the picture of current old age remains relatively positive. The fact that the picture is becoming more complicated and less stereotypically dichotomous (either miserable, lonely old age, or happy, vibrant old age) can only be seen as a step forward. We look forward to future editions and a still more nuanced picture.

REFERENCES

Ade-Ridder, L., & Brubaker, T. H. (1983). The quality of long-term marriages. In T. H. Brubaker (Ed.), *Family relationships in later life* (pp. 21–30). Beverly Hills, CA: Sage.

Albrect, S. L., Bahr, H. M., & Chadwick, B. A. (1979). Changing family and sex roles: An assessment of age differences. *Journal of Marriage and the Family, 41*, 41–50.

Altergott, K. (1988). Social action and interaction in later life: Aging in the United States. In K. Altergott (Ed.), *Daily life in later life: Comparative perspectives* (pp. 117–146). Newbury Park, CA: Sage.

Anderson, S. A., Russell, C. S., & Schumm, W. R. (1983). Perceived marital quality and family life-cycle categories: A further analysis. *Journal of Marriage and the Family, 45*, 127–139.

Bachrach, C. A. (1980). Childlessness and social isolation among the elderly. *Journal of Marriage and the Family, 42*, 627–637.

Ballweg, J. A. (1967). Resolution of conjugal role adjustment after retirement. *Journal of Marriage and the Family, 29*, 277–281.

Baltes, P. B. (1968). Longitudinal and cross-sectional sequences in the study of age and generation effects. *Human Development, 11*, 145–171.

Bell, R. A., Buerkell-Rothfuss, N. L., & Gore, K. E. (1987). "Did you bring the yarmulke for the cabbage patch kid?" The idiomatic communication of young lovers. *Human Communication Research, 14*, 47–67.

Blake, J. (1979). Is zero preferred? American attitudes toward childlessness in the 1970s. *Journal of Marriage and the Family, 41*, 245–257.

Blood, R. O., & Wolfe, D. M. (1960). *Husbands and wives*. Glencoe, IL: Free Press.

Brubaker, T. H. (1985). *Later life families*. Beverly Hills, CA: Sage.

Brubaker, T. H. (1990). Families in later life: A burgeoning research area. *Journal of Marriage and the Family, 52*, 959–981.

Brubaker, T. H., & Hennon, C. B. (1982). Responsibility for household tasks: Comparing dual-earners and dual-retired marriages. In M. Szinovacz (Ed.), *Women's retirement: Policy implications of recent research* (pp. 205–219). Beverly Hills, CA: Sage.

Burgess, E. W., & Cottrell, L. S. (1939). *Predicting success or failures in marriage*. New York: Prentice Hall.

Burman, B., & Margolin, G. (1992). Analysis of the association between marital relationships and health problems: An interactional perspective. *Psychological Bulletin, 112*, 39–63.

Burr, W. R. (1970). Satisfaction with various aspects of marriage over the life cycle: A random middle class sample. *Journal of Marriage and the Family, 32*, 29–37.

Cameron, O. (1970). The generation gap: Beliefs about sexuality and self-reported sexuality. *Developmental Psychology, 3*, 272.

Camp, P. L., & Ganong, L. H. (1997). Locus of control and marital satisfaction in long-term marriages. *Families in society: The journal of contemporary human services, 30*, 624–631.

Caplow, T., Bahr, H. M., Chadwick, B. A. Hill, R., & Williamson, M. H. (1982). *Middletown families*. Minneapolis: University of Minnesota Press.

Carlson, E. (1979). Divorce rate fluctuation as a cohort phenomenon. *Population Studies, 33,* 523–536.

Cartensen, L. L., Gottman, J. M., & Levenson, R. W. (1995). Emotional behavior in long-term marriage. *Psychology and Aging, 10,* 140–149.

Chayko, M. (1993). How you "act your age" when you watch TV. *Sociological Forum, 8,* 573–593.

Cooney, T. M., & Uhlenberg, P. (1990). The role of divorce in men's relations with their adult children after mid-life. *Journal of Marriage and the Family 52,* 677–688.

Coyne, J. C., Kahn, J., & Gottlieb, I. H. (1987). Depression. In T. Jacob (Ed.), *Psychopathology and the family* (pp. 509–533). New York: Plenum.

Damrosch, S. (1982). Nursing students' attitudes toward sexually active older persons. *The Gerontologist, 24,* 299–302.

Fengler, A. P. (1973). The effects of age and education on marital ideology. *Journal of Marriage and the Family, 35,* 264–271.

Fitting, M., Rabins, P., Lucas, M. J., & Eastham, J. (1986). Caregivers for dementia patients: A comparison of husbands and wives. *The Gerontologist, 26,* 248–252.

Fitzpatrick, M. A. (1988). *Between husbands and wives.* Newbury Park, CA: Sage.

Garza, J. M., & Dressel, P. L. (1983). Sexuality and later-life marriages. In T. H. Brubaker (Ed.), *Family relationships in later life* (pp. 91–108). Beverly Hills, CA: Sage.

George, L. K., & Weiler, S. J. (1981). Sexuality in middle and late life: The effects of age, cohort, and gender. *Archives of General Psychiatry, 38,* 919–923.

Gilford, R. (1984). Contrasts in marital satisfaction throughout old age: An exchange theory analysis. *Journal of Gerontology, 39,* 325–333.

Glass, T. A., & Seeman, T. E., Herzog, A. R. Change in productive activity in late adulthood: MacArthur studies of successful aging. *Journals of Gerontology: Series B, 50B,* S65–S76.

Glenn, N. D., & McLanahan, S. S. (1981). The effects of offspring on psychological well-being of older adults. *Journal of Marriage and the Family, 43,* 409–421.

Glenn, N. D. (1991). The recent trend in marital success in the United States. *Journal of Marriage and the Family, 53,* 261–270.

Glenn, N. D. (1998). The course of marital success and failure in five American 10-year marriage cohorts. *Journal of Marriage and the Family, 60,* 569–576.

Goldman, N., Koreman, S., & Weinstein, R. (1995). Marital status and health among the elderly. *Social Science & Medicine, 40,* 1717–1730.

Goldscheider, F. K. (1990). The aging of the gender revolution: What do we know and what do we need to know? *Research on Aging, 12,* 531–545.

Goodman, C. C. (1999). Reciprocity of social support in long-term marriage. *Journal of Mental Health and Aging, 5,* 341–357.

Gottman, J. M., & Levenson, R. W. (1992). Marital processes predictive of later dissolution: Behavior, physiology, and health. *Journal of Personality and Social Psychology, 63,* 221–233.

Gottman, J. M., & Notarius, C. I. (2000). Decade review: Observing marital interaction. *Journal of Marriage and the Family, 62,* 927–947.

Heaton, T. B. (1991). Time-related determinants of marital dissolution. *Journal of Marriage and the Family, 53,* 285–295.

Heckhausen, J., & Brim, O. G. (1997). Perceived problems for self and others: Self-protection by social downgrading throughout adulthood. *Psychology and Aging, 12,* 610–619.

Hennon, C. B. (1983). Divorce and the elderly: A neglected area of research. In T. H. Brubaker (Ed.), *Family relationships in later life* (pp. 149–172). Beverly Hills, CA: Sage.

Hicks, M. W., & Platt, M. (1970). Marital happiness and stability: A review of research in the sixties. *Journal of Marriage and the Family, 32,* 553–555.

Houseknecht, S. K. (1979). Childlessness and marital adjustment. *Journal of Marriage and the Family, 41,* 259–265.

Johnson, C. L. (1985). The impact of illness on late-life marriages. *Journal of Marriage and the Family, 47,* 165–172.

Johnson, M. P., Huston, T. L., Gaines, S. O., & Levinger, G. (1992). Patterns of married life among young couples. *Journal of Social and Personal Relationships, 9,* 343–364.

Keating, N. C., & Cole, P. (1980). What do I do with him 24 hours a day? Changes in the housewife role after retirement. *The Gerontologist, 20,* 84–89.

Keith, P. M. (1983). A comparison of the resources of parents and childless men and women in very old age. *Family Relations, 32,* 403–409.

Keith, P. M., (1986). The social context and resources of the unmarried in old age. *International Journal of Aging and Human Development, 23,* 81–96.

Keller, F. J., Eakes, E., Hinkle, D., & Hughston, G. A. (1978). Sexual behavior and guilt among women: A cross-generational comparison. *Journal of Sex and Marital Therapy, 4,* 259–265.

Kivett, V. R., & Lerner, R. M. (1980). Perspectives on the childless rural elderly: A comparative analysis. *The Gerontologist, 20,* 708–716.

Kovar, M. (1977). Elderly people: The population 65 years and over. In *Health: The United States 1976–1977* (U.S. Department of Education and Welfare Publication No. HRA 77-1232). Washington, DC: U.S. Government Printing Office.

Lachman, M. E., & Weaver, S. L. (1998). Sociodemographic variations in the sense of control by domain: Findings from the MacArthur Studies of Midlife. *Psychology and Aging, 13,* 553–562.

La Torre, R. P., & Kear, K. A. (1977). Attitudes towards sex in the aged. *Archives of Sexual Behavior, 6,* 203–213.

Lauer, R. H., Lauer, J. C., & Kerr, S. T. (1990). The long-term marriage: Perceptions of stability and satisfaction. *International Journal of Aging and Human Development, 31,* 189–195.

Lee, G. R., & Shehan, C. L. (1989). Retirement and marital satisfaction. *Journal of Gerontology, 44,* S226–S230.

Levenson, R. W., Cartensen, L. L., & Gottman, J. M. (1993). Long-term marriage: Age, gender, and satisfaction. *Psychology and Aging, 8,* 301–313.

Levenson, R. W., Cartensen, L. L., & Gottman, J. M. (1994). The influence of age and gender on affect, physiology, and their interrelations: A study of long-term marriage. *Journal of Personality and Social Psychology, 67,* 56–68.

Libman, E. (1989). Sociocultural and cognitive factors in aging and sexual expression: Conceptual and research issues. *Canadian Psychology, 30,* 560–566.

Locke, H. J. (1951). *Predicting adjustment in marriage: A comparison of a divorces and a happily married group.* New York: Holt.

Locke, H. J., & Wallace, K. M. (1959). Short marital adjustment and prediction tests: Their reliability and validity. *Marriage and Family Living, 21,* 251–255.

Luckey, E. B. (1966). Numbers of years married as related to personality perception and marital satisfaction. *Journal of Marriage and the Family, 28,* 44–48.

Malatesta, V. J., Chambless, D. L., Pollack, M., & Cantor, A. (1988). Widowhood, sexuality and aging: A life span analysis. *Journal of Sex & Marital Therapy, 14,* 49–62.

Millar, F. E., & Rogers, L. E. (1988). Power dynamics in marital relationships. In P. Noller & M. A. Fitzpatrick (Eds.), *Perspectives on marital interaction* (pp. 78–97). Clevedon, UK: Multilingual Matters.

Moen, P., Kim, J. E., & Hofmeister, H. (2001). Couples' work/retirement transitions, gender, and marital quality. *Social Psychology Quarterly, 64,* 55–71.

Murphy, G. J., Hudson, W. W., Cheung, P. P. L. (1980). Marital an sexual discord among older couples. *Social Work Research and Abstracts, 16,* 11–16.

Myers, S. M., & Booth, A. (1996). Men's retirement and marital quality. *Journal of Family Issues, 17,* 336–357.

Norton, A. J., & Moorman, J. E. (1987). Current trends in marriage and divorce among American women. *Journal of Marriage and the Family, 49,* 3–14.

Palmore, E. (1985). How to live longer and like it. *Journal of Applied Gerontology, 4*(2), 1–8.

Pfeiffer, E., & Davis, G. C. (1972). Determinants of sexual behavior in middle and old age. *Journal of the American Geriatric Society, 20,* 151–158.

Pienta, A. M., Hayward, M. D., & Jenkins, K. R. (2000). Health consequences of marriage for the retirement years. *Journal of Family Issues, 21,* 559–586.

Rands, M., & Levinger, G. (1979). Implicit theories of relationship: An intergenerational study. *Journal of Personality and Social Psychology, 37,* 645–661.

Rempel, J. (1985). Childless elderly: What are they missing? *Journal of Marriage and the Family, 47,* 343–348.

Rexroat, C., & Shehan, C. (1987). The family life cycle and spouses' time in housework. *Journal of Marriage and the Family, 49,* 737–750.

Roberts, W. L. (1979). Significant elements in the relationship of long-married couples. *International Journal of Aging and Human Development, 10,* 265–271.

Rogers, S. J., & Amato, P. R. (1997). Is marital quality declining? The evidence from two generations. *Social Forces, 75,* 1089–1100.

Rogers, S. J., & Amato, P. R. (2000). Have changes in gender relations affected marital quality? *Social Forces, 79,* 731–753.

Rollins, B., & Feldman, H. (1970). Marital satisfaction over the family life cycle. *Journal of Marriage and the Family, 32,* 20–28.

Ross, C. (1995). Reconceptualizing marital status as a continuum of attachment. *Journal of Marriage and the Family, 57,* 129–140.

Ryder, R. G. (1973). Longitudinal data relating marriage satisfaction and having a child. *Journal of Marriage and the Family, 35,* 750–755.

Schaie, K. W., & Hertzog, C. (1985). Measurement in the psychology of adulthood and aging. In J. E. Birren & K. W. Schaie (Eds.), *Handbook of the psychology of aging* (2nd ed., pp. 61–94). New York: Van Nostrand Reinhold.

Schiavi, R. C., Mandeli, J., & Schreiner-Engel, P. (1994). Sexual satisfaction in healthy aging men. *Journal of Sex & Marital Therapy, 20,* 3–13.

Sillars, A. L., & Wilmot, W. W. (1989). Marital communication across the life span. In J. Nussbaum (Ed.), *Lifespan communication* (pp. 225–253). Hillsdale, NJ: Lawrence Erlbaum Associates.

Spanier, G. B. (1976). Measuring dyadic adjustment: New scales for assessing the quality of marriage and similar dyads. *Journal of Marriage and the Family, 37,* 15–28.

Spanier, G. B., & Cole, C. L. (1975). Marital adjustment over the family life cycle: The issue of curvilinearity. *Journal Marriage and the Family, 37,* 263–278.

Sporakowski, M. J., & Hughston, G. A. (1978). Prescriptions for a happy marriage: Adjustments and satisfactions of couples married for 50 or more years. *The Family Coordinator, 27,* 321–327.

Stinnett, N., Carter, L. M., & Montgomery, J. E. (1972). Older persons' perceptions of their marriages. *Journal of Marriage and the Family, 34,* 665–670.

Stinnett, N., Collins, J., & Montgomery, J. E. (1970). Marital need satisfaction of older husbands and wives. *Journal of Marriage and the Family, 32,* 428–434.

Stone, R., Cafferata, G. L., & Sangl, J. (1987). Caregivers of the frail elderly: A national profile. *The Gerontologist, 27,* 616–626.

Suitor, J. J. (1991). Marital quality and satisfaction with the division of household labor across the family life cycle. *Journal of Marriage and the Family, 53,* 221–230.

Swenson, C. H., Eskew, R. W., & Kohlhepp, K. A. (1981). Stage of family life cycles, ego development, and the marriage relationship. *Journal of Marriage and the Family, 43,* 841–853.

Szinovacz, M. E. (1980). Female retirement: Effects on spousal roles and marital adjustment. *Journal of Family Issues, 16,* 423–440.

Szinovacz, M. E. (1996). Couples' employment/retirement patterns and perceptions of marital quality. *Research on Aging, 18,* 243–268.

Szinovacz, M. E. (2000). Changes in housework after retirement: A panel analysis. *Journal of Marriage and the Family, 62,* 78–92.

Szinovacz, M. E., & Schaffer, A. M. (2000). Effects of retirement on marital conflict tactics. *Journal of Family Issues, 21,* 367–389.

Treas, J., & Van Hilst, A. (1976). Marriage and remarriage rates among older Americans. *The Gerontologist, 16,* 132–136.

Tuckman, J., & Lorge, I. (1954). Old people's appraisal of adjustment over the life span. *Journal of Personality, 22,* 417–422.

Van Lear, A. (1992). Marital communication across the generations: Learning and rebellion, continuity and change. *Journal of Social and Personal Relationships, 9,* 103–123.

Verwoedt, A., Pfeiffer, E., & Wang, H. S. (1969). Sexual behavior in senescence. *Geriatrics, 24,* 137–154.

Weishaus, S., & Field, D. (1988). A half century of marriage: Continuity or change? *Journal of Marriage and the Family, 50,* 763–774.

Winn, R. L., & Newton, N. (1982). Sexuality in aging: A study of 106 cultures. *Archives of Sexual Behavior, 11,* 283–298.

Zarit, S., Todd, P., & Zarit, J. (1986). Subjective burden of husbands and wives as caregivers: A longitudinal study. *The Gerontologist, 26,* 260–266.

Zietlow, P. H., & Sillars, A. L. (1988). Life-stage differences in communication during marital conflicts. *Journal of Social and Personal Relationships, 5,* 223–245.

10

The Nature of Family Relationships Between and Within Generations: Relations Between Grandparents, Grandchildren, and Siblings in Later Life

Valerie Cryer McKay and R. Suzanne Caverly
California State University, Long Beach

INTRODUCTION

The life span perspective offers a view of life that is progressive and reflective, continuous and insightful. Congruent with this perspective, intergenerational relationships between grandparents and grandchildren, and intragenerational relationships between siblings, offer unique opportunities to examine the dynamic and communicative nature of these relationships in later life. In fact, more than four decades of research into the lives of older adults have provided considerable insight into this dynamic perspective of family relationships between and within generations. With this in mind, research has the potential to take new and creative directions in exploring the depth and richness of the human life experience. This chapter will offer a comprehensive review of research examining relationships between grandparents, grandchildren, and older adult siblings over the life course and consider the influences of gender, culture, and relevant socioeconomic factors fundamental to a feminist perspective. Integrating these perspectives offers an innovative framework by which to explore those aspects of human experience unique to communication and aging research and theory.

There are three basic assumptions intrinsic to feminist inquiry, assumptions that emerge from criticisms or limitations identified in traditional scientific models of social research (Millman & Kanter, 1987; Sherif, 1987). First, there is substantial evidence that social, economical, and cultural factors are integral to a comprehensive understanding of the nature of human life experience. Second, we now recognize that people of various genders and cultures inhabit diverse social worlds to the extent that generalizations inherent in traditional scientific models are limited in their applicability to these populations. Finally, the tendency for social scientists to pursue quantitative, rather than or in addition to qualitative, methods may have inhibited inquiry into less mainstream relationships such as those between grandparents, grandchildren, and siblings. Based on these premises, a review of past and current research literature exploring family relationships between and within generations in later life will provide the basis for recommending new directions for research within a framework of feminist inquiry and the life span perspective.

Perhaps the most viable argument for reevaluating traditional views of family life is the definition of "family" in our society today. As noted in the U.S. presidential elections of 1992, 1996, and again in 2000, the issue of "family values" raised questions regarding how we, as a society, define family, and consequently, what family unit was perpetuating which values. The traditional nuclear family comprises a comparably small percentage of the total number of "families" in the United States. More prevalent are the nontraditional family units made up of single-parent families (female head of household), grandparents parenting grandchildren, divorced or stepfamilies, extended family, and homosexual couples with biological or adopted children, to name a few. Integral to our examination of family relationships is the recognition that limitations in traditional scientific inquiry (i.e., utilizing the nuclear family as the unit of analysis) minimize generalization to nontraditional family forms.

Factors influencing our definitions of family and family relationships for older adults are many and varied. Not only are increased life expectancy and better health care offering individuals the opportunity to continue active and satisfying lives, but also the changing roles of women and men in our society continue to influence our conception of traditional and nontraditional family forms and relationships. For example, as women are increasingly more involved in career and educational goals, the convergence of life patterns in later years has lessened the social gap between women and men (Hagestad, 1987). "Three trends have been identified as contributing to such convergence: the decreasing life course involvement in work and parenting, the androgyny of later life, and the feminization of the population" (p. 421). It is not surprising, then, that the common perception of women as kin keepers occupied with maintaining family ties, child caregiving, and homemaking is rapidly changing, and men are becoming increasingly more involved in family relationships.

In contrast to the normative image of aging as decline, increased dependency, and senility, preventive health maintenance and health care provide older adults with the means to remain active and healthy well into their 70s and 80s (Hagestad, 1987). Likewise, many older adults continue to be active participants in their communities during their retirement. At least in terms of the grandparent–grandchild relationship, the image of "grandma" or "grandpa in a rocking chair" is, to a great extent, both unrealistic and outdated. The higher life expectancy for women ensures that they are more likely to be widowed in later life. The divergence from sex-stereotypical behavior that women are now experiencing in adulthood also guarantees that they are more likely to maintain friendships and active lifestyles for longer periods of time following the death of their spouse (Hagestad, 1987). Relationships with siblings are valued and often renewed in later years as the similarities in life experience and need for emotional support intensify (Atchley, 1991). For unmarried or widowed women, relationships between siblings, and specifically sisters, become more salient in later years (Connidis & Davies, 1992).

To comprehensively analyze family relations of the elderly, Brown (1990) recommended that a multicultural perspective must also be adopted. If it is assumed that our society is homogeneous, then a comparison between cultures within our society automatically imposes a model of "norm" and "deviation" consistent with positivistic views. In other words, if White culture is defined as the "norm," then by definition non-White cultures are "deviant" and fail to adhere to the standard definitions of family life. Consistent with the feminist view is recognition that our society is not homogeneous but heterogeneous, and efforts to characterize each culture within American society should recognize the uniqueness associated with each and not limit them by definition to normal and not normal. Accordingly, Bell, Bouie, and Baldwin (1990) argued that imposing the Euro-American view on various cultures within our society has resulted in the definition of relationships according to a Eurocentric rather than multicultural view. Thus, relationships in families of non-European origin endure the effects of social and psychological oppression for failing to adhere to what is accepted as normal in our society.

We expand our understanding of family in our society today by considering the influence of gender and culture and the pervasiveness of nontraditional family forms. No longer should research questions be limited to the investigation of predominantly demographic variables that are purported to affect or influence relationships between and within generations. No longer should it be assumed that the nuclear family is the appropriate unit of analysis, neither can we generalize from nuclear family to nontraditional family. No longer are notions of traditional gender roles acceptable, nor can we assume that men and women continue to adhere to gender stereotypes in relations between generations. No longer can we assume that family relationships do not vary by culture. On the contrary, the exploration of these unique and extraordinary family relationships encompasses new and uncommon dimensions that should become the focus of research into human experience. This is the premise on which this chapter is conceived.

THEN AND NOW: FOUR DECADES OF RESEARCH ON GRANDPARENTS AND GRANDCHILDREN

Early investigations into the nature of the grandparent–grandchild relationship focused primarily on the enjoyable aspects of the relationship for grandparents. Grandparents often conceived of their relations with grandchildren as "pleasure without responsibility" (Albrecht, 1954). In subsequent investigations, grandparents have been found to take a more participative orientation in their role, especially in situations related to divorce (Beal, 1979; Blau, 1984; Cogswell & Henry, 1995; Johnson, 1985; Wilson & DeShane, 1982), surrogate child care (Bell & Garner, 1996; Caputo, 1999; Hartfield, 1996; Updegraff, 1968; Woodworth, 1996), families of lower socioeconomic status (Clavan, 1978), and life in Black communities (Aston, 1997; Hays & Mindel, 1973; Hunter, 1997; Stack, 1974). Other studies have explored the nature and characteristics of the grandparent role and, specifically, the function of grandparents as adult role models for children. For example, in some families, grandparents are conceived of as mentors for grandchildren, offering advice on topics such as education and career (King & Elder, 1998a, 1998b; Roberto & Skoglund, 1996). However, grandparents' age, health status, gender, level of education, attitudes toward their life situations, and relations with their own children, as well as other social factors mediate the degree to which this aspect of the grandparental role can be fulfilled (King & Elder, 1998b; Kornhaber, 1985; Kornhaber & Woodward, 1981; Troll & Bengtson, 1979).

What is the effect of age on the role of grandparent? The perception of the role itself varies with the age at which adults become grandparents. As noted by Hagestad (1985), "in a society in which grandparents range in age from 30 to 110, and grandchildren range from newborns to retirees, we should not be surprised to find a wide variety of grandparenting styles" (p. 36). The onset of grandparenthood is not voluntary, and often adults are little prepared to accept the role of grandparent that has been traditionally characterized by old age (Johnson, 1983). Consequently, this unsuitable portrayal of traditional grandparenthood engenders feelings of ambivalence; in fact, age has been found to be a significant predictor of grandparents' participation in relations with grandchildren (Johnson, 1985). In an investigation comparing older and younger grandparents, younger grandparents (operationalized as under 65 years of age) were found to accept greater responsibility for child care, grandchild discipline, and the offer of advice more often than were older grandparents (Thomas, 1986). This was attributed to the fact that these individuals had recently been parents themselves. Older grandparents attributed their lack of participation to geographic distance, physical limitations, disinterest, and strained relations with their children (Neugarten & Weinstein, 1964, 1968). The wisdom that comes with age has also been found to influence the definition of the role for grandmothers. Older grandmothers reported that they knew more about their grandchildren, were better able to teach their grandchildren, and were less frustrated than women who became grandmothers at a younger age (Watson, 1997).

What are the effects of grandparents' health on relationships with grandchildren? Grandparents' health is often found to be a factor influencing the frequency of interaction with grandchildren; yet, it has only recently become the focus of intergenerational research. In a unique study of grandparents with Alzheimer's disease, Creasey and Jarvis (1989) assessed the effects of this disease on grandparents' relations with their grandchildren, and the child's relations with caregiving parents. The psychological effects of the disease are often devastating—both to the victim and his or her family. Research indicates that the stress involved in caring for an Alzheimer's victim is extreme; Alzheimer's disease is often a family crisis. In a three-generation family, the effects of stress on the caregiver (usually mother) are dynamic; children experience the lack of attention by the caregiver devoted to the disabled grandparent, anger at the lack of assistance provided by their father, and loss of a meaningful relationship with the grandparent. The objective of this research was to provide descriptive and prescriptive information to aid in counseling family members.

Is there a difference between men and women and their role as grandparent? Grandparent gender has frequently been utilized as a distinguishing variable between varying degrees of closeness and interaction with grandchildren; grandmothers reported more closeness, whereas grandfathers were more emotionally distant. These results have often been attributed to the expressive versus instrumental dimensions of traditional gender socialization (Hagestad, 1985). More recently, however, one study found no significant differences between grandmothers' and grandfathers' satisfaction with the grandparenting role. Grandparents disclosed that opportunities to observe their grandchildrens' development and share in their activities were the best predictors of role satisfaction (Peterson, 1999). In fact, a recent study of expectations associated with the grandparent role suggested that grandfathers felt more able to offer childrearing advice, whereas grandmothers reported greater satisfaction with their role overall (Somary & Stricker, 1998).

Research has also revealed social factors that influence the grandparent–grandchild relationship. For example, King and Elder (1998a) found that the level of the grandparents' education was directly related to the level of contact with grandchildren: Less educated grandparents were more likely to have frequent contact with grandchildren, rated the relationship quality higher, worried more about their grandchild's future, and had strong opinions about the importance of the role and family continuity. In contrast, grandparents with higher levels of education defined their role and responsibility differently; their level of activity depended on the grandchild's age and the quality of interaction focused more on their role as mentor. King and Elder (1995) also explored the characteristics of intergenerational relationships for rural farm and nonfarm families. It is interesting that the relationships between grandparents and grandchildren in both groups were dependent on the relations between parents and grandparents. Grandchildren from rural farm families also defined their relationships with grandparents by frequent contact and more closeness than did nonfarm families. Religiosity has also been found to be a factor in grandparent–grandchild relations (King & Elder, 1999). Specifically, results of this study suggest that religious grandparents are more involved in various aspects of family life. "Overall . . . religious grandparents are more likely to be enmeshed in social ties to others" (King & Elder, 1999, p. S317).

There are many functions that characterize the style with which grandparents enact their role. For some grandparents, the role "constitute[s] a source of biological renewal . . . and continuity" as they exchange information about their life experiences with grandchildren (Neugarten & Weinstein, 1968, p. 282). For others, their role as grandparent centers on resources: financial and advisory. Either way, grandparents often report a sense of satisfaction in their ability to contribute to their grandchildren's lives. Consistent with this view is the belief that grandparents and grandchildren often share similar values and attitudes, resulting in a high degree of personal satisfaction for grandparents (Bengtson, 1985; Neugarten & Weinstein, 1968; Thompson, Clark, & Gunn, 1985; Troll, 1971, 1972). These feelings are intensified by the strength of the intergenerational relationship and perceptions of emotional closeness (Brussoni & Boon, 1998; Holladay, Lackovich, & Lee, 1998). Frequency of interaction and perceptions of closeness in relations with grandchildren have often been used in studies characterizing styles of grandparenting (Cherlin & Furstenburg, 1985; Kivnick, 1985; Robertson, 1976); styles are also defined by grandparent age and the degree of care provided for grandchildren (Baydar & Brooks-Gunn, 1998). Extending the notion of grandparent styles, the diverse and symbolic aspects of grandparenting have also been revealed. The diversity associated with the grandparent role was described in terms of child care and recreational activities; symbolic aspects were portrayed as those behaviors that grandparents perform by "being there" for other family members by providing advice, financial assistance, or emotional support. Symbolically, grandparents provide a buffer against their childrens' fear of mortality. On their death, the second generation becomes the eldest (and perhaps grandparents themselves) and is forced to face the physical and mental changes associated with the onset of old age (Bengtson, 1985). When the parent dies first, however, grandparents are often the primary source of support and child care (Thompson, 1999).

The grandparent–grandchild relationship is often characterized by life experiences shared through stories and interpersonal discourse; this is accomplished by grandparents exchanging information about living and deceased family members, life experiences related to historical events, stories about parents, and stories of grandchildren in infancy and early childhood. Not surprisingly, research strongly suggests that through this role grandparents provide a source of identity development as grandchildren learn about their family "roots," function as a catalyst for improved relations with parents as the child realizes the human character of "mom" and "dad," and serve as a source of historical information as they disclose their personal experience, involvement in historical events, or both (Baranowski, 1982). Grandparents experience a sense of continuity in their own lives by sharing personal and family history (Atchley, 1991). Further, when grandchildren have the opportunity to interact with their grandparents, often their attitudes toward older adults and the aging process are positively influenced (Bekker & Taylor, 1966; Troll, 1971).

What are the effects of contemporary family situations and issues on the grandparent–grandchild relationship (e.g., divorce, drug and alcohol abuse, AIDS)? In a program of research investigating this intergenerational relationship following parents' divorce, Johnson (1983, 1985) concluded that grandparents are often a source of stability and security for grandchildren experiencing the divorce of their parents. Directly, grandparents provide child care and a source of social interaction; indirectly, they are a source of social support (Thompson, Tinsley, Scalora, & Parke, 1989). Gladstone (1988) concluded that the quantity and quality of advice and emotional support provided by grandmothers increases when given the opportunity to respond to grandchildrens' questions regarding their parents' divorce. This type of interaction provides grandchildren with a source of security during a relatively uncertain period of life. "Grandmothers' involvement in young grandchildren's care represents a powerful psychological experience and influences the relations between the grandmother and other members of the family" (Gattai & Musatti, 1999, p. 35). Derdeyn (1985) concluded that "continuing contact between grandparents and grandchildren is an invaluable part of the course of family life" (p. 285) that often provides grandchildren with a positive influence in an otherwise negative family environment. Apparently, our legal system has also recognized the value of this intergenerational relationship during the postdivorce period. Hartfield (1996) noted that "legislators have concluded that grandparents provide a unique type of nurturance and continuity required by their grandchildren in the midst of family crisis and upheaval" (p. 53). In fact, all 50 states have adopted laws regarding grandparent visitation and, although the law usually reviews individual family situations before granting visitation rights, standards are set according to what is in the "best interest of the child" ("Spotlight on the States," 2001).

Factors other than those stipulated by the law also influence postdivorce grandparent–grandchild interaction. For example, research revealed significant differences between maternal and paternal grandparents in the quantity and quality of postdivorce interaction with grandchildren. Not surprisingly, paternal grandparents must often make special efforts to maintain relations with grandchildren (Johnson & Barer, 1987); the matrilineal advantage in grandparent–grandchild relations is well documented (Chan & Elder, 2000).

In fact, Johnson (1988) noted that paternal grandmothers often experience a significant decrease in their involvement, whereas maternal grandmothers become more involved with grandchildren following a divorce. Given that women usually maintain custody of children following a divorce, visitation with grandchildren by maternal grandparents is facilitated. Visitation by paternal grandparents, however, is dependent on the nature of the divorce (hostile or amicable), mothers' relations with her in-laws during the marriage, and fathers' involvement with the children. Grandchild age has also been identified as a factor influencing postdivorce intergenerational relationships. Cooney and Smith (1996) found that "adult grandchildren from divorced families are more likely to initiate contacts with grandparents on their own" (p. 91). Counseling practitioners have only recently begun to explore the effects of the adult child's divorce on the grandparent–grandchild relationship within the context of mediation during family divorce (Kruk & Hall, 1995).

Of increasing concern in social science research is the increasing number of grandparents who accept the role of primary child caregiver when parents are incapable of enacting the parenting role. Although the reasons vary, in 1980, 2 million children in the United States were living with their grandparents or other relatives (Minkler, Roe, & Price, 1992). Currently, in the United States, there are over 3.9 million children under the age of 18 who live with their grandparents; of these, 1.3 million do not have a parent living with them (Bryson & Casper, 1999). Although statistics indicate that this phenomenon is occurring with increasing frequency in all socioeconomic and ethnic groups (American Association of Retired People), it is most prevalent in African American communities in which 12% of grandchildren live in households headed by grandparents. Within the Hispanic community, 5.8% of grandchildren live with surrogate grandparents, and 3.6% of the children in White families live with a family member other than their biological parents. Of fundamental concern are the physical, economic, and emotional conditions of these grandparents. Whereas most grandparents have completed their responsibilities as parents and are psychologically and physically ready to enter into retirement and their "golden years," these grandparents are again asked to accept the role of parent with little social or economic support.

An increasing number of studies have explored the problems and challenges associated with the surrogate parent role for grandparents. "As a consequence of their caregiving role, grandparents raising their grandchildren face a number of problems ranging from stress-related illnesses and social isolation to severe financial difficulties" (Minkler & Roe, 1996, p. 34). In an exploratory investigation of 71 African American grandmothers living in Oakland, California (and whose children were involved in drug use or rehabilitation), Minkler, Roe, and Price (1992) found that these grandmothers were likely to diminish the severity of their own health problems because they feared that their grandchildren would be placed in foster care if they were found to be physically incapable of rendering child care. Further, many of these grandmothers were engaged in either full-time or part-time employment, forcing them to secure additional child-care services. Characteristic of women in the African American community, many of these grandmothers were also involved in caring for other members of the family or community concurrently while caring for their grandchildren. These researchers conclude that practitioners, service providers, and policy makers should focus their efforts to include the social, economic, and physical needs of these grandparent caregivers.

In another study of grandchild caretaking, Burton (1992) found that joblessness, teenage pregnancy, single parenting, and illicit drug addiction are primary contributors to the increase in grandparent childrearing in recent years. Grandparents who parent their grandchildren can receive two types of support: informal support provided by the child's parents or other family members and formal support in terms of social services. However, in both cases, financial support is difficult to secure, and emotional support in any form is limited except in communities where support groups are available. Respondents in this study reported high levels of depression, anxiety, stress, and more frequent illness, contributing to their limited ability to provide maximum care; yet, these grandparents were willing to sacrifice their own needs so that the needs of their grandchild(ren) be met. Many feel isolated, frustrated, and angry; sometimes they feel ashamed (Woodworth, 1996). Burton (1992) argued that policy makers must consider the needs of these grandparents and enact legislation to provide economic support to those who accept the emotional and financial burdens associated with primary child care. Notably, sources of community support are increasing in relation to the number of grandparents accepting the role. For example, Project GUIDE (Grandparents United: Intergenerational Development Education) is "an innovative program for grandparents who are rearing their grandchildren [and] offers a comprehensive service program to help all family members cope with their stressful situation" (Jones & Kennedy, 1996, p. 636). Other national organizations, such as the American Association of Retired People, and regional associations such as Grandparents as Parents in Southern California provide support, information, and resources to empower grandparents parenting grandchildren.

Shifting our attention to grandchildren, research suggests that the grandparent–grandchild relationship varies both developmentally and characteristically with the ages of both grandparent and grandchild (Kahana & Kahana, 1970). In other words, younger grandchildren perceive the relationship differently than older grandchildren in much the same manner as older and younger grandparents differ in their perception of the grandparent role. Results vary, however, with regard to issues of grandchild age and characteristics of relationships with grandparents: Whereas some studies conclude that younger children interact more frequently with grandparents than do adolescents, others report a desire on the part of adolescents to engage in meaningful interaction with grandparents (Kornhaber & Woodward, 1981). Children who have the opportunity to meaningfully participate in relationships with grandparents also perceive the aging process differently from those who do not. For instance, research suggests that younger children (aged 4–8) who interact frequently with grandparents demonstrate less age prejudice toward older adults (Bekker & Taylor, 1966; Isaacs, 1986; Ivester & King, 1977; Thomas & Yamomoto, 1975). In contrast, children who have limited contact with older adults tend to attach negative stereotypes to older persons, in addition to expressing fear of their own aging (Burke, 1982), although this effect is less prevalent in children of older age groups (Luszcz, 1985).

In a study of adolescent children aged 9 to 13 years, three hypotheses were tested regarding grandchildren's relations with maternal grandparents. In situations of divorce, for example, results indicated that grandparents and adolescent grandchildren became more involved—emotionally and recreationally—following a divorce. This phenomenon was referred to as the latent function for grandparents in single-parent families. Moreover, given that adolescents were experiencing biological and social cognitive change and stress with the onset of puberty, they tended to form stronger attachments to parents; these attachments might also diminish strong negative emotions experienced during parents' divorce, that is, "stress buffer" relationships. Evidence indicated that these same attachments developed between grandmothers and granddaughters, grandsons and grandfathers, and grandsons and grandmothers (Clingempeel, Coylar, Brand, & Hetherington, 1992). The authors noted, however, that respondents were Caucasian, middle class, and, therefore, the results were limited in generalizability, because issues of cultural difference were not addressed.

For young adults, the grandparent–grandchild relationship is very important in providing grandchildren with a source of learning and emotional support (Hartshorne & Manaster, 1982; Robertson, 1976). Young adults also define their relationships with grandparents depending on the extent to which grandparents influenced various aspects of their lives, including engendering beliefs and values (Brussoni & Boon, 1998). In a comparative study of three adolescent age groups (aged 13–14, 15–16, and 17–18), Dellman-Jenkins, Papalia, and Lopez (1989) found that teenagers across all age groups voluntarily interacted with grandparents and engaged in moderate recreational activity (phone calls, walks, eating out, watching movies, etc). In a subsequent investigation, adolescent grandchildren reported that, although their relationships were not intimate and grandparents were not considered a primary source of assistance, they were attached to their grandparents and considered the relationship a foundation for identity formation (i.e., cultural and social learning; Creasey & Koblewski, 1991). Although the data obtained in this study were quantitatively analyzed, the authors advocated that open-ended experiential questions be asked with regard to mutuality of support between grandparents and grandchildren during age-related life events such as pursuit of education, marriage, and the death of a spouse (Creasy & Koblewski, 1991).

What happens as these grandchildren grow up? Mills (1999) explored the effects of grandchildrens' changing employment and marital and family status on relationships with grandparents, and, specifically, characteristics of solidarity and connection during periods of transition. "The model of role transition across the life course suggests a negative relationship between adult role acquisition and intergenerational solidarity" (p. 219). Consistent with these findings, Harwood and Lin (2000) discovered that intergenerational relationships are affected by major life transitions such as grandchildren moving away to college. These researchers revealed four major themes relevant to grandparents' relations with college-aged grandchildren: expressions of affiliation, pride, exchange, and feeling distant. Clearly, "transitions in the grandchildren's lives (e.g., moving to college) and the grandparents' lives (e.g., retirement) require continuous adaptation and development both individually and relationally" (Harwood & Lin, 2000, p. 44). Holladay, Lackovich, and Lee (1998) identified other turning points in relationships between grandmothers and granddaughters. Using a retrospective interviewing technique, granddaughters were asked to reflect on life episodes or turning points that significantly impacted their relationships with grandmothers. This project revealed that "decreases in closeness were associated with negative experiences with the grandmother, increases in geographic separation, and the transition to college, while increases in relational closeness resulted from decreases in geographic separation, engaging in shared activities, deaths or serious illnesses in the family, and family disruptions" (Holladay, Lackovich, & Lee, 1998, p. 287). Clearly, although some life transitions provide grandparents and grandchildren the opportunity to become closer and provide mutual support (e.g., divorce), others result in emotional and geographical distance that can have a significant impact on both the quality and quantity of intergenerational interaction.

Although few studies have investigated the influence of gender and culture on intergenerational family relationships, an exceptional study by Kennedy (1990) assessed college students' attitudes toward grandparents' expectations and role perceptions in relation to grandparent–grandchild gender, cultural background, and form. Results from this study contrasted even more with previous research in terms of quality and quantity of interaction. Accordingly

> Students in this study were somewhat more inclined than [students in earlier studies] to expect grandparents to be liaisons between them and parents, to be somebody they turn to for personal advice, to understand them when nobody else does, to be one who aids in their financial support, to be somebody who acts as a role model, or to be somebody whose occupation they might choose to imitate. (Kennedy, 1990, p. 47)

Not surprisingly, in this same study, granddaughters appeared to be more concerned about their relations with grandparents; the relational and emotional closeness and frequency of interaction reported by granddaughters were significantly higher than for grandsons. In addition, expectations for this intergenerational relationship varied between African American and Caucasian students. For African Americans, grandparents were expected to be there for everyday interaction and support; in contrast, Caucasian students did not expect grandparents to assume an everyday presence in their lives. The author advocated that continued research examine these cultural differences more thoroughly (Kennedy, 1990).

In response to this call, an increasing number of studies have focused on cultural aspects of intergenerational relationships: domestically and internationally. For example, consistent with the role of extended family in African American communities, Hunter (1997) found that working mothers were very reliant on grandmothers for child-care assistance and advice. Factors influencing this reliance included family closeness, the number of surviving generations (family lineage), residence, and geographic proximity. In a comparative study of mutual support between generations in Black and White families, Black grandmothers received more help from grandchildren and more requests for help from grandchildren than did White grandmothers. Factors influencing the opportunity for mutual support included geographic proximity and the grandchild's kin placement, gender, and race.

In a program of research exploring grandparent–grandchild relationships in African American families, in-depth interviews with men and women revealed information that served to discount these myths of the dysfunctional character of Black family life suggested by early research comparing Blacks and Whites (McKay, 1991, 1999). The interviews were conducted with both grandmother–granddaughter and grandfather–grandson biological pairs. Respondents were nonrandomly selected from a variety of social and economic backgrounds. When asked whether grandmothers considered themselves role models for their granddaughters, most replied positively; grandmothers used their life experiences as motivation for granddaughters to achieve higher educational goals than they themselves had achieved (McKay, 1999).

The socialization of religious beliefs was considered a primary responsibility; grandmothers also felt responsible for conveying both historical and personal aspects of African culture and heritage. One grandmother disclosed stories of her foremothers' life in slavery so that her granddaughter understood, and was proud of, the strength of character engendered in her family history. Of particular interest was the manner in which grandmothers advised granddaughters about relations with men. Although grandmothers did not discourage relationships, they advocated that granddaughters be independent and able to survive on their own. Some granddaughters conveyed confusion regarding their role in relations with men. One granddaughter indicated that she respected her grandmother for "shadowing" her grandfather: "I think if we could shadow our Black men a little better, but at the same time keep our own identity, the Black culture within itself would advance." Clearly, the dominant and matriarchal role of women in African American families was not evidenced in these interviews (McKay, 1999, p. 365).

Grandfathers were asked similar questions regarding their relationships with grandsons (McKay, 1991). As role models, grandfathers described their responsibilities as mentor, teacher, surrogate father figure, and teacher of work roles and ethics. As with grandmothers, a primary responsibility was to impart religious beliefs and family values. Grandfathers were also asked about the ease or difficulty associated with being an African American man in our society. Most acknowledged that racism is still present, but rather than being angry, they conveyed the importance of "passing on" a good attitude and advice in dealing with negative attitudes toward Blacks. Some grandfathers suggested that it is especially important to be proud of their African heritage, deal with change graciously, adhere to religious values, and stick with getting an education. Most of the grandfathers interviewed did not live close to their grandsons, and the pursuit of employment was the primary reason for moving away. This is not surprising given the employment statistics for African American men; the myth of the absence of men in Black families should be attributed to lack of available employment and economics as opposed to lack of concern for family responsibilities (McKay, 1991). The consequence of integrating gender and culture into this examination of the grandparent–grandchild relationship was to enrich our understanding of the African American family unit. This is only one example of the means by which inherent biases in traditional research can be revealed and refuted, our knowledge and awareness enhanced and enlightened. Utilizing the same framework to describe or explain family relations in myriad cultures, our understanding of history, heritage, and the essence of culture and gender is limitless and absent of the boundaries defined by stereotypes and cultural biases.

One of the fastest growing yet least investigated populations is the Mexican American community. Documented evidence indicates that Latinos "are facing a number of problems such as poverty, hunger, and a high dropout rate at school" (Strom, Buki, & Strom, 1997, p. 1), necessitating the need for research into their social, political, and economic status. Latino family structures are characterized by multiple generations living together, which raises issues of language and generational differences. Not surprisingly, Strom, Buki, and Strom (1997) found that "Spanish-speaking grandparents reported a greater need for information than English-speaking grandparents, and more frustration when dealing with adolescents than with younger children" (p. 1). These researchers advocate the need for programs to help these grandparents deal with language barriers and other domestic family issues associated with intergenerational residence.

Insight into homosexual family relationships is almost nonexistent in mainstream research literature. One exception is a recent study by Patterson, Hurt, and Mason (1998). The focus of their project was sexual orientation and family development, and, specifically, the role of extended family and friendship networks for lesbian mothers and their children. Results "counter stereotypes of such children as isolated from parents' families of origin" and speak to the behavioral and general well-being of children in lesbian families (Patterson et al., 1998, p. 390).

Researchers are just beginning to explore intergenerational family relationships on a global level. Focusing on grandparenthood as a developmental phase of life for Taiwanese grandparents, Strom, Strom, and Shen (1996) asked grandparents, parents, and grandchildren about aspects of being a grandparent and "realms of learning they should acquire to become more effective" (p. 1). Consistent with the role of grandparenting found in domestic research, grandparent and grandchild age, gender, frequency of interaction and care, and coresidency were found to be important qualities in defining this role and consequent relationships. Grandparents in Japan believe that their family status is eroding and, although they want to be influential, social policy prevents them from doing so. Not surprisingly, both the positive (regarding satisfaction, involvement, and teaching) and the negative (regarding difficulty, frustration, and being less informed) perceptions of their role were in stark contrast to perceptions of the parent and grandchild generations (Strom, Strom, & Collinsworth, 1995). A recent study of Finnish-Polish grandchildren revealed that they tend to focus on dimensions other than those found in extant domestic research. These grandchildren talked more about their feelings, differences and similarities in personalities, as well as emotional and intellectual skills learned from grandparents, and less about how frequently they interacted with grandparents (Hurme, 1997). Van Ranst, Verschueren, and Marcoen (1995) asked: Why do adolescents value their grandparents? This question was posed to adolescents and young adults in Flanders (Belgium). Grandchildren responded that grandparents provide affection, reassurance of worth, and are a source of reliable alliance in relations with parents. Maternal grandparents were generally perceived as closer and more important than paternal grandparents.

From early research into the grandparent–grandchild relationship to more recent investigations, the focus of study has progressed from simple questions analyzing demographic characteristics of the relationship and its participants to more complex relational questions. Heuristically, we are moving from research providing descriptive information to the exploration of interrelated issues and characteristics of this intergenerational relationship. Further, whereas comparatively few investigations have incorporated the concepts of gender and culture into a framework for analysis, many of the studies cited in this review suggest that these issues increasingly continue to become the focus of intergenerational research. What we know is that factors such as proximity, age, health, family roles, emotional closeness, and parents' marital status enhance or inhibit the development of close relations between grandparents and grandchildren. What we are just beginning to explore is how these factors vary by the gender or cultural characteristics of grandparent or grandchild—a more complex perspective than simply investigating differences between males and females in their relations with others. How do these relations vary with cultural groups? What can we learn about the nature of family ties in our society based on a new definition of "family"? How could our continued exploration of this intergenerational relationship and its dynamics contribute to the understanding and tolerance of aging in our society? Do instruments assessing intergenerational relationships translate meaningfully for use in multicultural research? These questions and more can be easily addressed within the feminist framework, which asserts that individual human experience is the key to understanding the nature of human relations.

The Sibling Relationship in Later Life

Only within the last two decades has the sibling relationship become the focus of research into the lives of older adults. Notably, 80% of the older people living today have living siblings; most people remain in contact with their siblings throughout their lives. These relationships are also receiving more attention, because of the presence of the baby-boom generation and its progress to adulthood; the number and variety of later-life sibling relations is steadily increasing. Many older people who are in their 80s still have at least one living sibling, over half of whom live within 100 miles, and 25% live in close geographic proximity. Sibling pairs living in close proximity usually see each other on a weekly basis (Atchley, 1991); others visit monthly or several times a year (Cicirelli, 1980). Burholt and Wenger (1998) found that relationships with siblings and children were especially salient to older adults 80 years of age or older. Why is this the case?

In a comprehensive analysis and review of sibling research, Cicirelli (1991) characterized the sibling relationship as one in which people share family history and intimate experiences. Further, "siblings share a common heritage, environment and childhood, which may lead to perspectives on life not shared by others" (Borland, 1989, p. 205). In a foundational work examining research about sibling relationships over the life span, Cicirelli (1995) offers theoretical, methodological, contextual, and topical recommendations for asking and answering questions about this lifelong relationship. Noting that sibling relationships have attributes in common with all interpersonal relationships, they are also unique. This relationship is usually the longest one that an individual will have throughout his or her lifetime. It is a relationship that is obtained by birth (ascribed) rather than sought out (e.g., friendships). In early childhood, the sibling relationship is characterized by daily contact in a wide array of forms, although in older adulthood mediated forms of communication are used to maintain it unless, as is sometimes the case, older adult siblings share a residence. Usually, sibling relationships are characterized by relative egalitarianism, although age, birth order, family size, and other characteristics can influence relations between siblings early or later in life. Given that siblings have a long history of both shared and unshared experiences, in later life these can become the foundation for empathy and vicariously shared experiences, "facilitating greater similarity" (Cicirelli, 1995, p. 3).

In fact, adopting a life span perspective to explore this intragenerational relationship informs both the questions that need to be asked and the methods by which those questions can be more suitably answered. Especially, "the fact that the sibling relationship is one of extraordinary longevity means that it is important to study this relationship not only in early childhood, but through all the life stages across the life span" (Cicirelli, 1995, p. 7). However, Cicirelli (1995) noted that researchers focusing on sibling relationships tend to study the relationship either in childhood or adulthood without using one area of research to inform another. Moreover, research has focused on sibling relationships at either end of the age continuum with little consideration of exploring sibling relations in young adulthood or characteristics of the relationship within or in comparison between cultures. "In sum, researchers from both parts of the life span need better communication to exchange ideas and stimulate on another's research work . . . [and] a greater collaborative effort to fill the gaps in existing knowledge of sibling relationships and to integrate this knowledge across the life span" (Cicirelli, 1995, p. 220). Although these considerations could inform a rigorous research program, biological sibling relationships within the traditional family may be increasingly few in number in parallel to the decreasing number of traditional families in our society. Perhaps the way(s) in which we conceptualize relations between siblings might evolve to consider cultural influences and nontraditional family forms.

How has previous research informed our understanding of sibling relationships in later life? Brother–sister and sister–sister relationships are usually extremely close; these relationships provide a sense of security in the lives of older people, and many feel comfortable asking siblings for assistance (Cicirelli, 1985), especially in terms of making major decisions and satisfying needs for transportation (Scott, 1983). As previously noted, women who are widowed or were never married maintain strong ties with their sisters (Hoyt & Babchuk, 1983; Rosenthal, 1985). In fact, in a study of companion and confidant networks of older persons, friends and sisters held a dominant presence in the networks of single older women (Connidis & Davies, 1992). Campbell, Connidis, and Davies (1999) extended this line of research to identify sibling social networks in older adulthood. Specifically, individual roles within four types of support networks (confidants, companions, emotional, and instrumental) were explored especially with regard to marital status. Using compositional analysis, network ties were described in terms of hierarchical compensatory, task specificity, and functional specificity, and the likelihood that siblings were included within the network itself. Findings indicated that older adult women who were either single, without children, or widowed had particularly involved sibling relationships that can best be described within the functional specificity model.

As noted by Cicirelli (1995), the sibling relationship has been characterized as egalitarian—one sibling has no prescribed power over another; the ascribed role is usually obtained by birth order. Although rivalry appears to be an inherent aspect of relations between young siblings, as they grow older, siblings usually become closer and rivalry decreases (Ross & Milgram, 1982), although some research suggests that relationships characterized by rivalry persist even into older adulthood (Bedford, 1998; Ross & Milgram, 1982). What are the implications of changing relational patterns between siblings over the life course? Conflicts arising in childhood are often healed when opportunities to reconcile (such as the death of a parent) are presented in later life (Schulman, 1999), except in relationships involving half siblings or stepsiblings (White & Riedmann, 1992). On the positive side, research suggests that sibling rivalry and conflict in childhood lay the groundwork for more social and emotional competence in adulthood (Bedford, 1998; Bedford, Volling, & Avioli, 2000). Not surprisingly, in addition to achieving and maintaining closer relationships in later life (especially between sisters) research has revealed that sibling cohabitation is considered a reasonable and comfortable alternative to institutionalization among elderly siblings, although some consider living with children or institutionalization preferable (Borland, 1989). In many cases, siblings depend on each other to engage in companionship activities such as recreation, regular visits, organizing family reunions, and, in rare cases, business ventures (Scott, 1983).

Unfortunately, research examining sibling relationships in later life is limited, and, as with the grandparent–grandchild relationship, most studies have focused on the descriptive characteristics of this bond within generations. More complex issues remain relatively uninvestigated. Exploration of this relationship within the framework of the life span perspective might reveal aspects of the dynamic nature of this intragenerational relationship. For instance, what is the motivation behind maintaining this relationship in varying degrees throughout the life course? What is the nature and consequence of shared life experience as a contributing factor to the sibling bonding process? What are the communication patterns between siblings over the life course, and what factors influence those patterns? How do gender and culture influence the variability in sibling relations? Whereas research to date has provided a foundation for understanding the characteristics of sibling relationships, future research should integrate both the exploration of individual human experience and the evolutionary nature of this relationship within generations and across the life span.

CONCLUSIONS AND IMPLICATIONS

The basic premise of this chapter purports that the consideration of gender, culture, and nontraditional family forms offer us a unique opportunity to better understand contemporary relationships between and within generations of family. Taking this one step further, both grandparent–grandchild and sibling relationships are long-term family relationships, and there exists a shared reality between participants: What better opportunity to explore the dynamic and communicative qualities of these relationships across the life span. This chapter advocates that, by integrating both the life span and feminist perspectives to create a framework for exploring these family relationships, we can more fully recognize the depth and richness of human experience these generations have to offer. Further, within this framework, the issues of gender and culture enhance the discovery of individual perceptions that influence the development of, and contribution to, our understanding of the unique qualities of each family bond.

Integrating gender and culture into the notion of shared experience, we find that research can take steps to advance our understanding by describing the quality, and ultimately the value, of ties between generations of family members. Moreover, "any attempt to bring theoretical clarity to human phenomena must be informed by an understanding of the sociocultural, economic, and political contexts in which the phenomena occur" (Dodson, 1981, p. 32). To illustrate this point, research focusing on intergenerational relations in African American families revealed that ties between women are stronger and more consistent than between men; Black men are perceived as being absent from the responsibilities of Black family life. African American womanhood is considered by some as dominant and matriarchal—deviant from models of Euro-American womanhood. To accept these notions is to fail to recognize the effects of culture and gender on human behavior (Sawyer, 1973). To go beyond these notions both expands and informs our understanding of family in our society today. In fact, Johnson (1999) found that Black men are no more likely than women to be isolated from their families and are likely to have close relationships with their sons, siblings, friends, and extended family.

The sibling relationship provides a unique and relatively unexplored context in which to investigate the effects of shared life experience, culturally defined family bonds, socialization and gender role learning, and the character of communicative interaction between persons of cohort generations within the framework advocated by this chapter. Whereas the life span perspective offers a view of life that is both progressive and reflective, the feminist perspective advocates the value of individual experience in the exploration of human nature. In combination, interviews with sibling pairs of various cultures and gender affiliation, or elderly brothers and sisters who have experienced the loss of a sibling relationship, provide us with singular insight into the longest relationship experienced over the life course.

Last, the issue of continuity functions to connect the grandparent–grandchild and sibling relationships within the life span perspective. By sharing information and events within these relationships, participants achieve a sense of continuity in their lives, continuity being realized by engaging in those strategies or activities for the purpose of maintaining and developing personal identity with, and understanding of, the past, present, and future. Continuity is viewed as "a healthy capacity to see inner change as connected to the individual's past," and the sense of self-satisfaction achieved is relevant not only to that individual but to others with whom he or she is associated (Atchley, 1991, p. 184). By reflecting on life events and experiences, an individual comes to better understand his or her own existence. Moreover, and consistent with the life span perspective, the individual can utilize new information for guidance in future endeavors or in advising the endeavors of others. In the process of telling one's story, then, he or she is better able to contemplate the nature and course of life events that were most important in their lives—or those events that have particular meaning for them. In this way, continuity contributes to both individual and relational development; for both grandparents and grandchildren, and older adult siblings, shared experience offers individuals the opportunity to reconnect with the past by reevaluating and reinterpreting life experiences in retrospect, and provide an opportunity to contemplate the present and, by bonding with others, the future.

In conclusion, this chapter has advocated the inception of a feminist view of the life span perspective of human development. Integrating these perspectives offers an innovative framework by which to explore those aspects of human experience that may be neglected by the imposition of traditional models of scientific inquiry. Focusing on the concept of continuity as a quality and consequence of human relations, grandparent, grandchild, and sibling relationships in older adulthood offer unique opportunities to explore the process of human bonding over the life course. With this in mind, research has the potential to take new and creative directions in exploring the depth and richness of the human life experience.

REFERENCES

Albrecht, R. (1954). The parental responsibilities of grandparents. *Marriage and Family Living, 16,* 201–204.

Aston, V. (1997). A study of mutual support between Black and White grandmothers and their adult grandchildren. *Journal of Gerontological Social Work, 26*(1–2), 87–100.

Atchley, R. C. (1991). Family, friends, and social support. In R. C. Atchley (Ed.), *Social forces and aging,* (pp. 150–152). Belmont, CA: Wadsworth.

Baranowski, M. D. (1982). Grandparent–adolescent relations: Beyond the nuclear family. *Adolescence, 17,* 375–384.

Baydar, N., & Brooks-Gunn, J. (1998). Profiles of grandmothers who help care for their grandchildren in the United States. *Family Relations, 47,* 385–393.

Beal, E. W. (1979). Children of divorce: A family systems perspective. *Journal of Social Issues, 35,* 140–154.

Bedford, V. H. (1998). Sibling relationship troubles and well-being in middle and old age. *Family Relations, 47,* 369–376.

Bedford, V. H., Volling, B. L., & Avioli, P. S. (2000). Positive consequences of sibling conflict in childhood and adulthood. *International Journal of Aging and Human Development, 51,* 53–69.

Bekker, L. D., & Taylor, C. (1966). Attitudes toward the aged in a multigenerational sample. *Journal of Gerontology, 21,* 115–118.

Bell, W., & Garner, J. (1996). Kincare. *Journal of Gerontological Social Work, 25*(1–2), 11–20.

Bell, Y. R., Bouie, C. L., & Baldwin, J. A. (1990). Afrocentric cultural consciousness and African-American male–female relationships. *Journal of Black Studies, 21*(2), 162–189.

Bengtson, V. L. (1985). Diversity and symbolism in grandparental roles. In V. L. Bengtson & J. F. Robertson (Eds.), *Grandparenthood* (pp. 11–26). Beverly Hills, CA: Sage.

Blau, T. H. (1984). An evaluative study of the role of the grandparent in the best interests of the child. *The American Journal of Family Therapy, 12,* 46–50.

Borland, D. C. (1989). The sibling relationship as a housing alternative to institutionalization in later life. In L. Ade-Ridder & D. B. Hennon (Eds.), *Lifestyles of the elderly: Diversity in relationships, health, and caregiving* (pp. 205–219). New York: Human Sciences Press.

Brown, A. S. (1990). The elderly in the family. In A. S. Brown (Ed.), *Social processes of aging and old age* (pp. 101–104). Englewood Cliffs, NJ: Prentice Hall.

Brussoni, M. J., & Boon, S. D. (1998). Grandparental impact in young adults' relationships with their closest grandparents: The role of relationship strength and emotional closeness. *International Journal of Aging and Human Development, 46,* 267–286.

Bryson, K., & Casper, L. M. (1999). Co-resident grandparents and grandchildren (US Census Bureau Publication No. P23–198; 1997 March Current Population Survey).

Burholt, V., & Wenger, G. C. (1998). Differences over time in older people's relationships with children and siblings. *Ageing and Society, 18,* 537–562.

Burke, J. L. (1982). Young children's attitudes and perceptions of older adults. *International Journal of Aging and Human Development, 14,* 205–222.

Burton, L. M. (1992). Black grandparents rearing children of drug-addicted parents: Stressors, outcomes, and social service needs. *The Gerontologist,* 744–751.

Campbell, L. D., Connidis, I. A., & Davies, L. (1999). Sibling ties in later life: A social network analysis. *Journal of Family Issues, 20*(1), 114–148.

Caputo, R. K. (1999). Grandmothers and coresident grandchildren. *Families in Society, 80*(2), 120–126.

Chan, C. G., & Elder, G. H. (2000). Matrilineal advantage in grandchild–grandparent relations. *The Gerontologist, 40*(2), 179–190.

Cherlin, A., & Furstenberg, F. F. (1985). Styles and strategies of grandparenting. In V. L. Bengtson & J. F. Robertson (Eds.), *Grandparenthood* (pp. 97–116). Beverly Hills, CA: Sage.

Cicirelli, V. G. (1980). Sibling influence in adulthood: A life span perspective. In L. W. Poon (Ed.), *Aging in the 1980's* (pp. 455–462). Washington, DC: American Psychological Association.

Cicirelli, V. G. (1985). The role of siblings as family caregivers. In W. J. Sauer & R. T. Coward (Eds.), *Social support networks and the care of the elderly: Theory, research, practice, and policy* (pp. 93–107). New York: Springer.

Cicirelli, V. G. (1991). Sibling relationships in adulthood. In S. K. Pfeifer & M. B. Sussman (Eds.), *Families: Intergenerational and generational connections* (Part 2, pp. 291–309). New York: Hayworth.

Cicirelli, V. G. (1995). *Sibling relationships across the life span.* New York: Plenum.

Clavan, S. (1978). The impact of social class and social trends on the role of grandparent. *The Family Coordinator, 27,* 351–358.

Clingempeel, W. G., Coylar, J. J., Brand, E., & Hetherington, E. M. (1992). Children's relationships with maternal grandparents: A longitudinal study of family structure and pubertal status effects. *Child Development, 63,* 1404–1422.

Cogswell, C., & Henry, C. S. (1995). Grandchildrens' perceptions of grandparental support in divorced and intact families. *Journal of Divorce and Remarriage, 23*(3–4), 127–150.

Connidis, I. A., & Davies, L. (1992). Confidants and companions: Choices in later life. *Journal of Gerontology, 47,* S115–S122.

Cooney, T. M., & Smith, L. A. (1996). Young adults' relations with grandparents following recent parental divorce. *Journal of Gerontology 51B (Series B: Psychological Sciences and Social Sciences),* S91–S95.

Creasey, G. L., & Jarvis, P. A. (1989). Grandparents with Alzheimer's disease: Effects of parental burden on grandchildren. *Family Therapy, 16,* 79–85.

Creasey, G. L., & Koblewski, P. J. (1991). Adolescent grandchildren's relationships with maternal and paternal grandmothers and grandfathers. *Journal of Adolescence, 14,* 373–387.

Dellman-Jenkins, M., Papalia, D., & Lopez, M. (1989). Teenagers' reported interaction with grandparents: Exploring the extent of alienation. In L. Ade-Ridder & D. B. Hennon (Eds.), *Lifestyles of the elderly: Diversity in relationships, health, and caregiving* (pp. 149–159). New York: Human Sciences Press.

Derdeyn, A. P. (1985). Grandparent visitation rights: Rendering family dissension more pronounced. *American Journal of Orthopsychiatry, 55,* 277–287.

Dodson, J. (1981). Conceptualizations of Black families. In H. P. McAdoo (Ed.), *Black families* (pp. 23–36). Beverly Hills, CA: Sage.

Gattai, F. B., & Musatti, T. (1999). Grandmothers' involvement in grandchildren's care: Attitudes, feelings, and emotions. *Family Relations, 48*(1), 35–42.

Gladstone, J. W. (1988). Perceived changes in grandmother–grandchild relations following a child's separation or divorce. *The Gerontologist, 28,* 66–72.

Hagestad, G. O. (1985). Continuity and connectedness. In V. L. Bengtson & J. F. Robertson (Eds.), *Grandparenthood* (pp. 31–48). Beverly Hills, CA: Sage.

Hagestad, G. O. (1987). Able elderly in the family context: Changes, chances, and challenges. *The Gerontologist, 27,* 417–422.

Hartfield, B. W. (1996). Legal recognition of the value of intergenerational nurturance: Grandparent visitation statutes in the 90s. *Generations, 20,* 53–56.

Hartshorne, T. S., & Manaster, G. J. (1982). The relationship with grandparents: Contact, importance, and role conception. *International Journal of Aging and Human Development, 15,* 233–255.

Harwood, J. (2000). Communicative predictors of solidarity in the grandparent–grandchild relationship. *Journal of Social and Personal Relationships, 17,* 743–766.

Harwood, J., & Lin, M. (2000). Affiliation, pride, exchange, and distance in grandparents' accounts of relationships with their college-aged children. *Journal of Communication, 50*(3), 31–47.

Hays, W. C., & Mindel, C. H. (1973). Extended kinship relations in Black and White families. *Journal of Marriage and the Family, 35,* 51–56.

Holladay, S., Lackovich, R., & Lee, M. (1998). (Re)constructing relationships with grandparents: A turning point analysis of grandaughters' relational development with maternal grandmothers. *International Journal of Aging and Human Development, 46,* 287–303.

Hoyt, D. R., & Babchuk, N. (1983). Adult kinship networks: The selective formation of intimate ties with kin. *Social Forces, 62,* 84–101.

Hunter, A. G. (1997). Counting on grandmothers: Black mothers' and fathers' reliance on grandmothers for grandparenting support. *Journal of Family Issues, 18*, 251–269.

Hurme, H. (1997). Cross-cultural differences in adolescents' perceptions of their grandparents. *International Journal of Aging and Human Development, 44*, 221–253.

Isaacs, L. W. (1986). The development of childrens' prejudice against the aged. *International Journal of Aging and Human Development, 23*, 175–194.

Ivester, C., & King, K. (1977). Attitudes of adolescents toward the aged. *The Gerontologist, 17*, 85–89.

Johnson, C. I. (1985). Grandparenting options in divorcing families: An anthropological perspective. In V. L. Bengtson & J. F. Robertson (Eds.), *Grandparenthood* (pp. 81–96). Beverly Hills, CA: Sage.

Johnson, C. I. (1988). Active and latent functions of grandparenting during the divorce process. *The Gerontologist, 28*, 185–196.

Johnson, C. I., & Barer, B. M. (1987). Marital instability and the changing kinship networks of grandparents. *The Gerontologist, 27*, 330–335.

Johnson, C. L. (1983). A cultural analysis of the grandmother. *Research on Aging, 5*, 546–468.

Johnson, C. L. (1999). Family life of older Black men. *Journal of Aging Studies, 13*(2), 145–160.

Jones, L., & Kennedy, J. (1996). Grandparents united: Intergenerational Development Education. *Child Welfare, 75*, 636–650.

Kahana, B., & Kahana, E. (1970). Grandparenthood from the perspective of the developing grandchild. *Developmental Psychology, 3*, 98–105.

Kennedy, G. E. (1990). College students' expectations of grandparent and grandchild role behaviors. *The Gerontologist, 30*, 43–48.

King, V., & Elder, G. H. (1995). American children view their grandparents: Linked lives across three rural generations. *Journal of Marriage and the Family, 57*, 165–178.

King, V., & Elder, G. H. (1998a). Education and grandparenting roles. *Research on Aging, 20*, 450–474.

King, V., & Elder, G. H. (1998b). Perceived self-efficacy and grandparenting. *Journal of Gerontology*, (Series B: Psychological Sciences and Social Sciences), *53*, S249–S257.

King, V., & Elder, G. H. (1999). Are religious grandparents more involved grandparents? *Journal of Gerontology*, (Series B: Psychological Sciences and Social Sciences), *54*, S317–S328.

Kivnick, H. Q. (1985). Grandparenthood and the "new social contract." In V. L. Bengtson & J. F. Robertson (Eds.), *Grandparenthood* (pp. 151–158). Beverly Hills, CA: Sage.

Kornhaber, A. (1985). Grandparenthood and the new social contract. In V. L. Bengtson & J. F. Robertson (Eds.), *Grandparenthood* (pp. 159–172). Beverly Hills, CA: Sage.

Kornhaber A., & Woodward, K. L. (1981). *Grandparents and grandchildren: The vital connection*. Garden City, NY: Anchor/Doubleday.

Kruk, E. and Hall, B. L. (1995). The disengagement of paternal grandparents subsequent to divorce. *Journal of Divorce and Remarriage, 23*(2), 131–147.

Luszcz, M. A. (1985). Characterizing adolescents, middle-aged, and elderly adults: Putting the elderly into perspective. *International Journal of Aging and Human Development, 22*, 105–122.

McKay, V. C. (1991, November). *Grandfathers and grandsons in African-American families: Imparting culture and tradition between generations of men*. Paper presented at the annual conference of the Speech Communication Association, Chicago.

McKay, V. C. (1999). Grandmothers and granddaughters in African-American families: Imparting cultural tradition and womanhood between generations of women. In H. Hamilton (Ed.), *Language and communication in old age: Multidisciplinary perspectives* (pp. 351–374). New York: Garland.

Millman, M., & Kanter, R. M. (1987). Introduction to another voice: Feminist perspectives on social life and social science. In S. Harding (Ed.), *Feminism and methodology* (pp. 29–38). Bloomington: Indiana University Press.

Mills, T. L. (1999). When grandchildren grow up: Role transition and family solidarity among baby boomer grandchildren and their grandparents. *Journal of Aging Studies, 13*(2), 219–239.

Minkler, M., & Roe, K. M. (1996). Grandparents as surrogate parents. *Generations, 20*, 34–38.

Minkler, M., Roe, K. M., & Price, M. (1992). The physical and emotional health of grandmothers raising grandchildren in the crack cocaine epidemic. *The Gerontologist, 32*, 752–761.

Neugarten, B. L., & Weinstein, K. K. (1964). The changing American grandparent. *Journal of Marriage and the Family, 26,* 199–204.

Neugarten, B. L., & Weinstein, K. K. (1968). The changing american grandparent. In B. L. Neugarten (Ed.), *Middle age and aging* (pp. 280–285). Chicago: University of Chicago Press.

Patterson, C. J., Hurt, S., & Mason, C. D. (1998). Families of the lesbian baby boom: Children's contact with grandparents and older adults. *American Journal of Orthopsychiatry, 68,* 390–399.

Peterson, C. C. (1999). Grandfathers' and grandmothers' satisfaction with the grandparenting role: Seeking new answers to old questions. *International Journal of Aging and Human Development, 49*(1), 61–78.

Roberto, K. A., & Skoglund, R. R. (1996). Interactions with grandparents and great-grandparents: A comparison of activities, influences, and relationships. *International Journal of Aging and Human Development, 43*(2), 107–117.

Robertson, J. F. (1976). Significance of grandparents: Perceptions of young adult grandchildren. *The Gerontologist, 16,* 137–140.

Rosenthal, C. J. (1985). Kinkeeping in the familial division of labor. *Journal of Marriage and the Family, 47,* 965–974.

Ross, H. G., & Milgram, J. I. (1982). Important variables in adult sibling relationships: A qualitative study. In M. E. Lamb & B. Sutton-Smith (Eds.), *Sibling relationships: Their nature and significance across the lifespan.* (pp. 225–249). Hillsdale, NJ: Lawrence Erlbaum Associates.

Sawyer, E. (1973). Methodological problems in studying socially deviant communities. In J. A. Ladner (Ed.), *Death of a White sociology* (pp. 38–54). New York: Random House.

Schulman, G. L. (1999). Siblings revisited: Old conflicts and new opportunities in later life. *Journal of Marital and Family Therapy, 25,* 517–524.

Scott, J. P. (1983). Siblings and other kin. In T. H. Brubaker (Ed.), *Family relationships in later life* (pp. 47–62). Beverly Hills, CA: Sage.

Sherif, C. W. (1987). Bias in psychology. In S. Harding (Ed.), *Feminism and methodology* (pp. 37–56). Bloomington: Indiana University Press.

Somary, K., & Stricker, G. (1998). Becoming a grandparent: A longitudinal study of expectations and early experiences as a function of sex and lineage. *The Gerontologist, 38,* 53–61.

Spotlight on the States. (2001, Spring). *Parenting grandchildren: A voice for grandparents* (newsletter published by the American Association of Retired People), p. 3.

Stack, C. B. (1974). Sex roles and survival strategies in an urban Black community. In M. Z. Rosaldo & L. Lamphere (Eds.), *Women, culture and society* (pp. 113–128). Stanford, CA: Stanford University Press.

Strom, R., Buki, L. P., & Strom, S. K. (1997). Intergenerational perceptions of English-speaking and Spanish-speaking Mexican-American grandparents. *International Journal of Aging and Human Development, 45*(1), 1–21.

Strom, R., Strom, S., & Collinsworth, P. (1995). Grandparents in Japan: A three-generational study. *International Journal of Aging and Human Development, 40*(3), 209–226.

Strom, R., Strom, S., & Shen, Y. (1996). Grandparents in Taiwan: A three-generational study. *International Journal of Aging and Human Development, 42*(1), 1–19.

Thomas, J. L. (1986). Age and sex differences in perceptions of grandparenting. *Journal of Gerontology, 41,* 417–423.

Thomas, E. C., & Yamamoto, K. (1975). Attitudes toward age: An exploration in school-age children. *International Journal of Aging and Human Development, 6,* 117–130.

Thompson, L., Clark, K., & Gunn, W. (1985). Developmental stage and perceptions of intergenerational continuity. *Journal of Marriage and the Family, 47,* 913–920.

Thompson, P. (1999). The role of grandparents when parents part or die: Some reflections on the mythical decline of the extended family. *Ageing and Society, 19,* 471–503.

Thompson, R. A., Tinsley, B. R., Scalora, M. J., & Parke, R. D. (1989). Grandparents' visitations rights: Legalizing the ties that bind. *American Psychologist, 44,* 1217–1222.

Troll, L. E. (1971). The family of later life: A decade review. *Journal of Marriage and the Family, 33,* 263–290.

Troll, L. E. (1972). Is parent–child conflict what we mean by the generation gap? *The Family Coordinator, 21,* 247–349.

Troll, L., & Bengtson, V. (1979). Generations in the family, In W. R. Burr, R. Hill, R. I. Nye, & I. L. Reiss (Eds.), *Contemporary theories about the family* (Vol. 1, pp. 127–161). New York: Free Press.

Updegraff, S. G. (1968). Changing role of the grandmother. *Journal of Home Economics, 60,* 177–180.

Van Ranst, N., Verschueren, K., & Marcoen, A. (1995). The meaning of grandparents as viewed by adolescent grandchildren: An empirical study in Belgium. *International Journal of Aging and Human Development, 41,* 311–324.

Watson, J. A. (1997). Grandmothering across the lifespan. *Journal of Gerontological Social Work, 28*(4), 45–62.

White, L. K., & Riedmann, A. (1992). When the Brady Bunch grows up: Step/half- and full-sibling relations in adulthood. *Journal of Marriage and the Family, 54,* 197–208.

Wilson, K. B., & Deshane, M. R. (1982). The legal rights of grandparents: A preliminary discussion. *The Gerontologist, 22,* 67–71.

Woodworth, R. S. (1996). You're not alone . . . you're one in a million. *Child Welfare, 75,* 619–635.

11

Friendships in Later Life

William K. Rawlins
Ohio University

Like most periods of the life course, friendships during later life involve both continuities and discontinuities with earlier patterns for individuals and social cohorts. Diverse friendship practices and experiences are apparent for several reasons. First, this developmental period potentially spans 40 or more years of human activity, with individuals exhibiting various opportunities, personal styles, decisions, and initiatives in conducting their social lives. Meanwhile, areas of diminishing or minimal individual control, such as persons' financial resources, social capabilities, health, and mobility, as well as their friends' proximity, abilities to interact, or mortality become salient at varying junctures. To a considerable degree, continuities in friendship practices reflect stability in the participants' capacities for interpersonal contact and enabling circumstances, whereas discontinuities arise from changes in these conditions, which frequently transcend individual choice (Rawlins, 1992; Roberto, 1997).

We need to recognize that the friendships of older adults exhibit as much diversity in life circumstances and outlooks as any other developmental period (Bury & Holme, 1990). In fact, later life probably should be viewed as a career in itself, a conception that emphasizes older persons' creative and conscious decisions and practices in purposefully shaping and adapting to their concrete exigencies and situations (Matthews, 1986a; Myerhoff & Simic, 1978). The period begins with options and imperatives relating to continuing or relinquishing full-time work, parenting, and neighborhood and civic responsibilities, often in the face of culturally imposed images of aging and human capabilities. It transitions to each person's ongoing pattern of individual and social activities shaped by the opportunities and constraints provided by one's health, financial resources, and social circumstances. Later on, the poignant and unique destiny of the oldest old is inevitable physical frailty and the lived recognition of impending death. Even so, as

Weisman (1991) observed, "... there is no age, stage, phase, or social gradient that is utterly free of vulnerability, or totally devoid of promising potential. To realize this fully means making our way through the thicket of preconceptions and prejudices. Unwarranted optimism, however, is not justified, nor is any judgment smacking of sermonizing" (p. 193).

What can be said about friendships during later life? Many of the same things that might be said about them throughout our lives (Rawlins, 1999). For example, the expectations older adults describe for their close friends are the same ones reported by diverse age groups beginning in late adolescence. That is, a close friend is someone to talk to, to depend on for practical and emotional assistance, and to enjoy spending time with (Rawlins, 1992). Further, older persons are typically not as isolated as many stereotypes hold; nor are they necessarily as lonely as was once thought (Altergott, 1985; Kovar, 1986; Mac Rae, 1992). We have come to recognize that individuals vary in their lifelong patterns of and needs for solitude, intimacy, and sociable interaction. Notwithstanding "objective" indices of isolation or time spent alone, persons vary in their subjective experiences of these patterns, typically because of the degree of choice they exercise in the matter and the extent to which their current situation continues or disrupts long-term tendencies (Chappell & Badger, 1989; Jerrome, 1983; Thompson & Heller, 1990). Finally, later-life friendships continue to reflect social stratification and economic disparities. Older persons tend to be friends with age contemporaries (though commensurable age is defined more broadly) who are similar to them in many ways (Heinemann, 1985; Peters & Kaiser, 1985; Rosow, 1970).

Though susceptible to pervasive social conditions and cultural patterning, all friendships are complex interactive achievements that remain highly contingent on friends' ongoing emotional needs, expectations, and initiatives, as well as their encompassing personal networks developed over the course of their lifetimes (Poulin, 1984). What is striking about friendship in later life are the continuities in its significance for personal well-being, its conditional yet protected and privileged status, and the gender alignments in its practices—all persisting in juxtaposition to the sometimes dramatic changes in its circumstances. We will examine the situated and communicative accomplishment and importance of later adult friendships in this chapter.

A DIALECTICAL PERSPECTIVE ON FRIENDSHIP

I employ a dialectical perspective in describing the varieties, tensions, and functions of friendship during later life (Rawlins, 1992). From a dialectical perspective, later-life friendships are dynamic, ongoing social achievements involving the constant interconnection and reciprocal influence of multiple individual, interpersonal, and social factors (Rawlins, 1999). The interaction of such friends continually generates and manages multiple contradictions and dialectical tensions, which arise over time within dyads and in light of how they are situated within enveloping social contexts. I view older adult friends as conscious, active selectors of possible choices from a field that is partially conceived by them, partially negotiated with others, and partially determined by social and natural factors outside of their purview. The choices persons continue to make throughout their lives in concrete circumstances simultaneously generate and restrict their options (Rawlins, 1992).

Both contextual and interactional dialectics interweave in shaping and reflecting patterns of friendship (Rawlins, 1989b). The contextual dialectics of the ideal and the real, and the private and the public derive from the place of friendship in Western culture. They describe cultural conceptions that frame and permeate interaction within specific friendships yet are conceivably subject to revision as a result of changes in everyday practices. The interactional dialectics of the freedom to be independent and the freedom to be dependent, affection and instrumentality, judgment and acceptance, and expressiveness and protectiveness are tensions emerging in the ongoing communication between friends.

The dialectic of the ideal and the real formulates the interplay between the ideals and expectations that are typically associated with friendship and the troublesome realities and unexpected rewards of actual friendships. As a celebrated dyadic relationship in Western society, close friendship involves certain ideal–typical characteristics (Brain, 1976; Rawlins, 1992). First, friendships are voluntary relationships; people choose and are chosen by their friends. So even though friendships may coincide with family ties, and work and neighborhood affiliations, you cannot compel people to be friends with each other. Second, friendship is a personal bond that is privately negotiated between particular individuals. People typically choose others as friends because of their unique personal qualities, not as representatives of a certain group or class of people. Third, a spirit of equality pervades friendship. Although individuals of differing status, ability, attractiveness, or age may become friends, some aspect of the relationship functions as a leveler. Friends tend to emphasize the personal attributes and styles of interaction that make them appear more or less equal to each other. By stressing equality, they minimize the risk of exploitation or indebtedness in the relationship (Fiebert & Fiebert, 1969; Kurth, 1970). Next, friendship reflects mutual involvement. The attachments of friendship arise from at least two individuals' collaboration in cultivating a shared social reality and history. The precise character and viability of specific friendships depends on whether the friends share the same stance toward their bond. Sustained friendships are ongoing, mutual achievements. Finally, friendship involves affection. Positive feeling, caring, and concern for the other, the touchstones of companionship, exist between friends.

Despite these positive characteristics ideally associated with close friendship, actual friendships vary considerably in the degree to which they embody them. The persistence and nature of particular friendships is notoriously contingent on the individuals' continued well-being and disposition toward the relationship, as well as the other requirements of their personal networks and social situation (Rawlins, 1994). For example, after young adulthood, most friendships end for reasons stemming from outside of the friendship, such as necessary geographical relocation because of employment requirements, rather than from problems occurring between friends (Rawlins, 1992; Rose, 1984). Persons may have considerably less control over who they are friends with than the ideals of friendship suggest, and for all their positive connotations, friendships can also cause confusion, exclusion, disappointment, and hurt feelings (Rawlins, 1998).

The dialectic of the private and the public addresses the tensions produced as the activities and decisions of friends weave in and out of public and private situations.

As mentioned, friendship is a privately negotiated and validated bond. This attribute distinguishes friendships from other relationships, such as those with family, which are warranted by blood ties and cultural expectations of kinship; or with coworkers or hired caregivers, which are stipulated by monetary contracts; or with a spouse, which is publicly sanctioned legally, religiously, or both. All of these relationships include fairly explicit duties and obligations that are recognized and regulated by the larger society. In contrast, the expectations of friendships are often tacitly developed and subject to minimal public enforcement, even though either friend may unilaterally withdraw affection under perceived unfavorable conditions (Wiseman, 1986).

Even so, the privately negotiated character of friendship permits persons such as family members, coworkers, caregivers, and spouses to become friends. This "double agency" of friendship allows it to serve personal integration to the extent that it fosters particularized caring between persons and fulfills individual needs, and to serve social purposes to the degree that it actively involves persons in larger social networks and publicly defined roles. But tensions can arise in situations where it is unclear whether the private expectations of friendship or the public expectations of the other role relationship should dictate the activities of the persons and their treatment of each other (High, 1990; Wolfsen, Barker, & Mitteness, 1990). How does being friends add to or reduce the expectations one has of a caregiver? A cousin? A daughter? A husband or wife? To what extent are friendships encouraged between structurally connected persons in certain settings as instruments of social control and cost cutting versus for the benefit of the particular individuals involved? The double agency of friendship provokes such questions (Rawlins, 1989a).

The interactional dialectic of the freedom to be independent and the freedom to be dependent further highlights the voluntaristic ethic underlying friendship in Western culture. In developing a friendship, persons concurrently grant to each other two freedoms: the privilege of living their own lives and making their own decisions without the other's input or interference, and the privilege of calling on the other for practical and emotional assistance (Rawlins, 1983a). Although these "conjunctive freedoms" function ironically to connect friends, they are not always easily reconciled, especially when friends' needs for either independence or dependence become excessive or unequal. How friends manage this dialectical tension combines in various ways with the others described here in accounting for much of the delicacy and vitality of friendships in later life (Rawlins, 1998).

The dialectic of affection and instrumentality involves caring for a friend as an end in itself versus caring for a friend as a means to an end. Persons of all ages want to feel that others' friendship for them is based primarily on the person they are, not on what they are able to do for or give to the other. Even so, helping each other is an important part of friendship; people feel good about helping friends and being able to call on friends. Both activities communicate affection. Potential problems develop, however, when individuals feel obligated to help friends or compelled to request too much assistance. In the first case, a person may begin to feel exploited or that their voluntary caring for their friend is becoming compulsory. Likewise, persons often resent feeling excessively indebted to friends; feeling too dependent on friends undermines the perceptions of equality and personal freedom that are important for individuals' dignity within friendships. From both friends' perspectives, extended asymmetrical instrumental assistance can strain the freely offered affection that lies at the heart of friendship in Western culture.

The dialectic of judgment and acceptance describes the dilemmas in friendship between providing objective appraisals of a friend's actions versus giving unconditional acceptance and support. People expect and desire acceptance from their friends; they value someone who likes them for who they are. Even so, individuals use certain standards in deciding who they will consider their friends, and together friends negotiate their expectations for each other's actions. Moreover, friends rely on the other's judgment in making decisions and in learning about proper conduct in new situations (Jerrome, 1984). Thus, even though people do not always enjoy being evaluated, a friend's honest input can communicate that someone is worthy of criticism. Good friends ackowledge each other's faults and limitations but still communicate overall acceptance of the other as a worthy person. In short, they often attempt to be judicious rather than judgmental and to provide evaluation carefully in ways that the other can handle, with a gentle manner, using discretion (Rawlins, 1998). Even so, the dialectic of judgment and acceptance frequently incites tensions among friends.

Finally, the dialectic of expressiveness and protectiveness addresses the opposing tendencies to speak openly with a friend and relate private thoughts and feelings, and the simultaneous need to restrain disclosures and commentary to preserve privacy and avoid burdening or hurting the friend. For many, having someone to confide in is a valued privilege of friendship, but it grows out of mutual trust that the other will protect the vulnerabilities that disclosing thoughts and feelings reveals. A friend protects one's privacy by retaining one's confidences and one's feelings by being discrete in discussing matters that are sensitive issues for one's friend (Rawlins, 1983b). The personal vulnerability occasioned by revealing personal information, and the corresponding responsibilities imposed on others not to misuse intimate knowledge of self, make confidence and trust problematic achievements throughout many individuals' lives. This dialectic has special salience in later life in light of connections that have been identified between having confidants and an older person's morale (Blau, 1973; Lowenthal & Haven, 1968). How participants manage these dialectics throughout adult life defines and reflects various types and degrees of friendship.

If one focuses on characteristic practices of friendships, two general modes are suggested, variously termed "communal and agentic friendships" (Rawlins, 1992; Wright, 1989), "primary and secondary friendships" (Adams, 1986), "friendships of commitment and convenience" (Fischer, Jackson, Stueve, Gerson, Jones, & Baldassare, 1977), and "confidants and companions" (Chappell & Badger, 1989; Connidis & Davies, 1990, 1992). Confidants approximate the ideals of close friendship in fulfilling individuals' needs for intimate communication, caring, subjective validation, and rather exclusive personal involvement with particular others (Rawlins, 1992). Such friends emphasize talking, as well as doing things together and for each other. These relationships are deepened by emotional attachment and empathy, diffuse mutual responsibilities and obligations, and are maintained through shared commitment and personal loyalty.

By comparison, companions reflect persons' needs for sociable interaction, group harmony, objective validation, and inclusive social involvement with a variety of others (Rawlins, 1992). Such friends emphasize joint activities and projects. Being more instrumentally oriented, these ties involve limited emotional investment, fairly explicit individual rights and notions of reciprocity, and are maintained as long as benefits exceed costs. It is not clear which of these modes is necessarily the most beneficial for persons during later life. In many respects, the answers to this question depend on individuals' needs and habitual participation in intimate and sociable relationships.

Circumstantial issues also affect how friendships are viewed in later life. When and under what conditions was the friendship formed? Is it a long-established bond extending back to childhood, adolescence, military service, college days, or one's primary parenting or working years? Were these especially happy or sad times? Was this friend a central or peripheral figure during that period? Or is the bond relatively recently established? How important and gratifying is time spent with this friend right now? Does the friend live nearby with regular contact, or is the person seldom seen despite being proximate? Does the friend live far away and communicate infrequently, or is contact with him or her an expected and meaningful part of one's social life? Is the friend also part of one's family, a daughter or son, sister or brother, husband or wife? Reflecting on one's "populated biography" of friendships (Matthews, 1986b) and current associates during later life therefore yields a potentially complicated array of perceived contributors to one's well-being and personal sense of identity. The term *friend* assumes multiple meanings in describing this range of persons (Patterson, Bettini, & Nussbaum, 1993).

Many older adults report that their closest friendships developed when they were younger and that these friends now live a considerable distance away (Adams, 1985–1986). These "old" friends are often "best" friends from a prior era who have invested the time and effort to sustain their bond over the years. Their separation in space and time has likely prevented them from routinely burdening each other or jeopardizing their jointly held favorable images (Hess, 1972). Such friends easily "pick up right where they left off" across the years, heartily enjoying the occasions spent together, exchanging letters and gifts, or talking on the phone. These lasting friendships reflect continual thoughts about each other, as well as attempts to sustain actual contact (Harre, 1977). The meaning-giving work includes ongoing assumptions of benevolence about the friend's actions, combined with assumptions of continuity in their importance to each other, sometimes despite contrary evidence. These friends receive the benefit of the doubt; for example, excuses are made for them if they have not been heard from for months or sometimes years. When they do write, call, or visit, these actions are often assigned more significance than analogous ones by relatives or other friends.

However, in the words of a 64-year-old woman, "Time and distance take their measure." The long run of adulthood involves ongoing intentional and unintentional trials and impediments, with friends being tested, falling by the wayside, drifting away, and losing touch. Consciously and unconsciously, friends are sorted out, with social calendars, role commitments, and career moves functioning both as facilitators and disqualifiers of friendships. Friends who have stood these "tests of time" and circumstances assume special status, and close friendships increase in stability across the life course (Brown, 1981).

As the years pass, the continuity of specific friendships documents the persistence of selves, images of selves, or both, as well as the concrete (or now imagined) social settings in which these persons were viable participants. These enduring friendships connect adults with still meaningful versions of their possibilities as human beings and rejected alternatives, transcending the finitude of "real time." As reservoirs of common histories and shared experiences, old friends are narrators and curators of the long-term coherence and significance of each other's lives. It is not surprising that well over three fourths of one sample of widows aged 50 or more years stated that no amount of effort could replace old friends (Lopata, 1973). In contrast, throughout life, new friends foster and reflect change and adaptation to altered conditions, though one implicitly compares them with one's "populated biography" of past relationships (Matthews, 1986b).

THE SOCIAL AND PERSONAL CONTEXTS OF AGING

Being "elderly" in any culture means undergoing biological aging processes in conjunction with perceptions by self and others that one is "old." But because persons vary in their pace and evidence of biological aging and their ability or willingness to view themselves as elderly—and cultures fluctuate in their conceptions and treatments of old age—its onset and expectations reflect continuous negotiations among individuals, age cohorts, and their social orders (Blau, 1973). Societies using functional definitions view old age as beginning when biological decline affects one's capacities to do work, thereby limiting one's endeavors and indicating "the end of active adult status" (Clark & Anderson, 1967, p. 6). By comparison, formal definitions arise from arbitrary assignments of symbolic import to specific events. This latter approach has prevailed in American culture, which defines old age temporally. Legislation stipulating initial eligibility for Social Security benefits (age 62) and the Age Discrimination Act regulating mandantory retirement (age 70 for most types of jobs) legally characterize old age as commencing sometime during and certainly by the end of a person's seventh decade (Clark & Anderson, 1967; Blau, 1973).

As throughout life, what is allowed and expected of older people is socially constructed and sanctioned. Until rather recently, it has been commonplace to assume that simply because a person is old, he or she is unwell, fragile, and in need of special attention and care (Matthews, 1986a). According to Mitchell and Acuff (1982), pervasive stereotypes such as "the elderly are generally sick, boring, lacking in sexual interest, mentally slower, withdrawn, unproductive, and so on" constitute "the stigma of old age" (p. 369). Actions reflecting such attitudes toward many older individuals strikingly recontextualize their experiences of self and others, with energetic and mentally astute older persons perceived and treated as deviants (Mitchell & Acuff, 1982). Such formal definitions and age-based preconceptions of older persons decreasingly correspond with functional realities in our culture because of medical advances and widespread improvements in lifestyle, diet, and exercise that extend physical and mental vitality. When these developments are coupled with retirement plans and health insurance policies that assure continued financial well-being, the possibilities of prolonged and active later lives multiply for these fortunately endowed persons. For example, some people may choose to work at gratifying jobs or to participate in professional or civic associations indefinitely, whereas others elect early retirement and vigorous travel and leisure. Thus, age norms are becoming highly flexible as a set of parameters, recognizing that images and actualities of aging are closely tied to historical moments and conditions, as well as the health practices and financial wherewithal of social cohorts.

Whereas cultural contexts shape the experiences of entire cohorts of older people in various ways, the concrete contingencies of physical health and financial assets interact with habitual patterns of intimacy and sociability and others' availability in fostering the individual's degree of personal well-being and social participation during later life. Continued good health is essential for contentment as one ages. Reviewing 30 years of relevant research, Larson (1978) concluded, "Among all the elements of an older person's life situation, health is the most strongly related to subjective well-being" (p. 112). Peters and Kaiser (1985) noted two effects of health status on relationships with friends and neighbors. First, a person's physical and mental condition affects his or her capacity to do things with friends. As long as individuals and their friends are in good health, they are likely to continue many of the same patterns of contact and activities they have enjoyed throughout their adult years. But, as strength and mobility wane, the elderly individual must increasingly rely on healthy and dedicated friends and nearby neighbors to initiate activities and maintain affiliations. When this occurs, the mutual process of defining friendships begins to change, with jointly offered comfort and assistance progressively giving way to unilateral help from concerned others (Peters & Kaiser, 1985).

Sadly, persons with chronic health problems experience constricted choices regarding their activities and greater social isolation (Stephens & Bernstein, 1984; Wenger, 1990). Diminished abilities to reciprocate may wither their pool of former friends; negative outlooks brought on by nagging illnesses may drive less devoted others away; and fading lucidity, speech, and vision may undermine the basic communication skills necessary for sustaining satisfactory face-to-face interactions with others (Adams, 1988; Retsinas & Garrity, 1985; Stephens & Bernstein, 1984). Such enforced solitude is especially disturbing because the support of friends positively affects a person's physical and psychological well-being, which recursively facilitate engagement in friendships in the first place (Ferraro, Mutran, & Barresi, 1984). With this simultaneous cause-and-effect association of friendship and health status, the negative spiral linking health problems and inhibited social life is indeed a gloomy prospect for some older individuals. Meanwhile, the poignant and unique destiny of the old is that eventually "the entire age cohort becomes physically frail" (Dono, Falbe, Kail, Litwak, Sherman, & Siegel, 1979, p. 409), and markedly increasing numbers of friends are lost because of death (Clark & Anderson, 1967; Hochschild, 1973; Matthews, 1986a).

Next to good health, adequate financial resources are crucial for dealing with the changing social circumstances and exigencies of aging (Larson, 1978; Morgan, 1988). Persons of higher socioeconomic status maintain larger collections of close friends and confidants, (Babchuk, 1978–1979), more functional friends (Cantor, 1979), and more friendship support (Ferraro, Mutran, & Barresi, 1984). Overall, affluent older people can afford to yield their institutional roles and make the transition to the ranks of the retired without seriously harming their morale (Lowenthal & Boler, 1965). In contrast, considerable research documents the stifled subjective well-being of older adults of lower socioeconomic status, which heightens their vulnerability to negative conditions such as poor health and isolation (Larson, 1978; Lopata, 1979).

Despite the unavoidable biological changes and likely limitations on sociality of extreme old age, many stereotypes of later life do not correspond with many people's ongoing management of its realities. In summarizing their study of financially secure and healthy 60- to 82-year-olds, Maas and Kuypers (1974) stated:

> Whether one considers women or men, and whether one examines their psychological capacities and orientations or their styles of living, most of these aging parents give no evidence of traveling a downhill course. By far most of these parents in their old age are involved in rewarding and diversely patterned lives. Most of them manifest high levels of coping capacities. (p. 215)

Babchuk, Peters, Hoyt, and Kaiser (1979) also described considerable involvement of older people, including persons over 80, in voluntary organizations.

People sustain customary and satisfying social participation as they revise and relinquish various social roles and responsibilities of middle adulthood. Some persons highlight their remaining options for involvement and cultivate other possibilities, notably friendships, by selectively investing their time and energy, by revising their definitions and expectations of friendships, or by adopting both approaches; other individuals are content with less active or gregarious lifestyles (Blau, 1973; Chown, 1981; Jerrome & Wenger, 1999). In short, individuals typically continue with practices and styles of friendship established earlier in their lives (Matthews, 1986a; Tesch, Whitbourne, & Nehrke, 1981), though increased freedom, and altered living arrangements during old age can also transform friendship patterns (Adams, 1985–1986, 1987; Hochschild, 1973).

ONGOING, GENDER-LINKED FRIENDSHIP PATTERNS AND CHANGING SOCIAL CONFIGURATIONS IN LATER ADULTHOOD

By and large, as their commitments to the parental responsibilities, occupational pressures, and civic obligations of middle adulthood diminish, the stage is set for healthy and financially secure older adults to engage in friendship as a primary endeavor (Brown, 1981). In gradually enacting this transition, individuals carry over various friendships, as well as habits and patterns of friendship from midlife into their later years (Clark & Anderson, 1967; Matthews, 1986a). Based on recollections obtained from over 60 interviews with older persons, Matthews (1986a, 1986b) describes in detail three primary friendship styles. Older persons exhibiting the "Independent" style refer more to circumstances than to specific individuals in describing friendships throughout their lives. Often resembling convenient, agentic friendships, their relationships are developed and maintained when situations allow them. As life structures change, so do these persons' main friends. Because the adults exhibiting this style primarily depend on the circumstantial availability of a pool of suitable others and not particular individuals to be friend, they are less susceptible to extreme grief over the loss of a given friend, which becomes commonplace during later life.

In contrast, the "Discerning" style reflects deep attachment to specific individuals, regardless of changing circumstances (Matthews, 1986a, 1986b). Often, such friends were made in or recalled from earlier periods in the life course (or both) and do not always survive into old age. Similar to committed, communal friendships and the old friends discussed above, adults remember developing few of these friendships in their lifetimes and view each as irreplaceble, becoming vulnerable to much sadness and loneliness if they lose any one of them (Matthews, 1983).

Finally, the "Acquisitives" preserve bonds with friends from their pasts and remain open to forming new friendships as situations permit (Matthews, 1986a, 1986b). As a result, they are likely to experience the personal integration and sense of continuity derived from enduring bonds, and the social integration and connectedness with present circumstances realized through developing what might be called "secondary friendships" (Adams, 1986) or "friendly relations" (Kurth, 1970). Matthews (1986b) did not discuss characteristic gender alignments in these styles, indicating through interview excerpts that both older women and men may exhibit them in recollecting their personal backgrounds and life experiences of friendship.

Although habitual styles of friendship remain salient, specific friendships continue to be negotiated within and contingent on each person's total configuration of social relations. If individuals have lived all or most of their lives in the same vicinity, they are likely to have a lifetime's array of friends and associates nearby to socialize with during later life. However, the prevailing middle-class pattern in Western culture finds older adults retiring in the community where their occupational careers stabilized, often several relocations away in time and space from where they were born and raised (Bellah, Madsen, Sullivan, Swidler, & Tipton, 1985). Still others make a "final move" to a retirement community, warmer climates, or both, requiring them once more to develop relationships with friends and neighbors. To the extent that married couples or individuals are accustomed to the interpersonal exigencies and routines of either staying put or relocating, granting good health and sufficient income, they are more likely to be practiced and reasonably satisfied with the social conditions their choices have produced.

Moreover, persons sustain and rearrange their life commitments differently depending on their preferred engagements in personal and social relationships. Overall, if individuals have been satisfied with limited social participation, preferring solitary pursuits and interaction with a small group of friends and family, they will probably strive contentedly to organize their interpersonal endeavors in later years accordingly, at least until they lose these favored others because of their relocations, disabilities, or deaths (Matthews, 1986b; Tesch et al., 1981). By comparison, persons who have always enjoyed considerable social participation in both sociable relations and close friendships are likely to configure their lives in ways fostering the continuation of these practices as long as they are able (Matthews, 1986b; Zborowski & Eyde, 1962).

In common with younger people, the elderly tend to be friends with age contemporaries with similar lifestyles, values, and experiences, the same gender and marital status, and residence in the same vicinity long enough for an attachment to develop (Heinemann, 1985; Peters & Kaiser, 1985; Rosow, 1970). Even so, age similarity is defined more broadly in later life than at earlier points, because the period known as "later life" potentially encompasses several decades. Thus, being an age peer has more to do with sharing similar functional capacities and restrictions, cohort membership and generational experiences, and relationship to the larger society than age per se (Peters & Kaiser, 1985; Rosow, 1970). Comparable age also increasingly involves a common sense of time, with the present shaped by an extensive past, limited future, and preparation for impending death (Marshall, 1975). However, as age peers die and available others noticeably decline in number during old-old age (75 years old and older), the elderly report more intergenerational friends (Armstrong, 1991; Brown, 1981; Usui, 1984).

Although many elderly people maintain some of their oldest and dearest friendships that are widely dispersed geographically and stem from both friends' changing addresses throughout their lives, practical restrictions foster more frequent interaction with and attachment to friends in their "immediate residential environment" (Adams, 1985–1986; Blieszner & Adams, 1998; Peters & Kaiser, 1985, p. 136). Spakes (1979) observed, "With increasing age, the respondents tended to report an increasing number of their friends they feel closest to living in the same community" (pp. 283–284). The heightened likelihood of older people with longer time in residence making friends and neighboring with nearby elderly, led Rosow (1970) among others to advocate age-segregated housing and retirement communities to facilitate their social activity and integration. Even so, a given individual's orientation toward sociablility, lifelong patterns of friendship, and degree of choice in living situation are likely to mediate his or her satisfaction with and inclination to form friendships in any housing arrangement (Rosow, 1970; Sherman, 1975).

As we have noticed since young adulthood, older persons primarily develop same-sex friendships; free-standing cross-sex friendships are uncommon in old age. The extent of their occurrence is rather consistently documented, with only one fifth of the women in an unmarried sample (Adams, 1985) and a similar ratio of married and widowed women (Babchuk & Anderson, 1989) reporting a male friend. By comparison, Powers and Bultena (1976) indicated less than one tenth of the women's and one third of the men's friends in their study as being cross-sex bonds. Clearly, the pronounced statistical minority of older males would seemingly make cross-sex friendships an important option for them. However, social suspicions and norms, as well as personal attributions of latent romantic or courtship interests, persist in later life and inhibit involvement in cross-sex friendships (Adams, 1985; Rawlins, 1982; van den Hoonaard, 1994). For the most part, cross-sex friendships are mediated by shared participation in an organization or club, friendship with a third party, or couple friendships (Adams, 1985; Lopata, 1979). Even so, Adams (1985) argued that these patterns of cross-sex friendship may also reflect generational differences in ideologies of gender, and recent work has reported more close cross-sex friends in later life (Armstrong, 1991; Jerrome & Wenger, 1999).

Friendship and Retirement

Keeping in mind the issues previously discussed, the effects of retirement on persons' social configurations depend on the centrality of work in arranging their earlier lives and therefore their friendships, their degree of liking for their occupations, coworkers, or both, the timing of their retirement vis-à-vis their favorite associates, their ability to sustain contact with admired coworkers after retirement, their other interests and opportunities for sociality, their preferred style of friendship, and the existence and character of their marital bond. I later discuss certain modal patterns and sources of variation associated with men and women.

As a statistical group, middle-class men derive a sense of continuity and identity from the limited contextual changes linked with well-ordered careers up until retirement (Bott, 1955; Hess, 1979; Maas & Kuypers, 1974; Wilensky, 1968). Typically, their wide networks of instrumental acquaintances and agentic friendships associated in midlife with enhancing careers and assuming positions of responsibility in their communities dwindle in later adulthood, with stabilized occupational and civic accomplishments. By the time they retire, they usually have a shrinking network of work-based acquaintances, a few closer friends developed during their middle years, and their wives as their best friends (Field, 1999; Fischer & Oliker, 1983). In this scenario, retirement separates men from their primary arena of social participation and most of their daily interaction with friends, and increases their dependence on their wives for friendship (Blau, 1973).

Although this pattern coheres with many findings regarding retirement-age males, a few matters modify the picture in given cases. First, the issue of "retirement age" is becoming progressively unclear. Even though money-making work is a fundamental organizer of most adult males' lives, men who dislike their occupations or coworkers increasingly have the option to retire early and configure their lives around part-time jobs, civic groups, personal interests, hobbies, or people that stimulate them more than working for their livelihood ever did. Thus, the statistically evident drop in the quantity of their relationships may be more than compensated for by their improved quality. Conversely, men in intrinsically satisfying occupations or with strong civic inclinations may continue to work or participate indefinitely, thereby retaining a pool of social contacts and friends. In either case, the timing of one's retirement relative to cherished work associates may facilitate or undermine continued association with them (Blau, 1961). Even if one remains working while others retire or vice versa, persons can always choose to devote the necessary time and effort to sustain mutual contact, just as they have with less available others throughout adulthood.

Despite these possibilities, the modal tendencies of males across life suggest that many men practice friendships of convenience with their work associates. When they retire from work and give up public endeavors, these unrehearsed breaches in their accustomed social routines greatly reduce their social encounters and confer special significance (and demands) on their families and lifelong patterns of close friendship with their wives (Blau, 1973; Powers & Bultena, 1976).

By comparison, women, especially working mothers, experience segmented, tentative and frequently interrupted careers across adulthood, which some authors argued prepare them to adapt to changing circumstances in later life (Hess, 1979; Maas & Kuypers, 1974). As the hampering of their friendship activities linked with the period of primary parenting responsibilities tapers off, women develop more acquaintances and friends than men do (Fischer & Oliker, 1983). It appears that the discontinuities and multiple contexts of their adult occupations can encourage them to make contacts and establish friendships with a variety of people in child-, community-, school-, religious-, and work-related settings. In addition to friendships with their husbands, women continue to cultivate other close friends and acquaintances (Field, 1999; Fischer & Oliker, 1983). In Roberto's and Kimboko's (1989, p. 16) study of people aged 60 years and older, though a majority of both genders reported "having close friends throughout their lives," the women demonstrated a greater tendency to maintain particular friendships over the entire life course than the men, who had begun their ongoing friendships during midlife. In short, women seem more inclined than men to make and maintain both communal and agentic friendships across public and private contexts in later adulthood (Fischer & Oliker, 1983).

Obviously, whether or not a particular woman demonstrates this modal pattern depends on her friendship style and practices, as well as how central money-making work and single minded careerism has been in organizing her life. To the extent that her adult endeavors resemble the modal female pattern of combining and juggling work, marriage, children, voluntary associations, and friendship, her retirement does not produce especially significant changes in her social activities (although her husband's may). She approaches later life with interactional habits and types of friendship that foster ongoing intimacy, as well as sociability. However, to the degree that a woman has emphatically organized her adult life around a profession or money-making career in the manner of the traditional male model, the gratifications of her occupation and coworkers, the timing of her retirement, her ability to stay in touch with appreciated associates, her avocations and chances for social interaction, her typical friendship style, and the presence and nature of her marital relationship become relevant concerns shaping her decisions about continuing to work or not. In summary, although retirement reduces the number of both men's and women's overall social contacts, the impact of this transition on their enjoyment and adjustment in later life depends on their customary interpersonal patterns both before and after they retire.

Friendship and Marriage

Key features of marriage viewed as a friendship during middle adulthood also endure into later life. First, husbands need their wives to confide in more than vice versa (Babchuk, 1978–1979; Chappell, 1983; Connidis & Davies, 1990; Keith, Hill, Goudy, & Powers, 1984). Whereas husbands mentioned their wives most, wives cited their husbands least as confidants in a sample of persons 60 years old or more (Lowenthal & Haven, 1968). In another study, wives' intimate friends accounted for almost as much of their overall interaction tallies as did their husbands (Powers & Bultena, 1976). Next, wives also serve as their husbands' principal links with neighbors, relatives, and other friends (Akiyama, Elliott, & Antonucci, 1996; Chappell, 1983). Reflecting our previous discussion of patterns of sociality, married older women do not consistently curtail their interactions with persons other than their husbands (Atchley, Pignatiello, & Shaw, 1979) and are more inclined to speak with relatives and close friends than are married older men (Kohen, 1983). They tend to remain connected with a variety of persons outside of their marriage. Finally, women assume the caregiving role with their aging husbands more than the reverse (Altergott, 1985; Chappell, 1983). Stoller (1990) reported, "Married women are institutionalized more frequently than married men, . . . so husbands may not be quite as dependable as wives in providing long-term care" (p. 229).

In view of these conditions, it is not surprising that older married women more often complain of loneliness than do older married men, perhaps when they are overburdened, prevented from interacting outside the marriage, or both (Blau, 1973). Older men are considerably more liable than women to engage in dating (Bulcroft & Bulcroft, 1991) and to marry in old age (Chown, 1981). They continue to view their wives as their best friends in later adulthood and to limit other close ties (Chappell, Segall, & Lewis, 1990). Yet as Hess (1979) observed, "While marriage is a mental and physical preservative for men, it is also an 'all the eggs in one basket' proposition, so that bereft of the one relationship, men have little else upon which to fall back except money and other material resources" (p. 505).

Although the marital friendship retains characteristic inequities and risks, it involves ongoing mutual benefits as well. Married persons report higher well-being than separated, divorced, or widowed individuals, though the ever-single exhibit comparable scores (providing further evidence for the satisfactions of continuous lifestyles; Larson, 1978). Married older people require less formal support and overall do not seem to lack companionship (Johnson, 1983; Kohen, 1983; Larson, Zuzaneck, & Mannelli, 1985) even in old-old age (Babchuk & Anderson, 1989). Finally and significantly, like other adult friendship patterns, couple friendships persist as a key basis for forming and maintaining interpersonal bonds with others (Babchuk, 1978–1979; Kohen, 1983).

Friendship and Widowhood

Older women are especially likely to lose their spouses to death in later life; after the age of 75, 70% of women are widows and 30% of men are widowers (Chown, 1981). Though researchers have reported lower rates of interaction with friends following widowhood (Blau, 1961; Lopata, 1973, 1979), recent studies emphasize the continuity in the close friendship patterns of older married women and widows. It includes their rates of interaction with friends and relatives (Atchley et al., 1979; Ferraro & Barresi, 1982; Petrowsky, 1976), number of intimate friends (Babchuk & Anderson, 1989; Roberto & Scott, 1984–1985), early formation and long duration of confidential close friendships, increased primary ties from work, stability of friendship network, and paucity of cross-sex friends (Babchuk & Anderson, 1989).

Although neither gender appears to be more socially isolated during widowhood than their married counterparts (Ferraro & Barresi, 1982; Petrowsky, 1976), men lose their foremost confidantes when their wives die. In contrast, widows are considerably more apt to mention at least one confidante (Barer, 1994; Lowenthal & Haven, 1968; Strain & Chappell, 1982). Lowenthal and Haven (1968) remarked, "Despite the fact that there were about twice as many widows as widowers in this sample, women were more likely to have a confidant than were men (69 percent, compared with 57 percent)" (p. 398). Based solely on this probability—reduced intimate dialogue with a caring other—it is easy to see why widowers complain of loneliness more than widows do (Blau, 1973).

Nevertheless, several changes in widows' friendship patterns are evident. Overall, contact with former couple friends diminishes markedly with the death of one's husband, initially reducing their total pool of friends (Heinemann, 1985; Lopata, 1979; van den Hoonaard, 1994). Meanwhile, widows progressively cultivate friendships with other single women (Arling, 1976; Heinemann, 1985), increase their daily contact with and receive more help from their close friends (Roberto & Scott, 1984–1985), confide more in their trusted friends (Babchuk & Anderson, 1989), and participate more in pleasurable and socially supportive activities with friends (Lopata, 1979). The composition of their friendship networks gradually alters in reflecting their marital status (as well as potential cohort effects). In one sample, nearly two thirds of the older widows's close friends were also widowed, with one third married; the opposite ratio characterized the friendships of married women (Roberto & Scott, 1984–1985).

Clearly, these heartening overall trends regarding widows' friendships are contingent on various factors. First, older women who retreat from society and center their lives around their husbands are at risk (Maas & Kuypers, 1974). As Lopata (1988) concluded, "In general, the greater a woman's dependence on the husband, or interdependence, the more every aspect of her life is disorganized when he dies" (p. 115). Next, resembling all older persons' capacities to adjust, husbands' deaths disrupt women's lives and friendship networks considerably more when they have less education and financial resources to help them cope and sustain voluntary bonds (Atchley, 1975; Atchley et al., 1979; Heinemann, 1985; Hess, 1979). Third, once again, an individual's initiative, self-confidence, and prevailing orientation toward friendships affects his or her chances of developing them (Heinemann, 1985; Jerrome, 1983; Lopata, 1979). Finally, limitations linked with chronic aging reduce the size of older widows' (age 75 and older) friendship networks, though not those of comparably older married women (Babchuk & Anderson, 1989).

Functions and Tensions of Friendship in Later Life

Friendships serve important and relatively specialized functions for older people for as long as they are able to actively engage in them in ways preserving their singular features. Thus, in considering its functional contributions to older persons' well-being, as well as possible tensions, it is useful to review friendship's ideal–typical attributes. First, friendship is a voluntary attachment that cannot be forced on anyone. Throughout life, people choose who they will treat and who they will allow to treat them as friends. Next, friendship involves a person-qua-person regard for the other (Suttles, 1970). Friends care about each other as specific individuals and not as members of a particular category or role. Third, friends view and deal with each other as equals. Accordingly, they seldom patronize or play up to each other, respecting the integrity and validity of their individual experiences and situations. Fourth and relatedly, friendships include mutual trust, support, and help. Over time, there are fairly symmetrical inputs into the relationship and to each other's welfare. Finally, friends feel and express shared and abiding affection. This set of features typically distinguishes friendships from other relationships, including those with most kin, in old age. But with constricted mobility and other incapacities, older people may have difficulty sustaining some of these qualities in their friendships, but also find them lacking in their family ties.

Socializing, Talking and Judging With Friends

Friendships persist as primary sources of enjoyment and pleasure for older people. Friends provide day-to-day companionship, as well as opportunities to talk, laugh, and have fun

(Adams, Blieszner, & de Vries, 2000; Crohan & Antonucci, 1989; Field, 1999; Gramer, Thomas, & Kendall, 1975; Mancini, 1980; Peters & Kaiser, 1985). Moreover, they link the older person to the larger community by encouraging participation in a variety of social activities (Arling, 1976; Peters & Kaiser, 1985; Sherman, 1975). Nussbaum and his colleagues (1989) observed, "Friends are likely to go shopping, to take a walk, to go to a ball game, or just to visit one another. These activities integrate the elderly individual into society" (p. 151). Noting married couples' self-sufficiency that approached "dysfunctional isolation" as they retreated from other social interaction and into their marital relationship, Johnson (1983) observed friends acting as intermediaries, drawing the spouses out and connecting them with various social agencies and collectivities. In conjunction with enjoyment and social integration, friends are also vital for relieving loneliness (Cantor, 1979; Heinemann, 1985; Mullins & Mushel, 1992; Perlman, 1988).

Older friends spend many of their moments together engaged in pleasurable and meaningful conversation (Field, 1999; Roberto, 1997). Moreover, sharing similar positions in the life cycle and perspectives on time, they often reminisce about periods when they both were younger or memorable events involving their spouses, careers, and children. Vivifying the past in this manner enriches the present, and longtime friends are especially liable to have helped create and to enjoy recalling such memories (Chown, 1981). Increasingly over the life course, talking with close friends confirms individuals in ways sensitive to the persistence of their valued personal characteristics and self-conceptions, despite unforeseen or unavoidable changes in circumstances, physical appearance, or actual capabilities. Moreover, this intimate acknowledgment comes from someone who responds voluntarily to the friend as a whole person, recognized and admired across time and a variety of social roles and predicaments. Consequently, their remarks bespeak a continuity, quality, and depth of relationship that few persons can emulate (Lemon, Bengtson, & Peterson, 1972).

This affectionate affirmation is a significant provision of friendship. Nearly one third of one sample of older women emphasized acceptance as a primary aspect of feeling close to someone (Adams, 1985–1986). Friends shield the older person from others' negative appraisals of his or her capabilities or worth as a person, which may be based on fair criteria or unwarranted prejudices (Heinemann, 1985). Likewise, friends help prevent negative self-evaluations (Mancini, 1980), warding off debilitating statements and thoughts about oneself as "old" (Bell, 1967), deterring identification with "age-related shortcomings" (Mitchell & Acuff, 1982), providing "status" (Candy, Troll, & Levy, 1981), and, overall, encouraging positive self-perceptions (Mancini, 1980; Mitchell & Acuff, 1982). As persons relinquish public opportunities to define themselves and demonstrate their competencies, they increasingly look to friends to validate them and to help formulate their self-evaluations (Atchley, 1977; Johnson, 1983; Shea, Thompson, & Blieszner, 1988; Sukowsky, 1977).

Consequently, the valued support and acceptance communicated by trusted close friends dialectically coexist with a reliance on them for frank judgments. Friends help to "maintain objectivity" about one's abilities, decisions, and dealings with others (Sukowsky, 1977) and to provide models of age-appropriate behaviors (Jerrome, 1986; Mitchell & Acuff, 1982). Candid exchanges help establish new standards, meanings, and interpretations of daily exigencies and problems. According to Francis (1981), difficulties with children are especially salient matters for discussion:

> Informants still discuss family problems with old friends who know their children well. They are able to assess their own status and the behavior of their children in comparison with that of their friends. They learn what is reasonable to expect from adult children, and they can adjust their expectations to conform to reality, rather than to an unrealistic ideal. This adjustment of expectations helps them to adapt their own behavior and to accept their changed status in their relations with children. (p. 93)

The counsel of friends conveys both judgment and acceptance by addressing the objective and possibly detrimental features of actions and situations, while still confirming and supporting each other's appropriate choices and inherent value as individuals. These practices of judging together and sharing advice distinguish old friends from new ones among older people (Shea et al., 1988).

In sensitively managing the dialectical tensions of expressiveness and protectiveness and judgment and acceptance in their ongoing interactions, many close friendships comprise relationships of confidence during old age, though women are more likely than men to reveal and discuss their concerns (Kohen, 1983; Roberto & Kimboko, 1989). In fact, three fifths of Adams' sample of older women considered "confiding behavior as a measure of closeness" in their friendships (1985–1986, p. 59). Although exceptions exist (Gupta & Korte, 1994), evidence repeatedly demonstrates that older persons' psychological well-being and morale are significantly associated with the quality of interaction characterizing their stable confidant relationships, not the overall quantity of their social encounters (Conner, Powers, & Bultena, 1979; Lowenthal & Haven, 1968; Strain & Chappell, 1982). Enjoying a relationship of confidence with at least one close friend may be enough to inhibit demoralization and prolong the older person's good mental health (Blau, 1973; Lowenthal & Haven, 1968).

Negotiating Independence and Assistance With Friends and Family

Most noninstitutionalized older people are self-sufficient and value their independence and privacy (Peters & Kaiser, 1985; Stoller & Earl, 1983; Wentowski, 1981). Reflecting our culture's preoccupation with individualism and self-reliance and its simultaneous tendency to underestimate and denigrate the capabilities and initiative of older persons, many of them are sensitive about their autonomy and reluctant to rely on others. Preserving their freedom to choose with whom, when, and whether or not they will be sociable is a key issue for them (Chown, 1981). Even married persons spend 40% of their time in solitude (Larson et al., 1985). Yet when persons must adjust to declining abilities to care for themselves, their lifelong patterns of independence, dependence, or interdependence influence the extent of others' availablity to them and how comfortable they feel about depending on and assisting others (Barer, 1994; Jonas, 1979; Wentowski, 1981).

When a voluntaristic ethic predominates, the limited obligations and flexible demands of friendship derive from and work to preserve both parties' freedom to be independent and their freedom to depend on each other (Rawlins, 1983a). Though friends remain able to call on each other for help and support in later life, they typically do so only in emergencies or for sporadic and limited assistance (Cantor, 1979). Enacting this policy accomplishes several individual and relational aims. First, by avoiding excessive dependence on their friends, they register their own continuing autonomy. Further, by not burdening their friends, they acknowledge their corresponding needs for independence, escape testing instrumentally the extent of affection and commitment upholding the friendship, and avoid setting precedents for assistance that they may not want or be able to deliver (Johnson, 1983). Meanwhile, occasionally relying on friends for necessary and appropriate help implicitly recognizes their caring, competence, and ability to assist. Such carefully managed participation enhances all parties' self-esteem and cultivates the voluntary, equal, mutual, and affective qualities of friendships (Bamford et al., 1998; Conner et al., 1979; Heinemann, 1985).

However, dialectical tensions may develop in friendships because one person provides too much assistance to the other or, conversely, makes undue demands. Crohan and Antonucci (1989) remarked, "Feeling needed is crucial to the elderly person's well-being. When older people define themselves as important to the welfare of a peer, their own ability to adapt to old age is enhanced" (p. 139). Hochschild (1973) described nurturant relationships where more capable persons care for clearly less fortunate others in mutually beneficial arrangements. Even so, both individuals must be reconciled to this form of relating; such complementarity can diminish friendships, because it highlights a lack of parity in the partners' personal resources and abilities and one person's dependence on the other (Ball & Whittington, 1995). In a sample of noninstitutionalized men and women aged 65 to 91 years old, Roberto and Scott (1986a) discovered that both equitably benefited and underbenefited individuals reported higher morale than did the overbenefited ones. In explaining why persons obtaining the same or less advantages than they offered friends felt better than those receiving more, the authors argued the following

> This inability to reciprocate undermines the older adult's sense of independence and self-worth, which may explain the greater anger expressed by those individuals who perceive themselves as receiving more from their friends than they were giving. The realization that one is less capable than in previous years to do for oneself is a stressful experience for older adults. (p. 246)

Although they recognize the enhanced self-esteem derived from helping less fortunate others, older persons are also well aware of the imposition and effort involved. They are often careful about how much they will inconvenience their friends, not wanting to incur obligations that they in turn may be unable to fulfil, nor to redefine the relationship as fundamentally instrumental in nature (Johnson, 1983; Nocon & Pearson, 2000). In most cases, they will turn to family members for long-range assistance or when they are more seriously or chronically ill (Cantor, 1979). Johnson (1983) reported that many older men and women recently discharged from a hospital orchestrated their encounters with friends, avoiding face-to-face contact until marked improvement in their health occurred, maintaining communication through letters and phone calls and preventing them from feeling pressured to do anything beyond the normal definition and expectations of their friendship (Johnson, 1983). Even so, in Johnson's (1983) study, friendships tended to dissolve when illnesses were too extended. Meanwhile, family members remained steadfast in their attention for as long as the former patient required it. The author stated:

> Nine months after hospitalization, the older people, who experienced an improvement in health and functional status, reported a decrease in contact with relatives and an increase in contact with friends. In contrast, those who remained highly impaired reported little change in contact and support from family members and a decline in support from friends [Johnson & Catalano, 1983]. This finding suggests that friendships have difficulty in standing the test of time if the relationship is also tested by illness and dependency. One can also conclude that friendships are quite specialized in the emotional or expressive domain, and in view of this voluntaristic character, are likely to break down when tested by the illness of one partner. (p. 113)

But several factors mediate who is liable to assist older persons in given circumstances. Dono et al. (1979) argued that potentially supportive others vary in terms of their proximity, sheer number, length and nature of commitment, physical capabilities, financial resources, and degree of caring. As long as older persons have able relatives nearby to rely on for aid, their friends are used minimally; yet when older adults are unmarried or lack available family members, friends and neighbors are likely to furnish support (Heinemann, 1985; Johnson, 1983; Stoller & Earl, 1983). Emotionally close friends who live nearby are particularly apt to help with a variety of tasks (Adams, 1984–1985). In most friendships, the potential strains of unbalanced reciprocity previously noted do not affect older persons' satisfaction with their best friends (Roberto & Scott, 1986b). As in communal friendships throughout life, older people are apparently more inclined to help and to receive help from their best friends without feeling burdened or uncomfortable; instrumental assistance emerges from and expresses affection. They usually view periods of inordinate demands as embedded in a long-term relationship of mutual reliance, whereas less close relationships involve more attention to fair exchange (Roberto & Scott, 1986b).

By and large, however, friendships are valued for companionship, and family members for instrumental help in later life (Bowling & Browne, 1991; Felton & Berry, 1992; Rook, 1989; Rook & Ituarte, 1999; Stoller & Earl, 1983; Wolfsen, Barker, & Mitteness, 1990). Although relationships with specific friends and relatives may share overlapping features and fulfill similar functions for older persons, they typically vary in important ways (Dono et al., 1979; Nocon & Pearson, 2000). Family relationships involve culturally sanctioned obligations (Brain, 1976); their ongoing enactment may be rooted in a sense of duty and responsibility for the older relative, not in the inherent pleasure of seeing him or her (Arling, 1976; Blau, 1973). Helpful kin usually are not peers (Lopata, 1988; Stoller & Earl, 1983). They differ in age and orientation toward life stages (Heinemann, 1985), with asymmetries in health status and personal resources generating lopsided capacities to reciprocate goods and services (Arling, 1976; Chappell, 1983). Because family members' perceived responsibilities may primarily dictate their sustained attention and regular visits, they may display positive concern but not necessarily liking or feeling close to the older person (Babchuk, 1978–1979; Heinemann, 1985). As a result, their patterns of contact may be neither mutually desired nor jointly fulfilling.

In contrast, friendships arise from voluntary attachments based on personal regard for another person and intrinsic enjoyment of his or her company. Friends are usually peers, sharing equivalent status based on their common membership in an age cohort, similar experiences of life course transitions, and comparable ability to assist each other (Arling, 1976; Heinemann, 1985; Jonas, 1979). The relationship rests on mutual choice and mutual need and continues as long as the partners feel shared affection. Even though they are more liable to dissolve than family bonds if circumstances, declining health, mobility or financial resources restrict interaction, either friend makes excessive demands, or they cannot preserve equal inputs over time, ongoing friendships are extremely important for older persons' self-esteem and continued enjoyment of life (Arling, 1976; Johnson, 1983). Across numerous studies, participating in friendships is more closely associated than family activity with high morale and psychological well-being (Arling, 1976; Johnson, 1983; Larson, 1978; Phillipson, 1997; Wolfsen et al., 1990; Wood & Robertson, 1978) and life satisfaction (Chappell, 1983; Edwards & Klemmack, 1973; Lemon et al., 1972; Pihlblad & Adams, 1972; Spakes, 1979). Despite their potential tensions and vulnerability to changes beyond either person's control, friendships help people feel good in later life.

CONCLUSION

In many ways, the patterns of solitude, intimacy, and sociability that persons negotiate within their evolving social configurations during mature adulthood continue into later life. Further, the boundaries marking individuals' transitions to old age itself are mutable and subject to the cultural constructions of older people prevailing at given historical moments interacting with their specific physical and mental capacities and financial resources. Accordingly, this chapter views later life as a career in itself, an extension of adult interpersonal and social practices involving active and conscious decisions by older people in shaping and managing their situated opportunities and impediments.

Nonetheless, friendships among older adults exhibit normative contours evident across the life course. Friends are close enough in age to share analogous generational experiences, and they are usually alike in race, gender, and marital and occupational status, all of which promote comparable lifestyles and values. Patterns of friendship related to activity in the workplace, participation in neighborhood, civic and professional associations, and recreation in groups of couples persist as long as these collectivities remain viable options. As persons voluntarily or involuntarily reduce or conclude their efforts as workers, association members, parents, or spouses, they may choose to change or sustain their degree of involvement in their other social endeavors. Conceivably, friendships endure as significant alternatives to pursue independently of diminished institutional activity and responsibilities throughout this period (Brown, 1981). Yet, once again, most persons' ongoing engagement in friendships reflects their previous practices and inclinations, given adequate health and finances.

For example, modal gender differences in friendship activity are clearly evident in later life. Men's social friends from professional and community settings taper off when they retire from these pursuits, yet they still depend on their wives for close friendship. They tend not to maintain their friends from youth and are less likely than women to have close friends or to replace ones made in adulthood as they are lost (Powers & Bultena, 1976; Wenger 1990). Though they report more frequent interpersonal contact than women do, it is confined to family and established friends (Powers & Bultena, 1976). In contrast, elderly women have more extensive ties with other people and are also liable to be more intimately concerned with and dependent on their friends than men; theirs is a diverse social landscape with enduring community involvement and amiable ties expanding their group of close friends (Armstrong, 1991; Babchuk, 1978–1979; Clark & Anderson, 1967; Roberto & Scott, 1986b; Spakes, 1979). They tend to preserve friendships from various points in their lives, losing their cherished friends only to death, and continue to acquire new close friends and sociable companions in later life (Armstrong, 1991; Zborowski & Eyde, 1962).

Sustaining personal integrity and autonomy, as well as self-chosen levels of social integration, presents increasingly salient challenges for older adults. On the one hand, there are the objective conditions for and practices of autonomous living and voluntary social involvement. These include utilitarian concerns, such as being able to provide for oneself and relevant loved ones, handle the routine tasks of personal hygiene, health care, and nutrition, maintain one's residence, and transport oneself to where one desires to go, such as shopping and social activities. On the other hand, there are the subjective feelings and perceptions of oneself as an autonomous and worthwhile being with ongoing relationships with others. These include such existential concerns as perceiving one's present life as meaningfully connected and continuous with one's prior moments, having fun, and feeling self-respect, dignity, and satisfaction with one's life as currently lived.

One must consider these two facets of autonomy in considering the value of friends. First, friends typically play minimal or specialized roles in the objective maintenance of older adults' autonomy; friends participate in helping networks but usually are not the mainstays of those networks (Nocon & Pearson, 2000). Friendship is based on freely offered affection, but this freedom from obligation often makes friends reluctant to impose on or burden each other. Moreover, friends are likely to be similar in age and functional ability (Bowling & Browne, 1991). Ironically, then, family members feel obligated to help kin with declining functional abilities, and it is largely their efforts that preserve autonomous living conditions for elderly persons. High (1990) observed, "Most important, the unavailability and absence of family members, especially adult children, to assume caregiving responsibilities is likely the strongest predictor of nursing home placement" (p. 278). By and large, the older one becomes, the greater the tendency to rely on family (Connidis & Davies, 1992).

For the most part, friendship appears to be a protected and privileged relationship during old age, reflecting both its vulnerable, voluntary basis and its specialized functions. Except for their time-tested and closest bonds, older people are reluctant to ask for too much instrumental assistance from friends. Family members, usually adult daughters, tend to provide material and service supports instead (Wilcox & Taber, 1991). However, despite their limitations, friends typically play vital roles in sustaining older persons' feelings of well-being and life satisfaction. Friends are uniquely valued to talk, reminisce, and judge with, and to keep confidences. They relieve loneliness, help with incidental needs, connect individuals to larger communities, and foster their ongoing enjoyment of life.

ACKNOWLEDGMENTS

Portions of this chapter have previously appeared in *Friendship Matters: Communication, Dialectics, and the Life Course,* by William K. Rawlins, (1992). I thank the publisher for permission to use this material.

REFERENCES

Adams, R. G. (1985). People would talk: Normative barriers to cross-sex friendships for elderly women. *The Gerontologist, 25,* 605–610.

Adams, R. G. (1985–1986). Emotional closeness and physical distance between friends: Implications for elderly women living in age-segregated and age-integrated settings. *International Journal of Aging and Human Development, 22,* 55–76.

Adams, R. G. (1986). Secondary friendship networks and psychological well-being among elderly women. *Activities, Adaptation and Aging, 8,* 59–72.

Adams, R. G. (1987). Patterns of network change: A longitudinal study of friendships of elderly women. *The Gerontologist, 27,* 222–227.

Adams, R. G. (1988). Which comes first: Poor psychological well-being or decreased friendship activity? *Activities, Adaptation and Aging, 12,* 27–41.

Adams, R. G., Blieszner, R., & de Vries, B. (2000). Definitions of friendship in the third age: Age, gender, and study location effects. *Journal of Aging Studies, 14,* 117–133.

Akiyama, H., Elliott, K., & Antonucci, T. (1996). Same-sex and cross-sex relationships. *Journal of Gerontology, 51B,* 374–382.

Altergott, K. (1985). Marriage, gender, and social relations in late life. In W. A. Peterson & J. Quadagno (Eds.), *Social bonds in later life* (pp. 51–70), Beverly Hills, CA: Sage.

Arling, G. (1976). The elderly widow and her family, neighbors and friends. *Journal of Marriage and the Family, 38,* 757–768.

Armstrong, M. J. (1991). Friends as a source of informal support for older women with physical disabilities. *Journal of Women and Aging, 3,* 63–83.

Atchley, R. C. (1975). Dimensions of widowhood in later life. *The Gerontologist, 15,* 176–178.

Atchley, R. C. (1977). *The social forces in later life* (2nd ed). Belmont, CA: Wadsworth.

Atchley, R. C., Pignatiello, L., & Shaw, E. C. (1979). Interactions with family and friends: Marital status and occupational differences among older women. *Research on Aging, 1,* 83–95.

Babchuk, N. (1978–1979). Aging and primary relations. *International Journal of Aging and Human Development, 9,* 137–151.

Babchuk, N., & Anderson, T. B. (1989). Older widows and married women: Their intimates and confidants. *International Journal of Aging and Human Development, 28,* 21–35.

Babchuk, N., Peters, G. R., Hoyt, D. R., & Kaiser, M. A. (1979). The voluntary associations of the aged. *Journal of Gerontology, 34,* 579–587.

Ball, M. M., & Whittington, F. J. (1995). *Surviving dependence: Voices of African American elders.* Amityville, NY: Baywood.

Bamford, C., Gregson, B., Farrow, G., Buck, D., Dowshell, T., McNamee, P., & Bond, J. (1998). Mental and physical frailty in older people: The costs and benefits of informal care. *Ageing and Society, 18,* 317–354.

Barer, B. M. (1994). Men and women aging differently. *International Journal of Aging and Human Development, 38,* 29–40.

Bell, T. (1967). The relationship between social involvement and feeling old among residents in homes for the aged. *Journal of Gerontology, 22,* 17–22.

Bellah, R. N., Madsen, R. Sullivan, W. M. Swidler, A., & Tipton, S. M. (1985). *Habits of the heart: Individualism and commitment in American life.* Berkeley: University of California Press.

Blau, Z. S. (1961). Structural constraints on friendships in old age. *American Sociological Review, 26,* 429–439.

Blau, Z. S. (1973). *Aging in a changing society.* New York:Watts.

Blieszner, R., & Adams, R. G. (1998). Problems with friends in old age. *Journal of Aging Studies, 12,* 223–238.

Bott, E. (1955). Urban families: Conjugal roles and social networks. *Human Relations, 8,* 345–384.

Bowling, A., & Browne, P. D. (1991). Social networks, health, and emotional well-being among the oldest old in London. *Journal of Gerontology, 46,* S20–S32.

Brain, R. (1976). *Friends and lovers.* New York: Basic Books.

Brown, B. B. (1981). A life-span approach to friendship: Age-related dimensions of an ageless relationship. In H. Z. Lopata & D. Maines (Eds.), *Research in the interweave of social roles: Friendship* (Vol. 2, pp. 23–50). Greenwich, CT: JAI.

Bulcroft, R. A., & Bulcroft, K. A. (1991). The nature and functions of dating in later life. *Research on Aging, 13,* 244–260.

Bury, M., & Holme, A. (1990). Quality of life and social support in the very old. *Journal of Aging Studies, 4,* 345–357.

Candy, S. G., Troll, L. E., & Levy, S. G. (1981). A developmental exploration of friendship functions in women. *Psychology of Women Quarterly, 5,* 456–472.

Cantor, M. H. (1979). Neighbors and friends: An overlooked resource in the informal support system. *Research on Aging, 1,* 434–463.

Chappell, N. L. (1983). Informal support networks among the elderly. *Research on Aging, 5,* 77–99.

Chappell, N. L., & Badger, M. (1989). Social isolation and well-being. *Journal of Gerontology, 44,* S169–S176.

Chappell, N. L., Segall, A., & Lewis, D. G. (1990). Gender and helping networks among day hospital and senior centre participants. *Canadian Journal on Aging, 9,* 220–233.

Chown, S. M. (1981). Friendship in old age. In S. Duck & R. Gilmour (Eds.), *Personal relationships 2: Developing personal relationships* (pp. 231–246). London: Academic Press.

Clark, M., & Anderson, B. G. (1967). *Culture and aging.* Springfield, IL: Thomas.

Conner, K. A., Powers, E. A., & Bultena, G. L. (1979). Social interpretation and life satisfaction: An empirical assessment of late-life patterns. *Journal of Gerontology, 34,* 116–121.

Connidis, I. A., & Davies, L. (1990). Confidantes and companions in later life: The place of family and friends. *Journal of Gerontology, 45,* S141–S149.

Connidis, I. A., & Davies, L. (1992). Confidants and companions: Choices in later life. *Journal of Gerontology, 47,* S115–S122.

Crohan, S. E., & Antonucci, T. C. (1989). Friends as a source of social support in old age. In R. G. Adams & R. Blieszner (Eds.), *Older adult friendship* (pp. 129–146). Newbury Park, CA: Sage.

Dono, J. E., Falbe, C. M., Kail, B. L., Litwak, R., Sherman, R. H., & Siegel, D. (1979). Primary groups in old age. *Research on Aging, 1,* 403–433.

Edwards, J. N., & Klemmack, D. L. (1973). Correlates of life satisfaction: A re-examination. *Journal of Gerontology, 28,* 497–502.

Felton, B. J., & Berry, C. A. (1992). Do the sources of the urban elderly's social support determine its psychological consequences? *Psychology and Aging, 7,* 89–97.

Ferraro, K. F. (1982). The impact of widowhood on the social relations of older persons. *Research on Aging, 4,* 227–247.

Ferraro, K. F., & Barresi, C. M. (1982). The impact of widowhood on the social relations of older persons. *Research on Aging, 4,* 227–247.

Ferraro, K. F., Mutran, E., & Barresi, C. M. (1984). Widowhood, health, and friendship support in later life. *Journal of Health and Social Behavior, 25,* 245–259.

Fiebert, M. S., & Fiebert, P. B. (1969). A conceptual guide to friendship formation. *Perceptual and Motor Skills, 28,* 383–390.

Field, D. (1999). Continuity and change in friendships in advanced old age: Findings from the Berkeley older generation study. *International Journal of Aging and Human Development, 48,* 325–346.

Fischer, C. S., Jackson, R. M., Stueve, C. A., Gerson, K. , Jones, L. M., & Baldassare, M. (1977). *Networks and places: Social relations in the urban setting.* New York: Free Press.

Fischer, C. S., & Oliker, S. J. (1983). A research note on friendship, gender, and the life cycle. *Social Forces, 62,* 124–133.

Francis, D. G. (1981). Adaptive strategies of the elderly in England and Ohio. In C. L. Fry (Ed.), *Dimensions: Aging, culture, and health* (pp. 85–107). New York: Praeger.

Gramer, E., Thomas, J., & Kendall, D. (1975). Determinants of friendship across the life span. In F. Rebelsky (Ed.), *Life: The continuous process* (pp. 336–345). New York: Knopf.

Gupta, V., & Korte, C. (1994). The effects of a confidant and a peer group on the well-being of single elders. *International Journal of Aging and Human Development, 39*, 293–302.

Harre, R. (1977). Friendship as an accomplishment: An ethogenic approach to social relationships. In S. Duck (Ed.), *Theory and practice in interpersonal attraction* (pp. 339–354). London: Academic Press.

Heinemann, G. D. (1985). Interdependence in informal support systems: The case of elderly, urban widows. In W. A. Peterson & J. Quadagno (Eds.), *Social bonds in later life: Aging and interdependence* (pp. 165–186). Beverly Hills, CA: Sage.

Hess, B. (1972). Friendship. In M. Riley, M. Johnson, & A. Foner (Eds.), *Aging and society: A sociology of age stratification* (Vol. 3, pp. 357–393). New York: Russell Sage Foundation.

Hess, B. B. (1979). Sex roles, friendships, and the life course. *Research on Aging, 1*, 494–515.

High, D. M. (1990). Old and alone: Surrogate health care decision-making for the elderly without families. *Journal of Aging Studies, 4*, 277–288.

Hochschild, A. R. (1973). *The unexpected community: Portrait of an old age subculture*. Berkeley: University of California Press.

Jerrome, D. (1983). Lonely women in a friendship club. *British Journal of Guidance and Counselling, 11*, 10–20.

Jerrome, D. (1984). Good company: The sociological implications of friendship. *The Sociological Review, 32*, 696–718.

Jerrome, D. (1986). Me Darby, you Joan! In C. Phillipson (Ed.), *Dependence and independence in old age* (pp. 348–358), London: Croom Helm.

Jerrome, D., & Wenger, G. C. (1999). Stability and change in late-life friendships. *Ageing and Society, 19*, 661–676.

Johnson, C. L. (1983). Fairweather friends and rainy day kin: An anthropological analysis of old age friendships in the United States. *Urban Anthropology, 12*, 103–123.

Johnson, C., & Catalano, D. (1983). A longitudinal study of family supports. *The Gerontologist, 23*, 612–618.

Jonas, K. (1979). Factors in development of community among elderly persons in age-segregated housing: Relationships between involvement in friendship roles within the community and external social roles. *Anthropological Quarterly, 52*, 29–38.

Keith, P. M., Hill, K., Goudy, W. J., & Powers, E. A. (1984). Confidants and well-being: A note on male friendship in old age. *The Gerontologist, 24*, 318–320.

Kohen, J. A. (1983). Old but not alone: Informal social supports among the elderly by marital status and sex. *The Gerontologist, 23*, 57–63.

Kovar, M. G. (1986). Aging in the eighties, age 65 years and over and living alone, contacts with family, friends, and neighbors. *National Center for Health Studies Advance Data, 116*, 3–11. Washington, DC: U.S. Department of Health and Human Services.

Kurth, S. B. (1970). Friendships and friendly relations. In G. J. McCall (Ed.), *Social relationships* (pp. 136–170). Chicago. Aldine.

Larson, R. (1978). Thirty years of research on the subjective well-being of older Americans. *Journal of Gerontology, 33*, 109–125.

Larson, R., Zuzanek, J., & Mannelli, R. (1985). Being alone versus being with people: Disengagement in the daily experience of older adults. *Journal of Gerontology, 40*, 375–381.

Lemon, B. W., Bengtson, V. L., & Peterson, J. A. (1972). An exploration of the activity theory of aging: Activity types and life satisfaction among in-movers to a retirement community. *Journal of Gerontology, 27*, 511–523.

Lopata, H. L. (1979). *Women as widows: Support systems*. New York: Elsevier.

Lopata, H. Z. (1973). Social relations of black and white widowed women in a northern metropolis. *American Journal of Sociology, 78*, 1003–1010.

Lopata, H. Z. (1988). Support systems of American urban widowhood. *Journal of Social Issues, 44*, 113–128.

Lowenthal, M. F., & Boler, D. (1965). Voluntary vs. involuntary social withdrawal. *Journal of Gerontology, 20*, 363–371.

Lowenthal, M. F., & Haven, C. (1968). Interaction and adaptation: Intimacy as a critical variable. In B. L. Neugarten (Ed.), *Middle age and aging* (pp. 390–400). Chicago: University of Chicago Press.

Maas, H. S., & Kuypers, J. A. (1974). *From thirty to seventy.* San Francisco: Jossey-Bass.

Mac Rae, H. (1992). Fictive kin as a component of the social networks of older people. *Research on Aging, 14,* 226–247.

Mancini, J. A. (1980). Friend interaction, competence, and morale in old age. *Research on Aging, 2,* 416–431.

Marshall, V. W. (1975). Socialization for impending death in a retirement village. *American Journal of Sociology, 8,* 1124–1144.

Matthews, S. H. (1983). Definitions of friendship and their consequences in old age. *Aging and Society, 3,* 141–155.

Matthews, S. H. (1986a). Friendships in old age: Biography and circumstance. In V. W. Marshall (Ed.), *Later life: The social psychology of aging* (pp. 233–269). Beverly Hills, CA: Sage.

Matthews, S. H. (1986b). *Friendships through the life course: Oral biographies in old age.* Beverly Hills, CA: Sage.

Mitchell, J., & Acuff, G. (1982). Family versus friends: Their relative importance as referent others to an aged population. *Sociological Spectrum, 2,* 367–385.

Morgan,, D. L. (1988). Age differences in social network participation. *Journal of Gerontology, 43,* S129–S137.

Mullins, L. C., & Mushel, M. (1992). The existence and emotional closeness of relationships with children, friends, and spouses. *Research on Aging, 14,* 448–470.

Myerhoff, B., & Simic, A. (1978). *Life's career—aging: Cultural variations on growing old.* Beverly Hills, CA: Sage.

Nocon, A., & Pearson, M. (2000). The role of friends and neighbors in providing support for older people. *Ageing and Society, 20,* 341–367.

Nussbaum, J. F., Thompson, T., & Robinson, J. D. (1989). *Communication and aging.* New York: Harper & Row.

Patterson, B. R., Bettini, L., & Nussbaum, J. F. (1993). The meaning of friendship across the life-span: Two studies. *Communication Quarterly, 41,* 145–160.

Perlman, D. (1988). Loneliness: A life-span, family perspective. In R. M. Milardo (Ed.), *Families and social networks* (pp. 190–220), Newbury Park, CA: Sage.

Peters, G. R., & Kaiser, M. A. (1985). The role of friends and neighbors in providing social support. In W. J. Sauer & R. T. Coward (Eds.), *Social support networks and the care of the elderly* (pp. 123–158). New York: Springer.

Petrowsky, M. (1976). Marital status, sex, and the social networks of the elderly. *Journal of Marriage and the Family, 5,* 749–756.

Phillipson, C. (1997). Social relationships in later life: A review of the research literature. *International Journal of Geriatric Psychiatry, 12,* 505–512.

Pihlblad, C., & Adams, D. L. (1972). Widowhood, social participation and life satisfaction. *International Journal of Aging and Human Development, 3,* 323–330.

Poulin, J. E. (1984). Age segregation and the interpersonal involvement and morale of the aged. *The Gerontologist, 24,* 266–269.

Powers, E. A., & Bultena G. L. (1976). Sex differences in intimate friendships of old age. *Journal of Marriage and the Family, 38,* 739–749.

Rawlins, W. K. (1982). Cross-sex friendship and the communicative management of sex-role expectations. *Communication Quarterly, 30,* 343–352.

Rawlins, W. K. (1983a). Negotiating close friendships: The dialectic of conjunctive freedoms. *Human Communication Research, 9,* 255–266.

Rawlins, W. K. (1983b). Openness as problematic in ongoing friendships: Two conversational dilemmas. *Communication Monographs, 50,* 1–13.

Rawlins, W. K. (1989a). Cultural double agency and the pursuit of friendship. *Cultural Dynamics, 2,* 28–40.

Rawlins, W. K. (1989b). A dialectical analysis of the tensions, functions and strategic challenges of communication in young adult friendships. In J. A. Anderson (Ed.), *Communication Yearbook 12* (pp. 157–189). Newbury, CA: Sage.

Rawlins, W. K. (1992). *Friendship matters: Communication, dialectics, and the life course.* Hawthorne, NY: DeGruyter.

Rawlins, W. K. (1994). Being there and growing apart: Sustaining friendships during adulthood. In D. J. Canary & L. Stafford (Eds.), *Communication and relational maintenance* (pp. 273–292). San Diego, CA: Academic Press.

Rawlins, W. K. (1998). Making meanings with friends. In R. L. Conville & E. Rogers (Eds.), *The meaning of "relationship" in interpersonal communication* (pp. 149–168). Westport, CT: Praeger.

Rawlins, W. K. (1999). Friendship. In D. Levinson, J. Ponzetti, & P. Jorgensen (Eds.), *Encyclopedia of human emotions* (pp. 280–285). New York: Macmillan Reference.

Retsinas, J., & Garrity, H. W. (1985). Nursing home friendships. *The Gerontologist, 25*, 376–381.

Roberto, K. (1997). Qualities of older women's friendships: Stable or volatile? *International Journal of Aging and Human Development, 44*, 1–14.

Roberto, K. A., & Kimboko, P. J. (1989). Friendships in later life: Definitions and maintenance patterns. *International Journal of Aging and Human Development, 28*, 9–19.

Roberto, K. A., & Scott, J. P. (1986a). Equity considerations in the friendships of older adults. *Journal of Gerontology, 41*, 241–274.

Roberto, K. A., & Scott, J. P. (1986b). Friendships of older men and women: Exchange patterns and satisfaction. *Psychology and Aging, 1*, 103–109.

Roberto, K. A., & Scott, J. P. (1984–1985). Friendship patterns among older women. *International Journal of Aging and Human Development, 19*, 1–10.

Rook, K. S. (1989). Strains in older adults friendships. In R. G. Adams & R. Blieszner (Eds.), *Older adult friendship* (pp. 166–194). Newbury Park, CA: Sage.

Rook, K. S., & Ituarte, P. H. G. (1999). Social control, social support, and companionship in older adults' family relationships and friendships. *Personal Relationships, 6*, 199–211.

Rose, S. M. (1984). How friendships end: Patterns among young adults. *Journal of Social and Personal Relationships, 1*, 267–277.

Rosow, I. (1970). Old people: Their friends and neighbors. *American Behavioral Scientist, 14*, 59–69.

Shea, L., Thompson, L., & Blieszner, R. (1988). Resources in older adults' old and new friendships. *Journal of Social and Personal Relationships, 5*, 83–96.

Sherman, S. R. (1975). Patterns of contacts for residents of age-segregated and age-integrated housing. *Journal of Gerontology, 30*, 103–107.

Spakes, P. R. (1979). Family, friendship, and community interaction as related to life satisfaction of the elderly. *Journal of Gerontological Social Work, 1*, 279–293.

Stephens, M. A. P., & Bernstein, M. D. (1984). Social support and well-being among residents of planned housing. *The Gerontologist, 24*, 144–148.

Stoller, E. P. (1990). Males as helpers: The role of sons, relatives, and friends. *The Gerontologist, 30*, 228–235.

Stoller, E. P., & Earl, L. L. (1983). Help with activities of everyday life: Sources of support for the noninstitutionalized elderly. *The Gerontologist, 23*, 64–70.

Strain, L. A., & Chappell, N. L. (1982). Confidants: Do they make a difference in quality of life? *Research on Aging, 4*, 479–502.

Sukosky, D. G. (1977). Sociological factors of friendship: Relevance for the aged. *Journal of Gerontological Nursing, 3*, 25–29.

Suttles, G. D. (1970). Friendship as social institution. In G. J. McCall (Ed.), *Social relationships* (pp. 95–135). Chicago: Aldine.

Tesch, S., Whitbourne, S. K., & Nehrke, M. F. (1981). Friendship, social interaction, and subjective well-being of older men in an institutional setting. *International Journal of Aging and Human Development, 13*, 317–327.

Thompson, M. G., & Heller, K. (1990). Facets of support related to well-being: Quantitative social isolation and perceived family support in a sample of elderly women. *Psychology and Aging, 5*, 535–544.

Usui, W. M. (1984). Homogeneity of friendship networks of elderly Blacks and Whites. *Journal of Gerontology, 39*, 350–356.

Van den Hoonaard, D. K. (1994). Paradise lost: Widowhood in a Florida retirement community. *Journal of Aging Studies, 8*, 121–132.

Weisman, A. D. (1991). Vulnerability and suicidality in the aged. *Journal of Geriatric Psychiatry, 24*, 191–201.

Wenger, G. C. (1990). The special role of friends and neighbors. *Journal of Aging Studies, 4*, 149–169.

Wentowski, G. J. (1981). Reciprocity and the coping strategies of older people: Cultural dimensions of network building. *The Gerontologist, 21*, 600–609.

Wilcox, J. A., & Taber, M. A. (1991). Informal helpers of elderly home care clients. *Health and Social Work, 16*, 258–265.

Wilensky, H. L. (1968). Orderly careers and social participation: The impact of work history on social integration in the middle mass. In B. L. Neugarten (Ed.), *Middle age and aging: A reader in social psychology* (pp. 321–340). Chicago & London: University of Chicago Press.

Wiseman, J. P. (1986). Friendship: Bonds and binds in a voluntary relationship. *Journal of Social and Personal Relationships, 3,* 191–211.

Wolfsen, C. R., Barker, J. C., & Mitteness, L. S. (1990). Personalization of formal social relationships by the elderly. *Research on Aging, 12,* 94–112.

Wood, V., & Robertson, J. F. (1978). Friendship and kinship interaction: Differential effect on the morale of the elderly. *Journal of Marriage and the Family, 40,* 367–375.

Wright, P. (1989). Gender differences in adults' same- and cross-gender friendships. In R. G. Adams & R. Blieszner (Eds.), *Older adult friendship* (pp. 197–221). Newbury Park, CA: Sage.

Zborowski, M., & Eyde, L. D. (1962). Aging and social participation. *Journal of Geronotology, 17,* 424–430.

IV

Organizational Communication

Bergstrom and Holmes have contributed a new chapter to this second edition of the handbook that concentrates on the contributions that organizational communication scholars have made to the study of aging. Although, as they point out, organizational communication is an area that attracts large numbers of scholars and their subsequent investigations, a focus on the possible affects of growing older at work has not received as much research attention as one might think. Bergstrom and Holmes situate their chapter within the changing demographics of the workplace. They contribute to the overall theme of the handbook by specifically discussing age as a social construction within organizations sometimes based on chronological age, sometimes based on functional age, and often based on what they call organizational age. Bergstrom and Holmes strongly suggest that future organizational communication scholarship must adopt a multidimensional, multiprocess view of aging and must consider how the organization and the individual relationships that transpire within the organization create their own aging realities.

Balazs begins her updated chapter (chap. 13) by showing how changes in socioeconomics and demography have caused marketing organizations to shift their focus "from the youth-oriented culture we have been to the aging culture we have become." The chapter focuses on how academic research (gerontology and geriatrics, sociology, economics, social psychology, and communication) is used to develop marketing strategies to serve members of society over 65. As in many other chapters, the link between communication and the formation of identity in aging is addressed here. The point of application here is that part of the role of communication research is said to involve enabling marketers to recognize "their role in shaping the expectations and behaviors of the older market that they strive to address."

Indeed, the issue of the terms used to describe old age by marketers is a potentially ageist force (Coupland & Coupland, 1990; Covey, 1988; Nuessel, 1984), but the euphemizing force of terms such as *the maturity segment, the ageless market,* and *gray America* is apparent, representing the discourse of institutions anxious not to alienate potential older customers. Balazs further indicates how the previous assumption by marketing practitioners that dramatic behavioral changes occur at a certain age has been displaced by the more recent rejection of chronological age as a consumer variable. On the one hand, she cites the notion of the New Age elderly, for whom age has become more a state of mind than a physical reality. In contrast to the negative stereotypes often associated with older people (see Hummert, Garstka, Ryan, & Bonnesen, this volume), this nomenclature and attendant ideology strongly suggest an empowered older population. On the other hand, there is evidence of marketers using models of deficit and impairment in planning communicative strategies of older customers.

Balazs argues that findings from today's seniors should not be extrapolated to older cohorts of the future. The recommendation (also made in other chapters) that longitudinal work should be done needs to be taken very seriously. This type of case study analysis would help us to track the effects of local and global life changes on individuals as members of cohort groups. There is also the implication that work on communication and aging needs to be an ongoing enterprise, reflecting changing beliefs and behaviors, not just of individuals, but also of societies as they come of age.

We would like to highlight some of the methodological points Balazs leaves us with. Her concern that the researcher should respect the individual respondent is of course relevant to all areas of social science research, but her preference for personal interviews matches our own experience in collecting data with older adults. There are at least two issues here. Older informants may have had limited educational opportunities, perhaps leading to a reluctance to participate in pencil-and-paper data-gathering designs. Also, they may have limited interactional opportunities, which puts a premium on spoken interaction of the sort that research can constitute. We have found one of the happy side effects of working on communication and aging to be the openness of older respondents to becoming involved in research activities, if studies are designed in a way sympathetic to their needs and wishes in the short term, as well as intended to be empowering in the longer term.

Bernard and Phillipson (chap. 14) orient to retirement (from paid work), which may be the longest and most stable period in our lives. Their chapter reviews the growth of retirement in the 20th century; assesses recent social trends, especially toward early retirement; and considers how retirement affects the social and leisure activities maintained by individuals. They examine the range of social and socioeconomic factors that first led to the establishment of retirement, and then early retirement, as aspects of individuals' changing expected or potential biographies. Again, we see a focus on choice rather than constraint, with the framing of retirement, and then early retirement, as aspects of individuals' changing expected or potential biographies. Again, we see a focus on choice rather than constraint, with the framing of retirement not as a lifestyle boundary marker but as a transition to be negotiated. Images of retirement, in turn, are described as currently being reconstructed, with concomitant uncertainty about and changes in the social position. The importance of the social categorization of individuals as representatives of age groups is quite clear in the case of legal and institutional definitions of old age resting on retirement from paid employment (as Rawlins, chap. 11, this volume, indicates), yet early retirees being too young to be called pensioners or senior citizens. Some interesting points of cross-cultural comparison arise between this discussion of marriage in retirement and Mars' and Fitzpatrick's review. Bernard and Phillipson review the two main foci of studies of retirement as leisure. The first is leisure time and the activities that constitute it, including levels of participation. The second is leisure as it relates to notions of adjustment, successful aging, and life satisfaction, which in turn becomes concerned with issues of identity in later life. Postretirement gender differences are revealed, in that women's lives remain more occupied, specifically, it seems, with domestic and family experiences. As the authors points out (and again there are clear links here with Mares' and Fitzpatrick's chapter), that what research has not yet shown us is whether the women and men themselves view this as significant.

REFERENCES

Coupland, N., & Coupland, J. (1990). Language and later life: The dischrony and decrement predicament. In H. Giles & P. Robinson (Eds.), *Handbook of language and social psychology* (pp. 451–468). Chichester, UK: Wiley.

Covey, H. C. (1988). Historical terminology used to represent older people. *The Gerontologist, 28,* 291–297.

Nuessel, F. (1984). Ageist language. *Maledicta, 8,* 17–28.

12

Organizational Communication and Aging: Age-Related Processes in Organizations

Mark J. Bergstrom
University of Utah

Michael E. Holmes
Ball State University

The inclusion of a chapter on organizational communication and aging in the second edition of this handbook is a response to the fact that individuals, organizations, societies, interdisciplinary researchers, and the popular press are becoming increasingly aware that the workplace is currently facing dramatic, and at times chaotic, change. Rapid changes and demographic imperatives force us to consider the dynamics and implications of an older workforce. Fortunately, the study of aging and work has a long history and rich corpus of studies from a variety of disciplines, albeit with few contributions from the communication discipline. Although organizational communication scholars have made sustained and significant contributions to our understanding of many organizational constructs of "difference" such as race, class, and gender, few organizational communication scholars investigate the age construct at work. This is unfortunate given the overwhelming evidence that aging impacts many, if not all, aspects of organizational life. Although age is sometimes noted as a variable in participant demographic sections of organizational communication research, along with gender and years of service in the organization, there is little attention to aging in its own right. We typically fail to reflect on what we're measuring when we collect chronological ages, or to consider the implications of the transformations involved when we talk of a sample's age distribution or define subgroups by 10-year groupings. For example, consider two defining mileposts of the state of organizational communication research: *The Handbook of Organizational Communication* (Jablin, Putnam, Roberts, & Porter, 1987) and *The New Handbook of Organizational Communication* (Jablin & Putnam, 2000). Neither handbook included "age" or "aging" in its index. Retirement is briefly mentioned in each volume as part of a developmental perspective on communication and organizational entry, assimilation, and exit (Jablin, 1987, 2000). In each case, however, retirement is treated as an age-neutral event of planned organizational exit or disengagement. Physical, psychological, social, and cultural facets of age and aging that might be related to retirement processes are ignored. Given that retirement is treated independently of aging, it is unsurprising that aging is also absent from theorizing about other subjects of organizational communication research.

Our goals in the present chapter are threefold: to selectively review the extensive demographic and social research on aging, to situate the study of aging and organizations as an important area of research for organizational communication, and to stimulate future studies of communication and aging in organizations. We hope that organizational communication and aging research will help ensure that a quality work life for older adults is available and productive; we also hope that attention to age and aging in organizations will provide communication scholars with a powerful site for the investigation of the social construction of aging and its interplay with the communicative enactment of organizations and organizing.

The work life of older workers is shaped by complex dynamics. Recent concern about the impact of aging baby boomers on the labor force and concern about Social Security stems from demographic arguments, and we begin our review by investigating the demographic exigency of aging in organizations. To appreciate more fully the complexity of aging in organizations and the implications of that complexity for our scholarship, we then unpack the age construct with a discussion of age, aging, and logics of explanation. In the United States, social policies like the Age Discrimination in Employment Act (ADEA) and the Americans with Disabilities Act (ADA) are designed to protect older workers. These policies, however, do not guarantee a full work life for older adults. The last section of our review describes potential challenges and obstacles that older workers may face in the organization.

DEMOGRAPHICS

Exploration of communication and aging in the workforce calls for attention, in particular, to the "older worker."[1] One of the most remarkable organizational changes in the last several decades is the "graying" of organizational membership, that is, a marked increase in the proportion of older workers. Recent projections from the Bureau of Labor Statistics note the enormous effects retiring baby-boomers could have on the workforce: "The current tight labor market situation could be exacerbated, hindering prospects for economic growth and putting a greater burden on those remaining in the workforce, perhaps forcing them to work longer hours" (Dohm, 2000, p. 17). Dohm suggested that these effects may be strongest in occupations with functions driven less by technological innovations, such as jobs in health and educational services, where workforce needs may go unmet unless older workers can be retrained or new workforce participants can be found. Before examining these effects in detail, it is important to note that even though

---

[1]Most of the "aging and organization" literature explicitly or implicitly focuses on formal organizations. This presumption is reflected in a tendency of authors to employ the terms *employee* and *worker* rather than the more generic "member" or "participant." Informal and noncommercial organizations are an important area for future research.

22 million people age 45 and older are projected to leave the workforce for retirement between 1998 and 2008, "there are encouraging signs that the labor force will not collapse in 20 years" (p. 25). Dohm posits that recent changes to Social Security, increased use of defined contribution pension plans, and a healthier older population who see work as beneficial may cause some individuals to delay retirement. In addition, the labor force may increase because of increased immigration and the entry of the baby-boom echo (the increased birth rate in the 1974–1994 period) into the workforce.

Researchers agree that chronological age is a flawed tool for characterizing aging, but reviewing demographic trends necessitates its use. For our purposes, the older worker, as defined by the U.S. Bureau of Labor Statistics, is one aged 55 and older and therefore past the key labor-force participation age range of 25 to 54. Although much attention has been given in recent years to the gender, racial, and ethnic diversity of American organizations, age diversity has been neglected despite predictions that "the aging of the U.S. workforce will be far more dramatic than its ethnic shifts" (Judy & D'Amico, 1997, p. 6). A host of demographic and policy pressures drive the changing age composition of the workforce in the United States (and similar trends in other industrial nations; cf. Gendell, 1998).

The aging of the workforce reflects demographic trends well-known to gerontologists and scholars of communication and aging but perhaps unfamiliar to organizational communication scholars. The general graying of the population reflects high birth rates after World War II and higher life expectancies associated with better health care, diets, and lifestyles. Baby-boomers, the 78 million people born in the United States between 1945 and 1965, are now reaching 55 and will begin to reach 65 in 2010 (creating, as some pundits have quipped, a "graybie boom"). The proportion of the U.S. population aged 65 and older was 12.6% in 2000. By 2020, almost 20% of the U.S. population will be 65 or older (Judy & D'Amico, 1997; Purcell, 2000). The impact of the aging boomer cohort is reflected in the demographics of the workforce. Purcell (2000) noted that boomers made up 55% of the age 25 to 54 population in the year 2000. According to Purcell, in the first 10 years of this century, the number of persons 25 to 54 will increase only 1.5 million, whereas the number of persons aged 55 to 64 will increase by 11.3 million, more than 47%. The workforce will continue to "gray" until around 2020; the trend will then dissipate as the oldest members of the boomer cohort retire in significant numbers (Sum & Fogg, 1990). In short, for the next several decades, the age profile of the workforce will undergo a dramatic shift to a greater number of older workers (Purcell, 2000).

Population demographics are not the only determinant of workforce age profiles. The size of a workforce age cohort also depends on its participation rate, defined as the proportion of potential workers who remain in the workforce. This factor may be less predictable because of opposing trends. These are (a) consistent decline in labor force participation with age (i.e., earlier retirement) over the last century (Sum & Fogg, 1990) and (b) recently increasing pressures for older workers to remain in the workforce.[2] The

---

[2]The exception to the decline is the participation of women in the 55 to 64 age group, which was expected to rise in the 1990s (Sum & Fogg, 1990). However, differences in workforce participation rates between men and women have consistently narrowed over the last several decades.

result of the conflicting pressures may be less predictability in retirement age (Judy & D'Amico, 1997) and heightened saliency of age in organizations.

The long-term trend toward earlier retirement is documented by the labor participation rates of men. In 1970, 83% of men 55 to 64 were in the labor force; this fell to 66% in 1995 (Judy & D'Amico, 1997). For those 64 and above, the participation rate in 1970 was 27%, and only 15% in 1995 (Judy & D'Amico, 1997). The nature of participation also changes with age: 83% of older workers (aged 55–64) work full-time; among those aged 65 and older, only 49% work full-time (Wiatrowski, 2001). It appears that increasingly "many men and women end their career jobs well before the 'normal' retirement age of 65" (Sum & Fogg, 1990, p. 43). The causes of this trend toward earlier retirement are many, including generous public and private pension benefits, low eligibility ages for full or partial benefits in many private pension plans, high marginal tax rates on retirement income, employer efforts to reduce labor costs through early retirement programs, and labor market shifts (Judy & D'Amico, 1997; Sum & Fogg, 1990; Wiatrowski, 2001).

It is possible, as Besl and Kale asserted, that "the long term decline in labor force participation at ages 55 and older has apparently leveled off in recent years" (1996, p. 19), but Gendell (1998) argued, on the basis of a study of workforce participation rates in the United States, Sweden, Germany, and Japan, that it is premature to conclude that the trend toward earlier retirement has ended. He found continuing declines in the average age of exit from the labor force for men and women in the four countries, but the magnitude and timing of the change varied.

Early retirement may not always be a matter of worker choice. The ADEA prohibits mandatory retirement in the United States, but it is important to recognize, as Herz and Rones (2000) noted, that "some retirements may not be strictly voluntary; rather, they may be in response to actual or probable job loss, or to a lack of acceptable job opportunities for older workers" (p. 14). They suggested that employer early retirement incentive programs, institutional barriers posed by Social Security and private pension rules, the undesirable nature of most part-time work, and age discrimination may encourage older workers to leave the workforce. The trend toward early retirement is even more surprising considering Morris and Bass' (1988) conclusion that "the physical and intellectual vigor of the elderly has so improved that the physiological and psychological age for removal from an economic role is probably better set at 75 than 65" (p. 4).

It is possible that economic pressures and policy shifts may diminish or reverse the trend toward ever earlier exit from the workforce. These pressures arise from workers' personal finance choices, shifts in the nature of pensions, cohort demographics, and government concerns about tax burdens posed by retirees. Some analysts suggested that "many aging baby boomers will be unable to afford to retire" (Judy & D'Amico, 1997, p. 73) because of their typically low rates of personal savings. Although most older adults assume they will retire around the age of 65, estimates indicate that up to 40% of adults will not have the resources necessary to do so (Sterns & Miklos, 1995). The decline of "defined benefit" pension plans may also contribute to the economic uncertainty of retirement (Besl & Kale, 1996).[3] Another contributing factor is that age cohorts are

---

[3] "Defined benefit pension plans" provide a set pension income determined by years of service, age of retirement, or both. These have increasingly given way to "defined contribution pension plans," where pension income depends on the financial performance of contributed funds (e.g., 401(k) plans).

increasingly well educated relative to earlier cohorts, and that the "opportunity cost" of labor force exit is greater. The participation rates of white-collar and professional older workers are consistently higher than those of blue-collar workers, and the trend away from industrial employment to white-collar and service industry employment also may diminish the importance of physical demands of work in the retirement decision (Judy & D'Amico, 1997). Government policy shifts may also discourage earlier retirement. The boomer demographics previously noted will result in a decline of the ratio of potential workers to potential retirees, from 4:1 around 2010 to 3:1 by 2020, and nearly 2:1 by 2030 (Judy & D'Amico, 1997). With fewer workers to support each retiree, the depletion of public funds for entitlements to older people looms as a real threat. One response in the United States has been to increase the age of eligibility for full Social Security benefits to 67 for those born in 1960 or later.

Older workers now have more choices regarding their exit from the workforce, including various forms of partial or phased retirement. Wiatrowski (2001) noted that retirees can be rehired as part-time, seasonal, temporary, or contract workers; may gradually reduce the numbers of hours worked; may take "trial retirement" leaves or absence; enter into job-sharing arrangements; or move to different jobs with reduced stress or fewer hours worked.

The demographics outlined above demonstrate the need to incorporate age constructs in organizational communication theorizing and research. Birth and death rate trends and opposing forces for early versus delayed exit from the workforce make age issues increasingly salient in organizational life. The demographics, however, beg the question of how to define age and aging in useful terms. Such definitions depend, ultimately, on a scholar's choice of a fundamental logic for explaining the social world.

AGE, AGING, AND LOGICS OF EXPLANATION

How "age" and "aging" are defined depends on the logic of one's preferred form of explanation of the social world: variance explanation or process explanation (Poole, Van de Ven, Dooley, & Holmes, 2000). These are not theories or methods per se, but rather are fundamental kinds of explanation exploited in theories and their associated research methods.[4]

Age as Variable Attribute

The logic of variance explanation goes hand in hand with the general linear model on which much social scientific research is based. Variance explanation describes the social world as composed of entities with attributes; these attributes interact to create outcomes. From this view, "age" is a variable attribute of an entity, such as a person or organization.

---

[4]The need for theories and research methods to share an underlying logic of explanation may be one reason for the emergence of theory–method complexes. Theories built on the logic of variance explanation posit outcomes created by the interaction of attributes of entities; traditional statistical methods for the analysis of variance are ideally suited for testing such theories. Theories built on process logic produce explanations based on change over time; therefore, time series analysis and interpretive methods are suited for theory testing or development.

The challenge for the researcher is to operationalize age and to determine its measure: chronological, biological, or psychosocial age.

If we assume a variance explanation logic, we define "age" as an attribute of persons and we explore the interaction of that attribute with other variables (such as financial status) in the production of outcomes such as health or satisfaction with retirement. We can combine an implicit process logic with our variance stance by defining an event (e.g., retirement) as an entity and applying age as an attribute of that entity (e.g., age at retirement). Age has been operationalized in a variety of ways by researchers applying a variance logic. For example, Ashbaugh and Fay (1987) noted that human resources researchers typically defined "older workers" by chronological age despite its obvious flaw: "Persons of the same chronological age often have very different physiological or mental ages. Many older individuals are youthful with regard to certain attributes and aged with regard to others." (Fries & Crapo, 1981, cited in Ashbaugh & Fay, 1987, p. 107).

If chronological age is a flawed measurement, how can we operationalize age? Sterns and Miklos (1995) provided an excellent review of five approaches to operationalizing age. Following Birren's (1959) proposal, they defined aging as "changes that occur in biological, psychological and social functioning through time" (p. 248). The three types are related to, but not the equivalent of, chronological age. They play a critical role in the study of aging in organizations. Biological age is defined as "an individual's position relative to his/her potential lifespan. Psychological age is indicated by the individual's capacity to adapt behavior to the demands of the environment. Social age refers to social norms and roles applied to an individual with respect to a culture or society" (Sterns & Miklos, 1995, pp. 248–249).

Chronological/Legal Age.   As noted above, older workers are most often defined on the basis of chronological age. Despite a lack of theoretical justification for defining an "older worker," Sterns and Miklos (1995) noted the implicit justification for age categories arises from legal definitions. The ADEA of 1967, amended in 1978 and 1986, was designed to protect workers over the age of 40 from age discrimination. Federal programs often use the Job Training Partnership Act and Older Americans Act age cutoff of 55 for defining the older worker. In contrast, the Americans with Disabilities Act (ADA) protects all workers, regardless of age. Employers must provide reasonable accommodations in testing and job duties (Sterns & Miklos, 1995). Thus, an older worker who has suffered a serious illness (e.g., a stroke) and desires to return to the workforce must be allowed to do so if she is able to perform the essential functions of the job with reasonable accommodations. Sterns and Miklos noted that many ailments associated with older adults are now classified as disabilities. The ADA is an example of legislation that protects all workers without age markers and avoids potential stereotypes that may accompany age categories.

Functional Age.   According to Stagner (1985), the term *functional age* was coined by McFarland (1953) as a way to emphasize the variability of older workers' rate of change in performance with age. Functional age indexes performance on a number of attributes related to job performance and defines individuals based on their performance as being either "younger" or "older" than their chronological age. For example, Sterns and Doverspike (1989) developed a performance-based functional age measure as an alternative to chronological age. This approach recognizes the wide variability in abilities and functioning for individuals at all ages.

Four factors were identified by Sterns and Miklos (1995) as potential sources of age-related changes that may or may not affect performance:

1. *Health status and aging.* Health problems increase gradually across age groups, and few differences are found between adjacent age groups.

2. *Physical capacity and aging.* Sterns and Miklos (1995) reported that physical capacity declines starting in middle age have been documented in virtually every index of physical activity. They noted, however, that the wide variation of decline and capacity in individuals, coupled with the unique demands of specific work tasks, make most declines relevant in only a small number of jobs. They used the example of decreased perceptual speed as potentially affecting a race car driver, although the same decline would likely have no effect on an executive role in an organization.

3. *Cognitive performance and age.* Studies of cognitive ability include learning, memory, intelligence, and speed. Sterns and Miklos (1995) noted that although any of these components may show age-related decline, any discussion of cognitive performance is complex because older workers may compensate with wisdom, expertise, and plasticity so that work-related outcomes are not affected by age-related changes. Most jobs, however, do not require maximal levels of processing, and Sterns and Miklos (1995) concluded that age is not useful in determining an individual's cognitive capacity for a given job.

4. *Age and job performance.* Their review concluded that there is no consistent relationship between age and job performance. In sum, although we may observe individual differences in functional age, the unique demands of a particular job and the compensatory mechanism employed by older adults may minimize the impact of functional age declines.

Psychosocial Age. This approach is based on definitions from "social perceptions, including age-typing of occupations, perceptions of the older worker, and the aging of knowledge, skills, and ability sets. How individuals perceive themselves and their careers at a given age may be congruent or incongruent with societal image" (Sterns & Miklos, 1995, p. 256). According to Sterns and Miklos, perceived attributes of older adults include being harder to train; less able to keep up with technological change, more accident prone, and less motivated, dependable, cooperative, conscientious, consistent, and knowledgeable. Job classification also brings age-related stereotypes for who is appropriate for a particular job, and expectations of performance may vary for older and younger workers. Organizational and technological changes may require older workers to update and learn new knowledge skills and abilities. Sterns and Miklos (1995) concluded that motivation is the most commonly studied factor in relation to updating behavior. Shearer and Steger (1975) found that lower levels of need for achievement and lower career expectations were associated with greater obsolescence.

Organizational Age. This approach studies aging in organizations through examination of age and tenure with an organization. Studies in this area also note that organizations age; in fact, an organization may be older than any of its members. The best example of work in this area is Lawrence's (1987) study. She concluded that organizational age norms provide a context that is useful in explaining judgments about older workers (cf. Lawrence, 1996; Zenger & Lawrence, 1989).

Lifespan Approach. This approach acknowledges the "possibility for behavioral change at any point in the life cycle" (Sterns & Miklos, 1995, p. 259) and recognizes substantial individual differences in aging. It therefore contests the possibility of an easily measured "age" attribute. Three sets of factors combine to form the unique status of an older adult worker. The first set resembles chronological age and includes normative age-graded biological determinants, environmental determinants, or both. The second set includes normative history-graded influences exhibiting a strong cohort effect. The third set is non-normative, including an individual's career and life changes, health, and stress-inducing events.

Studies using organizational and life span approaches to operationalizing age force us to consider further complexities that must be addressed in the study of work and aging. Two examples illustrate the scope of these complexities. Lawrence (1987) called attention to the fact that employees make decisions regarding everyday organizational events on the basis of age (which involve judgments based on psychological definitions of age), and organizations themselves use the age distributions of employees, which rely in part on chronological definitions of age. Her examples illuminate age effects, noting that indirect and direct organizational processes produce effects that are experienced by both organizations and individuals (Table 12.1).

TABLE 12.1
Lawrence's Typology of Age Effects

	What Process Produces the Outcome	
Who Experiences the Outcome	Direct	Indirect
Organization or group	An aerospace company wonders what will happen to management succession when retirements expose the recession-created gap in eligible 35- to 50-year-old managers.	A merger fails when the 40-year-old executives of one firm and the 65-year-old executives of a second firm do not "speak the same language."
Individual	A 45-year-old baseball player retires because he can no longer meet the physical demands of his profession.	A 38-year-old woman decides to reenter the workforce before she is seen as too old.

Note: From Lawrence (1987, p. 39).

Although her examples illustrate age effects that stem from individual aging processes, the organizational environment itself also produces age effects. The human factors perspective uses research to adapt the workplace environment to the worker, including adaptations of work for the aging worker (Sterns, Barrett, Czaja, & Barr, 1994). This work is similar to the interdisciplinary approach to aging at work advocated by Teiger (1994), Paoli (1994), and Marquié, Cau-Bareille, and Volkoff (1998). These approaches seek to improve productivity and the overall well-being of workers through the study of workplaces, work activities, and ergonomics. These authors focus on the importance of designs that serve workers of all ages.

Teiger (1994) recognized three different kinds of aging in the workplace. First, his conceptualization of *aging with regard to work* recognizes that organizations make age selections when hiring that are younger, about the same, or older than the age structures of the entire work population. This is explained by either work conditions (some jobs have stressors like posture and physical load, which seem not to support successful aging in that occupation). Other effects can be explained by history (high recruitment at a given time, then stabilization of the workforce). From this perspective, combining demographics with ergonomics allows an understanding of aging with regard to work. Second, his conceptualization of *aging through work* recognizes that there is differential aging, depending on job occupation. For example, sewing machine operators under time and accuracy constraints, with wages based on output, tend to leave this occupation at an early age. Further, most workers give up shift work and night work between the ages of 40 and 45. His conceptualization of *aging in work* calls attention to the possibility that organizations can maintain the mental and physical health of workers and develop their skills throughout their work life. For example, Davies and Sparrow (1985) found that in complex task performance, older engineers with greater experience performed better than young engineers, whereas younger engineers were more productive at simple tasks. Teiger (1994) proposed that by relieving time pressures and enabling the use of various strategies for task and work, performance of an older operator can be maintained. Such examples demonstrate the fluidity and complexity of the age construct considered as an attribute; we will now further complicate the notion of "age" by shifting from a variance to a process logic of explanation.

Aging as Process

Process explanation describes the social world in the logic of stories: as composed of continuous events, discontinuous events, or both, in meaningfully ordered sequences. The notion of "events" presumes that the continuous social process can be sectioned into bounded units (Abbott, 1984). Differences between such units define change over time. Differences in the position of units in the sequence allow the construction of chronologies of events. In process explanation, we assume that outcomes are created both by what happens and when it happens. Events have attributes, just as do entities, but chief among these attributes is location in an event sequence. Process explanation takes narrative form, whether a story in the traditional sense (e.g., ethnography) or in a more esoteric sense (e.g., time series analysis). The fundamental challenge for the process researcher is to define and operationalize events. Process explanation incorporates time, minimally at the level of relative ordering of events "before" and "after" each other and typically at the level of "real time" chronology and developmental cycles (the scope of a process from "start" to "end"). From this view, "the aging process" replaces "age" as a central concept. This logic shifts our focus to events that constitute aging and to the attributes of those events that are important to features and outcomes of the aging process: progression in chronological age, sequences of certain biological changes, psychological development, or sociohistorical context.

Attention to age in organizational communication scholarship should mean more than adding chronological age as a variable in our statistical models, or using arbitrary age groupings in 5- or 10-year spans (20–29, 30–39, etc.) to define subgroups for purposes of comparison. Just as organizational communication scholars have argued for focusing on organizing instead of organization (Weick, 1979, 1995), we advocate a multifaceted approach to aging in which age is only one facet. As demonstrated throughout this handbook, aging is a challengingly complex concept with physical, psychological, sociocultural, and sociohistorical elements. Organizations provide yet another context in which the meaning of aging is produced and reproduced, but at this point how the many facets of aging interact with complex organizational communication processes remains largely unexplored.

We suggest that to integrate age into organizational communication scholarship requires that we adopt a multidimensional, multiprocess view of aging and the organization–individual relationship. Process theorizing often begins with stage models, and that is where we focus. As Abbott (1984) noted, incorporating time turns the researcher's attention to the ordering of events in meaningful sequences; the questions often asked about event sequences are (a) whether there is a typical sequence or developmental path for a given social process and (b) whether there are families or clusters of typical sequences, and if so, what differentiates the paths through which a social process may unfold. We begin by briefly presenting models asserting typical sequences of individual development and of career stages, and then consider how we might integrate such models with each other and with other important facets of aging and organizations. We acknowledge that stage models have their weaknesses: They downplay individual differences and tend to smooth complex processes into simpler forms. To look only to general developmental stages and career stages applies a procrustean narrative to the diverse life stories of individuals. Stage models provide, nevertheless, a manageable framework for introducing time and sequence into our thinking about organizations and aging. When we understand patterns found across life histories, we can subsequently shift our attention to those dynamics that differentiate individual instances of the patterned histories.

Stage models presume that there are qualitative differences to be found within event sequences that distinguish portions of the sequences as distinct in structure, function or operation. Stage models therefore emphasize discontinuous change; discontinuity defines boundaries between stages or phases of relative stability. From a structuralist perspective, a stage model is driven by an internal engine and constrained by a coherent logic shaped by an end goal; context is not an inherent element in the model. A more developmental view frames stage models as descriptive tools; in such a view, change is not driven by teleology alone but is also shaped by contextual features and specific historical settings (Woolf, 1998). This developmental view is one we advocate; we would not claim, for example, that career stages in the Western world of the late-industrial and postindustrial period would hold in other contexts or settings.

Psychological Development and Aging.   There are many models of psychological development and aging; our discussion is informed by Woolf's review and synthesis (1998). She noted the complexity of aging and the need to consider life span development in light of the influence of "psychological, physiological, historical, sociological, and cultural factors" (p. 7) on the progression from childhood through adolescence, young adulthood, middle adulthood, and older adulthood.

Following Baltes, Reese, and Lipsitt (1980) and Sterns and Miklos (1995), Woolf (1998) called attention to three contextual influences that help account for both regularity and interindividual differences: normative age-graded influences, normative history-graded influences, and non-normative influences. According to Woolf, normative age-graded influences are those factors roughly correlated with chronological age because of biological or environmental forces. The former includes puberty, menopause, and other age-related physical influences; the latter includes the family life cycle, stages of education, and occupational development.

Normative history-graded influences are factors arising from political, social, and economic contexts that are largely shared within cultures, for example, wars, epidemics, and economic expansions and contractions. Of course, multiple cohorts can share experience of a set of history-graded influences, but the influence of that experience will be different in part because of chronological age. For example, the terrorist attacks of September 11, 2001, will have a different role in shaping worldviews and life experiences of members of the "Great Generation" and "Generation X." History-graded influences provide some of the basis for generalizations of generations, such as the different orientations toward financial security of those who came of age in the Great Depression and those who came of age in the economic expansion of the 1990's.

Non-normative influences are specific to individual experience, unlike age-graded and history-graded influences, which are shared within cultures. They represent the stock of unique experiences and influences that accumulate over the individual's life course. Woolf noted that "as each member of a cohort continuously experiences different non-normative life events throughout their [sic] life course, interindividual differences increase across the life-span" (1998, n. p.). Such interindividual differences are a reminder of the limitations of stage models—it is such differences that stage models neglect, and which make underlying patterns of stages and phases difficult to detect as the backdrop of individual differences.

Chronological development and age-graded, history-graded, and non-normative influences are not independent; rather, they are mutually influencing. This is evident if we compare these influences for two overlapping cohorts, as displayed in Fig. 12.1. History-graded events tend to have less developmental impact in the earliest and latest stages of life; therefore, the relative impact of such events may differ considerably across cohorts. Age-graded influences such as education and family timetables may shift over time and therefore influence cohorts differently. Such shifts in age-graded influences can be fueled by history-graded events, such as wars or economic cycles (Fig. 12.1).

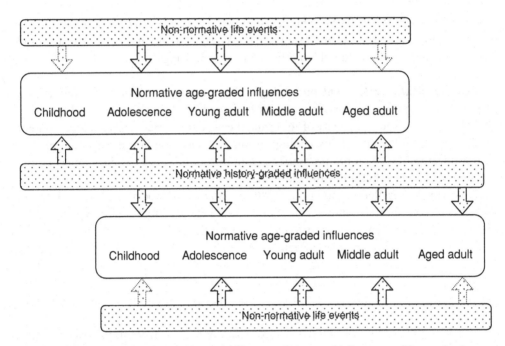

FIG. 12.1. Cohort-based similarities and differences of contextual influences on life span development (adapted from Woolf, 1998).

From this alignment we can see that age-graded influences work on both cohorts, though the evolution of such influences over time may alter their significance for different cohorts. The age grading of family life, for example, is less clearly defined now than several decades previously; being "off schedule" in childrearing therefore has different meanings and consequences over time. The cohorts in Fig. 12.1 share some history-graded influences but not others; the difference in chronological age of the cohorts will be an important determinant of the scope of shared experiences. However, age-graded influences will play a role in shaping different meanings for those events that are shared.

Career Theory.   The life span developmental perspective provided by Woolf (1998) is primarily a tool for understanding psychological development; it does not incorporate work history or the domain of organizational experiences except as elements of age-graded influences. Our next task is to explore the development of organizational life experiences as captured in the concept of career, "a sequence of positions occupied by a person during the course of the lifetime" (Super & Hall, 1978, p. 334). Our approach will be shaped by a concern for patterns over time, but several caveats must be noted. First, it is important to remember that organizational lives and career trajectories are shaped by a multitude of factors, not just the dynamics of life span development and career stages. For example, dispositional influences, social class, and technological change shape the course of careers (Sonnenfeld & Kotter, 1982). Second, we are concerned with patterns over the life span; this requires us to look beyond models limited implicitly or explicitly to a person's entry-to-exit experience within a particular organization (e.g., Jablin, 1987, 2000). Third, just as life span developmental theory tends to neglect organizational influences, career theory tends to ignore influences outside of the work sphere (Sonnenfeld & Kotter, 1982), and it will remain for us to integrate life span development and career development models.

Setting aside for now the many ways in which individual biographies may differ, as specified in the life span approach noted earlier, there remains considerable evidence for distinct occupational stages, or phases, in careers (Rush, Peacock, & Milkovich, 1980). Stage and life cycle career theories attempt to describe and explain these regularities. At the broadest level, we find the general tripartition of education, work, and retirement (the last a relatively recent addition; see Setterson & Hagestad, 1996). A typical model is offered by Hall (1976; see also Super, 1980). Hall suggests phases of career development with different themes or foci aligned, in part, with stages of psychological development, or "life tasks" (Fig. 12.2). These stages are broadly associated with chronological age through linking mechanisms of cultural age norms (Lawrence, 1987, 1996; Nydegger, 1986; Settersten & Hagestad, 1996) and ecological (i.e., economic and sociohistorical) influences on career development (Carroll, Haveman, & Swaminathan, 1992).

The career stages offered by Hall are (1) *exploration*, focused on precareer exploration and anticipatory socialization; (2) *establishment*, concerned with career trial, stabilization, and advancement and development within the career; (3) *maintenance*, or continuation along an established career path—alternately, there may be growth or stagnation; and (4) a terminal period of *decline* characterized by disengagement and exit from the career path. These stages progress concurrently, with developmental stages concerned with the formation of identity (adolescence), managing intimacy (young adulthood), providing generativity (serving the future good and/or guiding the next generation; middle adulthood), and achieving ego integrity or life satisfaction (older adulthood).

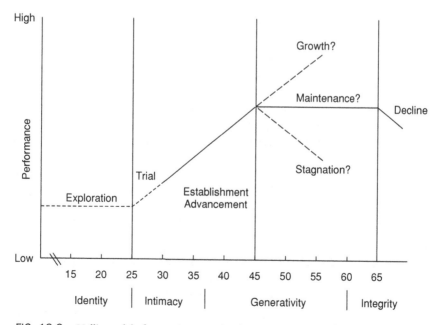

FIG. 12.2. Hall's model of stages in career development (adapted from Hall, 1976, p. 57).

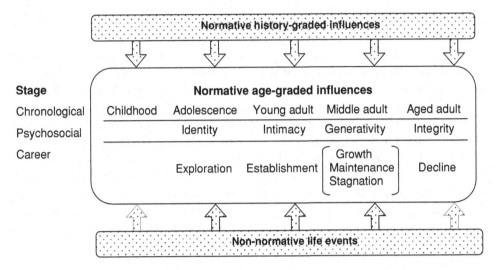

FIG. 12.3.   An integrated model of normative and nonnormative influences on psychological development and career development.

The relationship of psychological development, career development, age-graded and history-graded influences, and non-normative life events is displayed in Fig. 12.3. Given the complex interplay between these influences and domains, it is unsurprising that unitary stage models fail to capture the rich variety of individual experiences of aging and work. History-graded influences will feed cohort differences and contribute to ecological factors such as changing economic forces, technologically driven career opportunity shifts, and organizational age. Normative age-graded influences will also be subject to cohort differences: Contextual factors drive age distributions in organizations that in turn influence age norms, as well as influence broader patterns of life course development by shaping educational and family timetables. Non-normative life events will further differentiate individual developmental paths.

Clearly, communication and aging researchers interested in organizations, and organizational communication researchers interested in aging processes, face a daunting task when unpacking age and aging constructs. Recognizing one's preferred form of explanation of the social world and situating that logic in a multidimensional, multiprocess view of aging and organization–individual relationship should improve the quality of inquiry. We now turn to specific challenges and obstacles facing an older adult as a result of these aging processes. The symbolic nature of the age construct, and hence its roots in communicative action, will be evident as we consider the consequences of age—what it "means"—for older workers.

CHALLENGES AND OBSTACLES FACING OLDER WORKERS

There are many myths about older workers but little support for those myths in the literature. Stagner (1985) summed up the problem: "The problem of aging in industry is peculiarly complicated by the difference between perceived reality and 'objective' reality.

Most discrimination against aging employees is based on the erroneous perception of older persons as less capable, less efficient, less productive than their younger counterparts" (p. 789). In the following section, we expand on Peterson and Coberly's (1988) typology of myths regarding older workers. These are similar to myths facing older adults in general and provide a good sample of the numerous myths that may serve as challenges and obstacles facing older workers in the organization. Unfortunately, the real impact of these myths and stereotypes is that they have been shown to lead to discrimination against hiring people over the age of 50 (Stagner, 1985). He noted that discrimination is not limited solely to hiring practices; older adults are also discriminated against with regard to opportunities for training, performance appraisals, promotions, and management development programs, and may be the victims of early retirement policies.

Myths of Illness and Old Age

This myth characterizes older adult workers as suffering from generally poor health, loss of physical energy, and increased illness (Peterson & Coberly; 1988; Stagner, 1985). It leads to the assumption that older workers can not physically complete tasks, are often absent from work for health reasons, and do not have enough stamina for demanding tasks (Peterson & Coberly, 1988). In reality, those over 65 do have more sick days, more hospital stays, and more limited activity than do younger persons (Peterson & Coberly, 1988). Peterson and Coberly noted that although older adults are more likely to have chronic illness, this is somewhat counterbalanced by a decline in acute illness (cf. Atchley, 1985; Eisdorfer & Cohen, 1980; Ries, 1979). Although the five senses decline, older adults can and do compensate through experience, increased knowledge, and foresight. These declines seldom interfere with performance until individuals are in their 70s (Peterson & Coberly, 1988), and employers can help older workers compensate for losses through job redesign. One area of decline that may be particularly relevant for communication scholars is the ability to process higher frequency aural information. Stagner (1985) noted that in noisy environments, aural decline might produce noticeable conversational effects. Further, because there is so much heterogeneity in aging, Peterson and Coberly (1988) noted that for persons 60 to 80 years of age, "individuals may change rapidly during this time, and age alone becomes a very poor predictor of health, vitality, or performance" (p. 120).

Myths of Accidents and Old Age

This myth characterizes older adult workers as having higher accident rates, lost workdays, and increased medical and insurance costs (Peterson & Coberly, 1988; Rosen & Jerdee, 1976a; Stagner, 1985). The myth leads to the assumption that older workers are more expensive, create increased risks, and are a danger to themselves and coworkers. In reality, statistics from the U.S. Administration on Aging (1984; as cited in Peterson & Coberly, 1988) counter this myth. In 1981, workers 55 and older made up 13.6% of the workforce but accounted for only 9.7% of all accidents. Accidents that can be avoided by using judgment and experience decrease with age, whereas accidents requiring quick evasion increase with age (Birren, 1964). Although older workers have fewer mishaps, it is important to note they are off the job for a longer time per accident (Stagner, 1985). They are also more likely to suffer a fatal accident; workers over 65 have half as many accidents as younger workers but are nearly four times as likely to die from job-related causes (Moss, 1997).

Myths of Job Performance and Old Age

This myth characterizes older adult workers as having slower actions and performance, greater absence, and less commitment to the employer, leading to lower productivity (Peterson & Coberly, 1988; Rosen & Jerdee, 1976a, 1976b; Stagner, 1985). It leads to the assumption that older workers retire because they do not have the productivity of younger workers and that they are an economic cost to employers. In reality, the overall productivity of workers does not decrease with age at least up to the age of 65. Doering, Rhodes, and Schuster's (1983) review indicated no clear pattern of decline. In fact, although losses may be noticed if tasks require physical strength and rapid pace, evidence suggests that there may be increases in productivity when tasks require experience or the building of a clientele (Peterson & Coberly, 1988). Most studies of performance do not measure actual output; rather, studies generally rely on performance ratings. One of the biggest problems with performance rating studies is that they may reflect the stereotypes of the raters and not actual traits of the workers, and even experienced performance appraisers often disagree with one another (Stagner, 1985).

Myths of Learning and Change in Old Age

This myth characterizes older adult workers as suffering from rigidity of behaviors and irritability, inability to learn new skills, and as not being innovative (Peterson & Coberly, 1988; Rosen & Jerdee, 1976a, 1976b; Schrank & Waring, 1983; Stagner, 1985). It leads to the employers not promoting older adults, retraining them, or selecting them for challenging assignments. In reality, Schaie (1975) concluded that verbal intelligence changes little across the life span, at least until after age 65. Baugher (1978) concurred with Schaie and concluded that "the amount of age-related loss in general intelligence ability appears minimal and should rarely influence older persons' job performance" (p. 248). This myth is particularly damaging to older adults in times of rapid technological change, which increase the salience of adaptive learning.

Sterns, Barrett, Czaja, and Barr (1994) reported that although older workers have more difficulty acquiring computer skills, they can learn to use technologies. Age and prior experience with computers both predict performance. Thus, older adults who receive training and gain experience are more likely to improve performance. Although most executives indicate that retraining is available for older workers, Rosen and Jerdee's (1976b) findings indicated that this is not the case; in their study, older workers were rated as less motivated to keep up-to-date. It is interesting, that Chown (1972) found that an individual's rigidity was a better predictor of acquiring new skills than was chronological age. Thus, the need to update knowledge, skills, and abilities—and potential barriers to such updating—should be a concern for workers of all ages (Sterns et al., 1994).

This myth also supports the common stereotype that age is inversely related to innovation, that is, organizations must hire young people (new blood) to adapt to changing times and remain innovative. Schrank and Waring (1983) offered an interesting counter to this myth. They argued that it is too risky for newcomers to be innovative. New hires must learn new job duties and responsibilities, participate in the socialization process, and gain competence in their new organizational role. Schrank and Waring (1983) argued that innovation is more likely to come from a position of power, usually associated with organizational status and tenure. Thus, it may be that implicit organizational processes dictate that innovations spring from older workers.

Myths of Retraining in Old Age

This myth, similar to the myth of rigidity, characterizes older adult workers as having a reduced potential for retraining, job development, and skill upgrading (Bowers, 1952; Peterson & Coberly, 1988; Rosen & Jerdee, 1976a, 1976b; Stagner, 1985). It leads to the assumption that development costs for older workers are too high and the potential length of continued service for an older adult does not merit the cost. Peterson and Coberly (1988) countered this myth by noting, "Although it may take the older worker somewhat longer to learn the new skill or knowledge, experience has shown that the retrained worker is likely to be more dependable, have a better attendance record, stay on the job longer, and do as much work as the younger employee" (p. 125). Belbin's (1969) discovery method is an especially effective method for retraining older workers. The method involves the presentation of tasks and problems in such a way that enables "active discovery" on the part of the learner. Peterson and Coberly (1988) summarized the effectiveness of Belbin's and other similar methods: "The potential of the older worker for developing and improving skills can be met through creative approaches to training and employment policy. If utilized in an appropriate and knowledgeable manner, it will result in substantial value to both the individual and the employer" (p. 126).

CONCLUSION

Researchers applying a logic of explanation with age as variable attribute have an opportunity to integrate the literature previously presented. Three research programs pursued by scholars of communication and aging seem ideally suited as foundations for investigations of aging in organizations. Hummert and her colleagues have developed an approach to stereotype research that may be ideally suited for investigating older worker stereotypes (see Hummert, Shaner, & Garstka, 1995; and Hummert, Shaner, & Garstka, chap. 4, this volume, for excellent reviews). Two other approaches, the Communication Predicament of Aging (CPA) model (Harwood, Giles, Fox, Ryan, & Williams, 1993; Ryan, Giles, Bartolucci, & Henwood, 1986), and Communication Accommodation Theory (CAT; Coupland, Coupland, & Giles, 1991; Giles, Coupland, & Coupland, 1991) also offer valuable frameworks for investigating the interaction processes through which attitudes and stereotypes may lead to discrimination in organizations.

The common stereotypes or myths associated with older workers have not benefited from a programmatic, communication-theoretic investigation such as that undertaken by Hummert and her colleagues, yet there appears to a strong rationale for doing so. Although the stereotypes previously described were noted as being similar to stereotypes of older adults in general, the organizational context may merit additional investigation. Organizations provide a unique context for studying "older adults," in part because being "old" in an organization may, in some cases, be anywhere from 35 to 65 years of age (e.g., 35 in a new technology firm). Even the traditional definitions of older worker (typically 55 and older) are somewhat younger than the age of older adults studied in communication and aging research, and are best described as the young-old in conventional usage. For example, Hummert, Garstka, and Shaner (1997) used photographs of older adults representing three age groups, young-old (55–64 years), middle-old (65–74 years), and old-old (75 and older) in their investigation of facial cues.

Further, we expect the perceived traits of older workers (generated as the first step in organizing traits into stereotypes) will be similar to Hummert's (1990) typology, but may also contain additional traits specific to the organizational context. Schmidt and Boland (1986), for example, identified the trait of "doesn't like mandatory retirement" in the John Wayne Conservative stereotype, whereas Hummert's investigation did not address this trait.

Identifying the stereotypes of older workers is a necessary first step, but does little to explain actual interactions involving older workers in the organization. Researchers should, at a minimum, follow Hummert's (1994) lead and investigate how stereotypes influence beliefs about the communicative abilities of older workers. This work is particularly relevant given the positive and negative stereotypes of older adults in general and, more specifically, older workers. Although our previous discussion focused on negative stereotypes of the older worker, the literature also reports positive stereotypes of older workers. Bowers' (1952) ratings of older workers were higher than younger workers in attendance, steadiness, and conscientiousness, and Rosen and Jerdee (1976a) found ratings for older workers to be higher than younger workers on stability. Researchers need to investigate how negative and positive stereotypes are related to communication. Hummert's (1994) model addresses the importance of context; similarly, Lee and Clemons (1985) indicated the importance of studying context. They found "increasingly favorable employment decisions were made about older workers when (a) the situation did not require a choice between and older worker and a younger worker and (b) behaviorally stated performance information about the older worker was provided" (p. 787). Finally, Hummert's use of young, middle-aged, and older adult raters allowed her to conclude that older adults have more complex stereotypes of older people than do middle-aged and younger adults (Hummert, Shaner, & Garstka, 1995). This is particularly noteworthy given Stagner's (1985) claim that older adult performance appraisers have been found to rate other older workers more severely. Thus, future investigations of older worker stereotypes would have a useful and rich preliminary framework available in communication and aging research.

The CPA model may also be a useful model for explaining older worker interactions. In this model, communicators recognize old age cues, which activate negative stereotypes of the older worker. Modified speech to the older worker results in constrained opportunities for interaction and may subsequently reinforce negative stereotypes. As a result the older worker may experience psychological decline and loss of self-esteem, which could result in age-related changes in the older adult. This is a psychosocial instance of the process of "aging through work" previously noted.

The aim of CAT is to specify the strategies of convergence, divergence, and maintenance by examining how speakers modify their communication to reduce or increase the difference between speakers and their conversational partners (Coupland et al., 1991). Coupland et al. (1991) noted that when a speaker has relational goals for an interaction, she will select communication strategies attending to or anticipating the older adults' communicative characteristics. Given the evidence of negative stereotypes of older workers previously presented, it is clear that judgments about the older worker may be biased by negative stereotypes, which could lead to either overaccommodation or under-accommodation in an intergenerational context. These models offer strong theoretical frameworks for future investigations of older workers in organizations. Although they have been discussed here under an "age as variable attribute" logic, they could also serve as "age as process" research agendas. In fact, the Coupland et al. (1991) studies rely on discourse analytic methods that reflect process logic.

Researchers using "age as process" logic also have an opportunity to integrate the literature already presented. We have noted a variety of aging processes within organizations. Members age in chronological sense, they age with regard to their tenure with organization and tenure within a given position, and the membership of an organization itself grows older or younger in composition, and, of course, the organization itself ages. Research from Sterns and his colleagues (Sterns & Patchett, 1984, Sterns, Barrett, Czaja, & Barr, 1994) provided a potential model for organizational scholars interested in aging. Their model of older adult career development utilizes a life span approach that recognizes that behavior change processes can occur at any point in the life course. The model addresses training, updating, continued education, and retraining (Sterns et al., 1994). The model predicts that older adults' decisions regarding job or career changes, or to leave an organization, are based on attitudes toward mobility and experiences with previous career development activities. Although no clear relationship exists between age and obsolescence, organizational climate scholars should be interested in findings that indicate that organizational climate is a moderator in the prediction of updating. Age appears to influence updating in unsupportive climates, whereas expectancies influence updating in supportive climates (Sterns et al., 1994). It is important to point out that individual characteristics of the older worker appear to be the primary influence on decisions regarding updating; nevertheless, organizations may seek to improve these processes by implementing interventions that positively influence workers of all ages to seek updating.

Socialization scholars should continue to build on the groundwork established by Jablin (1987, 2000). Future socialization research should enhance and extend our understanding of socialization processes, with the inclusion of age and aging in the study of physical, psychological, social, and cultural facets related to organizational entry, assimilation, and exit.

Perhaps the most important area for future studies is the exploration of organization and organizing as sites for the social construction of age and aging. We must begin to recognize the importance of all aging processes. Cohort influences should be investigated, as they are likely to create and constrain communication. For example, the organization provides a context for members enacting cohort identities. The communicative context is different for a 50-year-old in an organization composed of largely 20- to 40-year-olds, compared to the communicative context for a 50-year-old in an organization composed largely of 45- to 65-year-olds. Organizational identity is constructed and embedded in a web of organizational relationships, and therefore the organization itself becomes a site for the exploration of the consequences of age-related identity constructions. Our review has focused primarily on traditional social scientific perspectives on aging and organizations; clearly, there are myriad issues to be addressed by the critical or cultural theorist as well. Critical scholars, in particular, have brought attention to gender, race, and sexual identity as contested social constructions and potential sites for the operation of marginalization. The critical (de)construction of age as another site for the contesting of power relationships should be of equal interest, especially as it may have complex and important relationships to other forms of oppression.

Older workers will have a tremendous impact on organizations through the next few decades. They will represent about the same proportion of the workforce, although their absolute workforce numbers will increase. Discrimination against older workers continues despite the clear evidence that they consistently perform at high levels on key areas of appraisal. Communication scholars must begin to investigate the interactive processes leading to these discriminatory practices if the significant number of older adults who would work if suitable jobs were available are to be afforded a productive, quality work life.

Age and aging are conspicuous in their absence in most organizational communication theorizing. It may be that we simply haven't gotten to them yet, and time itself will irresistibly add aging to the mix. The academic working class is not exempt from the age demographics noted earlier; we anticipate a steady increase in attention to aging in all areas of communication research, organizational communication included, fueled by the realization of an aging cohort of baby-boom scholars that they, too, grow old.

REFERENCES

Abbott, A. (1984). Event sequence and event duration: Colligation and measurement. *Historical Methods, 17*(4), 192–204.

Ashbaugh, D. L., & Fay, C. H. (1987). The threshold for aging in the workplace. *Research on Aging, 9*, 417–427.

Atchley, R. (1985). *Social forces in later life.* Belmont, CA: Wadsworth.

Baltes, P. B., Reese, H. W., & Lipsitt, L. P. (1980). Life-span developmental psychology. *Annual Review of Psychology, 31*, 65–110.

Barbato, C. A., & Feezel, J. D. (1987). The language of aging in different age groups. *The Gerontologist, 27*, 527–531.

Baugher, D. (1978). Is the older worker inherently incompetent? *Aging and Work, 1*, 243–250.

Belbin, R. M. (1969). *The discovery method: An international experiment in retraining.* Paris: Organization for Economic Cooperation and Development.

Besl, J. R., & Kale, B. D. (1996, June). Older workers in the 21st century: Active and educated, a case study. *Monthly Labor Review, 119*, 18–28.

Birren, J. E. (1959). Principles of research on aging. In J. E. Birren (Ed.), *Handbook of aging and the individual: Psychological and biological aspects,* (pp. 3–42). Chicago: University of Chicago Press.

Birren, J. E. (1964). *The psychology of aging.* Englewood Cliffs, NJ: Prentice Hall.

Bowers, W. H. (1952). An appraisal of worker characteristics as related to age. *Journal of Applied Psychology, 36,* 296–300.

Cantu-Weber, J. (1999, May–June). Harassment and discrimination: News stories show litigation on the rise. *Change, 31,* 38–45.

Carroll, G. R., Haveman, H., & Swaminathan, A. (1992). Careers in organizations: An ecological perspective. In D. L. Featherman, R. M. Lerner, & M. Perlmutter (Eds.), *Life-span development and behavior* (Vol. 11, pp. 111–144). Hillsdale, NJ: Lawrence Erlbaum Associates.

Chown, S. M. (1972). The effects of flexibility–rigidity and age on adaptability in job performance. *Industrial Gerontology, 13,* 105–121.

Coupland, N., Coupland, J., & Giles H. (1991). *Language, society & the elderly.* Oxford, UK: Blackwell.

Davies, D. R., & Sparrow, P. R. (1985). Age and work behavior. In N. Charness (Ed.), *Aging and human performance* (pp. 293–332). New York: Wiley.

Davis, J. A., & Tagiuri, R. (1988). Using life stage theory to manage work relationships. In H. Dennis (Ed.), *Fourteen steps in managing an aging work force* (pp. 123–137). Lexington, MA: Lexington Books.

Doering, M., Rhodes, S. R., & Schuster, M. (1983). *The aging worker: Research and recommendations.* Beverly Hills, CA: Sage.

Dohm, A. (2000, July). Gauging the labor force effects of retiring baby-boomers. *Monthly Labor Review,* 17–25. Washington, DC: U.S. Department of Labor.

Eisdorfer, C., & Cohen, D. (1980). The issue of biological and psychological deficits. In E. Borgatta & N. McCluskey (Eds.), *Aging and society: Current research and policy perspectives* (pp. 49–70). Beverly Hills, CA: Sage.

Frierson, G. C. (1997–1998). Sex plus age discrimination: Double jeopardy for older female employees. *Journal of Individual Employment Rights, 6*(3), 155–166.

Fries, S. F., & Crapo, L. M. (1981). *Vitality and aging.* San Francisco: W. H. Freeman.

Gendell, M. (1998, August). Trends in retirement age in four countries, 1965–95. *Monthly Labor Review, 121,* 20–30.

Giles, H., Coupland, N., & Coupland, J. (1991). *Contexts of accommodation: Developments in applied sociolinguistics.* Cambridge, UK: Cambridge University Press.

Hall, D. T. (1976). *Careers in organizations.* Pacific Palisades, CA: Goodyear.

Harwood, J., Giles, H., Fox, S., Ryan, E. B., & Williams, A. (1993). Patronizing young and elderly adults: Response strategies in a community setting. *Journal of Applied Communication Research, 21,* 211–226.

Herz, D. E. (1995, April). Work after early retirement: An increasing trend among men. *Monthly Labor Review, 118*(4), 13–20.

Herz, D. E., & Rones, P. L. (1989, April). Institutional barriers to employment of older workers. *Monthly Labor Review, 112*(4), pp. 14–21.

Herzog, A. R., & Morgan, J. N. (1992). Age and gender differences in the value of productive activities: Four different approaches. *Research on Aging, 14,* 169–198.

Hummert, M. L. (1990). Multiple stereotypes of elderly and young adults: A comparison of structure and evaluations. *Psychology and Aging, 5,* 182–193.

Hummert, M. L. (1994). Stereotypes of the elderly and patronizing speech. In M. L. Hummert, J. M. Wiemann, & J. F. Nussbaum (Eds.), *Interpersonal communication in older adulthood: Interdisciplinary theory and research* (pp. 162–184). Newbury Park, CA: Sage.

Hummert, M. L., Garstka, T. A., & Shaner, J. L. (1997). Stereotyping of older adults: The role of target facial cues and perceiver characteristics. *Psychology and Aging, 12,* 107–114.

Hummert, M. L., Shaner, J. L., & Garstka, T. A. (1995). Cognitive processes affecting communication with older adults: The case for stereotypes, attitudes, and beliefs about communication. In J. F. Nussbaum & J. Coupland (Eds.), *Handbook of communication and aging research* (pp. 105–131). Mahwah, NJ: Lawrence Erlbaum Associates.

Jablin, F. M. (1987). Organizational entry, assimilation, and exit. In F. M. Jablin, L. L. Putnam, K. H. Roberts, & L. W. Porter (Eds.), *Handbook of organizational communication* (pp. 679–740). Newbury Park, CA: Sage.

Jablin, F. M. (2000). Organizational entry, assimilation, and disengagement/exit. In F. M. Jablin & L. L. Putnam (Eds.), *The new handbook of organizational communication: Advances in theory, research, and methods* (pp. 732–818). Newbury Park, CA: Sage.

Jablin, F. M., & Putnam L. L. (2000). *The new handbook of organizational communication: Advances in theory, research, and methods.* Newbury Park, CA: Sage.

Jablin, F. M., Putnam, L. L., Roberts, K. H., & Porter, L. W. (1987). *Handbook of organizational communication: An interdisciplinary perspective.* Newbury Park, CA: Sage.

Judy, R. W., & D'Amico, C. (1997). *Workforce 2020: Work and workers in the 21st century.* Indianapolis, IN: Hudson Institute.

Lawrence, B. S. (1987). An organizational theory of age effects. In S. Bacharach & N. Di Tomasco (Eds.), *Research in the sociology of organizations:* (Vol. 5, pp. 37–71). Greenwich, CT: JAI.

Lawrence, B. S. (1996). Organizational age norms: Why is it so hard to know one when you see one? *The Gerontologist, 36,* 209–220.

Lee, J. A., & Clemons, T. (1985). Factors affecting employment decisions about older workers. *Journal of Applied Psychology, 70,* 785–788.

Marquié J. C., Cau-Bareille, D. P., & Volkoff, S. (1998). *Working with age.* Bristol, PA: Taylor & Francis.

McFarland, R. A. (1953). *Human factors in air transportation: Occupational health and safety.* New York: McGraw-Hill.

Morris, R., & Bass, S. A. (1988). Toward a new paradigm about work and age. In R. Morris & S. A. Bass (Eds.), *Retirement reconsidered: Economic and social roles for older people* (pp. 3–14). New York: Springer.

Moss, M. (1997, June 17). For older employees, on-the-job injuries are more often deadly. *Wall Street Journal.* pp. A1–A10.

Nydegger, C. N. (1986). Timetables and implicit theory. *American Behavioral Scientist, 29,* 710–729.

Paoli, P. (1994). Aging at work: A European perspective. In J. Snel & R. Cremer (Eds.), *Work and aging: A European perspective* (pp. 5–12). Bristol, PA: Taylor & Francis.

Peterson, D., & Coberly, S. (1988). The older worker: Myths and realities. In R. Morris & S. A. Bass (Eds.), *Retirement reconsidered: Economic and social roles for older people* (pp. 116–128). New York: Springer.

Poole, M. S., Van de Ven, A. H., Dooley, K., & Holmes, M. E. (2000). *Organizational change and innovation processes: Theory and methods for research.* New York: Oxford University Press.

Purcell, P. (2000, October). Older workers: Employment and retirement trends. *Monthly Labor Review, 123,* 19–30.

Ries, P. (1979). *Acute conditions: Incidence and associated disability, United States,* 1977–1978 (Vital and Health Statistics, Series 10, No. 132). Washington, DC: U.S. Government Printing Office.

Rosen, B., & Jerdee, T. H. (1976a). The influence of age stereotypes on managerial decisions. *Journal of Applied Psychology, 61,* 428–432.

Rosen, B., & Jerdee, T. H. (1976b). The nature of job-related age stereotypes. *Journal of Applied Psychology, 61,* 180–183.

Rush, J. C., Peacock, A. C., & Milkovich, G. T. (1980). Career stages: A partial test of Levinson's model of life / career stages. *Journal of Vocational Behavior, 16,* 347–359.

Ryan, E. B., Giles, H., Bartolucci, G., & Henwood, K. (1986). Psycholinguistic and social psychological components of communication by and with the elderly. *Language and Communication, 6,* 1–24.

Schaie, K. W. (1975). Age changes in adult intelligence. In D. S. Woodruff, & J. E. Birren (Eds.), *Aging: Scientific perspectives and social issues* (pp. 111–124). New York: Van Nostrand.

Schmidt, D. F., & Boland, S. M. (1986). Structure of perceptions of older adults: Evidence for multiple stereotypes. *Psychology and Aging, 1,* 255–260.

Schrank, H. T., & Waring, J. M. (1983). Aging and work organizations. In M. W. Riley, B. B. Hess, & K. Bond (Eds.), *Aging in society: Selected reviews of recent research* (pp. 53–69). Hillsdale, NJ: Lawrence Erlbaum Associates.

Settersten, R. A., Jr., & Hagestad, G. O. (1996). What's the latest? II. Cultural age deadlines for educational and work transitions. *The Gerontologist, 36,* 602–613.

Shearer, R., & Steger, J. (1975). Manpower obsolescence: A new definition and empirical investigation of personal variables. *Academy of Management Journal, 18*(2), 263–275.

Sonnenfeld, J., & Kotter, J. P. (1982). The maturation of career theory. *Human Relations, 35,* 19–46.

Stagner, R. (1985). Aging in industry. In J. E. Birren & K. W. Schaie (Eds.), *Handbook of the psychology of aging* (2nd ed. pp. 789–817). New York: Van Nostrand Reinhold.

Steinhauser, S. (1998, July). Age bias: Is your corporate culture in need of an overhaul? *HR Magazine, 43*(8), 86–91.

Sterns, H. L., Barrett, G. V., Czaja, S. J., & Barr, J. K. (1994). Issues in work and aging. *Journal of Applied Gerontology, 13,* 7–19.

Sterns, H. L., & Doverspike, D. (1989). Aging and the retraining and learning process in organizations. In I. Goldstein & R. Katzel (Eds.), *Age, health and employment* (pp. 229–332). San Francisco, CA: Jossey-Bass.

Sterns, H. L., & Miklos, S. M. (1995). Aging worker in a changing environment: Organizational and individual issues. *Journal of Vocational Behavior, 47,* 248–268.

Sterns, H. L., & Patchett, M. (1984). Technology and the aging adult: Career development and training. In P. R. Robinson & J. E. Birren (Eds.), *Aging and technology* (pp. 93–113). New York: Plenum.

Sum, A. M., & Fogg, W. N. (1990). Profile of the labor market for older workers. In P. B. Doeringer (Ed.), *Bridges to retirement: Older workers in a changing labor market* (pp. 33–63). Ithaca, NY: ILR.

Super, D. E. (1980). A life-span, life-space approach to career development. *Journal of Vocational Behavior, 16,* 282–298.

Super, D. E., & Hall, D. T. (1978). Career development: Exploration and planning. In M. R. Rosenzweig & L. W. Porter (Eds.), *Annual review of psychology* (pp. 333–372). Palo Alto, CA: Annual Reviews.

Teiger, C. (1994). We are all aging workers: For an interdisciplinary approach to aging at work. In J. Snel & R. Cremer (Eds.), *Work and aging: A European perspective* (pp. 65–84). Bristol, PA: Taylor & Francis.

Weick, K. (1979). *The social psychology of organizing* (2nd ed.). Reading, MA: Addison-Wesley.

Weick, K. (1995) *Sensemaking in organizations.* Thousand Oaks, CA: Sage.

Wiatrowski, W. J. (2001, April). Changing retirement age: Ups and downs. *Monthly Labor Review, 124,* 3–12.

Woolf, L. M. (1998). *Theoretical perspectives relevant to developmental psychology* [Online]. Available: http://www.webster.edu/~woolflm/theories.html

Zenger, T. R., & Lawrence, B. S. (1989). Organizational demography: The differential effects of age and tenure distributions on technical communication. *Academy of Management Journal, 32,* 353–376.

Marketing to Older Adults

Anne L. Balazs

Mississippi University for Women

The study of human aging is not restricted to gerontologists, developmental psychologists, and health care researchers. Interest in the elderly is far-reaching and not purely academic. An understanding of human behavior over the life span is critical to many for-profit and nonprofit businesses whose markets (and employees) are aging. Aging of the population has an economic impact that businesses of all sizes must monitor and respond to with new strategies (Miller & Kim, 1999). During the past half century, businesses have relied on the baby-boom as a market for its products and services. Although this is the largest single demographic segment the U.S. population has produced, it is by no means the only one. All Americans are living longer and fuller lives. Further, "boomers" (and likewise Generations X and Y) have inevitably aged and become moving targets, with different needs and wants. The baby-boom generation has been a profitable segment to cater to in the past, but marketers now realize that the boomers' parents are alive and well. This has caused marketing organizations to shift their focus from the youth-oriented culture we have been to the aging culture we have become (Hauser & Scarisbrick-Hauser, 1995).

BUSINESS MOTIVATIONS FOR INTEREST IN OLDER ADULTS

The practice of marketing has matured as well over the past few decades. The marketing concept has evolved from a distribution to a sales orientation, and more recently to a service and satisfaction focus and changed the way organizations respond to the needs of its markets. Indeed, in the past, there was little consideration for what people wanted and more emphasis on what could be produced. As the population grew and became more heterogenous and as other environmental variables impacted consumers, marketers began to segment their markets. Demographic variables are the simplest dimensions by which to categorize people, and these will always be used to describe a population. Age, income, household status, and education level are common variables used to identify and target a segment of consumers and limit the number of marketing strategies to be employed. Such demographic analysis led market researchers to take notice of the aging population and the size of the mature market.

In a quest to identify niches of opportunity and new markets to serve, businesses "discovered" the population over 65 years of age. Literally, dozens of articles in the popular business press have announced this fact over the years (for early examples, see Bartos, 1980; Block, 1974; Gelb, 1978; *U.S. News & World Report*, 1980). Here was a market, previously neglected, that was enormous in size with greater discretionary income than other markets. Marketers were willing and eager to serve this segment of the population and pursued older people with products and services to meet their needs and desires (Visvabharathy & Rink, 1984). In the last decade, more attention has been devoted to the "gold in the gray" (Hoey & Jugenheimer, 1997; Paul, 2001; Shaw, 1993).

The realization that the population is aging and that the older consumer represents the future of business has been the focus of more astute marketing organizations. Other, less adaptable firms have been reticent to target a market with which they are unfamiliar (or don't acknowledge as an important one), whereas others continue to pursue the markets that are appropriate to their product offerings. The narrow-mindedness of some firms will ultimately catch up with them. The marketplace has become increasingly global, and its consumers are aging universally ("More People Hit Old Age," 1993; van der Merwe, 1987).

This chapter will introduce the reader to the intellectual disciplines that support marketing research on aging and provide examples of how marketing organizations use this research to develop strategies and serve the mature segment of society. Successful efforts to promote products and services depend on a thorough understanding of older people and their needs.

DISCIPLINARY FOUNDATIONS OF MARKETING RESEARCH ON OLDER PEOPLE

Aging research, conducted by marketing organizations and academics, borrows from many disciplines. Gerontology and the more clinical discipline, geriatrics, are the bases for understanding the biological aspects of aging. These provide marketers with a knowledge of the physiological and psychological changes that accompany age and require adjustment and adaptation. Such knowledge assists businesses in designing products for assisted living; product labeling in the appropriate typeface, color, and size; store layouts that accommodate the older shopper; and services to maintain independence and allow for "aging in place," that is, remaining comfortably in one's home for as long as possible.

Demography and, by extension, sociology are the means by which marketers understand older adults as a generational cohort and in relation to other age groups. The tremendous growth of the older age groups (and particularly the "old-old") has caused as much excitement for demographers as the anomolous baby boom. The relationships between young and old are more complicated. Sociologists research these relationships and hypothesize about their effects on the status of the family, organizations, government, and other institutions. Marketers take note of these observations and structure their marketing plans and programs with aging family dynamics in mind.

In addition, sociologists and social gerontologists study social behavior. The normative roles played by each generation and especially by the older ones are being rewritten. Never before have older people been so vital, healthy, and active. Nor has there been the social freedom to play out their roles of spouse, parent, grandparent, retiree, widow, or volunteer in so many ways. Mores and restrictions on behavior are loosening up, and increasing numbers of older people are able to conduct their lives in ways they choose, rather than in ways they are assigned. Again, sociologists and social gerontologists observe these changes in behavior and offer theories on how these roles and role changes may be more fulfilling and satisfying. Marketers represent and support these new behaviors in advertising and through products, and bridge the gap between the theoretical and the practical.

Economic research provides the marketer with information on the spending and saving activities of older people. It also examines the extent to which older people represent a tax burden on the younger generations by demanding social services. The economic support required by retirees concerns younger workers, who face a rising deficit and increasing taxes. Older adults have become a powerful lobbying group, affecting economic legislation at the expense of other "special interest" groups who feel slighted next to such a bloc. There is, however, great variation in wealth in the older market, ranging from the "pension elite" (Longino, 1988) to those living at or below the poverty line. The structure of the older household, its income stream, and spending patterns are important clues to the marketer who is designing a new product line, pricing strategy, or promotional campaign.

Research in social psychology provides insight into the older individual's personality, attitudes, cognitive development, and conversely how others perceive and judge older adults. Life span development theory suggests that older adults are not merely young people "grown up." They are affected by the process of aging, psychologically and physiologically. This, in turn, affects their ability to interact with others and communicate their needs. Marketers must be sensitive to the changes that accompany aging and facilitate positive social interaction. This task will continue to challenge marketers who successfully serve older people now and eagerly await the next cohort.

The main objective of marketing is to promote exchange and thereby increase satisfaction. This goal cannot be accomplished without knowledge of the communication process. Communication is inherent in personal selling and the service encounter—advertising, packaging, and supplying product information. If the benefits of a product cannot be communicated effectively, no sale takes place and no benefits are received by the targeted market. Or, if the product's benefits are communicated in such a way as to offend the audience, the marketer has reduced the potential consumers' satisfaction. Further, the content of the message and the portrayal of the older consumer in marketing communications is a powerful force in creating images of older people and their "appropriate" behavior. Although the practice of marketing is supposed to be market driven, that is, based on consumer needs and wants, it also shapes consumer behavior. Consumers, both old and young, are affected by marketing messages and respond to them by changes in their purchasing, preference structure, and role enactment, to name but a few ways.

Communication research is vital to marketing practitioners interested in interpreting the needs of an older target market. Marketers must know how to relate informative, persuasive, and attractive messages to an audience that may process information more slowly, have different expectations of service, have physiological and cognitive deficiencies that affect their reception of the message, or face all of these difficulties. Marketers must also recognize their role in shaping the expectations and behavior of the older market that they strive to address. Marketing research draws from the social sciences to better apply marketing principles to older consumers. Businesses use this information to develop products and services to match the needs and wants of a heterogeneous market and to increase life satisfaction.

OBJECTIVE OF MARKETING TO OLDER ADULTS

Businesses manufacture, promote, and distribute products and services not simply with the hope of increased sales or profits. Indeed, if the products and services did not fulfill needs, they would not be demanded and fail in the marketplace. The marketing concept promotes exchange for the purpose of raising levels of satisfaction for all parties involved. Satisfaction with the product, service, information, or idea exchanged is explicit in this definition. Implicit in this exchange is a loftier goal of raising life satisfaction. Thus, an important objective in marketing to older consumers is increasing their life satisfaction through successful exchange.

Life satisfaction and quality of life considerations have been studied extensively in the gerontology literature (Larson, 1978; Palmore & Kivett, 1977; Pearlman & Uhlman, 1991), and marketers have borrowed these concepts in recent years (Bearden, Gustafson, & Mason, 1979). Cooper and Miaoulis (1988) claimed that the concept of life satisfaction will become more important as the number of older consumers grows. Cooper and Miaoulis (1988) reasoned that according to Maslow's theory of self-actualization, older people will have an interest in higher order needs because their basic needs would have been fulfilled. Concern with life satisfaction would take precedence over finding food, clothing, or shelter later in life. By virtue of their age and experience, older people may take greater pleasure in everyday activities and more time to reflect on their personal achievements. The review by Diener, Suh, Lucas, and Smith (1999) of 30 years of research on subjective well-being, drew the conclusion that life satisfaction of older people actually increases with age. In part, this may be because of adaptive strategies undertaken later in life in response to physiological constraints. Goals and preferences are adjusted to conform to the realities of the aging process. If this is indeed the case, then a regular monitoring of the older market's needs and wants is necessary.

The life satisfaction of older people partly depends on their consumption of appropriate and beneficial products and services to assist and simplify everyday activities. More important, opportunities to enhance self-esteem and self-fulfillment are needed. The incorporation of life satisfaction as a goal into existing marketing strategies will promote a more meaningful relationship with the aging consumer. Marketing strategies that incorporate the enhancement of life satisfaction as a criterion for their products and delivery of services (and their corresponding advertising messages) will be able to approach the senior market more effectively (Sherman & Cooper, 1988).

Life satisfaction, or how satisfied one is with one or many dimensions of life, is frequently used as a dependent variable. Quality of life is also a desired end state often measured objectively, through socioeconomic indicators, or subjectively, based on the individual's perceptions of the quality of various facets of life. These include experience with government, job, family, and leisure activities (Andrews & Withey, 1976). Campbell, Converse, and Rodgers (1976) use a different set of life domains that includes religion, marriage, and savings and investments. Recent criticism by Euler (1992) suggested that researchers have been measuring life satisfaction superficially. "Shallow" measures tap satisfaction with life circumstances, rather than deep, inner spiritual strength and fulfillment. Marketing researchers must assess the validity of scales and measures before borrowing from other fields of research.

Health has been found to be a significant predictor of life satisfaction and quality of life (Larson, 1978; Stolar, MacEntee, & Hill, 1992). If one is healthy, the opportunity to enjoy life is greater. The relationship between life satisfaction and health have been linked to the practice of marketing. According to Sirgy, Mentzer, Rahtz, and Meadow (1991), "the more the elderly feel satisfied, rather than dissatisfied, with health care services the greater the likelihood that they may evaluate their lives more positively" (p. 37). Therefore, satisfaction with the marketing exchange (in this case, a health care exchange) leads to greater overall satisfaction. (Note: Satisfaction with service delivery is measured in marketing practice by comparing consumers' expectations of various dimensions of service quality with their experience and by conducting a gap analysis. The resulting satisfaction "score" then depends on what was expected from the interaction in the first place and how the interaction was subsequently perceived. Also, the dimensions of service quality are often formalized in a scale, (for example, the popular SERVQUAL scale, and are neither arbitrary nor individually determined. See Zeithaml, Parasuraman, and Berry, 1990, for service quality theory and scale development.)

Marketing activities then strive to increase the life satisfaction of the elderly first and allow market forces to determine whether this goal has been accomplished. Increased market share, sales, and profit would indicate a job well done.

SEGMENTATION STRATEGIES

Sixty-five is the age most often cited as the onset of old age because of its association with retirement and receipt of Social Security benefits. However, many different reference points have been used to characterize old age, including 49+, 50, 55, 60, 62, and 65 (Tongren, 1988). According to the 2000 census, there are approximately 35 million people over the age of 65 in the United States. They have been called the maturity segment (Swartz & Stephens, 1984), the ageless market (Wolfe, 1987), the Gray (Tracy, 1985), and Gray America (Schewe, 1985). Segmentation strategies used by marketers begin with age, but there are many dimensions to the mature market (Lazer, 1986), determined in part by income, household characteristics, and behavioral patterns. The heterogeneous nature of a group as large as 35 million people is better addressed through responses to the marketers' offering, as opposed to a single objective measure (Moschis & Lee, 1997). Nonetheless, the availability of data often drives strategy development, and demographic data are relatively easy to collect.

Age

A common practice of marketers has been to underscore age when addressing an older audience. The maturing consumer market has been defined by chronological age even when it is not always appropriate (Schewe, 1988). It wasn't until the notion of "cognitive age" (or contextual age) became popular that marketers soon learned of their mistake. Older people often do not feel their chronological age. Instead, they tend to think of themselves as 10 to 15 years younger than they are. This would account for their rejection of advertising campaigns for products to relieve the signs, pains, or physical changes that accompany aging. Moschis (1991) agreed that the use of older models may be ineffective with cognitively "young" people who simply don't put themselves in the category of "old." Campanelli (1991) added that "the biggest trick in marketing to seniors . . . is in *not* mentioning their age" (p. 68).

Wolfe (1987) warned that "age-based strategies are inappropriate" (p. 28) and that marketers should pay more attention to what older consumers (aged 60 to 80) desire. Having satisfied their need for material possessions, older adults are interested in experiences rather than things. These "being experiences" allow them to achieve satisfaction more readily than by owning something. Examples of being experiences include taking a college course or an aerobics class, enjoying a evening walk or conversation with a friend, or getting involved in a church group.

The interest in being experiences is also related to comfort. Mature consumers are more interested in making their lives more comfortable, physically and psychologically. Thus, the older market has an extraordinary demand for home-based products and services (Campanelli, 1991). Home entertainment systems, security systems, landscaping services, and home health care products are popular with the older market. Schewe (1990, 1991) also cited comfort as a positioning tactic to be used by marketers.

Bone (1991) agreed that generalizing about chronological age has more drawbacks than advantages. After reviewing 33 separate studies on segmenting the older market, she identified five critical segmentation variables: discretionary income, health, activity level, discretionary time, and response to other people. Income levels range from the poor to the super affluent; health from poor (with 40% of the mature market in an unhealthy state because of chronic conditions) to excellent; activity level from limited to engaged in work, recreation, and social activities; discretionary time (as opposed to retirement status) from noninvolved to committed to activities; and response to other people ranging from sociable to seeking separation. The combinations of older adults possible with these variables represent an enormously diverse market.

Segmentation strategies are known for categorization schemes. Van der Merwe (1987) referred to people over 60 as GRAMPIES: Growing, Retired Active Monied People in an Excellent State. The GRAMPIES phenomenon is an international one, with 9% of the world population over 60 in 1990 and expected to climb to 12.6% in 2020. Marketers should not generalize the American aging experience to the rest of the world but recognize the significance of the cultural context.

Age and health have been combined with activity levels to classify older people. Pol, May, and Hartranft (1992) identified eight stages of aging for noninstitutionalized people aged 50 and above. (The eight stages were in 5-year increments until age 84. Thereafter, one category was formed for those 85 and older.) The first stages (50–54 and 55–59) showed relatively few differences in male/female ratio, health status, and work status. As the population reaches 60, a significant percent retire and the chances of being widowed nearly double. Throughout their 60s, most of those surveyed were in good to excellent health, and fewer (19.8%) were working. At age 70, 11% were living below the poverty line, only 10% were working, and increasing numbers of women were surviving and living alone. Half of those over 85 lived alone.[1] Two thirds were women, and 17% were impoverished. More than half had activity limitations (56.7%), and more than one fourth (28%) were living with relatives. The distinctions among older Americans raised by this study and others are important for marketers to realize lest they tend to group all persons over 50, 60, or 65 into one homogenous category.

Wealth and Employment Status

One of the attractions of the older market is the discretionary income they are purported to have (Kelly, 1992), and yet research indicates wide variation in income sources and levels. Longino and Crown (1991) remind marketers that although older Americans have lower incomes (40% of the income earned by 45- to 54-year-olds), they also have fewer expenses and smaller households. Many have completed mortgage payments and sent their children to college and thus have more discretionary income than the younger age groups. Today's mature consumers have a higher propensity to save than younger Americans, possibly because of their experience with the Great Depression (Lazer & Shaw, 1987), and are more creditworthy (Hoey & Jugenheimer, 1997). Marketers often hastily target the most affluent seniors, assuming this 20% of the older population accurately reflects the mature market. (See, for example Orr, 1994), who cites the 50-and-older market as one third of the U.S. population. Another 20% live below the poverty line and are likely to be female and non-White.

The older market needs to be broken down by household characteristics, as well as employment status to further delineate income potential (Waldrop, 1992). The striking difference between households with persons aged 65 and older and those younger is retirement status. Of married couple households aged 55 to 64, 63.9% had a household member in the workforce, versus 19.1% in the married couple household aged 65 to 74. These latter households depend more on Social Security income than the salary and wages earned by the preretired households. In fact, more than 90% of all households headed by someone over 65 receives Social Security benefits (Longino & Crown, 1991). (This is

---

[1]The average life expectancy has increased by 28 years in this century mainly due to lifestyle changes and public health efforts against infectious diseases. The large proportion of older widows is related to the life expectancy of women, which is greater for both White and Black females than it is for males of either race.

not expected in the future, as more baby-boomer women are working and contribute to retirement plans.)

Retirement has been suggested as a segmentation variable (Burnett, 1989), and yet marketers must have a more thorough understanding of retirement status than whether or not an individual is gainfully employed on a full-time basis. They must understand the role of retirement and the psychological adjustments made to cope with the role. A "successful" retirement is anticipated and planned for, resulting in greater life satisfaction at this stage in life.

Retirement status is a significant predictor of activity level and certain consumer behaviors. Burnett (1989) found that many newly retired males increased their participation in activities such as exercise, shopping, and entertaining. However, activity level was moderated by age, income level, and health status. Those retirees in poor health or with limited economic means were likely to drop activities such as returning an unsatisfactory product, joining a club, or dancing with a spouse. These moderating factors significantly affect the activity levels of retired males and must be accounted for when segmenting by retirement status.

The financial services industry is especially interested in mature affluent consumers, "since people over 55 own 77 percent of all financial assets" (Dychtwald, 1997, p. 274). Media messages and promotional campaigns abound to assist with retirement and estate planning, investment portfolio choices, and the challenges of a variable stock market. Whether the client is pre- or postretirement affects message, services, and investment strategy (Shaw, 1993).

Household Location

A cursory look at older adults in the United States would lead marketers to the states with the greatest number of retirees: California, Florida, and New York. When income is factored into the geodemographic landscape, other states stand out. In an earlier study, Longino (1988) estimated on a state-by-state basis the percentage of retirees who could be classified as "comfortably retired," with incomes double the poverty level, and the "pension elite," who have three sources of income: Social Security, assets, and pensions. The states with the highest concentrations of "comfortable" and "elite" retirees in their respective age groups are presented in Appendix A. A follow-up study by Longino (1994) focused on states with positive migration streams. These are listed in Appendix B.

Marketers also need to know the types of residences older people live in and the structure of their households. Contrary to popular belief, only 5% of those over 65 are institutionalized. The vast majority of older adults are "aging in place," that is, maintaining their own homes after retirement rather than moving. Their reasons for migrating are more often for a smaller, convenient home, a warmer climate, or a safer neighborhood. Residences that provide health care are projected to be a popular housing choice of the future (Riche, 1986). Other alternative housing arrangements for older people were explored by Rosenblum (1989). Prefabricated "echo" units (freestanding housing units on a family member's property), accessory apartments (additions to the family member's home), and home sharing (among independent older adults or on an intergenerational basis) are some ways for older people to remain close to family and

Appendix A
States With High Concentrations of Well-Off Older Adult

Comfortably Retired

Age Group	State	Percentage (of all retirees)
55–64	Connecticut	70.9%
	Hawaii	70.9%
	New Jersey	68.3%
65–74	Hawaii	64.0%
	Connecticut	62.7%
	New Jersey	61.6%
75+	Hawaii	57.5%
	Arizona	49.5%
	New Jersey	49.4%

Pension Elite

Age Group	State	Percentage (of all retirees)
55–64	New Hampshire	8.1%
	Florida	7.1%
	Wisconsin	6.8%
65–74	Connecticut	18.5%
	Washington	19.3%
	Pennsylvania	16.8%
75+	Connecticut	12.4%
	Vermont	12.1%
	Maryland	11.9%
	Oregon	11.9%
	Pennsylvania	11.9%

Note: Adapted from Longino (1988).

friends without the complete financial and emotional burden of living alone or in an uncomfortably large house.

Values and Lifestyles

As different cohorts join the ranks of older adults, they will present values and behaviors different from previous cohorts. The New Age elderly are a contemporary and distinct segment, according to Schiffman and Sherman (1991). Age, they suggested, has become more of a state of mind than a physical reality for many people. Those who are cognitively younger tend to experience greater life satisfaction and appreciate those products and services that allow them to do this. These older consumers are wiser by virtue of

Appendix B
Retiree Magnets
(States with the highest percentages of interstate
migrants aged 60 and over in 1990)

State	Percentage of Interstate Migrants Received in the Preceding 5-Year Period
1. Florida	23.9%
2. California	6.9
3. Arizona	5.2
4. Texas	4.1
5. North Carolina	3.4
6. Pennsylvania	3.0
7. New Jersey	2.6
8. Washington	2.5
9. Virginia	2.4
10. Georgia	2.3

Note: Adapted from Longino (1994).

experience and use their knowledge to their advantage. Twenty years ago, Schiffman (1972) recognized that older people who were less risk averse might be targeted for new product trial. Today, he finds that a less traditional and larger New Age group is similarly not fearful of product innovations. They are confident in their purchasing behavior, purchase new products selectively, and are more interested in inner growth and new experiences than physical possessions. This portrait of the New Age elderly is in sharp contrast to the stereotype often used of older people.

Likewise, Wolfe (1988) referred to this segment as the New Seniors, who are creative, wise, active, and concerned. They are interested in an experiential lifestyle and intellectual growth. The more experienced older consumers are misunderstood by most consumer researchers. Communicating with the New Seniors, Wolfe said, requires learning a new language.

Psychographic segmentation is another scheme used by marketers to identify homogenous groups of consumers with the same interests, attitudes, and lifestyles. A group of older people were subjected to a psychographic analysis by Sorce, Tyler, and Loomis (1989), which resulted in five lifestyle clusters: the Self-Reliants, Quiet Introverts, Family-Orienteds, Active Retirees, Young and Secures, and Solitaires. Strategic marketing suggestions were offered for each of these groups. Satisfaction would be increased if a product were designed and tailored to the lifestyle of the cluster. Others argued that not many differences are found between older and younger adults in terms of lifestyle (Moschis, 1991). Personality traits do not change over time, nor do they significantly affect the behavior of older people.

Moschis (1991) was critical of the segmentation strategies used by marketing researchers such as Sorce et al. (1989). First, he noted that "age is usually not a major factor in determining older consumers' responses to marketing activities" (p. 34). Second, the inclusion of other variables may still not accurately define different groups of older consumers. Moschis (1993) sought to address the limitations of previous research by developing "gerontographics" as a new segmentation model of older consumers. He used a combination of biological, social, and marketing variables to better describe the clusters within this heterogeneous market. Four new stable senior segments were found: Healthy Hermits, Ailing Outgoers, Frail Recluses, and Healthy Indulgers. Moschis continues to use this classification model to analyze the needs of the mature market.

Contradictions arise in fitting consumers into artificial categories. For example, in general, older people have more disposable income than younger people and are perceived as relatively well-off. However, they have higher medical care and insurance costs. They are considered to be thrifty and yet are in need of discounts. Fewer assumptions and more data on how older consumers allocate their income are needed. Marketers presuppose older adults' willingness to purchase a product or service and don't investigate actual spending. Segmentation strategies must be followed by analysis of spending patterns.

The segmentation process also must go beyond the two-dimensional model of "age × characteristic." Despite all the segmentation variables used by researchers, many studies are limited by the use of one model or one variable, such as activity level or attitude, to explain older consumer behavior. The older market is multidimensional. Reaching the older audience requires understanding older consumers' behaviors on a macro and micro level. On a social level, as one ages, different roles are assumed or discarded, affecting one's ability to enjoy life. On an interpersonal level, social interaction is affected by the ability to process and remember information. Marketers have examined older adults' social and cognitive changes and offer strategies for positive and attractive appeals.

Understanding Role Changes

Roles such as worker, spouse, and parent provide identification for most people. The loss of these roles, because of retirement or death, for example, is accompanied by stress and reduced life satisfaction (Elwell & Maltbie-Crannel, 1981). In addition, the roles often held by older people have negative stereotypes associated with them. The labels of "spinster," "widow," and "nursing home set" based on marital or household status conjure up weak or pathetic images. Unfortunately, the mass media has helped to perpetuate these negative characterizations (Hess, 1974).

Cox (1990) speculated that the roles for older adults in postindustrial societies will have a positive, freeing effect in the future. Older adults will have a greater choice of non–work-related roles available to them, and these roles will be more highly valued and respected by society at large. Schewe and Balazs (1990) commented specifically on the changing economic, social, and functional roles older people play. The stereotypes of retiree, grandparent, or widow(er) are no longer as restrictive as they once were. Marketers can help older people cope with role transitions and support these new roles by portraying older "actors" with an updated "wardrobe" and "props," in the forms of products and services.

Roles are defined also within a cultural context, and with the changing racial composition of the United States, interracial and international traditions affect how olders people behave. Longino and Earle (1996) reported on the increase in divorce and unwed pregnancies in the United States and the roles that grandparents and stepgrandparents play in these situations. Multigenerational households allow for new social and economic dynamics that defy stereotypes and impact purchasing patterns.

An early study by Phillips and Sternthal (1977) proposed that the role losses that occur later in life limit the older adults' exposure to other people and information. Older adults then rely on mass media more so than earlier in their lives. At the same time, the ability to learn and process information slows.

Understanding Cognitive Processing

Aging is certainly defined by physical change. The aging process also affects cognitive abilities, but not uniformly across all people. In general, older people experience a learning deficit, difficulty in entering information into memory or encoding, and an inability to process new information and information delivered at a semantic level (Roedder-John & Cole, 1986). A combination of factors (speech rate and complexity of message) have been found to have a significant effect on the processing time of older adults (Stine, Wingfield, & Poon, 1986). Marketers have borrowed psychological models of learning, recall, and memory to effectively address the older audience through the proper media. Spotts and Schewe (1989) used the Elaboration Likelihood Model to examine the various information-processing deficits in perception, learning, and memory that have been associated with the aging process. They suggested that physical factors (such as visual deficits), psychological factors (such as increased fear), and social factors (such as a change in values) affect the reception of messages. Messages must be structured to account for these deficiencies.

This information is helpful to marketers. The advertising medium better suited to overcome these cognitive deficiencies is television, according to Cole and Houston (1987). TV promotes the encoding of information by sensory-induced stimuli and avoids the ambiguity of semantics. TV would also allow for the high frequency of an ad, which increases recall for people of all ages (Stephens & Warrens, 1983–1984). However, the elderly are better able to process information if they control the pace of the communication, which would not be the case for television ads. Processing does not preclude susceptibility to misleading advertising. In one study (Gaeth & Heath, 1987), older adults were found to be more easily misled than younger adults by deceptive advertising, even when given the opportunity to review the ads and after being trained to recognize deceptive ads. Schewe (1989) suggested enhancing communication to older people by keeping the message simple, explicit, and familiar, using a rational appeal and choosing print media over broadcast media. Providing visual cues and aiding recall will help the older audience better process and remember the message. Also, an information ad with an emotional content will be more highly retained than an ad with an information format only (Goldberg, Gorn, Litvack, & Rosenblatt, 1985).

ELDERLY CONSUMER BEHAVIOR

Once the social, economic, and behavioral influences on the mature market are under-stood, this knowledge is used to develop appropriate and effective marketing strategies. These strategies must utilize previous marketing research involving older consumers and their interaction with marketing information, promotions, and the retail store environment.

Interaction With Media and Promotions

Research has shown that older adults are relatively greater consumers of news media than younger adults (Doolittle, 1979). Newspapers were the preferred news source for the young-old, television for the 67- to 74-year-olds, and the oldest seniors (75–93) were indifferent between the two. Burnett (1991) also found that affluent older adults were heavy users of newspapers and magazines. This is consistent with the research of Rahtz, Sirgy, and Meadow (1989), who found a negative relationship between income and television viewership and also education and television viewership. Further, a positive relation-ship exists between television orientation and low morale and television orientation and limited out-of-home activity. Perhaps print media are preferred because of their self-paced nature. Fewer TV commercials and more print advertising (at a relatively lower cost) would reach today's more affluent and active elderly. (See also Robinson, Skill, and Turner, chap. 17, this volume.)

The Portrayal of Older Adults in Advertising

Given the popularity of newsprint and magazines among an older audience, one would expect to find a frequent and positive portrayal of older adults in the media. Because content analyses first appeared in the academic literature, the research results were con-sistent. Older people were underrepresented in magazine advertising (Bramlett-Solomon & Wilson, 1989; Gantz, Gartenberg, & Rainbow, 1980; Peterson, 1992; Ursic, Ursic, & Ursic, 1986). This was especially true for older women. Older males were found to appear more frequently and in a work environment in advertising than were older females, who were more often shown in a social or family setting (Ursic, Ursic, & Ursic, 1986). Peterson (1992) also found the elderly to be depicted unfavorably, appearing as impaired, weak, or naive, whereas a positive portrayal of active, healthy older adults can be found in magazines targeted to older people (Kvasnicka, Beymer, & Perloff, 1982) and in newspapers (Buchholz & Bynum, 1982).

Underrepresentation also occurs in television advertising (Roy & Harwood, 1997), where older adults are given minor roles (Swayne & Greco, 1987). On a positive note, the role of advisor was frequently assigned to the older men and women in the ads versus a weaker representation. Older people were most often seen in transgenerational ads, that is, with multiple age groups present. And the setting was primarily in the home, with the most popular product class being food products. Roy and Harwood (1997) also found older people most often appearing with actors of all ages. They found fewer older models representing travel services and automobiles (which is contrary to a market-driven advertising campaign) and more present in commercials for financial services and retail chains. Although this is preferable to negative stereotyping, more positive and realistic advertising of older Americans, including more women as role models, is needed. It should be noted that this negligence of the older market is not limited to the United States Szmigin and Carrigan (2000a) and Carrigan and Szmigin (1999) found resistance to the use of older models in advertising and were pointed in their criticism of marketing practice in the United Kingdom: "Not only are most advertisers and marketers appearing disinterested in older consumers, but when they try to market products to them, they do so with tired, unimaginative efforts which serve only to alienate and offend the over-50s further" (p. 222).

There has been a significant use of older people in ads for liquor, automobiles, banks, and cameras (Ursic, Ursic, & Ursic, 1986). Greco's (1988) survey of advertising executives found the most suitable product classes for the use of older spokespeople were health and medicines, travel and vacations, and financial services. These would be product categories that older people would most likely purchase. They would identify with the spokesperson and judge him or her as more credible than younger models (Milliman & Erffmeyer, 1989–1990). Those product classes least suggested for use with older spokespeople were electronics and communications, sporting goods, and shampoo. The executives agreed that stereotyping of older adults does take place, but that they make effective spokespeople. Greco (1989) later acknowledged the effectiveness of older spokespersons for the older market, but qualified the older models' effectiveness for other target markets. Recent research and relaxed expectations may account for new directions in future advertising. Szmigin and Carrigan (2000a) suggested that today's New Age elderly (à la Schiffman & Sherman, 1991) are more comfortable with their successful aging and would be responsive to models of their own age in ads. Ads that depict older consumers in traditionally "young" settings such as theme parks or using technology are reflecting a growing reality of cross-generational patronage patterns (Paul, 2001).

Information-Seeking Behavior

Seniors like to gather as much information as possible before a purchase (Campanelli, 1991; Swartz & Stephens, 1984). Internal sources (such as past experience) and external sources (e.g., advertisements) of information are used by older people in their purchase decision making and affect new-product trial behavior (Schiffman, 1971). Klippel and Sweeney (1974) measured older participants' locus of control and risk preference and related it to uses of formal (e.g., print media) and informal (e.g., friends) information sources. Those with an external orientation (looking to others for validation) placed a higher value on informal sources of information.

Older people are not different from their younger counterparts in their choice of information sources (such as catalogs and point-of-purchase displays) for apparel purchase decisions. However, the uses of such information were different. For example, older people place more importance on guarantees and product and store information. This may stem from past purchase dissatisfaction. They also use salespeople, mass-media sources, and independent sources more so than the nonelderly (Lumpkin & Festervand, 1987–1988). Older consumers are used to the high customer-service levels once offered by department and specialty stores. They appreciate assistance from a salesclerk in selecting appropriate merchandise, providing fashion information, and offering personalized attention. (Unfortunately, retailers today rely less on salespeople and more on technology to cut costs and provide "service.") The newspaper, friends, and salespeople are also seen as more important sources of information to the fashion-conscious elderly (Greco, 1986). Older adults are actively seeking and using marketing information from various sources to make purchasing decisions, reduce risk, and avoid dissatisfaction.

Expert advice plays a role in other types of purchases as well. In a recent study of direct-to-consumer, or DTC, drug advertising, the majority of older respondents recalled having seen ads on TV and in magazines. However, when prompted for their sources of additional information, more than half consulted a pharmacist, and more than 40% communicated with friends or family members (Balazs, Yermolovich, & Zinkhan, 2000). The Internet was not a primary source of information, but a small percentage conducted further research online. Foley (2000–2001) reported that older consumers were less likely to comprehend the DTC ads, not recognizing they are for prescription drugs, and not likely to read the risks and summary information required in the ads. This creates a challenge for drug companies. According to the Food and Drug Administration, adults over the age of 65 buy 30% of all prescription drugs and 40% of all over-the-counter drugs (Williams, 1996). If these consumers are to take an active role in their health and treatment options, the information must be readily available to them in a format they can understand.

Response to Technology

Older people have been stereotyped as traditional, unable to learn, and resistant to new trends and technology. In the past decade, the reliance on technology of the economy as a whole and on individuals in business, education, entertainment, and leisure choices have led to remarkable changes in commerce, communication, and social interaction. In the early 1990s, more positive attitudes toward computer use were found among affluent, younger-old adults who recognized their potential in their lives (Festervand & Meinert, 1994). Overall, a tentative response was found for specific applications (such as home banking and shopping, communication, and entertainment) and the likelihood of using them. This may have been because of knowledge or financial gaps (the so-called Digital Divide) and lack of exposure to a innovative phenomenon. Oberndorf (1999) reported that mature adults (over age 51) account for 17% of the online audience and 9% of online sales. Szmigin and Carrigan (2000b) argued that indeed at the beginning of the new millenium, more older people were using technology deliberately to invest, shop, bank, learn, and communicate. Trocchia and Janda (2000) found the social rewards of e-mail and chat rooms were attractive to seniors, whereas limits on physical dexterity, the impersonal nature of the Web, and lack of understanding the technology were problematic. Social support and relationships with people of all ages are other motivations for Internet use (Trocchia & Janda, 2000; Wright, 2000). In a more technical study, Morrell et al. (2000) found that young-old adults were better able to learn and retain computer skills (specifically, an electronic bulletin board) and that age-related differences in cognitive processing affected computer performance as instructions and tasks became more complex. One motivation for using the Web turns out to be health-related. Older consumers are increasingly likely to resort to pharmaceutical Web sites for information (Balazs, Yermolovich, & Zinkhan, 2000). In her study of Swedish consumers, Maddox (1999) found that older participants rated such sites significantly higher than respondents under age 50 for obtaining a second opinion, choosing the correct drug, and gathering information to discuss with their physician.

This has great implications for the development of Internet sites, Web page design, and access routes to retail and noncommercial products and services. Many of the same recommendations for the printed page apply, such as increasing the size of type and pictorial images, ensuring a contrast of background and foreground colors, avoiding a complex or busy site with flashing graphics, and slowing the timing of sequentially delivered animation (Jakobi, 1999; Oberndorf, 1999). Considering the fact that younger adults are using technology regularly in the workplace and at home for shopping and entertainment, and children are familiar with a keyboard and mouse at the preschool levels, these differences will no doubt dissipate over time.

Patronage Patterns

Retail store shopping is the most frequently used mode of purchase by older adults (relative to mail order, door-to-door sales, or telephone order), but a great deal of anxiety and alienation is experienced (Barnes & Peters, 1982). Older consumers shop less frequently, oftentimes because of lack of convenient transportation and small household size. Quality merchandise, attractive prices, and store reputation are important to the older shopper (Lumpkin, Greenberg, & Goldstucker, 1985). Both dependent and self-reliant older adults are active store patrons, but the former group places greater importance on the physical aspects of the store (Lumpkin & Hunt, 1989). Older consumers place a greater importance on rest areas, convenient parking, and being close to a variety of other stores than do retailers (Balazs, 1994; Lumpkin & Hite, 1988).

Retailers are mistakenly placing more importance on fast checkout, credit, and having a wide variety of products. The older consumer goes to the marketplace in search of many things: product information, exercise, and social interaction (Lumpkin, 1985). Older people are "shopping" for fashion ideas, joining mall walking clubs, and meeting friends for lunch. However, they may have difficulty in getting to the shopping center; often do not find helpful personnel, rest areas, or rest rooms; and cannot find apparel that is fashionable and fits. Also, older consumers find current packaging difficult to open and the labeling difficult to read (Erickson, 1990). These problems lead older consumers to choose packaging that is easier to open and thus avoid brands they know are problematic. The use of color is important in designing labels. Use of flourescent colors, glossy paper, small print, and ornate typefaces are discouraged, whereas maximizing contrast, such as white on black, is recommended (Hunter, 1987; Silvenis, 1979). Elder-friendly packaging would not remind the users of their age or result in purchasing or usage mistakes. These are just some of the problems encountered by the older shopper. Manufacturers with older consumers must make products easier to recognize, open, and use. In fact, consumer goods manufacturers are learning that by designing products with the older consumer in mind, the younger market is not excluded ("Seniors Consult on Products for the Elderly," 1995; "Think Intuitively to Make Your Package Senior-Friendly," 2000). Increased market share and goodwill are likely results.

Many changes in the retail mix (controllable resources available to the retailer) have been suggested by researchers to meet the needs of the growing older market. However, few have been radical in scope, nor do they fully explore all the possibilities for an aging society. Many of the suggestions to assist older shoppers have been in the area of packaging design, signage, discounts, and customer service in general (Lambert, 1979; Schewe, 1985; Wilson, 1991). However, this market has not been served in other, more significant ways, such as specialty stores with attractive and comfortable apparel (Lumpkin & Greenberg, 1982), consumer education services such as product demonstrations and fashion shows, and opportunities to relax and enjoy the retail environment (with innovative food court design, lounges, or personalized services). Balazs (1994) found that elderly shoppers who identified themselves as recreational shoppers (who shopped more often and longer, and spent more) were more likely to respond positively to shopping mall specialty services such as a layaway plan, a post office branch, a gourmet food store, a variety store, a café/bakery, and hair styling services.

Complaint Behavior

One reason for limited action on the part of retailers may be that older consumers have been less likely to complain than the population as a whole (Bearden & Mason, 1979; Bernhardt, 1981). They are, however, as dissatisfied as younger consumers with certain services and attribute the problem to the vendor. Mature consumers often do not complain, because they feel their actions would be futile. La Forge used the theory of learned helplessness (LH) and the notion of uncontrollability to explain older-consumer complaint behavior. LH theory posits that people will become passive or give up trying if they routinely meet with failure. They consider the situation beyond their control, even if circumstances change. La Forge offered strategies to change feelings of uncontrollability to increase satisfaction. Other reasons for not taking action included the older adults unfamiliarity with in-store informational resources, Federal Trade Commission regulations, and store management policies. Increased consumer education on store return policies, product warranties, and exchange privileges would improve the relationship between vendor and older consumer.

Allen, Davis, Keesling, and Grazer (1992) characterized older adult segments as having different preferences for businesses handling their complaints. Older consumers evaluated 17 possible corporate responses to complaints. The consumers judged how likely they would be satisfied with the response and the importance of that response to them. Four distinct groups (Assentors, Credibles, Suspecters, and Agnostics) were identified through cluster analysis. Suspecters were more likely college educated and likely to complain. Half of the Assentors had incomes below $25,000 and were usually satisfied with the response to their complaints. The largest cluster, Credibles, would be satisfied with a written or verbal response from someone in an important position. Agnostics are skeptical and prone to complain and difficult to satisfy. The study, finding that a majority of older consumers do complain, disagreed with previous research. However, a prompt and courteous attempt to resolve a complaint by someone in a position of authority would be acceptable to most older consumers. More effort by retailers to solicit constructive criticism and feedback from older shoppers would result in more satisfied customers.

IMPLICATIONS FOR MARKETING RESEARCH AND PRACTICE

The interest in the older consumer has prompted a great deal of research by marketing academics and practitioners. Unfortunately, many attempts to communicate and serve this burgeoning market have failed (Szmigin & Carrigan, 2000a; Tongren, 1988). One reason for this is the assumption that the mature market is homogenous and that dramatic behavioral changes occur at a certain age (e.g., 65). Segmentation on a number of demographic and psychographic variables, such as attitudes toward unethical consumer practices (Vitell, Lumpkin, & Rawwas, 1991) or motivations for shopping (Lumpkin, 1985) would more accurately describe this heterogeneous group of mature consumers. Moschis (1991) founds fault with anecdotal research with no external validity, inferential conclusions that were not theoretically justified, and ambiguous empirical "evidence" used by researchers. Many of these problems arise in the research design and the methodology used by researchers.

Longitudinal cohort analysis has been suggested for research on the ever changing older consumer (Rentz, Reynolds, & Stout, 1983; Tongren, 1988). Cohort analysis measures the age, period, and cohort effects of the shifting age distribution. It is a more appropriate method than cross-sectional analysis when future consumption patterns are under study. Exter (1986) pointed out that chronological age, cohort membership, and the effects of history help shape the character of a generation. A single profile of older people is misleading. Marketing campaigns should be designed with the target audience and its particular experience in mind, rather than on an outdated profile of an age group. This would be one way to avoid extrapolating findings from today's older people to future older cohorts.

Special precautions should be taken when surveying an older sample (Beltramini, 1982). Health of the respondent and fear of strangers are two impediments to data collection. Approaching the older respondent demands patience, personalization, and respect from the interviewer. A personal interview is most effective with an older sample for clarifying responses, reducing distractions, and using probing techniques (Gruca & Schewe, 1992). Mail surveys are less confusing than telephone surveys, as they allow the respondent to pace their answers and process the information more carefully.

FUTURE DIRECTIONS IN MARKETING AND RESEARCH

The "graying of America" is a relatively new phenomenon, and businesses are only beginning to deal with one cohort of older adults. Future efforts at addressing and appealing to older consumers must account for a changing market. Baby boomers for example, are expected to have different values and be healthier and wealthier than their now, older parents. Also, the dynamic environment will prompt the need for different product and service offerings in the future. What has succeeded under present conditions may fail in a different social, political, and economic context. For example, the U.S. health care system will undergo revolutionary changes in the next few years. Firms offering health care services, insurance policies, and pharmaceuticals will undoubtedly face drastic reorganization. With new regulations and greater competition in this market, promotional campaigns will require a reeducation of consumers and a persuasive appeal based on real differentiation.

Fortunately, seniors are receiving increased attention from all types of researchers and businesses. Armed with this knowledge of senior consumer behavior, marketers will be able to sensitively address the needs of the mature market. The result should be more effective communication and delivery of appropriate products and services, which increase the life satisfaction of the older consumer.

REFERENCES

Allen, J., Davis, D., Keesling, G., & Grazer, W. (1992). Segmenting the mature market by characteristics of organizational response to complaint behavior. In R. P. Leone & V. Kumar (Eds.), *Enhancing knowledge development in marketing* (1992 AMA Educators' Proceedings, pp. 72–78). Chicago: American Marketing Association.

Andrews, F. M., & Withey, S. B. (1976). *Social indicators of well-being: America's perception of life quality.* New York: Plenum.

Balazs, A. L. (1994). The eldermall: Exploring new ways to position the aging retail shopping center for aging consumers. *Journal of Shopping Center Research, 1*(1), 39–59.

Balazs, A. L., Yermolovich, J., & Zinkhan, G. M. (2000). Direct to consumer advertising: An ad processing perspective. In Francessa Van Gorp Cooley (Ed.), *Marketing in a global economy: Proceedings of the international marketing educators' conference* (pp. 478–486), Buenos Aires, Argentina: American Marketing Association.

Barnes, N. G., & Peters, M. P. (1982, Fall). Modes of retail distribution: Views of the elderly. *Akron Business and Economic Review, 13*(3), 26–31.

Bartos, R. (1980, January–February). Over 49: The invisible consumer market. *Harvard Business Review, 58,* 140–148.

Bearden, W. O., Gustafson, A. W., & Mason, J. B. (1979). A path-analytic investigation of life satisfaction among elderly consumers. In W. L. Wilkie (Ed.), *Advances in consumer research* (Vol. 6, pp. 386–391). Ann Arbor, MI: Association for Consumer Research.

Bearden, W. O., & Mason, J. B. (1979). Elderly use of in-store information sources and dimensions of product satisfaction/dissatisfaction. *Journal of Retailing, 55*(1), 79–91.

Beltramini, R. F. (1982). Use special scaling techniques when researching senior market. *Marketing News, 15*(23), 9.

Bernhardt, K. L. (1981). Consumer problems and complaint actions of older Americans: A national view. *Journal of Retailing, 57*(3), 107–125.

Block, J. E. (1974). The aged consumer and the marketplace: A critical review. *Marquette Business Review, 18*(2), 73–81.

Bone, P. F. (1991). Identifying mature segments. *Journal of Consumer Marketing, 8*(4), 19–32.

Bramlett-Solomon, S., & Wilson, V. (1989, Spring). Images of the elderly in *Life* and *Ebony*, 1978–1987. *Journalism Quarterly,* 185–188.

Buchholz, M., & Bynum, J. E. (1982). Newspaper presentation of America's aged: A content analysis of image and role. *The Gerontologist, 22*(1), 83–88.

Burnett, J. J. (1989). Retirement versus age: Assessing the efficacy of retirement as a segmentation variable. *Journal of the Academy of Marketing Science, 17,* 333–343.

Burnett, J. J. (1991). Examining the media habits of the affluent elderly. *Journal of Advertising Research, 31*(5), 33–41.

Campanelli, M. (1991). The senior market: Rewriting the demographics and definitions. *Sales and Marketing Management, 143*(2), 63–69.

Campbell, A., Converse, P. E., & Rodgers, W. L. (1976). *The quality of american life*. New York: Russell Sage Foundation.

Carrigan, M., & Szmigin, I. (1999). In pursuit of youth: What's wrong with the older market? *Marketing Intelligence & Planning, 17*(5), 222–230.

Cole, C. A., & Houston, M. J. (1987, February). Encoding and media effects on consumer learning deficiencies in the elderly. *Journal of Marketing Research, 24*(1), 55–63.

Cooper, P. D., & Miaoulis, G. (1988). Altering corporate strategic criteria to reflect the changing environment: The role of life satisfaction and the growing senior market. *California Management Review, 31*(1), 87–97.

Cox, H. G. (1990). Roles for aged individuals in post-industrial societies. *International Journal of Aging and Human Development, 30*(1), 55–62.

Diener, E., Suh, E. M., Lucas, R. E., & Smith, H. L. (1999). Subjective well-being: Three decades of progress. *Psychological Bulletin, 125*, 276–302.

Doolittle, J. C. (1979). News media use by older adults. *Journalism Quarterly, 56*, 311–317, 345.

Dychtwald, M. K. (1997). Marketplace 2000: Riding the wave of population change. *Journal of Consumer Marketing, 14*, 271–275.

Elwell, F., & Maltbie-Crannell, A. D. (1981). The impact of role loss upon coping resources and life satisfaction of the elderly. *Journal of Gerontology, 36*, 223–232.

Erickson, G. (1990). Packaging for older consumers. *Packaging, 35*(13), 24–28.

Euler, B. L. (1992). A flaw in gerontological assessment: The weak relationship of elderly superficial life satisfaction to deep psychological well-being. *International Journal of Aging and Human Development, 34*, 299–310.

Exter, T. G. (1986, September). How to think about age. *American Demographics, 8*(9), 50–51.

Festervand, T. A., & Meinert, D. B. (1994). Older adults' attitudes toward and adoption of personal computers and computer-based lifestyle assistance. *Journal of Applied Business Research, 10*(2), 13–22.

Foley, L. (2000–2001). The medication information gap: Older consumers in the void between direct-to-consumer advertising and professional care. *Generations, 24*(4) 49–54.

Gaeth, G. J., & Heath, T. B. (1987, June). The cognitive processing of misleading advertising in young and old adults: Assessment and training. *Journal of Consumer Research, 14*, 43–54.

Gantz, W., Gartenberg, H. M., & Rainbow, C. K. (1980). Approaching invisibility: The portrayal of the elderly in magazine advertisements. *Journal of Communication, 30*(1), 56–60.

Gelb, B. D. (1978). Exploring the gray market segment. *Michigan State University Business Topics, 26*(2), 41–46.

Goldberg, M. E., Gorn, G. J., Litvack, D. S., & Rosenblatt, J. A. (1985). TV and the elderly: An experiment assessing the effects of informational and emotional commercials. In M. J. Houston & R. Lutz (Eds.), *Marketing communications—Theory and research* (pp. 215–219). Chicago: American Marketing Association.

Greco, A. J. (1986). The fashion-conscious elderly: A viable, but neglected market segment. *Journal of Consumer Marketing, 3*(4), 71–75.

Greco, A. J. (1988, June–July). The elderly as communicators: Perceptions of advertising practitioners. *Journal of Advertising Research, 28*, 39–46.

Greco, A. J. (1989). Representation of the elderly in advertising: Crisis or inconsequence? *Journal of Consumer Marketing, 6*(1), 37–44.

Gruca, T. S., & Schewe, C. D. (1992, September). Researching older consumers. *Marketing Research, 4*(3),18–24.

Hauser, W. J., & Scarisbrick-Hauser, A. (1995). From the fonts of knowledge: The partnership between older consumers and business. *Generations, 19*(1), 26–29.

Hess, B. B. (1974). Stereotypes of the aged. *Journal of Communication, 24*(4), 76–85.

Hoey, J., & Jugenheimer, D. (1997). Marketing to seniors: A meaningful and profitable segment. *Credit World, 85*(3), 21–23.

Hunter, B. T. (1987). Making food labels readable. *Consumers' Research Magazine, 70*(12), 8.

Jakobi, P. (1999). Using the World Wide Web as a teaching tool: Analyzing images of aging and the visual needs of an aging society. *Educational Gerontology, 25*, 581–592.

Kelly, M. E. (1992, May 18). Discounters grow wiser to seniors' spending potential. *Discount Store News*, 113–114.

Klippel, R. E., & Sweeney, T. W. (1974, April). The use of information sources by the aged consumer. *The Gerontologist, 14*(2), 163–166.

Kvasnicka, B., Beymer, B., & Perloff, R. M. (1982). Portrayal of the elderly in magazine advertisements. *Journalism Quarterly, 59,* 656–658.

La Forge, M. C. (1989). Learned helplessness as an explanation of elderly consumer complaint behavior. *Journal of Business Ethics, 8*(5), 359–366.

Lambert, Z. V. (1979). An investigation of older consumers' unmet needs and wants at the retail level. *Journal of Retailing, 55*(4), 35–57.

Larson, R. (1978). Thirty years of research on the subjective well-being of older americans. *Journal of Gerontology, 33*(1), 109–125.

Lazer, W. (1986). Dimensions of the mature market. *Journal of Consumer Marketing, 3*(3), 23–34.

Lazer, W., & Shaw, E. H. (1987, September). How older americans spend their money. *American Demographics, 9*(9), 36–41.

Longino, C. F., Jr. (1988, June). The comfortably retired and the pension elite. *American Demographics, 10*(6), 22–24.

Longino, C. F., Jr. (1994). From sunbelt to sunspots. *American Demographics, 16*(11), 22–28.

Longino, C. F., Jr., & Crown, W. H. (1991, August). Older americans: Rich or poor? *American Demographics, 12*(8), 48–52.

Longino, C. F., Jr., & Earle, J. R. (1996). Who are the grandparents at century's end? *Generations, 20*(1), 13–17.

Lumpkin, J. R. (1985). Shopping orientation segmentation of the elderly consumer. *Journal of the Academy of Marketing Science, 13,* 271–289.

Lumpkin, J. R., & Festervand, T. A. (1987–1988). Purchase information sources of the elderly. *Journal of Advertising Research, 27*(6), 31–44.

Lumpkin, J. R., & Greenberg, B. A. (1982). Apparel-shopping patterns of the elderly consumer. *Journal of Retailing, 58*(4), 68–89.

Lumpkin, J. R., Greenberg, B. A., & Goldstucker, J. L. (1985). Marketplace needs of the elderly: Determinant attributes and store choice. *Journal of Retailing, 61*(2), 75–105.

Lumpkin, J. R., & Hite, R. E. (1988). Retailers' offerings and elderly consumers' needs: Do retailers understand the elderly? *Journal of Business Research, 16,* 313–326.

Lumpkin, J. R., & Hunt, J.B. (1989). Mobility as an influence on retail patronage behavior of the elderly: Testing conventional wisdom. *Journal of the Academy of Marketing Science, 17*(1), 1–12.

Maddox, Lynda M. (1999). The use of pharmaceutical Web sites for prescription drug information and product requests." *Journal of Product & Brand Management,* Vol. 8, No. 6, 488–496.

Miller, N. J., & Kim, S. (1999). The importance of older consumers to small business survival: Evidence from rural Iowa. *Journal of Small Business Management, 37*(4), 1–14.

Milliman, R. E., & Erffmeyer, R. C. (1989–1990). Improving advertising aimed at seniors. *Journal of Advertising Research, 29*(6), 31–36.

More People Hit Old Age in Developing Nations. (1993, June 15). *The Wall Street Journal,* p. B1.

Morrell, R. W., Denise C. P., Christopher B. M., & Catherine L. K. (2000). "Effects of age and instructions on teaching older adults to use eldercomm, an electronic bulletin board system," Educational Gerontology, Vol. 26, Issue 3, (April/May) 221–235.

Moschis, G. P. (1991, Fall). Marketing to older adults: An overview and assessment of present knowledge and practice. *Journal of Consumer Marketing, 8*(4), 33–41.

Moschis, G. P. (1993). Gerontographics. *Journal of Consumer Marketing, 10*(3), 43–53.

Moschis, G. P., & Lee, E. (1997). Targeting the mature market: Opportunities and challenges. *Journal of Consumer Marketing, 14,* 282–293.

Oberndorf, S. (1999). Targeting those senior surfers. *Catalog Age, 16*(11), 53.

Orr, A. (1994). The $1 trillion senior market. *Target Marketing, 17*(8), 6–8, 10.

Palmore, E., & Kivett, V. (1977). Change in life satisfaction: A longitudinal study of persons aged 46–70. *Journal of Gerontology, 32,* 311–316.

Paul, N. C. (2001). Selling to seniors. *The Christian Science Monitor, 93*(78), 11.

Pearlman, R. A., & Uhlmann, R. F. (1991). Quality of life in elderly, chronically ill outpatients. *Journal of Gerontology: Medical Sciences, 46*(2), M31–M38.

Peterson, R. T. (1992). The depiction of senior citizens in magazine advertisements: A content analysis. *Journal of Business Ethics, 11,* 701–706.

Phillips, L. W., & Sternthal, B. (1977, November). Age differences in information processing: A perspective on the aged consumer. *Journal of Marketing Research, 14,* 444–457.

Pol, L. G., May, M. G., & Hartranft, F. R. (1992, August). Eight stages of Aging. *American Demographics, 14*(8), 54–57.

Rahtz, D. R., Sirgy, M. J., & Meadow, H. L. (1989). The elderly audience: Correlates of television orientation. *Journal of Advertising, 18*(3), 9–20.

Rentz, J. O., Reynolds, F. D., & Stout, R. G. (1983, February). Analyzing changing consumption patterns with cohort analysis. *Journal of Marketing Research, 20*(1), 12–20.

Riche, M. F. (1986, January). Retirement's lifestyle pioneers. *American Demographics, 8*(1), 42–44, 50–56.

Roedder-John, D., & Cole, C. A. (1986, December). Age differences in information processing: Understanding deficits in young and elderly consumers. *Journal of Consumer Research, 13,* 297–315.

Rosenblum, G. (1989, May). New ways to live together. *New Choices,* pp. 29–33.

Roy, A., & Harwood, J. (1997). Underrepresented, positively portrayed: Older adults in television commercials. *Journal of Applied Communication Research, 25,* 39–56.

Schewe, C. D. (1985). Gray America goes to market. *Business, 35*(2), 3–9.

Schewe, C. D. (1988). Marketing to our aging population: Responding to physiological changes. *Journal of Consumer Marketing, 5*(3), 61–73.

Schewe, C. D. (1989). Effective communication with our aging population. *Business Horizons, 32*(1), 19–25.

Schewe, C. D. (1990). Get in position for the older consumer. *American Demographics, 12*(6), 38–43.

Schewe, C. D. (1991). Strategically positioning your way into the aging marketplace. *Business Horizons, 34*(3), 59–66.

Schewe, C. D., & Balazs, A. L. (1990). Playing the part. *American Demographics, 12*(4), 24–30.

Schiffman, L. G. (1971, October). Sources of information for the elderly. *Journal of Advertising Research, 11*(5), 33–37.

Schiffman, L. G. (1972). Perceived risk in new product trial by elderly consumers. *Journal of Marketing Research, 9*(1), 106–108.

Schiffman, L. G., & Sherman, E. (1991). Value orientations of New-Age elderly: The coming of an ageless market. *Journal of Business Research, 22*(2), 187–194.

Seniors consult on products for the elderly. (1995). *Futurist, 29*(6), 42–43.

Seniors' spending: A big, new market. (1980, September 1). *U.S. News and World Report, 91*(10), pp. 51–52.

Shaw, R. (1993). Graying and green. *Mediaweek, 3*(29), 20–22, 24.

Sherman, E., & Cooper, P. (1988, March). Life satisfaction: The missing focus of marketing to seniors. *Journal of Health Care Marketing, 8*(1), 69–71.

Silvenis, S. (1979, October). Packaging for the elderly. *Modern Packaging, 52,* 38–39.

Sirgy, M. J., Mentzer, J. T., Rahtz, D. R., & Meadow, H. L. (1991). Satisfaction with health care services consumption and life satisfaction among the elderly. *Journal of Macromarketing, 2,* 24–39.

Sorce, P., Tyler, P. R., & Loomis, L. M. (1989). Lifestyles of older Americans. *Journal of Consumer Marketing, 6*(3), 53–63.

Spotts, H. E., & Schewe, C. D. (1989, September). Communicating with the elderly consumer: The growing health care challenge. *Journal of Health Care Marketing, 9*(3), 36–44.

Stephens, N., & Warrens, R. A. (1983–1984). Advertising frequency requirements for older adults. *Journal of Advertising Research, 23*(6), 23–32.

Stine, E. L., Wingfield, A., & Poon, L. W. (1986). How much and how fast: Rapid processing of spoken language in later adulthood. *Journal of Psychology and Aging, 1,* 303–311.

Stolar, G. E., MacEntee, M. I., & Hill, P. (1992). Seniors' assessment of their health and life satisfaction: The case for contextual evaluation. *International Journal of Aging and Human Development, 35,* 305–317.

Swartz, T. A., & Stephens, N. (1984). Information search for services: The maturity segment. In T. Kinnear (Ed.), *Advances in Consumer Research* (Vol. 11, pp. 244–249). Provo, UT: Association for Consumer Research.

Swayne, L. E., & Greco, A. J. (1987). The portrayal of older Americans in television commercials. *Journal of Advertising, 16*(1), 47–54.

Szmigin, I., & Carrigan, M. (2000a). Does advertising in the UK need older models? *Journal of Product & Brand Management, 9*(2) 128–143.

Szmigin, I., & Carrigan, M. (2000b). The older consumer as innovator: Does cognitive age hold the key? *Journal of Marketing Management, 16,* 505–527.

Think intuitively to make your package senior-friendly. (2000). *Brand Packaging, 4*(6), 48–51.

Tongren, H. N. (1988). Determinant behavior characteristics of older consumers. *Journal of Consumer Affairs, 22*(1), 136–157.

Tracy, E. J. (1985, October 14). The gold in the gray. *Fortune, 112*(8), 137–138.

Trocchia, P. J., & Janda. S. (2000). A phenomenological investigation of Internet usage among older individuals. *Journal of Consumer Marketing, 17,* 605–616.

Ursic, A. C., Ursic, M. L., & Ursic, V. L. (1986, June). A longitudinal study of the use of the elderly in magazine advertising. *Journal of Consumer Research, 13*(1), 131–133.

U.S. News and World Report (1980). "Seniors' spending: A Big, New Market," Vol. 91, No. 10, (Sept. 1), 51–52.

Van der Merwe, S. (1987). GRAMPIES: A new breed of consumers comes of age. *Business Horizons, 30*(6), 14–19.

Visvabharathy, G., & Rink, D. R. (1984). The elderly: Neglected business opportunities. *Journal of Consumer Marketing, 1*(4), 35–46.

Vitell, S. J., Lumpkin, J. R., & Rawwas, M. Y. A. (1991). Consumer ethics: An investigation of the ethical beliefs of elderly consumers. *Journal of Business Ethics, 10,* 365–375.

Waldrop, J. (1992). Old Money. *American Demographics, 14*(4), 24–32.

Williams, R. D. (1997, September–October). Medications and older adults. *FDA Consumer, 31*(6), 15–18.

Wilson, M. (1991, July). Sixty something. *Chain Store Age Executive,* pp. 30–33.

Wolfe, D. B. (1987). The ageless market. *American Demographics, 9*(7), 26–29, 55–56.

Wolfe, D. B. (1988, March). Learning to speak the language of the new senior. *Marketing Communications, 13*(3), 47–52.

Wright, K. (2000). Computer-mediated social support, older adults, and coping. *Journal of Communication, 50*(3), 100–118.

Zeithaml, V. A., Parasuraman, A., & Berry, L. L. (1990). *Delivering Service Quality.* New York: Free Press.

14

Retirement and Leisure

Miriam Bernard and Chris Phillipson
Keele University

INTRODUCTION

The period of old age is now characterized by two distinctive features: first, the growth in the proportion of the population aged 60 to 65 and over; second, the withdrawal of older people (especially men) from the labor market. In the case of Britain, in the late 19th century, nearly three fourths of men aged 65 and over were still in employment; by the early 1950s, this had fallen to one third; and by the end of the 1990s, just 3% of men were employed full-time (the figure for women was 1%).

These trends may be found in virtually all advanced industrialized societies (Kohli, Rein, Guillemard, & Gunsteren, 1991; Phillipson, 1998). For example, the ratio of employment to population for older men fell in virtually all Organisation of Economic and Cooperative Development (OECD) countries during the 1980s and 1990s, albeit with some upturn in economic activity for men in their 50s at the end of the decade (OECD, 2001). The situation of older women is more varied, with a mixture of experiences from country to country. However, it is the decrease in male employment that has been crucial over the past two decades, and that has led to the marked decline in labor force participation among older workers.

The purpose of this chapter is to move beyond simply describing the changes underpinning these developments. Rather, the focus will be on increasing our understanding of retirement as a social institution. This will involve, first, a brief overview of its growth in the 20th century; second, an assessment of recent social trends, especially in relation to the development of early retirement; third, a review of the experience of retirement; fourth, a consideration of the possibilities raised by retirement for the social and leisure activities maintained by individuals.

The emphasis of this chapter will be on viewing retirement as a socially constructed event, shaped by a number of competing influences within society. A significant role may be attributed to economic factors, these influencing both the growth of retirement, as well as perceptions about its desirability and value (Macnicol, 1998; Phillipson, 1982). Of added importance, however, is the value placed by individuals on what has come to be known as the third age, with ideas of activity and development replacing the traditional labels of decline and passivity associated with leaving work (Gilleard & Higgs, 2000; Laslett, 1989). An assessment of the variety of meanings associated with retirement will form a significant part of this chapter.

The Development of Retirement

Modern retirement policy is a product of the late 19th century, as large private companies and branches of the civil service adopted pension policies of different kinds. Subsequently, at key periods in the 20th century (often in periods of economic slump or through the impetus of war), pension coverage has extended across key groups within the working population (Hannah, 1986). In consequence, modern states have become responsible not only for the income maintenance of substantial sections of the older population, but also for the organization of the rules governing access to the various pathways into retirement (Kohli et al., 1991). In broader terms, these responsibilities reflect the linkages between retirement and modernity, with all stages of the life course being shaped by a more specialist division of labor combined with more extensive forms of state intervention (Blaikie & Macnicol, 1986; Macnicol, 1998).

During the 20th century, the spread of retirement and occupational pensions has been further increased by demands for greater efficiency and productivity in the workplace (Graebner, 1980; Phillipson, 1982). Fischer (1977) suggested that the growth of the factory system accelerated the process of retirement, with the development of assembly-line production hastening the displacement of older workers. However, as Quadagno (1982) and others noted, the issue is not simply one of technological change creating the right climate for the emergence of retirement.

First, older people probably entered the new industries at a slower rate than younger workers. The reverse side of this is that they tended to be clustered in industries subject to economic decline.

Second, retirement has to be seen in more global terms. For industrial capitalism, according to Graebner (1980), retirement provided a means of challenging security of tenure, or jobs for life. It was a reaction against the persistence of personal modes of behavior and provided a mechanism for discharging loyal workers. As an added bonus, it also assisted the stabilization of corporate hierarchies, creating a permanent flow of employees and guaranteeing promotion through the ranks (Hannah, 1986).

Third, as Graebner (1980) and Phillipson (1982) suggested, retirement has played an important role in periods of mass unemployment. The idea of older workers being surplus to labor requirements has been crucial to the development of pensions both in Britain (the 1925 pensions legislation) and America (the Social Security Act of 1935). Similarly, the growth of unemployment in the 1970s and 1980s stimulated early retirement policies in a number of countries (Laczko & Phillipson, 1991).

Early Exit From Work. An important trend in the United Kingdom and most European countries has been the declining age of exit from the labor force (Kohli, Rein, Guillemard, & Gunsteren, 1991). In the United Kingdom, among men aged 60 to 64, labor force participation declined from 82.9% in 1971 to 54.1% in 1991. By 1999, the rate had declined to below 50%. The proportion of men aged 50 to 64 neither in work nor looking for employment increased from 11% in 1976 to 27% by the end of the 1990s. In general, male employment rates now begin to decline from an earlier age: as early as age 50 according to one estimate (Campbell, 1999).

For women, employment patterns are more complex. In the period from the mid-1970s to the mid-1990s, there was a decrease in participation rates among nonmarried women in the 55 to 59 age group; for married women in this age group, however, the pattern appeared more stable, with rates fluctuating at around 53%. Overall, the evidence indicates that women as well as men are leaving work at younger ages across successive cohorts, even though cohort-specific employment rates are increasing.

The move away from paid work was accelerated by periods of high unemployment in the 1970s and 1980s. During the 1990s, with the move out of recession, the shift toward earlier retirement went into reverse, with modest increases in economic activity for men and women in their late 50s. This pattern is confirmed in comparative data from the OECD that looks at trends across a mix of OECD member countries (Tables 14.1 and 14.2). By 1999, 70% of men aged 55 to 59 in the United Kingdom were employed, representing a drop of six percentage points since 1983, but an increase of four points over the period since 1993. The trend for women aged 55 to 59 showed a gradual increase over the 16-year period, but with a sharper upturn in the period at the end of the 1990s (OECD, 2001).

In most countries, these changes illustrate modest upturns in economic activity among older workers. Together with cyclical recovery within the economy, some determining factors include

- Restriction of policies leading to early exit from work (e.g., initiatives to encourage those receiving disability benefits back into the workforce; de Vroom & Guillemard, 2002)
- Development of programs encouraging training and returning to work (e.g., in the United Kingdom, New Deal 50plus)
- Encouragement of gradual pathways to retirement (e.g., expansion of part-time work)
- Problems associated with the funding of occupational pensions schemes (e.g., the decline in final salary schemes).

Although these measures have slowed, and in some cases reversed, trends toward earlier retirement, the evidence suggests that many people in their 50s and early 60s will be reluctant to return to work (Phillipson, 2002). In the first place, the majority of those who leave the labor market become "economically inactive" rather than registered as "unemployed" and "looking for work." Just over one in two men indicate being long-term sick or disabled as the main reason for inactivity; for women, the reasons are split between long-term sick and caring at home (Barham, 2002). The majority of both men

TABLE 14.1
Employment/Population Ratios for Older Men
(Percentages, 1983–1999)

	Age	1983	1984	1985	1986	1987	1988	1989	1990	1991	1992	1993	1994	1995	1996	1997	1998	1999	*Variation (in percentage points)* 1992/1983	1999/1993
Canada	55–59	74	74	74	73	73	73	73	72	69	67	66	65	66	66	66	66	67	−7	2
	60–64	55	54	51	51	48	48	48	48	44	43	42	42	40	40	42	41	44	−12	1
Finland	55–59	61	60	59	60	57	56	58	62	58	52	49	48	46	50	50	51	54	−9	5
	60–64	39	36	36	34	31	29	29	30	29	25	22	21	22	23	23	23	23	−14	1
Germany	55–59	74	72	71	72	71	70	69	70	71	67	64	64	64	64	64	65	65	−7	1
	60–64	37	33	32	32	32	32	32	32	28	28	27	26	26	26	27	27	28	−9	1
Italy	55–59	71	69	68	67	68	66	64	66	66	61	62	59	55	55	52	51	52	−10	−10
	60–64	36	37	37	37	36	36	34	34	36	31	31	30	29	28	30	30	29	−5	−2
Japan	55–59	87	87	87	87	87	89	89	90	92	92	92	92	91	92	92	91	91	4	−2
	60–64	70	69	67	67	66	66	67	69	71	71	71	70	69	68	68	67	66	1	−4
Netherlands	55–59	65	—	64	—	62	63	64	64	61	61	60	59	58	59	62	65	67	−4	7
	60–64	34	—	29	—	28	25	22	22	22	21	21	21	20	20	21	24	27	−13	6
Sweden[a]	55–59	—	—	—	—	84	84	86	86	84	82	78	76	76	77	76	78	81	−2	3
	60–64	—	—	—	—	63	63	62	62	63	58	53	51	51	53	50	50	49	−4	−4
United Kingdom	55–59	76	74	75	73	71	73	73	75	74	69	66	67	66	68	69	69	70	−6	4
	60–64	52	51	50	49	49	49	50	49	49	47	45	45	45	45	48	46	47	−5	2
United States	55–59	76	76	76	75	77	77	77	77	75	75	74	74	75	75	76	76	76	−1	2
	60–64	54	53	53	53	53	52	53	54	52	51	51	50	51	52	53	54	53	−2	2

Source: OECD (2001).

[a] 2000 instead of 1999. Variations are calculated for years 1992/1987 and 2000/1993.

The table shows trends in the employment/population ratio for men in two older age groups. Taking the United Kingdom as an example, it shows that about 76% of men 55 to 59 were employed in 1983. By 1999, this had dropped to 70%. However, the period since 1993 has seen an increase of 3%.

TABLE 14.2

Employment/Population Ratios for Older Women
(Percentages, 1983–1999)

	Age	1983	1984	1985	1986	1987	1988	1989	1990	1991	1992	1993	1994	1995	1996	1997	1998	1999	Variation (in percentage points) 1992/1983	1999/1993
Canada	55–59	36	36	39	39	40	41	42	43	42	44	43	44	44	45	44	47	48	7	5
	60–64	23	23	22	22	23	23	21	23	22	22	22	23	22	21	23	24	25	−2	2
Finland	55–59	57	56	55	52	52	53	56	59	57	54	52	50	50	49	49	50	56	−3	5
	60–64	30	30	30	25	22	22	23	21	21	19	16	13	16	17	16	16	20	−11	4
Germany	55–59	37	35	34	34	34	34	34	35	37	36	37	37	40	42	43	44	44	0	8
	60–64	11	10	10	10	10	9	10	10	9	9	9	9	10	11	11	11	11	−2	3
Italy	55–59	19	19	19	20	19	20	19	19	21	17	19	19	19	20	21	22	22	−2	3
	60–64	9	11	10	9	10	10	9	10	10	8	8	8	7	8	8	7	8	−1	−1
Japan	55–59	50	50	50	49	50	50	51	53	55	55	56	55	56	57	57	57	57	5	1
	60–64	39	37	38	38	38	38	39	39	40	40	39	39	39	38	39	39	38	1	−1
Netherlands	55–59	17	—	18	—	21	22	21	23	22	25	26	27	28	29	32	30	32	8	6
	60–64	8	—	7	—	8	7	8	8	7	6	7	7	8	9	7	9	10	−1	3
Sweden[a]	55–59	—	—	—	—	77	78	77	78	78	77	75	75	73	74	75	75	76	0	1
	60–64	—	—	—	—	48	49	49	52	53	50	47	43	45	46	44	43	43	2	−4
United Kingdom	55–59	47	48	49	49	49	49	51	52	51	52	51	52	53	52	50	53	54	5	3
	60–64	20	20	18	18	18	19	22	22	23	23	24	25	25	26	24	23	24	3	1
United States	55–59	46	48	48	49	50	52	53	54	54	54	55	57	57	58	59	60	60	8	5
	60–64	32	32	32	32	32	33	35	35	34	35	36	36	37	37	39	38	38	3	2

Source: OECD (2001).

[a] 2000 instead of 1999. Variations are calculated for years 1992/1987 and 2000/1993.

The table shows trends in the employment/population ratio for women in two older age groups. Taking the United Kingdom as an example, it shows that about 47% of women 55 to 59 were working in 1983. By 1999, this had risen to 54%.

and women (around 70%) identify themselves as "not wanting a job." Barham's (2002) analysis of the economically inactive in the British Labour Force Survey suggested, at least in the case of older men, two different groups—neither of whom seem strong candidates for returning to the labor force: "One group appears to consist of voluntarily retired professional workers, who may well have occupational pension schemes enabling them to have an income before state pension age. A second group includes skilled or semi-skilled workers who have been made redundant and are now unable to work due to long-term sickness" (p. 307).

Another important trend, however, is the gradual acceptance of retiring ahead of state pension ages. One element here is the extent to which "early exit has . . . become part of the workers' 'normal biography' [and that] in many cases, expectations of early exit determine the life plan and strategic choices of employees . . . " (European Foundation for the Improvement of Living and Working Conditions, 1999, p. 10). McKay and Middleton (1998) found that successive age groups expect to retire at younger ages than their elders; working people aged 30 to 34 expected, on average, to retire at age 59, whereas the cohort aged 45 to 49 had an average expected retirement age close to 61. Further evidence is provided by the Employers Forum on Age (EFA, 2001) survey *Retirement in the 21st Century*, which explored attitudes to work and retirement among a sample of adults. The majority of respondents in this survey took the view that people should be allowed to retire by the time they are 60 at the very latest. Around a quarter of respondents felt that people should be able to retire before they are 50, and a similar proportion said that the age should be no more than 55 years.

Such findings are highly significant in the context of debates pressing for an upward revision of pension ages (see, e.g., Brooks, Regan, & Robinson, 2002). They also raise issues for governments encouraging a switch from an "early exit" to a "late exit" culture (de Vroom & Guillemard, 2002). Some of the strategies that are being canvassed here include the adoption of strategies of "active aging" (Cabinet Office, 2000), the promotion of "age diversity" (Department for Education and Employment, 2001), the encouragement of flexible retirement, and the removal of incentives to leave the labor force. Such policies, however, if they are to succeed, will need to take account of some of the broader changes affecting men and women in their 50s and 60s. The next section relates these to the literature on work and retirement before examining the implications of this material for the area of leisure and social participation.

Attitudes Toward Retirement.   The trends in labor force participation, outlined in the previous section, have produced a number of significant changes. Retirement (defined in terms of entry into a public old-age pension scheme) and withdrawal, or "exit," from the workforce, no longer coincide for increasing numbers of people. In this sense, it is misleading to view the fall in male participation rates as part of a general trend toward retirement. Retirement, as traditionally defined, is seen to come at a predictable point, accompanied (for men at least) by a pension provided from the state. In contrast, the retirement that developed from the 1970s did not come at a fixed point in the life course and was usually developed in isolation from systems of state pensions (Phillipson, 1998). Such developments reflect the emergence of a new phase in the history of retirement. In general terms, we can distinguish between two main periods:

- First, the gradual consolidation of retirement from the 1950s through to the late 1960s
- Second, the acceleration of early exit and complete withdrawal from work at age 60 (for women) and 65 (for men) in the period after 1970.

The contrast between the two is illustrated by the change from the middle to the end of the 20th century. In 1951 to 1960, the annualized labor force participation rate for 65 to 69-year-old men was 50%; for men over 70, the figure was 20%. The equivalent figures for the period between 1971 to 1980 were 24% and 8%, and for 1981 to 1990 were 14% and 5%. The first period can best be described in terms of the steady growth of retirement as a social and economic institution, with the expansion in occupational pensions, and the gradual acceptance of retirement as a major stage in the life course (Costa, 1998; Hannah, 1986; Macnicol, 1998). Sociologically, this period can be identified as one in which retirement was viewed as a (mostly) male phenomenon (and problem)—a phase subordinate in length and status to that of paid employment. Henretta (2003) summarized this aspect as follows:

> The decline in the role of traditional institutions, such as the family, in determining the pattern of individual lives, the growth of the welfare state, and the growing bureaucratization of the workplace created an "institutionalized" life course with rules based on age or its correlate, time. In the institutionalized life course, the age at which events occur became more predictable because their timing is strongly influenced by institutional rules. (p. 87)

The second phase of retirement, beginning from the late 1960s onward, was marked by a number of critical changes, these arising from the development of more flexible patterns of work and the emergence of high levels of unemployment. These produced what may be termed the reconstruction of middle and old age, with the identification of a "third age" in between the period of work (the second age) and the period of physical and mental decline (often referred to as the fourth age, Gilleard & Higgs, 2000). A characteristic feature of this new period of life is the ambiguity and flexibility of the boundaries between work at the lower end and the period of late old age at the upper end of the life course. Both had more complex periods of transition, with the ambiguity of "work ending" in the first period and the blurring of dependence and independence in the second (Schuller, 1989).

This structural transformation was further reinforced (during the 1980s and 1990s) by changes in attitudes and behaviors within the workplace. Scales and Scase (2001), in a report for the Economic and Social Research Council (ESRC), identified a changing work culture, where younger employees were planning careers of 25 to 30 years instead of 40 years duration. This development was related to (a) greater pressures within the workplace—increased job dissatisfaction, longer working hours, and greater stress—and (b) lifestyle choices that valued personal autonomy over higher material living standards. Both these factors strengthened (and themselves caused) moves out of the labor market, generating a new focus on early retirement lifestyles, especially among those from white-collar, managerial, and professional occupations.

The "push" factor relating to stresses in the workplace appears to have become an influential factor in shaping transitions after 50. Another ESRC study (Taylor, 2002), comparing changes in worker attitudes over an 8-year period, found a sharp fall in job satisfaction in respect of working hours among older employees over the age of 50. The study reported that

> The proportion in that age group who said they were completely or very satisfied with their working hours declined from 54 to 26 per cent between 1992 and 2000. The largest falls in job satisfaction over the number of hours worked was the most pronounced at both ends of the occupational grading—among senior managers and professionals and the unskilled and semi-skilled. It is also clear that people who hold university degrees or the educational equivalent experienced a sharper decline in their satisfaction with hours worked than those with fewer educational qualifications. Such facts point to a particular malaise among highly educated males. The disgruntled manager has joined the disgruntled manual worker, at least in complaints about the long hours culture. (p. 10)

Discontent with work may itself be expressed in life course terms, especially in relation to the individual's trajectory through large-scale organizations. Dannefer (cited in Elder & Johnson, 2003), for example, has put forward the hypothesis that the midlife crisis—in relation to work and other aspects of life—may be attributed in part to the nature of age-graded movement through bureaucratic hierarchies. The hierarchical nature of many organizations creates inequality, as members of the same entering cohort either are promoted or reach the limits of their career. Such events may create dissatisfaction with work, as well as foster a desire for early retirement.

Images and Experiences of Retirement.   These developments suggest new transitions affecting people as they move through their 50s and 60s. The changes affecting people in these age groups are especially interesting, because the traditional means of integration—in relation to work and the family—are much less apparent now than might have been the case 10 or 20 years ago. Employment is a less stable experience, with the rise of what has been described as "contingent work" and the decline of the traditional model of the work career (Beck, 2000). Family experiences are also changing, with a higher proportion of 50-year-olds living alone (around 13% of men in the U.K.), and an increase in those who have experienced a divorce (Scales & Scase, 2001).

The implications of these trends are likely to be at least twofold: first, new types of social integration will emerge for men and women moving through their 50s: work may become less important; personal networks and leisure experiences may matter more. Second, midlife will almost certainly become more varied, with the emergence of new forms of social exclusion. This is summarized by Scales and Scase (2001) as follows:

> As a result of changes in family forms, as well as broader patterns of social and economic restructuring, a socially homogeneous 50s age group has become fragmented into a number of diverse age groupings driven by different employment and other biographical experiences. The reconstitution of family forms and the more temporary nature of personal relationships

is leading to an increase in the number of 50-year-old men and women living alone. The outcome is personal lifestyles ranging from high degrees of social isolation and loneliness to a rich intensity of personal networks. (p. 7)

These developments have led to alterations in images associated with growing old and retirement. In the 1950s, the focus was on the idea of retirement as a social-psychological crisis, with the possibility of increased morbidity and mortality being cited in the research literature (Phillipson, 1993). This was seen as a consequence of the loss of work-based friendships and the reduction in status and self-esteem associated with old age. This approach was strengthened by sociological perspectives such as role theory, with its portrayal of retirement as a "roleless role" (Burgess, 1960).

By the 1970s, a new generation of studies was providing a more positive view of retirement. The potential for it to be experienced as a new stage in life is emphasized in this period, with the development of more active lifestyles being fostered in what was increasingly referred to as the third age (Laslett, 1989). Research (following pioneering work by Peter Townsend, Dorothy Wedderburn, and others) now acknowledged the crucial role of income in terms of its influence on the retirement experience. Atchley (1999), for example, conducted a series of studies in America through the late 1960s and 1970s, all of which pointed to the impact of reduced income, and the change in lifestyle associated with it, as the major factor in producing negative retirement attitudes. From his research, Atchley found little evidence for those with moderate incomes and above actually "missing work" to any great extent.

The analysis by Palmore, Burchett, Fillenbaum, George and Wallman (1985) of a number of American longitudinal data sets concluded that

1. Retirement at the normal age has little or no adverse effects on health for the average retiree. Some have health declines, but these are balanced by those who enjoy health improvement.
2. Retirement at the normal age has few substantial effects on activities, except for the obvious reduction in work and some compensating increase in solitary activities.
3. Retirement at the normal age has little or no effect on most attitudes for the average retiree. Some become more dissatisfied, but these are balanced by those who become more satisfied. (p. 167)

This research was conducted before the rapid falls in labor force participation of the 1980s and was based on interviews with cohorts of men and women who had managed to work up to (or beyond) either the minimum or maximum age limit for their occupation. It also tended to focus on retirement as a predominantly male concern and problem (Arber & Ginn, 1991; Bernard & Meade, 1993; Phillipson, 1999; Stone & Minkler, 1984).

From the late 1980s, the focus shifted toward the varied types of experiences affecting people in the transitions running up to and through retirement. In this period, research findings accumulated about the pressures associated with early retirement. The evidence suggested a distinction between, on the one hand, a minority of people (mainly men) choosing to retire early (partly through access to sufficient income) and who tended to be satisfied with their decision (McGoldrick & Cooper, 1989). On the other hand, a larger number were found to be leaving work ahead of state pension age, largely as a result of health problems (Bone et al., 1992; Disney, Grundy, & Johnson, 1997). This group appeared more likely to report uncertainty about the future and to experience problems of adjustment.

Another area of differentiation concerns the question of women in retirement. Women's transition to retirement is influenced by their position in the labor market and ideologies regarding their role as carers (Bernard & Meade, 1993). Early research argued that entering retirement was less stressful for women, partly because work had been, it was argued, a less important element in their lives and partly because of the maintenance of the role of housewife (Donahue, Orbach, & Pollack, 1960; Tibbitts, 1954). However, Dex (1985) suggested that the later research literature indicated similarities in the attitudes of men and women toward work. Women, she noted, have been found to be more instrumental than was thought to be the case (and men less so), and both experience variations through the life course in orientations to work (see also Ginn, Street, & Arber, 2001a).

Szinovacz (1991) noted that the discontinuous nature of women's work histories means that many women view their midlife as a special challenge and opportunity and may view the later phase of their work career in a different way than that of men. They are also less likely than men to have achieved their career goals at the time their spouses wish to retire or employers' retirement incentives encourage their husbands' retirement (Szinovacz & De Viney, 2000). Pension provision has also emerged as a major issue for women, with countries such as Britain and the United States providing what Ginn et al. (2001b, p. 233) referred to as "a particularly harsh pensions environment for many working-age women."

At the beginning of the 21st century, therefore, the literature on retirement has been attempting to bring together a range of divergent and often contradictory themes. On the one hand, differentiation and divisions within the retirement experience have almost certainly grown over the last quarter of the 20th century (Mann, 2001; Phillipson, 1998). This can be attributed to two main factors: first, increasing instabilities in pension provision; second, the greater diversity of routes and pathways that people take in moving from work to retirement. On the other hand, the idea of a "third age" lifestyle has also begun to take hold. Gilleard and Higgs (2000, p. 29) refer to the emergence of a distinct "postwork" stage within the life course, as well as " 'expanded' opportunities for consumption and the development of later life identities." Both points suggest, in fact, a "broadening" in the institution of retirement, with a more diverse mix of work, familial, caring, and leisure activities (Phillipson, 2002). The remainder of the chapter explores this theme, beginning first with the involvement of older people within the family.

Retirement and Social Relationships

A consistent observation from both American and British research is that retirement produces more intensive involvement within the family, for example, in grandparenting and caring roles of many kinds (Ade-Ridder & Hennon, 1989; Crawford, 1981; Cunningham-Burley, 1986; Phillipson, 1992). This is part of a broader finding that family life—for women and men—remains crucially important in later life, even if "intimacy at a distance" has tended to replace (in many communities) everyday face-to-face contact between parents and children (Phillipson, Bernard, Phillips, & Ogg, 2001).

Within the social world of retired people, the marital relationship now occupies a crucial role. This reflects the running together of a number of important factors. In the first place, society is adjusting to the phenomenon of the long-term marriage (defined here as marriages lasting 20 or more years). American research in the 1970s suggested that one marriage in five could expect to see its 50th anniversary. This contrasts sharply with the situation at the beginning of the 20th century, when most marriages were terminated during middle age by the death of a spouse (Sporakowski & Axleson, 1989).

The significance of these changes is reinforced by earlier retirement and changes in household composition. The former has meant an increase in the amount of time couples can choose to spend with each other, ahead of some of the familiar health changes associated with late old age. The latter is illustrated by the decline of coresidence between older parents and their adult children, a change that gathered momentum from the 1960s onward (Wall, 1992). Taken together, these transformations to both retirement and household structure will have profound consequences for leisure and other forms of social relationships and activities in later life.

Research in this area has identified a number of important issues affecting the lives of older couples. American studies suggest that gender-differentiation in household work may diminish after retirement, with greater participation by husbands in traditionally "feminine" tasks such as cleaning and shopping (Sporakowski & Axelson, 1989). Such participation, however, is usually defined as "helping" wives and therefore does not change the basic division of household tasks. British research by Mason (1987) reported older couples focusing their attention much more on the home, and that this in many ways represented an encroachment by men on what had formerly been viewed (often by default) as women's domain, or territory. Not only did husbands encroach on wives' personal time and space, but as Mason went on to indicate, their presence began to create different forms of domestic work for their wives.

Research in the 1960s and 1970s also suggested potential conflicts between husbands and wives in the period following retirement. Gilford (1984), drawing on this literature, suggested that retirement can bring out the negative aspects of a marriage, especially for women. She suggested (and Mason's 1987 exploratory study would appear to confirm), that women benefit less from their husbands' retirement than does the retiring husband. This negative view has been challenged by longitudinal studies, in particular those by Atchley and his colleagues (Atchley, 1992). Atchley concluded from his work that retirement has no positive or negative effect, either on individuals or on marriages. However, it can lead to problems for some and, for a larger number, the need to reorganize the division of labor within the home.

A more recent focus of research has been on differences between dual retiree and single retiree marriages (Szinovacz, Ekerdt, & Vinick, 1992). In the case of dual retirees, there are important issues to consider in terms of how the timing of the husband's and wife's retirement influences the marital relationship. The new dimension here is that there is an increasing need to see issues in terms of couples retiring, rather than the return of a dominant breadwinner into the home, as has been the overriding concern of existing research. The issue of the impact of early retirement on the marital relationship is another area that warrants further investigation (Cliff, 1989).

Paralleling these trends, it is also evident that other kinds of relationships people have will impact strongly on both leisure and retirement. It is to an examination of some of these relationships—notably those associated with friendship—that we now turn.

Friendship In Later Life.   In addition to issues that affect the family lives of older people, the search for meaning and fulfilment in later life is also closely bound up with the ability to create and sustain social relationships of other kinds. Relations with peers, and friendship in particular, have been shown by a number of writers on both sides of the Atlantic to be crucial to the aging process (see, e.g., Abrams, 1978; Hazan, 1980; Jerrome, 1981, 1984, 1993a; Rosow, 1975; Wenger, 1987). Despite the contention that friendship has received comparatively little attention from either the social sciences or gerontology (Chown, 1981; Jerrome, 1993b), a number of key themes can be identified in terms of its nature and how it is experienced. These include gender differences (and similarities), the influences of class and education on friendship formation, and whether relationships are dyadic or group based (Jamieson, 1998).

Gender differences in friendship are particularly marked and, over the years, research has increasingly pointed to the value of a "special relationship," or confidant(e), in adjusting to the stresses and strains of later life (Coleman, 1993). For women especially, this is also a life course issue in that the presence of a confidante or close friend has been found to be important in terms of social support, as well as in the maintenance of psychological well-being and mental health (Brown & Harris, 1978; Crohan & Antonucci, 1989; Larson, 1978). Moreover, rather than fulfilling this need for a confidante within the marriage relationship, women will tend to look to other women or an adult child for this kind of support (Blau, 1981; Jerrome, 1981). Men, by contrast, tend to name their wives as their main source of emotional support and as the only person they talk to about personal problems and difficulties. An examination of the family and community life of older people in three urban areas in England (Phillipson, Bernard, Phillips & Ogg, 2001) confirms some of these findings, showing that although the immediate (rather than the extended) family is still crucial in the social networks of older people, friends have a higher profile than they did 50 years ago and often play a substantial role in providing emotional support (see also Pahl, 2000).

It has also been the contention of certain writers that the nature and quality of friendship reveals marked gender differences. Women's friendships are said to be person oriented, emotionally richer than men's, and characterized by emotional support, intimacy, self-disclosure, and mutual assistance (Booth & Hess, 1974; Hess, 1979; Weiss & Lowenthal, 1975). Indeed, Wright (1982), reviewing the early literature in his classic article from 1982, suggested that it showed that "for men friendship tends to be a side-by-side relationship, with the partners mutually oriented to some external task or activity; while for women friendship tends to be a face-to-face relationship, with the partners mutually oriented to a personalized knowledge of and concern for one another" (p. 8). Although this generalization had later to be modified when more detailed and differentiated work began to show that men were also able to develop long-lasting and deep relationships, gender differences are still observed in terms of how friendships are developed and experienced over the life course. As Jerrome (1993b) argued, a particularly important way in which this is revealed is through talk: conversation being one of the main activities of female friendship from early childhood onward. Older men's friendships, on the other hand, still tend to be activity oriented, based on shared experiences, and are more likely to be confined to particular settings (such as pubs and the workplace) rather than be conducted in home settings.

Thus, friendship is self-evidently entwined with the other social roles that tend to shape social networks throughout life, and with influences such as educational levels and social class. A number of studies reveal that social class continues to exert a strong influence in later life, with middle-class people having more friends than working-class people (Allan, 1996; Phillipson, Bernard, Phillips, & Ogg, 2001). Middle-class older adults tend to express these friendships through dyadic relationships, through membership in various kinds of organizations, and by voluntary work (Jerrome, 1993b). Club or group membership is also an activity of working-class men and women.

A consideration of the group settings in which friendships are expressed points to two dilemmas concerning the nature of friendship and its relation to leisure in later life. First, the findings previously discussed are, of necessity, generalizations, and there is some conflicting evidence about both the extent and persistence of such differences over the life course and into old age. What seems to be emerging is a much more complex picture, stressing both the value of same-sex and cross-sex friendships, together with a need to explore the nature of friendship as it relates not only to other social roles, but also to other individual characteristics such as personality traits. For example, the Kansas City Studies of Adult Life, carried out by Bernice Neugarten (1964) and her colleagues, hold out the interesting possibility that, as we age, we become more androgynous, with women exhibiting increased autonomy and assertiveness, and men showing an increased capacity to express tenderness and compassion. This contention has been supported by more recent work (see, e.g., Sheehy, 1997) and holds out the possibility that confidant relationships in later life may not necessarily need to be with a friend of the same gender. Although hormonal influences (reduced testosterone levels in men and estrogen in women) may influence these developmental changes, social factors are also undoubtedly at work here: Economic, political, social, technological, and demographic factors are shaping and reshaping our gender roles and identities throughout life. Yet, although it may be tempting, as Rossi (1986) suggested, to take "comfort in the idea that the more androgynous qualities may become stable characteristics of men and women throughout their lives" (p. 132), we should not underestimate the time it will take, and the turmoil (both personal and social) it will entail, to effect such radical individual and social transformation.

The second dilemma concerns the nature of friendship and the meanings it has for people in later life, whether these be positive or problematic (Adams & Blieszner, 1998; Blieszner & Adams, 1998). Although many gerontologists might be interested in patterns of friendship among older adults, they have rarely systematically examined what older people actually do with their friends in the context of leisure (Adams, 1993). In her work in this area, Adams (1993) attempted to focus on the ways in which the meanings of the (social or leisure) activity and the relationships are related. By examining what older women do with friends, she attempts to understand how the form and content of the activity influence the development and maintenance of friendship. She concluded that "leisure activities both provide social structural contexts for friendship and are processes shared by friends" (p. 83). In so doing, she explicitly moved us into the realms of social activity, and then into what Kaplan (1979) argued is just one of the eight categories of "activity experiences" that contribute to the conception of leisure as a social role.

Consequently, we now turn our attention to a more detailed consideration of the possibilities raised by retirement for leisure and social activities. We shall assess the factors that influence leisure in retirement, exploring opportunities for the future and potential avenues for further research on this hitherto neglected issue (at least in a British context).

Leisure in Retirement

Given the social and demographic trends highlighted earlier in this chapter concerning the changing nature of old age in general and retirement in particular, it is evident that the cessation of paid work no longer necessitates retirement from an active life (Szinovacz, Ekerdt, & Vinick, 1992; Young & Schuller, 1991). Indeed, Kaplan (1979) estimated that, by the start of the 21st century and given a life expectancy of 85 years, we would roughly be spending the first 20 years or so in education; the next 30 years in work; and the last 35 years in retirement. Thus, retirement may in fact become the longest and most stable period in our lives: a period that is increasingly being promoted as "a time of opportunity for personal development, a time of freedom, of leisure and pleasure" (Bernard & Meade, 1993, p. 146).

Despite these transformations to later life, research has yet to really address the theoretical and empirical connections between leisure and aging, retirement, or both. More than a decade ago, it was observed that the situation in Britain was one "in which leisure scholars have paid scant attention to older people, and gerontologists for their part have yet to address the role of leisure" (Bernard, 1990, p. 14). This was echoed in an American context by Kelly (1993), who, in the introduction to his edited book on *Activity and Aging*, said

> Each perspective has developed theory and research that complements the other. With some exceptions, however, the two groups of scholars have not been in communication. . . . Social gerontologists have learned a great deal about relationships in later life as well as about certain age-designated environments and programs. Those studying leisure have learned a great deal about the forms and meanings of a wide range of activity but have seldom tied that knowledge to the realities of aging. (p. ix)

More recently still, the ghettoization and lack of intellectual dialogue between leisure studies and other fields has continued to exercise the minds of notable scholars on both sides of the Atlantic (Deem, 1999; Samdahl & Kelly, 1999). Indeed, the first of the new millennium issues of the *Journal of Leisure Research* was almost entirely devoted to soul-searching papers exploring the relevance of leisure research (Shaw, 2000; Witt, 2000) and its future (Samdahl, 2000), as well as the potential connections and parallels between theory and research in both leisure and gerontology (McGuire, 2000; Searle, 2000).

Given this continuing lack of any consistent dialogue between the two fields, our concern here is to try and draw out what we know about what leisure both consists of and might mean for people in retirement, and then to speculate on some of the possible directions for exploring these issues further. Thus, we now turn to consider some of the factors that influence leisure in this phase of the life course, drawing particular attention to gender dimensions and the still prevailing influence of work.

Work and Leisure—Poles Apart?   For decades, we have been used to opposing work and leisure. Work has been seen as the antithesis to relaxation, fun, amusement, and idleness, with recreation being literally, re-creation: re-creation for yet more work. Thus, it is hardly surprising that for many older people leisure has come to be associated with notions of "deservingness," of it being somehow "earned" through hard work (Mansvelt, 1997). Retirement within such an ideological framework is consequently problematic, especially for certain groups of older people who may never have "worked" in a traditional sense, for example, older women who may have devoted themselves to family raising and caring tasks of various kinds.

In addition, although the importance of work may have declined in temporal terms, its psychological impact is still considerable. The work ethic, and the "functional worth ethic," has tended to dominate many older people's lives. As Coleman (1993) argued, this "works against them in later life when lack of occupation seems akin to idleness" (p. 130). Our continued emphasis on juxtaposing work and leisure, in societies still dominated by the work ethic, means that equating retirement with leisure can reinforce negative views of older people as both functionless and unproductive (Dobbin, 1980). Retirement, as presently constituted, essentially means asking people to make a transition to a leisure role: a role that current generations of older people may find exceedingly difficult.

The difficulties of looking at retirement as leisure has meant that various approaches have been adopted to try to articulate what the experience of leisure is like in old age. Historically, we can identify two major foci: on leisure time and the activities that constitute it, and on leisure as it relates to notions of adjustment, successful aging, and life satisfaction (Kelly, 1990). With the former, research concerns have largely been taken up by issues such as delineating how much time we might have to spend on leisure, what activities are or are not defined as leisure, how levels of participation vary across different activities, and what roles might be ascribed to leisure in the absence of paid work. Many of the early studies were preoccupied with "the working man" and his leisure activities. Indeed, a major study on leisure around retirement conducted in Britain in the early 1980s was solely concerned with retired men (Long & Wimbush, 1985). One of its prime aims was to examine how leisure activities might be used to help men cope with the changes brought about by retirement from paid work. In this respect, it echoed both earlier empirical work and various theoretical constructs, which emphasized the different ways in which leisure might either compensate for the loss of work roles in terms of offering new opportunities for engagement and the expression of one's skills (Brehm, 1968; De Carlo, 1974), or in which it helps provide some of the necessary continuities in people's lives (Atchley, 1971, 1993; Long, 1989). However, as has been argued elsewhere (Bernard & Meade, 1993), these kinds of work substitution models do not really help us a great deal in terms of truly understanding the role and value of leisure in later life, particularly for women.

Similarly, research that has concentrated on participation levels and activities tends to give us a very restricted view of the potential of this phase of life. In fact, what we get is a picture of participation that is strongly and inversely related to age (Kelly, 1990). In Britain, the Carnegie Enquiry into the Third Age presented a depressing illustration of retirement as constituting an abundance of leisure in terms of time, but in which very small minorities of older people took part in any kinds of leisure activities, particularly those of an active nature beyond the home (Midwinter, 1992). The same picture of declining participation in outdoor activities with age still holds at the start of the 21st century, gardening among older men being the one notable exception (Office for National Statistics, 2002).

By contrast, sedentary and home-based activities constitute the bulk of the leisure time available to older men and women (as indeed they do for other age groups). Again, the Carnegie Enquiry found that three fifths of the leisure time of the average older woman, and more than half that of the older man, was taken up with sedentary activities such as television watching and listening to the radio (Midwinter, 1992). It is interesting that our own research showed that reading was in fact the most popular activity among our sample, ahead of gardening and television watching. These three activities together reflect changing opportunities and expectations among this generation and clearly attest to the significance of "home" for today's older people (Phillipson, Bernard, Phillips, & Ogg, 2001). However, it has to be said that conventional participation surveys still paint a rather gloomy picture of leisure in later life that is somewhat at odds with current images of retirement as a time of active leisure and personal fulfilment.

Historically, the second major focus of leisure research has concerned notions of adjustment, successful aging, and life satisfaction. This dates back to the seminal volume edited by Kleemeier (1961), entitled *Aging and Leisure*. In this volume, a number of the contributors explored what leisure itself means and how different leisure activities hold different meanings for different people. The contributions, some emanating from the Kansas City Studies of Adult Life, took us beyond sterile activity checklists and descriptive statistics about participation, into the realms of phenomenology, self-concepts, and the maintenance of identities in later life. From these and ensuing studies, we learned that engagement in leisure activity of all kinds is as important for older people as younger people, that it is a crucial dimension of well-being and functioning in later life (Everard, 1999), and that it is key to determining quality of life and life satisfaction (Lloyd & Auld, 2002). Indeed, despite the fact that we still lack any universally agreed on definitions of adjustment, or "successful aging" (Rowe & Kahn, 1997), substantial (if sometimes contradictory) evidence now exists of a generally positive relationship between engagement in various physical, social, and other leisure activities and the maintenance or improvement of one's ability to function competently into old age, together with improved levels of life satisfaction (Kelly, 1990). In the United States, the MacArthur Foundation Research Network on Successful Aging has shown that both physically demanding, as well as less strenuous, leisure activities are associated with better functioning (see, e.g., Seeman et al., 1995), whereas European studies, notably the Berlin Aging Study, clearly demonstrate how social and leisure activity contributes to a person's competence, and hence to a better quality of life in old age (Baltes, Mayr, Borchelt, Maas, & Wilms, 1993).

Moving leisure center stage in this way, looking at its multidimensionality and its contribution to the meaningful use of time in later life, holds out the possibility that it might actually be retirement itself, rather than leisure, that is the problematic element in the retirement and leisure equation. It also suggests that our present, and indeed future, concerns might more fruitfully be focussed on discussions about how we can minimize the impact of retirement on our leisure lives than on how leisure might compensate for loss of work through retirement.

Gendered Leisure. As indicated, approaches to the study of leisure over recent decades have tended to both emphasize its antithesis to work, and to be overly concerned with what activities do or do not constitute leisure. This, we would suggest, has obscured any substantive or meaningful discussion of the gender differences and similarities that might occur concerning the experience of leisure in later life. Only since the end of the 1980s have we seen the emergence of a number of studies that begin to counter the invisibility of leisure in the lives of aging women (see, e.g., Allen & Chin-Sang, 1990; Henderson, Bialeschki, Shaw, & Freysinger, 1989; Mowl & Towner, 1995; Siegenthaler & Vaughan, 1998; Stokowski, 2000; Szinovacz, 1992). In Britain, therefore, the information we do have is culled mainly from studies that have not primarily been focused on the leisure lives of older women and men, but which have looked at different, though related issues, like retirement experiences. Research from North America is rather more developed, though here, too, criticism has been leveled at its failure to grasp the complexity of factors that intersect to give meaning to the leisure and lifestyles of people in old age (Freysinger, 1993). With these provisos, what then do the leisure lives of men and women in the retirement years look like?

The picture of age-related decline in participation, with a concomitant increase in the sedentary activities previously noted, masks some notable gender differences in terms of the time, participation levels, and interests of older adults. In relation to time, it has been observed that "with the removal of paid work from the scenario of later life we might assume that women's and men's access to leisure time might converge" (Bernard & Meade, 1993, p. 157). However, this is patently not the case. Research has consistently shown that women have much less time for leisure than men, and that this difference persists into old age. In Britain, of the 105 waking hours available to us weekly, a retired man has 92 hours of "free time" a week, and a retired woman 75 hours (Central Statistical Office, 1990). Recent statistics still show that women's lives remain more occupied with domestic chores and family responsibilities, despite the changing roles of men and women (Office for National Statistics, 1999). Indeed, Mason's (1987) work, referred to earlier, illustrated the additional constraints put on women by the retirement of their husbands both in terms of increased domestic work and because of the pressure to facilitate and support his newfound leisure.

These findings echo Havighurst's (1961) work from the Kansas City Studies of Adult Life, which showed that women's leisure tended to be fitted in and around their domestic tasks. However, other research points to some convergence in domestic roles after retirement. In Young's and Schuller's (1991) study of retired men and women between the ages of 50 and 60 or 65, the men spent as much time on shopping as the women whose preserve it had once been. Men, though, tended to perform less domestic work of an ordinary routine nature, and to have more time for "passive leisure": watching television, reading, talking, and listening to the radio. Although they accepted that these differences existed, the authors contended that "they cannot be counted as substantial" (Young & Schuller, 1991, p. 138). However, what we do not know is whether the men and women concerned share this view.

For both women and men in later life, other related influences will also affect the time available for leisure. For example, some may have considerable caring roles for grandchildren or for relatives with disabilities, whereas others may have adult children living at home following divorce. Retirement then may not automatically herald a future of uninterrupted or autonomous leisure time.

Participation levels for a variety of both home-based and outside leisure activities also reveal patterns of differentiation and convergence. Our earlier discussion of friendship noted that club attendance was a common setting for the development and maintenance of such relationships. Jerrome (1993a) wrote: "Club-going is a way of life for a minority of retired people, mainly working-class white women . . . (it) is an activity of friendship. Club meetings are attended with friends: the object is less to make new ones than to reinforce existing friendly relations" (pp. 96–97). In fact, we know that in Britain, approximately 30% of older women join social clubs, and about 13% attend clubs that cater solely to older people (Midwinter, 1992). In addition, playing bingo is the only gambling activity in which women of all ages are more likely than men to participate, and it is the only gambling activity in which participation increases with age (Office for National Statistics, 2002).

Religion is also an important part of many older people's lives, with more older women than men involved in church and other organized activities. In addition to gender, participation is affected by other factors, such as race, social class, and health status. For example, a survey in Nottingham, England, revealed that churchgoing was the most common social–leisure activity amongst West Indian pensioners: 74% of older women attended weekly, compared with 46% of older men (Berry, Lee, & Griffiths, 1981). Similar findings have been reported from other communities.

Volunteering too is a long-established tradition in Britain, and here again we can observe gender differences in participation, with nearly 50% of third-age women (aged 50–74) taking part in voluntary activities, compared with 38% of men (Lynn & Davis Smith, 1991). Since the mid-1980s, cultural, demographic, and programmatic improvements have led to a growth of volunteering among most age groups, with the largest growth being among middle-aged and older people (Chambre, 1993). In fact, the 1997 *National Survey of Volunteering* (Davis Smith, 1998) showed that the participation rate for people aged 65 to 74 grew from 34% in 1991 to 45% in 1997, and from 25% to 35% for those aged 75 and over. Age, socioeconomic status, race, educational qualifications, and geographical location all affect patterns of participation with those still in paid work, who have higher incomes, are well educated, and have access to a car being the most likely to volunteer, whereas those who are widowed are the least likely to do so (Chambre, 1987, 1993; Davis-Smith, 1992, 1998; Knapp et al., 1996). Older people are also more likely than any other age group to agree that "doing voluntary work is a good thing . . . because it makes them feel they are contributing to society" (Office for National Statistics, 2002). The important role that volunteers and voluntary work play in community life was finally recognized when the United Nation established 2001 as the International Year of the Volunteer.

In contrast, British researchers have long observed that only small numbers of older people are engaged in educational activities in any formal sense, especially in higher education (Marshall, 2000; Midwinter, 1998). However, men over the age of 50 tend to be rather more involved than older women (see, e.g., the three national surveys conducted in the 1980s: Advisory Council for Adult and Continuing Education, 1982; Office of Population Censuses and Surveys, 1983, 1985; Schuller & Bostyn, 1992), although this pattern is reversed when one considers participation in growing self-help education groups such as the Universities of the Third Age. Here, older women outnumber men by a ratio of three to one (Midwinter, 1998), whereas across Europe, participation in education in all its forms plays a continuing and expanding role in the lives of older people (Midwinter, 1996; Norton, 1992; Tokarski, 1993).

We know, too, that participation levels in other forms of leisure and social activity demonstrate gender differences. Physical activities, sports, and other forms of outdoor recreation reveal very low levels of participation. In Britain, only 5% of women aged 60 and over take part in keep-fit activities such as yoga and aerobics, with even smaller percentages engaging in pursuits like swimming or cycling. American studies bear this out, reporting higher levels of involvement for men in exercise, sports, and outdoor recreation (Kelly, Steinkamp, & Kelly, et al., 1986). Social pursuits, too, such as visiting friends and relatives and being visited, show consistent gender-differentiated patterns. Jerrome's (1993a, 1993b) research for example, confirmed the importance of social relationships generally for older women, and intimate friendships in particular.

Explanations for these gender differences are many and varied, but what is evident from the research is that social and cultural norms and expectations influence the findings much more than do biological or physical differences between older men and women. The "image" of leisure is youthful, fit, active, and largely male. Combined with the use of women's sexuality to promote leisure, this creates a powerful ideology on how women in particular may choose to spend their leisure time. Participation in socially acceptable and respectable activities such as clubs, voluntary organizations, adult education classes, and visiting friends and relations is a reflection of these constraints. Many of these activities are also perceived as taking place in "safe environments." Consequently, this discussion now brings us to a consideration of what the future might hold in terms of the settings, opportunities, and research directions for leisure and retirement.

Future Prospects for Leisure and Retirement

What stands out from this examination of retirement and leisure are the ways in which their historical development and influencing factors mirror and parallel each other. In the conclusion to his book on activity and aging, Kelly (1993) wrote, from the standpoint of a leisure scholar, that we need to know a lot more about various aspects of leisure "activity" in later life:

> We need to know how it fits into the overall life patterns of older adults. We need to know more about contexts, provisions, barriers, facilitating factors, access, adaptation, and organization. We need to know more about continuities and discontinuities in identities,

values, risk, relationships, and inhibitions. We need to know more about what actually occurs in the process of activity, both in the discrete experience and in the longer term line of action. We need to know about contexts and resources as well as about attitudes and emotions. And we need to know more about the isolated, the withdrawn, and those who have been denied access and opportunity. (p. 265)

Substitute *retirement* for *activity* in this quotation and, although we might argue that we know much more about these dimensions of retirement, the proposition still rings true. What it also highlights is the fact that we are missing an explicit examination of the intersection of retirement and leisure that takes us beyond what people actually do and asks, more pertinently, why this should be so. Such an examination would need to be based, as Freysinger (1993) argued, on a broad understanding and appreciation of three areas:

- Knowledge of the politics of age, gender, race, class, health, and leisure
- Comprehension of the ways in which age, gender, race, and class interact to help construct human identity
- Recognition of the fact that relations of power and privilege are not static, but are continually changing and being renegotiated.

In broad terms, it should also be recognized that the basis for research has been drastically changed by a combination of developments in economic and social policy, and transformations in individual attitudes and behavior. Even up to the 1960s (especially in the case of Britain) the conjunction of leisure and retirement was somewhat awkward. The emphasis was on the rolelessness of the latter, and the lack of definition of the former. At the start of the 21st century, the situation has clearly changed, with macroeconomic polices consolidating the institutionalisation of retirement, and cultural changes loosening commitments toward paid work, especially in the latter half of the life course. Underlying this is greater flexibility in social definitions and images of aging. Older people have themselves been part of a significant movement that has challenged conventional age stereotypes, expressing their engagement with society in developing lifestyles built around new patterns of consumption, intensive involvement in educational and cultural activities, and developing campaigns with and on behalf of more deprived groups of pensioners (Bornat, 1998). For researchers, such developments offer important challenges in respect of further study. Three in particular may be identified.

First, we need to know more about the applicability of terms such as *leisure* and *retirement* for different cohorts of older people. The agenda here is to examine the socially constructed nature of these terms and to assess the way in which they are interpreted by different groups of older people. This is especially important in terms of the different ways in which people are responding to aging, with contrasts between those who view it as an opportunity for transforming their lives and others who are overwhelmed by the problems and crises it brings (Phillipson, Bernard, Philips, & Ogg, 2001).

Second, the relationship between leisure and retirement needs clarification. Traditionally, as we have argued, research has tended to focus on the importance of retirement in terms of shaping people's lives, with leisure being seen as filling in the space that work leaves behind. However, this somewhat crude model has limited value in terms of explaining and understanding how people might change in later life. An alternative approach may be that we need to give greater priority to life course perspectives in aging studies, examining both the interrelatedness of work and leisure, and the way in which attachments to each sphere may alter at different phases. Moving away from examining the problem of retirement toward the potential of leisure reorients research toward an exploration of growth and development in the later years (Bernard, 2001).

Third, a clearer focus needs to be given to both leisure settings and new patterns of consumption among older people. These are changing in different ways, reflecting increasing numbers of services, goods, and markets targeted at older people (Gilleard & Higgs, 2000). Again, the traditional model tends to emphasize older people passively responding to opportunities that evolve within the community. In contrast, an alternative view would emphasize the interaction between the emerging consciousness of age and the active involvement of older people in the development of different patterns of consumption. This raises significant opportunities for research to assess the variety of activities that are emerging as people challenge traditional definitions of retirement and the third age.

Overall, what is important to establish is that the ways in which we construct the notion of retirement is changing radically and, with it, our identities, outlooks, and potentials as retirees. At the beginning of the 20th century, the majority of people continued to work until the point of exhaustion or ill health set in. One hundred years later, the majority are leaving paid employment—in most industrialized countries—well before this point, with a rapid expansion in the number of years spent "in retirement." Yet, the key point to take from this chapter is the difficulty we still have of grasping the scale and rapidity of the changes affecting work, retirement, and leisure over the past century, and the impact these have on individual attitudes and behavior. The challenge for researchers is an exciting one: to document new lifestyles, institutions, and identities among the old. Understanding the way in which old age is being reconstructed, and its linkages to other areas of life and other institutions, is now a key task for social scientists to address.

REFERENCES

Abrams, M. (1978). *Beyond three-score and ten: A first report on a survey of the elderly.* Mitchum, UK: Age Concern Research Unit.

Adams, R. G. (1993). Activity as structure and process—Friendships of older adults. In J. R. Kelly (Ed.), *Activity and aging—Staying involved in later life* (pp. 73–85). Newbury Park, CA: Sage.

Adams, R. G., & Blieszner, R. (1998). Structural predictors of problematic friendship in later life, *Personal Relationships, 5,* 439–447.

Ade-Ridder, L., & Hennon, C. B. (1989). *Lifestyles of the elderly.* New York: Human Sciences Press.

Advisory Council for Adult and Continuing Education. (1982). *Adults: Their educational experience and needs.* Leicester, UK: Author.

Allan, G. (1996). *Kinship and friendship in modern Britain.* Oxford, UK: Oxford University Press.

Allen, K. R., & Chin-Sang, V. (1990). A lifetime of work: The context and meaning of leisure for aging Black women. *The Gerontologist, 30,* 734–740.

Arber, S., & Ginn, J. (1991). *Gender and later life.* London: Sage.

Atchley, R. C. (1971, Spring). Retirement and leisure participation: Continuity or crisis? *The Gerontologist 1,* 13–17.

Atchley, R. C. (1992). Retirement and marital satisfaction. In M. Szinovacz, D. J. Ekerdt, & B. H. Vinick (Eds.), *Families and retirement* (pp. 145–158). London: Sage.

Atchley, R. C. (1993). Continuity theory and the evolution of activity in later adulthood. In J. R. Kelly (Ed.), *Activity and aging—Staying involved in later life* (pp. 5–16). Newbury Park, CA: Sage.

Atchley, R. C. (1999). *Continuty and adaptation in aging.* Baltimore: The John Hopkins University Press.

Baltes, M. M., Mayr, U., Borchelt, M., Maas, I., & Wilms, H.-U. (1993). Everyday competence in old and very old age: An inter-disciplinary perspective [Special issue: The Berlin Aging Study]. *Ageing and Society, 13,* 657–680.

Barham, C. (2002, February). Economic inactivity and the labour market. *Labour Market Trends, 110, 69–77.*

Beck, U. (2000). *The brave new world of work.* Cambridge, UK: Polity Press.

Bernard, M. (1990). Leisure and older people: Constraints and opportunities. In A. Tomlinson (Ed.), *Leisure and the quality of life: Themes and issues* (pp. 14–30). Brighton, UK: Leisure Studies Association.

Bernard, M. (2001). Women ageing: Old lives, new challenges. *Education and Ageing, 16, 333–352.*

Bernard, M., & Meade, K. (1993). A third age lifestyle for older women? In M. Bernard & K. Meade (Eds.), *Women come of age: Perspectives on the lives of older women* (pp. 146–166). London: Arnold.

Berry, S., Lee, M., & Griffiths, S. (1981). *Report on a survey of West Indian pensioners in Nottingham* [mimeo]. Nottingham, UK: Nottingham Social Services Department.

Blaikie, A., & Macnicol, J. (1986). Ageing and social policy: A twentieth-century dilemma. In A. Warnes (Ed.), *Human ageing and later life: Multidisciplinary perspectives* (pp. 69–82). London: Arnold.

Blau, Z. S. (1981). *Aging in a changing society.* New York: Watts.

Blieszner, R., & Adams, R. G. (1998). Problems with friends in old age. *Journal of Aging Studies, 12, 223–238.*

Bone, M., Gregory, J., Gill, B., & Lader, D. (1992). *Retirement and retirement plans.* London: HMSO.

Booth, A., & Hess, E. (1974). Cross-sex friendships. *Journal of Marriage and the Family, 36,* 38–47.

Bornat, J. (1998). Pensioners organise: Hearing the voices of older people. In M. Bernard & J. Phillips (Eds.), *The social policy of old age: Moving into the 21st century* (pp. 183–199). London: Centre for Policy on Ageing.

Brehm, H. P. (1968). Sociology and aging: orientation and research. *The Gerontologist, 8,* 24–32.

Brooks, R., Regan, S., & Robinson, P. (2002). *A new contract for retirement.* London: Institute for Public Policy Research.

Brown, G. W., & Harris, T. (1978). *Social origins of depression: A study of psychiatric disorder in women.* London: Tavistock.

Burgess, W. (Ed.). (1960). *Aging in Western societies.* Chicago: University of Chicago Press.

Cabinet Office. (2000). *Winning the generation game.* London: H. M. Government, Performance and Innovation Unit.

Campbell, N. (1999). *The decline of employment among older people in Britain* (CASE Paper No. 19). London: London School of Economics.

Central Statistical Office. (1990). *Social Trends 20.* London: Her Majesty's Stationery Office.

Chambre, S. M. (1987). *Good deeds in old age: Volunteering by the new leisure class.* Lexington, MA: Lexington Books.

Chambre, S. M. (1993). Volunteerism by elders: Past trends and future prospects. *The Gerontologist, 33*(2), 221–228.

Chown, S. (1981). Friendship in old age. In S. Duck & R. Gilmour (Eds.), *Personal relationships* (Vol. 2, pp. 231–246). London: Academic Press.

Cliff, D. (1989). *Life after work: An investigation of men in early retirement.* Unpublished doctoral dissertation, University of Huddersfield.

Coleman, P. (1993). Adjustment in later Life. In J. Bond, P. Coleman, & S. Peace (Eds.), *Ageing in Society: An introduction to social gerontology* (2nd ed., pp. 97–132). London: Sage.

Costa, D. (1998). *The evolution of retirement.* Chicago: University of Chicago Press.

Crawford, M. (1981). Not disengaged: Grandparents in literature and reality. *Sociological Review, 29,* 499–519.

Crohan, S. E., & Antonucci, T. C. (1989). Friends as a source of social support in old age. In R. G. Adams & R. Blieszner (Eds.), *Older adult friendship* (pp. 129–146). Newbury Park, CA: Sage.

Cunningham-Burley, S. (1986). Becoming a grandparent. *Ageing and Society, 6,* 453–470.

Davis-Smith, J. (1992). *Volunteering: Widening horizons in the third age* (Research Paper No. 7, "The Carnegie Enquiry into the Third Age"). Dunfermline, UK: Carnegie U.K. Trust.

Davis-Smith, J. (1998). *The 1997 national survey of volunteering.* London: National Centre for Volunteering.

De Carlo, T. J. (1974). Recreation participation and successful aging. *Journal of Gerontology, 29,* 416–422.

De Vroom, B., & Guillemard, A. M. (2002). From externalisation to integration of older workers: Institutional changes at the end of the worklife. In J. G. Anderson & P. H. Jensen (Eds.), *Changing labour markets, welfare policies and citizenship* (pp. 183–209) Bristol, UK: Polity.

Dex, S. (1985). *The sexual divisions of work.* Brighton, UK: Wheatsheaf Books.

Deem, R. (1999). How to get out of the ghetto? Strategies for research on gender and leisure for the twenty-first century. *Leisure Studies, 18*(3), 161–177.

Department for Education and Employment. (2001). *Code of practice on age diversity.* London: Author.

Disney, G. E., Grundy & Johnson, P. (1997). *The dynamics of retirement.* London: Stationery Office.

Dobbin, I. (1980). *Leisure and the elderly.* Salford, UK: Centre for Leisure Studies.

Donahue, W., Orbach, H. L., & Pollack, O. (1960). Retirement: The emerging social pattern. In C. Tibbitts (Ed.), *Handbook of social gerontology* (pp. 330–406). Chicago: University of Chicago Press.

Elder, G., & Johnson, M. (2003). The life course and aging: Challenges, lessons and new directions. In R. Seltersten (Ed.), *Invitation to the life course* (pp. 49–81). Amityville, NY: Baywood.

Employers Forum on Age. (2001). *Retirement in the 21st century.* London: EFA.

European Foundation for the Improvement of Living and Working Conditions. (1999). *Active strategies for an ageing workforce: Conference report.* Dublin, Ireland: Author.

Everard, K. M. (1999). The relationship between reasons for activity and older adult well-being. *Journal of Applied Gerontology, 18,* 325–340.

Freysinger, V. J. (1993). The community, programs and opportunities—Population diversity. In J. R. Kelly (Ed.), *Activity and aging: Staying involved in later life* (pp. 211–230). Newbury Park, CA: Sage.

Gilford, R. (1986, Summer). Marriages in later life. *Generations,* (pp. 16–20).

Gilleard, C., & Higgs, P. (2000). *Cultures of ageing: Self, citizen and the body.* London: Pearson Education.

Ginn, J., Street, D., & Arber, S. (2001a). Cross-national trends in women's work. In J. Ginn, D. Street, & S. Arber (Eds.), *Women, work and pensions* (pp. 11–30). Buckingham, UK: Open University Press.

Ginn, J., Street, D., & Arber, S. (2001b). Women's pension outlook: Variations among liberal welfare states. In J. Ginn, D. Street, & S. Arber (Eds.), *Women, work and pensions* (pp. 216–235). Buckingham, UK: Open University Press.

Graebner, W. (1980). *A history of retirement: The meaning and function of an American institution, 1885–1978.* New Haven: Yale University Press.

Hannah, L. (1986). *Inventing retirement.* Cambridge, UK: Cambridge University Press.

Havighurst, R. J. (1961). The nature and values of meaningful free-time activity. In R. W. Kleemeier (Ed.), *Aging and leisure* (pp. 309–344). New York: Oxford University Press.

Hazan, H. (1980). *The limbo people: A study of the constitution of the time universe among the aged.* London: Routledge & Kegan Paul.

Henderson, K. A., Bialeschki, M. D., Shaw, S., & Freysinger, V. J. (1989). *A leisure of one's own: A feminist perspective on women's leisure.* Philadelphia: Venture.

Henretta, J. (2003). The life course perspective in work and retirement. In R. Schersten (Ed.), *Invitation to the life course* (pp. 85–105). Amityville, NY: Baywood.

Hess, B. (1979). Sex roles, friendships and the lifecourse. *Research on Aging, 1,* 494–515.

Jamieson, L. (1998). *Intimacy: Personal relationships in modern societies.* Cambridge, UK: Polity.

Jerrome, D. (1981). The significance of friendship for women in later life. *Ageing and Society, 1,* 175–197.

Jerrome, D. (1984). Good company: The sociological implications of friendship. *Sociological Review, 32,* 696–718.

Jerrome, D. (1993a). Intimacy and sexuality amongst older women. In M. Bernard & K. Meade (Eds.), *Women come of age: Perspectives on the lives of older women* (pp. 85–105). London: Arnold.

Jerrome, D. (1993b). Intimate relationships. In J. Bond, P. Coleman, & S. Peace (Eds.), *Ageing in society: An introduction to social gerontology* (2nd ed., pp. 226–254). London: Sage.

Kaplan, M. (1979). *Leisure: Lifestyle and lifespan—perspectives for gerontology.* Philadelphia: Saunders.

Kelly, J. R. (1990, Spring). Leisure and aging: A second agenda. *Society and Leisure, 13*(1), 145–167.

Kelly, J. R. (Ed.). (1993). *Activities and aging: Staying involved in later life.* Newbury Park, CA: Sage.

Kelly, J. R., Steinkamp, M., & Kelly, J. (1986). Later life leisure: How they play in Peoria. *The Gerontologist, 26,* 531–537.

Kleemeier, R. W. (Ed.). (1961). *Aging and leisure.* New York: Oxford University Press.

Knapp, M., Koutsogeorgopoulou, V., & Davis Smith, J. (1996). Volunteer participation in community care. *Policy and Politics, 24*(2), 171–92.

Kohli M., Rein, M., Guillemard, A.-M., & Gunsteren, H. (Eds.). (1991). *Time for retirement: Comparative studies of early exit from the labour force.* Cambridge, UK: Cambridge University Press.

Lackzo, F., & Phillipson, C. (1991). *Changing work and retirement.* Milton Keynes, UK: Open University Press.

Larson, R. (1978). Thirty years of research on the subjective well-being of older Americans. *Journal of Gerontology, 33,* 109–125.

Laslett, P. (1989). *A fresh map of life.* London: Weidenfeld & Nicholson.

Lloyd, K. M., & Auld, C. J. (2002). The role of leisure in determining quality of life: Issues of content and measurement. *Social Indicators Research, 57,* 43–71.

Long, J. (1989). A part to play: men experiencing leisure through retirement. In B. Bytheway, T. Keil, P. Allat, & A. Bryman (Eds.), *Becoming and being old: Sociological approaches to later life* (pp. 55–71). London: Sage.

Long, J., & Wimbush, E. (1985). *Continuity and change: Leisure around retirement.* London: Sports Council and Economic and Social Research Council.

Lynn, P., & Davis Smith, J. (1991). *The 1991 national survey of voluntary activity in the U.K.* (Voluntary Action Research Second Series, Paper No. 1). Berkhamsted, UK: Volunteer Centre U.K.

Macnicol, J. (1998). *The politics of retirement in Britain 1878–1948.* Cambridge, UK: Cambridge University Press.

Mann, K. (2001). *Approaching retirement.* Bristol, UK: Polity.

Mansvelt, J. (1997). Working at leisure: Critical geographies of ageing. *Area, 29,* 289–298.

Marshall, P. (2000). Older women undergraduates: Choices and challenges. In M. Bernard, J. Phillips, L. Machin, & V. Harding Davies (Eds.), *Women ageing: Changing identities, challenging myths* (pp. 93–109). London: Routledge.

Mason, J. (1987). A bed of roses? Women, marriage and inequality in later life. In P. Allat, T. Keil, A. Bryman, & B. Bytheway (Eds.), *Women and the life cycle* (pp. 90–106). London: Macmillan.

McGoldrick, A., & Cooper, C. (1989). *Early retirement.* Aldershot, UK: Gower.

McGuire, F. (2000). What do we know? Not much: the state of leisure and aging research. *Journal of Leisure Research, 32*(1), 97–100.

McKay, S., & Middleton, S. (1998). *Characteristics of older women.* Centre for Research in Social Policy Research Report No. 45, Loughborough University. London: The Stationery Office.

Midwinter, E. (1992). *Leisure: New opportunities in the third age.* Dunfermline, UK: Carnegie United Kingdom Trust.

Midwinter, E. (1996). *Thriving people: The growth and prospects for the U3A in the UK.* London: Third Age Trust.

Midwinter, E. (1998). Age and education. In M. Bernard & J. Phillips (Eds.), *The social policy of old age: Moving into the 21st century* (pp. 40–55). London: Centre for Policy on Ageing.

Mowl, G., & Towner, J. (1995). Women, gender, leisure and place: Towards a more "humanistic" geography of women's leisure. *Leisure Studies, 14,* 102–116.

Neugarten, B. L. (Ed.). (1964). *Personality in middle and later life.* New York: Atherton.

Norton, D. (1992). Social provision for older people in Europe—In education and leisure. In L. Davies (Ed.), *The coming of age in Europe: Older people in the European community* (pp. 38–62). London: Age Concern England.

Office for National Statistics. (1999). *National statistics omnibus survey* [Online]. Available: http://www.statistics.gov.uk

Office for National Statistics. (2002). *Social Trends No 32.* London: HMSO.

Office of Population Censuses and Surveys. (1983). *General household survey, 1981.* London: HMSO.

Office of Population Censuses and Surveys. (1985). *General household survey, 1983.* London: HMSO.

Organisation for Economic Co-operation and Development. (2001). *Ageing and income: Financial resources and retirement in 9 OECD countries.* Paris: Author.

Pahl, R. (2000). *On friendship.* Cambridge, UK: Polity.

Palmore, E. G., Burchett, B., Fillenbaum, G., George, L., & Wallman, L. (1985). *Retirement: Causes and consequences.* New York: Springer.

Phillipson, C. (1982). *Capitalism and the construction of old age.* London: Macmillan.

Phillipson, C. (1992). Family care of the elderly in Great Britain. In J. Kosberg (Ed.), *Family care of the elderly* (pp. 252–270). London: Sage.

Phillipson, C. (1993). The sociology of retirement. In J. Bond, P. Coleman, & S. L. Peace (Eds.), *Ageing in society: An introduction to social gerontology* (pp. 180–199). London: Sage.

Phillipson, C. (1998). *Reconstructing old age.* London: Sage.

Phillipson, C. (1999). The social construction of retirement: Perspectives from critical theory and political economy. In M. Minkler & C. Estes (Eds.), *Critical gerontology: Perspectives from political and moral economy* (pp. 315–328). New York: Baywood.

Phillipson, C. (2002). *Transitions from work to retirement.* Bristol: Policy Press.

Phillipson, C., Bernard, M., Phillips, J., & Ogg, J. (2001). *The family and community life of older people: Social networks and social support in three urban areas.* London: Routledge.

Quadagno, J. (1982). *Ageing in early industrial society: Work, family and social policy in 19th century England.* London: Academic Press.

Rosow, I. (1975). *Socialization to old age.* Berkeley: University of California Press.

Rossi, A. S. (1986). Sex and gender in the aging society. In A. Pifer & L. Bronte (Eds.), *Our aging society: Paradox and promise* (pp. 111–139). New York: Norton.

Rowe, J. W., & Kahn, R. L. (1997). Successful aging. *The Gerontologist, 37,* 433–440.

Samdahl, D. M. (2000). Reflections on the future of leisure studies. *Journal of Leisure Research, 32*(1), 125–128.

Samdahl, D. M., & Kelly, J. J. (1999). Speaking only to ourselves? Citation analysis of *Journal of Leisure Research* and *Leisure Sciences. Journal of Leisure Research, 31*(2), 171–180.

Scales, J., & Scase, R. (2001). *Fit at fifty.* Swindon, UK: Economic Research and Social Research Council.

Schuller, T. (1989). Work-ending: Employment and ambiguity in later life. In B. Bytheway, T. Keil, P. Allat, & A. Bryman (Eds.), *Becoming and being old* (pp. 41–54). London: Sage.

Schuller, T., & Bostyn, A. M. (1992). *Learning: Education, training and information in the third age* (Research Paper No. 3). Dunfermline, UK: Carnegie U.K. Trust.

Searle, M. S. (2000). Is leisure theory needed for leisure studies? *Journal of Leisure Research, 32*(1), 138–142.

Seeman, T. E., Berkman, L. F., Charpentier, P. A., Blazer, D. G., Albert, M. S., & Tinetti, M. E. (1995). Behavioural and psychosocial predictors of physical performance: MacArthur studies of successful aging. *Journal of Gerontology: Medical Sciences, 50A,* M177–M183.

Shaw, S. M. (2000). If our research is relevant, why is nobody listening? *Journal of Leisure Research, 32*(1), 147–151.

Sheehy, G. (1997). *New passages.* London: HarperCollins.

Siegenthaler, K. L., & Vaughan, J. (1998). Older women in retirement communities: Perceptions of recreation and leisure. *Leisure Sciences, 20,* 53–66.

Sporakowski, M., & Axelson, L. (1989). Long-term marriages: A critical review. In L. Ade-Ridder & C. Hennon (Eds.), *Lifestyles of the elderly* (pp. 9–28). New York: Human Sciences Press.

Stokowski, P. A. (2000). Exploring gender. *Journal of Leisure Research, 32*(1), 161–165.

Stone, R., & Minkler, M. (1984). The socio-political context of women's retirement. In M. Minkler & C. Estes (Eds.), *Political economy of aging* (pp. 225–238). New York: Human Sciences Press.

Szinovacz, M. (1991). Women and retirement. In B. Hess & E. Markson (Eds.), *Growing old in America* (pp. 293–304). New York: Transaction.

Szinovacz, M. (1992). Leisure in retirement: Gender differences in limitations and opportunities. *World Leisure and Recreation, 34*(1), 14–17.

Szinovacz, M., & De Viney, D. J. (2000). Marital characteristics and retirement decision. *Research on Aging, 22,* 470–489.

Szinovacz, M., Ekerdt, D. J., & Vinick, B. H. (Eds.). (1992). *Families and retirement.* Newbury Park, CA: Sage.

Taylor, R. (2002). *Britain's world of work: Myth and realities.* Swindon, UK: Economic and Social Research Council.

Tibbitts, C. (1954). Retirement problems in American society. *American Journal of Sociology, 59,* 301–308.

Tokarski, W. (1993). Later life activity from European perspectives. In J. R. Kelly (Ed.), *Activity and aging: Staying involved in later life* (pp. 60–67). Newbury Park, CA: Sage.

Wall, R. (1992). Relationships between the generations in British families past and present. In C. Marsh & S. Arber (Eds.), *Families and households* (pp. 63–85). London: Macmillan.

Weiss, L., & Lowenthal, M. F. (1975). Life course perspective on friendship. In M. F. Lowenthal, M. Thurnher, & D. Chiriboga (Eds.), *Four stages of life* (pp. 48–61). San Francisco: Jossey-Bass.

Wenger, G. C. (1987). *Relationships in old age: Inside support networks* (Third report of a follow-up study of older adults in northern Wales, the Centre for Social Policy Research and Development). Bangor, UK: University College of North Wales.

Witt, P. A. (2000). If leisure research is to matter II. *Journal of Leisure Research, 32*(1), 186–189.

Wright, P. H. (1982). Men's friendships and women's friendships and the alleged inferiority of the latter. *Sex Roles, 8,* 1–20.

Young, M., & Schuller, T. (1991). *Life after work: The arrival of the ageless society.* London: HarperCollins.

V

Political and Mass Communication

Holladay and Coombs (chap. 15) turn to the political power of older adults, and in particular the role of communication in the creation and exercise of political power. Their analysis is particularly relevant to the United States and the United Kingdom. First, they focus on the influence of the media in shaping the political power of older adults. Second, the actual political participation of older adults is analyzed. Third, the access to power and the possible influence of aging interest groups is discussed. Finally, the future political power of older adults, with special attention to the aging baby boomers, is highlighted.

Holladay and Coombs cite research that has in fact shown that the attitudes of the seniors in the United States do not differ markedly from those of the young on most issues, even on those pertaining directly to seniors' concerns. Official figures show that there is also economic variation, but still a great deal of poverty, among seniors, displacing the myth of an affluent, but greedy, older population depriving the young of necessary financial resources. The case that the successful senior interest groups are an easy but inappropriate target for the problems of the young is vociferously argued here. We must question the myth of intergenerational conflict. A future approach the authors recommend is the use of case studies involving the exercise of senior political power and its influence on specific policy decisions. Holladay and Coombs make an excellent point concerning the current and future use of new technologies within the political process. The current negative stereotype that older adults are afraid of the Internet and will never adapt to e-mail is simply not consistent with the current facts and will be far from fact within the aging baby-boomer cohort.

Kaid and Garner (chap. 16) consider how aging concerns have been conveyed through political candidates' television campaign strategies in the United States. In this cultural context, there is a growing acknowledgment of the increase in numbers of the older population, and of their political commitment, coupled with the fact that the mass medium of choice for seniors is television (see also Robinson, Skill, & Turner, chap. 17, this volume). An important mode of political communication is the short spot commercial, a mode only currently used in the United Kingdom. The authors cite strong research evidence for the effectiveness of such advertising, but very little research previous to the authors' own, with colleagues, presented here, had directly considered ways in which seniors are addressed in campaign advertising.

Content analysis of the U.S. presidential campaign ads dating from 1960 to 2000 show that the number of advertisements addressing issues defined as related to older adults (generally, health care and social welfare issues) has changed drastically. Indeed, there was a complete absence in the 1992 elections. Individual campaigns and, more generally, political parties have portrayed the experience of being an older adult more often positively than negatively. Nevertheless, portrayals have been highly stereotypical, with positive portrayals, including senior citizens being seen talking to candidates about political issues, and negatively, including portrayals of older people as dependent and as victims. As Kaid and Garner point out, rarely are seniors portrayed as "contributing members of society who are working to live productive, independent lives." The authors suggest that seniors' problems are being used as fear appeals or as the target of optimistic reassurance, without their concerns being genuinely addressed (see Glendenning, chap. 21 this volume). Change, however, can be documented in the most recent (2000) presidential election in the United States. George W. Bush's spots may indicate a change in attitude by the Republican Party toward older adults. These particular political ads gave older adults a more positive image, as well as a more positive role.

Robinson, Skill, and Turner stress the importance of media usage by seniors, not only in terms of their high proportion of leisure time spent in this way. This is also an activity that impacts on other behaviors, interpersonal interaction, and perceptions of aging and social reality. The nature of portrayals of seniors as a social group is also potentially influential in these three areas, because of responses to such images by seniors and their interlocutors.

Self-reports on television viewing, radio listening, and newspaper and magazine reading indicate time spent and levels of preference, divided here into age groups 18 to 34, 55 to 64, and over 65. These data led Robinson, Skill, and Turner, and others here, to warn future researchers against treating seniors as a homogeneous group in the light of the socioeconomic and gender differences in media usage.

Studies of television portrayals of older characters (here, 65 and over) are reviewed. Media representations of older people are significant not only because of the power of the media to portray images of aging to all age groups (see Hepworth, chap. 1, this volume), but also because of the preferences expressed by older adults for watching television programs showing characters of their own age group. Generally speaking, the studies reported seem to show that the very young (usually, children and young teens) and the elderly (persons over 65) are underrepresented in comparison with census data, that males outnumber females, and that older characters have minor roles. The authors' own study replicates earlier findings. Robinson, Skill, and Turner make the point that "because TV shows are essentially narratives . . . the way stories are communicated tells us a great deal about the storyteller and the audience." Some research work presented here finds older audience members actively using TV characters as bases for self-social comparison, with unhappy viewers looking for characters to provide negative comparisons and avoiding positive comparisons with their own situation.

REFERENCE

Fowler, R. (1991). *Language in the news: Discourse and ideology in the press.* London: Routledge.

15

The Political Power of Seniors

Sherry J. Holladay and W. Timothy Coombs
Eastern Illinois University

INTRODUCTION

Media accounts of the political power of seniors commonly portray them as dominating the policy-making process. The following quotes are typical of media descriptions of the power of senior interest groups: "There may be a more feared special interest organization in Washington than AARP. But no one has the same capacity to fill congressional mailbags or jam the switchboards as does this 33-million member big business" (Hall, 1995, p. 4); senior interest groups are "the 800-pound gorilla" of politics (Mufson, 1990, p. A1); "Championing the cause of the elderly is one of Washington's growth industries" (Anderson & Van Atta, 1990, p. C19).

Groups representing seniors are described as nearly omnipotent and striking fear in the hearts of politicians. Seniors themselves frequently are faulted for exercising too much political power: "The elderly have become so powerful that very few politicians have the nerve to tangle with them" (Dychtwald, 1999, p. 24).

This sample of quotations from the print media is alarming. If the quotes reflect reality, then why aren't we living in a world dominated by the interests and agendas of seniors? Closer scrutiny reveals distortions in these media portrayals of seniors and their interest groups. In an effort to debunk these exaggerated images of the power of seniors, this chapter discusses the political power of seniors by examining the political landscape in the United States and the United Kingdom. First, we focus on the influence of the media in shaping the public's perceptions of the political power of seniors. The concept of framing is discussed to demonstrate how media portrayals influence people's perceptions of the political power of seniors. Although seniors are portrayed as a powerful voting bloc, their impact on the political process is limited.

Second, the actual political participation of older adults is reviewed. Factors related to political activity are identified. Research on their political behavior indicates that seniors vote at higher rates than the rest of the population, but their voting patterns differ from common media images. Political activity should not be confused with political power.

Next, the policy-making process is discussed to demonstrate the role and true political power of senior interest groups in the United States and United Kingdom. Although these interest groups represent the interests of seniors, their power and influence is limited to certain arenas.

Fourth, the future of the political power of seniors is discussed, including the future of their interest groups, the anticipated impact of the aging of the baby-boom generation, the oldest members of which begin turning 65 in less than a decade, and the potential role of the Internet in their political activity.

Fifth, suggestions for future research into the political power of seniors and their interest groups are forwarded. Future research could help demystify the impact of the senior population on politics in the United States and United Kingdom.

MEDIA PORTRAYALS OF SENIORS' POLITICAL POWER

To a great extent, the public is dependent on the news media for information about political issues (Iyengar, 1989; Linksy, 1986). The way in which the media cover policy issues pertaining to seniors, the "grey vote," and activities of senior interest groups influences the way the public perceives the political power of seniors. In this way, our images of the political influence of seniors are, to a large extent, a "media effect." The actual political activity and impact of older adults differs from common stereotypes.

Media Frames

To understand how the media influences public perceptions of political issues, it is instructive to cover the concept of framing. Framing refers to the way in which media portray events, issues, or people. Because events, issues, and people can be presented in multiple ways, reporters must organize information in a way that facilitates audience members' sense making and "makes a good story." They follow news values that emphasize factors like conflict, the impact on and proximity to viewers or readers, human interest, timeliness, and prominence (Treadwell & Treadwell, 2000). Because these news values are shared among the media, news stories tend to be covered in similar ways by different media organizations (Bantz, 1995).

The frames used by the media are powerful, because they can influence people's perceptions, public opinion, and policy decisions (Iyengar, 1989, 1990; Ryan, 1991). For example, depicting seniors as "greedy geezers," the "grey peril," or the "deserving poor," or presenting conflicting views about budget issues as "intergenerational conflict" or "age wars," both dramatizes and oversimplifies the issues, and provides vivid, memorable images. Because of common news values, these frames are likely to be repeated in different news stories, thus reinforcing the images presented to consumers.

In the United States, media coverage of the political power of seniors tends to perpetuate the image of a powerful voting bloc. The myth of the grey vote entails the depiction of seniors as a large, homogeneous, self-interested voting bloc. For example, quotations from public figures often refer to politicians' fear of senior interest groups and senior voters, serving to perpetuate the image of the power of these voters. Seniors "quickly become a unified power block whenever their interests are threatened" (Dychtwald, 1999, p. 23). Representatives from senior interest groups such as the American Association of Retired Persons (AARP) in the United States and Age Concern in the United Kingdom are eager to act as information sources for news coverage of issues pertaining to older adults. Their advocacy role entails persuading people that older adults are a force that cannot be ignored. In their role as an information subsidy (Gandy, 1992), these interest groups frequently are referenced in news reports, increasing the visibility and perceived importance of these organizations, their causes, and their constituency. They may be victims of their own success in that their prominence in the news makes them convenient targets of criticism (Coombs & Holladay, 1995).

To understand how the political activity of seniors came to be framed in less than positive ways in the United States, it is helpful to examine the types of frames and news values typically applied by the news media. In their review of the framing literature, Valkenburg, Semetko, and De Vreese (1999) observed that news is commonly framed various ways, by using: (1) conflict, (2) attribution of responsibility, and (3) economic consequences for the audience. Although a single frame may dominate a news story, more than one frame may be integrated within the same news report. In the case of the stories concerning the political power of seniors, the conflict frame is frequently used in conjunction with the attributing responsibility frame and the economic consequences frame.

The "conflict" frame portrays groups or people in conflict, thus adding drama to the news story. The coverage of seniors' role in American politics tends to be conflict driven, with the conflict emerging frequently in the coverage of senior policy issues. Typically, the interests of the young and the old are depicted as being in conflict (Binstock, 1992c; Kingson, Hirshorn, & Cornman, 1986; Phillipson, 1991; Wisensale, 1988). Budget battles may be framed as seniors win–children lose. References to "intergenerational conflict," "intergenerational justice," or "age wars" typify the conflict frame. The extent to which intergenerational conflict actually exists is not questioned. The terms provide a convenient symbol for depicting a complex issue.

The intergenerational conflict frame suggests that the old and young are in conflict because of scarce resources. In this frame, the demographic aging of society is viewed as a problem, because it threatens economic growth and burdens the working population (Walker, 1990). Writers argued that the interests of the old are supported without regard for younger generations, and this situation will lead to economic disaster when the baby boomers reach old age (Fairle, 1988; Longman, 1985, 1986; Pifer, 1986). They argued that children are suffering as an increasingly larger portion of the budget is devoted to seniors (Richman & Stagner, 1986; Smith, 1992).

Frames that blame the "greedy geezers" for consuming more than their fair share of the U.S. budget also reflect the "attributing responsibility" frame. Here, institutions, especially AARP and, more generally, the "senior lobby" and seniors individuals, are blamed for creating budget problems and preventing change that would benefit other needy groups and the population as a whole. Senior interest groups may be portrayed as making unreasonable demands and exercising inordinate influence among policy makers. This frame also implicates the conflict frame by suggesting that the responsibility for the conflict between generations lies with seniors and their representatives. For example, one writer claimed, "The elderly are currently the most malign influence on our politics, the beneficiaries of the most outrageous affirmative-action scam in the country . . . " (Steyn, 2000, p. 56), thus blaming economic woes on seniors. The lack of policy-making action often is attributed to seniors: "Veteran members of Congress are wary of the potential backlash if they attempt to tinker with Social Security" (Rosenblatt, 1993, p. A1) and "Fear of the seniors' lobby has largely held congress hostage when it comes to even discussing Social Security reform—which serves the country badly since the program will bankrupt the federal government if it isn't reined in" (Chavez, 1995, p. 13A). The greedy geezer frame involves portraying seniors as powerful, affluent, and motivated by self-interest to pursue age-based entitlements, irrespective of their effects on other generations (Minkler, 1986; Walker, 1986, 1990).

The third type of frame, the "economic consequences" frame, focuses on the economic consequences of the issue for the audience. The consequences may be portrayed as affecting a group, an institution, an individual, a region of the country, or other aspects of society. This frame is evidenced when stories reference how younger generations in the United States will be forced to bear the burden of affluent seniors' retirement, and how more younger people believe in unidentified flying objects (UFOs) than the availability of Social Security in their own old age. Predictions about when Social Security will go bankrupt often are included in this frame. The underlying idea is that seniors' unwillingness to make concessions will have dire economic consequences for future generations.

The framing of stories pertaining to pensioners in the United Kingdom (aged 65 and older for men and 60 and older for women) differs from that in the United States. Although the "demographic time bomb" metaphor is used commonly in both the United States and the United Kingdom (Vincent, 1999), the voting of pensioners in the United Kingdom has aroused less acrimony than in the United States. Vincent (1999) argued that political parties in the United Kingdom do not feel pressured to attend to the interests of pensioners. When political parties court the older vote, they typically focus on pension and welfare issues. Older voters do not seem to carry the same political clout as their counterparts in the United States. Political parties are not portrayed as giving in to the special interest groups representing pensioners.

Although Vincent (1999) reported that pensioners in the United Kingdom are generally portrayed as the "deserving poor" and as "in need" (p. 73), other researchers (Minkler, 1986; Walker, 1990) have suggested that politicians have been redefining older adults as more affluent and are attempting to use them as scapegoats. Their interest groups must walk a line between cultivating a positive image of pensioners and also seeing that needed social and medical services are provided to them.

In sum, the frames provided in media coverage influence viewers' and readers' perspectives on the issues involving the political power of older adults. Frames emphasizing conflicting interests between generations, the power and greed of the grey vote, and the economic impact of their political power are common in the United States, and senior interest groups in the United Kingdom are cognizant of the negative image of "grey power" in the United States and hope to avoid the stigma. Valkenburg et al. (1999) found that frames affected readers' thoughts and concluded that "news media can have the capacity not only to tell the public *what* issues to think about but also *how* to think about them" (p. 567). Media frames can distort and oversimplify reality in the interest of a "good story." Misconceptions about the political power of seniors are created and perpetuated. Typically, the frames do not focus on a significant part of the story—the minimal impact of senior voters on the actual policy-making process in the United States. The next section discusses the political involvement of seniors that provides the fodder for the media myths.

POLITICAL INVOLVEMENT OF SENIORS

As we have seen, fears regarding the political power of seniors and their interest groups have dominated media frames in the United States. In spite of the fact that the media tend to portray seniors as a large, monolithic voting bloc, research has demonstrated that they are diverse in their political attitudes and group identifications (Binstock, 1990, 1992b, 1992c; Hudson & Strate, 1985; Rhodebeck, 1993). As Binstock (1992b) observed, there is no reason to expect that seniors become homogeneous in their political behavior and self-interests as they grow old. They do not possess an age-consciousness that would lead them to bloc vote (Cutler, 1977). Nevertheless, news stories frequently mention how courting the "senior vote" is critical to winning elections (e.g., MacManus & Shields, 2000).

In the 1970s and 1980s, advocates for older adults in the United States and the United Kingdom were successful in portraying them as needy and deserving of aid (Binstock, 1992a). This portrayal, coupled with media coverage of the age group, may have contributed to the view that seniors are a homogeneous group (Binstock, 1992a). In the United Kingdom, the image of seniors as the "deserving poor," combined with less emphasis on their voting power, has enabled them to benefit from a more sympathetic image in the press.

Although the popular press has sounded warnings about the power of the grey vote and politicians have complained about their power, how accurate are these depictions of the political power of seniors? An examination of political attitudes and voting behavior illuminates the issue. The following discussion outlines findings related to the political activities of older adults in the United States and the United Kingdom.

Registration and Voting

One aspect of the media portrayals of the political activity of seniors that is accurate is the fact that they are more likely to be registered to vote and to vote than members of the younger generations (Alwin, 1998; Jennings & Markus, 1988; Stanley & Niemi, 1992; Verba & Nie, 1972). Voter registration is a necessary precursor to voting, and older adults are more likely to be registered than any other age group (Binstock, 2000). Among the total U.S. population, only 69.5% reported they were registered voters in 2000. Among those aged 65 and older, 78.4% were registered. About 75.4% of those aged 45 to 64 and

67.1% of those 25 to 44 were registered to vote in 2000. Only 50.7% of those aged 18 to 24 were registered (U.S. Bureau of the Census, 2002).

In the 2000 U.S. presidential election, only 59.5% of the voting age population voted. Among those aged 65 and over, 69.6% voted, representing 20% of the total U.S. vote. About 67.8% of those aged 45 to 64 and 56.1% of those 25 to 44 voted, representing about 35% and 37% of the total vote. Those aged 18 to 24 had the lowest voting participation rate, with only 36.1% voting, representing 7.8% of the total vote (U.S. Bureau of the Census, 2002).

Hence, in the United States, seniors are more likely to be registered to vote and exercise their vote. Although the age groups vote at different rates, older voters seem to vote similarly to other age groups (Binstock, 1997). Seniors may vote in greater numbers, but this does not mean that all members of the 65-and-over age group vote in the same way (Binstock & Day, 1996). As we will see, their voting preferences are diverse.

In contrast to the United States, in the United Kingdom, 95% of people are registered to vote (Howarth, 1999). As in the United States, greater numbers of older voters in the United Kingdom turn out for elections than younger voters. Pensioners (age 60 for women and 65 for men) are twice as likely to vote as are those 18 to 24 (Mortimore, 2000). Worcester (2001) reported preelection poll results indicating that 82% of those aged 65 and older, and 83% of those 55 to 64, compared to 63% of those 18 to 24, planned to vote in the 2001 election. A poll conducted prior to the 1997 election indicated that 68% of people over age 55 planned to vote, whereas, only 29% of those aged 18 to 24 intended to vote (as cited in "Values," 1999). In both the United States and the United Kingdom, older people vote at higher rates than other age groups.

Other Political Activity

Older adults are more interested in and knowledgeable about politics than younger generations and tend to follow the news more closely (Binstock, 2000; Binstock & Day, 1996; Jennings & Markus, 1988). It may be that they are more interested, because they feel they have more of stake in government programs from which they benefit (Binstock & Day, 1996; Day, 1990)

As they age, people are more likely to be politically active, especially in terms of political activities like voting (Jennings & Markus, 1988; Peterson & Somit, 1994). AARP's survey of political involvement indicated that older adults are more active than younger age groups in political activities such as participating in campaigns, working on local, state, or national issues, and voting in elections. Research has suggested that more passive forms of political involvement, like voting and contacting public officials, persist into later life (Binstock & Day, 1996; Jennings & Markus, 1988). More active political involvement, such as working on campaigns, tends to decrease with age (Verba & Nie, 1972). Older voters also express greater identification with their political parties than younger voters (Binstock, 1997, 2000; Binstock & Day, 1996). In both the United States and the United Kingdom, older voters tend to be more conservative than younger voters (Mortimore, 2000; Worcester, 1999).

Although they vote in greater numbers, research has suggested that seniors in the United States feel less able to affect the political process than do other age groups (Jennings

& Markus, 1988). The relatively low feelings of political efficacy may stem from their years of experience with the political process, which tends to support incrementalism rather than major change.

Other factors associated with participation in the political process include health and education. Older adults who are healthier are more likely to follow politics (Peterson & Somit, 1994), and people who are better educated are more politically active (Binstock & Day, 1996; Peterson & Somit, 1994).

Policy Preferences

The actual policy preferences of seniors are as varied as within other age groups. Research in the United States suggests there are no discernible differences between younger, middle-aged, and older people in political attitudes and behavior (Binstock & Day, 1996). All age groups consistently support programs for seniors (Conner, 1992; Cook, 1996; Day, 1990; Gilliland & Havir, 1990; Jacobs, 1990; Tropman, 1987). Rhodebeck (1993) found that younger age groups were more supportive than older adults of increased spending on programs that benefit seniors. Through the 1980s, seniors expressed a greater desire for increased spending on programs that would directly benefit them and less supportive of programs that would benefit all age groups. Being older, personally experiencing more financial difficulties, and holding more liberal views seems related to seniors' support of policies that benefit them. Hence, economic status and ideological views have a greater effect on policy preferences than does age (Rhodebeck, 1993).

In spite of media frames depicting them as "greedy geezers," seniors tend to be perceived as more deserving of public support than other groups, and opinion polls show that the public does not approve of cuts in age-based entitlements (Gilliland & Havir, 1990; Hudson & Strate, 1985). For instance, Americans of all ages are concerned about Social Security. Public support for programs like Social Security and health care increased in the 1980s (Rhodebeck, 1993). American voters reported that health care, an issue that cuts across generations and includes concerns such as Medicare, patients' rights, and prescription drug coverage for older adults, was an important issue in the 2000 presidential election (Blendon, Benson, Brodie, Altman, & James, 2000). About 80% of Americans in one survey reported that, in the 2000 presidential election, the candidates' positions on Social Security would have an extremely important or very important influence on their voting (Hinds, 2000). Old as well as young benefit from government-funded programs like Social Security and Medicare (Deets, 1991; Kingson, Hirshorn & Cornman, 1986; Minkler, 1986). Support for these programs also stems from the fact that younger generations benefit from the programs, because they make it possible for seniors to be more financially independent from their families (Kingson et al., 1986; Minkler, 1986).

The idea of intergenerational support may resonate more strongly with people than the intergenerational conflict frame popularized by the media and some politicians. It may be the case that people perceive more connections than conflicts between generations. For example, media attention has been devoted to the difficulties experienced by the "sandwich generation," people who are simultaneously caregiving for both children and parents. These caregivers may experience financial strains, as well as strains on their time, but nonetheless experience feelings of filial responsibility. People who are not currently assisting older parents may be likely to envision a time in the future when they will be called on to provide physical as well as financial assistance. The media's coverage of the strains of the sandwich generation may be sufficient to cause them to be sympathetic to programs for seniors.

When the media covers seniors issues, it seems natural that viewers' thoughts would turn to their own senior parents and relatives. The media and politicians often accuse seniors of voting only in terms of their own self-interest. However, what they seem to neglect is the fact that younger generations may see programs for seniors as benefiting their own self-interests as well. As noted by Greene and Marty (1999)

> Few adult children and grandchildren would stand by idly and see their forebears suffer from penury or untreated illness without feeling compelled to make significant financial and other personal sacrifices on their behalf. So, programs like Social Security and Medicare provide large insurance benefits to members of the workforce generation by providing a way of pooling the risk that they might individually face in substantial and in-kind responsibilities for older relatives who might fall into poverty and/or need expensive health care. (p. 646)

The programs help ensure their parents' financial independence, thus lessening their parents' dependence on them. In addition, their parents' accumulation of wealth may benefit them in the future through inheritances.

Greene and Marty (1999) argued for the "generational investment" paradigm, which acknowledges that programs like Medicare and Social Security are mechanisms that provide returns to older people for the investments made earlier in life—paying taxes, raising children, and so forth. Through investing in human capital (i.e., the well-being of younger generations), people are essentially engaging in "societal savings," which benefit the young as well as the old. In this way, the "generational accounts" are balanced.

This discussion of the political activity of seniors demonstrates that they currently do not possess the age consciousness that would lead them to vote as a bloc (Cutler, 1977). Many older adults do not identify as strongly with "seniors" as they do with other groups that are based on characteristics other than age, especially because old age may be stigmatized (Day, 1990). Day (1990) speculated that seniors will develop a group consciousness only when they become dissatisfied or frustrated. The economic, social, racial, and ethnic diversity of seniors discourages them from developing a sense of unity based on age (Binstock, 1983, 1992b, 1992c; Hudson & Strate, 1985; Rhodebeck, 1993; Walker, 1986, 1990). As demonstrated through the enactment and subsequent repeal of the Medicare Catastrophic Coverage Act of 1988, seniors do not think and act as one mind (Crystal, 1990; Rhodebeck, 1993; Williamson, 1998).

SENIOR INTEREST GROUPS

The previous sections have explained how the political power of seniors portrayed in the media is a distorted fiction. So what, then, is reality? Political power is a very complicated subject to assess correctly. This section provides a framework for assessing the political power of senior interest groups. It begins by explaining how to evaluate the strength of interest groups and then moves to a discussion of political power itself and evaluations of the political power of various senior interest groups in the United States and the United Kingdom.

Interest Group Strength

The United States and United Kingdom are pluralistic political systems based on the premise that concerns compete with one another. The idea is that good decisions are made through a public discussion of ideas. Concerns do not exist on their own; interest groups have developed to support and to push concerns. Interest groups are collections of individuals and organizations that link together to actively promote their concerns and values in the policy-making process (*Online Dictionary of Social Science*, 2001). Senior concerns are no exception; interest groups have formed to promote them in the United States and the United Kingdom. Hence, when we speak of the power of seniors, we are really talking about the power of senior interest groups. Because interest groups compete with one another, we often talk about the relative strength of each. *Fortune* magazine deemed AARP the strongest interest group in the United States in three of its four yearly surveys. The National Rifle Association bumped AARP to second in the 2001 list of the top lobbying groups (Birnbaum, 2001). It would be fair to say that AARP is considered a policy elite, a group that controls resources and influences policies (Von Beyme, 1996). The question becomes: What gives an interest group strength allowing it to become an elite? The strength of an interest group is a function of three factors: (1) power, (2) legitimacy, and (3) urgency (Mitchell, Agle, & Wood, 1997). We will define each of these factors and see how they relate to primary senior interest groups in the United States and the United Kingdom.

Power

Power is defined as the ability to get a person or group to do something they otherwise would not do (Mitchell, Agle, & Wood, 1997). An interest group has power if it can get politicians to do things they would not ordinarily do, such as introduce or vote for a particular piece of legislation. Power, in turn, is a function of access and mobilization. To change a politician's behavior and interest, you must have access to the politician. AARP has fairly easy access to U.S. politicians. Founded in 1968, AARP has over 34 million members, all of whom are 50 years of age and older and that are a mix of retirees and those near retirement. Numbers are power in politics, because the numbers reflect potential votes. It follows that politicians listen when AARP calls or visits. In addition, AARP is considered an expert on subjects related to aging. As a result, when legislation involving senior appears, AARP representatives are asked for their input and appear to testify about the legislation. AARP is a regular part of policy decisions involving seniors and a bona fide political elite.

The United Seniors Association (USA), formed in 1991, is a relatively new, conservative senior group devoted to lower taxes, smaller government, and a robust free enterprise system ("About USA," n.d.). With 550,000 members, USA is much smaller than

AARP and much less integrated into the policy-making system. USA is a political elite "wanna-be."

Age Concern, in the United Kingdom, is the largest charitable organization devoted to senior concerns ("Quick Facts," n.d.). Formerly the Committee for the Welfare of the Aged and then the National Old People's Welfare Committee, Age Concern traces its roots back to 1940 and the harmful effects of World War II on seniors. Age Concern transformed itself into an interest group by attempting to influence public policy in the 1970s. Although Age Concern does not support political candidates or parties, it accesses the policy-making process by acting as policy advisors on political issues related to aging (Pratt, 1993). The Association of Retired and Persons over 50 (ARP/O50), another U.K. senior interest group, is modeling itself after AARP. ARP/O50 has a membership of over 80,000. ARP/O50 also positions itself as a voice for seniors in the policy-making process ("Join," n.d.). Like Age Concern, neither have the membership numbers of AARP. However, ARP/O50 noted that it is ahead of AARP's numbers when AARP was its age, and observes that there is a pool of over 19 million people in the United Kingdom over 50 who are potential members ("About Us," n.d.).

Grassroots lobbying is when constituents, or voters, contact politicians about a concern. Grassroots lobbying is very powerful; when voters speak, politicians listen (Ryan, 1991; Wittenberg & Wittenberg, 1990). Effective grassroots lobbying requires the ability to get members of an interest group to take action to support the interest group's concerns. Large numbers are meaningless if the interest group's members are passive and do not actively support its political agenda. When a large number of constituents contact a politician, the politician will most likely attend to that concern. The grass roots must be rallied to be effective. Activating the grass roots involves three steps: (1) awareness of a concern, (2) reasons to take action, and (3) information that allows a person to take action—mobilizing information (Coombs & Cutbirth, 1998; Keefer, 1993). One, people do not act on a concern if they do not know about it. An interest group must be able to create awareness of a concern. Two, people must have some reason to take action—the concern must be relevant to them. The interest group should link the concern to the personal concerns of its membership. Three, people must have mobilizing information—know how, when and where to act. E-mail addresses, phone numbers, and locations for rallies are all examples of mobilizing information. People cannot act if they do not know how to do so (Coombs & Cutbirth, 1998).

Through e-mail and land mail, AARP can generate constituency pressure quickly. The AARP's Web pages note the organization's ability to generate thousands of constituent responses rapidly ("AARP FAQ," n.d.). An interest group is powerful when it can mobilize a large number of its members quickly. AARP can do exactly that. AARP has quick access to its member base, and this access has shifted from land mail to e-mail as technology altered the communication landscape of politics. AARP has links to its members, and those members are willing to take action. Moreover, AARP uses the Internet to link its members to politicians. The AARP site contains a page where people can write and send a letter to his or her members of Congress. The site directs the message by examining the members' postal address; the postal address then signals what member of the House and which senators should receive the message ("Legislative Issues," n.d.; "You Make," n.d.). Granted, when AARP puts out a call to action, thousands rather than millions respond, but thousands are very powerful in grassroots lobbying (Wittenberg & Wittenberg, 1990).

USA offers similar political provisions on its Web site. A member can sign up for e-mail alerts about senior concerns and easily contact officials in the legislative, executive, or even judicial branches. For instance, a member can go to the "Contact Congress" Web page and submit his or her address. The member can even select prewritten text for a number of issues or create his or her own message. With a few clicks of the computer, the message is sent to members of Congress ("About USA," n.d.). USA also has 778 grassroots leaders. These people volunteer to make others aware of concerns and to urge them to contact government officials ("Grassroots," n.d.).

Age Concern also uses its Web site to activate members. People can join the e-campaigner network, which alerts them to concerns and connects them to politicians. By visiting the Web site, a member can simply click on a link and be taken to a Web page where they can write a message to a politician. Again using the address as the key, the computer finds the relevant member of Parliament (MP). Age Concern goes a step further by developing a Web page for contacting prospective Parliament members during an election ("Get Involved," n.d.). Political involvement is increased when it is made easier, and the Internet makes contacting politicians very easy.

ARP/O50 does not have a Web page devoted to grassroots lobbying. However, ARP/O50 offers a bimonthly publication for informing members (the other three aged interest groups have print publications, too) and nearly 200 Friendship Centres. Friendship Centres are primarily social, offering members a place to relax, play cards, or bowl ten pin. However, the ARP/O50 Web site also notes that it is a place to talk about political concerns ("Join," n.d.). It follows that Friendship Centres could be used to make members aware of issues, develop concern for the issue, and to provide mobilizing information.

Power draws together a number of elements: number of members of an interest group, access to the interest group members, and access to politicians. Ideally, an interest group uses these three factors to connect its constituents and members to politicians. Through access to members, interest groups can activate the grass roots. The activation connects the members to the politicians. The significance of the activation is amplified by numbers. Access to politicians reinforces the grassroots efforts, and the grassroots messages increase the credibility of the interest group leaders. The power potential of senior interest groups is high. Currently, only AARP has tapped this potential to become a political elite.

Legitimacy

Legitimacy refers to whether or not people perceive a concern as proper or appropriate. Politicians and other people are more likely to support a concern if it is legitimate. People do not want to support a concern reflective of the lunatic fringe (Coombs, 1992). Senior concerns carry legitimacy (Vincent, 1999). Heath care, cost of living, discrimination, retirement, and abuse are all legitimate concerns. Moreover, people usually have older relatives or know they will one day older, so they can connect with older people's concerns even if they themselves are not seniors. If a group builds its identity around a legitimate issue and articulates its arguments rationally, it will be perceived as legitimate. The legitimacy potential is strong for all senior interest groups both in the United States and the United Kingdom. Only USA has limited appeal because of its very conservative agenda.

Urgency

Urgency refers to the need for action or commitment from the interest group members. AARP uses its grassroots lobbying and history to build urgency. People contacting politicians reflects a commitment to the concern. AARP has history of commitment to aging issues. Hence, its involvement with a concern is deep-seated rather than a passing fancy. As noted earlier, Age Concern, USA, and ARP/O50 demonstrate a similar urgency by promoting grassroots lobbying. This urgency potential is strong for senior interest groups.

Summary

AARP is at the pinnacle of interest group strength in the United States. It has a very large and politically active membership and the means to activate it for grassroots lobbying. AARP leaders are themselves ingrained into the policy-making process as expert witnesses for government hearings. The concerns promoted by AARP are generally considered legitimate, and AARP's commitment to aging concerns raises the urgency factor. We have painted a picture of AARP as a strong interest group, because it has all three factors of strength: power, legitimacy, and urgency. In fact, ARP/O50 in the United Kingdom, and USA in the United States, openly model themselves after AARP in terms of grassroots lobbying. Age Concern's long history gives it easy access to politicians in the United Kingdom, which complement its grassroots lobbying efforts. Although Age Concern, ARP/O50, and USA have legitimacy and urgency, they lack the power AARP derives from its massive political base and political access. AARP is what many other senior interest groups hope to become—an extremely strong interest group. Note, we have not said that AARP has strong political power, as many in the media are want to proclaim. As an interest group, the political power of even AARP is bounded. Our forthcoming discussion of political power will expand on that point. We must contextualize interest group strength to understand its true political power.

POLITICAL POWER: A CLEARER PICTURE

The impression created in the United States is that seniors are a very powerful political force. This conclusion is derived from a limited view of their political power. We need to examine the broader meaning and context of political power to truly assess the power of seniors. For our purposes, political power is the ability to create public policy, laws, and regulations. Political power has two components, influence and decision making. Influence is the ability to shape decisions by persuading the policy makers. Decision making involves saying yes or no to a policy. Seniors have strong influence, in a limited policy arena, and weak decision making. We will elaborate on each of these components as a way of unpacking the true political power of seniors.

Influence

To understand influence, we need to consider how policy is made in pluralist political systems. There are three basic stages to policy making: (1) placement of the policy agenda, (2) selection of policy options, and (3) implementation of the policy option (Almond & Powell, 1984; Eyestone, 1978; Ripley, 1987). The power of a political entity, such as seniors, can vary from stage to stage. We should note that when influencing policy, we are taking about seniors as members of interest groups. In the United States and the world, AARP is the largest and most vocal senior interest group, so it will be used to illustrate discussion of influence. The implications for other senior interest groups are the same as those for AARP.

A concern receives attention only after it enters the policy agenda. Many important issues and concerns are never acted on, because they cannot break through to the policy agenda (Manheim, 1987; Rogers & Dearing, 1988). Once on the political agenda, politicians will consider various options for resolving a concern. An interest group must ensure that the politicians take action on their concern and select its preferred policy option. For example, the insurance industry wants to ban the use of handheld mobile phones while driving. It is not enough to get the concern on the agenda. Politicians must take action—make a decision on the concern—and the outcome must be the one desired by the insurance industry. A concern can linger and die on the political agenda if the concern is discussed without action being taken on it. The House of Representatives might discuss the use of mobile phones while driving but never vote on a proposed law. Or the politicians can select an undesirable policy option. Instead of banning mobile phones, a stronger warning statement about the dangers of driving while using a mobile phone is placed in the information consumers receive about mobile phones. In sum, an interest group has power when it can place concerns on the political agenda, see that action is taken, and persuade politicians to accept their preferred policy option.

AARP is considered one of the strongest interest groups in U.S. politics, so we will use it to illustrate the influence of senior interest groups. Senior concerns are easy to place on the political agenda. This ease is a function of political champions, grassroots lobbying, and media coverage. A political champion is a someone on the inside who will support and push for your concern. AARP has found no lack of champions to support its causes. In general, people support senior concerns, so politicians are happy to champion popular concerns among voters. AARP has a system, including e-mail alerts, that can rally the grass roots. When voters say a concern is important, politicians take notice. Expressions of public opinion are a powerful influence on the policy agenda (Cassara, 1998; Cobb & Elder, 1972). Media coverage is another form of influence. An interest group can use the media to shape the public debate by presenting the group's message in news stories (Terkildsen, Schnell, & Lang, 1998). AARP is the eighth most cited interest group in the U.S. newspapers, averaging over 400 stories per year (Fountain, 2001). When AARP wants a concern to appear on the policy agenda, it usually does.

Although senior concerns appear regularly on the political agenda, they are not always acted upon. Prescription drug prices provide an excellent illustration. Politicians in the United States, including President George W. Bush when he campaigned, have talked about the problem of high drug costs for seniors. However, talk without action does not address the concern. Even with its strong grassroots presence, AARP has been unable to get action on prescription drug prices, and politicians do not always select the policy options desired by AARP. Early attempts by the Clinton administration to reform health insurance would have benefited seniors, but Congress did not support them.

In contrast, AARP is extremely successful at protecting Social Security and Medicare. Popular wisdom suggests it is political suicide in the United States to tamper significantly with either. For instance, we have seen an increase in the age for receiving benefits in the United States but little beyond that. AARP found itself in hot water with seniors when it supported Medicare changes in the 1990s. The protection of Social Security and Medicare is more a function of the power of the status quo than the power of seniors. It is harder to create political change than it is to stop it (Cobb & Elder, 1972; Ryan, Swanson, & Buchholz, 1987). Hence, AARP can easily stop advances on established policies such as Social Security and Medicare, but its ability to have politicians take actions and to accept its policy options is overstated. Politicians do not roll over for this "massive" voting block each time the AARP trots out a concern. AARP's successes are of the easy variety. It has not fared as well in the tough policy battles. Their victories have involved policies that are noncontroversial and enjoy widespread popular support such as protecting funding for social security.

Decision Making

The second component of political power is decision making. Seniors have opportunities to make decisions, because they are voters. Voters can decide ballot initiatives—policy decisions that are placed before voters. In the United States, ballot initiatives appear on the state and local level. In contrast, Switzerland often has national-level initiatives. In the United States, interest groups are involved in decision making but on a limited scale, although interest groups do get involved when voters are asked to decide. In the 2000 election, California and Michigan had ballot initiatives about school vouchers, a system that would allow a family to take the tax money it would have paid to support its local school district—a voucher for a set amount of money—and to use the money to pay for private school fees. A variety of interest groups came out to support and oppose each side (Chase, 2000). Ballot initiatives are the exception rather than the rule in political decision making—most decisions are made by politicians, not the voters.

As discussed in the previous section, seniors are viewed as a powerful voting bloc, because a higher percentage of seniors register to vote and vote than any other segment of the population. Moreover, the number of seniors and their percentage of the population in the United States and the United Kingdom is growing. The implication is that this massive bloc of active voters imposes its will by voting for its own issues and defeating others. There are two serious flaws to this perspective. First, seniors are not a monolithic voting bloc (Binstock, 2000; Vincent, 1999). Like any age group, seniors have diverse interests and concerns. USA formed because it disagreed with AARP's liberal agenda ("About USA," n.d.). Second, seniors support a variety of concerns, not just their own interests. AARP has been active in supporting school initiatives, and research data shows that senior voters support the concerns of others ("U.S. Department of Education," 1999). The decision-making power of seniors is limited, because so few ballot initiatives appear and seniors are not a monolithic voting bloc in the United States or the United Kingdom (Binstock, 2000; Vincent, 1999).

Summary

The political power of senior interest groups is more media myth than reality. AARP is a political elite at or near the top of the interest groups in terms of influence. USA, Age Concern, and ARP/O50 are using similar strategies to tap into this influence. The key is that senior interest groups have a large constituency that is politically active. Their members are more likely to contact a politician and to vote than other age groups, making them a serious grassroots threat. The members can be activated, and they do vote—this will draw the attention of politicians and get senior concerns on the policy agenda. However, all interest groups are limited in political power. Even AARP cannot force politicians to act on a concern or choose its desired resolution. If AARP could, the United States would have seen action on the high cost of prescription drugs and health care long ago. Further, senior interest groups are not the unified voting bloc the media portrays. As a result, they do not control the outcomes of the limited number of ballot initiatives that appear in the United States and United Kingdom. When we look at the totality of political power, we realize that senior interest groups have limited power but strong influence over the policy agenda. AARP provides an example of how senior interest groups can utilize their resources to become strong interest groups. But even strong interest groups have their limits.

THE FUTURE OF SENIOR POLITICS

Senior Interest Groups

What challenges face senior interest groups in the future? Although AARP is well established, other interest groups are attempting to emulate its strategies. AARP has a very broad-based, diverse membership (those aged 50 and up). It will find it increasingly difficult to meet the needs of such a diverse constituency. It changed its name to AARP from the American Association of Retired Persons as part of a change in its identity to appeal to the "younger old"—baby boomers nearing retirement age who do not see themselves as "old." In doing so, it may have alienated its older members.

Other senior interest groups are likely to compete with AARP on specific issues. Because of their economic, social, and ethnic diversity, AARP members may not embrace the organization's stand on all issues. AARP needs to balance the diverse and sometimes conflicting demands of its constituencies. Alternative organizations like USA may appeal to older voters who object to AARP's agenda. USA and other smaller organizations may find "niche markets" by emphasizing specific concerns, such as conservative agendas and the preservation of Social Security.

It seems likely that senior interest groups in the United States and United Kingdom will continue to promote intergenerational programs and alliances to combat media frames depicting them as self-serving. For example, AARP cosponsored Student Vote 2000, a program designed to help about 2.1 million students and their parents consider issues important to them and their families. AARP's voter education campaign was directed toward people of all ages and provided information about candidates and issues, sponsored forums, and encouraged voting. The intergenerational program attempted to deal with the image of the dominant older voter. In the United Kingdom, Age Concern's Millennium Debate of the Age was a project involving all age groups. It was designed to create cooperation between generations and develop awareness of the implications of demographic changes in Great Britain and to recommend policy initiatives related to these demographic realities.

Senior interest groups will continue to be active in promoting awareness of the way in which their favored issues affect people of all ages. People of all ages benefit from programs like Social Security and Medicare. The high cost of health insurance and prescription drugs affects everyone. Most people will have senior parents and relatives, and senior interest groups can highlight how their agendas are family friendly. Forging intergenerational links may help senior interest groups preserve their legitimacy and prevent their opponents from using unfair criticism to take resources from seniors.

The Aging Generation of Baby Boomers

The baby-boom generation, because of its sheer numbers and diversity, has the potential to alter how aged interest groups operate and the issues they promote. Like the current group of seniors, the baby boomers (born between 1946 and 1964) are not a homogeneous group ("Baby Boomers," n.d.). Baby boomers are expected to be more economically and socially diverse than current seniors (Williamson, 1998). The U.S. boomers will begin turning 65 in 2011 and will swell the senior proportion of the population to 20% by 2030 ("Estimates," n.d.).

Baby boomers are described as better educated than previous generations, a characteristic that may lead them to be more politically active. Money may be a problem for some boomers who emphasized "consumerism" and did not save or invest money like previous generations. Even though they tended to be workaholics in the 1980s, it is likely that some will experience some financial difficulties in old age. The early boomers who experienced greater opportunities are likely to be better off financially than the later boomers who experienced a less lucrative job market and less favorable purchasing power. Boomers also tend to have a sense of "entitlement" and are self-indulgent (Smith & Clurman, 1997), factors that may lead them to become dissatisfied with their economic circumstances in later life. However, their identification with a "youth culture" may make them reluctant to identify with "old age" and senior issues (Day, 1990).

Baby boomers seem to have become somewhat more conservative over time (Alwin, 1998; Binstock, 2000; Williamson, 1998). Fewer identify themselves as Independents or Democrats, and more describe themselves as Republicans (Alwin, 1998). If the boomers embrace conservative agendas, government-supported programs may be at risk. Although we commonly associate the 1960s, early 1970s, and the college-aged boomers with liberalism and political activism, boomers are unlikely to become activists in old age. In spite of media coverage that indicated that activism was widespread, relatively small proportions of the young boomers participated in the protests (Williamson, 1998). Because of their diversity, it will be difficult for a single issue to capture their attention.

Williamson (1998) predicted that baby boomers are much more likely to engage in "checkbook activism" to support their causes rather than personal activism. This is likely to finance lobbyists and media campaigns, and is likely to be more important than participation at political events. However, if senior baby boomers become frustrated with government-supported programs and their financial security and unify and focus on a single issue, like Social Security, their large numbers could make them a political force.

Technology and Future Political Activity of Seniors

In the future, the Internet is expected to play an even greater role in the political arena. Although the Internet was heralded as ushering in a new era in democracy, it has fallen short of those optimistic predictions. As use of the Internet expands and politicians and interest groups determine how to reap benefits from its use, the Internet's role in politics will expand.

Future generations will continue to become more comfortable using the Internet for a variety of functions. One survey showed that nearly 67% of Americans use the Internet at home, work, or school ("UCLA Center for Communication Policy," 2000). Another indicated that 45% of adults in the United States had used online services in the past 30 days (Interep, 2000). People who use the Internet and other online services are more likely to vote, be above average in income, and be better educated than the rest of the population (Rash, 1997). These three factors are associated with greater political involvement. Research suggests that baby boomers and current older adults go online for news and information (Smith & Clurman, 1997). As a general information source, the Internet ranks ahead of television, radio, and magazines. Only books and newspapers were ranked ahead of the Internet ("UCLA Center for Communication Policy," 2000). More than two thirds of the Internet users in the survey reported that the Internet is "important" or "extremely important" as an information source. Its importance as a news source is expected to grow as more people become familiar with the technology.

Estimates indicate that 15% of people aged 65 and older are Internet users and average about 8 hours each week online ("Nua Internet Survey," 2000). This rate of use puts them ahead of college students, adults, and teens. This survey estimates that by 2003, 27% of Americans over 55 will be actively using the Internet.

How do people use the Internet to obtain information about political issues? The University of California–Los Angeles study found that the Internet is viewed as offering an important information resource about politics. About 45% of the Internet users in the survey agreed, and about 21% disagreed, with the statement, "by using the Internet, people like you can better understand politics." However, survey respondents were more skeptical of the Internet's influence on politics and government, with only about 29% of users agreeing with the statement, "By using the Internet, people like you can have more political power." About 25% of users agreed with the statement, "By using the Internet, people like you have more say about what the government does." Nearly 28% of the users agreed that "by using the Internet, public officials will care more about what people like you think."

Developing technologies will help interested users collect the information they desire and participate in political activities. Politicians as well as voters will find the Internet to be a useful tool. The Internet is used to send press releases, mobilize members, create online groups, and collect data through online polling. Politicians prefer the Internet when they want to disseminate information without it being filtered. A Web site could be helpful when candidates believe that reporters are not clearly articulating their stances or providing enough depth (Birnbaum, 2000; Fromm, 2000; Rash, 1997). The Internet is relatively inexpensive for candidates to use when compared to budgets from TV, print, and radio ads (Fromm, 2000; Rash, 1997). The Internet makes it easy to make a campaign contribution (Birnbaum, 2000; Fromm, 2000). Users also may have an opportunity to interact directly with politicians and access policy-related information (Birnbaum, 2000).

Although only about 20% of Americans currently get political information from the Internet, these users are likely to be highly motivated and may be capable of swinging the outcome in low-turnout elections (Birnbaum, 2000). In the United States, the campaigns of Jesse Ventura (governor of Minnesota) and John McCain (presidential candidate) demonstrated that the Internet can be used effectively to mobilize volunteers and collect campaign contributions (Birnbaum, 2000; Fromm, 2000). Candidates who support senior issues may find that seniors are eager to help their causes and have the leisure time to devote to political activity. Senior interest groups can also use their Internet presence to facilitate political involvement. According to AARP's Web site, AARP/VOTE, the purpose of this portion of the Web site is "to educate and involve voters on issues of concern to older Americans and the community at large." Senior interest groups have been at the forefront of integrating the Internet as part of their political strategy. However, they cannot abandon their print and interpersonal strategies for fear of alienating some constituencies. Although Internet use is relatively high, those with less education and lower incomes are less likely to use the technology. The Internet provides an opportunity for older adult users who are interested in more passive political activities to access political information and participate in the political process through grassroots lobbying.

FUTURE RESEARCH

Future research should be concerned with both media portrayals and the actual political behaviors of senior interest groups and voters. The media's use of the intergenerational conflict frame should be examined longitudinally. This frame seemed to be more dominant in the late 1980s and early 1990s. If the use of the conflict frame has declined, can it be explained by the efforts of senior interest groups to build intergenerational linkages?

Aged interest groups have attracted a great deal of attention from politicians and the media. How are the activities of senior interest groups portrayed in discussions of reform for Social Security and Medicare? Conservatives consistently use AARP as a scapegoat for failures to reform the programs. They portray AARP as a symbol of senior interest groups and a roadblock to any change, even constructive change. Closer examination of attempts to reform Social Security and Medicare should be undertaken to reveal the veracity of conservative's complaints. Only through such examinations can we illuminate the impact these groups have on proposed reforms. We must determine if they are impediments to any change or selective in their efforts, intervening only when their constituency is threatened. How is the legitimacy of senior interest groups affected by their support of certain issues?

Researchers should look at the growth and success of the various interest groups. Are the memberships of niche groups or those with a more broad-based appeal growing at a faster rate? What appeals to constituents seem to be more effective? What does it mean to be "successful" as a senior interest group? What indexes could be used to gauge the success of senior interest groups? For example, success could be examined in terms of the number of times the groups have served as media sources (their visibility). It also could be assessed in terms of the groups' abilities to position their issues on the policy agenda and to prompt politicians to take action on them. Success also could be measured in terms of the group's ability to mobilize its members.

A case study analysis could be used to evaluate the success of AARP's attempts to alter its identity and appeal to relatively younger older adults: Did its strategies attract new members while maintaining old members?

The political activity of senior adults also warrants investigation. Although we have alluded to the potential of the Internet to stimulate political activity, we do not know how Internet users search for and use the political information available on the Web. What types of information do senior voters find most helpful? To what extent are surfers using Web sites for grassroots organization?

The current generation of seniors has been accused of being "greedy geezers." As the large baby-boom generation ages and begins to qualify for age-based benefits, to what extent are they portrayed negatively? Writers have foreshadowed increasing resentment toward seniors among younger voters. Will the actions of the boomers intensify perceptions of them being "greedy elders"?

As demonstrated by this brief inventory of areas for future research, investigations of the political power and activity of senior interest groups and older adults, along with the media reports of their activities, provide opportunities to separate fact from fiction.

CONCLUSION

In conclusion, this chapter has examined the common (mis)perceptions and reality of the political power of seniors in the United States and United Kingdom to demonstrate how the activities of the media, older voters, and their interest groups influence how we conceive of the political power of seniors. Media coverage in the United States perpetuates the view of seniors as a self-interested, monolithic voting bloc. Reports of the political participation of older adults emphasize their voting rates but downplay their diversity of interests and identifications. Intergenerational dependencies also are neglected. The role of senior interest groups in the policy-making and decision-making processes was discussed to demonstrate that although they are influential in setting the policy agenda, their actual decision-making and policy-making power is limited. The future of senior politics was discussed to illustrate how senior interest groups, baby boomers, and the Internet are likely to affect the political power of older adults. Last, directions for future research were presented to guide scholars in their investigations into the political power of seniors.

REFERENCES

Age Concern. (n.d.). *Get involved with age concern* [Online]. Retrieved June 14, 2001. Available: http://www.ageconcern.org.uk/SiteArchitek/news.nsf/html/4HGPYX?OpenDocument&style=Get_involved

Age Concern. (n.d.). *Join the age concern e-campaign network* [Online]. Retrieved June 14, 2001. Available: http://www.ageconcern.org.uk/SiteArchitek/news.nsf/html/4J2J5K?Open Document&style=news

Age Concern. (n.d. c). *Quick facts* [Online]. Retrieved June 14, 2001. Available: http://www.ageconcern.org.uk/SiteArchitek/campaignalerts.nsf/campaignalertreg?OpenForm

AARP. (n.d. a). *AARP FAQ* [Online]. Retrieved June 14, 2001. Available: http://aarp.org/aarpfaq.html

AARP. (n.d. b). *Baby boomers envision their retirement: An AARP segmentation analysis* [Online]. Retrieved June 14, 2001. Available: http://research.aarp.org/econ/boomer_seg_1.html

AARP. (n.d. c). *Legislative issues* [Online]. Retrieved June 14, 2001. Available: http://www.aarp.org/legislativeguide

Administration on Aging. (n.d.). *Estimates and projections of the older population by age group: 1990–2050* [Online]. Retrieved June 21, 2001. Available: http://www.aoa.dhhs.gov/aoa/stats/prog1tbl.html

Age Concern. (1999). *Values and attitudes in an ageing society.* Glasgow: Bell & Bain Ltd.

Almond, G. A., & Powell, G. B., Jr. (1984). *Comparative politics today: A world view* (3rd ed.). Boston: Little, Brown.

Alwin, D. F. (1998). The political impact of the baby boom: Are there persistent generational differences in political beliefs and behavior? *Generations, 22*(1), 46–54.

Anderson, J., & Van Atta, D. (1990, April 10). Lobbyists tap into "gray power." *Washington Post,* p. C19.

ARP/O50. (n.d.). *About us* [Online]. Retrieved June 14, 2001. Available: http://www.arp.org.uk/about-social.html

Bantz, C. (1995). Organizing and enactment: Karl Weick and the production of news. In S. R. Corman, S. P. Banks, C. R. Bantz, & M. E. Mayer (Eds.), *Foundations of organizational communication: A reader* (2nd ed., pp. 151–160). White Plains, NY: Longman.

Binstock, R. H. (1983). The aged as scapegoat. *The Gerontologist, 23,* 136–143.

Binstock, R. H. (1990). The politics and economics of aging and diversity. In S. A. Bass, E. A. Kutza, & F. M. Torres-Gil (Eds.), *Diversity in aging: Challenges facing planners and policy-makers in the 1990s* (pp. 73–99). Glenview, IL: Scott, Foresman.

Binstock, R. H. (1992a). Another form of "elderly bashing." *Journal of Health Politics, Policy and Law, 17,* 269–272.

Binstock, R. H. (1992b). Older voters and the 1992 presidential election. *The Gerontologist, 32,* 601–606.

Binstock, R. H. (1992c). The oldest old and "intergenerational equity." In R. M. Suzman, D. P. Willis, & K. G. Manton (Eds.), *The oldest old* (pp. 394–417). New York: Oxford University Press.

Binstock, R. H. (1997). The 1996 election: Older voters and implications for policies on aging. *The Gerontologist, 37,* 15–19.

Binstock, R. H. (2000). Older people and voting participation: Past and future. *The Gerontologist, 40,* 18–31.

Binstock, R. H., & Day, C. L. (1996). Aging and politics. In R. H. Binstock & L. K. George (Eds.), *Handbook of aging and the social sciences* (4th ed., pp. 362–387). San Diego, CA: Academic Press.

Birnbaum, J. H. (2000, March 6). Politicking on the Internet. *Fortune, 141,* 84–86.

Birnbaum, J. H. (2001, May 28). Fat and happy in D.C. *Fortune, 143,* p. 94–5, 97–100.

Blendon, R. J., Benson, J., Brodie, M., Altman, D., & James, M. (2000). Health care in the upcoming 2000 election. *Health Affairs, 19*(4), 210–221.

Cassara, C. (1998). U.S. newspaper coverage of human rights in Latin America, 1975–1982: Exploring President Carter's agenda-building influence. *Journalism and Mass Communication Quarterly, 75,* 478–486.

Chase, B. (2000, Nov. 16). Vouchers fail to impress. *USA Today,* p. 18A.

Chavez, L. (1995, June 28). Simpson prods congress to stand up to AARP. *USA Today,* p. A13.

Cobb, R. W., & Elder, C. D. (1972). *Participation in American politics: The dynamics of agenda-building*. Baltimore: Johns Hopkins University Press.

Conner, K. A. (1992). *Aging America: Issues facing an aging society*. Englewood Cliffs, NJ: Prentice Hall.

Cook, F. L. (1996). Public support for programs for older Americans: Continuities amidst threats of discontinuities. In V. L. Bengtson (Ed.), *Adulthood and aging: Research on continuities and discontinuities* (pp. 327–349). New York: Springer.

Coombs, W. T. (1992). *The RBOC's quest for competition: A case study of "values" in public policy argument*. Paper presented at the annual meeting of the Speech Communication Association, Chicago.

Coombs, W. T., & Cutbirth, C. W. (1998). Mediated political communication, the Internet, and the new knowledge elites: Prospects and portents. *Telematics and Informatics, 15,* 203–218.

Coombs, W. T., & Holladay, S. J. (1995). The emerging political power of the elderly. In J. F. Nussbaum & J. Coupland (Eds.), *Handbook of communication and aging research* (pp. 317–341). Mahwah, NJ: Lawrence Erlbaum Associates.

Crystal, S. (1990). Health economics, old-age politics, and the catastrophic Medicare debate. *Journal of Gerontological Social Work, 15*(3/4), 21–31.

Cutler, N. E. (1977). Demographic, social, psychological, and political factors in the politics of aging: A foundation for research in "political gerontology." *American Political Science Review, 71,* 1011–1025.

Day, C. L. (1990). *What older Americans think: Interest groups and aging policy*. Princeton, NJ: Princeton University Press.

Deets, H. B. (1991, April–May). Young and old, we are one people. *Modern Maturity*, p. 9.

Dychtwald, K. (1999). *Age power: How the 21st century will be ruled by the new old*. New York: Tarcher/Putnam.

Eyestone, R. (1978). *From social issues to public policy*. New York: Wiley.

Fairlie, H. (1988, March 28). Talkin' 'bout my generation. *The New Republic, 198*(13), 19–22.

Federal Election Commission. (n.d. a). *Voter registration and turnout by age, gender & race 1998* [Online]. Retrieved June 27, 2001. Available: http://www.fec.gov/pages/98demog.htm

Federal Election Commission. (n.d. b). *Voter registration and turnout in federal elections by age, 1972–1996* [Online]. Retrieved June 27, 2001. Available: http://www.fec.gov/pages/agedemog.htm

Fountain, K. (2001). *Political advocacy groups: Newspaper citations methodology* [Online]. Retrieved June 27, 2001. Available: http://www.csuchico.edu/~kcfount/citations.html

Fromm, E. (2000, May 15). suppl. Digital Marketing). E-savvy politics. *Marketing Magazine, 105*(19), 22–23.

Gandy, O. H., Jr. (1992). Public relations and public policy: The structuration of dominance in the information age. In E. L. Toth & R. L. Heath (Eds.), *Rhetorical and critical approaches to public relations* (pp. 131–164). Hillsdale, NJ: Lawrence Erlbaum Associates.

Gilliland, N., & Havir, L. (1990). Public opinion and long-term care policy. In D. E. Biegel & A. Blum (Eds.), *Aging and caregiving: Theory, research, and policy* (pp. 242–253). Newbury Park, CA: Sage.

Greene, V. L., & Marty, K. (1999). Generational investment and social insurance for the elderly: Balancing the accounts. *The Gerontologist, 39,* 645–647.

Hall, J. (1995, June 22). Senators dodge AARP hearings. *The Tampa Tribune*, p. 4.

Hinds, M. D. (2000, September). Voter 2000. *American Demographics, 22*(9), 20–21.

Howarth, G. (1999, October). Final report of Working Party on electoral procedures.

Hudson, R. B., & Strate, J. (1985). Aging and political systems. In R. H. Binstock & E. Shanas (Eds.), *Handbook of aging and the social sciences* (2nd ed., pp. 554–585). New York: Van Nostrand Reinhold.

Interep. (2000, December 13). *Interep research: Extreme peaks begin to moderate as adult on-line usage approaches 50% threshold nationwide* [Online]. Retrieved June 14, 2001. Available: http://www.interep.com/Tpress.asp

Iyengar, S. (1989). How citizens think about national issues: A matter of responsibility. *American Journal of Political Science, 33,* 878–900.

Iyengar, S. (1990). Framing responsibility for political issues: The case of poverty. *Political Behavior, 12,* 19–40.

Jacobs, B. (1990). Aging and politics. In R. H. Binstock & L. K. George (Eds.), *Handbook of aging and the social sciences* (3rd ed., pp. 349–361). San Diego, CA: Academic Press.

Jennings, M. K., & Markus, G. B. (1988). Political involvement in the later years: A longitudinal survey. *American Journal of Political Science, 32,* 302–316.

Keefer, J. D. (1993). The news media's failure to facilitate citizen participation in the congressional policy making process. *Journalism Quarterly, 70,* 412–424.

Kingson, E. R., Hirshorn, B. A., & Cornman, J. M. (1986). *Ties that bind: The interdependence of generations.* Washington, DC: Seven Locks.

Linsky, M. (1986). *Impact: How the press affects federal policy making.* New York: Norton.

Longman, P. (1985, June). Justice between generations. *Atlantic Monthly,* pp. 73–81.

Longman, P. (1986, January–February). Age wars: The coming battle between young and old. *The Futurist,* pp. 8–11.

MacManus, S. A., & Shields, K. H. (2000). *Targeting senior voters: Campaign outreach to elders and others with special needs.* Lanham, MD: Rowman & Littlefield.

Manheim, J. B. (1987). A model of agenda dynamics. In M. L. McLaughlin (Ed.), *Communication Yearbook 10* (pp. 499–516). New York: Sage.

Minkler, M. (1986). "Generational equity" and the new victim-blaming: An emerging public policy issue. *International Journal of Health Services, 16,* 539–551.

Mitchell, R. K., Agle, R. A., & Wood, D. J. (1997). Toward a theory of stakeholder identification and salience: Defining the principle of who and what really counts. *Academy of Management Review, 22,* 853–886.

Mortimore, R. (2000, August 11). *Grey power* [Online]. Retrieved June 27, 2001. Available: http://www.mori.com/digest/2000/c000811.shtml

Mufson, S. (1990, October 15). Older voters drive budget. *Washington Post,* pp. A1, A6.

Nua. (2000, September 21). *Nau internet surveys: Senior citizens to embrace the Web* [Online]. Retrieved June 29, 2001. From http://www.nua.ie/surveys/index.cgi?f=VS&art_id=905356057&rel=true

On line Dictionary of Social Science. (2001). http://datadump.icaap.org/cgi-bin/glossary/SocialDict/

Peterson, S. A., & Somit, A. (1994). *The political behavior of older Americans.* New York: Garland.

Phillipson, C. (1991). Inter-generational relations: Conflict or consensus in the 21st century. *Policy and Politics, 19*(1), 27–36.

Pifer, A. (1986). The public response to population aging. *Daedalus, 115,* 373–395.

Pratt, H. J. (1993). *Gray agendas: Interest groups and public pensions in Canada, Britain, and the United States.* Ann Arbor: University of Michigan Press.

Rash, W., Jr. (1997). *Politics on the nets: Wiring the political process.* New York: Freeman.

Rhodebeck, L. A. (1993). The politics of greed? Political preferences among the elderly. *Journal of Politics, 55,* 342–364.

Richman, H. A., & Stagner, M. W. (1986). Children: Treasured resource or forgotten minority? In A. Pifer & L. Bronte (Eds.), *Our aging society: Paradox and promise* (pp. 161–179). New York: Norton.

Ripley, R. B. (1987). *Policy analysis in political science* (2nd ed.). Chicago: Nelson-Hall.

Rogers, E. M., & Dearing, J. W. (1988). Agenda-setting research: Where has it been, where is it going? In J. A. Anderson (Ed.), *Communication Yearbook 11* (pp. 555–594). Beverly Hills, CA: Sage.

Rosenblatt, R. A. (1993, February 16). 8 million elderly may be a tough sell for Clinton. *Los Angeles Times,* pp. A1, A16.

Ryan, C. (1991). *Prime time activism: Media strategies for grass roots organizing.* Boston: South End.

Ryan, M. H., Swanson, C. L., & Buchholz, R. A. (1987). *Corporate strategy, public policy and the Fortune 500.* Cambridge, MA: Basil Blackwell.

Smith, J. W., & Clurman, A. (1997). *Rocking the ages: The Yankelovich report on generational marketing.* New York: HarperBusiness.

Smith, L. (1992, January 13). The tyranny of America's old. *Fortune,* pp. 68–72.

Stanley, H. W., & Niemi, R. G. (1992). *Vital statistics on American politics* (2nd ed.). Washington, DC: Congressional Quarterly Press.

Steyn, M. (2000, October 23). Gray dawn. *National Review, 52*(20), 59–61.

Terkildsen, N., Schnell, F. I., & Lang, C. (1998). Interest groups, the media, and policy debate formation: An analysis of message structure, rhetoric, and source cues. *Political Communication, 15,* 45–61.

Treadwell, D., & Treadwell, J. B. (2000). *Public relations writing: Principles in practice.* Boston: Allyn & Bacon.

Tropman, J. E. (1987). *Public policy opinion and the elderly, 1952–1978.* New York: Greenwood.

UCLA Center for Communication Policy (2000, November). *Surveying the digital future* [Online]. Retrieved June 5, 2001. Available: www.ccp.ucla.edu

U.S. Bureau of the Census. (2002, February). *Current population reports: Voting and registration in the election of November 2000.* (Series P20-542). Washington, DC: U.S. Government Printing Office.

U.S. Department of Education, American Institute of Architects. (1999, October). AARP call for federal support for school construction. *ALA Washington News, 51*(10), 5.

USA. (n.d. a). *About USA* [Online]. Retrieved June 14, 2001. Available: http://www.unitedseniors.org/about/

USA. (n.d. b). *Grassroots action guide* [Online]. Retrieved June 14, 2001. Available: http://www.unitedseniors.org/grassroots/actionguide.cfm

Valkenburg, P. M., Semetko, H. A., & De Vreese, C. H. (1999). The effects of news frames on readers' thoughts and recall. *Communication Research, 26,* 550–569.

Verba, S., & Nie, N. H. (1972). *Participation in America: Political democracy and social equality.* New York: Harper & Row.

Vincent, J. A. (1999). *Power, politics and old age.* Philadelphia: Open University Press.

Von Beyme, K. (1996). The concept of political class: A new dimension of research on elites. *Western European Politics, 19*(1), 68–88.

Walker, A. (1986). Pensions and the production of poverty in old age. In C. Phillipson & A. Walker (Eds.), *Ageing and Social Policy* (pp. 184–216). London: Gower.

Walker, A. (1990). The economic "burden" of ageing and the prospect of intergenerational conflict. *Ageing and Society, 10,* 377–396.

Williamson, J. B. (1998). Political activism and the aging of the baby boom. *Generations, 22*(1), 55–59.

Wisensale, S. K. (1988). Generational equity and intergenerational politics. *The Gerontologist, 28,* 773–778.

Wittenberg, E., & Wittenberg, E. (1990). *How to win in Washington: Very practical advice about lobbying, the grassroots and the media.* Cambridge, MA: Basil Blackwell.

Worcester, R. M. (1999, February 19). *Grey power: The changing face* [Online]. Retrieved June 5, 2001. Available: http://www.mori.com/pubinfo/social.shtml

Worcester, R. M. (2000, November 27). *Challenges of the demographic shift* [Online]. Retrieved June 5, 2001. Available: http://www.mori.com/pubinfo/articles.shtml

Worcester, R. M. (2001, 19 May). *Grey power and class voting* [Online]. Retrieved June 27, 2001. Available: http://www.mori.com/pubinfo/rmw-ep-010519.shtml

16

The Portrayal of Older Adults
in Political Advertising

Lynda Lee Kaid
University of Florida

Jane Garner
University of Oklahoma

Given the growth of the senior population over the past several decades, it is not surprising that increased importance has been attached to older adults as a voting bloc. Such a substantial portion of the voting-age population has tremendous potential to influence the political process in the United States, raising questions about how political candidates attempt to attract seniors during political campaigns. This chapter addresses that question by considering how aging concerns have been conveyed through the candidate's primary form of political campaigning—political television advertising.

OLDER ADULTS AS VOTERS

Although the size of the aging population has been increasing in recent decades, early voting behavior theorists discounted the notion that older adults would become a major political force (Campbell, 1971; Hudson & Binstock, 1976). More recently, however, scholars have espoused the possibility that older adults could become a more cohesive, organized constituency (Powell, 1985).

There are several reasons why politicians give weight to the concerns of older adults in the campaign process. First, older adults make up a substantial and increasing percentage of the voting population (Williamson, Evans, & Powell, 1982). The population of older adults in the United States is growing at a rapid rate, outpacing the growth in younger age groups, as the baby-boom generation begins to be classified as an aging population (Taylor, 1995). Second, research has shown that seniors vote more frequently than do younger members of the American population (Straits, 1990; Uhlaner, 1989). Third, the senior population has shown a particular willingness to engage in and win political battles where specific issues, such as Social Security benefits and health care, are involved (Dye, Zeigler, & Lichter, 1992; Palmore, 1999).

MEDIA AND OLDER ADULTS

Given the political force of the older voter, it seems natural to consider the campaign mechanisms that political candidates would select for communicating political choices. For politicians, this decision is not a difficult one. Researchers have clearly shown that the mass medium of choice for seniors is television. Seniors watch more television than any other age group (Robinson, 1989), although the portrayal of older adults in television programming is not always a positive one (Nussbaum, Thompson, & Robinson, 1989; Robinson, 1998).

Both positive and negative images of older people are seen in the mass media, demonstrating the presence of both "positive ageism" and "negative ageism" (Palmore, 1999). Product advertisers on television generally portray older adults positively (Roy & Harwood, 1997), whereas print advertisements in U.S. magazines show a decreasing number of positive stereotypes and an increasing number of negative stereotypes (Miller, Miller, McKibbin, & Pettys, 1999). Older people are also underrepresented in both televised and print ads compared to their percentage of the total U.S. population (Miller et al., 1999; Roy & Harwood, 1997). These effects are stronger for women and minorities (Roy & Harwood, 1997). Unrealistic portrayals of older persons in televised ads has been shown at rates lower than their population percentage (Hiemstra, Goodman, Middlemiss, Vosco, & Ziegler, 1983). Bell (1992) found positive images of older persons as characters in prime-time television programs and a decrease in negative stereotypes. Atchley (1997) suggested that more representative images of older persons are becoming a part of television programming.

Although seniors seem to show a preference for political information garnered via television news and public affairs programming (Bower, 1973; Powell, 1985; Rubin & Rubin, 1982), the older portions of the population are also heavy consumers of other types of television programming, including game shows, variety shows, and situation comedies (O'Keefe & Reid, 1989; Robinson, 1989).

The conclusion from such research findings is clear. Although politicians must concern themselves with how they are presented to seniors via news and public affairs programming, the high general consumption of television by seniors also provides the opportunity for political candidates to reach them via political television advertising.

THE IMPORTANCE OF POLITICAL ADVERTISING

Whatever the group being addressed, there is no question that television advertising has become the dominant method of communication between presidential candidates and voters in the United States (Devlin, 2001; Kaid, 1999; Kaid & Johnston, 2001). Candidates choose the short spot commercial, because it reaches large percentages of the population, allows the candidate to present an uncontrolled and uncensored message, and has been shown to have direct cognitive, affective, and behavioral effects on voters (Kaid, 1981, 1999).

The ability of televised political spots to serve the first purpose—reaching large segments of the population—needs no elaboration. However, as with all communication, the possibility of selective exposure, particularly on the basis of political party affiliation, may be questioned. In this regard, however, political television spots have proven particularly effective for politicians. Research has shown consistently and convincingly that political television ads overcome partisan selective exposure (Atkin, Bowen, Nayman, & Sheinkopf, 1973). In the case of seniors, research has verified that they receive higher exposure to television ads than other voting groups (Patterson & McClure, 1976).

The second concern, that the candidate have control of the message, is also easily met by political advertising. The candidate or a chosen surrogate is usually the source of political ad messages. The candidate pays, and the candidate controls the content; television stations are prohibited from altering or regulating the content of political spots (Johnson-Cartee & Copeland, 1989; Kaid, 1981, 2000). Consequently, candidates are free to determine the message and format used to communicate with voters.

Finally, research has increasingly shown that television spots do have effects on voters. Ads have cognitive effects by increasing voter knowledge about issues and candidate image characteristics (Cundy, 1986; Kaid & Sanders, 1978; Patterson & McClure, 1976). Affective responses to ads have also been demonstrated by researchers who have found that political spots can cause changes in candidate evaluations (Garramone, 1985; Garramone, Atkin, Pinkleton, & Cole, 1990; Kaid, 1997, 2001; Kaid & Boydston, 1987; Kaid, Leland, & Whitney, 1992; Kaid & Sanders, 1978). Direct effects on voting behavior have been less easily established, but several researchers have found strong relationships between aspects of political spot ads and voting intentions (Kaid & Sanders, 1978; Lang, 1991; Thorson, Christ, & Caywood, 1991).

Given the strong research evidence for political advertising effects, it is not surprising that political candidates have put their money where the action is—in political advertising. Over the past several decades, the percentage of campaign monies devoted to political television advertising has climbed to new highs. In the 1988 presidential campaign, for instance, George Bush and Michael Dukakis, bolstered by political party expenditures, spent almost $80 million to produce and buy time for television spots (Devlin, 1989). These expenditures were easily topped in the 1992 and 1996 presidential campaigns (Devlin, 1993, 1997), and all-time highs were set in 2000, when Al Gore and George W. Bush and their national political parties spent over $200 million on campaign advertising (Devlin, 2001).

However, despite the importance of older adults as a voting group, research has not directly considered the ways in which older adults are addressed in campaign advertising. The dominance of advertising as a political message format raises questions about how often these voters and their concerns are presented in national campaign advertising, how often seniors actually appear in political spots, what specific issues are addressed, and how seniors are portrayed in the spots.

METHODOLOGY

Content analysis was used to investigate the role of seniors in national political advertising. Extending on earlier analyses of the presentation of seniors in political advertising (Kaid & Garner, 1995), the researchers first screened television ads used in the general election campaigns of American presidential candidates from 1960 through 2000. This initial ad sample consisted of 1,191 ads provided by the Political Commercial Archive at the University of Oklahoma, the world's largest collection of political radio and television commercials. Using a computerized database of these ads (Kaid, Haynes, & Rand, 1996), the researchers selected all presidential campaign ads from 1960 through 2000 in which seniors or issues directly related to them were addressed. Issues used to identify "senior-related" spots included Social Security, Medicare, "fixed incomes," and similar topics. Individual examinations of the ads produced by the database indicated that of the 1,191 total ads, 158 (or 13%) contained appeals related to older adults in some way. This sample of 158 ads was then subjected to content analysis.

The categories used for the content analysis were developed by the researchers specifically for this project. In addition to a category for candidate and year, categories included (1) political party of the candidate, (2) presence and number of older individuals in the ad, (3) gender of older adults present in the ad, (4) presence of fear appeals, (5) focus of the ad (negative or positive),[1] (6) portrayal of seniors in the ad (positive, negative, or neutral).[2]

A codebook was developed for the categories, and graduate students were trained to act as coders for the project. During the training sessions, coders identified and discussed problems with the coding instrument and codebook, and revisions were made before final coding was initiated. To test for intercoder reliability, the coders coded a subset of 10% of the sample. Using Holsti's formula,[3] the average intercoder reliability across all categories was +.85. The individual category reliability ranged from +.72 to +.96. Reliability across categories for most ads was at least +.87. Following the training sessions and the assessment of intercoder reliability, the ads were divided among the coders for final coding.

---

[1]In coding the focus of the commercial, a commercial was coded as positive if it focused on good qualities or actions, setting forth the candidate's position with positively stated goals and positions. A spot was coded as negative if it focused on undesirable aspects of the issue, particularly if it attacked the opposition party or candidate's stance on an senior-related issue. A neutral spot took neither a positive nor negative stance on the senior-related issue. Ads in the neutral category typically included announcer references to issues, with no elaboration or information that implied an evaluation of the concept presented.

[2]The portrayal of seniors was coded as positive if seniors in the ad were shown in a good, productive manner. For instance, if they were portrayed as self-sufficient or in good health, the ad would be coded as positive. However, an ad was coded as negative if the seniors were portrayed as victims, unhealthy, or without control of their lives. In neutral ads, no positive or negative presentation seemed apparent.

[3]The formula used to calculate intercoder reliability is that given in North, Holsti, Zaninovich, and Zinnes (1963). It is given for two coders and can be modified for any number of coders:

$$R = \frac{2(C_{1,2})}{C_1 + C_2}$$

$C_{1,2}$ = number of category assignments both coders agree on, and $C_1 + C_2$ = total category assignments made by both coders.

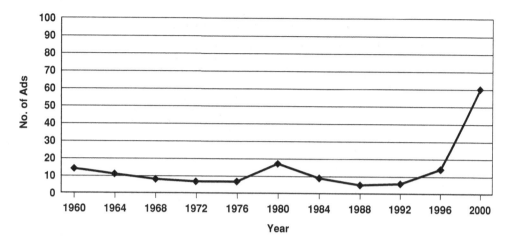

FIG. 16.1.   Trends in presidential ads with senior concerns.

The exploratory nature of the study and the nominal nature of the variables suggested that descriptive statistical analyses would be appropriate. Consequently, the results reported later primarily consist of frequencies, supplemented by breakdowns by the political party affiliations of the candidates. Because the content analysis was performed on the universe of ads addressing elderly concerns and not on a sample of the ads, no statistical testing was appropriate.

RESULTS

One initial finding that resulted from the screening of the total presidential ad sample bears repeating here. Despite the substantial size and importance of seniors as voters, only 13% of the television spots used in presidential campaigns from 1960 through 2000 addressed issues related to seniors.

Perhaps even more interesting has been the trend over time in the treatment of senior issues in presidential campaign advertising. As Fig. 16.1 shows, with the exception of a sharp upturn in 1980, the number of commercials addressing senior concerns declined drastically from 1960 until the last two election cycles. In fact, this decline has been so sharp in recent years that the number of political spots addressing seniors dropped to almost none in 1992. However, a slight increase in 1996 preceded a dramatic resurgence in the 2000 election.

Specific Candidates and Seniors

In the early years of campaign advertising, no presidential candidate gave greater direct attention to seniors in his spots than did John F. Kennedy. Table 16.1 shows that Kennedy's 1960 campaign spots included 14 different spots that addressed such issues. Several spots were direct "head-on" appeals in which Kennedy speaks with intensity about the plight of seniors and their need for medical assistance. In one spot, Eleanor Roosevelt awkwardly

TABLE 16.1
Candidate Usage of Television Spot Ads Related to
Seniors: Presidential Campaigns, 1960–2000 (*N* = 158)

Candidate		Number of Spots
1960	Kennedy	14
1960	Nixon	0
1964	Goldwater	3
1964	Johnson	8
1968	Nixon	1
1968	Humphrey	7
1972	McGovern	5
1972	Nixon	2
1976	Carter	2
1976	Ford	5
1980	Carter	6
1980	Reagan	11
1984	Mondale	9
1984	Reagan	0
1988	Bush	2
1988	Dukakis	3
1992	Bush	2
1992	Clinton	4
1996	Clinton	8
1996	Dole	6
2000	Gore	44
2000	Bush	16

endorses Kennedy, because of his stands on Social Security and social welfare issues. In another stilted testimonial, Dr. Benjamin Spock uses similar issues as a reason to weigh in on Kennedy's side of the campaign. However, seniors themselves rarely appear in Kennedy's spots, although in one Kennedy message an older couple (identified only as the McNamaras) sit silently, neither talking nor moving as Kennedy talks. Kennedy's unsuccessful opponent, Richard Nixon, on the other hand, addressed not a single spot to the plight of the senior population. In fact, over the course his three presidential campaigns and hundreds of spots, Nixon addressed senior concerns only in three (one in 1968 and two in 1972).

Kennedy's successor, Lyndon Johnson also addressed several spots (*n* = 8) to seniors' concerns. Many of these were negative spots in which he attacked his opponent Barry Goldwater's stance on Social Security. In more than one spot, an announcer recounts Goldwater's alleged anti–Social Security statements and concludes with a visual destruction of a Social Security card while the voiceover firmly states that Goldwater's plans would "wreck" or destroy Social Security. Another ad on Goldwater's policy "flip-flops" insists that Goldwater has favored a "voluntary Social Security" system. Goldwater was forced to respond. In one spot, he used a question-and-answer format to insist that he "favors a sound Social Security system" and wants to preserve the value of the dollars placed in it. In another ad, Senator Margaret Chase Smith tries to correct misimpressions about Goldwater's record on Social Security.

In 1968, Hubert Humphrey used seven of his television spots to make appeals to seniors. Humphrey was quick to take credit for his sponsorship of Medicare, and his spots had a much more positive focus than did Johnson's. Nixon's single 1968 ad focusing on seniors was much more negative, portraying older persons as "forgotten Americans" who "eke out a living." By 1972, however, Nixon had decided that things had completely turned around for the older Americans. He used his two senior-oriented spots to show them being useful and productive, enjoying increased Social Security benefits and independent lifestyles.

In general, however, the 1970s showed a decline in spots using senior themes, although George McGovern should be credited with spots that seemed to imply a need to listen to seniors themselves. In several of his ads, known for their cinema verité production style, McGovern actually appears with large numbers of older citizens and appears to be really listening and responding to their concerns. However, despite this tendency to treat seniors as worthy of concern, many McGovern ads also insert negative stereotypical references to them. These inserts were probably designed to combat Nixon's presumptions that no problems existed for seniors. Such strategies on McGovern's part were typical of the trend throughout all of the ads for the Democratic Party—to use seniors' concerns to promote sympathy by portraying itself as the only one concerned about seniors as "forgotten," "living on minimum incomes," "suffering," "lonely," "old, poor, and hungry," and trying to "make a few dollars stretch."

In 1976, Jimmy Carter took a much more negative tone, accusing Gerald Ford of voting against Medicare and reducing food stamps for seniors. Although Carter used only two senior-focused ads in 1976, their negative focus made a strong statement about his perception of seniors as victims. The still photos and slides used in these ads included pictures of seniors looking "pitiful" and conveying a stereotypical image of them as helpless. In contrast, Ford's five ads about seniors emphasized a more positive message, suggesting that there were problems for seniors but that these could be addressed in constructive ways. A repeated message assured seniors that there were ways to avoid the serious problems of health care, that "there's no reason for somebody to go broke just to get well."

It was not until the 1980 campaign that senior issues seemed to capture substantial attention again, with Ronald Reagan's 1980 campaign including eleven spots about such issues. Until the 2000 election, the Reagan 1980 campaign had the highest number of senior-related spots of any presidential campaign after John F. Kennedy's 1960 race. Many of these spots were devoted to Reagan's attempts to defend statements he had made about cutting Social Security. Reagan made several spots with speech excerpts and direct head-on defenses of his positions. Even Nancy Reagan chimed in with spots in which she defended her husband against Carter's attacks on Reagan's positions.

This latter finding is particularly interesting when, as Table 16.1 indicates, Reagan's coverage of this issue in his 1984 race dropped to zero. Even the fact that his challenger, Walter Mondale, gave considerable attention to seniors in nine of his spots did not spur

Reagan to provide any coverage in 1984. Perhaps the concerns of seniors had no place in the beautiful and progressive picture of America painted in Reagan's 1984 ads. However, the content of the Mondale ads was nothing for seniors to cheer about. Older women in particular were portrayed negatively in Mondale's ads. They were shown as helpless and whining about how bad things were.

The 1980s saw a continuing decline in ad emphasis on seniors. Bush and Dukakis barely mentioned them in 1988, although Dukakis did suggest in one spot that Americans should try to "imagine a place where people can grow old without being made to feel like a burden. . . ." Bush's 1988 senior ads were the result of responses to attacks made by Dukakis. Bush quickly pointed out that Dukakis was "cruel to scare people with cuts" and that there was "no place for partisan politics with Social Security." In 1992, seniors and their concerns were similarly missing from the spots of both Bush and Clinton.

The 1996 campaign saw a slight increase in attention to seniors, with Clinton addressing the concerns of older adults in eight of his spots and Dole portraying them in six ads. Clinton's approach was mainly a negative one, typical of his overall attack campaign in 1996. He focused his ads about older adults on accusations toward Dole's positions on Social Security and Medicare. His ads generally portrayed the older citizen in a negative light—old, decrepit, stumbling along in walkers, or frustrated with their high medical bills. Dole, considered a senior citizen himself by many voters, found himself on the defensive, trying to refute Clinton's portrayal of him as someone who would wreck Social Security and rob senior citizens of their benefits.

It was the 2000 campaign, however, that gave older adults a starring role in the advertising air war. Al Gore devoted 44 ads, 38% of his spots, to issues that involved older adults. George W. Bush was not far behind, addressing these concerns in 16 (35%) of his spots. For the first time since John F. Kennedy's 1960 campaign, the discussion of policy issues that related to senior citizens took on major importance. There was, however, a major difference in the tone of the two candidates' ads in 2000. Like his predecessor Bill Clinton, Gore's spots were primarily negative in tone, attacking and accusing Bush of plans to harm senior citizens. A good example of Gore's portrayal of older adult issues was "Siding," a spot in which Bush is accused of "forcing seniors into HMOs" and "leaving millions with no help." The ad features negative images of seniors—a frail hand and a woman moving slowly in a walker.

The 2000 Bush spots were overall more positive in tone. His ads addressed issues similar to Gore's—health care and prescription drugs—but the images of older adults were less depressing. In an ad called "Surplus," for instance, Bush discussed the budget surplus and pledged to "protect and strengthen Social Security" while the ad showed images of senior citizens playing with younger family members. In "Hard Things," Bush's mention of Social Security is accompanied by positive images of senior citizens singing and playing the piano.

Ad Characteristics

It is apparent from Table 16.1, and summarized in Table 16.2, that Democrats are much more likely than Republicans to address the concerns of older Americans. Of the 158 ads in the sample, 70% were sponsored by Democratic presidential candidates. This is not

TABLE 16.2
Characteristics of Presidential Ads Related
to Seniors, 1960–2000 (*N* = 158)

Party	
Democrat	110 (70%)
Republican	49 (30%)
Incumbent	34 (22%)
Older Person Present	91 (58%)
Gender of Older Person Present in Ad	
Female	28 (18%)
Male	6 (4%)
Both male and female	57 (36%)
None present	67 (42%)
Fear Appeal in Ad	30 (19%)
Focus of Ad	
Positive	79 (50%)
Negative	79 (50%)
Portrayal of Older Person	
Positive	49 (31%)
Negative	11 (7%)
Neutral	19 (12%)
Not applicable	79 (50%)

a surprising finding, because throughout most of the past four decades, the Democratic Party has been associated with spending programs to provide assistance to specialized groups, particularly Medicare, increased Social Security benefits, and other such programs. Therefore, it is natural that the Democrats would seek to solidify their support among the groups benefited by such programs.

Table 16.2 also shows that incumbent presidential candidates are much less likely than challengers to address senior issues in their spots. Only about 22% of the ads addressing seniors were sponsored by incumbents.

Overall, the ads addressing senior issues were split between positive and negative ads. Exactly 50% of the 158 ads were predominantly focused on the sponsoring candidate; the other 50% focused on attacking the opponent.

Portrayal of Seniors in Political Spots

One mark of the significance accorded to seniors is how often older persons actually appear in the ads that address their issues. By this measure, candidates did not accord seniors much significance until the 1996 and 2000 campaigns. Older persons appeared in less than half the spots in this sample (38%) from 1960 through 1992. However, when the 1996 and 2000 ads were included, the percentage of ads in which older individuals appeared rose to 58%.

In tune with reality, when an older person did appear in an ad, the gender of the person was more often female than male (see Table 16.2), although it was more likely that a combination of both males and females would appear. The spots in which older persons appeared often included several older persons. Several spots contained as many as 10 women or men. McGovern's 1972 spots in which he appeared with groups of older citizens, and Ford's 1976 ads in which he promised senior groups that he would work for catastrophic health insurance, are good examples of spots in which multiple male and female senior citizens appeared.

The setting of an ad is another indication of how seniors are being treated and addressed in the ad. In this sample, many of the settings were nursing or retirement homes. Candidate offices or formal settings were the place to see seniors in about one fourth of the ads. Candidates were also anxious to show seniors as part of the audience at their campaign rallies or gatherings, but only small numbers of older adults were shown in private homes or in work settings. The message was clear that the older citizen can expect to live a marginal life, not often involved in everyday experiences of working-class America.

Another consideration in the content of the ads analyzed was whether a fear appeal was being used. Table 16.2 shows that almost one fifth of the ads addressed to older citizens used a fear appeal of some kind. Johnson's 1964 negative ads on Social Security are a good example of such appeals, because these ads attempted to frighten older citizens into voting against Barry Goldwater. Similar tactics have been used in subsequent presidential campaigns.

Finally, there is the overall concern of whether seniors were portrayed in a positive or negative light. Table 16.2 shows that when older citizens or their issues were identified specifically in the spots, they were more often portrayed positively (31%) than negatively (7%). Positive portrayals in the spots included senior citizens discussing issues and concerns with candidates. Negative portrayals include the perception of older persons as victims and dependent on assistance programs such as food stamps. Older adults were also often portrayed as nonactive adults. In the 1996 and 2000 samples, for instance, older individuals were usually portrayed as leading a sedentary lifestyle. They were shown sitting, reading, or watching television in 39% of the ads during these election years.

The issues included in the spot ads related to seniors were predictable ones. Not surprisingly, the most frequently addressed policy issue was Social Security. Other frequently mentioned topics included health care and Medicare and the general provision of government services. The election of 2000 brought Social Security to the forefront. In a 2000 spot, George W. Bush pledged to defend Social Security, stating "no changes, no reductions, no way." Gore's many negative spots included attacks on George W. Bush, including Bush's statement about Social Security "like it's some kind of government program," as well as Gore's promise to keep Social Security funds in "a lock box."

Political Parties and Political Spots Related to Older Citizens

Interesting comparisons can be made in the handling of seniors' concerns by the two major political parties. As mentioned, Democrats were much more likely than Republicans

TABLE 16.3
Political Party Differences in Presidential Spots Related
to Seniors

	Democrat (n = 110)	Republican (n = 48)
Older Person Present in Ad	72%	69%
Focus of Ad		
Positive	41	69
Negative	59	31
Portrayal of Older Person		
Positive	27	48
Negative	25	4
Neutral/NA	58	58
Fear Appeals	23	10

to use television ads to address seniors' concerns in presidential campaigns. As Table 16.3 shows, Democrats also were slightly more likely to show older persons in their ads.

Perhaps the most interesting difference in the partisan treatment of older adults in spot ads is the difference in ad focus between the parties. Republicans were much more likely than Democrats to appeal to seniors in ads by presenting positive images of older persons and their concerns. Over two thirds (69%) of Republican ads were positive. Democrats, however, used negative ads 59% of the time, a percentage of negative ads that is much higher than the average for all political spots across this same time span (Kaid & Johnston, 2001). Negative ads presented images and references to seniors as victims, poor, hungry, and without dignity.

Not only did Republicans address seniors' concerns in more positive ads, they were also more likely to present a positive portrayal of seniors. Table 16.3 indicates that 48% of Republican ads displayed a positive portrayal of seniors, compared to only 27% of Democratic ads. Democrats were also much more likely to use fear appeals in their ads than were Republicans.

CONCLUSION

Overall, these findings point to some interesting conclusions about the use of political advertising to appeal to the elderly. A particularly significant finding is the trend in increased frequency of appeals to older citizens in the 2000 election. At a time when the size of the older voting bloc is continuing to grow, presidential candidates seem to have finally recognized the need to increase their efforts to communicate directly with older adult voters. The near absence of seniors concerns in the 1988 and 1992 presidential ads is particularly puzzling, because it cannot be explained by an absence of issue concerns related to the elderly. For instance, a major initiative of the Clinton administration was health care, a particular concern of seniors. During political discussions, however, distinctions are made regarding medicare and health care in general.

It is also interesting to note that seniors themselves are becoming increasing players in these ads. In earlier election cycles, from 1960 to 1992, seniors were not major players in ads that address their concerns, because older persons appeared in less than 40% of these ads (Kaid & Garner, 1995). However, the percentage of visible older adults rose to 58% when 1996 and 2000 were included in the sample. When older adults do appear, they are often pictured in nursing homes and in staged political settings, rarely in private homes or outdoor or natural settings. Although seniors are sometimes pictured positively, these are stereotypical images that continue the perception of the aging person as often helpless and dependent on the government.

The finding that Democrats cover seniors' concerns more often than Republicans is not unexpected, but it was surprising that Republican ads were more often positive and presented a more positive portrayal of seniors. This finding, however, may be a natural result of the tendency of the Democratic Party to use fear appeals to make its case to seniors, setting up criticism of the opposition party by accusing Republicans of supporting policies designed to damage Social Security and health care benefits and disadvantage seniors. Such partisan portrayals may be understandable from a political campaign communication standpoint, but they send an unfortunate signal to seniors that they are merely pawns to be used as political capital in the power game of politics.

It is possible that George W. Bush's spots from 2000 mark a change in the attitude of Republican candidates to older adults. As mentioned earlier, Bush's campaign gave the older voter a more positive role and a more positive visual image. The images of older people in Bush's spots are happy and vibrant, standing in front of a white background. In "New Americans" and "Hard Things," the image of a man spinning and throwing his hat around interspersed with more realistic shots of activities of older people highlights the variations in use of images of seniors. Bush also used scenes of an older man playing softball and a couple loading a travel trailer to show active and outdoor involvement.

Overall, then, these findings about the role of seniors in candidate-controlled presidential communication for the past four decades make a significant statement about the perceptions that candidates and parties have about seniors and their place in the political system. First of all, the finding that only 13% of all presidential ads from 1960 through 2000 even addressed senior concerns is a significant indication that older citizens are not perceived as important political constituents. The message was loud and clear—despite their increasing significance as a voting bloc, older citizens were not seen by either party as important enough to justify significant attention in the primary medium of modern campaign communication. The 2000 election, however, offered new hope that the older voter might be taken seriously, with both candidates addressing significantly higher percentages of their spot ads to the older adult voter. It remains to be seen if this trend is sustained in future elections.

Second, over the past four decades, when seniors have been portrayed in ads, the nature of their presentation was telling. Although half the ads addressed to seniors were positive in overall focus, the images and stereotypes present in the visual aspect were especially significant. Older people were often presented as helpless victims, dependent on the government and Social Security, often in pitiful straits, and in constant fear of reduced circumstances. Rarely were they persons portrayed as contributing members of society who were working to live productive, independent lives. It is interesting, however, that Republicans were much more likely than Democrats to present seniors in a positive light. This difference is probably because of the Republican philosophy of emphasizing individualism and less need for government intervention for social programs. These partisan definitions of what it means to be an older person imply a great deal about the differing views of seniors held by American political parties.

All too often, the result of these differing partisan definitions is that seniors become a microcosmic battleground between the parties, with Democrats using their problems as fear appeals to attack their Republican opponents and the Republican Party responding with optimistic reassurances. Neither side, however, has taken senior concerns seriously enough to use their primary communication source (political ads) as a way of introducing new or innovative policy initiatives that might genuinely address senior concerns.

This research also suggests other avenues of research on political advertising and older citizens. A continuation of such research in future elections is, of course, necessary to continue to track the trends in candidate use of political ads to appeal to seniors. Future content analyses might also address in more detail the types of emotional appeals, particularly fear appeals, that are being used. Research is also needed to determine the approach taken by other candidates at other electoral levels, particularly in races for the U.S. Senate and U.S. House of Representatives. These findings might also serve as the basis for experimental studies that could determine the effectiveness of certain types of appeals and help to ascertain how older voters react to various types of appeals and strategies used in political ads.

The importance of such future research cannot be overestimated. Given the life span communication perspective on seniors (Nussbaum, 1989), it is clear that they continue to develop and evolve socially and politically. Politicians must learn to understand this developmental process if they are to communicate successfully with older voters. Likewise, older voters can ensure a place for themselves and their interests in the political system only if they reach a better understanding of the approaches and strategies of political candidates.

ACKNOWLEDGMENTS

The authors wish to express their appreciation to Eric Bull for assistance with the analysis of these ads.

REFERENCES

Atchley, R. (1997). *Social forces and aging*. Belmont, CA: Wadsworth.

Atkin, C. K., Bowen, L., Nayman, O. B., & Sheinkopf, K. G. (1973). Quality versus quantity in televised political ads. *Public Opinion Quarterly, 37,* 209–224.

Bell, J. (1992). In search of a discourse on aging: The elderly on television. *The Gerontologist, 32,* 305–311.

Bower, R. (1973). *Television and the public*. New York: Holt, Rinehart, & Winston.

Campbell, A. (1971). Politics through the life cycle. *The Gerontologist,* 112–117.

Cundy, D. T. (1986). Political commercials and candidate image. In L. L. Kaid, D. Nimmo, & K. R. Sanders (Eds.), *New perspectives on political advertising* (pp. 210–234). Carbondale: Southern Illinois University Press.

Devlin, L. P. (1989). Contrasts in presidential campaign commercials of 1988. *American Behavioral Scientist, 32,* 389–414.

Devlin, L. P. (1993). Contrasts in presidential campaign commercials of 1992. *American Behavioral Scientist, 37,* 272–290.

Devlin, L. P. (1997). Contrast in presidential campaign commercials of 1996. *American Behavioral Scientist, 40,* 1058–1084.

Devlin, L. P. (2001). Contrast in presidential campaign commercials of 2000. *American Behavioral Scientist, 44,* 2338–2369.

Dye, T. R., Zeigler, H., & Lichter, S. R. (1992). *American politics in the media age* (4th ed.). Pacific Grove, CA: Brooks/Cole.

Garramone, G. M. (1985). Effects of negative political advertising: The role of sponsor and rebuttal. *Journal of Broadcasting and Electronic Media, 29,* 147–159.

Garramone, G. M., Atkin, C. K., Pinkleton B., & Cole, R. T. (1990). Effects of negative political advertising on the political process. *Journal of Broadcasting and Electronic Media, 34,* 299–311.

Hiemstra, R., Goodman, M., Middlemiss, M. A., Vosco, R., & Ziegler, N. (1983). How older persons are portrayed in television advertising: Implications for educators. *Educational Gerontology, 9,* 111–122.

Hudson, R. B., & Binstock, R. H. (1976). Political systems and aging. In R. H. Binstock & E. Shanas (Eds.), *Handbook of aging and the social sciences.* (pp. 369–400). New York: Van Nostrand Reinhold.

Johnson-Cartee, K., & Copeland, G. (1989). *Negative political advertising.* Hillsdale, NJ: Lawrence Erlbaum Associates.

Kaid, L. L. (1981). Political advertising. In K. R. Sanders & D. Nimmo (Eds.), *Handbook of political communication* (pp. 249–271). Beverly Hills, CA: Sage.

Kaid, L. L. (1997). Effects of the television spots on images of Dole and Clinton. *American Behavioral Scientist, 40,* 1085–1094.

Kaid, L. L. (1999). Political advertising: A summary of research findings. In B. Newman (Ed.), *The handbook of political marketing* (pp. 423–438). Thousand Oaks, CA: Sage.

Kaid, L. L. (2000). Ethics in political advertising. In R. E. Denton, Jr. (Ed.), *Political communication ethics* (pp. 146–177). Westport, CT: Praeger.

Kaid, L. L. (2001). Technodistortions and effects of the 2000 political advertising. *American Behavioral Scientist, 44,* 2370–2378.

Kaid, L. L., & Boydston, J. (1987). An experimental study of the effectiveness of negative political advertisements. *Communication Quarterly, 35,* 193–201.

Kaid, L. L., & Garner, J. (1995). Political advertising and the elderly. In J. F. Nussbaum & J. Coupland (Eds.), *Handbook of communication and aging research* (pp. 343–357). Mahwah, NJ: Lawrence Erlbaum Associates.

Kaid, L. L., Haynes, K. J. M., & Rand, C. (1996). *The Political Commercial Archive: A catalog and guide to the collection.* Norman, OK: Political Communication Center.

Kaid, L. L., & Johnston, A. (2001). *Videostyle in presidential campaigns: Style and content of televised political advertising.* Westport, CT: Praeger.

Kaid, L. L., & Johnston, A. (1991). Negative versus positive television advertising in U.S. presidential campaigns, 1960–1988. *Journal of Communication, 41,* 53–64.

Kaid, L. L., Leland, C., & Whitney, S. (1992). The impact of televised political ads: Evoking viewer responses in the 1988 presidential campaign. *Southern Communication Journal, 57,* 285–295.

Kaid, L. L., & Sanders, K. R. (1978). Political television commercials: An experimental study of type and length. *Communication Research, 5,* 57–70.

Lang, A. (1991). Emotion, formal features, and memory for televised political advertisements. In F. Biocca (Ed.), *Television and political advertising: Vol. 1. Psychological processes* (pp. 221–243). Hillsdale, NJ: Lawrence Erlbaum Associates.

Miller, P. N., Miller, D. W., McKibbin, E. M., & Pettys, G. L. (1999). Stereotypes of the elderly in magazine advertisements 1956–1996. *International Journal of Aging and Human Development, 49,* 319–337.

North, R. C., Holsti, O., Zaninovich, M. G., & Zinnes, D. A. (1963). *Content analysis: A handbook with applications for the study of international crisis.* Evanston, IL: Northwestern University Press.

Nussbaum, J. (1989). Life-span communication: An introduction. In J. Nussbaum (Ed.), *Life-span communication: Normative processes* (pp. 1–4). Hillsdale, NJ: Lawrence Erlbaum Associates.

Nussbaum, J. F., Thompson, T., & Robinson, J. D. (1989). *Communication and aging.* New York: Harper & Row.

O'Keefe, G. J., & Reid, K. (1989, November). *Individual and interpersonal predictors of media use among elderly persons.* Paper presented at the Speech Communication Association Convention, San Francisco.

Palmore, E. B. (1999). *Ageism: Negative and positive* (2nd ed.). New York: Springer.

Patterson, T. E., & McClure, R. D. (1976). *The unseeing eye.* New York: Putnam.

Powell, L. (1985). Mass media as legitimizers of control. In J. A. Shindell & L. Evans (Eds.), *Aging and public policy: Social control or social justice* (pp. 180–205). Springfield, IL: Thomas.

Robinson, J. D. (1989). Mass media and the elderly: A uses and dependency interpretation. In J. Nussbaum (Ed.), *Life-span communication: Normative processes* (pp. 319–337). Hillsdale, NJ: Lawrence Erlbaum Associates.

Robinson, T. E., II. (1998). *Portraying older people in advertising: Magazines, television and newspapers.* New York: Garland.

Roy, A., & Harwood, J. (1997). Underrepresented, positively portrayed: Older adults in television commercials. *Journal of Applied Communication Research, 25,* 39–56.

Rubin, A. M., & Rubin, R. B. (1982). Older persons' viewing patterns and motivations. *Communication Research, 9,* 287–313.

Straits, B. C. (1990). The social context of voter turnout. *Public Opinion Quarterly, 54,* 64–73.

Taylor, S. C. (1995, November–December). Prime time's big sleep. *Modern Maturity,* 38–43.

Thorson, E., Christ, W. G., & Caywood, C. L. (1991). Selling candidates like tubes of toothpaste: Is the comparison apt? In F. Biocca (Ed.), *Television and political advertising: Vol. 1. Psychological processes* (pp. 145–172). Hillsdale, NJ: Lawrence Erlbaum Associates.

Uhlaner, C. J. (1989). Turnout in recent American presidential elections. *Political Behavior, 11,* 57–79.

Williamson, J. B., Evans, L., & Powell, L. A. (1982). *The politics of aging: Power and policy.* Springfield, IL: Charles C. Thomas Pub.

Media Usage Patterns and Portrayals of Seniors

James D. Robinson and Tom Skill
University of Dayton

Jeanine W. Turner
Georgetown University

Like other age groups, the elderly spend a great deal of time with the mass media. They spend more than 40% of their leisure time watching television, reading, going to the movies, listening to the radio, and listening to music (Spring, 1993). Although any behavior that consumes so much time should be of concern to social scientists, mass media usage is a particularly important leisure activity, because it affects other leisure behaviors (Robinson, 1972, 1981), impacts interpersonal interaction (Bleise, 1986; Faber, Brown, & McLeod, 1979; Lull, 1980), and also affects perceptions of aging and social reality (Davis, 1984; Gerbner, Gross, Signorielli, & Morgan, 1980).

In this chapter, we describe the media usage of older adults and compare these patterns with other adult age cohorts. Much of the previous research has treated people 65 and older as if they were completely homogeneous and, consequently, many of those findings are somewhat misleading. We attempt to clarify these misconceptions and identify where differences within the older population exist.

In addition, we review the extant literature that has examined how older people have been portrayed on television. By carefully reviewing this literature, patterns in the frequency and nature of such portrayals can be identified. Examination of this literature indicates that changes in portrayals have not occurred since the 1970s.

TELEVISION CONSUMPTION IN THE UNITED STATES

In 1999, 98.2% of all homes in the United States had one or more TV sets, 99% of those homes had at least one color set, and over 76% had two or more sets in working order (*The World Almanac*, 2001). Currently, cable TV lies in or near about 90% of all homes (*Plunket's, Entertainment and Media Industry Almanac*, 1999), and 68% of all homes with televisions subscribe to basic cable service (*The World Almanac and Book of Facts*, 2001) at an average cost of $342 a year (*Plunkett's*, 1999). In addition, 32% of all U.S. homes pay for a premium cable TV package (*The World Almanac and Book of Facts*, 2001). Currently, about 85% of all homes have at least one videocassette recorder (VCR; *The World Almanac and Book of Facts*, 2001), and the average U.S. consumer spends about $79 a year on home video (*Plunket's*, 1999). Blockbuster Video estimates that each customer spends about $120 per year on video rentals, and Blockbuster is the largest video rental chain in the United States, with 29% of the rental market (Villa, 2001).

The 99.4 million or so U.S. homes with TV have them turned on about 7 hours a day (*The World Almanac and Book of Facts*, 1991) and consume almost 50% of the leisure time of adult Americans (Robinson & Godbey, 1999). In fact, Americans spend more of their leisure time watching TV than any other single activity (Robinson & Godbey, 1999).

Not only do adults spend a great deal of time watching TV, they also report that it is a very important part of their daily lives. Hart Research Associates (1993) reported that 25% of all Americans would not give up TV for a million dollars, and another 20% would not give it up for less than a million dollars. It is interesting to note that in a recent Harris Poll, 23% of the respondents indicated their favorite leisure time activity was watching television (Gallop-Goodman, 2001).

Although nearly half of the people surveyed reported they would not give up TV for less than a million dollars, TV is rated as being less enjoyable than most other leisure activities (Spring, 1993). However, for people with few alternatives (e.g., people living on poverty or alone), television is considered to be an extremely important part of their daily lives (Hart Research Associates, 1993).

Television Usage and Older Adults

Television viewing increases over the adult life span (Bogart, 1972; Bower, 1973; Chaffee & Wilson, 1975; Harris & Associates, 1975; Hoar, 1961; Simmons Market Research Bureau, 1991, 1997; Steiner, 1963; Time Buying Services, 1986; *The World Almanac and Books of Facts*, 2001), and older viewers watch more television than any other age cohort, including children (Simmons Market Research Bureau, 1991, 1997; *The World Almanac and Books of Facts*, 2001). Television has been and continues to be the most frequently reported daily activity for older Americans (De Grazia, 1961; Robinson & Godbey, 1999; Schramm, 1969; Spring, 1993; Time Buying Services, 1986), and viewing levels are even higher among the widowed and retired older adults (Davis & Kubey, 1982; Kubey, 1980). This pattern has recently been shown to hold true with Chinese American audiences (Allison & Geiger, 1993) and audiences in Germany (Grajczyk & Zollner, 1998).

It is difficult to estimate exactly how much time people spend watching TV. Nielsen suggests that the set is on about, 1 hours per week in the typical household. Adults spend about 4 hours per day actually watching TV (Kubey & Csikszentmihalyi, 1990; *The World Almanac and Book of Facts*, 2001). The disparity often found in such estimates stem from differences in the way "adults" and "TV consumption" is conceptualized and measured. Almost every estimate of adult TV consumption is between 2.5 and 4 hours per day for people 18 years of age and older. As you can see in the Nielsen (2000) data below, television viewing habits vary with age and gender (see Table 17.1). Viewing habits

TABLE 17.1
Weekly TV Viewing by Age Cohort Over Three Years

Age & Gender	October 1999	November 1998	November 1997
Women 55+	41:20	42:00	41:50
Women 25–54	30:35	30:35	31:45
Women 18–24	21:30	22:11	25:22
Men 55+	36:28	36:47	36:17
Men 25–54	27:33	27:53	28:44
Men 18–24	20:10	19:29	20:30

also vary by income level and time of year. Generally, the higher the income level, the less television viewed. Households also watch significantly more television in the winter when the shows are new and the weather is bad in much of the country.

The average adult has about 41 hours of leisure time each week, and older people generally have more free time than the typical adult. Thus, it is not surprising that the elderly spend more time watching TV than the average adult and may spend as much as 6 hours a day tuning in (Danowski, 1975; Time Buying Services, 1986; *The World Almanac and Book and Facts*, 2001).

Of course, the elderly do not watch more television than younger adults solely because of their age. It is generally accepted that older people watch more television because they have the time and opportunity. As Bower (1985) suggested, "A person's age, sex, race, education, income make little difference—everyone views television about the same amount except when prevented from doing so by external factors, like work" (pp. 40–41). This is of particular salience because TV viewing tends to reduce the amount of time spent reading magazines and newspapers, as well as listening to the radio and attending movies (Robinson, 1972, 1981).

TV Content Preferences of Older Adults

Although older adults are less likely than other adult age cohorts to be cable subscribers, when they do have cable they are more likely to watch community access programming than most other age cohorts. Community access programming often consists of local school board and city council meetings broadcast as a community service. Such programming does not generally attract a very large audience, but retired older adults watch more community access programming than any age cohort except the 25 to 34 age group, and they are generally more satisfied with such programming than the other age cohorts (Atkins & LaRose, 1991).

The fact that viewers 65 and older watch community access programming is consistent with previous research that suggested older people prefer to watch the news, documentaries, and public affairs programming over other genres of shows (Bower, 1973;

Davis, 1971; Davis & Westbrook, 1985; Goodman, 1990; Korzenny & Neuendorf, 1980; Meyersohn, 1961; Phillips & Sternthal, 1977; Simmons Market Research Bureau, 1991, 1997; Steiner, 1963; Wenner, 1976). Because viewers over 65 constitute the single largest group of television news viewers (Davis & Westbrook, 1985; Simmons Market Research Bureau, 1991, 1997), the conclusion that the older adults prefer informative content over entertainment-oriented content appears to be highly defensible.

Examination of relatively current programming preferences indicates that older adults still enjoy informative programming and watch at rates higher than younger age cohorts. The Simmons Market Research Bureau (1997) data presented here asked respondents if they watched the news the previous night, and it is clear that older adults were more likely to watch early evening prime-time network news than the 18 to 34 and 55 to 64 age groups ($x2 = 15.497 = 4 \leq .004$, two tailed; see Table 17.2) Just under 83% of the elderly, 48.07% of the 55 to 64 age cohort, and 59.49% of the 18 to 34 age group reported watching the early evening prime-time network on the night before the survey. This trend holds true with weekend news viewing (see Table 17.3) and other news programs such as *MacNeil/Lehrer NewsHour* (see Table 17.4).

In addition, older adults report paying more attention to the news programs that they watch. This was true on the early evening network news broadcasts, as well as the *MacNeil/Lehrer Newshour* and *Nightly Business Report* (see Table 17.5).

Examination of the viewing patterns of the elderly is only one way to ascertain their preferences in programming. Using self-report measures, Goodman (1990) found that older adult males and females rate news and public affairs programming as their favorite. Older males report sporting events as their second favorite type of programming, and older adult females prefer educational programming. Elderly males reported their third favorite type of programming to be educational, whereas elderly females ranked dramas as their third favorite program type and sporting events as their least favorite type of programming.

Mundorf and Brownell (1990) also examined TV show preferences of older adults. They asked them to rank their favorite shows, identify their motivations for viewing,

TABLE 17.2
Network News Viewing by Age Cohort

Viewed Last Night	65+	55–64	18–34
ABC World News Tonight	25.06%	16.36%	18.52%
CBS Nightly News	31.73%	16.03%	18.36%
NBC Nightly News	25.89%	15.68%	22.61%
Combined News Viewing	82.68%	48.07%	59.49%

Chi-square = 15.497; $df = 4$, $p < .004$, two-tailed.
Data from Simmons Market Research Bureau (1997).

TABLE 17.3
Weekend Network News Viewing by Age Cohort

Viewed Last Night	65+	55–64	18–34
ABC World News Tonight Saturday	29.25%	16.75%	16.9%
ABC World News Tonight Sunday	27.65%	16.33%	17.13%
CBS Evening News Saturday	30.27%	16.38%	14.55%
CBS Evening News Sunday	28.29%	13.18%	16.88%
NBC Nightly News Saturday	29.57%	15.46%	18.43%
NBC Nightly News Sunday	31.58%	15.78%	17.06%

Data from Simmons Market Research Bureau (1997).

TABLE 17.4
Frequency of *MacNeil/Lehrer* Viewing by Age Cohort

Frequency of Viewing MacNeil/Lehrer	65+	55–64	18–34
Once a week	23.92%	11.29%	21.69%
Twice a week	22.43%	12.24%	15.91%
Three times a week	22.26%	18.78%	16.79%
Four times a week	32.24%	19.95%	13.37%
Five times a week	52.95%	14.59%	6.29%

Chi-square $= 69.077$; $df = 8$, $p < .001$, two-tailed.
Data from Simmons Market Research Bureau (1997).

and to name their favorite TV characters. Not surprisingly, the self-reported program preferences are generally consistent with the viewing patterns reported here.

When asked why they watch, however, older adults reported that their primary motivation was entertainment. Similarly, almost 70% of the college students surveyed also reported watching TV to be entertained. Although for most viewers the desire to be informed was the second most common usage of TV, this motivation was dwarfed by the desire to be entertained. Research into these motivations is sorely needed.

TABLE 17.5
Full Attention Viewing of News Programs by Age Cohort

Full Attention Viewing	65+	55–64	18–34
ABC World News Tonight	32.42%	16.53%	16.08%
CBS Nightly News	40.28%	12.98%	14.65%
NBC Nightly News	33.34%	15.6%	19.36%
MacNeil/Lehrer NewsHour	55.0%	13.12%	5.95%
Nightly Business Report	43.5%	15.78%	19.42%

Chi-square $= 24.287$; $df = 8$, $p < .002$, two-tailed.
Data from Simmons Market Research Bureau (1997).

Previous research has also suggested that older adults enjoy travelogues, game shows (Bower, 1973; Danowski, 1975; Davis, 1971), and soap operas (Barton, 1977; Mundorf & Brownell, 1990). Examination of the 1997 viewing patterns indicates that older people continue to watch these types of programs. For example, they watched more syndicated quiz and audience participation shows, more daytime quiz and audience participation shows, and more general variety programs than the younger age cohorts. The younger cohorts were much more likely to watch syndicated and prime-time situation comedies, and true crime and police dramas than their elderly counterparts (Simmons Market Research Bureau, 1997).

In terms of soap operas, Mundorf and Brownell (1990) reported that almost 91% of the college-age females and 60% of elderly females reported watching soaps on a regular or occasional basis, but only 18.8% of elderly males reported watching soaps. It will be interesting to see if those college age males continue to report watching soaps throughout the life span.

Mundorf and Brownell (1990) also asked older viewers and college students to list their favorite characters and found, generally, that older adults do prefer elderly characters and programs that contain elderly characters, and college students generally prefer programs featuring younger stars. This is consistent with previous research (Harris & Associates, 1975; Meyersohn, 1961; Parker, Berry, & Smythe, 1955). It is also consistent with research by Mares and Cantor (1992), who provided experimental evidence that people prefer programs with characters about their own age.

An examination of 1991 and 1997 viewing patterns (Simmons Market Research Bureau, 1991, 1997) indicates that this pattern appears to have continued. For example, in 1991 *Golden Girls* and *Murder, She Wrote*, were the two most often watched fictional TV programs for the elderly audience and both contained older characters. Although these two programs were quite popular with other age cohorts, they were much more popular in the older age groups. In the 1997 data (see Table 17.6), it is also clear that the favorite programs of the elderly and the 55- to-64 age cohort contain actors or characters that are similar in age. For example, *Walker, Texas Ranger* star Chuck Norris was 57 at the time, *Diagnosis Murder* star Dick Van Dyke was 72, *Cosby Show* star Bill Cosby was 60.

TABLE 17.6
Top Seven Weekly Television Programs by Age Cohort (1997)

Program Ranks, Adults 65+		65+	55–64	18–34
1	60 Minutes	30.94%	23.86%	6.1%
2	Touched by an Angel	24.0%	21.57%	8.88%
3	Walker, Texas Ranger	18.39%	17.43%	8.71%
4	20/20	17.85%	18.52%	10.87%
5	Cosby	17.0%	15.58%	7.18%
6	Dr. Quinn, Medicine Woman	15.33%	11.87%	5.2%
7	Diagnosis Murder	15.14%	12.41%	2.54%

Program Ranks, Adults 55–64				
1	60 Minutes	30.94%	23.86%	6.1%
2	Touched by an Angel	24.0%	21.57%	8.88%
3	20/20	17.85%	18.52%	10.87%
4	Walker, Texas Ranger	18.39%	17.43%	8.71%
5	Home Improvement	14.93%	18.08%	27.58%
6	Seinfeld	11.6%	15.77%	31.57%
7	Cosby	17.0%	15.58%	7.18%

Program Ranks, Adults 18–34				
1	Seinfeld	11.6%	15.77%	31.57%
2	Home Improvement	14.93%	18.08%	27.58%
3	Friends	5.36%	7.06%	22.1%
4	Simpsons	1.08%	4.03%	21.43%
5	E.R.	7.05%	11.07%	17.85%
6	X-Files	2.27%	7.34%	17.01%
7	Mad About You	2.27%	7.34%	16.82%

This is clearly different from the typical age of central characters in the shows most often watched by younger adults (e.g., *Seinfeld* and *Friends*).

Burnett (1991) examined the media usage of older adults and found that there were significant differences in program preferences attributable to gender and income. Specifically, he reported that affluent males preferred *Family Ties*, PBS, premium cable, and CNN, whereas the less affluent elderly watched more prime-time moves, *Monday*

Night Football, religious programs, and late-night reruns. Much like their male counterparts, the affluent elderly females watched *The Cosby Show*, PBS, premium cable, and CNN, and the less affluent elderly females preferred to watch prime-time movies, late-night reruns, religious programs, and soap operas. These findings are again generally supported by the viewership data from 1997 (Simmons Market Research Bureau, 1997). It is interesting to note that although religious television programs are popular with less affluent males and females, older adults do not substitute religious radio or television programs for actual church attendance as they grow older or become less able to get to church (Hays, Landerman, Blazer, Koenig, Carroll, & Musick, 1998). As Abelman (1987) pointed out, the motivation for religious television program viewing is typically dissatisfaction with secular or nonreligious programming and not religious instruction.

Radio Usage by Older Adults

Radio usage generally decreases with age (Simmons Market Research Bureau, 1991, 1999; Time Buying Services, 1986), with typical elderly adults listening between 1 and 2.5 hours a day (Beyer & Woods, 1963; Danowski, 1975; Parker & Paisley, 1966; Time Buying Services, 1986). Simmons Market Research Bureau data from 1991 and 1999 indicates that the elderly are significantly less likely to use the radio on a typical day than their younger counterparts. In the 1997 data set, 61.42% of older adults reported listening to the radio during the previous week. In the 55- to-64 age cohort, 71.52% had listened, and in the 18-to-34 age group, nearly 91% had listened to the radio in the week preceding the survey.

Believe it or not, more U.S. households have radios than have televisions. About 99% of all U.S. homes contain at least one radio (*New York Times Almanac*, 1999), and the average home had 5.6 radios in 1990 (*Universal Almanac*, 1991). In addition to the 576.5 million radios in homes, over 209 million cars, trucks, vans, and motor homes also have radios (*New York Times Almanac*, 1999). Because over 40% of all radio listening occurs outside the home (Beville, 1985), the number of car radios is of particular importance. Because most people listen to the radio while they are doing something else (e.g., driving to the store or working), age differences in radio usage may be, in part, attributable to the amount of time spent in the car and at work and not age.

About 95.3% of all Americans 12 or older listen to the radio regularly, for an average of 3 hours and 20 minutes each workday (*New York Times Almanac*, 1999). Zorn (1987) reported that nearly 70% of the adults surveyed reported that they pay as much or more attention to the news as they do to music when they listen to the radio. When listening to radio news, people listen for weather information, local news, regional news, and school closings (Zorn, 1987). Males, unlike females, also listen to sports on the radio.

Radio Programming Preferences by Older Adults

There were approximately 10,716 radio stations in the United States in the year 2000. Just under 21% of those stations were country music format stations. The second most common format was adult contemporary music (14.5%), and the third most common U.S. radio format was news/talk/business/sports (13.3%). The top six most common formats for radio stations can be seen in Table 17.7.

TABLE 17.7
Radio Station Formats by Percentage of Total
U.S. Radio Stations

Rank	Format	Percentage of Total U.S. Radio Stations
1	Country	20.9%
2	Adult contemporary	14.5%
3	News/talk/business/sports	13.3%
4	Oldies/classic hits	10.6%
5	Religious	10.4%
6	Rock	7.7%

Data from Simmons Market Research Bureau. (1997).

When older adults listen to the radio, they prefer new/talk/business/sports, country music, adult contemporary music, news, nostalgia, and religious programming (Simmons Market Research Bureau, 1997). The adult contemporary (AC) format grew out of the "middle of the road" (MOR) format and includes music that may be best described as soft rock. MOR radio might include songs by Frank Sinatra or Barbara Streisand, whereas AC might contain a song by Diana Ross or a ballad by the Kinks. In addition, in MOR radio, the deejay is more likely to be the focus of the program and spend a good deal of the time talking. In AC radio, the music is the focus and little talk, sports, or news surrounds it. In the vernacular of the trade, "More music—less talk."

An examination of radio usage suggests that older adults are no more likely to listen to talk or news programming than the 55- to-64 age cohort, but both are more likely to listen to talk or news than the 18- to-34 age group. The 18- to-34 group, on the other hand, are much more likely to listen to music). Generally speaking, older adults listen to daytime AM radio and very little FM radio (Phillips & Sternthal, 1977; Schiffman, 1971; Schreiber & Boyd, 1980). It is interesting to note, however, that older adults are no more likely to listen to religious programming than the other age cohorts (Simmons Market Research Bureau, 1997).

Burnett (1991) found differences in radio usage within the elderly population attributable to income. Specifically, he found that affluent elderly men and women prefer easy listening music, whereas less affluent males like to listen to country/western music, religious or gospel music, and sports. The less affluent elderly females enjoy the same music as their male counterparts, but do not listen to sports on the radio.

Reading and Older Adults

Spring (1993) reported that Americans spend about 2.5 hours a week reading. If you count the time spent reading as a secondary activity (e.g., reading and watching television or reading and eating), Americans spend almost exactly 3 hours per week reading (Robinson

& Godbey, 1999). Although the overall time spent reading has decreased since the 1960s, most of the decline can be attributed to the decrease in time spent reading a daily newspaper. In fact, the amount of time spent reading books and magazines has actually increased over that time span (Robinson & Godbey, 1999). Also keep in mind that the figure of 3 hours represents leisure time spent reading. All reading done for work or school is not included in the 3 hours.

Generally, women read more than men, the affluent more than the less well off, and college graduates read twice as much as their counterparts with a grade school education (Robinson & Godbey, 1999). Although adults may only spend about 3 hours a week reading, they spend more leisure time reading than any other activity except watching television and socializing (Robinson & Godbey, 1999). In fact, the amount of time spent reading books, magazines, and newspapers is growing closer to the amount of time spent in socializing over time. It is interesting to note that conversational time is divided nearly equally between conversations at home, telephone conversations, and correspondence / interpersonal communication.

Although it is generally agreed that the time spent reading decreases across the life span (Gordon, Gaitz, & Scott, 1976; Harris & Associates, 1975; McEnvoy & Vincent, 1980), much of the precipitous decline in reading previously reported can be attributed to the high incidence of illiteracy and physical factors such as loss of sight (Scales & Briggs, 1987). Ngandu and O'Rourke (1980) reported that 26% of older adults surveyed had given up reading or did not read because of eye difficulties. Although some older adults compensate by using magnifying glasses, large-print books, or audiotapes (Duncan & Goggin, 1981), others do not. Avid readers generally remain avid readers across the life span (Smith, 1993).

Adults 55 to 64, however, spend nearly three times more time reading than adults 18 to 24 (Robinson & Godbey, 1999). Reading expenditures are also higher for the 55- to 64-year-old age cohort than any other age cohort. In 1995, the average individual consumer spent about $81 on books, $50 on daily newspapers, and $36 on magazines (Plunkett's, Entertainment and Media Industry Almanac, 1999). The percentage of money spent, although increasing with household income, remains relatively stable, at about 0.5% of the household budget.

Magazine and newspaper readers are generally motivated by the desire to increase their knowledge, whereas book readers are more likely to read for pleasure. The research on book readership indicates that people read books to escape reality and learn about themselves, but not to fulfill social or interpersonal needs. In general, people only employ the media to fulfill their social needs when interpersonal opportunities are unavailable.

Newspaper Readership and Older Adults

Although audience perceptions of the educational value, coverage, and efficiency of the newspaper have decreased over the years (Bower, 1985), in 1994, 50% of all adults read a paper everyday, nearly 73% of all adults 18 and older read a newspaper at least a few times a week, and nearly 87% read a newspaper at least once a week (National Opinion Research Center, 1997).

Newspaper readership is significantly and positively related to age, education, and income (Doolittle, 1979; National Opinion Research Center, 1997; Schramm & White, 1947), as well as visual acuity and the perceived salience of the news (Adams & Groen, 1975; Salisbury, 1981). It is negatively related to TV viewing (Mediamark Research, 1991; National Opinion Research Center, 1997; *Universal Almanac*, 1991). Although newspaper readership has dropped from 62% of all reading in 1965 to about 34% of the time spent reading in 1985 (Robinson & Godbey, 1999), newspaper readership steadily increases across the life span (ANPA, News Research Bulletin [ANPA], 1973; Burgoon & Burgoon, 1980; Chaffee & Wilson, 1975) until the age of 70 (Chaffee & Wilson, 1975; Doolittle, 1979; National Opinion Research Center, 1997). This decline in newspaper readership has been attributed to vision problems associated with aging (Salisbury, 1981). Visual acuity begins to decline gradually from about the age of 40, and by the age of 70, few individuals have normal eyesight before correction (Botwinick, 1977). In addition, nearly 25% of the elderly have cataracts (Botwinick, 1977; Fozard, Wolf, Bell, McFarland, & Podolosky, 1977)—another reason time spent reading tends to decline with age.

Examination of the General Social Survey indicates that throughout the 1970s, 1980s, and 1990s, the vast majority of older adults read a daily newspaper. Just over 30% of adults 18 to 24 read a daily newspaper. In the 25- to-54 age cohort, 44.6% read a daily newspaper. The rate of daily newspaper reading jumped to 57.1% of the adults 55 to 64, and 75.3% of adults between 55 and 70 reported reading a newspaper daily (National Opinion Research Center, 1997).

Burnett (1991) also found newspaper readership to vary among income levels of elderly adults. Specifically, he found that affluent elderly males read the news section, business section, travel section, and magazine section of the newspaper more than their less affluent male counterparts. In addition, the affluent elderly male was more likely to read *USA Today* and *The Wall Street Journal* than elderly adults with a more moderate income. Elderly females do not read the sports section, business section, or *The Wall Street Journal*, but affluent elderly females are more likely to read the news section, food section, lifestyle section, and travel section than their less affluent counterparts. They are also more likely to read *USA Today*, but less likely to read the advertising supplement than less affluent elderly female newspaper readers.

Magazine Readership and Older Adults

Unlike newspaper readership, which tends to increase with age, magazine readership tends to decline with age (ANPA, 1973; Harris & Associates, 1975; Phillips & Sternthal, 1977; Schiffman, 1971; Schreiber & Boyd, 1980; Time Buying Services, 1986) and drops off dramatically at the age of 70 (Chaffee & Wilson, 1975). Danowski (1975) reported that the elderly spend about 30 minutes a day reading magazines, but almost all other estimates are lower. About 12 minutes a day is commonly reported for the amount of time older adults typically spend reading magazines (Robinson & Godbey, 1999). Such estimates are less useful, however, because the percentage of older people reading magazines is not that large, and consequently those who do read magazines tend to spend more time reading them than the estimate suggests (Burnett, 1991; Simmons Market Research Bureau, 1991, 1999) and enjoy reading magazines a good deal (Robinson & Godbey, 1999; Spring, 1993).

The Simmons Market Research Bureau (1999) data asked respondents what magazines they had read or looked into in the past six months and provided insight into the magazine reading habits of older adults. The top-rated magazines for the elderly are *Reader's Digest, People, Better Homes & Garden, Family Circle, Modern Maturity, National Geographic, Woman's Day, Good Housekeeping, TV Guide*, and *Newsweek*. This obviously reflects the large number of women in that age cohort. The top-ranked magazines for the 55-to-64 age cohort are quite similar to those of the 65+ age cohort. Only *Modern Maturity* was dropped from the list and was replaced by *Time*. In the 18-to-34 age cohort, magazine titles change dramatically. *Family Circle, Modern Maturity, National Geographic, Woman's Day, and Good Housekeeping* are less likely to be read, and *Time, Sports Illustrated, Entertainment Weekly, Cosmopolitan*, and *The National Enquirer* are more likely to be read. The percentages reported in Table 17.8 indicate the percentage of respondents who had read or looked into the magazine within the past 6 months.

Mundorf and Brownell (1990) compared magazine readership of college students with older adults and found differences in magazine preferences. In addition, gender differences help account for some of the apparent age differences in preference. For example, almost 94% of the elderly males read *Sports Illustrated*. However, whereas 29% of the female college students read *Sports Illustrated*, only about 2.5% of the elderly females read it on a regular or occasional basis. Mundorf and Brownell (1990) reported that *TV Guide* is the only magazine widely read by both age cohorts. Examination of the Simmons Market

TABLE 17.8
Top Ten Magazines Ranked by Percentiles of Readers by
Age Cohort

Magazine	% Elderly Readers	Readers 55–64	Readers 18–34
Reader's Digest	54.53%	46.93%	30.91%
People	40.66%	45.62%	59.5%
Better Homes & Gardens	38.71%	40.95%	33.19%
Family Circle	33.66%	36.22%	
Modern Maturity	33.65%		
National Geographic	32.43%	29.51%	
Woman's Day	32.32%	35.54%	
Good Housekeeping	31.31%	32.14%	
TV Guide	30.87%	33.63%	49.4%
Newsweek	27.43%	30.04%	33.94%
Time		30.99%	36.87%
Sports Illustrated			38.39%
National Enquirer			33.24%
Cosmopolitan			27.56%
Entertainment Weekly			29.21%

Note: Data from Simmons Market Research Bureau (1999).

TABLE 17.9
Top Ten Magazine Preferences for Elderly Males and Females

Magazine	% Elderly Female Readers	% Elderly Male Readers
Reader's Digest	54.07%	55.18%
Better Homes & Gardens	48.68%	24.66%
Family Circle	46.67%	
People	45.58%	33.73%
Woman's Day	45.34%	
Good Housekeeping	41.05%	
Modern Maturity	36.11%	30.18%
McCalls	34.25%	
TV Guide	33.74%	26.82%
National Geographic	27.76%	39.02%
Newsweek		33.43%
U.S. News & World Report		30.48%
Time		29.79%
Sports Illustrated		26.38%

Note: Data from Simmons Market Research Bureau (1999).

Research Bureau (1999) data by gender provided support for Mundorf's and Brownwell's (1990) contention that gender and not age may be the explanation for content preference differences in older adults. See Table 17.9 for a list of the Top 10 Magazine preferences for males & females.

Burnett (1991) also examined the magazine readership of older adults and found that the affluent elderly read many more magazines than their less affluent counterparts. The affluent and older male readers were more likely to read *Business Week, Newsweek, Time, The New Yorker, U.S. News & World Report, Forbes, Fortune, Money,* and *National Geographic* than the less affluent elderly males. The only magazine that moderate-income elderly males were more likely to read than the affluent elderly males was *Field & Stream.* There were no differences in the likelihood of reading *People, Sports Illustrated, TV Guide, Reader's Digest* or tabloids such as *The National Enquirer.*

Moderate-income females were more likely to read *Family Circle,* and more affluent elderly females were more likely to read *Newsweek, Time, U.S. News & World Report, Money,* and *National Geographic* than their less affluent female counterparts.

Enjoyment and Media Use

In an extensive study of leisure activities, Spring (1993) examined Americans' use of time and found that the average adult has about 41 hours of leisure time per week. Robinson and Godbey (1999) reported that the average adult had 39.6 hours of leisure time in 1985. Robinson and Godbey (1999) reported data from a number of leisure studies, including the American's Use of Time studies funded by the National Science Foundation.

TABLE 17.10
Percentage of Time Spent on Leisure Activities

Rank	Activity	Percentage of Time
1	Television viewing	30%
2	Socializing	8.0%
3	Reading	5.0%
4	Do-it-yourself projects	4.0%
5	Shopping	3.0%
6	Outdoor activities	2.6%
7	Hobbies	1.6%
8	Vacations	1.3%
9	Gardening	1.3%
10	Golf	1.1%
11	Team sports	1.0%
12	Sewing/knitting	0.9%
13	Eating out	0.89%
14	Movies	0.89%
15	Swimming	0.77%
16	Fishing	0.75%
17	Self-improvement	0.71%
18	Yard/home maintenance	0.70%
19	Eating at home	0.70%
20	Religious activities	0.64%
21	Listening to the radio	0.04%
22	Spectator sports	0.03%
23	Cooking/baking	0.03%
24	Parlor sports/games	0.03%
25	Volunteer work	0.03%
26	Walking	0.03%
27	Listen to tapes/CDs	0.02%
28	Other	33.94%

Note: Data from Spring (1993).

Within that data is a detailed analysis of not only how much time people spend (see Table 17.10) but how much enjoyment they derive from their leisure activities (see Table 17.11).

Using a scale from 0 to 10, Robinson (1993) found that people enjoy interpersonal interaction (chatting with family and friends) more than they enjoy watching television or listening to the radio, but not as much as they enjoy attending movies or watching videos and about as much as they like reading magazines. Television viewing was rated 7.8—higher than letter writing and telephone conversations with loved ones but still lower than most other leisure activities. Listening to the radio was the least enjoyable media-based leisure activity, and attending movies was the most enjoyable media-based leisure activity.

TABLE 17.11
Audience Self-Report Estimates of Enjoyment
of Communication Activities

Rank	Activity	Enjoyment
1	Having sex	9.3
2	Watching videos	8.3
3	Socializing with friends	8.2
4	Reading magazines	8.2
5	Socializing with family	8.0
6	Television viewing	7.8
7	Reading newspapers	7.8
8	Listening to the radio	7.3
9	Taking car to repair shop	4.6

Note: Data from Robinson (1993). Differences of .05 are significant.

Although the data from Robinson (1993) suggests that adults enjoy attending movies, older adults do not attend very frequently. McGloin's (2001) analysis of movie attendance figures from 1995 and 1999 demonstrated that adults 60 and older comprise 20% of the U.S. population but only 8% of total moviegoers during that time frame. Adults 50 to 59 attend at about the same rate. Teens 16 to 20 comprised 9% of the U.S. population and 20% of moviegoers in 1999—in fact, the 12-to-24 age cohort represents 22% of the U.S. population and 41% of the total moviegoing audience in 1999. If you exclude the 16-to-20 age cohort, film attendance increases from the age of 12 to 40 and then begins to decline. It is interesting to note that in a recent Harris Poll, 6% of the respondents indicated that their favorite leisure time activity is going to the movies (Gallop-Goodman, 2001).

Computer Usage and Older Adults

Recent figures indicate that in 1998, 42.1% of all U.S. households contained a personal computer (PC; *World Almanac*, 2001). This is an increase of 18% since 1994. Not surprisingly, nearly 80% of all households with an annual income of $75,000 or higher have a PC, and only 21.2% of all homes with an annual income of $15,000 to $20,000 had a PC in 1998. When the head of household was 55 years of age or older, the percentage of homes with a PC was 25.8%—by far the smallest percentage of all other age cohorts. In addition, when the head of household had a college degree, the home was much more likely to contain a PC. Just under 69% of all homes with a college graduate as the head of household had a PC, and the percentage of homes with a PC was less than 16% in the least well-educated families (*World Almanac*, 2001).

Similarly, the number of homes with Internet access has increased dramatically in recent years. In 1998, 26.2% of all U.S. households had Internet access (*World Almanac*, 2001). Percentages of homes with Internet access vary with education level, race, and region of the country, as well as with the age of the household head (*World Almanac*, 2001). Just over 14% of all homes headed by an adult 55 years of age or older had Internet access in 1998 (*World Almanac*, 2001).

Although the percentage of homes with computers may seem low, keep in mind that only about 10% of all U.S. homes had a computer in the middle of the 1980s and that number had grown to 35% by 1995 (Robinson & Godbey, 1999). In a survey of computer usage, Robinson and Godbey (1999) reported that the average daily use of computers was about 34 minutes per day for those with a computer. It is interesting to note that computer usage is positively related to other media usage. In fact the more one uses the computer the more time they spend reading newspapers, magazines, and books (Robinson & Godbey, 1999). In the case of television, however, the relationship is negative. People who do not have a computer or never use their computer spend more time watching television than those who use their computers (Robinson & Godbey, 1999).

In an effort to provide older adults with the skills they need to use personal computers, SeniorNet and other such organizations have been born. SeniorNet is a nonprofit organization with over 210 learning centers and over 39,000 members across the U. S. created to help older adults gain some measure of computer literacy. SeniorNet members are 50 years of age and older and dedicated to providing other older adults access to computer technology and computer training in a variety of areas including desktop publishing, word processing, Internet access, e-mailing and financial planning. SeniorNet is the result of a research project funded in 1986 by the Markle Foundation to determine whether telecommunications and computers enhance the lives of older adults. Since then, elderly SeniorNet volunteers have taught well over 100,000 adults 50 years of age and older a variety of computer skills.

In addition to training and access, SeniorNet also operates two discussion groups that seniors can access through American Online (keyword: SeniorNet) and the World Wide Web. These discussions cover a wide variety of topics and are open to the public. Membership in SeniorNet is not a requirement for participation—all seniors are welcome to participate. On the Web, the discussion boards are called the SeniorNet RoundTables and can be found at http://www.seniornet.org.

The E-telephone survey of adults 55 years of age and older was completed in November 1995 and underwritten by a grant from the Intel Corporation. The results of the survey clearly demonstrate that many seniors have adopted this technology and that increasing numbers have become rabid users of computers.

In one of the few in-depth, qualitative case studies of home-based recreational and personal computer usage, Ito, Adler, Linde, Mynatt, and O'Day (2001) examined the role SeniorNet plays in the lives of its members. In the 1-year-long ethnographic investigation, Ito et al. (2001) found that older adults were indeed able to use the Internet in ways that were meaningful. Older adults reported using SeniorNet as a place for social interaction, gaining information, recreation, and entertainment, as well as an opportunity to gain social support. Most significant, however, is the finding that older adults use SeniorNet for social inclusion, identity maintenance and strengthening, networking, and overcoming fear of technology or alienation from it. Although the SeniorNet participants did use the Internet for information retrieval and entertainment, the primary benefits were more social in nature.

Media Usage Conclusions

From the research reported here, it is clear that older adults use the mass media a great deal and that the mass media plays an important role in the lives of seniors. Although it is clear that seniors do have a strong preference for informational media content, they still prefer to be entertained when they use the media.

It is also clear, however, that the elderly are not a homogeneous group that uses the mass media in the same proportions or manner. Certainly, senior males use the mass media differently than senior females, and affluent seniors use the media differently than their less affluent counterparts. Future research needs to carefully consider differences within senior elderly population to avoid lumping seniors into a single group.

As much of the research reviewed in this chapter indicates, gender differences and income differences must be considered in future media usage studies. In addition, mobility and marital status and psychological variables such as loneliness, self-esteem, and social integration are other factors that merit consideration in future research.

Although the elderly spend the majority of their leisure time with television, exactly what they will find in terms of characters is another question entirely. The second section of this chapter will focus on how television has portrayed elderly characters throughout the history of television.

TV Portrayals of Older Adults

Over the past 35 years, a number of studies have examined how the elderly are portrayed on television. In a recent review of this literature, Robinson and Skill (1995) concluded that characters 65 years of age and older are significantly underrepresented on television. Most investigations found that characters 65 years of age and older represent less than 5% of the television population (Ansello, 1978; Arnoff, 1974; Bell, 1992; Downing, 1974; Elliott, 1984; Gerbner, Gross, Signorielli, & Morgan, 1980; Greenberg, Korzenny, & Atkin, 1980; Levinson, 1973; Northcott, 1975; Robinson & Skill, 1995; Signorielli & Gerbner, 1978). In fact, only Peterson (1973) reported finding significantly more elderly characters. She reported that 13% of the characters in her sample of 30 prime-time network programs were 65 years of age or older.

These percentages hold true in prime time (Arnoff, 1974; Gerbner, et al., 1980; Greenberg et al., 1980; Northcott, 1975; Robinson & Skill, 1995), in network programming throughout the day (Harris & Feinberg, 1977), game shows (Harris & Feinberg, 1977), Saturday morning programming (Gerbner, Gross, Signorielli, & Morgan, 1980; Greenberg et al., 1980) and cartoons (Levinson, 1973). The percentage of older characters only increases slightly—to 8% or 9%—in daytime serial programs or soap operas (Cassata, Anderson, & Skill, 1980; Elliott, 1984).

When older characters do appear on television they tend to be male (Arnoff, 1974; Elliott, 1984; Gerbner et al., 1980; Greenberg et al., 1980; Northcott, 1975; Robinson & Skill, 1995; Signorielli, 1982). Female characters occupy only about 30% of the roles (Arnoff, 1974; Gerbner et al., 1980; Greenberg et al., 1980; Northcott, 1975) and tend not to play serious roles or be treated with respect (Gerbner et al., 1980). Older women are often portrayed as being eccentric or lacking common sense (Gerbner et al., 1980), and elderly women tend to be of lower socioeconomic status than elderly men (Greenberg et al., 1980; Robinson & Skill, 1995).

The vast majority of older characters are White (Greenberg et al., 1980; Robinson & Skill, 1995), have no discernible religious affiliation (Robinson & Skill, 1995), more likely to be cast as widows than other age cohorts (Robinson & Skill, 1995), and unlikely to be cast in central roles (Robinson & Skill, 1995).

Roy and Harwood (1997) conducted an extensive study of television commercials and found that, again, older people are significantly underrepresented on evening network programming. Elderly males outnumber elderly females on these commercials, and the vast majority of the elderly were White. When appearing on commercials, however, the elderly were portrayed in a positive light. Although Roy and Harwood (1997) voiced some concern over the possibility of subtle agist stereotyping, the older characters were portrayed as strong, active, and happy, and consistent with the findings of Atkins, Jenkins, and Perkins (1991) and Swayne and Greco (1987). Roy and Harwood (1997) pointed out that this positive portrayal may be a manifestation of stereotyping and the desire to have products and services paired with favorable attitude objects. Research in this area is sorely needed.

Print Portrayals of Older Adults

A number of researchers have examined the way older people are portrayed in the various print media. Vasil and Waas (1993) reviewed this literature and concluded that the elderly are underrepresented in print, much like they are underrepresented on television. Specifically, the elderly are seldom depicted in newspaper advertisements—about 1% of the space is devoted to them (Bucholz & Bynum, 1982), magazine advertisements—less than 3% of the ads contained elderly models (England, Kuhn, & Gardner, 1981; Gantz, Gartenberg & Rainbow, 1980; Hollenshead & Ingersoll, 1982), magazine cartoons—4.3% of the characters (Smith, 1979), children's magazines—5.6% of the stories contain elderly characters (Almerico & Filmer, 1988), children's literature—3.3% of the characters (Barnum, 1977; Janelli, 1988), and Basal readers—5.6% of the characters (Robin, 1977).

In nearly every study, older women were even more underrepresented than older men, and the majority of the portrayals were negative (Vasil & Wass, 1993). In most studies, there was evidence of age stereotyping and generally the characters were cast in minor roles (Vasil & Wass, 1993).

Media Portrayals: Conclusions

The results of these studies suggest that overall, very little has changed over the years. Seniors continue to be infrequently seen on television, and when they do appear, they occupy lead roles at about one half the rate of all other age groups. Hacker (1951) suggested that fictional portrayals are an indication of a groups' social status. Given this perspective, there is clearly a great deal of room for improvement in the way older people are depicted on television, as there is room for improvement in the print arena.

However, if television writers are trying to attract audiences of a particular demographic composition that are appealing to advertisers, it makes sense that they would create stories and characters that they believe will attract and hold those viewers. Gitlin (1983) suggested that writers and producers create programs that are consistent with the "main contours of popular culture" (p. 25), because they are trying to attract a broad audience. Industry programmers attempt to understand the viewing interests of the target audience by employing research tools such as focus groups, surveys, Stanton-Lazarsfeld Program Analyzers, and Nielsen ratings in an effort to test the appeal of the stories they tell. The writers and network executives do not rely on empirical methods alone when trying to create programs. Industry axioms and superstitions, past programming successes, and gut feelings of writers, directors, and producers also guide television programming (Gitlin, 1983).

Regardless of whether the information used by the writers and directors comes from their own personal feelings about what the target audience will like, past show successes, or a detailed analysis of the demographic group being targeted, the programs represent the writers' best judgments about what the target audience will watch. As CBS programmer Herman Keld explained, "We always look for programming that appeals to everybody, but that kind of program doesn't exist, really. So the networks tend to program for their core audiences" (Gitlin, 1983, p. 57). It is clear that at this time, they do not believe the prime demographic market will watch elderly characters on TV.

Consequently, the way groups are portrayed on television is not so much an indication of how that group is viewed by society as it is an indication of how the writers believe the target audience views that group. Because TV programs are essentially narratives (cf. Fisher, 1989), the way stories are communicated tells us a great deal about the storyteller and the audience.

In describing the narrative form, Gallie (1964) pointed out that skillful story tellers do not provide explicit explanations within the story: "Characters must be presented and described in general terms, so that we can know them as types and are interested in them as individuals; but it is in the latter spirit that we follow their actions" (p. 23). Clearly, Gallie (1964) was describing characters that are central enough to the plot to actually develop over the course of the program. Less central or peripheral characters that do not develop over the course of the episode may provide a more interesting picture of the target audience than do central characters—more interesting in the sense that the peripheral characters represent the writers' conception of what the target audience believes about a social group unencumbered by central or focal story lines and the accepted stereotypical characteristics of that group.

In outlining Peripheral Imagery Theory, Robinson and Skill (1995) suggested that because peripheral characters are portrayed in general or stereotypical terms, examination of peripheral character portrayals on successful shows may provide insight into the attitudes and stereotypes of the target audience. In addition, the peripheral characters provide a context for understanding the central characters and provide the audience with the information they need to follow the story. It would be interesting for future research to examine whether or not shows that successfully tap into audience stereotypes are more successful than programs that present peripheral characters in a manner that is inconsistent with audience attitudes toward those social groups or individuals.

From the research describing media usage patterns by older people, it is obvious that they are a heterogeneous audience that use the media for a variety of different reasons. Bleise (1986) found that older people use the mass media in at least 10 ways:

1. To supplement or substitute for interpersonal interactions
2. To gather content for interpersonal interactions
3. To form and/or reinforce self-perceptions and to gather information about societal perceptions of various groups of people
4. To learn appropriate behaviors (including age-appropriate behavior)
5. For intellectual stimulation and challenge
6. As a less costly substitute for other media
7. For networking and mutual support
8. For self-improvement
9. For entertainment
10. For company and safety. (p. 575)

There are undoubtedly other reasons older people watch television. Mares and Cantor (1992) conducted an experiment and found older viewers use the mass media to alter their mood. Testing contrasting predictions from Zillman's (1988) mood management model and Festinger's (1954) theory of social comparison, they found that viewers can and do use TV characters as comparison others. These comparisons yield information about the relative abilities and opinions of audience members, as well as information that can alter their affective states.

For example, Mares and Cantor (1992) found that audience members do prefer to watch programs containing characters that are similar (in terms of age) to themselves. Further, they found that lonely and unhappy viewers rated programs that depicted the elderly in a negative fashion more favorably than they rated programs that portrayed them as being happy and socially integrated. Mares and Cantor also reported that people who are unhappy actually want to find programs with characters similar to themselves and in circumstances that are worse than their own. Consistent with social comparison theory, Mares and Cantor (1982) suggested that such comparisons actually raise audience levels of affect through ego enhancement. Similarly, unhappy, depressed, and lonely audience members may actually avoid programs with positive portrayals of similar others, because they do not want to receive the ego-threatening information such comparisons would yield.

Mares and Cantor were discussing audience use of central characters. Peripheral characters seem less likely to be useful in the social comparison process because audience members need information about specific opinions and abilities to engage in social comparisons. Future theorizing of media effects needs to examine the different roles that central and peripheral characters may play in that process.

The elderly are a diverse, heterogeneous group that must be studied as such. A starting point in this rather complex process might be through the enhancement of elderly images in the mass media. Diverse portrayals of elderly characters may help improve societal attitudes toward the elderly, as well as provide the opportunity for mood tone improvement via self-enhancing social comparisons.

Providing positive and negative portrayals of the elderly in all media will afford audience members of all ages the opportunity to increase their knowledge about the elderly and aging, improve their attitudes toward the elderly and aging, and provide viewers with comparison others to improve their affective states through social, or perhaps parasocial, comparisons.

REFERENCES

Abelman, R. (1987). Religious television and uses and gratifications. *Journal of Broadcasting and Electronic Media, 31*, 93–107.

Adams, M., & Groen, R. (1975). Media habits and preferences of the elderly. *Journal of Leisurability 2*, 25–30.

Allison, M. T., & Geiger, C. W. (1993). Nature of leisure activities among the Chinese–American elderly. *Leisure Sciences, 15*, 309–319.

Almerico, G. M., & Filimer, H. T. (1998). Portrayal of older characters in children's magazines. *Educational Gerontology, 14*, 15–31.

American Newspaper Publishers Association. (1973, April 26). *News and editorial content and readership of the daily newspaper* (ANPA News Research Bulletin). Washington, DC: Author.

Ansello, E. F. (1978, November). *Broadcast images: The elderly woman in Television.* Paper presented at the annual scientific meeting of the Gerontological Society, Dallas, TX.

Arnoff, C. (1974). Old age in prime time. *Journal of Communication, 24*(4), 86–87.

Atkin, D., & La Rose, R. (1991). Cable access: Market concerns amidst the marketplace of ideas. *Journalism Quarterly, 68*, 354–362.

Atkins, T. V., Jenkins, M. C., & Perkins, M. H. (1991). Portrayal of persons in television commercials age 50 and older. *Psychology: A Journal of Human Behavior, 28*, 30–37.

Barnum, P. W. (1997). Discrimination against the aged in young children's literature. *Elementary School Journal, 77*, 301–306.

Barton, R. (1977). Soap operas provide meaningful communication for the elderly. *Feedback, 19*, 5–8.

Bell, J. (1992). In search of a discourse on aging: The elderly on television. *The Gerontologist, 32*, 305–311.

Beville, H. M. (1985). *Audience ratings.* Hillsdale, NJ: Lawrence Erlbaum Associates.

Beyer, G., & Woods, M. (1963). *Living and activity patterns of the aged* (Research Report No. 6). Ithaca, NY: Cornell University Center for Housing and Environmental Studies.

Bleise, N. W. (1986). Media in the rocking chair: Media uses and functions among the elderly. In G. Gumpert & R. Cathcart (Eds.), *Intermedia: Interpersonal communication in a media world* (pp. 573–582). New York: Oxford University Press.

Bogart, L. (1972). *The age of television.* New York: Ungar.

Botwinick, J. (1977). Intellectual abilities. In J. Birren & K. Schaie (Eds.), *Handbook of the psychology of aging* (pp. 580–605). New York: Van Nostrand Reinhold.

Bower, R. T. (1973). *Television and the public.* New York: Holt, Rinehart & Winston.

Bower, R. T. (1985). *The changing television audience in America.* New York: Columbia University Press.

Broadcasting & Cable. (1993, May 31). 40 top cable programs (pp. 18–19). New York: Cahners.

Brunner, B. (Ed.). (1998). *Information please almanac.* Boston: Houghton Mifflin.

Buchholz, M., & Bynum, J. E. (1982). Newspaper presentation of America's aged: A content analysis of image and role. *The Gerontologist, 22*, 83–87.

Burgoon, J., & Burgoon, M. (1980). Predictors of newspaper readership. *Journalism Quarterly, 57*, 589–596.

Burnett, J. J. (1991). Examining the media habits of the affluent elderly. *Journal of Advertising Research, 31*(5), 33–41.

Cassata, M., Anderson, P., & Skill, T. (1980). The older adult in daytime serial drama. *Journal of Communication, 30*, 48–49.

Chaffee, S., & Wilson D. (1975, August). *Adult life cycle changes in mass media usage.* Paper presented at the annual meeting of the Association for Education in Journalism, Ottawa, Canada.

Dail, P. (1988). Prime-time television portrayals of older adults in the context of family life. *The Gerontologist, 28*, 700–706.

Danowski, J. (1975, November). *Informational aging: Interpersonal and mass communication patterns at a retirement community.* Paper presented at the annual meeting of the Gerontological Society, Louisville, KY.

Davis, R. H. (1971). Television and the older adult. *Journal of Broadcasting, 15*, 153–159.

Davis, R. H. (1984). TV's boycott of old age. *Aging, 346*, 12–17.

Davis, R. H., & Kubey, R. W. (1982). Growing old on television and with television. In D. Pearl, L. Bouthilet, & J. Lazar (Eds.), *Television and behavior: Ten years of scientific progress and implications for the eighties* (Vol. 2, pp. 201–208). Washington, DC: U.S. Department of Health and Human Services.

Davis, R. H., & Westbrook, G. J. (1985). Television in the lives of the elderly: Attitudes and opinions. *Journal of Broadcasting and Electronic Media, 29*, 209–214.

De Grazia, S. (1961). The uses of time. In R. Kleemeier (Ed.), *Aging and leisure* (pp. 113–153), New York: Oxford University Press.

Doolittle, J. (1997). News media use by older adults. *Journalism Quarterly, 56*(2), 311–345.

Downing, M. (1974). Heroine of the daytime serial. *Journal of Communication, 24*, 130–137.

Duncan, P. H., & Goggin, W. F. (1981, December). *Reading habits, patterns, and interests of older adult readers.* Paper presented at the annual meeting of the American Reading Forum, Sarasota, FL. (ERIC Document Reproduction Service No. ED 209 654).

Elliott, J. A. (1984). The daytime television drama portrayal of older adults. *The Gerontologist, 24*, 628–633.

England, P., Kuhn, A., & Gardner, T. (1981). The ages of men and women in magazine advertisements. *Journalism Quarterly, 58*, 468–471.

Faber, R. J., Brown, J. D., & McLeod, J. M. (1979). Coming of age in the global village: Television and adolescence. In E. Wartella (Ed.), *Children communicating: Media and development of thought, speech, and understanding* (pp. 215–249). Beverly Hills, CA: Sage.

Festinger, L. (1954). A theory of social comparison processes. *Human relations, 7*, 117–140.

Fisher, W. (1989). *Human communication as narration: Toward a philosophy of reason, value, and action.* Columbia, SC: University of South Carolina Press.

Fozard, J. L., Wolf, E., Bell, B., McFarland, R. A. & Podolsky, S. (1977), "Visual Perception and Communication," in J. E. Birren and K. W. Schale (Eds.), *Handbook of Psychology of Ageing* (pp. 497–534). New York: Van Nostrand Reinhold.

Gallie, W. B. (1964). *Philosophy and the historical understanding.* New York: Shocker.

Gallop-Goodman, C. (2001, March). Time off. *American Demographics, 24*.

Gantz, W., Gartenberg, H., & Rainbow, C. (1980). Approaching invisibility: The portrayal of the elderly in magazine advertisements. *Journals of Communication, 30*, 56–60.

Gerbner, G., Gross, L., Signorielli, N., & Morgan, M. (1980). Aging with television: Images on television drama and conceptions of social reality. *Journal of Communication, 30*(1), 37–48.

Gitlin, T. (1983). *Inside prime time.* New York: Pantheon.

Goodman, R. I. (1990). Television news viewing by older adults. *Journalism Quarterly, 67*(1), 137–141.

Gordon, C., Gaitz, C., & Scott, J. (1976). Leisure and lives: Personal expressivity across the life span. In R. Binstock & E. Shanas (Eds.), *Handbook of aging and the social sciences* (pp. 310–341). New York: Van Nostrand Reinhold.

Grajczyk, A., & Zollner, O. (1998). How older people watch television: Telemetric data on the TV use in Germany in 1996. *Gerontology, 44*, 176–181.

Greenberg, B. S. (1986). Minorities and the mass media. In J. Bryant & D. Zillman (Eds.), *Perspectives on media effects* (pp. 165–188). Hillsdale, NJ: Lawrence Erlbaum Associates.

Greenberg, B. S., Korzenny, F., & Atkin, C. K. (1980). Trends in the portrayal of the elderly. In B. S. Greenberg (Ed.), *Life on television: Content analysis of U.S. TV drama* (pp. 23–33). Norwood, NJ: Ablex.

Hacker, H. (1951). Woman as a minority group. *Social Forces, 30*, 39–44.

Harris & Associates, Inc. (1975). *The myth and reality of aging in America.* Washington, DC: National Council on Aging.

Harris, A., & Feinberg, J. (1977). Television and aging: Is what you see what you get? *The Gerontologist, 17*, 464–468.

Harris, R. J. (1989). *A cognitive psychology of mass communication*. Hillsdale, NJ: Lawrence Erlbaum Associates.

Hart Research Associates. (1993). How viewers feel about television. *American Demographics, 15*(3), 18.

Harwood, J. (1997). Viewing age: Lifespan identity and television viewing choices. *Journal of Broadcasting & Electronic Media, 41*, 203–213.

Hays, J., Landerman, L., Blazer, D., Koenig, H., Carroll, J., & Musick, M. (1998). Aging, health, and the electronic church. *Journal of Aging and Health, 10*, 458–482.

Hoar, J. (1961). A study of free time activities of 200 aged persons. *Sociology and Social Work, 45*, 157–163.

Hollenshead, C., & Ingersoll, B. (1982). Middle-aged and older women in print advertisements. *Educational Gerontology, 8*, 25–41.

Ito, M., Adler, A., Linde, C., Mynatt, E., & O'Day, V. (2001). Final report: Broadening Access: Research for Diverse Network Communities [Online, NSF No. 9712414]. Available: http://www.seniornet.org/research/9911.shtml

Janelli, L. M. (1988). Depictions of grandparents in children's literature. *Educational Gerontology, 14*, 193–202.

Korzenny, F., & Neuendorf, K. (1980). Television viewing and the self concept of the elderly. *Journal of Communication, 30*, 71–80.

Kubey, R. (1980). Television and aging: Past, present, & future. *The Gerontologist, 20*, 16–35.

Kubey, R., & Csikszentmihalyi, M. (1990). *Television and the quality of life: How viewing shapes everyday experience*. Hillsdale, NJ: Lawrence Erlbaum Associates.

Levinson, R. (1973). From Olive Oyle to Sweet Poly Purebread: Sex role stereotypes and televised cartoons. *Journal of Popular Culture, 9*, 561–572.

Lull, J. (1980). Family communication patterns and the social uses of television. *Communication Research, 7*, 319–334.

Mares, M., & Cantor, J. (1992). Elderly viewers' responses to televised portrayals of old age: Empathy and mood management versus social comparison. *Communication Research, 19*, 459–478.

McEnvoy, G., & Vincent, C. (1980). Who reads and why? *Journal of Communication, 30*(1), 134–140.

McGloin, T. (2001, May). On with the show. *American Demographics*, 54–55.

Mediamark Research, Inc. (1991). *Audio & video equipment & leisure activities*. New York: Author.

Meyersohn, R. (1961). A critical examination of commercial entertainment. In R. Kleemeier (Ed.), *Aging and leisure* (pp. 258–279). New York: Oxford University Press.

Mundorf, N., & Brownell, W. (1990). Media preferences of older and younger adults. *The Gerontologist, 30*, 685–692.

National Opinion Research Center. (1993). *General social survey* [Machine-readable data file]. Chicago: National Opinion Research Center.

National Opinion Research Center. (1997). *General social survey* [Machine-readable data file]. Chicago: National Opinion Research Center.

New York Times Almanac. (1999). New York: Penguin Books.

Ngandu, K. M., & O'Rourke, B. (1980, November). *Reading attitudes, habits, interests, and motivations of the elderly*. Paper presented at the annual meeting of the College Reading Association, Boston.

Northcott, H. C. (1975). Too young, too old—Age in the world of television. *The Gerontologist, 15*, 184–186.

Parker, E. B., Berry, D., & Smythe, D. (1955). *The television–radio audience and religion*. New York: Harper & Row.

Parker, E., & Paisley, W. (1966). *Patterns of adult information seeking*. Stanford, CA: Stanford University, Institute for Communication Research.

Petersen, M. (1973). The visibility and image of old people on television. *Journalism Quarterly, 50*, 569–573.

Phillips, L. W., & Sternthal, B. (1977). Age differences in information processing: A perspective on the aged consumer. *Journal of Marketing Research, 12*, 444–457.

Plunkett's Entertainment and Media Industry Almanac. (1999). Houston, TX: Plunkett Research.

Robin, E. P. (1977). Old age in elementary school readers. *Educational Gerontology, 2*, 275–292.

Robinson, J. P. (1972). Toward defining the functions of television. In E. Rubinstein, G. Comstock, & J. Murray (Eds.), *Television and social behavior: Vol. 4. Television in day-to-day life: Patterns of use* (pp. 568–603). Washington, DC: U.S. Government Printing Office.

Robinson, J. P. (1981). Television and leisure: A new scenario. *Journal of Communication, 31*, 120–130.

Robinson, J. P. (1993). As we like it. *American Demographics, 15*(2), 44–51.

Robinson, J., & Godbey, G. (1999). *Time for life: The surprising ways Americans use their time.* University Park: Pennsylvania State University Press.

Robinson, J., & Skill, T. (1995). The invisible generation: Portrayals of the elderly on prime-time television. *Communication Reports, 8*(2), 111–119.

Roy, A., & Harwood, J. (1997). Underrepresented, positively portrayed: Older adults in television commercials. *Journal of Applied Communication Research, 25*(1), 39–56.

Salisbury, P. (1981). Older adults as older readers: Newspaper readership after age 65. *Newspaper Research Journal, 3*(1), 38.

Scales, A. M., & Briggs, S. A. (1987). Reading habits of elderly adults: Implications for instruction. *Educational Gerontology, 13,* 521–532.

Schiffman, L. G. (1971). Sources of information for the elderly. *Journal of Advertising Research, 11*(5), 33–37.

Schramm, W. (1969). Aging and mass communication. In M. Riley, J. Riley, & M. Johnson (Eds.), *Aging and society: Vol 2. Aging and the professions* (pp. 352–375). New York: Russell Sage: Foundation.

Schramm, W., & White, D. M. (1947). Age, education, and economic status: Factors in newspaper reading. *Journalism Quarterly, 26,* 149–159.

Schreiber, E. S., & Boyd, D. A. (1980). How the elderly perceive television. *Journal of Communication, 30*(4), 61–70.

Signorielli, N. (1982). Marital status in television drama: A case of reduced options. *Journal of Broadcasting, 26*(2), 585–597.

Signorielli, N., & Gerbner, G. (1978). The image of the elderly in prime time television drama. *Gererations, 3,* 10–11.

Simmons Market Research Bureau. (1991). *The 1990 study of media and markets.* New York: Author.

Simmons Market Research Bureau (1997). *The 1996 study of media and markets.* New York: Author.

Smith, M. C. (1976). Portrayal of the elderly in prescription durg advertising. *The Gerontologist, 16,* 329–334.

Smith, M. C. (1993). The reading abilities and practices of older adults. *Educational Gerontology, 19,* 417–432.

Spring, J. (1993). Seven days of play. *American Demographics, 15*(3), 50–55.

Steiner, G. (1963). *The people look at television.* New York: Knopf.

Swayne, L. E., & Greco, A. J. (1987). The portrayal of older Americans in television commercials. *Journal of Advertising, 16,* 47–54.

Time Buying Services. (1986). *TV Dimensions '86.* New York: Author.

Universal Almanac. (1991). Harrisburg, VA: Banta.

Vasil, L., & Wass, H. (1993). Portrayal of the elderly in the media: A literature review and implications for educational gerontologists. *Educational Gerontology, 19,* 71–85.

Villa, J. (2001, May 31), *Blockbuster Launches DirecTV PPV Service* [Online]. Available: http://www.videostoremag.com/news/html/industry_article.cfm?article_id=1262

Wall Street Journal Almanac. (1998). New York: Ballantine.

Walsh, W. (1987). Going international: Foreign leisure. *American Demographics, 9*(2), 60.

Wenner, L. (1976). Functional analysis of TV viewing for older adults. *Journal of Broadcasting, 20,* 77–88.

World Almanac and Book of Facts. (2001). Mahwah, NJ: World Almanac Books.

Zillman, D. (1988). Mood management through communication choices. *American Behavioral Scientist, 31,* 327–340.

Zorn, E. (1987, December). Radio news: Alive and struggling—Debt, deregulation, and info-bits. *Washington Journalism Review 9*(10), 16–18.

VI

Health Communication

The continued growth of our older population, and the continued frequency with which older people visit their physicians, makes Thompson, Robinson, and Beisecker's topic (chap. 18) particularly salient. As they point out, older adults are relatively powerless, given predictable asymmetries of age, gender, and ethnicity in patient–practitioner encounters. Most of the existing research has focused on the communicative behaviors of patients and their doctors separately, but there is a growing appreciation and discussion of health care interaction as a jointly achieved phenomenon (Coupland, chap. 3, this volume). After a review of the research that contrasts young and older patients' interactions with physicians, and analyses of older patient–physician interactions in their own right, the authors highlight some contradictory findings (some finding positive and some negative aspects in and outcomes from interactions). These may be partly accounted for by different behaviors being manifested by geriatricians and other types of physicians. The review incorporates a wide sweep of studies in areas such as levels of information giving, control of the conversation, involvement in medical decision making, perceptions of control, and gender differences. Having looked at work on outcome variables such as patient satisfaction and patient compliance with treatment regimens, the authors suggest that the impact of communication variables on the health status of older patients is an area ripe for research. One distinctive feature of physician–older patient consultations is the frequency with which the patient is accompanied on a visit to the physician. In addition, the authors point to the increase in telemedicine and the unique impact telemedicine has on the older population.

The communication experiences of institutionalized older adults (in the United Kingdom, comprising only 5% of the population over 65 but about 20% of those over 85) are the focus of Grainger's chapter (chap. 19). Grainger's concern is that although there is literature on this population in the nursing, sociological, and psychological disciplines, little of it pays more than scant attention to communication. Grainger's own work, reported here in brief, is based on ethnographically grounded discourse analysis of talk between residents and patients in long-stay hospital wards.

The literature is reviewed under three themes: absence of talk, task-oriented talk, and dependency-including talk. It is not mainly the patient's own communication inadequacy but features of the institution that seem to militate against adequate and meaningful interaction. This results in communication-impaired environments, not the least occurring because of the high level of effort for low return that nurses associate with talking to older residents. All studies reviewed found that the majority (sometimes the vast majority) of talk is focused on the goals of the care task as opposed to interpersonal or relational goals. Studies on goals in discourse (see Tracy & Coupland, 1990) often stress the presence of multiple goal discourse, with, for example, relational goals assisting the pursuance of task goals (Coupland, Robinson, & Coupland, 1994). But the problem that Grainger is outlining stems from the conflict between such goals when tasks are too many for too few, which results in the deindividualization of older carees. In effect, talk is sometimes seen as interfering with care routines. Nurses' coping strategies include forms of evasion and suppression, for example, of patient's expressed troubles.

The extent to which institutionalized older adults have been made powerless is most clear in the section on dependency-inducing talk (which can make connections with studies on dependency, cited in Coleman's and O'Hanlon's chapter). As Grainger points out, Lanceley's (1985) characterization of nurses' language as controlling underrepresents frequent hedging or mitigating strategies, achieved through politeness formulas, for example. Power here is implied rather then directly enforced, and perhaps the more insidious for that. What nurses are doing is not using powerful forms of talk (or power behind discourse). Fairclough (1989) saw power behind discourse (which may well be quite widely a factor in young-to-elderly talk) as far more pervasive and indicative of imbalances of rights and obligations in societal exchanges generally.

Communication in a long-stay residential context is undoubtedly fraught with problems, where even attempts to nurture may be perceived as condescending. Grainger's description casts both the carees and the nurses as victims of impoverished communication opportunities. Nussbaum, Pecchioni, Robinson, and Thompson (2001) refer to the communication climate of long-term care facilities in the United States as "interactive starvation." Grainger recommends that intervention could benefit from more studies with the kind of critical sociolinguistic perspective she has taken, in which the solution to observed communication problems reflects the dialectical relationship between local interpersonal discourse processes and macroinstitutional and social structure.

Wright and Query (chap. 20) have contributed a new chapter to the handbook discussing the benefits and limitations of computer-mediated support networks for older adults. A generally accepted myth of our society is that older adults do not utilize new technologies and even may intentionally spurn any advance requiring new learning (See the final section of this handbook for a discussion of learning across the life span). Older adults are not only being affected by the new technologies, but are playing a very active role in the utilization of new technologies to gather health information and to join support groups in an attempt to remain physically, psychologically, and spiritually healthy. Numerous unique characteristics of interpersonal communication with the use of computer-mediated transmissions are advantageous to older adults who may be less mobile and who are seeking to maintain an active, social lifestyle. However, the limitations of computer-mediated support groups may be intensified within an older population.

REFERENCES

Coupland, J., Robinson, J. D., & Coupland, N. (1994). Frame negotiations in doctor–elderly patient interactions. *Discourse and Society, 5*(1), 89–124.

Fairclough, N. (1989). *Language and power*. London: Longman.

Fowler, R. (1985). Power. In T. van Dijk (Ed.), *Handbook of discourse analysis, 4*, 61–82. London: Academic Press.

Giles, H., Coupland, N., & Wiemann, J. (1992). "Talk is cheap" but "my word is my bond": Beliefs about talk. In K. Bolton & H. Kwok (Eds.), *Sociolinguistics today: Interactional perspectives* (pp. 218–243). London: Routledge.

Holmes, J. (1992). *Introduction to sociaolingusitics*. London: Longman.

Lanceley, A. (1985). Use of controling language in the rehabilitation of the elderly. *Journal of Advanced Nursing, 10*, 125–135.

Nussbaum, J. F., Pecchioni, L., Robinson, J. D., & Thompson, T. (2001). *Communication and aging*. Mahwah, NJ: Lawrence Erlbaum Associates.

Tracy, K., & Coupland, N. (1990). (Eds.). *Multiple goals in discourse*. Clevedon, UK: Multilingual Matters.

The Older Patient–Physician Interaction

Teresa L. Thompson and James D. Robinson
University of Dayton

Analee E. Beisecker*
University of Kansas Medical Center

The topic of how older individuals interact with their physicians is a particularly appropriate one for inclusion in this volume in light of the frequency with which older people utilize health care. During 1997, people in the United States between 65 and 74 averaged 5.5 office visits to a physician, and those 75 years of age and older averaged 6.5 visits (U.S. Census Bureau, 1999). This represents 24.3% of all physician office visits for the year 1997 (U.S. Census Bureau, 1999). Of course, not all physician–older patient interactions occur in the physician's office. During 1996, adults 65 years of age and older averaged just under 12 patient–physician contacts. Table 18.1 provides some indication of how older adults compare with other age cohorts in terms of their frquency of physician contacts.

Older adults are also more likely to spend time in a hospital than younger adults. The American Geriatric Society estimates that older adults are hospitalized three times more often than younger adults, and they stay in the hospital nearly twice as long after being admitted (Pritchett, 2000). Manton and Suzman (1992) suggested that the number of physician visits by people aged 65 to 74 is expected to nearly double between 1980 and 2040, increasing from 100 million visits to 186 million visits; and physician visits by those aged 75 and over are expected to quadruple, increasing from 66 million in 1980 to 115 million in 2000, 144 million in 2020, and 241 million in 2040. If, as anticipated, people continue to live longer, it is likely that this will be accompanied by increases in the duration and prevalence of chronic diseases among older people, especially for those over 85 years of age (Manton & Soldo, 1992). Thus, older people spend a significant portion of their time interacting with physicians, and these interactions have important consequences for the quality of life of older patients.

---

*deceased

TABLE 18.1
Annual Physician–Patient Contacts by Age Cohort

Year	18–24 Years of Age	25–44 Years of Age	45–64 Years of Age	65 Years of Age and Older
1985	4.2 visits per year	4.9 visits per year	6.1 visits per year	8.3 visits per year
1990	4.3 visits per year	5.1 visits per year	6.4 visits per year	9.2 visits per year
1995	3.9 visits per year	5.2 visits per year	7.1 visits per year	11.1 visits per year
1996	4.1 visits per year	4.9 visits per year	7.2 visits per year	11.7 visits per year

This chapter reviews the research on older patient–physician communication by focusing on the empirical work published since Haug's and Ory's (1987) thorough review of the topic. These studies have been, for the most part, quantitative field studies. Some of the more recent work has entailed the observation of actual interactions; most earlier work used self-report questionnaires or interviews. A small amount of the research has been qualitative or ethnographic (cf. Koch, Webb, & Williams, 1995, and a few other pieces cited throughout this review). Most of the early research was conducted in the United States, but the recent research has taken place in a variety of cultures. Recent writings have begun to emphasize the need to attend to cultural, language, and ethnic differences in regard to older patients (Adler, McGraw, & McKinlay, 1998; Seijo, Gomez, & Freidenberg, 1995; Wood, 1989).

We begin by discussing the differences between physician–patient interactions when the patient is older versus when the patient is younger, with a special emphasis on physiological barriers to communication. We then move on to a review of some descriptive research on older patient–physician interaction. Turning to a focus on some more specific variables operating within this context, we discuss such concepts as information transmittal, control issues, provider–patient agreement, and gender differences, all of which have received some research attention in recent years. Important outcome variables will then be examined, including satisfaction, compliance, and health status. Finally, we discuss three variables not always considered in the interpersonal health communication research that have special significance for older adults: the influence of a companion during physician–patient interaction, telemedicine, and advanced directives. The conclusion offers some suggestions regarding the application of theoretical perspectives to research in this area.

INTERACTIONS INVOLVING OLDER VERSUS
YOUNGER PATIENTS

There are many ways in which all physician–patient interactions are similar regardless of the age of the patient (for a review of the research on physician–patient interaction in general, see Thompson & Parrott, 2002). There are probably more ways in which physician–patient interaction is the same regardless of age than there are ways in which interaction differs depending on the age of the patient. However, research has identified some of the ways in which physician–patient interaction does differ with the age of the patient, and these differences are worth noting. Although, as discussed later in this chapter, it is most fruitful to look at talk as something that is jointly achieved, past research has typically focused separately on the behavior of patients and providers. According to this research, some of these differences are apparently because of older patients behaving differently than younger patients, and some are because of physicians treating older patients differently from younger patients. Others, however, appear to be because of the interaction of those two factors.

Several physiological factors affecting older people may complicate physician–patient interaction. These include sensory deficits, cognitive impairment, functional limitations, and multiple medical problems, many of which probably cannot be cured (Adelman, Greene, & Charon, 1991). Hearing impairment is, of course, more common in older patients (Desselle, 1999), and research has examined the communicative strategies used by hearing-impaired older adults to manage interaction (Stephens, Jaworsk, Lewis, & Aslan, 1999), the impact of hearing impairment on health care interaction (Witte & Kuzel, 2000), and the role that care providers can play in removing communication barriers associated with hearing impairment (Lindblade & McDonald, 1995). Many of these physiological factors are associated with unpleasant vocalizations in some older adults (Clavel, 1999). Even apart from hearing loss, language comprehension may deteriorate with age (Pichora-Fuller, 1997). Conversational interaction style alters with increasing age, with individuals beyond 75 conveying information with reduced efficiency (Mackenzie, 2000). Decision making also changes with aging (Yates & Patalano, 1999).

Communication changes associated with aging are particularly noticeable in the case of individuals with Alzheimer's disease (AD; Orange, Molly, Lever, Darzins, & Ganesan, 1994). Alzheimer's patients tend to use fewer statements and longer utterances, ask more questions, and list fewer essential and optional steps when communicating a procedure (Ripich, Carpenter, & Ziol, 1997). They typically speak in shorter turns, rely on more nonverbal responses, use more requestive and fewer assertive statements, and utter more unintelligible statements (Ripich, Vertes, Whitehouse, Fulton, & Ekelman, 1991). The caregivers of AD patients indicate that the patients engage in more repetition of questions and have difficulty both following conversations in a group and keeping a conversation going (Powell, Hale, & Bayer, 1995). Gesturing also changes as a result of Alzheimer's disease, with Alzheimer's patients using more referentially ambiguous gestures, fewer gestures referring to metaphorical as opposed to concrete contents, and fewer conceptually complex gestures involving both arms (Glosser, Wiley, & Barnoski, 1998). Other people speak in shorter turns when interacting with someone with senile dementia of the Alzheimer's type (Ripich et al., 1991), although effectiveness in both interpretting the communication of Alzheimer's patients (Acton, Mayhew, Hopkins, & Yauk, 1999) and professional support for caregiver coping (Vernooij-Dassen & Lamers, 1997) have been empirically demonstrated. Age is also a factor in the disclosure of dementia to the patient (Heal & Husband, 1998).

Assessing pain is problematic in any patient who is nonverbal or voiceless, but is more common in older patients (Simons & Malabar, 1995). Voicelessness also negatively impacts other aspects of interaction (Happ, 2000). Reliance on verbal communication is impacted by psychiatric issues as well (Haley, 1996; Lesser, Beller, & Harmon, 1997).

Older patients also have had different health care experiences than younger patients, because of exposure to economic depressions, war, and lack of access to education, all of which may affect their attitudes toward self-care and skepticism about the efficacy of medicine (Haug & Ory, 1987). There are other attitudinal differences between older and younger patients. For instance, older women are more accepting of mammography. Despite stereotypical conceptions, older patients do not demand a formal mode of address from their physician (Tiernan, White, Henry, Murphy, Twomey, & Hyland, 1993). Older persons report higher frequencies of preventive health actions, are more compliant with treatment regimens, and are more vigilant and responsive to health threats, but they are more likely to attribute symptoms to age rather than to illness (Ory, Abeles, & Lipman, 1992).

Older and younger patients' communicative behaviors have also been compared. Younger patients have been found to ask questions more frequently and with more precision, talk about their problems more, give information in more detail, and be more assertive (Adelman et al., 1991; Breemhaar, Visser, & Kleijnen, 1990; Greene, Adelman, Charon, & Friedmann, 1989). There is disagreement in the literature in regard to question asking, however. Svarstad (1976) found that older patients ask more questions, but Beisecker and Beisecker (1990) qualified this by concluding that older patients ask more questions only in longer interactions, and the questions tended to arise well into the interaction rather than in early parts of it. Additional research should explore such issues as the kinds of questions found in longer interactions and whether they are elicited by the care provider.

Several authors have noted that negative attitudes of physicians toward older patients can affect interaction (Adelman et al., 1991). Physicians spend less time with older patients, even though those older individuals are slower to undress and provide information (Haug & Ory, 1987). Older patients also receive less information in general (Street, 1991b) and about their medication in particular (Wiederholt, Clarridge, & Svarstad, 1992), although this may be because of differences in patient elicitation of such information. However, physicians who work with older patients more frequently, as reflected in Medicare billings, have been observed to spend as long with their older patients as with their younger ones (Radecki, Kane, Solomon, Mendenhall, & Beck, 1988). Ageism on the part of physicians has been inferred by researchers (Greene, Adelman, Charon, & Hoffman, 1986) as reflected in findings that physicians are more condescending, abrupt, and indifferent with older patients (Adelman et al., 1991), and are more egalitarian, patient, engaged, respectful, and likely to share decision making with younger patients (Adelman, Greene, Charon, & Friedmann, 1990). The studies by Adelman et al. (1990, 1991) used the 36-category Multi-Dimensional Interactional Analysis (MDIA) instrument developed specifically for coding audiotapes of physician–older patient interactions. The instrument allows for the coding of the frequencies of various behaviors and considers the content of the talk and the quality or success of the interaction. However Coupland and Coupland (1993) found that, in a U.K. geriatric clinic, most physicians enacted an antiageist discourse even when patients made ageist remarks.

Hall's, Roter's, and Katz's (1988) meta-analysis of 41 studies concluded that older patients receive more information, more total communication, more questions about drugs, more courtesy, and less tension release than younger patients. Older patients are also more satisfied with their health care. This is consistent with the finding of Breemhaar et al. (1990), that older patients judge the medical information they receive as clearer and more adequate.

Physicians provide better information to younger patients on physician-initiated issues and are more supportive of younger patients on patient-raised issues (Adelman et al., 1991). Similarly, physicians are more likely to raise psychosocial issues with younger patients than with older patients (Greene, Hoffman, Charon, & Adelman, 1987), although discussion of psychosocial issues is even more important with older patients (Greene & Adelman, 1996). Doctors also respond significantly better to psychosocial topics raised by younger patients. Overall, physicians are more responsive to younger patients regardless of topic (Greene et al., 1987).

The difference in discussion of psychosocial topics noted in the study by Greene et al. (1987) may be an interactive one. The authors noted that younger patients may talk about more psychosocial topics, because physicians first initiate such topics. Similarly, the lack of responsiveness of physicians to psychosocial concerns raised by older patients may have discouraged patients from further initiation of these concerns.

Beyond this, there is more agreement or concordance between younger patients and their doctors (Adelman et al., 1991; Greene, Adelman, Charon, & Friedmann, 1989) and a more egalitarian sharing of interests (Greene et al., 1987) than is found with older patients. Haug and Ory (1987) also noted that older people find it more difficult to interact with physicians on an egalitarian basis.

INTERACTION ANALYSES

The preceding section provided some insight into the nature of older patient–physician interaction by contrasting it with interactions between physicians and younger patients. Additional insight can be yielded by examining some studies that have undertaken detailed descriptive analysis of encounters between older patients and their physicians, although critics of research on provider–older patient interaction have noted that much research tends to ignore the contribution of the patient to the interaction (Jarrett & Payne, 1995) and does not actively examine the effectiveness of interventions (McCormick, Inui, & Roter, 1996). Building on some of the research previously mentioned, Adelman, Greene, Charon, and Friedmann (1992) coded the content of older patient–physician interactions using the MDIA system. They found that physicians and patients were equally likely to raise medical topics for discussion. Psychosocial subjects accounted for 11.4% of physician-initiated topics and 25.7% of patient-initiated topics. Physicians brought up almost two thirds of all topics raised during the visit. About 70% of medical topics and personal habits topics were initiated by physicians, whereas 55% of psychosocial topics were raised by patients. Patients were also more likely than physicians to initiate discussion of the physician–patient relationship. For every content area, physicians responded more positively and with more detailed information when they had initiated the topic. The authors concluded that these results indicate that older patients may experience difficulty getting their concerns addressed by physicians, because the physicians frequently did not respond with much information to issues raised by the patients.

Similar issues were raised in recent research by Rost and Frankel (1993). Their goal was to determine whether older diabetic patients' concerns are not addressed in medical encounters, because the patients choose not to raise them or because they intend to raise them but do not. Results indicated that an average of 27% of the patients' current problems were not addressed, because the patient did not raise them. More than half of the patients in this study had at least one important medical problem that was never raised; 60% had at least one important psychosocial problem that was not raised. The problem identified by the patient as most important was not the problem that 70% of patients introduced initially. It is possible, of course, that the placement of the topic in the discussion is not an important variable, because such characteristics as whether the topic is embarrassing, threatening, or painful to discuss also play a role. Rost and Frankel also found, however, that patients who participated in a previsit interview conducted by a researcher asking them about their problems raised an average of 7.6 additional problems they had not included in their initial agenda for the visit. This finding is consistent with other research indicating that patients can be encouraged to raise relevant issues and ask questions of physicians (Feeser & Thompson, 1993; Greenfield, Kaplan, & Ware, 1985; Robinson & Whitfield, 1985; Roter, 1977). Ethnographic research indicated that patients are more likely to perceive openness in care providers if the providers respond to patient needs; maintain eye contact, smile, explain procedures; do helpful, little things for patients; use time flexibly and efficiently; and use a gentle touch (Scott, 1997). Grainger's (1993) discourse analysis of interaction in nursing homes found that discourse modes function to construct institutional definitions of reality that give priority to the physical nursing task to the detriment of patients' relational needs. She argued that this contributes to the poor quality of life in most nursing homes. In regard to the caregiving decisions made by frail elders and their families, Agee and Blanton (2000) reported that care providers are more likely to assume an "educator" (p. 313) role than to function as a listener.

Finally, Coupland and Coupland (1993; see also Coupland, Coupland, & Robinson, 1992) provided some provocative data demonstrating that both patients and physicians are involved in the negotiation of the interaction. Their findings indicate that a holistic perspective on the health of older people is manifest through physician talk within a geriatric clinic by providing emphasis on sociorelational as well as medical issues. These findings, in combination with those of Radecki, Kane, Solomon, Mendenhall, and Beck (1988), indicating that nonageist behaviors on the part of physicians who see many older patients, may lead us to question generalizing the findings discussed within all medical settings. There seem to be some distinct differences between physicians specializing in geriatric medicine and other types of physicians.

INFORMATION

Beyond the issues raised in the studies previously discussed, some research has focused specifically on information transmittal between patients and physicians. The quality of diagnosis is dependent, in part, on the information shared by the patient. In addition, research has also associated increased information giving on the part of the physician with increased compliance, patient recall and understanding, patient satisfaction

(Hall et al., 1988), decreased levels of concern in patients (Arntson, Makoul, Pendleton, & Schofield, 1989) and improved health status (Kaplan, Greenfield, & Ware, 1989a, 1989b). Despite these positive effects of physician information giving, most physicians spend a lot more time seeking information than they do giving it (Roter, Hall, & Katz, 1988). Most older patients and their families have an incomplete understanding of the health information they receive from the emergency room, but do not voice complaints about their experiences (Majerovitz et al., 1997). Although patients report needing more information about their medications and side effects, they receive it only if they introduce the topic (Smith, 1998; Smith, Cunnngham, & Hale, 1994). Increased information giving by the patient is associated with poorer health status (Kaplan et al., 1989a) but with greater adherence to prescription regimens if the patient volunteers information, as well as responds to questions (Rost, Carter, & Inui, 1989).

The amount of information communicated by physicians to patients has been found to be determined, in part, by the patient's communicative style, particularly patient question asking and expressions of affect. More specifically, patient question asking predicts the communication of diagnostic health information from physicians, patient question asking and expressions of concern predict treatment information, whereas patient anxiety predicts information received about medical procedures (Street, 1991b).

Although, as was mentioned earlier, older patients have been found to not receive as much information as younger patients, research has indicated that older patients do desire information (Beisecker, 1988; Haug & Ory, 1987; Rogers et al., 2000). This desire for information is indicated not only by self-report data, but also by the information-seeking comments made by patients (Beisecker & Beisecker, 1990). In the Beisecker and Beisecker (1990) study, information-seeking comments included direct questions by the patient to gain information, patient initiation of a topic to gain information, and asking the doctor for clarification. Situational variables such as length of interaction, diagnosis, and reason for visit are stronger determinants of patterns of information seeking than are attitudinal or sociodemographic characteristics (Beisecker & Beisecker, 1990). This desire to receive information has also been substantiated in other research. Blanchard, Labrecque, Ruckdeschel, and Blanchard (1988) reported that 92% of older cancer patients in their research preferred to receive all information—good or bad.

INVOLVEMENT, PARTICIPATION, AND CONTROL

The notion of patient involvement, participation, and control has become an important one in the research on physician–patient interaction in light of the growing trend toward what has been labeled a consumeristic attitude on the part of patients (Beisecker, 1988, 1990; Beisecker & Beisecker, 1993) and, in some cases, an accommodating response on the part of physicians (Roter & Hall, 1989; Roter et al., 1988). Even though, as noted earlier, older patients desire information, they frequently have not been found to want as much involvement in medical decision making as do younger patients (Adelman et al., 1991; Beisecker, 1988; Haug & Ory, 1987). This difference is not only an attitudinal one, it is also indicated by less assertive and controlling behaviors such as topic changes or direct challenges to the doctor on the part of older patients (Beisecker, 1990). Lack of a desire for participation is particularly noticeable in older, sicker men (Beisecker, 1990; Blanchard, Labrecque, Ruckdeschel, & Blanchard, 1988). This is consistent with the finding that physicians are less likely to share decision making with older patients (Greene et al., 1989), although Street (1991b) found physicians engaging in more partnership building, as indicated by allowing input, with older patients. Patient assertiveness can also be increased with intervention (Adler et al., 1998).

Patients who do desire involvement in medical decision making typically perceive that they have received different information than do those who do not desire such control. Patients desiring input in decision making more often perceive that the physician has discussed future tests or treatments, discussed test or treatment results, gave good news, discussed discharge, and discussed side effects. Those who prefer to leave decisions to physicians more often perceive that the doctor has discussed the family's role in patient care (Blanchard et al., 1988).

As noted in the research on information transmittal, physicians tend to dominate conversations with patients, especially during the history-taking portion of the interview (Rost et al., 1989). Even when patients do voice a comparable number of utterances, they are at a disadvantage in getting their viewpoint across when most of their utterances are in response to physicians' questions (Rost et al., 1989), unless those questions include, "Is there anything else you'd like to ask or tell me about?"

Whether the physician or the patient controls an interaction or controls decision making has been shown to have some consequences for both parties. For instance, the more the physician dominates a conversation, the less the patient is satisfied with the interaction (Bertakis, Roter, & Putnam, 1991). Patients preferring to participate in decision making perceive themselves as more involved in the interaction and are less satisfied with the medical encounter (Blanchard et al., 1988). Patients who are more controlling during the office visit report fewer days lost from work, fewer health problems, and fewer limitations in their activities, and rate their health more favorably at a follow-up (Kaplan et al., 1989a). Similarly, patients who desire greater involvement in their health care and perceive that they have more control recover more quickly, as evidenced by data that they are released from the hospital more quickly (Mahler & Kulik, 1990). Overall, Coe (1987) concluded that communication styles that reduce doctor–patient status differences and incorporate the patient's viewpoint are associated with positive outcomes.

Based on a review of nursing care for older people, Davies, Laker, and Elils (1997) identified several institutional indicators that are associated with attempts to promote patient autonomy and independence. These include: systems of care promoting comprehensive individualized assessment and multidisciplinary care planning; encouraging patients to participate in decisions about their care; patterns of communication that deemphasize exertion of power and control over patients; and modifications to the environment to allow independence and minimize risk.

In addition to the research that has looked at patient involvement or control as a variable of the interaction, some research has looked at control as a personality variable by focusing on health locus of control (HLC) in patients. Health locus of control is a self-report measure of the extent to which people feel their health status is determined by their own behaviors (internal) or forces external to themselves (fate, doctor, etc.; Rodin, 1986; Rotter, 1966). Older patients seem more likely to show an external locus of control, and an external HLC is related to patient question asking and discussing problems (Breemhaar et al., 1990). Breaking this concept down more specifically, however, Arntson et al. (1989) looked at three dimensions of HLC: internal HLC, fate HLC, and doctor HLC. Each of these dimensions operates independently; a high score on one does not necessarily correlate with high or low scores on the other dimensions. Older patients were more likely to rate themselves high on doctor HLC or fate HLC. Patients with high doctor HLC or fate HLC felt their physicians did not respond well during the encounter. Patients with high internal HLC, doctor HLC, or fate HLC reported that their physicians encouraged significantly more patient involvement than those low on these dimensions.

AGREEMENT OR CONCORDANCE

Some research has compared physician and patient perceptions of control within the interaction. Patients (average age of 64) were more likely than were their physicians to rate relationships as partnerships rather than as physician controlled. Physician–patient agreement regarding control in the interaction was no better than expected by chance, although agreement was higher when status differences were narrower (Anderson & Zimmerman, 1993).

Noting that the sociodemographic characteristics of age, gender, and race are often mismatched in the older patient–physician relationship, suggesting that patients are often less powerful in these interactions and that patients and physicians are not homophilous, Adelman et al. (1991) looked at agreement between doctors and patients on the main goals and primary medical problems of the encounter. They found significantly less concordance in encounters involving older patients than in those involving younger patients. This is even more true with regard to medical topics than for psychosocial topics and personal habits (Greene et al., 1989). Thorne and Robinson (1989) also noted the relevance of conflicting perspectives of doctors and patients on many dimensions, leading to unmet expectations and shattered trust. Similarly, Rose, Bowman, and Kresevic (2000) documented a lack of agreement between family caregivers and health care providers of older individuals about the patients' health conditions, needs to stay healthy, and problems in self-care. The quality of the relationship between the physician and the older patient, and the medical progress of the patient, depend on the congruence of aims between the interactants and the complementarity of their roles (Jones, O'Neill, Waterman, & Webb, 1997). Indeed, patient selection of cancer treatment and subsequent patient longevity are related to physician–patient agreement on chances of survival (Baumrucker, 1998). In those cases where physicians do not communicate their estimates of likely patient survival, patients are more likely to choose treatment options that result in earlier death. Patient decisions about life-prolonging care, however, are impacted by interaction with all persons in the clinical setting who know the patient's treatment wishes, not just the physician (Shidler, 1998)

Although rarely a topic of study in the research on communication between older individuals and their physicians, we know at least a little about gender differences in this context. The paucity of such research is unfortunate, considering the notable gender differences that have been uncovered in research on patient–physician interaction that does not focus on older patients (Gabbard-Alley, 1995), including the negative stereotypes that physicians may hold of female patients, such as the tendency toward hysteria and a belief that "female" complaints should not be taken seriously (Bernstein & Kane, 1981), and the fact that most females, regardless of age, typically visit physicians more frequently than do males (Arntson et al., 1989). Some research, of course, has reported no gender differences in the variables that have been examined. The research that has uncovered gender differences in older patients indicates the following:

1. Females report that their doctors give them more responsibility for their own health care and that they are less concerned than are men following the consultation (Arntson et al., 1989).
2. Significantly more females rate their internal HLC as low (Arntson et al., 1989).
3. Age and gender interact, such that women under 65 report higher familiarity on a paper-and-pencil measure with their doctors than do males, whereas after age 65 males perceive higher familiarity (Arntson et al., 1989).
4. Gender match between doctors and patients fosters rapport and facilitates disclosure (Haug & Ory, 1987).
5. Male patients tend to be more opinionated and to receive more partnership-building statements that request or allow input from physicians (Street, 1991b).
6. More women than men report receiving no information from physicians regarding their medication (Wiederholt, Clarridge, & Svarstad, 1992).
7. Negative stereotypes of women in general and older women in particular inhibit accurate reporting of some symptoms by female patients (Root, 1987).

Because women predominate in the older population, further research is needed to document gender differences in the interaction and behaviors attributable to doctor–patient gender concordance.

OUTCOME VARIABLES

In addition to determining the characteristics of provider–patient interaction, health communication research frequently looks at the outcomes of various interaction variables to determine their impact and to ascertain whether they make a difference in the health care context. We now move to a discussion of such outcome variables.

Satisfaction

The most frequently studied outcome variable in research on provider–patient relations in general has traditionally been satisfaction. The same is true of research on older patient–physician communication. The preponderance of research in this area is likely because of the ease of obtaining patient self-reports of satisfaction, rather than to the importance of satisfaction in the health care context. Some have criticized reliance on satisfaction as a key outcome variable, noting that patients are sometimes satisfied even with inadequate health care (Kaplan et al., 1989b), and older patients typically express a high degree of satisfaction with their medical care and the providers who deliver it (Anderson, Rakowski, & Hickey, 1988).

Patient satisfaction is, of course, a determinant of such behaviors as leaving one doctor for another, which is frequently called doctor shopping (Kasteler, Kane, Olsen, & Thetford, 1976), disenrolling from health insurance plans (Davies, Ware, & Brook, 1986), ignoring medical advice (Hayes-Bautista, 1976; Haynes, Taylor, & Sackett, 1979), and the initiation of malpractice suits (Ware et al., 1978). More important, satisfaction is sometimes related to health status or how healthy a patient is (Adelman et al., 1991). Hall, Milburn, and Epstein (1993) examined these variables utilizing regression analyses and found that health status predicts satisfaction rather than satisfaction predicting health status. These researchers speculated that medical care providers play a role in producing patient satisfaction, in that providers like healthier patients and may verbally and nonverbally express this affect to their patients. This good feeling may have been returned to the providers through high patient-satisfaction scores. This interpretation is consistent with the findings of Roter and Hall (1989), that physician affect is related to patient satisfaction, and of Anderson et al. (1988), that how a medical resident assesses an older patient's health status is positively related to satisfaction levels in the care provider.

Some have argued that satisfaction has not been adequately studied in the older patient (Adelman et al., 1991). As stated previously, it has been determined that older individuals are typically satisfied with their encounters with health care providers (Anderson et al., 1988; Anderson & Zimmerman, 1993; Blanchard, Labrecque, Ruckdeschel, & Blanchard, 1990; Breemhaar et al., 1990; Hall et al., 1988; Haug & Ory, 1987; Like & Zyzanski, 1987). Both lower levels of patient education and higher physician ratings of a partnership within the relationship are related to higher levels of satisfaction (Anderson & Zimmerman, 1993; Hall et al., 1988).

Other behavioral and communicative variables are also related to the satisfaction of older patients. Using the Roter interaction analysis instrument, Bertakis et al. (1991) concluded that how satisfied a patient is with a physician's medical skills and communication regarding psychosocial topics are the best predictors of overall patient satisfaction. They found that physician question asking about biomedical topics, patient talk about biomedical topics, and physician domination of talk are all negatively related to satisfaction (Bertakis et al., 1991). Patients are more satisfied when they perceive that their needs have been addressed by the physician, that they have received emotional support from him or her, that the physician has discussed treatment (Blanchard et al., 1990), when they feel that the medical information they have received is adequate (Breemhaar et al., 1990), and when they have received more information from the physician (Hall et al., 1988; Roter & Hall, 1989). Longer interactions and more statements indicating approval and confirmation also predict satisfaction (Hall et al., 1988; Like & Zyzanski, 1987). Validation and confirmation have also been advocated in the case of advanced dementia (Benjamin, 1999).

Other research has found that patients who prefer to participate in decision making are less satisfied (Blanchard et al., 1988), perhaps because they are not allowed the level of participation they would prefer. Some nonverbal behaviors also predict patient satisfaction. Greater satisfaction is associated with more nonverbal immediacy and attention from the physician (Hall et al., 1988). Nonverbal immediacy is communicated through such behaviors as forward leans, direct eye contact, more direct-body orientation, and less interpersonal distance. Hall et al.'s (1988) meta-analysis reported that, of all provider variables, only question asking is not related to patient satisfaction (see also Roter & Hall, 1989). In a rather thorough interaction analysis that then related communicative behaviors to satisfaction, Greene, Adelman, Friedmann, and Charon (1994) found higher patient satisfaction with increased physician supportiveness, shared laughter, opportunities to be questioned and share information about their own agenda items, and provision of some physician structuring through the wording of initial questions in the negative. Thus, it appears that physician behaviors are important determinants of patient satisfaction. These provider behaviors can be improved by training (Caris-Verhallen, Kerkstra, Bensing, & Grypdonck, 2000). Video Interaction Analysis training improves provision of information to patients, reliance on open-ended questions, and ratings of warmth, involvement, and patronization (Caris-Verhallen et al., 2000)

Patients who are more seriously ill, however, report lower levels of satisfaction (Labrecque, Blanchard, Ruckdeschel, & Blanchard, 1991) as do those patients whose desires and expectations are not met (Like & Zyzanski, 1987). Sicker patients are more often accompanied by family members, of course, but Labrecque et al. (1991) found that satisfaction is related to the physician's actual behavior rather than to the presence of a family member during the medical encounter. We discuss the presence of a patient's companion shortly.

Satisfaction has also been studied in health care providers, although rarely in physicians who interact with older patients. Anderson et al. (1988) noted that residents who interact with older patients are typically satisfied with these interactions, and that attitudes toward the older adults and the expected benefits of health education are predictors of these levels of satisfaction. There was little similarity in paired patient and resident satisfaction ratings, however. This indicates different satisfaction levels in the patients and the medical residents who are treating those patients.

Patient Compliance and Adherence

Although lack of patient compliance with the treatment regimen is a common complaint among care providers and is frequently studied in the general health communication research (see Thompson & Parrott, 2002, for a summary of this research), it has not been studied among older adults nearly as frequently as it has been studied within the broader population. This is somewhat surprising, given the frequency with which older people take medication. Salzman (1995) suggested that over 86% of the adults between 65 and 69 take medication, and 93.4% of those in the 85+ cohort take one or more medications. In nursing homes, medication usage rates may be as high as 97% (Salzman, 1995). Noncompliance has been found to be greater with multiple medications and complex regimens, situations frequent for older patients (Whittington, 1982) and often exacerbated by consultations with multiple physicians (Penner, 1988). Aproximately 10% of older people take a medication that was not prescribed for them, 20% take medication that should not be included in their current treatment regimen, and as many as 40% discontinue taking medicines without their physicians' awareness (Salzman, 1995). Intentional dosage changes or underuse of medication is also common with older patients (Haug & Ory, 1987).

It has been noted that compliance problems of older patients may be related to physician communication (Adelman et al., 1991) and that a doctor's autocratic behavior may erode trust and cause patients to conceal noncompliance (Haug & Ory, 1987). Coe (1987) content analyzed tape recordings of older patient–physician interactions to determine the educational strategies used by the physicians to promote compliance with medical regimens. This research found that the physicians engaged the accompanying relatives as medication supervisors, adapted their language to the perceived level of understanding of the patient and relative, used medication reminders or aids, and seemed sensitive to the patient's input in adapting the medication schedule to the patient's activities. Such strategies are necessary to ensure that patients and family members fully understand the prescribed routine and the reasons it is necessary. Much nonadherence to treatment regimens, especially among older patients, is not willful noncompliance but simply misunderstanding of instructions.

Both patient and physician communicative behaviors are related to patient adherence with medication regimens. Rost et al. (1989) found positive relationships between adherence and both patient volunteering of information and patient question answering during the initial history-taking portion of the medical encounter, and, during the actual exam, negative relationships between adherence and patient provision of information or behaviors sought by the physician. These findings were especially strong when the recommendations were solely for new medications. Part of the information that patients perceived as relevant in the medical context involved small talk and thus included getting to know the doctor as a person.

A growing number of physicians and other health care practioners are using written patient information, brochures, and leaflets to inform their patients about their illness, treatment or both. Although much of this research has not focused on older patients, the results have obvious applicability. It is clear from this literature that patients want more information about their disease and treatment (Donovan, 1991; Kay & Punchak, 1988) and more written patient information or health care literature (Donovan & Blake, 1992). The written instructions must be visually appealing (Meade & Smith, 1991), legible or easy to read (Albert & Chadwick, 1992), and easy to understand (Dixon & Park, 1990; Larson & Schumacher, 1992; Ley, 1982; Priestly, Campbell, Valentine, Denison, & Buller, 1992; Tackett Stephens, 1992) to be effective. Ease of understanding is of particular concern in the older patient population, because lower levels of literacy in this patient cohort have been observed (Jackson, Davis, Murphey, Bairnsfather, & George, 1994; Jolly, Scott, Feied, & Sanford, 1993).

Vignos, Parker, and Thompson (1976) and others suggested that written information should be used in conjunction with oral instructions. The written instructions serve to reinforce the verbal message (Caughey, 1989; Dunkelman, 1979; Vignos, et al., 1976), can increase patient knowledge or recall of information (Gauld, 1981; Sandler, Mitchell, Fellows, & Garner, 1989; Wilkinson, Tylden-Pattenson, Gould, & Wood, 1981), and may increase patient compliance (Hawe & Higgins, 1990; Myers & Calvart, 1984). Compliance rates, however, are not always increased when patients are provided written information (Gauld, 1981). It is interesting to note that in one of the few studies to examine the impact of written instructions and older patients, the rate of compliance was increased, but patient knowledge did not increase (Hawe & Higgins, 1990). Taylor and Cameron

(2000) suggested that whenever patients are discharged from the emergency room, written patient discharge information should also be provided. They suggested the use of structured, preformatted written instructions. This is of particular importance given that many people are unable to recall physician instructions after 5 minutes (Raynor, Booth, & Blenkinsopp, 1993).

Health Status

Although, as noted earlier, such variables as satisfaction and compliance are related to patient health, how healthy a patient is as an outcome of communication has been rarely studied in older patients. This is important because one of the unique advantages of health communication as an area of study is the opportunity to examine the impact of various aspects of communication on patient health. From studies of the general population, we do know that some communication variables are related to such health variables as blood pressure, postoperative vomiting, speed of recovery, blood sugar control, functional status and the need to rely on pain medication (see Thompson & Parrott, 2002, or Kaplan et al., 1989a, 1989b, for a summary of this research). Focusing specifically on older adults, research tells us only that sicker males prefer to leave medical decisions to physicians (Blanchard et al., 1988). However, the Hall et al. (1993) study, reported earlier, indicating a link from health to satisfaction rather than from satisfaction to health, also utilized an older population. Thus, it appears that the impact of communication variables on the health status of older persons is an area ripe for research.

SPECIAL ISSUES

The Impact of a Companion

Most research on patient–care provider interaction has focused exclusively on that dyad. Recently, investigators have begun examining triadic health care encounters. They have been interested in the interpersonal dynamics involved in interactions among a physician, a patient, and an accompanying friend or relative of the patient. This type of interaction is much more likely to occur when the patient is older than when the patient is a younger adult; the companion is most commonly a family member (a spouse or an adult child) and is frequently the family caregiver (Adelman, Greene, & Charon, 1987; Beisecker, 1989; Glasser, Prohaska, & Roska, 1992).

The presence of such a companion is likely to change the nature of the interaction, including allowing the formation of coalitions if the companion functions as an advocate of the patient or an antagonist of either the patient or the care provider (Adelman, Greene, & Charon, 1987; Coe, 1987). These companions frequently question the doctor on behalf of the patient (Beisecker, 1989). There is some indication that companions take time away from the patient, because there is no difference in the length of interactions with or without companions during physician visits by rehabilitation medicine patients (Beisecker, 1989), although Labrecque et al.'s (1991) study of oncology patients found that interactions involving companions lasted longer. Companions initiate more comments to doctors than doctors do to them, indicating that companions respond to or initiate comments when doctors are not addressing them (Beisecker, 1989). The communication of the companion is most likely to occur during the history-taking and physician feedback portions of the encounter. Beisecker also identified three roles played by companions: (a) watchdog, verifying information for both patient and physician; (b) significant other, providing feedback regarding the appropriateness of various behaviors within the medical interaction to both patient and physician; and (c) surrogate patient, answering questions directed to the patient.

These findings are consistent with those of Coe (1987), who noted that the companion acted as an interpreter to elaborate the patient's complaint to the physician or explain to the patient how the physician's orders could be carried out at home. Companions also acted as negotiators with physicians, offering alternative suggestions or compromises. In this study, physicians tended to encourage the companions, who were all relatives, to become caretakers of the patients.

In a study of cancer patients, Labrecque et al. (1991) determined that physicians discuss future treatment more often when a family member is present, but only with a patient who is more seriously ill. Physicians discuss the patient's current medical status more often when another member of the family is present, regardless of symptoms. Physicians are more likely to greet patients when another member of the family is there. Overall, Labrecque et al. (1991) concluded that physicians provide more information when a family member is present.

A series of studies by Glasser and colleagues have also examined the influence of a companion on the physician–older patient encounter. Glasser et al. (1992) looked at differences between those patients who do and do not bring companions and the companions' influence on the decision to seek care. However, the 1992 study did not focus on the actual interaction, as was done by Glasser, Rubin, and Dickover (1989). The 1989 study indicated that caregivers desire more explanation of the diagnosis and future expectations from physicians. Caregivers of Alzheimer's patients wanted the doctors to better explain the disease and what they could expect to happen. Over 50% of the family caregivers were dissatisfied or very dissatisfied with support offered by physicians. Glasser et al. (1990) later suggested that physicians need to be especially attentive to the information and support needs of family caregivers when the patient experiences a dementing disorder, such as Alzheimer's disease. These concerns are also reflected in the findings of Haley, Clair, and Saulsberry (1992), that family caregivers of Alzheimer's patients are not highly satisfied with the information they receive from physicians ($M = 2.39$, with $0 =$ very dissatisfied, $4 =$ very satisfied).

Other studies of family caregivers of Alzheimer's patients indicate that family members are more concerned about input from physicians in the early stages of caregiving and have many negative reactions to physicians, although specialists are mentioned more positively. Family caregivers respond more positively to those physicians with whom they are able to communicate in more depth and breadth and who show sensitivity to the caregiver's feelings and responsiveness to the caregiver's viewpoint (Morgan & Zhao, 1993). Morgan and Zhao also noted that family caregivers play an important role in determining how much a patient will comply with a treatment regimen, and frequently feel that physicians overmedicate patients to control undesirable patient behavior, sedating patients so much that they were not able to relate to their family members at all. Silliman (1989) echoed many of these concerns and urged separate meetings with caregivers and, as patient–family member relations deteriorate, active intervention on the part of the physician to help both parties deal with changes. Given Burns' and Mulley's (1992) findings that only 28% of the ophtalmology patients in their study administered eyedrops to themselves, it seems that the relationship between the older individual, a companion, and the physician are often intertwined with compliance.

Telemedicine

In addition to studying the role of companions, researchers have begun investigating the impact of telemedicine on patient–physician interaction. Telemedicine is the use of any telecommunications device interposed between patients and physicians. As far back as 1917, the telegraph was used to treat patients (Kropff & Grigsby, 1999), but today telemedicine programs use telephones, fax machines, and videoconferencing equipment to bring patients and health care providers together. Research by Mekhjian, Turner, Gailiun, and McCain (1999) suggests that patients are satisfied with telemedical consultations on both the information exchange or content level, as well as the relational level. Turner et al. (2001) found that in a telemedical consultation, physicians are less likely to provide patient education information than in face-to-face office visits. They also found that physicians are less likely to engage in "small talk" in videoconferenced medical consultations than when they meet the same patients face-to-face.

Although telemedicine is currently being used with patients of all ages, it is of particular interest to the older population, because of the tremendous shortage of physicians with a specialization in geriatric medicine. King (1999) suggested that there will be approximately 7,000 qualified geriatricians in the year 2000, although there is a need for 22,000 to 23,000. Because 21% of all health care expenditures are accrued after the age of 65 (Mustard, Kaufert, Kozyrsky, & Mayer, 1998) and diminished levels of mobility make going to the doctor more difficult, it makes sense to bring such a technology to the older patient. This is particularly true in light of the dramatic increases—a nearly 60% increase between the years of 1990 and 1996 (Meyer, 1997)—in the number of home health care agencies. The effectiveness of home health care can be dramatically improved if telemedicine programs can be incorporated into those systems of health care delivery.

Research has shown that telemedicine can be a cost-effective method of delivering health care in the short run (Barbaro, Bartolini, & Bernaducci, 1997; Kaye, 1997), as well as in the long run. Short run studies compare costs of medical treatments such as the cost of home visits versus telemedical visits. Kaye (1997) found telemedical visits to be less expensive ($73 vs. $94), and these costs will decrease dramatically as the cost of the videoconferencing equipment decreases. Jones (1996) argued that in the long run, the benefits of getting patients to physicians faster will allow the detection of problems more quickly, decrease the need for emergency treatments and transportation, and reduce nursing home placements in many cases.

Telemedicine programs have traditionally been used to improve accessibility to health care and health information. They have been used with older patients to improve the monitoring of basic health parameters such as blood pressure and temperature (Allen,

Roman, Cox, & Caldwell, 1996), as well as cancer patients (Allen & Hayes, 1994), diabetes patients (Owens, 1997), and older adults with heart conditions (Barbaro, Bartolini, & Bernaducci, 1997; Roth, Carthy, & Benedek, 1997). In addition, telemedicine programs have been used in both emergency situations to provide specialized medical care (Ekert, 1997) and in situations where emergency treatment personnel can be dispatched to treat patients with chronic heart problems (Roth et al., 1997).

Telemedicine programs could also be used to provide mental health treatment for older adults (Baer et al., 1995; Ball, 1996; Brandt, Spencer, & Folstein, 1988; Burke, Roccaforte, Wengel, Conley, & Potter, 1995; Lanska, 1993; Montani et al., 1997) in addition to providing health care information to elderly patients and care providers of the elderly (Ball, 1996).

Advance Directives

As older patients grow closer to the end of their lives, patient–physician communication becomes more difficult and at times impossible. Diminished capacity, the involvement of other family members, discrepancies in values and attitudes toward life and health care, and a lack of understanding of medical terms and treatments can all contribute to the patient's inability to communicate with his or her physician. The Patient Self-Determination Act was passed in 1990 as a vehicle to ensure that patients are able to communicate their preferences for end-of-life care. Through the use of advanced directives, patients are able to inform their physician of the kinds of care and treatments they prefer and to avoid futile care in the event they can no longer communicate those wishes directly. Although definitions of "futile care" vary, generally, treatments with a 1% or less chance of increasing patient survival for 2 months are deemed futile.

Advanced directives are legal documents patients generate before end-of-life care is needed. These directives fall into two general classes of directives: instructional directives and proxy directives. A proxy directive is a durable power of attorney that identifies who should make medical decisions for patients in the event they can no longer make such decisions for themselves. Instructional directives are also known as living wills and provide the physician with instructions about the type and circumstances in which particular types of treatments are acceptable or unacceptable to the patient. Shmerling, Bedell, Lilenfeld, and Delbanco (1988) reported that only about 3% of the nonimpaired older adults surveyed reported having treatment limitation discussions with their physician, and such discussions between physicians and other family members occurred only at a rate of about 12%. Gamble, MacDonald, and Lichstein (1991) found that somewhere between 4% and 24% of all Americans have actually completed an advance directive. Research suggests that 70% of elderly patients feel the physician should discuss end-of-life treatment issues such as the use of CPR before the patient becomes ill (Shmerling et al., 1988).

Older adults generally do not bring up the topic of advanced directives and prefer their physician to initiate such conversations (Reilly et al., 1994; Rubin, Strull, Fialkow, Weiss, & Lo, 1994). In cases where the physician suggests treatments should be withdrawn, 45% of the patients and their families agree with little or no delay, and 90% concur with the physician within 5 days (Prendergast & Luce, 1997). The rate of physician-initiated advance directive discussions can be increased through the use of automated reminding systems (Emanuel, Barry, Stoeckle, Ettelson, & Emanuel, 1991; Lurie, Pheley, Miles, & Bannick-Mohrland, 1992).

Although seemingly a good idea, research suggests that advanced directives have not been particularly effective (Schneiderman, Kronick, Kaplan, Anderson, & Langer, 1992; Teno, Stevens, Spernak, & Lynn, 1998; Teno et al., 1997) for a variety of reasons. Physicians are often unaware of the existence of the advance directive (Hanson, Tulsky, & Danis, 1997; Layson, Adelman, Wallach, Pfeifer, Johnston, McNutt, 1994; Teno et al., 1997), there is much equivocality in the language of such directives (Eisendrath & Josen, 1983), and patients have so little experience or expertise that the decisions they make in advance are arguably not well informed (Brett, 1991).

In fact, advance directives have not been shown to increase patient–physician discussions of end-of-life treatment (Teno et al., 1997), decrease medical costs (Teno, Lynn, Connors, Wenger, Phillips, Alzola, Murphy, Desbiens, & Knaus, 1997) or effect medical decisions and treatment (Schneiderman, Kronick, Kaplan, Anderson, & Langer, 1992; Teno et al., 1997; Teno, Lynn, Connors, Wenger, Phillips, Alzola, Murphy, Desbiens, & Knaus, 1997). Widespread adoption and refinement of advanced directives, however, is believed to be a step in the right direction. Without such an innovation, patients would be unable to ensure their health care decisions last a lifetime.

CONCLUSION

All of the issues discussed in this chapter warrant further examination. More insight into characteristics of older patient–physician interaction is needed, but researchers also need to go beyond description, to more completely understand communication processes and to look at important outcomes that derive from such interactions. Research in this area has great potential if the "so what" question can be answered by tying characteristics of the interaction to pragmatic concerns. Indeed, it is difficult to obtain entry to medical settings or funding for health communication research without tying interaction variables to significant outcomes. Further, intervention attempts, utilizing communication interventions to affect outcomes, have yet to undergo much research investigation.

Few measurement issues have been discussed in the literature on older patients and their physicians. It has been noted, however, that older patients' perceptions of physician behavior are better predictors of patient satisfaction than are the actual physician behaviors (Blanchard et al., 1990). This raises a number of research questions and concerns over intervention, as behaviors can be more easily changed than perceptions.

In addition, researchers such as Coupland, Coupland, and Robinson (1992) have demonstrated the utility of applying microanalytic techniques to this research area. Their discourse analyses of older patient–physician interactions have significantly extended our knowledge base.

Another pressing need in this area is to make sense of separate studies and disparate findings by providing a theoretical framework. Few theories of aging or communication theories have yet been applied to this context, although perspectives such as constructivism (Burleson, 1984; Delia, O'Keefe, & O'Keefe, 1982) might prove illuminating. Constructivism is one of the few communication perspectives that has been applied developmentally, but until now only with younger populations. Traditional medical models, such as the biomedical model and the sick role model, have been applied to older patient–physician interaction (see Haug, 1981; Ory et al., 1992), although their emphasis on paternalism has frequently been seen as counterproductive as medicine has moved more toward a biopsychosocial model. Woodward and Wallston (1987) used the concept of self-efficacy, derived from social learning theory, to demonstrate that the amount of impact people perceive they could have on particular health conditions or health-related behaviors moderates their desire for control. Beisecker (1988, 1989) used role theory to explain the communication behaviors of patients. Coupland, Robinson, and Coupland (1994) used the notion of multiple goals in discourse to examine the negotiation of the medical interaction. Finally, Waitzkin (1991), drawing on many theoretical perspectives, provided a theoretical structure to "plan how contextual social issues affect doctor–patient encounters."

Other researchers have begun to develop new theoretical frameworks to facilitate understanding of medical interactions. Roter's and Hall's (1989; see also Hall et al., 1988) discussion of reciprocity is such an example. This framework looks at the reciprocal influence of patient and physician behavior on each other. Reciprocity has also been applied by Thorne and Robinson (1989) to the development of trust in alliances between physicians and chronically ill patients. A similar but more elaborate approach using communication accommodation theory has been provided by Giles and others. This perspective takes reciprocity further by looking not only at the mutual influence of two communicators, but also at how one's behavior accommodates the other. Accommodation theory has been applied to health care interactions (Street, 1991a), intergenerational interactions (Coupland, Coupland, Giles, & Henwood, 1988), and older people in health care (Williams, Giles, Coupland, Dalby, & Manasse, 1990), and has promise for our understanding of older patient–physician interactions. More recently, Barker and Giles (2003; see also Barker, Giles, & Harwood, chap. 6, this volume) have brought together the communicative predicament model of Ryan, Giles, Bartolucci, and Henwood (1986) and the communicative enhancement model of aging of Ryan, Meredith, MacLean and Orange (1995) into a new framework that shows promise for an understanding of older patient–provider interaction.

Frederikson (1993) has also offered an integrative model for medical consultation based on an information-processing perspective. The model attempts to synthesize diverse findings on provider–patient interaction by focusing on how patients process medical information. This framework, too, has direct applications to older adults and their interactions with physicians. Although they did not focus on the medical encounter, Leventhal, Leventhal, and Schaefer (1992) advanced a "commonsense model," which they applied to older adults. This model utilizes a systems framework that views individuals as active problem solvers whose instrumental actions are a product of their perception, intellectual understanding, and emotional response to a health threat and the coping procedures used to regulate both the threat and their emotional response to it. Showing great promise is Babrow's (1992) problematic integration theory (see Hines, Babrow, Badzek, & Moss, 1997, for an application to older patients, especially in regard to end-of-life decisions), which provides a rather sophisticated look at the process of decision making and communication.

There is obviously a need to develop theoretically grounded research projects from an aging perspective and to test the assumptions of diverse theoretical perspectives utilizing data from physician–older patient encounters. In addition, much more development and testing of interventions is needed before significant change can occur within this area (McCormick et al., 1996). It appears, then, that students of older patient–physician communication face a multitude of challenges and questions as they continue to describe the nature and outcomes of such interactions. It is hoped that this review has prompted attention to these issues.

REFERENCES

Acton, G. J., Mayhew, P. A., Hopkins, B. A., & Yauk, S. (1999). Comunicating with individuals with dementia: The impaired person's perspective. *Journal of Gerontological Nursing, 25*(2), 6–13.

Adelman, R. D., Greene, M. G., & Charon, R. (1987). The physician–elderly patient–companion triad in the medical encounter: The development of a conceptual framework and research agenda. *The Gerontologist, 27*, 729–734.

Adelman, R. D., Greene, M. G., & Charon, R. (1991). Issues in physician–elderly patient interaction. *Ageing and Society, 2*, 127–148.

Adelman, R. D., Greene, M. G., Charon, R., & Friedmann, E. (1990). Issues in the physician–geriatric patient relationship. In H. Giles, N. Coupland, & J. M. Wiemann (Eds.), *Communication, health and the elderly* (pp. 126–134). London: Manchester University Press.

Adelman, R. D., Greene, M. G., Charon, R., & Friedmann, E. (1992). The content of physician and elderly patient interaction in the medical primary care encounter. *Communication Research, 19*, 370–380.

Adler, S. R., McGraw, S. A., & McKinlay, J. B. (1998). Patient assertiveness in ethnically diverse older women with breast cancer: Challenging stereotypes of the elderly. *Journal of Aging Studies, 12*, 331–350.

Agee, A., & Blanton, P. W. (2000). Service providers' modes of interacting with frail elders and their families: Understanding the context for caregiving decisions. *Journal of Aging Studies, 14*, 313–333.

Albert, T., & Chadwick, S. (1992). How readable are practice leaflets? *British Medical Journal, 305*, 1266–1268.

Allen, A., & Hayes, J. (1994). Patient satisfaction with telemedicine in a rural clinic. *American Journal of Public Health, 84*, 1693.

Allen, A., Roman, L., Cox, R., & Caldwell, B. (1996). Home health visits using cable television network: User satisfaction. *Journal of Telemedicine and Telecare, 2*(Suppl. 1), 92–94.

Anderson, L., Rakowski, W., & Hickey, T. (1988). Satisfaction with clinical encounters among residents and geriatric patients. *Journal of Medical Education, 63*, 447–455.

Anderson, L. A., & Zimmerman, M. A. (1993). Patient and physician perceptions of their relationship and patient satisfaction: A study of chronic disease management. *Patient Education and Counseling, 20*(1), 27–36.

Arntson, P., Makoul, G., Pendleton, D., & Schofield, T. (1989). Patients' perceptions of medical encounters in Great Britain: Variations with health loci of control and sociodemographic factors. *Health Communication, 1*, 75–95.

Babrow, A. S. (1992). Communication and problematic integration: Understanding diverging probability and value, ambiguity, ambivalence, and impossibility. *Communication Theory, 2*, 95–130.

Baer, L., Jacobs, D. G., Cukor, P., O'Laughlen, J., Coyle, J. T., & Magruder, K. M. (1995). Automated telephone screening survey for depression. *Journal of American Medical Association, 273*, 1943–1944.

Ball, C. J. (1996). Telepsychiatry: The potential for communications technology in old age psychiatry. *Journal of Telemedicine and Telecare, 2*(1), 117–118.

Barbaro, V., Bartolini, P., & Bernaducci, R. (1997). A portable unit for remote monitoring of pacemaker patients. *Journal of Telemedicine and Telecare, 3*(1), 96–102.

Barker, V., & Giles, H. (2003). Integrating the communicative predicament and enhancement of aging models: The case of older Native Americans. *Health Communication, 15*, 255–275.

Baumrucker, S. J. (1998). An obvious call for better communication. *The American Journal of Hospice & Palliative Care, 15*, 253–254.

Beisecker, A. E. (1988). Aging and the desire for information and input in medical decisions: Patient consumerism in medical encounters. *The Gerontologist, 28*, 330–335.

Beisecker, A. E. (1989). The influence of a companion on the doctor–elderly patient interaction. *Health Communication, 1*, 55–70.

Beisecker, A. E. (1990, November). *The older patient's companion*. Paper presented at the annual meeting of the Gerontological Society of America, Boston.

Beisecker, A. E., & Beisecker, T. D. (1990). Patient information-seeking behaviors when communicating with doctors. *Medical Care, 28*, 19–28.

Beisecker, A. E., & Beisecker, T. D. (1993). Using metaphors to characterize doctor–patient relationships: Paternalism versus consumerism. *Health Communication, 5*, 41–58.

Benjamin, B. J. (1999). Validation: A communication alternative. In L. Volicer & L. Bloom-Charette (Eds.), *Enhancing the quality of life in advanced dementia* (pp. 107–125). Philadelphia: Taylor & Francis.

Bernstein, B., & Kane, R. (1981). Physicians' attitudes toward female patients. *Medical Care, 19*, 600.

Bertakis, K. D., Roter, D., & Putnam, S. M. (1991). The relationship of physician medical interview style to patient satisfaction. *Journal of Family Practice, 32*, 175–181.

Blanchard, C. G., Labrecque, M. S., Ruckdeschel, J. C., & Blanchard, E. B. (1988). Information and decision-making preferences of hospitalized adult cancer patients. *Social Science and Medicine, 27*, 1139–1145.

Blanchard, C. G., Labrecque, M. S., Ruckdeschel, J. C., & Blanchard, E. B. (1990). Physician behaviors, patient perceptions, and patient characteristics as predictors of satisfaction of hospitalized adult cancer patients. *Cancer, 65*, 186–192.

Brandt, J., Spencer, M., & Folstein, M. (1988). The telephone interview for cognitive status. *Neuropsychiatry, Neuropsychology, and Behavioural Neurology, 1*, 111–117.

Breemhaar, B., Visser, A., & Kleijnen, J. (1990). Perceptions and behavior among elderly hospital patients: Description and explanation of age differences in satisfaction, knowledge, emotions and behaviour. *Social Science and Medicine, 31*, 1377–1384.

Burleson, B. R. (1984). Age, social-cognitive development, and the use of comforting strategies. *Communication Monographs, 51*, 140–153.

Brett, A. S. (1991). Limitations of listing specific medical interventions in advance directives. *Journal of the American Medical Association, 266*, 825–828.

Burke, W. J., Roccaforte, W. H., Wengel, S. P., Conley, D. M., & Potter, J. F. (1995). The reliability and validity of the Geriatric Depression Rating Scale administered by telephone. *Journal of the American Geriatric Society, 43*, 674–679.

Burns, E. & Mulley, G. P. (1992). Practical problems with eye-drops among elderly ophtalmology out-patients. *Age Ageing, 21*, 168–170.

Caris-Verhallen, W. M. C. M., Kerkstra, A., Bensing, J. M., Grypdonck, M. H. F. (2000). Effects of video interaction analysis training on nurse–patient communication in the care of the elderly. *Patient Education & Counseling, 39*, 91–103.

Caughey, D. (1989). Patient communication and effective rheumatology. *Journal of Rheumatology, 16*, 5–6.

Clavel, D. S. (1999). Vocalizations among cognitively impaired elders. *Geriatric Nursing, 20*(2), 90–93.

Coe, R. M. (1987). Communication and medical care outcomes: Analysis of conversations between doctors and elderly patients. In R. A. Ward & S. S. Tobin (Eds.), *Health in Aging* (pp. 180–193). New York: Springer.

Coupland, J., Coupland, N., Giles, H., & Henwood, K. (1988). Accommodating the elderly: Invoking and extending a theory. *Language in Society, 17*, 1–41,

Coupland, J., Coupland, N., & Robinson, J. D. (1992). "How are you?" Negotiating phatic communion. *Language in Society, 21*, 207–230.

Coupland, J., Robinson, J. D., & Coupland, N. (1994). Frame negotiation in doctor–elderly patient consultations. *Discourse and Society.*

Coupland, N., & Coupland, J. (1993). Discourses of ageism and anti-ageism. *Journal of Aging Studies, 7*, 279–301.

Davies, A. R., Ware, J. E., Jr., & Brook, R. H. (1986). Consumer acceptance of prepaid and fee-for-service medical care: Results from a randomized controlled trial. *Health Services Research, 21*, 429–452.

Davies, S., Laker, S., & Ellis, L. (1997). Promoting autonomy and independence for older people within nursing practice: A literature review. *Journal of Advanced Nursing, 26*, 408–417.

Delia, J. G., O'Keefe, B. J., & OKeefe, D. J. (1982). The constructivist approach to communication. In F. E. X. Dance (Ed.), *Human communication theory* (pp. 147–191). New York: Harper & Row.

Deselle, D. D. (1999). Improving the quality of communication between health care professionals and their hearing-impaired elderly patients. *Annals of Long-Term Care, 7*, 418–420.

Dixon, E., & Park, R. (1990). Do patients understand written health information? *Nursing Outlook, 38*(6), 278–281.

Donovan, J. (1991). Patient education and the consultation: The importance of lay beliefs. *Annals of Rheumatic Diseases, 50*, 418–421.

Donovan, J., & Blake, D. (1992). Patient compliance: Deviance or reasoned decision making? *Social Science Medicine, 34*(5), 507–513.

Dunkelman, H. (1979). Patients' knowledge of their condition and treatment, how it might be improved. *British Medical Journal, 2*, 311–314.

Eisendrath, S. J., & Jonsen, A. R. (1983). The living will: Help or hinderance? *Journal of the American Medical Association, 249*, 2054–2058.

Emanuel, L. L., Barry, M. J., Stoeckle, J. D., Ettelson, L. M., & Emanuel, E. J. (1991). Advance directives for medical care: A case for greater use. *New England Journal of Medicine, 324*, 889–895.

Ekert, T. (1997). High-quality television links for home-based support for the elderly. *Journal of Telemedicine and Telecare, 3*(Suppl. 1), 26–28.

Feeser, T., & Thompson, T. L. (1993). Study suggests method of increasing patient involvement in physician–patient interactions. *Cosmetic Dermatology, 6*(9), 51–55.

Fox, S. A., & Stein, J. A. (1991). The effect of physician–patient communication on mammography utilization by different ethnic groups. *Medical Care, 29*, 1065–1082.

Frederikson, L. G. (1993). Development of an integrative model for medical consultation. *Health Communication, 5*, 225–237.

Gabbard-Alley, A. S. (1995). Health communication and gender: A review and critique. *Health Communication, 7*, 35–56.

Gamble, E. R., MacDonald, P. J., & Lichstein, P. R. (1991). Knowledge, attitudes, and behavior of elderly persons regarding living wills. *Archives of Internal Medicine, 151*, 277–280.

Gauld, V. A. (1981). Written advice: Compliance and recall. *Journal Royal College of General Practitioners, 31*, 553–556.

Glasser, M., Prohaska, T., & Roska, J. (1992). The role of the family in medical care-seeking decisions of older adults. *Family and Community Medicine, 15*, 59–70.

Glasser, M., Rubin, S., & Dickover, M. (1989, July–August). Caregiver views of help from the physician. *The American Journal of Alzheimer's Care and Related Disorders and Research, 4*(4), 4–11.

Glasser, M., Rubin, S., & Dickover, M. (1990). The caregiver role: Review of family caregiver–physician relations and dementing disorders. In S. M. Stahl (Ed.), *The legacy of longevity—Health and healthcare in later life* (pp. 321–337). London: Sage.

Glosser, G., Wiley, M. J., & Barnoski, E. J. (1998). Gestural communication in Alzheimer's disease. *Journal of Clinical & Experimental Neuropsychology, 20*, 1–13.

Grainger, K. (1993). "That's a lovely bath dear": Reality construction in the discourse of elderly care. *Journal of Aging Studies, 7*, 247–262.

Greene, M. G., & Adelman, R. (1996). Psychosocial factors in older patients' medical encounters. *Research on Aging, 18*, 84–102.

Greene, M., Adelman, R., Charon, R., & Friedmann, E. (1989). Concordance between physicians and their older and younger patients in the primary care medical encounter. *The Gerontologist, 29*, 808–813.

Greene, M. G., Adelman, R., Charon, R., & Hoffman, S. (1986). Ageism in the medical encounter: An exploratory study of the doctor–elderly patient relationship. *Language and Communication, 6*, 113–124.

Greene, M. G., Adelman, R. D., Friedmann, E., & Charon, R. (1994). Older patient satisfaction with communication during an initial medical encounter. *Social Science & Medicine, 38*, 1279–1288.

Greene, M. G., Hoffman, S., Charon, R., & Adelman, R. (1987). Psychosocial concerns in the medical encounter. A comparison of the interactions of doctors with their old and young patients. *The Gerontologist, 27,* 164–168.

Greenfield, S., Kaplan, S., & Ware, J. E. (1985). Expanding patient involvement in care. *Annals of Internal Medicine, 102,* 520–528.

Haley, W. E. (1996). The medical context of psychotherapy with the elderly. In S. H. Zarit & B. G. Knight (Eds.), *A guide to psychotherapy & aging.* Washington, DC: APA.

Haley, W. E., Clair, J. M., & Saulsberry, K. (1992). Family caregiver satisfaction with medical care of their demented relatives. *The Gerontologist, 32,* 219–226.

Hall, J. A., Milburn, M. A., & Epstein, A. M. (1993). A causal model of health status and satisfaction with medical care. *Medical Care, 31,* 84–94.

Hall, J. A., Roter, D. L., & Katz, N. R. (1988). Meta-analysis of correlates of provider behavior in medical encounters. *Medical Care, 26,* 657–675.

Hanson, L. C., Tulsky, J. A. & Danis, M. (1997). Can clinical interventions change care at the end of life? *Annals of Internal Medicine, 126,* 381–388.

Happ, M. B. (2000). Interpretation of nonvocal behavior and the meaning of voicelessness in critical care. *Social Science & Medicine, 50,* 1247–1255.

Haug, M. R. (1981). *Elderly patients and their doctors.* New York: Springer.

Haug, M. R., & Ory, M. G. (1987). Issues in elderly patient–provider interactions. *Research on Aging, 9,* 3–44.

Hawe, P., & Higgins, G. (1990). Can medication education improve the drug compliance of the elderly? Evaluation of an in hospital programme. *Patient Education and Counseling, 16,* 151–160.

Hayes-Bautista, D. E. (1976). Modifying the treatment: Patient compliance, patient control and medical care. *Social Science and Medicine, 10,* 233–238.

Haynes, R. B., Taylor, D. W., & Sackett, D. L. (1979). *Compliance in health care.* Baltimore: Johns Hopkins University Press.

Heal, H. C., & Husband, H. J. (1998). Disclosing a diagnosis of dementia: Is age a factor? *Aging & Mental Health, 2,* 144–150.

Hines, S. C., Babrow, A. S., Badzek, L., & Moss, A. H. (1997). Communication and problematic integration in end-of-life decisions: Dialysis decisions among the elderly. *Health Communication, 9,* 199–218.

Jackson, R. H., Davis, T. C., Murphy, P., Bairnsfather, L., & George, R. (1994). Reading deficiencies in older patients. *The American Journal of the Medical Sciences, 308*(2), 79–82.

Jarrett, N., & Payne, S. (1995). A selective review of the literature on nurse–patient communication: Has the patient's contribution been neglected? *Journal of Advanced Nursing, 22,* 72–78.

Jones, E. (1996). New horizons beckon to home health industry. *Telemedicine Journal, 3,* 243–246.

Jolly, B. T., Scott, J. L., Feied, C. F., & Sanford, S. M. (1993). Functional illiteracy among emergency department patients: A preliminary study. *Annals of Emergency Medicine, 22,* 573–578.

Jones, M., O'Neil, P., Waterman, H., & Webb, C. (1997). Building a relationship: Communications and relationships between staff and stroke patients on a rehabilitation ward. *Journal of Advanced Nursing, 26,* 101–110.

Kaplan, S. H., Greenfield, S., & Ware, J. E., Jr. (1989a). Assessing the effects of physician–patient interactions on the outcomes of chronic disease. *Medical Care, 27*(Suppl. 3), S110–127.

Kaplan, S. H., Greenfield, S., & Ware, J. E. (1989b). Impact of the doctor–patient relationship on the outcomes of chronic disease. In M. Stewart & D. Roter (Eds.), *Communicating with medical patients* (pp. 228–245). Newbury Park, CA: Sage.

Kasteler, J., Kane, R., Olsen, D., & Thetford, C. (1976). Issues underlying the prevalence of "doctor-shopping" behavior. *Journal of Health and Social Behavior, 17,* 328.

Kay, E. A., & Punchak, S. S. (1988). Patient understanding of the causes and medical treatment of rheumatoid arthritis. *British Journal of Rheumatology, 27,* 396–398.

Kaye, L. W. (1997). Telemedicine: Extension to home care? *Telemedicine Journal, 3,* 243–246.

King, M. (September 27, 1999). *Geriatrics find a new popularity* [Online]. Available: seattletimes.com. http://seattletimes.nwsource.com/news/health-science/html98/docs_19990927.html

Koch, T., Webb, C., & Williams, A. M. (1995). Listening to the voices of older patients: An existential-phenomenological approach to quality assurance. *Journal of Clinical Nursing, 4,* 185–193.

Kropff, N. P., & Grigsby, R. K. (1999). Telemedicine for older adults. *Home Health Care Services Quarterly, 17*(4), 1–11.

Labrecque, M. S., Blanchard, C. G., Ruckdeschel, J. C., & Blanchard, E. B. (1991). The impact of family presence on the physician–cancer patient interaction. *Social Science and Medicine, 33,* 1253–1261.

Lanska, D. J., et al. (1993). Telephone assisted mental state. *Dementia, 4,* 117–119.

Larson, I., & Schumacher, H. R. (1992). Comparison of literacy levels of patients in a VA arthritis centre with the reading level required by educational mateials. *Arthritis Care Research, 5*(1), 13–16.

Layson, R. T., Adelman, H. M., Wallach, P. M., Pfeifer, M. P., Johnston, S., McNutt, R. A. (1994). Discussions about the use of life-sustaining treatments: A literature review of physicians' and patients' attitudes and practices. *Journal of Clinical Ethics, 5,* 195–203.

Lesser, J. M., Beller, S., & Harmon, H. (1997). Communication with the elderly psychiatric outpatient. *Texas Medicine, 93*(11), 56–59.

Leventhal, H., Leventhal, E. A., & Schaefer, P. M. (1992). Vigilant coping and health behavior. In M. G. Ory, R. P. Abeles, & P. D. Lipman (Eds.), *Aging, health and behavior* (pp. 109–140). Newbury Park, CA: Sage.

Ley, P. (1982). Satisfaction, compliance, and communication. *British Journal of Clinical Psychology, 21,* 241–254.

Like, R., & Zyzanski, S. J. (1987). Patient satisfaction with the clinical encounter: Social psychological determinants. *Social Science and Medicine, 24,* 351–357.

Lindblade, D. D., & McDonald, M. (1995). Removing communication barriers for the hearing-impaired elderly. *MEDSURG Nursing, 4,* 379–385.

Lurie, N., Pheley, A. M., Miles, S. H., & Bannick-Mohrland, S. (1992). Attitudes towards discussion of life-sustaining treatments in extended care facility patients. *Journal of American Geriatric Society, 40,* 1205–1208.

Mackenzie, C. (2000). Adult spoken discourse: The influences of age and education. *International Journal of Language & Communication Disorders, 35,* 269–285.

Mahler, H. I., & Kulik, J. A. (1990). Preferences for health care involvement, perceived control and surgical recovery: A prospective study. *Social Science and Medicine, 31,* 743–751.

Majerovitz, S. D., Greene, M. G., Adelman, R. D., Brody, G. M., Leber, K., & Healy, S. W. (1997). Older patients' understanding of medical information in the Emergency Department. *Health Communication, 9,* 237–252.

Manton, K. G., & Soldo, B. J. (1992). Disability and mortality among the oldest old: Implications for current and future health and long-term-care service needs. In R. M. Suzman, D. P . Willis, & K. G. Manton (Eds.), *The oldest old* (pp. 199–250). New York: Oxford University Press.

Manton, K. G., & Suzman, R. (1992). Forecasting health and functioning in aging societies: Implications for health care and staffing needs. In M. G. Ory, R. P. Abeles, & P. D. Lipman (Eds.), *Aging, health and behavior* (pp. 327–357). Newbury Park, CA: Sage.

McCormick, W. C., Inui, T. S., & Roter, D. L. (1996). Interventions in physician–elderly patient interaction. *Research on Aging, 18,* 103–136.

Meade, C., & Smith, C. (1991). Readability formulas: Cautions and criteria. *Patient Education and Counseling, 17,* 429–434.

Mekhjian, H., Turner, J. W., Gailiun, M., & McCain, T. (1999). Patient satisfaction with telemedicine in a prison environment: A matter of context. *Journal of Telemedicine and Telecare, 5,* 55–61.

Meyer, H. (1997). Home (care) improvement. *Hospitals and Health Networks, 71*(8), 40–41.

Montani, C., Billaud, N., Tyrrell, J., Fluchaire, I., Malterre, C., Lauvernay, N., Couturier, P., & Franco, A. (1997). Psychological impact of a remote psychometric consultation with hospitalized elderly people. *Journal of Telemedicine and Telecare, 3*(3), 140–145.

Morgan, D. L., & Zhao, P. Z. (1993). The doctor–caregiver relationship: Managing the care of family members with Alzheimer's disease. *Qualitative Health Research, 2,* 133–164.

Mustard, C. A., Kaufert, P., Kozyrsky, A., & Mayer, T. (1998). Sex differences in the use of health care services. *New England Journal of Medicine, 328,* 1678–1683.

Myers, E. D., & Calvart, E. J. (1984). Information, compliance, and side effects: A study of patients on anti-depressant medication. *British Journal of Clinical Pharmacology, 17,* 221–225.

Orange, J. B., Molly, D. W., Lever, J. A., Darzins, P., Ganesan, C.R. (1994). Alzheimer's disease: Physician-patient communication. *Canadian Family Physician, 40,* 1160–1168.

Ory, M. G., Abeles, R. P., & Lipman, P. (1992). *Aging, health, and behavior.* Newbury Park, CA: Sage.

Owens, D. R. (1997). Telemedicine in screening and monitoring of diabetic eye disease. *Journal of Telemedicine and Telecare, 3*(Suppl. 1), 89.

Penner, M. (1988). Patient-to-physician communication on medication use. *The Journal of Family Practice, 26*(1), 82–84.

Pichora-Fuller, M. (1997). Language comprehension in older listeners. *Journal of Speech-Language Pathology and Audiology, 21,* 125–142.

Powell, J. A., Hale, M. A., & Bayer, A. J. (1995). Symptoms of communication breakdown in dementia: Carers' perceptions. *European Journal of Disorders of Communication, 30,* 65–75.

Prendergast, T. J., & Luce, J. M. (1997). Increasing incidence of withholding and withdrawal of life support from the critically ill. *American Journal of Respiratory Critical Care Medicine, 155,* 15–20.

Priestly, K. A., Campbell, C., Valentine, C. B., Denison, D. M., & Buller, N. P. (1992). Are patient consent forms for research protocols easy to read? *British Medical Journal, 305,* 1263–1264.

Pritchett, L. M. (2000). New age of medicine [Online]. *Business first: The Weekly Business Newspaper of Greater Louisville.* Retrieved June 23, 2000. Available: http://louisville.bcentral.com/louisville/stories/2000/06/26/focus4.html

Radecki, S., Kane, R., Solomon, D., Mendenhall, R., & Beck, J. (1988). Do physicians spend less time with older patients? *Journal of the American Geriatrics Society, 36,* 713–718.

Raynor, D. K., Booth, J. G., & Blenkinsopp, A. (1993). Effects of computer generated reminder charts on patients' compliance with drug regimens, *British Medical Journal, 306,* 1158–1161.

Reilly, B. M., Magnussen, C. R., Ross, J., Ash, J., Papa, L., & Wagner, M. (1994). Can we talk? Inpatient discussions about advance directives in a community hospital. Attending physicians' attitudes, their inpatients' wishes, and reported experience. *Archives of Internal Medicine, 154,* 2299–2308.

Ripich, D. N., Carpenter, B. D., & Ziol, E. W. (1997). Procedural discourse of men and women with Alzheimer's disease: A longitudinal study with clinical implications. *American Journal of Alzheimer's Disease, 12,* 258–271.

Ripich, D. N., Vertes, D., Whitehouse, P., Fulton, S., & Ekelman, B. (1991). Turn-taking and speech act patterns in the discourse of senile dementia of the Alzheimer's type patients. *Brain & Language, 40,* 330–343.

Robinson, E. J., & Whitfield, M. J. (1985). Improving the efficiency of patients' comprehension monitoring: A way of increasing patients' participation in general practice consultations. *Social Science and Medicine, 21,* 915–919.

Rodin, J. (1986). Aging and health: Effects of the sense of control. *Science, 233,* 1271–1276.

Rogers, A. E., Addington-Hall, J. M., Abery, A. J., McCoy, A. S. M., Bulpitt, C., Coats, A. J. S., & Gibbs, J. S. R. (2000). Knowledge and communication difficulties for patients with chronic heart failure: Qualitative study. *British Medical Journal, 321,* 605–607.

Root, M. J. (1987, July–August). Women and their health care providers: A matter of communication. *Public Health Reports Supplement, 6,* 152–155.

Rose, J. H., Bowman, K. F., & Kresevic, D. (2000). Nurse versus family caregiver perspectives on hospitalized older patients: An exploratory study of agreement at admission and discharge. *Health Communication, 12,* 63–80.

Rost, K., Carter, W., & Inui, T. (1989). Introduction of information during the initial medical visit: Consequences for patient follow-through with physician recommendations for medication. *Social Science and Medicine, 28,* 315–321.

Rost, K., & Frankel, R. (1993). The introduction of the older patient's problems in the medical visit. *Journal of Health and Aging, 5,* 387–401.

Roter, D. L. (1977). Patient participation in the patient–provider interaction: The effects of patient question asking on the quality of interaction, satisfaction and compliance. *Health Education Monographs, 5,* 281–315.

Roter, D. L., & Hall, J. A. (1989). Studies of physician–patient interaction. *Annual Review of Public Health, 10,* 163–180.

Roter, D. L., Hall, J. A., & Katz, N. R. (1988). Patient–physician communication: A descriptive summary of the literature. *Patient Education and Counseling, 12,* 99–119.

Rotter, J. B. (1966). Generalized expectancies for internal versus external control of reinforcement. *Psychological Monographs, 80*(Whole No. 609).

Roth, A., Carthy, Z., & Benedek, M. (1997). Telemedicine in emergency home care—the "Shahal" experience. *Journal of Telemedicine and Telecare, 3*(Suppl. 1), 58–59.

Rubin, S. M., Strull, W. M., Fialkow, M. F., Weiss, S. J., & Lo, B. (1994). Increasing the completion of durable power of attorney for health care. *Journal of the American Medical Association, 271,* 209–212.

Ryan, E., Giles, H., Bartolucci, G., & Henwood, K. (1986). Psycholinguistic and social psychological components of communication by and with the elderly. *Language & Communication, 6,* 1–24.

Ryan, E. B., Meredith, S. D., MacLean, M. J., & Orange, J. B. (1995). Changing the way we talk with elders: Promoting health using the communiation enhacement model. *International Journal of Aging & Human Development, 41,* 89–107.

Sandler, D. A., Mitchell, J. R. A., Fellows, A., & Garner, S. T. (1989). Is an information booklet for patients leaving hospital helpful and useful? *British Medical Journal, 288,* 915–919.

Salzman, C. (1995). Medication compliance in the elderly. *Journal of Clinical Psychiatry, 56,* 18–22.

Schneiderman, L. J., Kronick, R., Kaplan, R. M., Anderson, J. P., & Langer, R. D. (1992). Effects of offering advance directives on medical treatments and costs. *Annals of Internal Medicine, 117,* 599–606.

Scott, A. (1997). Patient perceptions of openness in nurses: A strategic ethnography. *The Journal of Theory Construction & Testing, 1,* 40–45.

Seijo, R., Gomez, H., & Freidenberg, J. (1995). Language as a communication barrier in medical care for hispanic patients. In A. M. Padilla (Ed.), *Hispanic psychology* (pp. 169–181). Thousand Oaks, CA: Sage.

Shidler, S. (1998). A systemic perspective of life-prolonging treatment decision making. *Qualitative Health Research, 8,* 254–269.

Shmerling, R. H., Bedell, S. E., Lilienfeld, A., & Delbanco, T. L. (1988) Discussing cardiopulmonary resuscitation: A study of elderly outpatients. *Journal of General Internal Medicine, 3,* 317–321.

Silliman, R. (1989). Caring for the frail older patient: The doctor–patient–family caregiver relationship. *Journal of General Internal Medicine, 4,* 237–241.

Simons, W., & Malabar, R. (1995). Assessing pain in elderly patients who cannot respond verbally. *Journal of Advanced Nursing, 22,* 663–669.

Smith, D. H. (1998). Interviews with elderly patients about side effects. *Health Communication, 10,* 199–210.

Smith, D. H., Cunningham, K. G., & Hale, W. E. (1994). Communication about medicines: Perceptions of the ambulatory elderly. *Health Communication, 6,* 281–296.

Stephens, S. D. G., Jaworski, A., Lewis, P., & Aslan, S. (1999). An analysis of the communication tactics used by hearing-impaired adults. *British Journal of Audiology, 33,* 17–27.

Street, R. L. (1991a). Accommodation in medical consultation. In H. Giles, J. Coupland, & N. Coupland (Eds.), *Contexts of accommodation* (pp. 131–156). Cambridge, UK: Cambridge University Press.

Street, R. L., Jr. (1991b). Information-giving in medical consultations: The influence of patients' communicative styles and personal characteristics. *Social Science and Medicine, 32,* 541–545.

Svarstad, B. L. (1976). Physician–patient communication and patient conformity with medical advice. In D. Mechanic (Ed.), *The growth of bureaucratic medicine* (p. 220). New York: Wiley.

Tackett Stephens, S. (1992). Patient education materials: Are they readable? *Oncology Nursing Forum, 19*(1), 83–85.

Taylor, D. M., & Cameron, P. A. (2000). Discharge instructions for emergency department patients: What should we provide. *Journal of Accidental Emergency Medicine, 17*(2), 86–90.

Teno, J. M., Lynn, J., Connors, A. F., Wenger, N., Phillips, R. S., Alzola, C., Murphy, D. P., Desbiens, N., & Knaus, W. A. (1997) The illusion of end-of-life resource savings with advance directives. (1997). *Journal of American Geriatric Society, 45,* 513–518.

Teno, J. M, Lynn, J., Phillips, R. S., Murphy, D., Youngner, S. J., & Bellamy, P. (1994). Do formal advance directives affect resuscitation decisions and the use of resources for seriously ill patients? *Journal of Clinical Ethics, 5,* 23–30.

Teno, J. M., Lynn, J., Wenger, N., Phillips, R. S., Murphy, D. P., Connors, A. F., Desbiens, N., Fulkerson, W., Bellamy, P., & Knaus, W. A. (1997). Advance directives for seriously ill hospitalized patients: Effectiveness with the Patient Self-Determination Act and the SUPPORT intervention. *Journal of the American Geriatric Society, 45,* 500–507.

Teno, J. M., Stevens, M., Spernak, S., & Lynn, J. (1998). Role of written advance directives in decision making: insights from qualitative and quantitative data. *Journal of General Internal Medicine, 13,* 439–446.

Thompson, T. L., & Parrott, R. (2002). Interpersonal communication and health care. In M. L. Knapp, J. Daly, & G. R. Miller (Eds.), *Handbook of interpersonal communication* (3rd ed., pp. 680–725). Newbury Park, CA: Sage.

Thorne, S. E., & Robinson, C. A. (1989). Guarded alliances: Health care relationships in chronic illness. *Image—Journal of Nursing Scholarship, 21,* 153–157.

Tiernan, E., White, S., Henry, C., Murphy, K., Twomey, C., & Hyland, M. (1993). Do elderly patients mind how doctors address them? *Irish Medical Journal, 86*(2), 73.

Turner, J. W., Robinson, J. D., Alaou, A., Winchester, J., Neustadtl, A., Levine, B. A., Collman, J., & Mun, S. K. (2001, May). *Media attitudes versus media use: Understanding the contribution of context to the communication environment.* Paper presented at the annual meeting of the International Communication Association, Washington, DC.

U.S. Census Bureau. (1999). *Statistical abstract of the United States: The national data book* (119th ed.). Washington, DC: U.S. Government Printing Office.

U.S. Senate Special Committee on Aging. (1991). *Aging America—Trends and projections* (U.S. Department of Health and Human Services No. FCOA 91-28001). Washington, DC: U.S. Government Printing Office.

Vernooij-Dassen, M. J., & Lamers, C. (1997). Activation of care-giver coping processes through professional support. In B. M. Miesen & G. M. M. Jones (Eds.), *Care-giving in dementia* (Vol. 2, pp. 178–190). New York: Routledge.

Vignos, P. J., Parker, W. T., & Thompson, H. M. (1976). Evaluation of a clinical education programme for patients with RA. *Journal of Rheumatology, 3,* 155–165.

Waitzkin, H. (1991). *The politics of medical encounters—How patients and doctors deal with social problems.* New Haven, CT: Yale University Press.

Ware, J. E., Jr., Davies, A. R., Kane, R. L., et al. (1978). *Effects of differences in quality of care on patient satisfaction and behavioral intentions: An experimental simulation.* Paper presented at the annual meeting of the Research in Medical Education, New Orleans, LA.

Whittington, F. (1982). Misuse of legal drugs and compliance with prescription directions. In *Drugs and the elderly adult.* Rockville, MD: National Institute on Drug Abuse Research Issues.

Wiederholt, J. B., Clarridge, B. R., & Svarstad, B. L. (1992). Verbal consultation regarding prescription drugs: Findings from a statewide study. *Medical Care, 30,* 159–173.

Williams, A., Giles, H., Coupland, N., Dalby, M., & Manasse, H. (1990). The communicative contexts of elderly social support and health: A theoretical model. *Health Communication, 2,* 123–144.

Wilkinson, P., Tylden-Pattenson, L., Gould, J., & Wood, P. (1981). Comparative assessment of two booklets about rheumatoid arthritis, intended for use by patients. *Health Education Journal, 40*(3), 84–88.

Witte, T. N., & Kuzel, A. J. (2000). Elderly deaf patients' health care experiences. *The Journal of the American Board of Family Practice, 13,* 17–22.

Wood, J. B. (1989). Communicating with older adults in health care settings: Cultural and ethnic considerations. *Educational Gerontology, 15,* 351–362.

Woodward, N., & Wallston, S. (1987). Age and health care beliefs: Self-efficacy as a mediator of low desire for control. *Psychology and Aging, 2,* 3–8.

Yates, J. F., & Patalano, A. L. (1999). Decision making and aging. In D. C. Park, R. W. Morrell, & K. Shifren (Eds.), *Processing of medical information in aging patients* (pp. 31–54). Mahwah, NJ: Lawrence Erlbaum Associates.

Communication and the Institutionalized Elderly

Karen Grainger

Sheffield Hallam University

The study of institutional life and institutional care of older adults is undoubtedly an important area of social gerontology. Although only a minority of older people experience long-term institutional living (in the United Kingdom, about 5% of the total population over 65; Royal Commission on Long Term care, 1999), it is nevertheless the case that the "very old" section of the population (those over 85) is growing more rapidly than any other: By 2050 it will be three times the current figures (Royal Commission on Long Term Care, 1999). About 20% of the over 85s are likely to need institutional care (Henwood, 1990). Quality of care in residential institutions for older adults is therefore of increasing interest and relevance to those involved in the care of the dependent older adults. It is now well-recognized in the social gerontology and geriatric nursing literature that quality of older adult care is in large part down to the carer–caree relationship (see, e.g., Brown, Nolan, & Davies, 2001; Marr, 1996; Nussbaum, 1990; Ory & Bond, 1989), and the communication process itself is seen by some to be central to this relationship (e.g., Caris-Verhallen, Kerkstra, & Bensing, 1997). Nussbaum (1990) stated that "communication researchers should study the messages which are exchanged within the nursing home" (p. 168). Discourse analysis provides, arguably, the best approach to such a research task. Van Dijk (1985) stated that "Discourse analysis provides us with rather powerful, while subtle and precise, insights to pinpoint the everyday manifestations and displays of social problems in communication and interaction . . . " (Vol. 4, p. 7). However, in the applied geriatric and gerontological literature, the detailed analysis of talk is only just beginning to be valued (e.g., Bryan & Maxim, 1998; Erber, 1994).

In reviewing the literature in this area, I have found that studies of the living conditions of older adults in long-term care institutions are not difficult to find in the nursing, sociological, and social-psychological literature. A number of these are reviewed in this chapter. However, despite the obvious importance of language to social and psychological welfare in this environment, as in any other (Giles, Williams, & Coupland, 1990), few of these studies give more than a global and superficial description of communication styles, paying little attention to the detailed processes of discourse. Caris-Verhallen et al. (1997) have found that although many nursing studies document the amount of nurse–patient interaction, they do not describe the "interactive nature of communication" (p. 931). There are some exceptions to this, for example, Gibb and O'Brien (1990; cited in Caris-Verhallen et al., 1997), which is a conversation analysis of interactions recorded during routine nursing procedures, and Hewison (1995; cited in Caris-Verhallen et al., 1997), which uses participant observation and shows how nurses use language to exert power over patients.

My own research is another exception, being a discourse study of nurse–patient interaction in two British long-stay geriatric hospitals. Ethnographic and discourse data were gathered in the form of field observations, interviews with nurses, and tape recordings of naturally occurring interactions between nurses and residents during the performance of the main daily care tasks. All data were gathered with the permission of participants and the relevant authorities over a period of 10 months in 1987. In reviewing the literature, then, I shall give an account of the main themes that arise in relation to communication in the institutional context, and also refer to my discourse-based study in which an analysis of the carer–older adult relationship is given through the micro-analysis of discourse as it occurs in the macro-institutional and social context.

Discussions of communication with the institutionalized older adults are mainly aimed at, and written either by, practitioners (chiefly in nursing and speech therapy) or those with a more theoretical orientation, aimed at the academic (social scientific) community. The former tend to have a more "hands on," interventionist approach, whereas the latter are more deeply analytical of the social processes taking place. Many nursing studies are based on tape-recorded or observed interactions but lack the necessary linguistic expertise with which to carry out discourse analyses or to provide satisfactory explanations based on theoretical frameworks (e.g., Lanceley, 1985; Thomas, 1994; cited in Caris-Verhallen et al., 1997). And ethnographic studies with more theoretical accounts of institutional life tend not to provide examples of actual verbal interaction to support their claims about communication (e.g., Evers, 1981). Nevertheless, a number of insights can be gained from the existing literature, and I discuss these later under the headings of the three main themes that emerge from the studies I have looked at. These themes are absence of talk, task-oriented talk, and dependency-inducing talk.

ABSENCE OF TALK

A basic, but important, observation on the quality of life of institutionalized older adults is that there is a noticeable absence of talk between carers and residents, or among residents only. Jones and Jones (1985) found that only 850 words were spoken by nursing staff to 36 residents during 72 hours of taping. This is partly attributable to the fact that a number of residents suffer from neurological or physiological communication disorders such as dysphasia, dementia, and dysfluency (Gravell, 1988; Meikle & Holley, 1991), but

Gravell (1988) stated that physical disorders are compounded by other factors such as low staff expectations of residents' mental faculties or ability to communicate. Gravell (1988) also cited some basic environmental barriers to communication in the institutional setting, such as the television being on too loud for much of the day and the arrangement of residents' chairs around the walls so that they cannot face each other to converse. Gravell (1988), citing Lubinski (1984), stated that many institutions are "communication impaired environments": Residents are given no stimulation and therefore no reason to talk; there is little or no privacy, and staff do not value communication with patients. Lubinski, Morrison, and Rigrodsky (1981) surveyed resident attitudes to interaction and found that residents were inhibited about talking, partly because they were very selective about their conversational partner, and partly because of lack of privacy, adoption of the patient role, and inactivity. Meikle and Holley (1991) stated that for older people in nursing institutions, the hospital ward environment can be disorienting, resulting in patients' withdrawal and silence. Further, it is claimed, all their physical needs are anticipated, giving patients little reason to communicate.

Commonly cited as a reason for limited communication on the part of staff is lack of time. Meikle and Holley (1991) stated, "It is often very difficult for professional staff to fulfill the communication needs of patients because staff shortages result in tremendous stresses and often there is not time for a nurse, for example, to sit and just have a chat" (p. 158). Niewenhuis (1989) wrote, "Nurses may wish to offer them more stimulation and conversation, but lack of staff (and hence lack of time) may prevent them from doing this to their own satisfaction" (p. 34). In my own research, also, nurses reported that lack of time prevented them from talking with residents (Grainger, 1993a), but I observed that even when nurses appeared to have spare time, they preferred to talk to each other rather than with residents. Nolan, Grant, and Nolan (1995) supported this observation: Despite nurses' claims that communicating with patients is an important part of their job, levels of nurse–patient interaction remain low. Marr (1996) captured the perspective of the young nurse thus:

> If we had to communicate, we wanted it to be as easy as possible! Someone who was our own age, independent and usually the opposite sex, offered the best chance for an amusing and interesting exchange. We had things in common to discuss, perhaps shared memories and a mutual understanding of each other's experiences. Rarely would we seek out a patient who was old. Older people, generally, were more difficult to talk to and required more time and effort on our part. (p. 142)

It seems, then, that talk with the patients involves some effort and does not come as easily as talk with colleagues (see also Caris-Verhallen et al., 1997). In some cases, this may have been because the residents had physiological restrictions on their communicative abilities and in other cases because patients simply made poor conversationalists: They were passive and withdrawn and had little to talk about. Allen's and Turner's (1991) study provided some empirical evidence for this in that nurses were less likely to interact with physically dependent patients.

In terms of communication accommodation theory, older patients are underaccommodative (Coupland, Coupland, Giles, & Henwood, 1988). That is, the quality of their talk is inadequate relative to the needs of the nurses (cf. Coupland, Coupland, & Giles, 1991). Some studies, as in my own, also report very little interaction between residents. Allen-Burge, Burgio, Bourgois, Sims, and Nunnikhoven (2001) attributed this partly to the residents impaired sensory and cognitive functioning, but also to the "impoverished social ecology of nursing homes" (Allen-Burge et al., 2001, p. 214), as Gravell noted.

There is no doubt, then, that institutionalized older adults tend to leave a communicatively impoverished life, if only because of the absence of talk, and much of the clinical literature is concerned with outlining schemes of intervention that attempt to motivate residents to interact with each other. Proposed interventions include staff training to raise awareness and to encourage nurses to stimulate residents to interact (Allen & Turner, 1991; Allen-Burge et al., 2001; Niewenhuis, 1989); changes in the environment and layout of the ward (Niewenhuis, 1989); increased activities on the ward (Niewenhuis, 1989); and therapy, such as reality orientation, reminiscence therapy (Gravell, 1988), and group therapy (Meikle & Holley, 1991). Although valuable as far as they go, such remedial programs only address the problem at a superficial level and seem to rely largely on the resident altering her or his behaviour as a result of particular therapies. Indeed, Allen and Turner (1991) found that a program of intervention made little difference to the amount of nurse–patient interaction, and Heaven and Maguire (1996) concluded that communication skills training of nurses did not lead to noticeable improvements in communication with cancer patients. (The later section, "Implications and Applications" of the Research, gives a more detailed critique of schemes of intervention.)

The quantitative observation of the absence of talk is clearly only part of the picture. For a full critical account of institutional communication, we must also look at the quality of interaction that does occur. Only then can meaningful remedial action be proposed. Thus, the knowledge that there is a paucity of social interaction in nursing homes for seniors should serve as a part of an entire picture in which any talk that does take place becomes all the more meaningful because of its rarity.

TASK-ORIENTED TALK

The most common qualitative observation about communication in old age institutions is that most of the carer–older adult interaction that occurs is centred around particular care tasks. Jones and Jones (1985) tape-recorded nurses and patients on a long-stay geriatric ward and identified the number of "words," "commands," "statements," "questions," and "answers." It is not clear how these categorizations were decided (are they grammatical or pragmatic "questions"? how do "statements" differ from "answers"?), but their findings nevertheless suggest an emphasis on task-oriented talk. They conclude that commands related to the care task were the "primary form of communication" (p. 13). In addition, Seers (1986) found that 64% of nurse–elderly interactions on a geriatric ward over a 7-day period were task oriented. Wells (1980) found that 75% of all communication between nurses and patients on long-stay geriatric wards occurred while nurses were occupied with a physical care task, and that all talk centred on the task. Wells divided talk between nurses and patients into procedural and personal talk, with a third category of "procedural/personal" talk. She found that over 50% of talk was procedural, and about

25% was mixed procedural and personal, with the remainder being solely personal. Ethnographic studies make similar claims. Fairhurst (1978) asserted that most verbal interaction between patients and nurses in a nursing home is talk that is "aimed to ensure the smooth execution of tasks." Evers (1981) noted that the residents of long-stay geriatric wards are treated as work objects. The staff–resident interaction is mostly task-centred as a result of the institutional pressure on nurses to get certain things done.

More recent work by Koch and Webb (1996) confirmed that geriatric nursing care remains dominated by the biomedical model of ageing, resulting in the dominance of the set care routine. However, there is one study that is inconsistent with these findings: Caris-Verhallen, Kerkstra, van der Heijden, and Bensing (1998) found that when home carers' and insititutional carers' communication with patients is compared, the institutional carers' talk is relatively high in "socioemotional talk." They suggested that this may be because most of the patients in the study had been in the hospital for over a year and so there was less need for task-related communication. As their study essentially involved categorizing and quantifying different types of utterances, it is impossible to judge whether the process of interaction was in fact "directed at the establishment of a relationship" (p. 105), as they claim. It may be that on closer inspection, apparently affective talk, such as joking, was actually in the service of task-oriented talk, as shown in the later discussion.

Lack of adequate training is often given as the reason for such a limited style of communication. Allen and Turner (1991) cited Spence (1985), saying that "part of the reason for the low level of interactions between staff and residents may be due to the absence from nurse education of adequate communication skills training" (p. 1172). Seers (1986) claimed that communication is not given priority and that nurses should be encouraged to view interacting with patients more often as more important. Other explanations delve a little deeper and are critical of the institutional regime. Wells (1980) stated that "nursing work routines created irrational and impossible work goals which tend to put pressure on the nurse to accomplish their work as quickly as possible" (p. 87). She observed that "individual patient preference or even necessary variation in care appeared to be obstructive to the goal, which was completion of the routine" (p. 92). Evers (1981) asserted that geriatric nurses experience conflict between the need to treat patients as people and to treat them as work objects, but that "routinisation and treatment of patients as work objects too often prevails ..." (p. 114).

These explanations begin to examine the problem of communication in geriatric care in terms of the institutional goals and roles of nursing staff, and they hint at a more complex solution than simply improving staff training. Indeed, Wells (1980) stated that "admonishing such nurses to improve care for the elderly and work harder is unlikely to improve care for the elderly ... there is no simple way to deal with years of educational and administrative neglect in geriatric wards" (p. 94). Koch's, Webb's, and Williams' (1995) study attested to the fact that, despite a wealth of research calling for changes in geriatric care since the early 1980s, it has changed little in 16 years. They argued that this lack of change is because of societal ageism.

A combined discourse and ethnographic study such as mine (Grainger, 1993a) provided graphic and powerful support for the claims of Wells and Evers. Examples of interactions from tape-recorded data, showing procedural and personal talk, demonstrate how the conflict between procedural (task-related) and personal (relational) goals that Evers talked about surfaces in nurse–patient interactions. I argued, like Evers, that nurses have conflicting relational and task-oriented goals (see Grainger, 1993b), but that the principal institutionally endorsed aim of task completion overrides relational goals. This is demonstrated through the detailed analysis of discourse in process. The following fragment of an interaction between a nurse and a patient on a long-stay geriatric ward serves to demonstrate this point:

Nurse:	Morning, Albert.[1]
Patient:	Morning, nurse.
Nurse:	How are you this morning?
Patient:	Not good.
Nurse:	Mm?
Patient:	Not too good.
Nurse:	Not too good? What's wrong?
Patient:	I don't know.
Nurse:	Mm?
Patient:	A lot of things wrong.
Nurse:	Oh. (2.0) I'm going to get you up now and have some breakfast (1.0), have some nice porridge.
Patient:	Oh, yes. (nurse)
Nurse:	Yes?
Patient:	That'll be lovely.
Nurse:	That'll be lovely good. (13.0) Get you some clothes ready first . . .

The extract shows nurse and patient engaging in ostensibly relational discourse at the beginning of the routine. The nurse offers a greeting and asks about the well-being of the patient. But when the patient responds nonphatically (see Malinowski, 1923; Coupland, Coupland, & Robinson, 1992) by saying, "Not too good," this threatens the continuance of the task. To delve into the patient's problems would take time, concentration, and skill. So, after initially inviting the patient to expand on his assertion ("What's wrong?"), the nurse brings the patient's focus back to the routine ("I'm going to get you up now and have some breakfast"). This is despite the patient's assertion that there are "A lot of things wrong." Thus, the nurse subordinates the investigation of the patient's feelings to the primary reason for the interaction—to get the patient washed and dressed for the day (see Grainger, 1993b, for a full discussion).

The analysis, although giving support for Wells' and Evers' observations, in fact goes further and claims that relation-oriented talk between patients and nurses is, in practice, in the service of task-oriented talk, and thus serves the primary goal of the institution, which is the custodial, physical care of its residents. It is argued, through discourse analysis, that relation-oriented talk, although it fulfills the caring–personal requirements of the nursing identity that Evers talks about, actually merely facilitates the orientation

---

[1]Transcription conventions used here are derived from those used by Jefferson (Atkinson & Heritage, 1984). Double brackets indicate unclear speech; numbers in brackets refer to the length of a pause in seconds. All names are fictitious.

to task completion. In fact, I found it useful to separate nurses' relation-oriented talk into that which sustains institutional roles and that which allows speakers to step outside their institutional role. I called these "nurturing discourse" and "personal discourse," respectively. Nurturing discourse occurs alongside procedural talk (which in my analysis is called "routine management discourse") and has affectionate and intimate overtones, which are conveyed via a soothing tone of voice, the use of endearing terms of address, such as "my love," the uttering of reassuring messages to the patient, for example, "Alright" and "There you are," and by verbally checking on the patient's welfare, such as asking, "okay?" These examples are all found in the following extract:

(The patient is about to go in the bath; the nurses are undressing her and taking the dressing off her leg.)

Patient:	. . . You're hurting
	[　　]
Nurse 1:	All right, Faith.
Nurse 2:	(softly) There you are.
Patient:	What are you doing to me?
Nurse 2:	We're going to take your dressing off now, my love,' cause you're going to go in the bath, OK?
Patient:	I've got some dressing on my legs, have I ? (2.0) Don't take all that dressing off. It hurts too much.
Nurse 1:	All right, Faith.

Personal discourse, however, is usually sequentially separate from talk focused on the task and, unlike nurturing discourse, allows the participants to background their institutionally prescribed roles. In the following fragment of talk, patient and nurse engage in a short sequence about holidays before returning to the business of the routine:

(The patient has just been put into the bath.)

Nurse:	How's that?
Patient:	Lovely (.) (beautiful)
Nurse:	All right?
Patient:	Yeah.
Nurse:	You going on holiday anywhere this year?
Patient:	No.
Nurse:	I am.
Patient:	Where you going?
Nurse:	I dunno (.) Going to England.
Patient:	(laughs)
Nurse:	Escape from Wales.
Patient:	Where's your home?
Nurse:	Sorry?
Patient:	Where's your home?
Nurse:	Here (.). Cardiff (1.5), right? Alan, I'm gonna wash your hair, okay?

From my analysis of extracts such as these, then, a more complex but arguably more explanatory picture of nurse–patient interaction emerges, in which nurses engage in talk that reflects multiple roles, but all of which serve the institutional goals of task achievement. Even personal discourse, it is argued, indirectly contributes to the instrumental goal of the institution by orienting to, and redressing, possible face threats (Brown & Levinson, 1978, 1987; Goffman, 1967) incurred by the imposition of the routine (Grainger, 1993a, 1993b). I also argue that patient talk, as well as nurse talk, may be both task oriented and relational. Routine management discourse (which is largely task oriented) and personal discourse occur in their talk but, whereas nurses sustain part of their professional identity with nurturing discourse (enabling task completion without face loss), patients engage in sick–dependent discourse that reflects a needy, dependent identity for patients that justifies and possibly invites nurturing discourse. In the following extract, the nurse and patient together construct the patient's identity as one who is sick. The nurse's sympathetic responses here validate the theme of the patient's medical dependency and sustain her own role as an altruistic service provider.

(The nurse is about to give the patient an insulin injection.)

Patient: . . . This one leg is very bad. (.) Ooh.
                              [   ]
Nurse: I know (.) ((it's sore)) (.)
                 It's sore on your (buttock).
Patient: It is sore, love (.) No, don't put it (the injection)
                 in there, love, it's too painful.
Nurse: No, I won't (.) I'll do it in your arm.
                              [   ]
Patient: Oh, yes (.)
                 It's agony. Honestly, it is (.) agonizing.

In sum, this analysis sees the communication between institutionalized older people and their carers in terms of institutional goals and participant roles. These goals and roles are often conflicting and result in communication problems. One problem is that the concentration on task completion leads nurses to respond to patient requests and complaints using "evasive techniques" (Wells, 1980). Wells observed that if a patient made a request that could interfere with task completion, the nurse may have referred to "unspecified people" or "unspecified times." She gave the example of a patient who complained of being given cabbage instead of potato for lunch. The nurse responded by saying, "They'll bring you the potato." Wells stated that evasion or ignoring patients' desires is a "defence against the myriad of conflicting demands" (p. 116). Grainger, Atkinson, and Coupland (1990) referred to a similar phenomenon—that of "deflecting" patients' expressions of discomfort or unhappiness. It is argued that nurses display a number of verbal strategies to avoid dealing with patients' talk about troubles (Jefferson 1984a, 1984b), which include ignoring the trouble, contradicting the patient's perception of the trouble (for example, by claiming it to be warm when a patient complains of being cold), referring to a higher authority (for example, by suggesting the patient should consult a more senior nurse or a doctor), or by making light of the trouble. The following extract from the data illustrates some of these strategies:

(The patient is having a bath. She is sitting in the bath on the seat of the hoist, which is used to lift patients in and out of the bath.)

Patient: Oh, it [is beginning to] hurt this bad leg now under here.
Nurse: Mm?
Patient: It hurts this bad leg now.
Nurse: Right, are we ready to get out?
                 (1.5)
Patient: Mm.

The complaint is taken up again a few moments later:

Patient: Ooh (I do hate these blessed things.)
Nurse: What you done? (2.0) Are you moaning? (1.5) Are you moaning?
                                                    [   ]
Patient: [2 syllables]
                 Yes, it's [paining me right up the top] here.
Nurse: Is it?
Patient: Where it's broke.
Nurse: Oh, by there? (.) It's hurting, is it? (1.0) Ah, dear (.) You're falling apart, you
                 are.
Patient: Oh, it's very painful, you know.
Nurse: (with exaggerated intonation) Is it? (1.0) Tell Mary when she . . .
                                                    [   ]
Patient: Well, you . . .
Nurse: . . . comes back.
Patient: . . . know it is.
Nurse: Tell Mary when she comes back (.), right?
Patient: Who?
Nurse: That your leg's been paining you.
Patient: Oh.
Nurse: Did you see the doctor yesterday? (1.0) Did you see him?

In these interchanges, we see the nurse ignoring the trouble (the fourth line down in the extract, when the nurse responds with "right, are we ready to get out?"), making light of the trouble ("You're falling apart, you are") and referring to higher authorities (the staff nurse, Mary, and the doctor). These "deflection" strategies, I suggest, occur when the troubles are not amenable to a plausible solution and when addressing the problem threatens to interfere with the furtherance of the physical care task being conducted. The realization of the deflection strategies reflects the communicative dilemma that nurses face in dealing with troubles talk. If they do not at least superficially attend to the trouble, their identities as carers are put at risk. But to fully engage with the patient about their trouble may also mean putting their efficient, task-oriented role in abeyance. This is the conflict that Evers (1981) has noted. She claimed it arises from the ideological goal of nursing to treat patients as people, on the one hand, and the practice of nursing on the other, which is highly routinized and subject to hierarchical control. Thus, the cheerful manner that characterizes the popular stereotype of nurses' speech may reflect an attempt to adhere to an imposed routine while maintaining a caring identity. But it is likely that there is more at issue here than the facility to continue with the routine. Nurses lack the specialized training and expertise to respond effectively to deeply felt emotional or physical stresses. Thus, deflection may be seen as a damage-limiting strategy, representing the least threatening alternative to both troubles teller and troubles recipient. In many cases, the data show that by focussing on the routine, the nurse can more easily "escape" from a trouble that she feels ill equipped to deal with. The following exchange provides an example:

Patient: Only wish I could sleep better at night, that's my only trouble.
Nurse: Mm (.) Stand up, love.

Given that nurses' training is largely task oriented and lacks any substantial component in communication and counseling (Clark, 1981), their tendency to focus on the physical care task may signal an inability to know what else to do. It is probably worth noting here that even hospice nurses, trained to deal with patients facing death, are reported to use evasive strategies when patients raise emotionally difficult topics (Booth, Maguire, Butterworth, & Hillier, 1996). We can speculate, then, that responding to troubles talk can be a frustrating experience for the carers in these situations, arousing, as it does, role conflicts and interactional dilemmas. But for patients, the consequences of evasive or deflective strategies must be even more pernicious. Interactions with staff members constitute the majority of their daily interactional experiences, and yet their communicative agendas repeatedly go unaddressed: Their troubles are consistently avoided, made light of, or dismissed. This, it seems to me, cannot fail to have damaging effects on the physical and mental health of the older recipients of long-term care. The implications for remedial policies must involve far-reaching institutional changes. These are discussed in the "Implications and Applications" section following a review of the third major group of findings around the quality of carer–older patient interaction.

DEPENDENCY-INDUCING TALK

It is an inherent irony of hospitals, nursing homes, and other "total" institutions (see Goffman, 1961) that, even though the ultimate aim is often said to be to enable the patient, resident, or inmate to become an independently functioning member of society, many of the practices and characteristics of institutions in fact induce dependency. Dependency-inducing behavior is in part displayed through verbal interaction, and research has found that caregivers' expectations of dependency in institutionalized older adults is apparent in their speech to the care receivers. Baltes, Wahl, and Reichert (1991, cited in Ryan, Hummert, & Boich, 1995) argued that one reason for this is that older residents have learned that one of the few ways they can elicit social interaction is by increasing their dependence on staff. Ashburn and Gordon (1981) found that speech addressed to older residents of a rest home had many of the features of speech addressed to young children (less complex, shorter utterances, more redundancy, more interrogatives). Staff members were more likely to use this kind of speech than non–staff members, and it was more likely to be used to those residents considered to be most dependent. Ashburn and Gordon (1981) therefore concluded that the use of the "baby talk" register depended both on the role of the speaker and on what the speaker perceived the communicative competence of their interlocutor to be.

Caporael and colleagues have also investigated the way caregivers talk to institutionalized older adults. Caporael (1981) found that a substantial amount of talk addressed to care receivers was found to have the paralinguistic features of baby talk (BT). It was further discovered that caregivers employed this mode of talk regardless of the perceived dependency of the addressee. A subsequent study (Caporael, Lucaszewski, & Culbertson, 1983) found that caregivers seemed to have a stereotyped expectation that BT would be preferred and better understood by residents regardless of actual level of dependency. The authors concluded, like Ashburn and Gordon (1981), that this is a speech register (i.e., a situation-dependent style) that is adopted in this environment and that is not adjusted according to addressee.

Caporael (1981) also identified an alternative mode of carers' talk to older adults, which she called non–baby talk (NBT). NBT is described as talk that occurs between caregiver and care receiver but does not have the prosody of BT. It is surmised that this constitutes an institutional register that, like BT, denotes and promotes dependency in institutionalized older adults and, unlike BT, does not convey affection. However, little information is given as to the content and quality of this style of address and, from Caporael's account, it is difficult to see how its communicative function differs from BT. A later study (Caporael & Culbertson, 1986) found that, although the paralinguistic characteristics of BT and NBT differ, the verbal content of NBT is very similar to BT. This would tend to suggest that BT is not usefully labeled as a discrete category. It would suggest that in fact no absolute distinction between the two modes of talk exists, and that to describe carers' talk in terms of separate registers overlooks the possibility that they are slightly different surface realizations of a communication accommodation strategy that is subject to contextual and psychosocial influences.

In my view, the occurrence of BT and other dependency-inducing behaviors in the nursing home setting is better explained by a model of interpersonal communication that is applicable to dynamic interaction analyzed in context. Communication accommodation theory (previously mentioned) may provide such an explanation. The theory claims that individuals attune their speech either toward or away from the speech style of the addressee depending on whether they have cooperative or antagonistic intents. In some cases, speakers may react, not to the addressee's actual speech style, but to a stereotypical expectation of the addressee's speech performance, interpretive competence, or communication needs or wishes. According to this framework, BT can usefully be seen as part of an interactive process that is a reaction to perceived dependency in older people and includes the use of directive and regulatory speech. This view is supported by Ryan et al. (1995), who reviewed the literature on baby talk and overly directive talk to elders, calling it collectively "patronizing communication." They define it as "overaccommodation in communication with older adults based on stereotyped expectations of incompetence and dependence" and may involve "underaccommodation to the individual needs of the recipient" (p. 145). This assessement captures the fact that different styles of talk (such as giving orders and baby talk) may be motivated by the same (often ageist) psychosocial factors and that all convey a lack of respect to the recipient.

Lanceley's work (1985) also argued that stereotyping of older patients is responsible for nurses' speech style on geriatric wards. She said that nurses' use of controlling language to older patients keeps patients "frozen in existing stereotyped roles." This produces the wrong climate for rehabilitating the elderly in that it does not promote independence. Basing her claims on casual observation, anecdote, and previous research, Lanceley provides a list of features of nurse talk to the elderly. The features she identifies include the use of "I versus we," as in "We'll just stand you up," which is described as a refusal on the part of the nurse to treat the patient as an individual; the use of "softeners" such as *just* (as in "I just want you to lie on your side"); conditionals (for example, "That would be difficult"); tag questions (as in "You will have some lunch, won't you?"); and a prevalence of modal verbs, such as *should*, *ought*, and *must*. Some of Lanceley's observations are in line with Goffman's claims about total institutions (Goffman, 1961). She noted that nurses often talk about the patient to a third person in the patient's presence and that elderly patients tend to be addressed by their first names. These are said to be tokens of the impersonal and disrespectful attitude of the staff member toward the patient.

Fairhurst (1981), whose study is an ethnographic description of communication styles at a nursing home for older adults, also identifies naming practices as a feature of nurses' talk to the institutionalized elderly. She states that in some institutions, patients acquire nicknames such as Gran or Pop. As with first naming, the use of such informal address terms conveys an affection and intimacy that, if not actually felt, and if undesired, will be perceived as disrespect. First naming, though, is not inherently disrespectful. It could also be said to convey an interest in the well-being of the individual and is therefore not necessarily a negative feature (see Ryan & Cole, 1990). Whether or not first naming is perceived as disrespectful by the recipient will depend on a number of contextual factors, such as perceived social distance and the social status of the interlocutors relative to each other. Where a solidary relationship cannot be assumed (as between nurses and patients) first naming may be viewed negatively by the older person. Within the framework of communication accommodation theory, inappropriate first naming is an instance of miscommunication (Coupland, Nussbaum, & Coupland, 1991), in which first naming is an interpersonal control strategy motivated by the speaker's perception of the social or institutional roles of the interlocutors. Ryan and Cole (1990), and subsequently, Ryan, Hummert, and Boich (1995), pointed out that carers of the elderly may find it difficult to communicate the correct balance between a nurturing and a respectful attitude, because the two are, to some extent, incompatible. In the cases of the address terminology identified by Lanceley (1985) and Fairhurst, nurses may opt for nurturing styles at the expense of the patients' dignity.

Fairhurst further identifies a category of "superlative talk," defined as comments that highly praise a patient's performance at doing a fairly ordinary task. Thus, a nurse might say a patient's painting done in occupational therapy is "marvellous" when in fact it is no better than average. The implication here is that the staff have low expectations of the capabilities of their elderly patients, so that anything they can achieve, no matter how small, is considered praiseworthy. This sort of talk, which may be well intentioned but arises out of a stereotyped idea of the addressee's competence, is another example of the overaccommodation strategy previously mentioned (Coupland et al., 1988). As with first naming, such expressions of approval (often accompanied by a pat on the back or the arm, according to Fairhurst) may either be perceived as condescending or as nurturing by the recipient, and again illustrates what Ryan and Cole (1990) called the carers' "communication challenge" (p. 187), and what Ryan, Hummert, and Boich (1995) call the "Communication Predicament": how to convey solicitude without patronizing the patient.

Persuasive talk, or compliance gaining, forms another subcategory of dependency-inducing communication strategies. These consist of nurses' verbal attempts at getting the patient to cooperate with specific activities. Fairhurst (1981) described persuasive talk as a reflection of the patient's loss of autonomy and the adoption of the "nurse knows best" attitude, which goes with the taking on of the sick role in hospital. She describes a sequence of interactions in which a patient is being persuaded to take some medicine. The forms of persuasion start with the general message that "It's good for you" and move on to an urging to "be good"; it then becomes a sort of emotional blackmail ("You'll get me into trouble") and finally turns into direct threats ("You won't get any dinner"), and even physical force—the nurse holds the patient's nose and pours the liquid down her throat. In my own data, compliance gaining is discussed in the light of Brown's and Levinson's theory of politeness, (1987) which seems to provide a partial explanation for the use of certain persuasive strategies in this context. That is to say, the form of compliance gaining or persuasive messages can be attributed to the interlocutor's desire to maintain their own and their addressee's face during the interaction. According to the theory, then, an utterance such as "It's good for you" suggests that the speaker has the addressee's wants in mind and is therefore orienting to their positive face wants (see Brown & Levinson 1978, 1987). My analysis suggests, however, that although personal face-saving may be going on in interactions such as these, nurses are also orienting to their own professional face, which involves the need to complete the task at hand, regardless of the wishes of the patient. This can explain why face-saving strategies (such as the one just discussed) occur alongside face-threatening strategies such as threats and physical force. It is the nurse's professional face that is paramount in these interactions, and politeness strategies are merely part of this, giving the impression of interactional sensitivity (for a full discussion, see Grainger, 1990).

Discourse analyses of communicative strategies that are backed by social psychological and sociolinguistic theory, such as those described on compliance gaining and overaccommodative talk, complement the other observational or experimental studies by Lanceley (1985), Caporeal at al. and Fairhurst. Together, they make a valuable contribution to our knowledge of the dependency-inducing nature of long-term care for older adults and can begin to provide explanations for the communicative behaviors observed. Any interventions affecting policy and practice in that area will therefore be better informed. The following section discusses how these, and the other findings discussed in this chapter, could be applied to institutional settings to improve the quality of care for older adults.

IMPLICATIONS AND APPLICATIONS OF THE RESEARCH

The foregoing review of recently available literature demonstrates that researchers from diverse fields of inquiry are at least in agreement that communication between dependent older adults and their formal carers is impoverished and in need of improvement. Older residents are subjected to loss of personal identity and self-esteem, loss of control over their own environment, and lack of meaningful social contact. For their part, the nurses who care for them are in the impossible position of having to meet institutional needs by objectifying the patient, while still maintaining at least a façade of personalized care. As some authors have observed (e.g., Denham, 1991; Evers, 1981; Gravell, 1988; Koch & Webb, 1996; Wells, 1980), this situation can be attributed to social and medical expectations of chronically sick older adults and their carers, the routinized and task-oriented institutional regimen, and the physical environment of care. On this basis, it would seem sensible to recommend a course of action where institutional changes are introduced. This is sometimes advocated in the literature (e.g., Evers, 1981; Koch, Webb, & Williams, 1995; Millard, 1991), and I have already mentioned some specific recommendations for remedial programs of care. But given the insights that can be gained from ethnographic and discourse studies of relationships and interaction in these contexts, it seems that these attempts at intervention could benefit from a critical sociolinguistic perspective in which the "solution" to observed communication problems reflects the dialectal relationship between local interpersonal discourse processes and macro-institutional and social structure. If fundamental and long-lasting changes are going to be made, then research in this area should be applied at three levels: the social, the institutional, and the interpersonal. To take the last of these first, interpersonal relationships between nurses and patients could benefit from training in communication skills. As previously mentioned, this is something that is already recognized in the clinical literature (e.g., Allen & Turner, 1991; Bryan & Maxim, 1998; Lanceley, 1985; Meikle & Holley, 1991; Seers, 1986; Caris-Verhallen et al., 2000).

My own discourse study reinforces these attempts to put communication skills high on the agenda for geriatric nursing training and can provide data and analyses through which carers can be made aware of interactional processes that are potentially alienating for patients (e.g., deflection). Indeed, recent investigations into various forms of communication skills training have found that engaging nurses in the micro-analysis of videotaped interactions can be beneficial (Bryan & Maxim, 1998; Caris-Verhallen, Kerkstra, Bensing, & Grypdonck, 2000). There is danger in assuming that intervention can rest at this level, however. The research has found that there are severe pressures on nurses doing this job. These pressures are not simply the lack of time, but pressures much less tangible than this, concerning lack of prestige and the previously discussed conflicts inherent in the institutional culture, in which "caring" must exist alongside regimented efficiency. In this case, then, it seems unfair to expect the nurses to cope with the additional load of implementing good communication practices when the caring environment does not facilitate this (see Clark, Hopper, & Jesson, 1991; Koch, Webb, & Williams, 1995). Much of the nursing literature seems to voice such an expectation, however. In concluding her article on nurses' use of controlling language to older people, Lanceley (1985) exhorted nurses to "use language so that elderly patients will not be deceived, confused, kept ignorant nor frozen in existing stereotyped roles" (p. 133). Salvage (1985) was critical of such a stance, however, and said that "the advice . . . tends to skate over or ignore the larger issues which act as communication blocks, whether they are of an institutional or more personal kind . . . " (p. 168). Heaven and Maguire (1996) questioned the effectiveness of communication training that concentrates on the acquisition of basic skills but does not address nurses' perceptions of their own efficacy, their beliefs and attitudes, and their need for professional support.

Thus, although intervention at the level of individual communication styles has its place, it is too simplistic if it ignores the influence of contextual factors on discourse processes. Further, to place the onus of responsibility with those who have the least power in a hierarchical system (such as that operating in a hospital or nursing home) allows those in authority to shift the blame from their own shoulders. It should be recognized that much can also be done at the institutional level to facilitate improved communication.

Perhaps the most important move would be for long-term care of older adults to take place in an environment in which the status of "caring" (versus curing) is elevated to the level of a valued occupation and skill. Thus, those who nurse chronically sick older adults should be well trained (including training in communication and counseling) and better paid. This could have the effect of giving staff pride and satisfaction in their work, which would affect their treatment of patients and residents. Given the time, expertise, and permission to look after patients' emotional needs, the need to deflect patients' talk about troubles may be reduced or even removed completely. Similarly, as part of a caring ethos, it is the patient as a person—not the needs of the institution—who should dictate modes of care (see Brown et al., 2001). Thus, the daily institutional routine would ideally be abandoned in favor of an individual routine decided by the patient or resident. In this situation, the need for coercive and manipulative compliance-gaining strategies could be much reduced.

These suggestions for institutional adjustments are not without precedent. There exist a few experimental projects that have set up alternative long-stay nursing establishments for older people (Glossop, 1991; McGregor, 1991; Millard, 1991; Sander, 1991). These claim considerable success in improving the quality of life for dependent older adults by permitting flexibility of routine, promoting the autonomy of the resident, and by giving special training to the staff. However, for this style of care to become generalized to all those who require long-term nursing care, there must be a change of priorities in resourcing and a change of the political will of social planners. This brings me to the level of changes in social attitudes. If, as a society, we had higher expectations and higher opinions of older people, we could not morally countenance the abandonment of chronically sick older people to the neglect of long-stay hospitals and residential homes such as the ones researched in the studies reviewed in this chapter. Seymour and Hansen (2001) argued that "The assumption that illness and death are somehow 'natural' for older people . . . has to be challenged . . . " and that "Palliative care . . . must engage on a broadly political base with a 'moral order', so that the experience of dying for older people is seen as no less worthy than that of younger people." (p. 118). It is to be hoped that research done in this area, and more like it in the future, will at least help to raise the collective social consciousness so that we no longer, as a society "punish the elderly for being sick" (Millard, 1991).

To summarize, research into communication in institutions for the care of older adults could be applied to the improvement of individual interactions, to the relaxing of institutional regimes, and to the changing of social attitudes. These levels of context are not as separable as may be implied here, though. A change in interactive practices at the "hands-on" level can influence more general social attitudes toward care of the older adults, whereas a change in social attitudes would lead to the implementation of antiageist social policies. We cannot intervene at one level (personal, institutional, or social) and expect this to be the entire solution. Carers who attempts to practice communication skills in which they have been trained may be unsuccessful if they are not supported in their efforts by the wider context; they may be forced to return to "coping strategies" (such as deflection) if professional conflicts are not resolved.

CONCLUSION

The literature reviewed for this chapter has necessarily been from diverse disciplines, because at present there does not exist a coherent tradition of research that examines communication with institutionalized older adults. Indeed, such an enterprise ideally requires a multidisciplinary approach that looks at both micro-processes of interaction and macro-environmental influences on talk. Nevertheless, we can conclude from the extant literature that talk between formal carers and the frail elderly is generally poor in both quantity and quality, and that little has changed in this respect for at least the last decade. Residents of nursing homes and long-stay hospitals have little opportunity for meaningful and fulfilling conversation and as a result the quality of their lives suffers. I have argued that the remedy to this situation is not easily arrived at without changes in social attitudes, institutional practices, and carer–caree relationships. But improvements are more likely and more possible once it is recognized that the quality of verbal interaction plays a central part in the provision of effective care and, perhaps more important, once the skill of "caring" itself is recognized (see Brown et al. 2001). A combination of macro-ethnographic descriptions of institutional environments and micro analyses of carer–resident interactions gives us invaluable access to the quality of life in homes for older adults. It is to be hoped that future research will encompass both these approaches.

REFERENCES

Allen, C. I., & Turner, P. S. (1991). The effect of an intervention programme on interactions on a continuing care ward for older people. *Journal of Advanced Nursing, 16,* 1172–1177.

Allen-Burge, R., Burgio, L., Bourgois, M., Sims, R., & Nunnikhoven, J. (2001). Increasing communication among nursing home residents. *Journal of Clinical Geropsychology, 7*(3), 213–230.

Ashburn, G., & Gordon, A. (1981). Features of a simplified register in speech to elderly conversationalists. *International Journal of Psycholinguistics, 8,* 7–31.

Atkinson, J., & Heritage, J. (Eds.). (1984). *Structures of social action: Studies in conversation analysis.* Cambridge, UK: Cambridge University Press.

Baltes, M., Wahl, H.-W., & Reichert, M. (1991). Successful aging in long-term care institutions. *Annual Review of Gerontology and Geriatrics, 11,* 311–338.

Booth, K., Maguire, P., Butterworth, T., & Hillier, V. (1996). Perceived professional support and the use of blocking behaviours by hospice nurses. *Journal of Advanced Nursing, 24,* 522–527.

Brown, J., Nolan, M., & Davies, S. (2001). Who's the expert? Redefining lay and professional relationships. In M. Nolan, S. Davies, & G. Grant (Eds.), *Working with older people and their families* (pp. 19–32) Buckingham, UK: Open University Press.

Brown, P., & Levinson, S.C. (1978). Universals on language usage: Politeness phenomena. In E. N. Goody (Ed.), *Questions and politeness* (pp. 56–289) Cambridge, UK: Cambridge University Press.

Brown, P., & Levinson, S. C. (1987). *Politeness: Some universals in language use.* Cambridge, UK: Cambridge University Press.

Bryan, K., & Maxim, J. (1998). Enabling care staff to relate to older communication disabled people. *Language and Communication Disorders, 33*(Suppl.), 121–125.

Caporael, L. R. (1981). The paralanguage of caregiving: Baby talk to the institutionalised aged. *Journal of Personality and Social Psychology, 40,* 876–884.

Caporael, L., & Culbertson, G. H. (1986). Verbal response modes of baby talk and other speech at institutions for the aged. *Language and Communication, 6*(1/2), 99–112.

Caporael, L., Lucaszewski, M. P., & Culbertson, G. H. (1983). Secondary babytalk: Judgements by institutionalized elderly and their caregivers. *Journal of Personality and Social Psychology, 44,* 746–754.

Caris-Verhallen, W., Kerkstra, A., & Bensing, J. (1997). The role of communication in nursing care for elderly people: A review of the literature. *Journal of Advanced Nursing, 25,* 915–933

Caris-Verhallen, W., Kerkstra, A., Bensing, J., & Grypdonck, M. (2000). Effects of video interaction analysis training on nurse–patient communication in the care of the elderly. *Patient Education and Counseling, 39,* 91–103.

Caris-Verhallen, W., Kerkstra, A., van der Heijden, P., & Bensing, J. (1998). Nurse–elderly patient communication in home care and institutional care: an explorative study. *International Journal of Nursing Studies, 35,* 95–108.

Clark, J. M. (1981). Communication in nursing. *Nursing Times, 77,* 12–18.

Clark, J. M., Hopper, L., & Jesson, A. (1991, February 20). Progression to counselling. *Nursing Times,* pp. 41–43.

Coupland, J., Coupland, N., & Robinson, J. (1992). "How are you?": Negotiating phatic communion. *Language in Society, 21,* 207–230.

Coupland, J., Nussbaum, J., & Coupland, N. (1991). The reproduction of aging and agism in intergenerational talk. In N. Coupland, H. Giles, & J. Wiemann (Eds.), *"Miscommunication" and problematic talk* (pp. 85–102). London: Sage.

Coupland, N., Coupland, J., & Giles, H. (1991). *Language, society and the elderly: Discourse, identity and ageing.* Oxford, UK: Blackwell.

Coupland, N., Coupland, J., Giles, H., & Henwood, K. (1988). Accommodating the elderly: Invoking and extending a theory. *Language in Society, 17*(1), 1–42.

Denham, M. (1991). The elderly in continuing care units. In M. Denham (Ed.), *Care of the long-stay elderly patient* (pp. 3–11) London: Chapman & Hall.

Erber, N. (1994). Conversation as therapy for older adults in residential care: The case for intervention. *European Journal of Disorders of Communication, 29,* 269–278.

Evers, H. (1981). Care or custody? The experiences of women patients in long-stay geriatric wards. In B. Hutter & B. Williams (Eds.), *Controlling women: The normal and the deviant* (pp. 108–130). London: Croom Helm.

Fairhurst, E. (1978, September) *Talk and the elderly in institutions.* Paper presented at the annual conference of the British Society of Social and Behavioural Gerontology, Edinburgh, Scotland.

Fairhurst, E. (1981). *A sociological study of the rehabilitation of the elderly in an urban hospital.* Unpublished doctoral dissertation, University of Leeds.

Freer, C. (1988). Old myths: Frequent misconceptions about the elderly. In N. Wells & C. Freer (Eds), *The ageing population: Burden or challenge?* (pp. 3–16) New York: Macmillan.

Gibb, H., & O'Brien, B. (1990). Jokes and reassurance are not enough: Ways in which nurses relate through conversation with elderly clients. *Journal of Advanced Nursing, 15,* 1389–1401.

Giles, H., Williams, A., & Coupland, N. (1990). Communication, health and the elderly: Frameworks agenda and a model. In H. Giles, N. Coupland, & J. Wiemann (Eds.), *Communication, health and the elderly* (Proceedings of the Fulbright Colloquium, 1988, pp. 1–28.) Manchester, UK: Manchester University Press.

Glossop, E. S. (1991). Roxbourne Hospital. In M. Denham (Ed.), *Care of the long-stay elderly patient* (pp. 3–11) London: Chapman & Hall.

Goffman, E. (1961). *Asylums*. Harmondsworth, UK: Penguin.

Goffman, E. (1967). *Interaction ritual: Essays on face to face behaviour*. New York: Pantheon.

Grainger, K. (1990). Care and control: Interactional management in nursing the elderly. In R. Clark, N. Fairclough, R. Ivanic, N. McLeod, J. Thomas, & P. Meara (Eds.), *Language and power* (pp. 147–157) London: Centre for Information on Language Teaching and Research.

Grainger, K. (1993a). *The discourse of elderly care*. Unpublished doctoral dissertation, University of Wales, Cardiff.

Grainger, K. (1993b). "That's a lovely bath dear": Reality construction in the discourse of elderly care. *Journal of Aging Studies, 7*, 247–263.

Grainger, K. (2002). Politeness or impoliteness? Verbal play on the hospital ward. In *Linguistic politeness and context* (Working Papers on the Web, Vol. 3). Sheffield Hallam University. www.shu.ac.uk/wpw

Grainger, K., Atkinson, K., & Coupland, N. (1990). Responding to the elderly: Troubles talk in the caring context. In H. Giles, N. Coupland, & J. Wiemann (Eds.), *Communication, health and the elderly* (Proceedings of the Fulbright Colloquium, 1988, pp. 192–212) Manchester, UK: Manchester University Press.

Gravell, R. (1988). *Communication problems in elderly people: Practical approaches to management*. London: Croom Helm.

Heaven, C., & Maguire, P. (1996). Training hospice nurses to elicit patient concerns. Journal of Advanced Nursing 23, 280–286.

Henwood, M. (1990). No sense of urgency. In E. McEwan (Ed.), *Age: The unrecognised discrimination* (pp. 43–57) London: Age Concern.

Hewison, A. (1995). Nurses' power in interactions with patients. *Journal of Advanced Nursing, 21*, 75–82.

Jefferson, G. (1984a). On the organisation of laughter in talk about troubles. In J. Atkinson & J. Heritage (Eds.), *Structures of social action: Studies in conversation analysis* (pp. 346–369) Cambridge, UK: Cambridge University Press.

Jefferson, G. (1984b). On "stepwise transition" from talk about a "trouble" to inappropriately next-positioned matters. In J. Atkinson & J. Heritage (Eds.), *Structures of social action: Studies in conversation analysis* (pp. 191–222) Cambridge, UK: Cambridge University Press.

Jones, D. C., & Jones, G. (1985). Communication between nursing staff and institutionalized elderly. *Perspectives, 9*(3), 12–14.

Koch, T., & Webb, C. (1996). The biomedical constuction of ageing: implications for nursing care of older people. *Journal of Advanced Nursing, 23*, 954–959.

Koch, T., Webb, C., & Williams, A. (1995). Listening to the voices of older patients: An existential-phenomonological approach to quality assurance. *Journal of Clinical Nursing, 4*, 185–193.

Lanceley, A. (1985). Use of controlling language in the rehabilitation of the elderly. *Journal of Advanced Nursing, 10*, 125–135.

Lubinski, R., Morrison, E. B., & Rigrodsky, S. (1981). Perception of spoken communication by elderly chronically ill patients in an institutional setting. *Journal of Speech and Hearing Disorders, 46*, 405–412.

Lubinski, R. (1984). Environmental considerations for the institutionalised demented patient. Paper at NYSSHA convention.

Malinowski, B. (1923). The problem of meaning in primitive languages. In the supplement to C. K. Ogden & I. A. Richards (Eds.), *The meaning of meaning* (pp. 146–152) London: Routledge & Kegan Paul.

Marr, J. (1996). Communication. In P. Ford & H. Heath (Eds.), *Older people and nursing* (pp. 142–153) Oxford, UK: Butterworth-Heinemann.

McGregor, H. (1991). The Anmer Lodge project. In M. Denham (Ed.), *Care of the long-stay elderly patient* (pp. 275–282) London: Chapman & Hall.

Meikle, M., & Holley, S. (1991) Communication with patients in residence. In M. Denham (Ed.), *Care of the long-stay elderly patient* (pp. 149–160) London: Chapman & Hall.

Millard, P. H. (1991). The Bolingbroke Hospital long-term care project. In M. Denham (Ed.), *Care of the long-stay elderly patient* (pp. 283–298) London: Chapman & Hall.

Niewenhuis, R. (1989). Breaking the speech barrier. *Nursing Times, 85*(15), 34–36.

Nolan, M, Davies, S & Grant, G. (Eds.). (2001) *Working with older people and their families*. Buckingham, UK: Open University Press.

Nolan, M., Grant, G., & Nolan, J. (1995). Busy doing nothing:activity and interaction levels amongst differing populations of elderly patients. *Journal of Advanced Nursing, 22,* 528–538.

Nussbaum, J. (1990). Communication and the nursing home environment: Survivability as a function of resident–nursing staff affinity. In H. Giles, N. Coupland, & J. Wiemann (Eds.), *Communication, health and the elderly* (Proceedings of the Fulbright Colloquium, 1988, pp. 155–171) Manchester, UK: Manchester University Press.

Ory, M. G., & Bond, K. (Eds.). (1989). *Aging and health care: Social science and policy perspectives.* London: Routledge.

Royal Commission on Long Term Care. (1999). *With respect to old age: Long term care: Rights and responsibilities.* London: Her Majesty's Stationery Office.

Ryan, E. B., & Cole, R. (1990) Perceptions of interpersonal communication with elders: Implications for health professionals. In H. Giles, N. Coupland, & J. Wiemann (Eds.), *Communication, health and the elderly* (Proceedings of the Fulbright Colloquium, 1988, pp. 172–191) Manchester, UK: Manchester University Press.

Ryan, E. B., Hummert, M. L., & Boich, L. (1995). Communication predicaments of aging: Patronizing behavior towards older adults. *Journal of Language and Social Psychology, 14*(1–2), 144–166.

Sander, R. (1991). Jubilee House. In M. Denham (Ed.), *Care of the long-stay elderly patient* (pp. 299–310) London: Chapman & Hall.

Seers, C. (1986). Talking to the elderly and its relevance to care. *Nursing Times, 82*(1), 51–54.

Seymour, J., & Hanson, E. (2001). Palliative care and older people. In M. Nolan, S. Davies & G. Grant (Eds.), *Working with older people and their families* (pp. 99–119) Buckingham, UK: Open University Press.

Spence, F. (1985). The continuing problem of nurse–patient communication with particular reference to low self-esteem in the elderly patient. *Nursing Review 3,* 4–5.

Van Dijk, T. A. (Ed.). (1985) *Handbook of discourse analysis* (Vol. 4). London: Academic Press.

Thomas, L. H. (1994). A comparison of the verbal interactions of qualified nurses and nursing auxiliaries in primary, team and functional nursing wards. *International Journal of Nursing Studies, 31*(3), 231–244.

Wells, T. J. (1980). *Problems in geriatric nursing care.* Edinburgh, Scotland: Churchill Livingstone.

Online Support and Older Adults: A Theoretical Examination of Benefits and Limitations of Computer-Mediated Support Networks for Older Adults and Possible Health Outcomes

Kevin B. Wright
University of Memphis

James L. Query
University of Houston

In the last decade, our society witnessed an explosive growth in computer-mediated communication (CMC), because of the proliferation of computer use and the expansion of the Internet. These advances in technology have increased our ability to develop and maintain new networks of interpersonal relationships that would be difficult to form in the face-to-face world (Wellman, 1997; Wood & Smith, 2001). During the same time period, we have also seen a rapid increase in the number of older adults (age 50 and older), because of the aging of the baby-boomer cohort and health and medical developments that have expanded the average life span (Nussbaum, Thompson, Robinson, & Pecchioni, 2000). In the year 2000, there were an estimated 35 million people age 65 or older in the United States, accounting for almost 13% of the total population. By 2011, the baby-boom generation will begin to turn 65, whereas by 2030, it is projected that one in five people will be age 65 or older (Federal Interagency Forum on Aging Related Statistics, 2000). During the next decade, the use of computer technology and computer-mediated communication, and the number of older adult Internet users within the United States are projected to increase steadily (Fox et al., 2001).

Health communication researchers and other social scientists have long been interested in the association between interpersonal relationships and health, particularly the effects of social support on specific mental and physical health outcomes. These include a person's ability to cope with stressful situations and the impact of reduced stress on physical symptoms related to chronic health conditions (Albrecht & Adelman, 1987; Antonucci, 1990; Cohen, 1988; Cohen, Gottlieb, & Underwood, 2000; Krause, 1990; Pierce, Sarason, & Sarason, 1996).

In recent years, researchers interested in communication and aging have focused a good deal of attention on the impact of interpersonal relationships and social support on the mental and physical well-being of older individuals (Krause, 1990; Nussbaum, 1994; Query & James, 1989; Rawlins, 1995; Rook, 1995; Wright, 2000b). This interest has been spurred in part by preliminary empirical studies that suggest interpersonal relationships can have a positive effect on the mental and physical health of older adults (Query & Kreps, 1996; Query & Wright, 2003) and may ultimately help reduce morbidity and mortality rates among people within this segment of the population (see Schulz & Beach, 1999).

However, these studies also imply that there is considerable variability among older individuals in terms of the types of relationships in which they are involved, their interpersonal skills, their reactions to supportive attempts, and the ways in which people adjust mentally and physically to stressful situations. Successful adjustment to stressful situations, as well as a general sense of well-being, are thought to be affected by numerous aspects of interpersonal relationships (Burleson, 1990; Cutrona & Cole, 2000; Query & Flint, 1996; Rook, 1995). These variables, along with many other individual differences, such as coping styles, conceptions of health, trait and environmental factors, and severity of physical health problems, all contribute to the complex relationship between interpersonal relationships and health (Query & Flint, 1996; Query & James, 1989; Query & Kreps, 1996; Schwarzer & Leppin, 1991).

Despite nearly 4 million older adults who are currently using the Internet, and a projected increase in Internet use among this segment of the population over the next decade (Fox et al., 2001), few communication researchers have discussed the impact of computer-mediated relationships on the health of older people (for a recent exception, see Query & Wright, 2003). In particular, there has been little focus on these relationships as potentially new social support networks for older adults, or the differences between computer-mediated relationships and face-to-face relationships, and the implications of these relationships on the mental and physical well-being of older individuals.

This chapter discusses some of the possible benefits and limitations of computer-mediated support networks for older adults, as well as possible health implications for older people using the Internet for support and companionship within interpersonal relationships. Toward that end, it examines current research in the areas of interpersonal relationships, social support and health outcomes for older people, and unique features of computer-mediated relationships as social support networks. In addition, it attempts to link theory from these diverse areas of research and to explain their implications for older adults who are currently using this technology and for those who will likely be using it in the near future.

OLDER ADULTS AND CMC USAGE

The focus of research on interpersonal relationships and health among older people has been centered primarily on traditional relationships in the face-to-face world. Yet a growing number of older adults are turning to the Internet as a medium for developing new relationships and maintaining relationships between family members and friends who may be separated geographically (Furlong, 1989; SeniorNet, 2000; Wright, 2000a).

Current and Projected Internet Use by Older Individuals

Although the current number of seniors (aged 65 and older) using the Internet for interpersonal relationships represents a relatively small percentage of this segment of the population (about 5% of all Internet users), people over the age of 50 have been identified as one of the fastest growing segments of the population who are accessing the Internet. Moreover, the number of these individuals is projected to grow significantly over the next 5 to 10 years (National Telecommunications and Information Administration, 2000).

It is predicted that a so-called gray tsunami phenomenon will likely take place during this time period as younger cohorts, who are already using the Internet for interpersonal relationships, make the transition into older adulthood (Fox et al., 2001). There are several arguments supporting this prediction. First, Internet users between the ages of 50 and 64 are currently one of the largest groups of individuals using this technology. However, because most surveys of Internet usage have defined seniors as 65 and older, many of the statistics appear to be influenced by a cohort effect. The majority of individuals in recent surveys report being introduced to computer technology at work, and the Internet has only seen widespread usage for about a decade (Fox et al., 2001). It is likely that many seniors who have been in retirement since the late 1980s and early 1990s missed the rapid growth of the Internet in the workplace while they were actively involved in their careers.

Second, computer ownership among older adults has significantly increased in the past few years. According to a 1998 survey of older computer users, one fourth of the respondents mentioned that they had bought a computer within the last year, and 60% of the sample reported they used their computer on a daily basis (Charles Schwab, Inc., 1998).

Finally, because activities such as "surfing the Net" are social phenomena more ingrained into the popular culture and lifestyles of younger generations, and because many older people are living on fixed incomes, it is not surprising that there are only a small number of seniors using computer-mediated communication. However, given the rapid growth of computer technology and number of baby boomers who will soon reach age 65, the upcoming cohort of seniors is expected to become the heaviest users of the Internet and computer-mediated communication (Fox et al., 2001).

To provide a glimpse of how rapidly Internet usage patterns change, the National Telecommunications and Information Administration reported a 58% increase in overall Internet usage in the 2-year period between 1998 and 2000 (National Telecommunications and Information Administration, 2000). In short, although older adults currently may not be engaged in computer-mediated communication on the Internet, they most likely will be in the very near future.

Profile of Current Older Internet Users. According to Fox et al. (2001), older adults who are currently using the Internet tend to be well educated, affluent, and male. Seventy-six percent of the seniors in the Fox et al. (2001) study had at least some college education, and one fourth of the respondents had an annual household income of more than $75,000. Adler (1996), in a survey of 700 older computer users, found that 7% of individuals with less than a high school education had a computer, whereas 53% of people with a college degree had a computer. In addition, Adler (1996) found that older men were more likely to use the Internet for communication than were older women.

Although older men may have been the first to adopt the Internet, more recent data indicates that the number of older women using this technology may be increasing. More than half of all senior citizens who obtained access to the Internet from March 2000 to December 2000 were female (Rice, 2001). Moreover, in terms of all Internet users, women are more likely than men to search for health-related information, including the use of online support groups (Rice, 2001).

Current senior Internet users tend to differ from their non-Internet using peers in other ways. Sixty-eight percent of current users say they would miss the Internet if they could no longer use it, whereas 81% of nonusers say they have no intention of going online. In addition, among nonusers, the oldest segments of the aging population appear to be the most resistant to adopting this technology (Fox et al., 2001). Older adults who are more experienced with the Internet are more likely than less experienced peers to engage in a variety of online activities, spend more time online each day, and to have visited state, local, and federal Web sites (Fox et al., 2001).

Although seniors who currently use the Internet represent only a small share of users, they are more likely to spend time online each day, use the Internet for non-work-related purposes, and to use this technology to maintain interpersonal relationships than other groups (Fox et al., 2001). In terms of specific uses of the Internet, the most popular activity among seniors is e-mail and using the Internet as a research tool (Charles Schwab, Inc., 1998; Fox et al., 2001). More than half of the seniors who frequently use the Internet access search engines for conducting online research on a variety of topics (Charles Schwab, Inc., 1998). The most popular activities, however, appear to be seeking information about hobbies, health, and news (Fox et al., 2001).

Online Communities for Older Adults

Although there have been relatively few studies of online communities specifically for older adults, several researchers have examined SeniorNet, one of the first online communities for this segment of the population (established in 1986). Furlong (1989) found that SeniorNet participants enjoyed "an opportunity to meet people with similar interests and to share not only information, but also communication on emotional and social issues that are particularly relevant to older adults" (p. 145). SeniorNet members also reported engaging in companionship relationships centered around mutually shared interests, including health-related issues.

A recent SeniorNet member survey found that the majority of participants (1001 respondents aged 50 and older) reported using the Internet primarily as a means to keep in touch with family and friends and to access information about various topics of interest. Most people said they used the Internet (including SeniorNet) between 10 to 19 hours a week, and that most of the respondents were female (SeniorNet, 2000).

In other recent work, Wright (2000a) found that SeniorNet participants frequently promote the SeniorNet Web site as a useful source of social support within messages posted on a variety of discussion group bulletin boards. Many participants in SeniorNet groups reported that other group members often served as "surrogate family," and that they found it easier to discuss some sensitive topics with their friends on SeniorNet than with close friends or family members. For example, 18% of the sample reported that some issues, such as caring for a spouse, were often easier to discuss with other caregivers. SeniorNet forums provided an opportunity to "obtain advice about the problem, moral support, encouragement, and sympathy . . . and an opportunity to complain about family members to people outside of their immediate network" (Wright, 2000a, p. 39).

In a related study, Wright (2000b) found that SeniorNet members who spent more time communicating with others online were more satisfied with their online support network, whereas individuals who spent less time communicating with others online were more satisfied with their face-to-face support network. SeniorNet members were found to seek companionship online more than support for specific problems, and there was a significant negative correlation between the number of hours participants interacted with others online and their perceived stress scores.

Although SeniorNet is one of the more popular Internet Web sites exclusively tailored for older adults, there are many other Web sites and IRC chat rooms available to seniors, including Worldwide Seniors (http://www.wwseniors.com/), Baby Boomer Bistro (http://www.babyboomerbistro.org.uk/), Retire.net (http://www.retire.net/html/chat.html), and the AARP Web site (http://www.aarp.org/), to name a few. These communities offer researchers access to large numbers of seniors who currently use the Internet; however, there are probably many more older adults who are likely members of other types of online communities dealing with issues regarding almost every imaginable interest rather than age-related concerns. For example, most online newsgroup portals and health-related Web sites include a forum dealing with senior issues, as well as topics that may attract people from multiple age groups.

INTERPERSONAL RELATIONSHIPS, SOCIAL SUPPORT, AND HEALTH OUTCOMES FOR OLDER ADULTS

Older Adults, Interpersonal Relationships, and Social Support

Perceptions of social support are frequently affected by the relationship an older adult has with a support provider (Kahn & Antonucci, 1980). These perceptions may influence appraisals of social support and whether or not the supportive attempt will ultimately have a positive or negative effect on a person (Cutrona & Cole, 2000; La Gaipa, 1990). When contrasting social support within older adult friendship relationships with family relationships, interpersonal theorists have argued that several unique features of friendship may affect perceptions of social support (Hegelson & Cohen, 1996; Kahn & Antonucci, 1980; Nussbaum, 1985; Rawlins, 1995; Rook, 1995).

First, social support within friendship relationships has been associated with greater psychological well-being and successful adaptation to the aging process than other types of relationships, including those with family members (Arling, 1976; Beckman, 1981; Nussbaum et al., 2000; Rook, 1995). Friendships are thought to often be better suited for providing emotional support, because of the notion that there are fewer role obligations in friendship relationships than in family relationships.

According to Nussbaum (1994), "Friendship is a relationship uniquely suited to provide emotional support to individuals within the relationship. This support may be linked to the nonobligatory nature of friendship" (p. 211). For example, if a member of our family becomes ill, we may feel obligated to help this person even if we don't want to. However, role obligations are less stringent within friendship relationships, and they tend to be more voluntary in nature. Moreover, friendship relationships are less likely to be affected by relational roles of dominance or submission (power) than are family relationships, which may negatively impact perceptions of social support. According to Nussbaum (1994), within older adults friendships, "Social support is freely given, with very little of the expectation of support often associated with the family" (p. 213).

Friends not only provide support during times of crisis, but they also may help to elevate a person's mood and lower his or her stress level through companionship. Rook (1995) asserted that companionship may be motivated more by the desire for positive affect and stimulation through mutually enjoyed activities rather than as a reaction to stressful situations, and companionship may help to elevate an older individual's overall mood on a daily basis. The heightened mood associated with companionship is thought to enhance psychological well-being by having a direct effect on stress levels, whereas support may be sought by older people during times of crises and serve as a buffer to stressful situations (Rook, 1987, 1995).

In brief, friendship relationships appear to have positive implications for the psychological and physical well-being of older adults. Online support networks may offer the opportunity for older people to extend the size and diversity of their friendship networks, and these relationships may be called on for companionship or support.

Social Support and Health Outcomes Among Older Adults: Recent Developments

As individuals mature and complete their life odysseys, they confront a formidable array of stressors, ranging from the passing of friends and family members (weak and strong ties), role reversals, an increased likelihood of contracting many types of cancer, and an increased risk for late-onset Alzheimer's disease. Unfortunately, when many of these events occur, large numbers of older adults are also contending with reduced personal wealth levels. As Query and associates have argued, a symbolic crisis is often triggered during such trials and tribulations (Query, 1987; Query & Flint, 1996; Query & Kreps, 1996). A similar perspective was advanced by Barnlund (1968), Good (1977), Good and Good (1981), Kreps (1988a), and Mechanic (1983). In particular, the older person may undergo an irrevocable loss of self or identity, may misdefine particular stressors in ways that block seeking social support or medical assistance (see Kreps, 1988a), change interaction patterns in unpredictable and often bizarre ways (see Query & Flint, 1996), or experience a dramatic and lasting loss of self-esteem because of bodily changes. For example, certain treatment regimens for various types of cancer, such as chemotherapy or colostomies, can create pervasive feelings of disgust and self-hatred among older adults, their families, as well as their general and primary caregivers (Wortman & Dunkel-Schetter, 1979). These powerful emotions frequently forge "new" identities among individuals living with cancer that discourage meaningful interaction with significant others and helping populations.

Health communication and life span communication specialists can help reduce the likelihood and intensity of symbolic crises among older adults by examining the communication behaviors that help shape the form and quality of social support processes, and by assessing their impacts on health status. Such an analysis should also provide an impetus for the development of training programs targeting older adults, their families, and helping populations. These interventions would be designed to assist individuals in developing communication skills integral to the reduction of equivocality and the development of therapeutic exchanges, as well as those competencies related to the effective mobilization and utilization of supportive social structures such as support, mutual aid, and self-help groups, and social networks (Kreps, 1988a, 1988b; Query & Kreps, 1996). Cutrona and Cole (2000) identified nine mechanisms for increasing support among naturally occurring networks. These include: "increase understanding; change attitudes; improve interaction skills; increase communication; coordinate responsibilities; strengthen bonds with positive network members; weaken bonds with destructive network members; remove structural barriers to support; and provide support to network members" (Cutrona & Cole, 2000, p. 282).

A series of studies by Query and associates (Query, 1987, 1990; Query & James, 1989; Query & Kreps, 1993, 1996; Query, Kreps, Arneson, & Caso, 2001; Query & Wright, 2003) among older adults have provided partial support for adopting many of the preceding mechanisms. In essence, this line of research has demonstrated that communication competence levels among older adults (e.g., older individuals residing in retirement communities, older adult caregivers for individuals with Alzheimer's disease, older adults living with cancer, as well as their lay caregivers and healthy peers) influence key health outcomes, such as social support satisfaction and adaptive coping levels. The bulk of the preceding studies were field based, addressed Bowers' (1969) Distress Relief Recommendation and were grounded theoretically in Kreps' (1988b) relational health communication model and Fisher's (1984, 1985, 1987) narrative paradigm, or both of the latter models.

To further inform the sound development of future interventions, that target the social support networks of older adults, longitudinal and multiple methodological research designs must be implemented. Face-to-face and online environments should be assessed to provide a rich comparative database. These investigations would attempt to track the daily message strategies evaluating social support provision and reception and ascertain their influence on key health outcomes.

COMPUTER-MEDIATED RELATIONSHIPS AS SOCIAL SUPPORT NETWORKS

Computer-Mediated Communication

Computer-mediated communication refers to "how human behaviors are maintained or altered by exchange of information through machines" (Wood & Smith, 2001, p. 4). Messages are typically exchanged through various technological means that facilitate CMC, including individual e-mail, bulletin board systems (BBS), real-time Internet chat rooms (IRC), and listserves—e-mail lists dealing with specific topics. Information, however, is also disseminated through Web sites and online libraries throughout the World Wide Web.

Communication researchers have identified a variety of ways that CMC differs from face-to-face communication and how it may affect interpersonal relationships and perceptions of relational partners (Rice & Love, 1987; Walther, 1992, 1996; Walther & Burgoon, 1992). Reduced nonverbal cues within the computer-mediated environment, the ability to communication in asynchronous (not in real time) and synchronous (real time) formats, and an increased ability for optimal self-presentation may all affect the nature of interpersonal relationships (Walther, 1996).

Early research on computer-mediated communication tended to focus on negative aspects for relationships, such as the ability to engage in deception more frequently and the tendency to engage in undesirable social behaviors, such as making rude or inappropriate remarks online and hiding behind one's anonymity (Sproull & Kiesler, 1986). However, later work began to examine more positive features of the medium, such as the ability to transcend social status cues that may inhibit the development of relationships and the opportunity for a diversity of ideas to be communicated (Walther, 1996; Wood & Smith, 2001).

Despite evidence that computer-mediated channels may affect our perceptions of interpersonal relationships, communication scholars are only beginning to understand the effects of this medium on relational outcomes, such as the effects of social support on mental and physical health. Yet some idea of how supportive relationships may function within this environment is emerging from recent work on computer-mediated support networks.

Computer-Mediated Social Support Networks

Many of the findings from empirical studies of computer-mediated support groups, including studies of online communities for older adults, shed light on ways in which computer networks may function as support networks (Alexy, 2000; Braithwaite, Waldron, & Finn, 1999; Finn, 1999; Smith, 1998; Winzelberg, 1997; Wright, 2000a, 2000b, 2000c). In addition, these studies suggest differences between computer support networks and more traditional sources of social support, as well as possible relationships between computer-mediated support and health outcomes. Although studies of computer-mediated support groups have given communication researchers insight into these new types of support networks, few studies have linked empirical findings to existing theory in the areas of social support and computer-mediated communication. The following sections attempt to link empirical findings from research on computer-mediated support groups to various theories of social support and computer-mediated communication, as well as possible implications for health outcomes.

Computer-Mediated Support Networks as Weak-Tie Networks

"Weak tie" network theory in the social support and social network literature may provide some explanation for certain aspects of computer-mediated support groups (Adelman, Parks, & Albrecht, 1987; Granovetter, 1973, 1982). The term *weak ties* refers to the relationships we engage in that differ in terms of intimacy and frequency of interaction from close relationships. Weak-tie relationships are typically individuals with whom we communicate on a daily basis, but we are not necessarily close to them (Granovetter, 1973). Prior to the Internet, weak tie networks for most people consisted of neighbors, service providers, and other individuals a person could turn to during times of stress when closer ties (e.g., friends and family members) were unavailable. The ability of the Internet to connect people with large networks of people has greatly expanded the number of relationships that could become potential weak-tie networks for social support.

According to Adelman, et al. (1987), weak-tie networks serve several functions, including access to diverse information, and the facilitation of disclosure of risky topics, or topics perceived to have a negative social stigma. However, these authors also discuss several limitations of weak ties as sources for social support. The ways in which computer-mediated support groups might serve these functions for older adults and some of their negative aspects are discussed next.

Diverse Information. One function of weak-tie networks identified by Adelman, et al. (1987) is access to diverse points of view and information that may not be available within closer ties. According to Wellman (1997), in the computer-mediated environment, weak-tie networks tend to be more heterogeneous than closer networks, and "weak ties are usually better connected to other, more diverse social circles, and hence are more apt to be sources of new information" (p. 189). Closer ties tend to be more homogenous because of the tendency of people to form relationships with others based on proximity and demographic, background, and attitudinal similarity (Adelman, et al. 1987). Wellman (1997) also stated, "The relative lack of social presence on-line fosters relationships with people who have more diverse social characteristics than might normally be encountered in person" (p. 191). Social presence is limited by reduced nonverbal cues in the computer-mediated environment (such as physical and social cues), and "this allows relationships to develop on the basis of shared interests rather than to be stunted at the onset by differences in social status." (Wellman, 1997, p. 191).

Moreover, computer networks can increase the range of social networks by facilitating more relationships than would be possible in the face-to-face world, because these networks transcend spatial and temporal boundaries, and because they are a less expensive form of communication than other current alternatives (Flaherty, Perse, & Rubin, 1998; Weinberg, Schmale, Uken, & Wessel, 1995). As a result, computer-mediated support groups may bring diverse individuals together, whereas the format of these groups allows each posting to the group to be read by all members (which gives participants access to multiple sources of information and diverse viewpoints about issues).

There is some empirical evidence suggesting that computer-mediated support groups for older adults exhibit this function of weak-tie networks. Wright (1999, 2000a) found that participants using the SeniorNet community enjoyed discussing family issues with nonfamily network members, because they were able to find individuals who were interested in similar issues, but who also had much different backgrounds and experiences with the issues. This allowed for multiple viewpoints to be expressed by people in online discussions, and participants reported enjoying the opportunity to encounter ways of solving problems that they had never considered.

Wellman (1997), regarding computer-mediated networks, mentioned, "the relative lack of social presence on-line fosters relationships with people who have more diverse social characteristics than might normally be encountered in person" (p. 191). King and Moreggi (1998) contended that fewer status cues in the online context may level the playing field for people who may be from different socioeconomic groups. Other researchers have found that online networks can increase the diversity of relationships individuals form, because these relationships are developed on the basis of shared interests and they are less likely than face-to-face relationships to be inhibited by physical cues such as appearance (Hiltz & Turoff, 1993; Lea & Spears, 1995).

These findings may have implications for older adults who could possibly increase the diversity of people from different age groups with whom they interact in the computer-mediated world, especially because intergenerational relationships are often difficult to form because of appearance cues, communication styles, and negative perceptions of out-groups (Harwood, Giles, & Ryan, 1995). Computer networks also may extend an older person's social network to include younger individuals who can offer different insights about issues than might be found within one's own age group.

Reduced Stigma and the Ability to Disclose Information About the Self Safely.
Weak-tie network theory might also partially explain how computer-mediated support groups offer a forum for people who feel stigmatized by their problems, including negative perceptions about one's identity or the sensitive nature of certain topics, such as health concerns. Adelman, et al. (1987) contended that communicating about sensitive issues is less risky in weak-tie networks, because relational partners are less likely to know members of one's closer tie network, and information that is revealed through self-disclosure is less likely to make it back to closer ties. Some researchers have found that people are often more willing to discuss sensitive topics with others online than they would in other contexts (Mickelson, 1997; Wright, 2000a). In addition, online support groups may be helpful for people who experience apprehension in other contexts, because of highly visible appearance cues (King & Moreggi, 1998).

The sense of safety is in part because of to the anonymity of online communication, which gives people an opportunity to talk about their problems with others dealing with the same issues without all the complications of face-to-face relationships (Wallace, 1999). Researchers dealing with the relationship between stigma and social support have found that as the visibility of a person's problem increases, social support from family and friends often decreases (Cluck & Cline, 1989; Kiecolt-Glaser, Dyer, & Shuttleworth, 1988; Turner, Hayes, & Coates, 1993). For older adults who do not want to burden their closer ties with problems, they may find online support networks to be an alternative when discussing sensitive issues, or if they perceive that they will be perceived negatively for bringing up certain topics.

According to Wood and Smith (2001), "Participants in online exchanges have been found to disclose more about their conditions, probably because they do not sense being as readily judged by any recipients of their messages, given the lack of nonverbal cues to indicate disapproval or disappointment." (p. 102). Wallace (1999) added that people might also want to remain anonymous "to voice their complaints, test out bizarre ideas and identities, ask questions that might reveal their stupidity, or engage in behavior they prefer others would not know about" (p. 240).

Wright (2000b), in a study of computer-mediated support groups for people dealing with health-related issues (e.g., substance abuse problems, eating disorders, cancer, and mental illness), found that the most frequently mentioned advantage of these groups was the perception that there was less stigma attached to one's illness or condition by other online support group members, because of the anonymity of the medium.

In addition, Wright (2000a) found that older adults enjoyed using SeniorNet, because of the ability to obtain feedback about ideas that they felt their closer ties might perceive negatively. For example, participants mentioned enjoying the ability to discuss issues surrounding caring for a spouse who has Alzheimer's disease with someone outside of their family, especially when people wanted to complain about problems associated with caregiving. According to the participants in this study, the Internet provided a network of individuals with whom they could vent their frustrations about caregiving without offending or burdening members of their family.

Negative Aspects of Weak Ties. Although some benefits of computer-mediated support can be explained by weak-tie support network theory, there are obvious disadvantages to weak ties. For example, although reduced social obligations among weak ties have been found to influence positive support functions, they can also cause a number of problems for relationships. One of the problems that computer-mediated communication researchers have identified in online groups is the presence of hostile messages, because of the lack of social presence within this environment (Walther, 1996). Because there are fewer nonverbal cues associated with computer-mediated communication, some people may engage in negative behaviors, such as hostility or deception, because minimal relational obligations means that there is little risk involved (Preece, 1998).

Another problem with computer-mediated support that can be attributed to characteristics of weak-tie networks is the difficulty of forming long-term relationships with people. Wright (2000b) found that online support group participants often found it difficult to seek support from people they had previously interacted with in online support groups, and this was perceived by participants to be one of the major disadvantages of using these groups. Some reasons why it is difficult to locate individuals in online support groups is because of the sporadic use of the group and the tendency for people to change e-mail address or online pseudonyms. In addition, some individuals turn to online support during a specific crisis period, but they will often refrain from using the groups at other times. This can make it difficult for an older individual to contact a specific person when seeking support.

Similarity, Empathy, and Support

Although the Internet may facilitate the development of a more diverse network of relationships than the face-to-face world, computer-mediated communities for older adults may also help to bring people together who are similar in terms of background, demographic characteristics, and attitudes. Because of the ability of the Internet to transcend geographic and temporal space, individuals can form relational networks among people with common similarities regarding very specific concerns. Unlike face-to-face support networks, individuals in the computer-mediated world can conveniently find people dealing with similar issues, despite the fact that they may be thousands of miles away. This may be an advantage to older individuals with limited face-to-face support networks or people who find it difficult to use face-to-face support groups in their community, because of mobility problems.

Perceptions of similarity among relational partners are an important part of relationships, and they influence perceptions of attraction and credibility (Wood, 2000). Similarity is often a key component in the social support process, particularly when individuals communicate empathy or the ability to communicate an understanding of another person's perspective when providing support (Adelman & Frey, 1997; Cline, 1999; Cluck & Cline, 1989). According to Cluck and Cline (1989), "sharing similar crises creates empathic understanding" (p. 313) and may influence positive appraisals of supportive communication.

Communication scholars and other social scientists have long been interested in the study of similarity between relational partners and relationship development (Burleson, 1998; Burleson, Samter, & Lucchetti, 1992; Byrne, 1971; Duck, 1973; Newcomb, 1961). Demographic, attitude, background, and communicative similarity, like other relational perceptions, can be communicated within the computer-mediated environment through a text-based medium, although it may take more time for perceptions to develop than in the face-to-face world (Walther, 1996; Walther & Burgoon, 1992). Early researchers who looked at the relationship between similarity and relational development found that perceptions of similarity are often more important than actual similarities (Newcomb, 1961). This may be significant within the computer-mediated environment, because the reduction of nonverbal cues limits a person's ability to assess actual similarities.

One reason why people may be drawn to online support communities is because they can locate others who might have similar problems or experiences (Braithwaite, Waldron, & Finn, 1999; Campbell & Wright, 2002; Miller & Gergen, 1998; Wright, 2000a, 2000b). Finding other individuals who share experiences online allows people to discuss fears, ask factual questions, and discuss common experiences with their peers. Online groups provide an alternative source of information and support that is typically obtained from a professional, such as a physician or therapist, or a family member (Preece, 1999).

Being able to talk about shared problems can be "cathartic, a vicarious learning experience, and a good way of sharing ideas and of gaining new information" (Preece & Ghozati, 2001, p. 242). According to Preece (1999), people who come from similar backgrounds (i.e., the same family or culture) or who share similar experiences tend to exhibit more empathy toward each other than complete strangers. Levenson and Ruef (1992) claim that the term *empathy* has at least three different meanings. It can mean knowing what another person is feeling, sensing what another person is feeling, or responding compassionately to another person's distress (p. 234).

According to Ickes (1997), our ability to empathize affects the way in which we communicate with others and how comfortable others feel when communicating with us. Our ability to comfort others and communicate deep concern about their problems is an important aspect of providing emotional support, and the ability of an individual to empathize with others has been found to be a crucial skill when comforting people during times of stress (Burleson, 1990). Similarity between individuals may help to facilitate empathy and the construction of appropriate comforting messages.

In terms of how empathy is communicated within online relationships, Preece and Ghozati (2001) found that the majority of messages conveyed within a survey of 100 online support groups were empathic. Participants in these groups viewed the ability to communicate empathy as an important aspect of the support process, and they found it relatively easy to convey empathy despite the limitations of the medium.

In a similar study, Wright (2000b) examined 20 online support groups dealing with various health concerns and found a positive correlation between participant perceptions of similarity and both support network size and network support satisfaction among participants. In addition, perceptions of similarity were related to perceptions of credibility when discussing problems online and perceptions of support satisfaction.

However, Wallace (1999) argued that a negative aspect of locating similar others online is that polarization can occur further exaggerating extreme opinions. She explained, "We could quickly acquire and exaggerated perception of the rightness of our views because we found others who not only agreed with us, but who are even further out on the attitudinal limb" (p. 79).

Changes in Supportive Communication Because of Computer-Mediated Context

Other unique features of computer-mediated support networks can be attributed to characteristics of the medium itself and their influence on perceptions of senders and receivers of supportive messages.

The research on CMC has identified several features of the medium that may affect interpersonal communication. Although reduced nonverbal cues in this environment often cause problems during the initial stages of relationship development, as the number of messages accumulate between relational partners, people learn to compensate for these limitations through the creative use of text-based communication (Matheson & Zanna, 1988; Walther, 1996; Walther & Burgoon, 1992). However, many aspects of computer-mediated communication add to the complexity of relational communication, and these are not very well understood by communication scholars.

For example, Walther (1996) introduced the idea of "hyperpersonal interaction," or a type of relational communication that is facilitated by the features of the computer-mediated environment. Computer-mediated communication allows people to prepare their messages more mindfully. Yet this can also lead to greater manipulation of self-presentation when computer-mediated messages are produced, and possibly more skepticism about the credibility of the source among recipients of computer-mediated messages.

Online group members often develop idealized perceptions of the people with whom they are interacting. Walther (1996) mentioned that because of limited information, participants in online relationships may "fill in the blanks" when it comes to forming perceptions of others. In other words, participants in online relationships frequently develop unrealistic images of their relational partners by projecting images of the partner based on schemas developed in other contexts and idealizing their communicative abilities (e.g., perceiving the person to be more supportive than he or she really is, a better listener to problems, etc.). Walther (1996) also claimed that idealized perceptions are perpetuated in the feedback cycle through an "intensification loop," where confirming messages of each partner reinforce the behavior of the other.

In terms of online support, some limited evidence suggests that the phenomenon of hyperpersonal interaction can influence perceptions of online community participants, as well as perceptions of the support offered within these groups. For example, Wright (2000a), in a qualitative study of an online community for older adults, found that a relatively large number of participants reported that other people in their online support network were perceived as being closer than even members of their own immediate family. Participants mentioned that the people they turned to for online support understood their problems better than non-Internet supporters, despite the fact that they had never met members of their online support network face-to-face.

Although these highly idealized perceptions may actually increase relational satisfaction (Walther, 1996), it is unknown precisely how they influence support appraisals. Future research is needed to assess whether limited nonverbal information in computer-mediated support groups actually influences positive perceptions of those who are providing support and the type of support that is offered. If older adults have idealized perceptions of the support they receive or its provider in this environment, there is a danger that they may perceive the support as more beneficial than other sources of support. If this phenomenon influences people to reduce their reliance on family members and friends (who could better provide certain types of support, such as tangible assistance), then this may limit supportive behaviors or information that can potentially help an older person ameliorate stress during a time of crisis.

Theoretical Implications and Directions for Future Research

The purpose of this chapter was to examine possible benefits and limitations of computer-mediated networks as sources of social support for older individuals, particularly in terms of health outcomes. Although the areas of social support and health, communication, aging, and computer-mediated communication represent diverse areas of interest in the communication discipline, communication theory from research in these areas may shed light on some potential advantages of using computer-mediated communication for social support, as well as many problems with these networks.

Benefits of computer-mediated support networks include access to diverse information and relationships, the ability for seniors to disclose sensitive information in a relatively risk-free environment, and the opportunity to find individuals dealing with similar problems or concerns in a convenient way. Relationships formed through online communities may be an additional source of social support for older adults during times of stress or crisis, particularly when face-to-face friends and family members are not available. In addition, the Internet may facilitate the maintenance of relationships among close friends and family members who are separated by geography. Because there is a paucity of research on interpersonal processes such as relational maintenance through computer-mediated communication, future research would benefit by examining the types of strategies older people use to maintain long-distance relationships via the Internet.

Older adults may benefit in terms of health outcomes by engaging in online relationships even when they are not actively seeking social support for some crisis. The literature on companionship (Rook, 1995) suggests that even relationships that are formed on the Internet based on shared activities and mutual interests may have a positive effect on the psychological well-being of older individuals.

Features of computer-mediated networks may also allow older individuals to engage in different types of supportive relationships (e.g., weak ties) and gain the support benefits that these networks provide. The ability to interact with weak ties allows older people to communicate with others about issues that they may not choose to discuss within closer relationships. In addition, the reduction of nonverbal cues may help seniors develop relationships that have traditionally been hampered by appearance and paralinguistic cues because of the negative stereotypes of older people, and this may facilitate intergenerational relationships based on attitudinal similarities, as opposed to demographic and background similarities. Future research would benefit from examining the extent to which older people engage in intergenerational relationships online.

The ability of the medium to bring distant voices together in an effort to find specific others who are willing to discuss similar issues is one of the greatest strengths of CMC. Online communities for older people, as well as specific forums within these communities, allow older people to find relationships that are tailored to their interests. This specificity may allow people to express empathy better when providing social support, because of the increased similarity among participants. Future research should focus on computer-mediated communication features that promote similarity and their relationship to empathy and other types of support. For example, it would be helpful to assess whether increased similarity between online partners affects perceptions of informational support or other types of assistance.

However, the limitations of the medium might lead to idealized perceptions of relational partners or the inability to locate others quickly during times of heightened stress. In addition, it is unclear how a generation of individuals who were socialized to seek and provide support within face-to-face contexts perceives the quality of computer-mediated social support versus support offered face-to-face. Preliminary evidence from empirical studies suggests that seniors have generally positive perceptions of online relationships (Furlong, 1989; Wright, 2000a, 2000c). However, more research needs to investigate differences between older individuals' perceptions of computer-mediated support and support in more traditional contexts.

Some broader concerns about older people using computer-mediated support networks include issues of access and diversity. At the present time, a relatively small percentage of seniors are using the Internet for interpersonal relationships and social support. Older adults who are using the Internet tend to be relatively well educated and affluent. One disturbing finding from Internet user profile surveys is that less educated and affluent older people, as well as older individuals representing minority groups, are not using the Internet (Adler, 1996; Fox et al., 2001). In addition, the oldest cohort of the aging population (80 and above) appear to be relatively nonexistent on the Internet. Again, this may be largely because of a cohort effect, and in the future we may see more individuals representing this age group using the technology. More research needs to be conducted that examines possible benefits of the Internet for older individuals besides the capability of the medium to increase the size of a person's support network, including the ability to learn new skills or further develop lifelong interests. Older individuals also may benefit from more applied research studies, particularly interventions that examine health-related effects associated with training older individuals to use the technology for support and information about various concerns.

As the number of older adults continues to rise, an already overtaxed and hierarchical health care system may become incapable of meeting accompanying challenges. One possible vehicle to help confront many of these expected obstacles is the integration of technology into the homes and everyday lives of older adults. As noted earlier in the chapter, however, access and personal preference barriers often block technological adoption and use. Rather than consider these as insurmountable forces, it is imperative that health communication and life span communication scholars, as well as practitioners, reframe these "problems" into opportunities for promoting the wellness of many older adults. Ample support for this reconceptualization is provided by Street, Gold, and Manning (1997), who cogently argue that "Interactive technology is perhaps the most promising medium for achieving health promotion initiatives" (p. xii). Whitten and Gregg (2001) further documented the promise of technological innovations in an incisive chapter explicating the development, effects, and implications of telemedicine among older adults. A concerted regional initiative should thus be developed and implemented, underwritten by federal and state monies, as well as private foundations, to develop partnerships with technological firms designed to increase access and lessen personal biases against such technology. Such interventions could then be evaluated in longitudinal, multimethodological, quasi-experimental designs using pre- and posttest measures assessing a variety of health outcomes that shape individuals' quality of life (see Street & Rimal, 1997). Such an undertaking will undoubtedly be daunting; however, potential rewards, such as improved levels of wellness among large numbers of older adults, easily justify accepting such a momentous challenge.

A final issue surrounding research on older adults and computer-mediated support networks is concerns about methodology. Although many surveys of older-adult computer usage have employed large random samples (usually phone interviews), most studies of older adults and the Internet, and studies of online support communities, have relied on convenience samples and online surveys of older Internet users. Although these studies shed some light on the communication patterns of seniors using this technology and the nature of computer-mediated social support, they have several limitations.

First, because of the anonymity of the Internet, it can be difficult to know if respondents to online surveys are misrepresenting themselves. However, the problem of participant misrepresentation in survey research is not limited to online surveys, because people can misrepresent themselves in mailed surveys, phone surveys, and other ways. Researchers can also use each respondent's Internet protocol (IP) address, which is a unique number assigned to each computer along with demographic information to keep track of participants and the potential for multiple response.

In terms of avoiding convenience samples, it is possible for researchers to get e-mail lists of community members from Web community administrators to form a sampling frame for random sampling procedures. Some organizations such as SeniorNet keep e-mail records of every individual who registers to use the community, and administrators may be willing to e-mail an advertisement for a study to all registered participants. However, because only certain individuals may be willing or able to respond to the survey, research may end up with a self-selected sample that may be systematically different from the rest of the population.

The study of computer-mediated support networks for older adults and their relationship to health outcomes is a challenging area on many levels. However, given the trend toward most age groups in society adopting this technology, it is important for communication scholars and other social scientists to assess the impact of this technology on interpersonal relationships, social support, and health outcomes.

REFERENCES

Adelman, M. B., & Frey, L. R. (1997). *The fragile community: Living together with AIDS*. Mahwah, NJ: Lawrence Erlbaum Associates.

Adelman, M. B., Parks, M. R., & Albrecht, T. L. (1987). Beyond close relationships: Support in weak ties. In T. L. Albrecht & M. B. Adelman (Eds.), *Communicating social support* (pp. 126–147). Newbury Park, CA: Sage.

Adler, J. (1996). *Older adults and computer use* [Online]. Available: http://www.seniornet.org/research/survey2.html

Albrecht, T. L., & Adelman, M. B. (1987). Communicating social support: A theoretical perspective. In T. L. Albrecht & M. B. Adelman (Eds.), *Communicating social support* (pp. 18–39). Newbury Park, CA: Sage.

Alexy, E. M. (2000). Computers and caregiving: Reaching out and redesigning interventions for homebound older adults and caregivers. *Holistic Nursing Practice, 14*, 60–66.

Antonucci, T. C. (1990). Social supports and social relationships. In R. H. Binstock & L. K. George (Eds.), *Handbook of aging and the social sciences* (pp. 205–226). San Diego, CA: Academic Press.

Arling, G. (1976). The elderly widow and her family, neighbors, and friends. *Journal of Marriage and the Family, 38*, 757–768.

Arneson, P. A., & Query, J. L. (2001). The case study: Revitalizing a nonprofit health organization. In S. L. Herndon & G. L. Kreps (Eds.), *Qualitative research: Applications in organizational life* (2nd ed., pp. 151–169). Cresskill, NJ: Hampton.

Barnlund, D. (1968). Introduction: Therapeutic communication. In D. Barnlund (Ed.), *Interpersonal communication: Survey and studies* (pp. 612–645). Boston: Houghton Mifflin.

Beckman, L. J. (1981). Effects of social interaction and children's relative inputs on older women's psychological well-being. *Journal of Personality and Social Psychology, 41*, 1075–1086.

Bowers, J. W. (1969). Implications of the New Orleans conference recommendations from the perspective of behavioral scholarship. In R. J. Kibler & L. J. Barker (Eds.), *Conceptual frontiers in speech communication* (pp. 184–189). New York: Speech Association of America.

Braithwaite, D. O., Waldron, V. R., & Finn, J. (1999). Communication of social support in computer-mediated groups for people with disabilities. *Health Communication, 11*, 123–151.

Burleson, B. R. (1998). Similarities in social skills, interpersonal attraction, and the development of personal relationships. In J. S. Trent (Ed.), *Communication: Views from the helm of the twenty-first century* (pp. 77–84). Boston: Allyn & Bacon.

Burleson, B. R. (1990). Comforting as social support: Relational consequences of supportive behaviors. In S. Duck & R. C. Silver (Eds.), *Personal relationships and social support* (pp. 66–82). Newbury Park, CA: Sage.

Burleson, B. R., Samter, W., & Lucchetti, A. E. (1992). Similarity in communication values as a predictor of friendship choices: Studies of friends and best friends. *Southern Communication Journal, 57*, 260–276.

Byrne, D. (1971). *The attraction paradigm*. New York: Academic Press.

Campbell, K., & Wright, K. B. (2002). On-line support groups: An investigation of relationships among source credibility, dimensions of relational communication, and perceptions of emotional support. *Communication Research Reports, 19*, 183–193.

Charles Schwab, Inc. (1998). *Research on seniors' computer and Internet usage: A report of a national survey* [Online]. Available: http://www.seniornet.org/php/default.php?PageID=5474&Version=0&Font=0

Cline, R. J. (1999). Communication within social support groups. In L. R. Frey, D. S. Gouran, & M. S. Poole (Eds.), *Handbook of group communication theory and research* (pp. 516–538). Thousand Oaks, CA: Sage.

Cluck, G. G., & Cline, R. J. (1989). The circle of others: Self-help groups for the bereaved. *Communication Quarterly, 34,* 306–325.

Cohen, S. (1988). Psychosocial models of the role of support in the etiology of physical disease. *Health Psychology, 7,* 269–297.

Cohen, S., Gottlieb, B. H., & Underwood, L. G. (2000). Social relationships and health. In S. Cohen, L. G. Underwood & B. H. Gottlieb (Eds.), *Social support measurement and intervention: A guide for health and social scientists* (pp. 3–25). New York: Oxford University Press.

Cutrona, C. E., & Cole, V. (2000). Optimizing support in the natural network. In S. Cohen, L. G. Underwood, & B. H. Gottlieb (Eds.), *Social support measurement and intervention: A guide for health and social scientists* (pp. 278–308). New York: Oxford University Press.

Duck, S. (1973). Interpersonal communication in developing acquaintance. In G. R. Miller (Ed.), *Explorations in interpersonal communication* (pp. 127–148). Beverly Hills, CA: Sage.

Federal Interagency Forum on Aging Related Statistics. (2000). *Older Americans 2000: Key indicators of well-being* [Online]. Available: http://www.agingstats.gov/chartbook2000/population.html

Finn, J. (1999). An exploration of helping processes in an on-line self-help group focusing on issues of disability. *Health and Social Work, 24,* 220–240.

Fisher, W. R. (1984). Narration as a human communication paradigm: The case of public moral argument. *Communication Monographs, 51,* 1–22.

Fisher, W. R. (1985). The narrative paradigm: An elaboration. *Communication Monographs, 52,* 347–367.

Fisher, W. R. (1987). *Human communication as narration: Toward a philosophy of reason, value and action.* Columbia: University of South Carolina Press.

Flaherty, L. M., Pearce, K. J., & Rubin, R. R. (1998). Internet and face-to-face communication: Not functional alternatives. *Communication Quarterly, 46,* 250–268.

Fox, S., Rainie, L., Larsen, E., Horrigan, J., Lenhart, A., Spooner, T., & Carter, C. (2001). *The Pew Internet and American life project* [Online]. Available: http://www.perinternet.org/

Furlong, M. S. (1989). An electronic community for older adults: The SeniorNet network. *Journal of Communication, 39,* 145–153.

Good, B. S. (1977). The heart of what's the matter: The semantics of illness in Iran. *Culture, Medicine, and Psychiatry, 1,* 25–29.

Good, B. S., & Good, M. D. (1981). The meaning of symptoms: A cultural hermeneutic model for clinical practice. In L. Eisenberg & A. Kleinman (Eds.), *The relevance of social science for medicine* (pp. 165–196). Dordrecht, the Netherlands: Reidel.

Granovetter, M. S. (1973). The strength of weak ties. *American Journal of Sociology, 78,* 1360–1380.

Granovetter, M. S. (1982). The strength of weak ties: A network theory revisited. In P. V. Marsden, & N. Lin (Eds.), *Social structure and network analysis* (pp. 105–130). Newbury Park, CA: Sage.

Harwood, J., Giles, H., & Ryan, E. B. (1995). Aging, communication, and intergroup theory: Social identity and intergenerational communication. In J. F. Nussbaum & J. Coupland (Eds.), *Handbook of communication and aging research* (pp. 133–159). Mahwah, NJ: Erlbaum.

Hegelson, V. S., & Cohen, S. (1996). Social support and adjustment to cancer: Reconciling descriptive, correlational, and interventional research. *Health Psychology, 15,* 135–148.

Hiltz, S. R., & Turoff, M. (1993). *The network nation* (2nd ed.). Cambridge, MA: MIT Press.

Ickes, W. (Ed.). (1997). *Empathic accuracy.* New York: Guilford.

Kahn, R. L., & Antonucci, T. C. (1980). Convoys over the life course: Attachment, roles, and social support. In P. B. Baltes & O. Brim (Eds.), *Life-span development and behavior* (Vol. 3, pp. 253–286). New York: Academic Press.

Kiecolt-Glaser, J. K., Dyer, C. S., & Shuttleworth, E. C. (1988). Upsetting social interactions and distress among Alzheimer's disease family care-givers. *American Journal of Community Psychology, 16,* 825–837.

King, S. A., & Moreggi, D. (1998). Internet therapy and self-help groups—the pros and cons. In J. Gackenbach (Ed.), *Psychology and the Internet: Intrapersonal, interpersonal, and transpersonal implications* (pp. 77–109). San Diego, CA: Academic Press.

Krause, N. (1990). Stress, support, and well-being in later life: Focusing on salient social roles. In M. A. Stephens, J. H. Crowther, S. E. Hobfoll, & D. L. Tennenbaum (Eds.), *Stress and coping in later-life families* (pp. 71–97). New York: Hemisphere.

Kreps, G. L. (1988a). The pervasive role of information in health and health care: Implications for health communication policy. In J. A. Anderson (Ed.), *Communication Yearbook 11* (pp. 238–276). Newbury Park, CA: Sage.

Kreps, G. L. (1988b). Relational communication in health care. *Southern Speech Communication Journal, 53,* 344–359.

La Gaipa, J. J. (1990). Interactive coping: The negative effects of informal support systems. In S. Duck & R. Silver (Eds.), *Personal relationships and social support* (pp. 122–139). Newbury Park, CA: Sage.

Lea, M., & Spears, R. (1995). Love at first byte? Building personal relationships over computer networks. In J. T. Wood & S. Duck (Eds.), *Understudies relationships: Off the beaten track* (pp. 197–233). Thousand Oaks, CA: Sage.

Levenson, R. W., & Ruef, A. M. (1992). Empathy: A physiological substrate. *Journal of Personality and Social Psychology, 63,* 234–246.

Matheson, K., & Zanna, M. P. (1988). The impact of computer-mediated communication on self-awareness. *Computers in Human Behavior, 4,* 221–233.

Mechanic, D. (Ed.). (1983). *Handbook of health, health care, and the health professions.* New York: Free Press.

Mickelson, K. D. (1997). Seeking social support: Parents in electronic support groups. In S. Kiesler (Ed.), *Culture of the Internet* (pp. 157–178). Mahwah, NJ: Lawrence Erlbaum Associates.

Miller, J. K., & Gergen, K. J. (1998). Life on the line: The therapeutic potentials of computer-mediated conversation. *Journal of Marital & Family Therapy, 24,* 189–202.

National Telecommunications and Information Administration. (2000). *Falling through the Net: Toward digital inclusion* [Online]. Washington, DC: United States Commerce Department. Available: http://www.search.ntia.doc.gov./pdf/fttn00.pdf

Newcomb, T. M. (1961). *The acquaintance process.* New York: Holt, Rinehart & Winston.

Nussbaum, J. F. (1985). Successful aging: A communicative model. *Communication Quarterly, 33,* 262–269.

Nussbaum, J. F. (1994). Friendship in older adulthood. In M. L. Hummert, J. M. Wiemann, & J. F. Nussbaum (Eds.), *Interpersonal communication in older adulthood* (pp. 209–225). Thousand Oaks, CA: Sage.

Nussbaum, J. F., Thompson, T., Robinson, J. D., & Pecchioni, L. L. (2000). *Communication and aging.* Mahwah, NJ: Lawrence Erlbaum Associates.

Pierce, G. R., Sarason, I. G., & Sarason, B. R. (1996). Coping and social support. In M. Zeidner & N. S. Endler (Eds.), *Handbook of coping* (pp. 434–451). New York: Wiley.

Preece, J. (1998). Empathetic communities: Reaching out across the Web. *Interactions, 2,* 32–43.

Preece, J. (1999). Empathetic communities: Balancing emotional and factual communication. *Interacting with Computers: The Interdisciplinary Journal of Human–Computer Interaction, 12,* 63–77.

Preece, J. J., & Ghozati, K. (2001). Experiencing empathy on-line. In R. E. Rice & J. E. Katz (Eds.), *The Internet and health communication: Experiences and expectations* (pp. 237–260). Thousand Oaks, CA: Sage.

Query, J. L. (1987, March). *A field test of the relationship between interpersonal communication competence, number of social supports, and satisfaction with the social support received by an elderly support group.* Unpublished master's thesis, Ohio University.

Query, J. L. (1990, June). *A field assessment of the relationships among interpersonal communication competence, social support, and depression among caregivers for individuals with Alzheimer's disease.* Unpublished doctoral dissertation, Ohio University.

Query, J. L., & Arneson, P. (1993, November). *Implications of the caregiving role during Alzheimer's disease: A narrative analysis.* Paper presented at the annual meeting of the Speech Communication Association, Miami, FL.

Query, J. L., & Flint, L. J. (1996). The caregiving relationship. In N. Vanzetti & S. Duck (Eds.), *A lifetime of relationships* (pp. 455–483). Pacific Grove, CA: Brooks/Cole.

Query, J. L., & James, A. C. (1989). The relationship between interpersonal competence and social support among elderly support groups in retirement communities. *Health Communication, 1,* 165–184.

Query, J. L., & Kreps, G. L. (1993). Using the critical incident method to evaluate and enhance organizational effectiveness. In S. L. Fish & G. L. Kreps (Eds.), *Qualitative research: Applications in organizational communication* (pp. 63–77). Cresskill, NJ: Hampton.

Query, J. L., & Kreps, G. L. (1996). Testing a health communication model among caregivers for individuals with Alzheimer's disease. *Journal of Health Psychology, 1,* 335–351.

Query, J. L., Kreps, G. L., Arneson, P. A., & Caso, N. S. (2001). Towards helping organizations manage interaction: The theoretical and pragmatic merits of the Critical Incident Technique. In S. L. Herndon & G. L. Kreps (Eds.), *Qualitative research: Applications in organizational life* (2nd ed., pp. 91–119). Cresskill, NJ: Hampton.

Query, J. L., & Wright, K. B. (2003). Communication competence, social support, and health outcomes for older cancer patients. *Health Communication, 15,* 203–218.

Rawlins, W. K. (1995). Friendships in later life. In J. F. Nussbaum & J. Coupland (Eds.), *Handbook of communication and aging research* (pp. 227–257). Mahwah, NJ: Lawrence Erlbaum Associates.

Rice, R. E. (2001). The Internet and Health Communication: A framework of experiences. In R. E. Rice & J. E. Katz (eds.), *The Internet and Health Communication* (pp. 5–46). Thousand Oaks, CA: Sage.

Rice, R. E., & Love, G. (1987). Electronic emotion: Socioemotional content in a computer-mediated communication network. *Communication Research, 14*(1), 85–108.

Rook, K. S. (1995). Support, companionship, and control in older adults' social networks: Implications for well-being. In J. F. Nussbaum & J. Coupland (Eds.), *Handbook of communication and aging research* (pp. 437–463). Mahwah, NJ: Lawrence Erlbaum Associates.

Rook, K. S. (1987). Social support versus companionship: Effects on life stress, loneliness, and evaluations by others. *Journal of Personality and Social Psychology, 52,* 1132–1147.

Schulz, R., & Beach, S. R. (1999). Caregiving as a risk factor for mortality: The caregiver health effects study. *Journal of the American Medical Association, 282,* 2215–2219.

Schwarzer, R., & Leppin, A. (1991). Social support and health: A theoretical and empirical overview. *Journal of Social and Personal Relationships, 8,* 99–127.

SeniorNet. (2000). *SeniorNet survey about the Internet, April 2000* [Online]. Available: http:// www.seniornet. org/php/default.php?PageID=5472&Version= 0&Font=0

Smith, J. (1998). "Internet patients" turn to support groups to guide medical decisions. *Journal of the National Cancer Institute, 90,* 1695–1696.

Sproull, L., & Kiesler, S. (1986). Reducing social context cues. Electronic mail in organizational communication. *Management Science, 32,* 1492–1512.

Street, R. L., Jr., Gold, W. R., & Manning, T. (1997). Preface. In R. L. Street, Jr., W. R. Gold, & T. Manning (Eds.), *Health promotion and interactive technology: Theoretical applications and future directions* (pp. xi–xxiv). Mahwah, NJ: Lawrence Erlbaum Associates.

Street, R. L., Jr., & Rimal, R. N. (1997). Health promotion and interactive technology: A conceptual foundation. In R. L. Street, Jr., W. R. Gold, & T. Manning (Eds.), *Health promotion and interactive technology: Theoretical applications and future directions* (pp. 1–18). Mahwah, NJ: Lawrence Erlbaum Associates.

Turner, H. A., Hays, R. B., & Coates, T. J. (1993). Determinants of social support among gay men: The context of AIDS. *Journal of Health and Social Behavior, 34,* 37–53.

Wallace, P. (1999). *The psychology of the internet.* Cambridge, UK: Cambridge University Press.

Walther, J. B. (1992). Interpersonal effects in computer-mediated interaction: A relational perspective. *Communication Research, 19,* 50–88.

Walther, J. B. (1996). Computer-mediated communication: Impersonal, interpersonal, and hyperpersonal interaction. *Communication Research, 23,* 3–43.

Walther, J. B., & Burgoon, J. K. (1992). Relational communication in computer-mediated interaction. *Human Communication Research, 19,* 50–88.

Weinberg, N., Schmale, J. D., Uken, J., & Wessel, K. (1995). Computer-mediated support groups. *Social Work With Groups, 17,* 43–55.

Wellman, B. (1997). An electronic group is virtually a social network. In S. Kiesler (Ed.), *Culture of the Internet* (pp. 179–205). Mahwah, NJ: Lawrence Erlbaum Associates.

Whitten, P., & Gregg, J. L. (2001). Telemedicine: Using telecommunication technologies to deliver health services to older adults. In M. L. Hummert & J. F. Nussbaum (Eds.), *Aging, communication, and health: Linking research and practice for successful aging* (pp. 3–22). Mahwah, NJ: Lawrence Erlbaum Associates.

Winzelberg, A. (1997). The analysis of an electronic support group for individuals with eating disorders. *Computers in Human Behavior, 13,* 393–407.

Wood, J. T. (2000). *Relational communication: Continuity and change in personal relationships* (2nd ed.). Belmont, CA: Wadsworth.

Wood, A. F., & Smith, M. J. (2001). *On-line communication: Linking technology, identity, and culture.* Mahwah, NJ: Lawrence Erlbaum Associates.

Wortman, C., & Dunkel-Schetter, D. (1979). Interpersonal relationships and cancer: A theoretical perspective. *Journal of Social Issues, 35,* 120–155.

Wright, K. B. (1999). *Computer-mediated social support, older adults, and coping.* Unpublished doctoral dissertation, University of Oklahoma.

Wright, K. B. (2000a). The communication of social support within an on-line community for older adults: A qualitative analysis of the SeniorNet community. *Qualitative Research Reports in Communication, 1,* 33–43.

Wright, K. B. (2000b). Computer-mediated social support, older adults, and coping. *Journal of Communication, 50,* 100–118.

Wright, K. B. (2000c). Perceptions of on-line support providers: An examination of perceived homophily, source credibility, communication and social support within on-line support groups. *Communication Quarterly, 48,* 44–59.

VII

Senior Adult Education

Glendenning's chapter (chap. 21) works to deny the stereotypes of inevitable cognitive decline with aging, not least by highlighting the methodological flaws in the cross-sectional comparative research that established this claim. He shows how more recent longitudinal research on individuals challenged these earlier findings, and that decline can be attributable to contextual factors and, indeed, can be reversed. His emphasis is on the need to match the lifelong human potential for cognitive development with education that takes place throughout one's whole life. This initiative is in theory enabled by increased life span, early retirement, midlife redundancy, and generally increased leisure time. Yet in practice, as Glendenning shows us, continuing education across the entirety of the life span is a long way from becoming widespread. Various factors contribute to this situation: from older people's negative experience of schooling, to poor provision of adult education, poor marketing of that which exists, and, in particular, attitudes questioning the utility of educating seniors. This last factor Glendenning sees as the crux of the matter. The existential problem, whereby the increase in life expectancy is completely unmatched by clear public understanding of the purpose of the meaning of lengthened old age, lies at the heart of this book's concerns. Widely prevalent ageist public and personal discourses, such as one being too old to learn, are produced within societies that regularly marginalize the rights and needs of older adults, as well as their potential abilities and contributions.

The particular perspective brought to bear on this review is that of critical educational gerontology, which wrangles with the need to afford older learners control and empowerment. Glendenning explains how educational experience is socially constructed, both in terms of the way that learning experiences are organized and the versions of knowledge taught. He proposes an emancipating approach to educating older adults based on the notion that individuals should be given the opportunity to control their own thinking and learning, not least regarding their own position in society. In this framework, education is not a matter of passing knowledge from teachers to students, but a process of group dialogue. Thus informed, older people can then, it is argued, help younger people to learn what it is like to grow old.

Baringer, Kundrat, and Nussbaum (chap. 22) have contributed a new chapter to this handbook that complements the dialogic approach put forth by Glendenning. Instructional communication scholars have long studied the importance of communication within the formal classroom. The great majority of this research has investigated the communication behaviors of teachers in traditional classrooms that are directly or indirectly linked to cognitive, affective, or behavioral outcomes. As previously mentioned, these studies have reinforced the notion of the classroom as a one-way, static communication environment. Baringer, Kundrat, and Nussbaum attempt to point to the existing knowledge within instructional communication as a good starting point with which to understand classroom communication, but point to the challenges and new opportunities that all educators must confront when attempting to educate older adults.

Education for Older Adults: Lifelong Learning, Empowerment, and Social Change

Frank Glendenning
Newcastle Under Lyme

AGING AND INTELLIGENCE

There is now general agreement that both physical and mental activity are essential ingredients for quality of life as we grow older (Belsky, 1988, 1990; Dychtwald, 1990; Evans, Goldacre, Hodkinson, Lamb, & Savory, 1992; Groombridge, 1989; Shaw, 1991). It has, however, been part of the general stereotype about old age that intellectual decline is inevitable and conflicting voices are heard about the psychology of aging. These issues need some clarification.

There is increasing support for the belief that cognitive decline is not inevitable and as Schaie (1990) pointed out, there is promising evidence in the Seattle longitudinal study that in cases where intellectual decline has been shown to exist, it is possible through carefully planned instruction in strategies at the ability level, to reverse the process, and 40% of those who declined significantly over 14 years were returned to their predecline level. The message about inevitable decline, no doubt expected in the early days partly because of the biomedical model, has caused a great deal of harm in supporting the stereotype of aging.

In the late 1960s, the evidence relating to psychological development in early and middle adult life was virtually nonexistent. There was, however, a considerable body of evidence relating to the later years. It was affirmed in the literature that anyone living to retirement age could expect to experience rapid decline in almost every intellectual ability. However, in a sense, this was regarded as good news, because psychologists had previously believed through intelligence testing that decline in mental abilities appeared as early as the age of 25 years for females and after 30 for males. In the 1960s and before, studies had been dominated by the employment of cross-sectional designs. These were based on the principle whereby equal numbers of different age cohorts (e.g., 20–35, 40–55, and 65+) had their performances on psychological or other measures compared at a single point in time. The focus on older cohorts, for example, was employed when norms (or average levels of performance) for younger people were known and researchers were expected to group and subsequently compare the cohorts on the basis of socioeconomic class, gender distribution, and past levels of formal education, ignoring the presence of sociohistorical change. In the 1960s, it was assumed that these were developmental studies, that they provided evidence of the qualitative and quantitative changes in intellectual ability that occurred merely as a result of chronological aging during the life course. Thus, a 70-year-old would be compared on this basis with a 20-year-old, and any variations were attributed to developmental change. Using norms disguised the fact that some in the older cohorts performed as well as some younger subjects and better than many.

To establish these norms in the first place, large statistical samples were required, and these were provided by schools. When psychologists wished to establish what the cognitive abilities of older people were, they also required a large statistical sample. They found this sample among institutionalized elders, rather than among older people who lived in their own homes or in the community. In addition to these cross-sectional designs, there was also an alternative in longitudinal studies. Here, a cohort was followed over a long period of time to measure the intellectual ability of individuals within that cohort as they grew older. Among the difficulties encountered in these studies was a tendency for participants to become accustomed to taking tests over a period of years and for the more able to return for measurements over the years with less coaxing. Nevertheless, results from longitudinal studies did challenge cross-sectional findings.

To refine these research designs, Schaie and his associates developed sequential studies, which by combining cross-sectional and longitudinal techniques, enabled a cross-section of cohorts to be measured and for the same procedure to be repeated at longitudinal intervals. As a result, it has become possible to distinguish between age differences and age change, and to describe the undulating pattern that emerged as "plasticity" (Labouvie-Vief, 1985; Schaie, 1990). This undulating model assumes that when there is evidence of decline in intellectual performance, with appropriate intervention, the decline can be reversed, as mentioned earlier. This suggests that apparent decline is the result of the environmental context in which it takes place. The implication is that there is human potential for cognitive development throughout the whole of life, and whether people's abilities develop or decline as they grow older is not connected with chronological aging.

It is necessary also to refer to another aspect of intelligence. Stuart-Hamilton (1991) noted that early in the century, every type of intellectual task (verbal, numerical, or visuospatial), was believed to be dictated by general intellectual capacity. Most researchers now reject this rigid definition by arguing that intelligence is composed of several interrelated skills. What these are is still open to debate, but Stuart-Hamilton cited Horn and Cattell (1967), who identified two of these special skills, calling them *crystallized intelligence* and *fluid intelligence*. The first refers to wisdom; the second wit.

The participants Horn's and Cattell's study were mostly under 50 and showed little decline in crystallized intelligence, though there was an age-related decline in fluid skills. Schaie (1979) found no decline in crystallized skills, but did not find an appreciable decline in fluid intelligence until the mid-60s. Rabbitt (1984) studied a group of 600 volunteers in their 50s, 60s, and 70s. He also found no significant differences in crystallized intelligence among the three age groups and 10% to 15% of the 70-year-olds were high scorers in the fluid intelligence tests.

What are the implications of this? According to Stuart-Hamilton, they appear to be that crystallized intelligence does not decline and may increase with age and that fluid intelligence does decrease during adulthood. This decrement has been widely reported and is generally accepted. Crystallized intelligence measures the amount of knowledge a person has acquired during his or her lifetime. Fluid intelligence is the ability to solve problems for which there are no solutions derivable from formal education or cultural practices.

As we discuss aging and intelligence, we are referring not to a homogeneous group of people, but to "a very diverse group . . . [who are] seeking to maintain a variety of competent skills and behaviors in various intellectual, artistic, social and political domains of their lives" (Kimmel, 1990, p. 194). Nevertheless, the level of formal education attained in the past by those over 65 years of age plays a significant part in achieving test scores (Belsky, 1990). In the United States, for example, only half of the over-65 population in 1986 were high school graduates, compared with three fourths in the 25 to 65 age group (Belsky, 1990). In Britain currently, nearly 70% of the over 50 population left school at or before 15 years of age, compared with less than 20% for the 20 to 29 age group (Schuller & Bostyn, 1992). Stuart-Hamilton (1991) drew attention to the connection between a decline in fluid intelligence and the amount of full-time schooling received by the person concerned.

As we have seen, we are able to use interventionist techniques to reduce the learning difficulties of some older learners. Shea (1985) suggested that the most successful model is to enable older learners to see someone using a successful technique. As we have moved to an acceptance that over 65s make more use of crystallized rather than fluid intelligence and therefore operate more at a concrete level, it would seem that verbal instructions may be too abstract for older learners, who may have an inability to hold too much in their short-term memory. There is some evidence that older students experience anxiety when they return to learning, and this can be lessened by enabling them to study in a relaxed atmosphere. Some may be experiencing difficulties with sight, hearing impairment, or physical discomfort as well. Responses to new learning may prove to be slower than in younger age groups, so older students need to work at their own pace. This is a major reason why the Open University in Britain, with its distance learning programs, has proved to be so popular with older students.

An investigation some years ago at the Open University demonstrated that "a greater proportion of older students (in the 60–64 age group) obtain marks in the lower pass range of 40–69 percent than the under 60s" (Clennell, 1984, p. 62). In this particular survey, no analysis of reasons for this success was attempted. A later survey of students over 60 in Belgium, Britain, France, and Germany (Clennell, 1990) was based on nearly 4,500 responses to a questionnaire. The evidence showed that manual and low-paid workers were poorly represented, most of these students had a continuous involvement in education, and the majority had experienced disruption in their secondary education as a result of economic and political events in the 1930s, World War II, and the difficulties of the immediate postwar years. They were highly motivated, well organized, hardworking, and adaptable, and between 1982 and 1988 the pass rate for over-60s was similar to that for younger undergraduates.

THE CONTEXT

The context in which I am discussing senior adult education, or education for older adults, is that the increase in life expectancy is one of the most significant achievements of the 20th century. If we take Britain as an example, there were more than half a million people over 75 years of age in 1901 and 57,000 over 85. By the national census in 1981, the numbers had increased to over 3 million and just over half a million, respectively. More than one-third of the population is over 60. In the United States, there were less than 1 million people over 75 in 1901 and 100,000 over 85. By 1988, there were nearly 12.5 million over 75, and nearly 2 million of those were over 85. Global predictions suggest that the over 60s will increase from 416 million in 1985 to 1.1 billion in 2025. Never before have so many people been living so much longer. This raises social policy issues as to how the increased demands on health and social services are to be met and how society plans for such an aging population worldwide.

There are absolutely no precedents. We are inexperienced in caring and planning for so many older people. Few trained social workers in Britain specialize in work with older people. Very many professional carers (e.g., home carers, care assistants, social work assistants) have had little or no training in the care of older people. Those who are trained tend to have a health background (doctors, nurses, health visitors), thus advancing models of care that emphasize the pathological aspects of aging, rather than dealing with normal aging. In this respect, gerontological education (the teaching of gerontology) needs to be undergirded by a strong examination of the stereotypes of aging. It is profoundly necessary for an understanding of the process of normal human aging to become part of the total educational curriculum, so that with education taking place throughout the life course, everyone from childhood onward will be reminded continually that their "quality of life in the later years is a product of a lifetime" (Bond & Coleman, 1990, p. 288).

The pace of social change during this century has been remarkable, and the results of technological developments impinge on us daily. Some of these results have a negative effect. The faster pace of work can increase stress. Family patterns are changing. Patterns of employment have altered radically. Workers are leaving full-time paid work much earlier. Long-term unemployment in midlife has become normative for many millions of people in the developed postindustrial societies. How are they to manage this transition in a positive way and how are they to spend their time for the next 30 years or more?

On the positive side, much of the physical workload has been reduced. Communication systems have changed out of all recognition as a result of electronics and computer technology. With increased life expectancy, facilities for independent living have increased, assisting frail and older people. Leisure time has increased, but there is uncertainty as to whether additional resources to meet this new development will materialize. Attitudes to old age, however, continue to be negative and people's self-conceptions as they grow older are molded by what they see around them, reinforcing a sense of marginalization, a theme to which I will return.

When the U.N. World Assembly on Aging met in Vienna in 1982, with its Western-dominated thinking, it recommended to member states that they develop a bewildering infrastructure of services to meet the new demographic situation, including day centers and day hospitals, outpatient clinics, domestic services, appropriate institutional care, health screening, geriatric clinics, and family and volunteer involvement. It also stressed that education was a basic human right and should be available for all older people. How all this was to come to fruition, particularly in the less developed regions, was not established. The 1991 Declaration on the Rights of Older Persons (Nusberg, 1991) went over similar ground and repeated the U.N.'s conviction that education and training were a basic human right and would enhance literacy, facilitate employment, and encourage informed planning. Tout (1992), however, gave a more realistic appraisal of the opportunities and problems presented by older people in the Third World. He placed emphasis on the reorientation of older people in situations of rapid social and technological change. He wrote of the isolation of the older person who continues to work in a partially vacated agricultural village and the older person without education and technical ability who lives alone in an urban setting. Both need a revised means of ensuring subsistence. Such a situation is in marked contrast to the experiences of adult educators in developed countries.

Of the 9 million people in Britain who are entitled to a state retirement pension, the majority left school at 14 or 15 or younger. For them, schooling and education in general have little relevance. Having been taught by rote and having suffered or witnessed corporal punishment at school 50 years ago was for the majority a negative experience (Abrams, 1981). Such experience, as they grew older, was unlikely to offset the frequently dull image of adult education, which is still regarded as the prerogative of the middle and professional classes (Ward & Taylor, 1986), in a society characterized by inequality (Walker, 1991; Walker & Walker, 1987; Ward & Taylor, 1986). Certainly, some may wish to explore educational opportunities to make good the educational debt they are owed by society, by studying participants of their choice or by catching up on new ideas, but it seems likely that they will be in a minority.

Nevertheless, education is about social change, and this is as true for older people as it is for the younger generations. Moreover, the social tensions experienced by many older people in postindustrial society are related to the social construction of old age, and education can enable many of them to explore the implications of this for their own lives.

SENIOR ADULT EDUCATION AND PUBLIC RESOURCES

Those who have responded to the educational needs of older adults since the 1970s have discovered that policy makers have paid only lip service to the international movement for education for older adults and in remarkably few countries has public funding been made available.

France has been one of the exceptions, because of the 1971 education laws. After the student uprising in 1968, it was decreed that French universities must be open to all. Legislation in 1971 brought into effect a levy of 1% of the salary bill of all firms with more than 10 employees as a contribution toward the national lifelong education program in both the public and the independent sector. One result has been the development since 1973 of the Universités du Troisième Age (U3A), the concept of which has now spread to many countries throughout the world. The presuppositions of the very first U3A in Toulouse were

- To contribute to the raising of the standard of living of elderly people by health building activities, sociocultural activities, and research.
- To contribute to the improvement of living conditions of elderly people through multidisciplinary research . . . and by the dissemination of information.
- To help private and public services and business through cooperative activities in training, information, and applied research. (Phillipson, 1983)

Here is a vision of education, for and about older people, as an agent of social change.

Another exception has been educational development in China, a country with a population of nearly 100 million over 60. During the last decade, as a result of the educational reforms of 1979, there has been a considerable attempt to make provision for the senior age group and an endeavor to find how they can spend their late years in a way that is meaningful to both themselves and society. By 1988, nearly 1,000 universities and schools for the "aged" had been created, with 130,000 enrolments (Du Zicai & Liu Pingsheng, 1989). Since 1988, education for Chinese elders has spread rapidly, well beyond these figures, and in 1991, Shanghai alone was estimating 200,000 students by the year 2000, and Wuhan projected 44,000 students. These programs offer courses in new agricultural methods, foreign languages (especially English), social development and reform, keeping fit and health care, calligraphy, and subjects promoting "the socio-cultural life of the elderly rather than education for jobs" (Du Zicai & Cheng Yuan, 1992, pp. 2–7).

These attitudes toward education for older people in France and China are different from those in the United States and in Britain, for example, where the tendency is to see education simply as a means of improving the quality of life of older people. There have been unsupported claims that education meets older people's real needs, which was one of the themes of the 1981 White House Conference on Aging, that education keeps people out of institutional care (Peterson, 1983; Weinstock, 1978), and that it favorably affects the health and performance of older people (Jones, 1982). But this is not the whole story. All these positions have depended on anecdotal evidence. But Moody has, since 1976, been asking intermittently and strenuously, why should we promote late-life learning (Moody, 1976, 1986, 1987, 1988)? For some, this has now become the critical question, because unless it can be answered, there is little likelihood that we will, in the majority of countries, progress the issue of economic resources for late-life learning programs. As Moody (1987) put it: "As the mood of liberal hopes cooled (in the 1980s) there was little understanding of the new political and economic environment in which adult education programs would exist in the foreseeable future" (p. 197). He went on to cite four reasons or problems that presented themselves: (a) the economic problem—high unemployment among older workers and lack of retraining for them; (b) the social service problem—excessive dependency on social services that cannot be met for financial reasons and lack of staff training; (c) the political problem—the lack of empowerment of older people and their inability to participate in decision making that affects them; and (d) the existential problem, whereby the increase in life expectancy is completely unmatched by any clear public understanding of the purpose or meaning of lengthening old age.

Indeed, as this chapter was being written, more than a century of British liberal adult education was slowly but surely being negated because of the educational dogmas now being legitimized by the New Right. Liberal adult education in Britain has for over 100 years offered courses in the humanities and the sciences, designated as nonvocational courses (expressive education or learning for learning's sake). This has traditionally been partially funded by the government, until legislation was introduced in 1988 leading to the end of direct funding for liberal adult education and the gradual contraction of such provision. Part-time adult student fees, for example, rose 15.7% in 1991–1992 and many courses have to be self-financing if they are to continue. By the end of the decade, it seems unlikely that in the majority of the universities in England and Wales any vestige of part-time liberal adult education will remain. Although there is evidence of money being more readily available to support training and skills-centered courses, it is likely that the whole adult education sector in Britain faces obliteration in due course.

With the marginalization of older people in society, and after more than 20 years of striving in both the United States and in Britain to establish a place for education for older adults on the social policy agenda, there seems to be sufficient evidence to demonstrate that this exercise is fruitless and education for older people has only slogan status. It is not on the political agenda, nor will it be in the foreseeable future. The Mondale Bill on Lifelong Learning proposed in 1976 never received basic U.S. federal funding. The British government, when approached by a deputation, refused to fund senior adult education in 1984 and in 1993, the Under Secretary of State for Further and Higher Education informed a recent U3A deputation "that direct funding [for U3A] from central government was unlikely" (*Third Age News*, 1993, p. 1).

An exception to the rule in senior adult education is the Pre-Retirement Association of Great Britain and Northern Ireland (PRA), which has been in acceptance of government grant aid since 1983. Its primary task is to maintain professional standards in relation to preretirement education through a code of good practice and the building up of a register of professionally trained tutors. The PRA has distanced itself from the traditional courses of the 1970s and 1980s that were characterized by and overloaded with the giving of information. It has moved instead into the provision of courses on positive change management, with the study of a raft of issues that are related to changing circumstances in later life, including the use of time, the experience of financial change, health needs, and changes in the patterns of relationships.

With adult education under attack in Britain, it is not surprising that senior adult education is being ignored. We saw the growth of social consciousness in the 1940s, 1950s, and 1960s in Britain, in the 1950s and 1960s in the United States, and in the 1960s and 1970s in France and Germany, but this proved to be short-lived. The free market approach to the economy has resulted in the reduced functioning of central or federal governments in a variety of matters that relate to the well-being of members of society. The welfare state is being systematically dismantled on the grounds of its increasing cost.

Nevertheless, there appears to be evidence that the number of older adults engaging in educational activities is increasing. In the United States, the Louis Harris National

Survey of 1975 reported that 2% of older people were enrolled in courses. This figure had grown to 5% by 1981 (Harris & Associates, 1981). In Britain, in 1980, Abrams (1980) found that 2% of older people participated in formal adult education at any time during a 12-month period. The same results were found by Midwinter (1982, p. 19). However, there has been a considerable increase, as shown in the Schuller and Bostyn (1992) report to the Carnegie U.K. Trust. The 2% for the over-60s changed to 9% for the over-65s and 14% of those between 55 and 64 years of age. Schuller and Bostyn estimated that some 755,000 over-50s participate in formal education and 675,000 in training opportunities. This total of nearly 1.5 million over 50s in some kind of formal education or training course compares with a further 1.5 million who are involved in informal learning (by which the authors mean systematically finding out about something (Schuller & Bostyn, 1992). These indications are then that participants in formal and informal senior adult education who are over 50 amount to about 18% of the population, and those in formal education and training amount to about 9%. In view of what has been said previously about the rapidly changing situation in the adult education sector in Britain, the future level of enrollments remains uncertain.

THE NATURE OF LATE-LIFE LEARNING

Whether late-life learning comes through the private sector, as with Elderhostel or the Senior Center Humanities Program in the United States, Schools and Colleges for Seniors in Australia, and Saga Holidays in Britain, or whether it comes through the voluntary sector with various manifestations of self-help programs, including the U3A, or whether it comes through statutory provision, education for older adults can make a considerable contribution to social change.

Phillipson (1983), for example, suggested that "education is a means of clarifying social and political rights in old age" (p. 19). Groombridge (1982) wrote

> The first task of education is to arouse self-awareness, rather than to provide content, to enhance the consciousness of the elderly in relation to themselves and to their social setting, to strengthen their self-esteem and to encourage their questioning or hidden aspirations. (p. 318)

Allman (1984) suggested that

> If through encouraging learning in the later years, we see the added opportunity for real-locating some value to people's mental labour thereby affecting the quality and meaning of their lives, then we can argue that the way in which learning and thinking is organised is extremely important. The organisation of those learning experiences must allow the individual to regain control over what is produced or created. (p. 87)

Older students then must be left in control of their own thinking. Phillipson, Groombridge, and Allman were all really suggesting that education viewed in this way leads to empowerment and this in itself can be dangerous to those who favor a structured curriculum that transmits socially produced knowledge. The relation between knowledge, power, and control is one that requires critical study, especially in relation to older adults, many of whom may be realizing for the first time that the knowledge they have acquired over the years has been socially constructed.

There has been a tendency throughout the debate about education for older adults to treat them as a homogeneous group. This is an error, because we are in reality describing more than one generation. Even if we start with persons over 60, rather than over 50 as some do, we can still be speaking of people who will live another 20 or more years and whose backgrounds and experiences may be profoundly different, in relation to education, social class, gender, and ethnicity. In addition, there was a tendency in the 1970s and 1980s in both Britain and the United States for some adult educators to affirm that education was self-evidently good, could improve the quality of life of older people, and even helped to diminish the effects of aging. Older people were perceived as needing assistance to cope with the years of retirement from full-time paid work. McClusky's (1974) hierarchy of needs—coping, expressive, contributive, influential, and transcendence needs—is a good example and it has been discussed by a number of commentators, both favorably and critically (Battersby, 1985; Glendenning, 1985, 1992; Lowy & O'Connor, 1986; Peterson, 1983). Battersby (1985) challenged what he perceived as "the middle-class notions which perpetuate 'medicinal' images of education for older people" (p. 78), where their needs can be remedied by educational programs. The vocabulary of adult education also treats older people as if they were a disadvantaged group along with those with mental illness, blindness, and deafness (Phillipson, 1983). The result has been that older people have been marginalized in relation to educational policy.

Radcliffe (1982), writing from a Canadian perspective, warned: served? Why do we need education for older adults? What are the power relationships involved in our work with older people? Where do the ideas that we embody in our work come from historically? How did we come to adopt them? Why do we continue to endorse them? How do these ideas influence the way in which we relate to older people? Such questions have been asked very rarely over the years. Indeed, Moody (1987) concluded that "we have more educational opportunities for older people but no better understanding about why and opportunities should exist in the first place" (p. 6). The normal response has been that self-evidently education is a good thing for all people, a claim based entirely on anecdotal evidence.

Moody (1988) suggested, just as Phillipson (1983) and Allman (1984) had done in Britain, that what was required was "a rationale more in tune with the political and economic realities of the years ahead" (p. 198). This would involve the retraining of older workers (suggesting the importance of lifelong education for economic survival), education in coping skills (to offset a public service strategy of dependency), a move toward self-reliance and self-sufficiency, education for citizenship and empowerment (there have been few efforts to encourage serious education or training for citizen participation among older people), and liberal arts education. Moody, however, pointed out that although education for self-fulfillment and expressive education have a powerful claim as a humanistic ideal, it has a weak claim on public resources. Moody (1987) concluded that all these programs for older people remain "far outside the mainstream of higher education . . . (and are) confined to an elite group of elders who have been fortunate enough to have access to learning opportunities" (p. 6).

The retraining of older workers, which Moody mentioned, is necessary to update obsolescent knowledge and skills. In general, we need to be more flexible about retraining, which will be required increasingly in the years to come in the postindustrial societies. Workers will change jobs much more frequently as work patterns change. There was much discussion in the 1980s about the need to retrain older workers, strongly supported by researchers in the International Labor Organisation (Plett & Lester, 1991) and by Waldman and Avolio (1983), who demonstrated that work performance shows little deterioration with increasing age. Older workers were advised by Plett and Lester (1991): "Older workers interested in continuing their working life and in keeping up with technological changes and a chance for the good jobs must assume some of the responsibility for getting the skills and knowledge training they need to remain competitive in an ever-changing world economy" (p. 13).

The implication of this is that the traditional notion of a lifetime career is likely to be superseded by a series of work opportunities, for each of which fresh training will be required. But such proposals were not heeded by the vast majority of British employers, who believed that the employment of older workers was less efficient and too expensive, given that older workers would remain in employment for a shorter time than younger workers (Laczko & Phillipson, 1991).

Although there has been a strong tendency in postindustrial societies to emphasize the importance of instrumental education (vocational education resulting in practical results) over expressive education (learning for learning's sake and self-fulfillment), opportunities for training have not been extended automatically to those over 50 years of age. We are then back in the area of age discrimination in relation to all forms of education for older adults. The critical task is to convince policy makers that learning of all kinds introduces us to a wide range of activities, making it possible for older adults to take charge of their own development (Allman, 1984; Freire, 1972; Moody, 1987). Our inability to achieve this, and to convince policy makers that every human being has the right to achieve their own potential, has forced older people in many parts of the world to take matters into their own hands and we have seen the formation of the self-help education for older adults movement, a topic to which I return later.

Because of the marginalization of older people, they tend to be robbed of their self-worth by being denied a meaningful role in society. Moody (1978) pointed out how Freire learned from his work in education for literacy among Brazilian peasants. There was a persistent block in their learning to read as a result of a deep-rooted negative self-image and a feeling that they could not learn or acquire the "high culture" of literacy. What they did know from their experience they felt was also worthless. The parallels with many older people are significant. Many older people say: "I'm too old to learn," or are bewildered by the technological revolution: "My experience is obsolete and worthless." Moody (1978) responded: "These false messages cannot be overcome simply by putting older people in a classroom and filling them up with new information" (p. 41; cf. Coupland & Coupland, 1993, p. 281).

Elsewhere, Moody (1988) stressed the significance of older people recovering "their identity as culture bearers and culture creators" (p. 259). In saying this, he reminded us of the importance of storytelling and reminiscence, which in traditional cultures contributed to positive respect and dignity for old age. Reminiscence is about placing value on memories. It can help to preserve a sense of identity. Indeed, reminiscence is now regarded "as the 'new' means of preserving mental functioning in old age" (Coleman, 1993, p. 158). But not all elderly people find reminiscence a positive experience. Some find reminiscence helpful because it enables them to find continuity in life. For others, it can be painful to recall unhappy events from the past. Coleman believed it is essential to pay due attention to each person's situation and experience before we embark on the reminiscence process. The majority of the evidence about the value of reminiscence is anecdotal, but reports are consistently positive. Greater participation and socialization and improvement in self-esteem and behavior are all stressed.

POPULAR CULTURE, THE MEDIA, AND EDUCATION

In the postindustrial societies, observers have for years been drawing our attention to the weakening of traditional forms of popular culture. With retirement, leisure time increases, and with so-called early retirement, opportunities for leisure can be seen to extend for 30 or 40 years. Television, radio, and the press have invaded our homes and begun to erode the boundaries between public and private life. Old people spend more time watching television than any other age group. In fact, the most recent figures available in Britain (Schuller & Bostyn, 1992) suggest that the over-65s on average watch television for 36.5 hours a week. This in turn suggests that basic weekly working hours have been almost replaced by hours of watching television, to fill the emptiness of social life and to act as a bridge between the end of full-time work and the end of life. Moody (1988) responded to such a situation by remarking that "television represents the domination of late-life leisure through one-way communication that reduces the last stage of life to silence" (p. 242). It is obvious that older people with a good educational background are likely to be selective and use the educational and learning opportunities that radio and television provide to reinforce and sustain their own cultural capital. But for many, the pervasive, latent effect of the negative images of old age on the screen, or over the air, are subconsciously and continually reinforced by the rapid, one-way, undifferentiated images and stereotypical language that accompanies them. This confirms many older people in their sense of marginalization.

All this is a far cry from those cultures that revere the old and see old age as a time for self-fulfillment, reflection, and the handing down of wisdom. Many aspects of rapid technological change may subvert such a situation, and when this occurs, what was once an essential part of collective memory is eroded. Continuity between the generations is damaged. The concentration by so much of media presentation on youth culture and premiddle age intrudes on the relationships between old and young.

Images in the historical memory of old and young differ, as do their experience of the life course. Those born before the invention of jet engine, computers, electronics, space travel, and mass worldwide travel and before the existence of multiethnic societies, especially in Europe, have witnessed change on a hitherto unknown and unprecedented scale. But it is possible for those who are so inclined to take advantage of information technology, for example, learning how to adapt it to their own personal needs.

What is certainly true is that since the 1970s in some postindustrial societies, there has been a steady growth of educational and cultural programs designed specifically for older people or designed on an intergenerational pattern, conforming more comfortably to mainstream adult education. Both patterns exist and both are valid. The dangers of isolation implied by separate provision for older people are real enough, but they must be balanced by the fact that older people often wish to spend time with their peers. The times when intergenerational programs take place may be inconvenient and those who are experiencing anxiety at returning to a learning situation after many years may find it easier to join groups of people of similar age or experience. Moreover, the collective term *older people* does not mean that they are a homogeneous group. As I said earlier, in terms of age, they span at least one generation. Some of these educational programs are provided on a commercial basis as has been already noted, particularly in the United States, Australia, and Britain; others are part of mainstream adult education study programs, but a great deal has developed on a self-help basis as well.

EDUCATION FOR OLDER ADULTS IN THE SELF-HELP MODE

The self-help movement for senior adult education is now a worldwide phenomenon. In very many cases, educational institutions have not provided appropriate opportunities and older people have taken the matter into their own hands, driven to do this because of unsuitable course design, high fees, inadequate facilities, unsuitable premises, timing of classes, and transportation. Self-help is not a new concept. Its historical roots have been examined in numbers of studies including Katz and Bender (1976), and Pancoast, Parker, and Froland (1983), and also specifically in relation to education (Erskine & Phillipson, 1980; Glendenning, 1985; Midwinter, 1984; Moody, 1988) There is a general pattern of mutual aid groups being set up for a particular purpose to bring about personal or social change. In relation to older people, self-help groups provide an increasing confirmation that older people are becoming more vocal in their endeavor to clarify their needs and attain their goals.

Certainly, there has been discussion about the implication of self-help in the face of the statutory authorities refusing or being unable to provide particular services and in effect avoiding social change when social change is desirable. The effect of the policies of the New Right will ensure that this is increasingly the case. It may be said then with some discomfort in this respect, but with congratulatory recognition as well, that self-help is here to stay. Moody (1988) estimated that in the United States in the mid-1980s, there were half a million self-help groups with more than 15 million people involved in a wide variety of activities. A proportion of these would be dedicated to education for older adults, involving a variety of academic subjects, as well as groups to study coping skills, peer health care, paralegal matters, or practical skills training for those who are bereaved, divorced, or separated, especially older women. An obvious example is the success of U3A, which has already been mentioned and has spread from France in different forms to many countries, including Argentina, Australia, Belgium, Britain, Canada, Finland, Germany, Mexico, Poland, Slovenia, Spain, Sweden, the United States, and elsewhere.

The first U3A was founded in Toulouse, France, in 1973. At the outset, the U3A groups were university oriented, and the term *Third Age* referred to the third stage of life (after childhood and work)—retirement from full-time paid work and prior to the "Fourth Age" of dependency. As the years have passed, U3As, both in France and elsewhere, have abandoned their early homogeneity and have developed as nonprofit companies or self-help charitable organizations, concerned chiefly with organizing study groups in the humanities for retired people. There are over 200 groups in Britain and in excess of 40,000 members. Other self-help groups outside U3A have also developed for those older people who, for a variety of reasons, do not wish to participate in mainstream adult education courses.

The existence of self-help education raises issues regarding standard setting and quality control. There is no rigorous policy of building evaluation into these self-help (education for older adults) groups. Unless there are clear aims and objectives, and stated professional educational standards, then the self-help groups run the risk of being accused of offering educational events that are substandard, even though they use tutors who are known to be experts, are retired themselves, and, at the most, receive only their expenses. It is surprising that the concept has not received more research attention since the first U3A was founded.

The rehearsal of the success of education for older adults is always anecdotal. Older learners, it is said, tend not to be interested in degrees and qualifications. In Britain, there is little reliable evidence to prove or disprove this view. The Open University research report on "Older Students in Adult Education" (Clennell, 1987) found that of its non–Open University respondents, who were over 60, only 14% were taking courses that led to an examination. There was a similar finding in a more recent report (Clennell, 1993): "The main pattern of learning activity is continuing education for personal interest" (p. 8).

Several thousand students over 60 are enrolled for degree courses at the Open University, and it is clear from experience (rather than evaluation) that learning in later life does enable older people to enhance the meaning of their lives through interactive learning occasions, which often lead to practical outcomes and social activities. They should be enabled to do this with the best tools available.

THE AIM AND PURPOSE OF EDUCATION IN LATER LIFE

The difficulty that continues to face adult educators who are concerned with education for older adults in Britain and North America is that of arguing that education for self-fulfillment should be provided by public resources when policy makers do not see the value and insist that it must be self-financing. At the same time, the practical, vocational type of course does not easily commend itself either. Moody (1988) suggested that if we can enlarge our vision of education for older adults with

> An image of education in late life as the systematic development of human capacities, if we can design programs to address problems of increasing literacy, improving health care, maintaining nutrition, retraining people for jobs, enlisting the talents of volunteers, helping younger generations, and contributing to the common good of society, then there is far less trouble selling the idea . . . to legislators, to departments of higher education, to the taxpayers and to the public. (p. 212)

The rhetoric is fine, but education for older people is still looked on as a frill, as inessential, and not worth spending money on.

For this reason, it is worth asking, because I have already suggested that education for older adults is not even on the political agenda, whether even Moody's approach goes far enough and whether in fact we do not need to examine in its entirety what passes for education for older adults or senior adult education. Programs in general are aimed at middle-class older adults who are financially secure and who have already benefited from the educational system. The nuances that accompany this type of education, for example, the language used and spoken or the type of instruction given, have been those that have alienated certain groups from education for many years. Learning any subject involves learning a language that is appropriate to that subject. Language also is learning social relations—how to perform properly and acceptably—so that we can reach our social destination. The type of instruction given is often tied to a set course and as often as not tied to particular textbooks. Knowledge is what the teacher teaches or what is in the books the teacher requires students to read. For example, the selectivity and interpretation of events in history textbooks is designed to create a conformity in the national presentation of history. In other words, we need to acknowledge that there is hegemonic control over what we learn and believe.

It is important then to ask, and to have answered, the questions: Whose interests are being served by this educational process? Are we saying that education should be perceived as a commodity that someone else has, or that is acquired through the medium of a lecture or a seminar? Are we saying that education should be perceived as being what someone else teaches you, what you learn from a book or from an activity someone else designs? Education should be about self-awareness and gaining power over one's life. Education may also be about a number of other things. It may be a vehicle for retraining or orientation to technological change; it may relate to self-fulfillment and the reinforcement of a sense of purpose and identity; and it should be an important mechanism for individual and group empowerment. Such philosophical elements are missing from much of the conventional wisdom about the nature and purpose of education in later life (Battersby & Glendenning, 1992).

THE NEED TO SHIFT THE PARADIGMS

Glendenning and Battersby (1990) proposed a paradigm shift away from the functionalist approach where older people are seen as a social problem to a sociopolitical framework that examines society's treatment of older people within the context of the economy and the state. Educational gerontology is currently shackled by this functionalist paradigm.

Battersby and Glendenning (1992) pointed to Estes (1986) as a starting point. She suggested that there was a need to develop an understanding of how economy and society in advanced capitalism impacted on older people and created enormous variations in their lifestyles and culture. She also proposed the need for an analysis of how the role of the state, the conditions of the labor market, race, class, and gender create age divisions in society: "the status and resources of the elderly, and even the experience of old age itself, are conditioned by one's location in the social structure" (Estes, 1986, p. 121).

The implication of this is, first, that we need to disentangle the complex sociological and economic elements that have contributed to the marginalization of older people by society and the structured dependency within which society has encapsulated them. Later, Estes (1991) reiterated her position: "Social policy for the old will continue to be a major battle ground on which the social struggles presently engulfing capitalist states are fought" (p. 68).

Second, it is necessary to move from educational gerontology to what Glendenning and Battersby (1990) called *critical educational gerontology* (CEG). Critical educational gerontology does not have its roots within the functionalist paradigm, but in critical theory. First of all, it challenges the notion that education for older people is neutral. By locating education for older people within a sociopolitical framework, CEG seeks to disclose the falseness of conventional paradigms that are conservative in outlook and in practice lead to the further domestication of older people, as they learn a structured adaptation to their environment. It aims to highlight whose interests might be served by educational policy and provision in later life. Overtly, CEG attempts to acknowledge and expose the feelings of frustration, lack of power, and helplessness experienced by older people, as they feel their independence and autonomy escaping from their control whether for economic or health reasons, or for reasons of structured dependency. CEG would attempt to uncover the factors that prevent older people from taking control of, and influencing, those decisions that crucially affect their lives. CEG is certainly concerned with the moral dimension of education and the transformation of those conditions that promote the disempowerment of older people. CEG concentrates also on "Why do what you are doing?" rather than "How do you do what you are doing?" questions and stresses the need of the individual for autonomy, freedom, and creativity. This would involve questioning the place of power and control in the pursuit of knowledge in later life and take into account the importance of life experience as a learning process (Battersby & Glendenning, 1992).

Third, the principles of CEG include such concepts as emancipation, empowerment, transformation, social and hegemonic control, and consciousness raising (Freire's term is *conscientization*). Allman (1984), like Moody, drew attention to the work of Freire showing how he provided deep insight into the way in which traditional forms of education are oppressive, relying on what he always called "the banking concept," whereby knowledge is simply deposited in the minds of learners. In turn, this domesticates the learner, preventing him or her from realizing his or her full potential. But if people are given the opportunity to control their own thinking and learning, significant transformations can take place in both the individual and in society as a result of this liberating form of education, which is currently not available to the majority of older adults (Glendenning, 1997).

Fourth, CEG is predicated on the notion of praxis—theory based on action. This becomes the practice of CEG, which Glendenning and Battersby (1990) called *critical gerogogy* (CG). Allman (1984) relied on a CG approach in which she discussed the importance of dialogue in educational practice with older adults:

> Dialogue requires people to be engaged in the process of questioning their existing knowl-
> edge. It also requires them to have a genuine desire to share group members' meanings,

ideas and feelings. Therefore dialogue presupposes attitudes such as mutual respect and trust of each member for every other member, so that genuine equality exists within the group experience. (p. 85)

To achieve this dialogue and facilitate communication, openness, trust, and commitment must develop within the interpersonal relationships of the group. The meanings, ideas, and feelings communicated by each individual require investigation. The group works collectively to explore each person's thinking, and this helps individuals to question or reflect on what they know or think they know. Considered in this way, "learning experiences during the later years of life have a potential to do more for an older person than simply providing mental activity.... Self-help group learning provides an opportunity for individuals to realise their full developmental potential" (Allman, 1984, p. 86).

Central to CG also should be an attempt to unsettle the complacency or the apathy that older people feel about their social conditions in the community and in wider society. They have been conditioned to believe that they are an insignificant force, if a force at all, and certainly powerless to effect any social change. But central to CG are the educational activities for older people that promote critical reflection and action and contribute to consciousness raising. Of all educational activities, as I have suggested, reminiscence work is perhaps the most interesting example of how educational activity can be an enabling process that emancipates older people, helps them to gain power over their lives, promotes personal growth and self-awareness, and enables them to assert themselves (Battersby & Glendenning, 1992). Considered this way, CEG and CG have the potential to reshape the current thinking and practice concerned with education in later life.

EDUCATION FOR OLDER PEOPLE AND SOCIAL CHANGE

When an individual faces threats to his or her survival because of low income, ill health, or inadequate housing, it is clear that access to education could also enhance coping skills. The key to this is the strengthening of an individual's sense of personal control and mastery over her or his own life situation (Lowy & O'Connor, 1986; Ruiz, 1993). Such empowerment enables individuals to understand themselves and the world around them better. It even calls into question the way in which policy markers arrive at their decisions. Individuals in society should not be seen merely as recipients of services. Midwinter (1992) suggested that "older people are uncritical consumers from the standpoint of fighting for their rights and demanding value for money, and when it comes to criticism of either commercial or public providers, this age group is the most uncomplaining of all" (p. 8). This statement underlines the significance of the why questions.

Older adults are a largely untapped source of potential. Society needs to realize that it can make a significant contribution to productivity with education programs for retired people. In the report for the National Council on Aging, Harris & Associates (1981) revealed that 75% of retired people would have preferred to be in part-time or full-time work. This underlines the earlier argument for programs that help older adults update their skills or learn new skills for new jobs. Realism, of course, accepts that unless employers abandon their stereotype of older workers, progress will not be made, but it would make sense to test out such a program in practice.

A major way in which older people can promote social change is through the democratic process. Not only do they form a significant proportion of the electorate, but society has much to gain and to learn if their participation in democratic processes is based on informed knowledge, which education can provide. Again, with the graying of society, how can existing and new policies and institutions serve the needs of the increasing numbers of older adults? Only by learning what it is like to grow old, and seeking the practical cooperation of those for whom policies and services are intended. Social change implies conflict, which in turn may require mediation. Society's institutions and services exist for the benefit of the members of a particular society. For this reason, we need to be able to identify a sense of purpose for that society. It is this that postindustrial society demonstrably lacks, a deprivation that owes much to our inability to think in life-developmental terms and our failure to listen to life experiences. Lifelong learning in its fundamental sense could help to redress the balance (Glendenning, 2000).

ACKNOWLEDGMENT

The author acknowledges his indebtedness to Professor David Battersby, Dean of the Faculty of Health Studies, Charles Sturt University, New South Wales, Australia, with whom he has discussed these issues for some years.

REFERENCES

Abrams, M. (1980). Education and the elderly. In *Research perspectives on ageing* (pp. 1–11). Mitcham, UK: Age Concern Research Unit.

Abrams, M. (1981). Education in later life. In *Research perspectives on ageing* (pp. 1–19). Mitcham, UK: Age Concern Research Unit.

Allman, P. (1984). Self-help learning and its relevance for learning and development. In E. Midwinter (Ed.), *Mutual aid universities* (pp. 72–89). London: Croom Helm.

Battersby, D. (1985). Education in later life. What does it mean? *Convergence, 18*(1–2), 75–81.

Battersby, D., & Glendenning, F. (1992). Reconstructing education for older adults: An elaboration of the statement of first principles. *Australian Journal of Adult and Community Education, 32,* 115–121.

Belsky, J. K. (1988). *Here tomorrow: Making the most of life after fifty.* New York: Ballantine.

Belsky, J. K. (1990). *The psychology of aging* (2nd ed.). Pacific Grove, CA: Brooks/Cole.

Black Report. (1982). In P. Townsend & N. Davidson (Eds.), *Inequalities in health* (pp. 39–233). Harmondsworth, UK: Penguin.

Bond, J., & Coleman, P. G. (1990). *Ageing and society: An introduction to social gerontology.* London: Sage.

Clennell, S. (Ed.). (1984). *Older students in the Open University.* Milton Keynes, UK: Open University, Regional Academic Services.

Clennell, S. (Ed.). (1987). *Older students in adult education.* Milton Keynes, UK: Open University, Regional Academic Services.

Clennell, S. (Ed.). (1990). *Older students in Europe.* Milton Keynes, UK: Open University, Regional Academic Services.

Clennell, S. (Ed.). (1993). *Older students and employability*. Milton Keynes, UK: Open University, Regional Academic Services.

Coleman, P. G. (1993). *Ageing and reminiscence processes*. Chichester, UK: Wiley.

Coupland, N., & Coupland, J. (1993). Discourses of ageism and anti-ageism. *Journal of Aging Studies, 7*(3), 279–301.

De Zicai, & Liu Pingsheng. June (1989, July). *A research into the laws governing education for aged people in China*. Paper delivered at the World Congress of Gerontology, Acapulco, Mexico.

Du Zicai, & Cheng Yuan. (1992). *Newsletter No. 1*. Wuhan, China: Institute of Research on Elderly Education, Wuhan University for the Aged.

Dychtwald, K. (1990). *Age wave*. New York: Bantam.

Erskine, A., & Phillipson, C. (1980). Self-help, education and older people. In F. Glendenning (Ed.), *Outreach education and the elders: Theory and practice* (pp. 103–113). Stoke-on-Trent, UK: University of Keele, Beth Johnson Foundation and Department of Adult Education.

Estes, C. (1986). The politics of ageing in America. *Aging and Society, 6*(2), 121–134.

Estes, C. (1991). Retrospective view of C. Phillipson's "Capitalism and the construction of old age (1982)." *Ageing and Society, 11*(1), 67–68.

Evans, J. G., Goldacre, M. J., Hodkinson, M., Lamb, S., & Savory, M. (1992). *Health: Abilities and wellbeing in the third age* (Research Paper No. 9). Dunfermline, UK: Carnegie U.K. Trust.

Freire, P. (1972). *Pedagogy of the oppressed*. Harmondsworth, U.K.: Penguin.

Glendenning, F. (Ed.). (1985). *Educational gerontology: International perspectives*. London: Croom Helm.

Glendenning, F. (1992). Educational gerontology and gerogogy: A critical perspective. In C. Berdes, A. A. Zych, & G. D. Dawson (Eds.), *Geragogics: European research in gerontological education and educational gerontology* (pp. 5–21). New York: Haworth.

Glendenning, F. (1997). Why educational gerontology is not yet established as a field of study: Some critical implications. *Education and Ageing*.

Glendenning, F. (2000). *Teaching and learning in later life: Theoretical implications*. Aldershot, U.K.: Ashgate.

Glendenning, F., & Battersby, D. (1990). Educational gerontology and education for older adults: A statement of first principles. *Australian Journal of Adult and Community Education, 30*(1), 38–44.

Groombridge, B. (1982). Learning, education and later life. *Adult Education, 54*, 314–327.

Groombridge, B. (1989). Education and later life. In A. M. Warnes (Ed.), *Human ageing and later life* (pp. 178–191). London: Arnold.

Harris & Associates. (1975). *The myth and reality of aging in America*. Washington, DC: National Council on the Aging.

Harris & Associates. (1981). *Aging in the eighties: America in transition*. Washington, DC: National Council on the Aging.

Horn, J. L., & Cattell, R. B. (1967). Age differences in fluid and crystallised intelligence. *Acta Psychologia, 26*, 107–129.

Jones, S. (1982). *Learning and meta-learning with special reference to education for the elders*. Unpublished doctoral dissertation, University of London.

Katz, A. H., & Bender, E. (1976). *The strength in us*. New York: Watts.

Kimmel, D. C. (1990). *Adulthood and aging* (3rd ed.). New York: Wiley.

Labouvie-Vief, G. (1985). Intelligence and cognition. In J. E. Birren & K. W. Schaie (Eds.), *Handbook of the psychology of aging* (2nd ed., pp. 500–530). New York: Van Nostrand Reinhold.

Laczko, F., & Phillipson, C. (1991).*Changing work and retirement*. Milton Keynes, UK: Open University Press.

Lowy, L., & O'Connor, D. (1986). *Why education in the later years?* Lexington, MA: Heath.

McClusky, H. Y. (1974). Education for aging: The scope of the field and perspectives for the future. In S. Grabowski & W. D. Mason (Eds.), *Learning for aging*. Washington, DC: Adult Education Association of the USA.

Midwinter, E. (1982). *Age is opportunity*. London: Centre for Policy on Aging.

Midwinter, E. (Ed.). (1984). *Mutual aid universities*. London: Croom Helm.

Midwinter, E. (1992). *Citizenship: From ageism to participation*. (Research Paper No. 8). Dunfermline, UK: Carnegie U.K. Trust.

Moody, H. R. (1976). Philosophical presuppositions of education for old age. *Educational Gerontology, 1*(1), 1–16.

Moody, H. R. (1978). Education and the life cycle: A philosophy of aging. In R. H. Sherron & D. B. Lumsden (Eds.), *Introduction to educational gerontology* (pp. 31–47). Washington DC: Hemisphere.

Moody, H. R. (1987). Why worry about education for older adults? *Generations, 12*(2), 5–9.

Moody, H. R. (1988). *Abundance of life: Human development policies for an aging society.* New York: Columbia University Press.

Nusberg, C. (1991). UN takes action on principles for older persons. *Ageing International, 18*(1), 3–6.

Pancoast, D. L., Parker, P., & Froland C. (Eds.). (1983). *Rediscovering self-help: Its role in social care.* Beverly Hills, CA: Sage.

Peterson, D. A. (1983). *Facilitating education for older learners.* San Francisco: Jossey-Bass.

Phillipson, C. (1983). Education and the older learner: Current developments and initiatives. In S. Johnston & C. Phillipson (Eds.), *Older learners: The challenge to adult education* (pp. 19–30). London: Bedford Square.

Plett, P. C., & Lester, B. T. (1991). *Training for older people.* Geneva, Switzerland: International Labour Organisation.

Rabbitt, P. M. A. (1984). Memory impairment in the elderly. In P. E. Bebbington & R. Jacoby (Eds.), *Psychiatric disorders in the elderly* (pp. 101–119). London: Mental Health Foundation.

Radcliffe, D. (1982). Third age education: Questions that need asking. *Educational Gerontology 8*, 311–323.

Ruiz, C. (1993). Facing a short future (Special issue). *Ageing International, 20*(1), 5.

Schaie, K. W. (1979). The primary mental abilities in adulthood: An exploration in the development of psychometric intelligence. In P. B. Baltes & O. G. Brim (Eds.), *Life-span development and behavior* (Vol. 2, pp. 67–115). New York: Academic Press.

Schaie, K. W. (1990). Intellectual development in adulthood. In J. E. Birren & K. W. Schaie (Eds.), *Handbook of the psychology of aging* (3rd ed., pp. 291–309). San Diego, CA: Academic Press.

Schuller, T., & Bostyn, A. M. (1992). *Learning: Education, training and information in the third age* (Research Paper No. 3). Dunfermline, UK: Carnegie U.K. Trust.

Shaw, M. W. (1991). *The challenge of ageing* (2nd ed.) Melbourne, Australia: Churchill Livingstone.

Shea, P. (1985). The later years of lifelong learning. In F. Glendenning (Ed.), *Educational gerontology: International perspectives* (pp. 58–80). London: Croom Helm.

Stuart-Hamilton, I. (1991). *The psychology of ageing.* London: Kingsley.

Third Age News. (1993, Summer) Education minister sympathetic to needs of U3A. *30*. London: Third Age Trust.

Tout, K. (1992). Debt, duty, and demands in third world ageing: The role of education. *Journal of Educational Gerontology, 7*(1), 25–35.

Townsend, P. (1981). The structured dependency of the elderly: Creation of social policy in the twentieth century. *Ageing and Society, 1*(1), 5–28. U. N. Department of International Economic and Social Affair. (1986). *Periodical on Aging.*

Waldman, D. S., & Avolio, B. S. (1983). *Enjoy old age.* New York: Norton.

Walker, A. (1991). *Social and economic policies and older people: National report—United Kingdom.* Brussels, Belgium: Observatory of the Commission of European Communities.

Walker, A., & Walker, C. (Eds.). (1987). *The growing divide: A social audit 1979–87.* London: Child Poverty Action Group.

Ward, K., & Taylor, R. (1986). *Adult education and the working class: Education for the missing millions.* London: Croom Helm.

Weinstock, R. (1978). *The graying of the campus.* New York: Educational Facilities Laboratories.

Instructional Communication and Older Adults

Doreen K. Baringer, Amanda L. Kundrat* and Jon F. Nussbaum
The Pennsylvania State University

LEARNING AS A LIFELONG PROCESS

As we age, we give thought and attention to the more noticeable changes that occur across the life span. These changes may include physical aspects, such as the graying of our hair, the addition of wrinkles, and random aches and pains. We notice changes in popular culture, such as musical genres, transportation trends, and the introduction of new technologies. In addition, the changes that occur with regard to our financial concerns are also noticeable—as teenagers we may have saved our money for the newest 8-track or stereo, whereas now we concern ourselves with our children and, perhaps, preparing for our retirement.

Less obvious to us may be the more invisible changes that take place, such as modifications in our motivations, shifts in relational dynamics, and the adoption of innovations. Some of the subtlest changes may be those that involve the acquisition, use, and dissemination of knowledge. We gain knowledge through the process of learning. The ongoing developmental process of learning is not restricted to the formal classroom context but is experienced in our everyday activities. Learning how to surf the Internet, refinish a cabinet, or read a newspaper are all types of learning experiences. Learning is necessary to the human condition so that we may exist and thrive in an ever changing environment. As Edwards and Usher (2001) so eloquently stated, "[Learning] is not, as many would see it, a secure ground on which to stand but is better understood as a process of constant traveling that is never completed and where destinations are always uncertain and constantly changing" (p. 284).

---

*deceased

Although years of formal education have shaped our learning experiences, so too do the interactions and experiences we engage in on a daily basis. Whether these events involve listening to a college-aged child discuss his or her own classroom experiences, learning how to create a Web page, or learning how to play a new game, we learn as we live. These daily occurrences are opportunities for growth, which enable us to revise, update, or reteach information we had never experienced, learned incorrectly, or have long forgotten. These experiences shape our attitudes, beliefs, and values about context-specific information, and influence our interpersonal interactions with those with whom we communicate. Teachers and students can learn from the older learners in the classroom, and these intergenerational interactions can inform and enhance our traditional classrooms.

Information gained within and outside of the classroom can directly and indirectly influence our abilities to age successfully. An older adult returning to the classroom not only benefits from the experience, but the younger students and instructors in the classroom benefit as well. Older adults bring with them information and experiences that their younger classmates do not have. The presence and participation of these older adults may influence the instruction and communication that takes place in the classroom. To aid instructors in the teaching of older learners, this chapter provides insight into the various changes that take place as we age and how these changes must be considered in the process of enabling older students to learn. We begin this chapter by reviewing the motivations for older adults to return to formal schooling. Next, we review psychological and physical aspects of older adults that affect learning ability. The final and most substantial portion of this chapter is dedicated to informing teachers as to how they can aid older adults in maximizing their learning potential. We discuss teaching in the classroom context with suggestions of effective methods, inclusive of the uses of media and technology, for educating and sustaining the motivations of older learners. We discuss classroom relational concerns that may influence the communication and dynamic of the classroom; these concerns are power and compliance, immediacy, and communication apprehension. Finally, we suggest techniques and strategies that can be used effectively by instructors in the classroom context to enhance learning for adult students.

MOTIVATION TO LEARN

Millette and Gorham (2001) define motivation is a "force or drive that influences behavior to achieve a desired outcome" (p. 141). Individuals are compelled for various reasons, inclusive of subject matter, and these motivations may change across the life span. With respect to formal education, younger individuals are more homogeneous than older individuals in their reasons for learning in college. They are often motivated to learn because of pressure from parents, competition with classmates, and the consequences of failure (Knowles, 1984). Traditional-aged college students may attend an institution of higher education because it is the norm. They observe their friends applying to colleges, believe their parents expect them to attend, or believe that they need to attain a higher degree to be able to pursue the career or economic bracket they desire. They may be influenced by any or all of these factors.

Older adults, on the other hand, are motivated to return to college when they feel a need to know or do something or to improve their lives (Knowles, 1984), whether for survival, such as retaining a job or gaining a promotion, or for self-actualization. They may be especially motivated at turning points in their lives (Knowles, 1984). Such turning points may include life changes such as divorce, children leaving home, and retirement. John (1988) gave various reasons why older adults may wish to learn in a formal setting. These include contributing to society, enhancing quality of life, improving mental and physical health, reducing economic problems (by remaining active and self-sufficient for a longer amount of time), and providing society with creative products. For example, one who wants to inspire others and benefit society simultaneously may enroll in an adult literacy–training course to teach others how to read. Another example is a person who wants to learn about staying physically fit. She or he may attend a twice a week aqua-aerobic class at a local YMCA. There are as many reasons for learning as there are individuals.

When recognizing motivations to learn, attention needs to be drawn to the agist tendency, yet mistake, of treating older individuals as a homogeneous group (Glendenning, 1995). Individuals from various cohorts (50s, 60s, 70s and over) are often categorized together as "older adults." In fact, the U.S. Bureau of the Census (2000) reported group statistics for those currently in college by using age categories such as 14 to 17, 18 to 19, 20 to 21, 23 to 25, 29 to 32, and then those over 35. This last group may include six decades of individuals that are treated as one. The mistake of treating older individuals as a homogeneous group is magnified when one considers the various contextual ages of those individuals in different age groups. Contextual age best reflects what a person has experienced during his or her chronological years of life. It is a "transactional view of aging that incorporates physiological, psychological, social, and communication influences on life-position" (Rubin & Rubin, 1986, p. 30). As life experiences shape individuals, and as individuals in generations become older, the more different from one another they may become. Thus, older adults are a heterogeneous group who have considerably more diverse backgrounds than do younger individuals (Graham, 1989).

This diversity in older individuals again reminds us of the various motivational factors that this "homogeneous" group has. The distinction that "older adults" are more heterogeneous than they are treated is not denying that there are cohort effects. Those of similar chronological ages do share similar experiences with those who witnessed the same political events, just as they share the technology of their time, as well as music and fashion preferences. Nonetheless, it is important for older individuals to be treated individually, as they do have diverse life experiences. When older adults return to school, they may or may not have attended an institute of higher learning. Even for those who have experienced higher learning, education and its institutions have continually changed throughout the decades. In addition, these individuals often have held jobs or maintained a career for years and have lived through unique experiences, which have shaped their interests or sparked a desire to explore various subjects.

The various motivations to learn are related to the learning orientations of individuals; just as motivations change over life, so may learning orientations. These two elements, motivations and orientations, are different for older adults than for younger individuals (Knowles, 1984). Knowles (1984) indicated that differences in learning are because younger individuals are interested in mastering subject matter. This can be seen in the desire to pass the class to advance to another class and ultimately in attaining a degree, enabling one to pursue a career. He contrasted this to the orientations of adults, who may be oriented to learn because of something that happened in their lives or a task they need to accomplish. For example, we recently read a newspaper article about a student who received a scholarship from the American Institute of Professional Geologists. After losing his job because of downsizing, he enrolled in university classes at the age of 32 to fulfill his lifelong dream of becoming a geologist. Others who may have dropped out in their teens or early 20s may have feelings of regret and the desire to finish what they once began. Teachers return to the classroom as students to continue learning and to maintain certification.

Houle (1961) identified three specific orientations that produce the desire to learn. These categorical orientations are goal, activity, and learning. Cross (1981) described the characteristics that accompany these various motivational orientations. Goal-oriented learners strive for achievement of concrete objectives, such as the ability to use more technology. An older adult goal-oriented learner may have encountered a need to understand technology in the workforce or else to more easily communicate with family members. The second learning orientation that Cross (1981) discussed are activity-oriented learners. These learners use the educational environment as an opportunity to interact with others. Individuals may be striving to have a new activity where they can meet new people. The third learning orientation that Cross (1981) identified is the learning-oriented individual. These individuals aspire to gain knowledge to develop and enhance their worlds. They learn for the sake of learning. Perhaps they were involved in a conversation that prompted a desire for them to learn more about a subject or had always had an interest in a subject they had never had time to explore. Smith (1990) described the learning orientation as being a lifelong, developmental process that expands an individual's attitudes, understanding, and skills. This enables individuals to become better learners, may cultivate critical thinking and enable spiritual growth, and is enhanced through formal and informal schooling. The more we engage in learning, the more enlightened we become. Formal and self-directed learning has the opportunity to enable these processes through structure and self-direction. Hately (1987–1988) declared that "through reaching out, people come to understand themselves in new ways, by adding information to their life contexts, by embracing a larger world, geographically, historically, intellectually" (p.42). In other words, learning enables us to understand our worlds through our search for meaning. This psychological process, related to our motivations, ultimately affects spiritual well-being (however one defines it), a concept that correlates strongly with physical health (Hately, 1987–1988).

As discussed, motivations to learn vary between younger individuals and older adults. Older adults' motivations are heterogeneous, as individuals considered "older adults" have diverse life experiences. In addition, motivations to learn change across the life span. Just as motivations change with age, the psychological and physical capabilities of individuals change as well.

PSYCHOLOGICAL AND PHYSICAL ASPECTS

We now turn our attention to psychological and physical aspects that may influence and regulate the ability to learn. There are age-related psychological changes, such as those in working memory, processing speed, name retrieval, and intelligence (Belsky, 1990; Glendenning & Stuart-Hamilton, 1995; Nussbaum, Hummert, Williams, & Harwood, 1996; Nussbaum, Pecchioni, Robinson, & Thompson, 2000; Salthouse, 1988; Schaie & Willis, 1996) that can sometimes make learning more challenging for older adults. These challenges are expanded by age-related physical changes such as a reduction in reaction time, vision, and hearing that takes place as we age beyond 30 years (Cross, 1981; Nussbaum et al., 1996, 2000).

Decline in working memory and processing speed can affect an individual's ability to receive and produce messages, which hinders the ability of older individuals to learn (Nussbaum et al., 1996). Although memory problems can occur in either long-term (tertiary) or short-term (secondary) memory systems, or possibly both, the meanings of words and ideas remain intact for older adults (Nussbaum et al., 1996; Wlodkowski, 1999). Although it may take longer for older students to process complex sentences and think of appropriate words when encoding messages than it would for younger students, they can still accomplish such tasks. An older adult may have trouble remembering names yet vividly remember an experience from childhood. The ability for older adults to memorize lists or highly complex subject matter decreases, perhaps because of declines in processing speed (Wlodkowski, 1999).

Older adults more often use organizational strategies when processing information, as opposed to younger students who place and remember unrelated words into categories (Schaie & Willis, 1996). A report on metacognitive differences (Justice & Dornan, 2001) informs us that older learners self-reported that they are more likely to use higher level cognitive study strategies, hyperprocessing, and the generating of constructive information than are younger adults. These applied, sophisticated strategies may be beneficial in that they may enable increased comprehension and integration of information. Hyperprocessing involves "extra processing of difficult or challenging materials" such as taking extra notes for material older learners perceive as difficult. Generation of constructed information involves "elaborating, reorganizing, or integrating information" (Justice & Dornan, 2001, p. 240). Older learners are also more likely than younger adults to do memory and comprehension checks to determine that they understand information (Justice & Dornan, 2001). For example, when learning a new theory, an older adult may jot down extra notes for some of the concepts he or she finds difficult. He or she may also think about it for an hour after class, reread the notes before going to bed, mentally or verbally elaborate on the concepts, and apply the information to past and present experiences to comprehend the theory more fully.

With regards to relevance, older adults are also able to process relevant material more adequately than material that is meaningless (Wlodkowski, 1999). For example, a primary point one of the authors was taught in an adult-literacy training course was to select reading materials that are of interest to the person learning to read. Adults who are learning to read are more likely to continue and see the benefits of the program if the materials are relevant to their needs and goals. Strategies that older adults do not spontaneously employ for remembering information are mnemonic devices, or classification strategies. It is important for teachers to know that when instructed to categorize unfamiliar words to remember them, older adults' performances improve considerably (Schaie & Willis, 1996).

Age may be an advantage in learning, because self-awareness has had the opportunity to develop over a longer period of time and may be the reason that older adults are found to reach performance criterion in fewer tries than younger people (Peterson, 1983). However, the need for more time to complete tasks is a disadvantage for the older student, especially when retrieval of information is necessary (Peterson, 1983). Much of this process can be linked to students' listening behaviors, which are influenced by previous knowledge and experiences. How students listen influences how well (or not so well) students remember (see Chesebro, 2002). "Interference" occurs when previous knowledge and the material being presented conflict (Peterson, 1983). Lower scores from older students may have nothing to do with cognitive abilities, but from the cohort in which she or he is a member. If a task requires categorizing and the older adult has not been instructed to do so, he or she may answer questions involving the task incorrectly. The adult may have a lower educational level, less supportive learning environment, or lack of previous educational challenges (Wlodkowski, 1999).

Older learners also suffer from greater test anxiety than do younger students. Some of this anxiety can be attributed to the debilitating fear that one's "mind is not what it used to be," thus becoming a barrier to recalling information (Belsky, 1990, p.130). The perception of memory problems may be a barrier to learning (Justice & Dornan, 2001). If older adults perceive that they will not test well, this may prove to be a self-fulfilling prophecy, resulting in an actual decreased test performance. In the next section of this chapter, we will discuss strategies and provide suggestions for assisting adult learners, but for now we turn to some of the physical changes of aging that may affect adult learners.

The physical changes that take place as we age can also affect adult learning. Competencies in formal classrooms are often based on abilities that focus or at least assume certain physical abilities. These may include mobility, vision, hearing, and health. Motor skills decline across the life span, causing us to perform tasks at a slower rate. However, it is important to note that another explanation for this tendency of older adults to work at a slower rate is their attempt to minimize potential error (Birren, 1996). For example, they make take an additional 30 minutes or longer to carefully review an assignment before submitting it in class. This contributes to their preference for more feedback on tasks than younger adults require (Peterson, 1983). An excellent example of the need for feedback is portrayed in the 1983 film *Educating Rita*, in which Rita, an older learner played by Julie Walters, consistently requests detailed feedback on her essays from her English professor, Frank (played by Michael Caine). Rita wants to know the "what" and "why" that contributes to poorly written essays, and, more important, "how" her essays can be improved. Although the older learner in this film had no mobility problems that hindered learning, nonetheless, her desire for explicated feedback took time. Extra time would likely have been needed if she had mobility problems.

In real life classrooms, decreased mobility of older students may contribute to learning difficulties. Some students may experience decreased mobility because of declines in vision and hearing. As a general guideline, vision is at peak performance at the age of 18, after which it begins to decline until the age of 40. Decline becomes even sharper between the ages of 45 and 55, after which it once again returns to a slow but steady rate of decline (Kidd, 1973). Loss of vision is the result of our eyes beginning to decrease in elasticity and transparency, and of our pupils becoming smaller and not reacting as they did when we were younger (Cross, 1981; Nussbaum et al., 2000). Other ailments such as glaucoma or cataracts hinder older adults and need to be considered with the adult learner. Hearing loss, which gradually declines until the age of 65 or 70, also contributes to declines in mobility because of increased problems with pitch, volume, and rate (Cross, 1981; Kidd, 1973). This decline could also occur earlier in the life span as the result of working in high-noise environments (i.e., bridge construction, day care, etc.). Hearing problems are related to deciphering problems, as rapid speed and high pitch become difficult to translate (Wlodkowski, 1999).

In addition to vision and hearing impairments, the overall health of a student will contribute to his or her ability to learn. Researchers have reported that physical declines are also attributable to lifestyle choices (i.e., diet, exercise, types of accidents, and possible vices, Peterson, 1983) and chronic illness (Belsky, 1990). Physical losses contribute to cognitive inability in the processing of information and, as a consequence, may result in the need for more time to react and process information (Belsky, 1990; Cross, 1981).

Vision, hearing, anxiety, and health all have the ability to affect communication and learning in the classroom, but are by no means exhaustive contributors to declines in mobility (e.g., self-concept, etc.). Employment and relationships with others can also affect communication and hinder learning in the classroom. Older students who may be simultaneously caring for older parents and younger children may have less time to devote to studies, which may result in decreased learning or participation in class discussions. Likewise, outside pressures from a job, such as being on time for a meeting, may detract the learner's attention and perhaps force him or her to miss class time. With such responsibilities, older learners' mental and physical resources may be additionally depleted.

More psychological aspects related to aging include self-esteem and self-efficacy. Past learning experiences can decrease participation in class discussions and learning. Negative experiences with former teachers, or past classroom atmospheres that were strict in nature, may cause students to feel anxious or reticent. The perception of large gaps in knowledge may induce older learners to worry about performance issues and contribute to their fears of participation. In addition, older students who lack self-efficacy may have more learning difficulties; if they do not believe they have the skills or abilities to perform assigned tasks, they may be less likely to attempt them or possibly drop out of the class. There are steps, however, that instructors can take to make older learners more comfortable in a classroom. We will now focus our attention on strategies and suggestions for communication in the classroom.

INSTRUCTIONAL COMMUNICATION

Pedagogy, a term frequently introduced to first-time instructors for the college classroom, is the basic study of how to teach. Instructors who study this "art and science of teaching" learn such things as how to structure and present lessons, solicit student feedback, encourage student participation, and assess and critique student performance (Walkin, 1990, p. 175). Two additional terms, not usually introduced to instructors, are *andragogy* and *geragogy*. *Andragogy*, first defined by Knowles (1980), is "the art and science of helping adults learn" (p. 6). This idea of different ages having different learning preferences, advanced by Knowles, was further developed when the term *geragogy*, defined as the "process involved in stimulating and helping the elderly persons to learn" (John, 1988, p. 12), was introduced. The terms *pedagogy*, *andragogy*, and *geragogy* call to instructors' attention that a student's learning preference, orientation to learning, readiness and motivation to learn, and preferred teaching techniques may change throughout the life span. As a result, instructors need to be aware that for instruction to be effective, they may need to adapt their teaching to benefit older adults.

Motivation

Teaching methods can encourage, maintain, or hinder the motivations of older learners (Wlodkowski, 1999). Hence, special attention should be devoted to learning precisely how to continue and sustain the motivation of older individuals. Andragogy differs from pedagogy with the assumption that as individuals age they become more psychologically independent. Pedagogical conjectures are realistic for young, dependent individuals. However, as we grow older, our experiences become a resource to draw on. As we age, our sense of self becomes more independent and we begin to feel more responsibility for our decisions. Knowles (1984) stated that younger students are dependent on the instructor for learning, but older adults learn best if they are encouraged to be more self-directing. Such self-directing is related to enabling older learners to draw on life experiences to enhance their learning, and to encouraging them to reflect on what they want to learn (Knowles, 1984). Knowles (1984) drew attention to the idea that "young children derive their self-identity largely from external definers . . . as they mature, they increasingly define themselves in terms of the experiences they have had. . . . To children, experience is something that happens to them; to adults, their experience is *who they are* . . . in any situation in which adults' experience is ignored or devalued, they perceive this as not rejecting just their experience, but rejecting them as persons" (p. 60, italics in original).

Further, the reasons why older adults want to learn more about a given subject is motivated more intrinsically and is less dependent on perceptions of what an educator may want them to learn. These motivations are present whether we are consciously or unconsciously aware of them. Older adults returning to the classroom may face a challenge to their independence and control. Individuals grow to be independent in life, but as adults return to the classroom, they may fall back on the educational experience of their youth. This change to being dependent may cause an internal conflict for returning adults who feel the need to be independent. To foster the control older learners may be consciously or unconsciously lacking, they should be encouraged to take responsibility for their education by choosing methods and resources by which to learn, and such methods should be included in the evaluation of their performance (Knowles, 1990). Increased learning is the result of encouraging older adults to set their own learning goals and evaluation criteria (Patterson & Pegg, 1999).

Instructors can stress the importance of personal goal setting at the beginning of the semester and encourage it through activities such as making a personal statement of what the student wishes to get out of the class. A copy of this personal statement can be given to the instructor, who would then be aware of the goals of the older adults. Through this personal statement, students can be given the opportunity to define how they wish to evaluate themselves. The criteria for grade evaluation may already be determined; however, this exercise acknowledges to students that people have various goals and that all are valid. The student can thus be in control by evaluating his or her learning goal, enabling him or her to be more motivated during the course.

Earlier in this chapter, we noted that adults can return to school for a range of reasons. To explore how the different motivations may manifest themselves in the students' goal setting, imagine that there are two individuals: one motivated by a promotion if she or he attains a college degree and another motivated to explore a topic that has always interested him or her. These individuals may have different goals in the classroom. The individual present to attain a college degree may have the goal of passing the course with a C, just learning enough to be able to do so. The other one, present to explore the material, may have the goal of understanding the material. These two very different goals can affect the instruction of the course. A student who is attending class just for the C and is not particularly interested in the material may not participate in class discussion, whereas someone who finds the material intriguing may ask many questions, contribute to lively discussions, or perhaps bring newspaper clippings to share with classmates. Also, with regard to evaluation, a younger student may ask the instructor, "What do I need to score on the final to keep a C in this class?" An older student may ask, "Can you make recommendations of other books that address this topic? It may provide me with more information in understanding the concepts." We have used these examples for explanatory purposes and are aware that there are also younger students in our classes who may ask the latter question. Nevertheless, the main issue we wish to convey is that different motivations and goals influence communication that occurs in the classroom (Knowles, 1984). The older adult who wants to learn the course material for self-actualization may ask more questions of the instructor than the student who only wants to remember the information until the final exam. This influence of motivations and goals reinforces our earlier discussion of self-direction. We suggest that instructors create a multidimensional learning environment within the classroom that will enable both younger and older students to flourish.

Wlodkowski's (1999) discussion and suggestions of his culturally responsive motivational framework is especially relevant for older students. Although age groups are not generally thought of as separate cultures per se, there are issues that are more relevant to older than younger adults. Wlodkowski's (1999) framework, which consists of four motivational conditions that positively influence learning, provides a useful structure in any attempt to enhance motivation in older learners. A first suggestion grounded in this framework is to establish an atmosphere of inclusion where students respect and connect to each other. This requires a collaborative environment in which students share their experiences, express concerns, and discuss their expectations, and this can be done by placing students in small groups. A second strategy in Wlodkowski's framework is that of developing attitude, which can be accomplished by asking students what they want to learn to promote interest in the material. The final two steps in the framework are those of enhancing meaning and engendering competence. Enhancing meaning is accomplished by asking students to form questions and ideas related to the material. Engendering competence is accomplished by asking students to complete self-assessments—reflecting and writing about their perceived learning (Wlodkowski, 1999). These ideas are complementary and reinforce our earlier discussion of Knowles' (1984) explication of andragogy and how older learners benefit from self-direction.

Life Experience

Younger learners have fewer life experiences than do older individuals, and so lectures, assigned readings, and audiovisual demonstrations may be highly effective with this group (Knowles, 1984). These methods of instruction provide information and examples to students who may otherwise have no experience. Older adults, however, have a wider variety of life experiences that can be incorporated into the classroom. Instructors can draw on the life experiences of adult students in a number of effective ways: by allowing them to share examples of their experiences with the class, and encouraging them to think about how those examples relate to class information. For example, a 51-year-old retired marine officer who has returned to higher education may have the potential to contribute much to an organizational class from his or her past experiences. Describing for a class how the Marines, as an organizational unit, have changed within the last two decades can provide younger students with a firsthand account of how organizations and the people who form them change over time. There would be political reasons and an individual perspective as a unit member. Such stories would tie in well with the course material, and may facilitate a lively and possibly even humorous discussion. The benefit of such a contribution may be that even when an example and discussion are forgotten, other students may remember the "flavor" of the material.

Instructors can also help adult learners by using other life experiences that the older student may have had in the "real world." Peterson (1983) suggested that we present course material in concrete ways that are personally meaningful for our older students. Information that is related to real-world experiences and personally relevant is more easily understood and remembered. Because older adults are accustomed to real-world experiences, the older students may have more difficulty with abstract components but have unlimited patience for concrete elements. For example, students who are older may be unwilling or unable to deal competently with abstract components such as shapes or colors, but have more patience for concrete elements such as vegetables or specific beverages. For example, when describing an object as octagonal and red in color, one could also say it resembled the shape of a stop sign. Associating the information with a real-world item makes it more quickly and easily understood. Teachers should use real-world scenarios or stories as examples as often as possible rather than consistently explicate in the abstract. Rather than explaining a theory only in the abstract, one should apply it to a situation that students can easily understand and identify. For example, rather than explaining uncertainty reduction only in abstract terms, one could also provide a personal or public story that may enable older students to understand and remember the theory.

When using real-world examples, instructors need to be aware of the fact that many older adults spend more time socializing or communicating with their peers than with younger adults. As a result, they are likely to be less familiar with the jargon of pop culture. Therefore, instructors should be sensitive to this possibility and use examples and stories to which older adults can relate, drawing from their life experiences. For example, although younger students may identify with the use of the Wall Flowers in an example, older students may identify with performers such as Bob Dylan. To older students, the "Wall Flowers" may conjure up images of homely, single women at a high school dance rather than a band. An instructor can use this example competently by including enough background information (i.e., informing students that the band is called the Wall Flowers) so that older students are also able to follow the story.

In addition to using familiar icons and life experiences, it is important for instructors to be aware of language choice when instructing older adults as compared with younger adults. In the same way that popular culture changes, language changes over time (John, 1988). For example, use of the term *phat* in the context of something "being cool" would likely cause an older adult learner to misunderstand the conversation, as they are accustomed to the word *fat* meaning being physically large. Say, for example, that an 18-year-old student gives a presentation about an island she visited during winter break. She shows the class a picture of herself on the beach with her friend. A student in the class remarks, "Wow . . . really phat." An older student may perceive the student's remark as extremely rude, interpreting *phat* for *fat*. As an extreme example, this illustrates changes in vernacular that may pose communicative misunderstandings for older learners. In addition to life experiences, attention to wording, and icons, instructors need to be cognizant of the extra time older adults require to grasp, ponder, and complete assignments associated with subject material.

Learning and Instructional Pace

Peterson (1983) reported that research on older adults indicated that teachers should instruct at a slower pace and, when possible, allow for self-paced instruction in an effort to provide older learners with the time they need to comprehend the material. Instructional material can be decreased and presentations can be constructed to offer more clarity, specificity, and depth to enable older students to grasp information. Instead of presenting two theories to a class in a given lecture period, an instructor can present only one and allow for more specificity and depth to enhance understanding and comprehension. This will allow the students more time to process the material and ask questions about it. The second theory can then be assigned as a self-paced chapter on which more specificity and depth can be provided in discussion during the following class meeting. Instructors should also provide a considerable amount of feedback to older adults. Written feedback allows the older adult to review ways to improve and better understand material at their own pace. When feedback is given in conjunction with alternative approaches, it prevents older learners from continuing to repeat unsuccessful learning techniques, a trap into which they sometimes fall. Types of feedback that would be useful to older adults are providing examples, explaining why an answer or technique is incorrect and offering an improved strategy, providing ways in which older adults can organize information, and showing how the material relates or links to other ideas.

Teachers must also keep in mind the psychological and physical challenges that students in their classes may have and tailor their lectures to make the information easier for them to obtain. Older adults who were once visual learners often develop a preference for auditory learning (Van Wynen, 2001). This means that in the past an older adult may have preferred to see or read information to understand it, but their preference has changed to hearing the information or listening to stories they find more interesting. Another reason for this preference for auditory learning is that students who have visual challenges may find it easier to learn the material. Providing handouts will also assist students with vision impairments. Note, however, that in the same classroom there could also be another student with auditory challenges. Instructors should thus present the material using more than one method whenever it is logical and possible to do so. In addition to instructional pace learning styles, and methods to help with possible difficulties related to older age, educators should also assess the students in their classes to determine levels of technological abilities.

Media and Technology

Earlier in this chapter, it was noted that learning is not only a lifelong process; it transpires in various contexts, including contexts outside of the classroom. Robinson, Skill, and Turner (see chapter in this volume) discuss the fact that older individuals spend a great deal of time consuming mass media. These media include television, radio, reading, music, film, and more recently, the Internet. Although television is the most widely used medium for older adults, personal computer use is on the rise. Attainment of information through such a medium, as well as knowledge about such a medium, adds to older individuals' life experiences and thus has implications for the classroom. We now take a look at how technology can enhance learning and then examine what this new trend allows instructors to do in the classroom.

A national survey commissioned by Charles Schwab and Co. (1998) for SeniorNet reported that since 1995, the number of older adults who spend more than 10 hours weekly online has more than doubled, from 6% to 15%. The findings report that the top two reasons seniors use the Internet is for e-mail and research purposes, respectively. In another SeniorNet survey (2000), older adults also reported using the Internet to stay "current with news and events" as a third reason. This study also reported that older adults overwhelmingly access the Internet by using their own computer, and that they mostly learned to use the Internet by teaching themselves, getting help from a relative, or taking a computer class.

Cody, Dunn, Hoppin, and Wendt (1999) reported that a group of older adults who were trained on how to surf the Internet and use e-mail and chat rooms reported increased feelings of social support, connectivity, and reduced technology-related anxiety. The challenges provided by learning to use computer technology promote perceptions of support and the feeling of being young and mentally alert. Further, those who had learned to use the Internet also reported more positive attitudes toward aging (Cody et al., 1999). Even though older adults prefer mass media for entertainment purposes, as mentioned earlier by Robinson, Skill and Turner (chapter in this book), the use of these media contributes to their knowledge base. In addition to the media, we also learn lessons from living life on a day-to-day basis. Recognizing the implications of a medium's use, we next discuss the uses of media and technology in the classroom. Following this, strategies to aid teachers in communicating with older adults in the classroom are be offered.

Web-based dialogues can be quite complementary to face-to-face classroom instruction (King, 2001). For example, King (2001) stated that students reported their contributions to Web boards as "based on deeper thought about questions, lengthier consideration, and greater analysis/critical thinking" (p. 346). This forum may provide the older learner with more time and comfort in which to participate in discussions. These students may also be provided with more opportunity to direct the discussions, sharing knowledge and skills drawn from life experience and previous learning. For those who are reticent, the opportunity to more easily engage in discussions is provided through Web-based interaction. Web board participation enables both older and younger students to improve writing and reading comprehension skills and feel a sense of accomplishment in using technology.

The use of Web boards may be extremely beneficial to older individuals. Although Web boards have primarily been used in distance education classes, there are benefits to using them as a supplementary aid to the traditional classroom (King, 2001). King (2001) informed us that students reported they thought longer, more reflectively, and more critically in regard to their Web board contributions than they did to class discussions. Because Web boards are posted for a duration of time, students are able to pause and reflect on classmate comments, have greater access to each other, and take a more active role in the direction of discussions. In addition, reticent and less articulate students may also have the opportunity to participate in discussions.

Web boards encourage collaboration among instructors and their students and also improve reading and writing skills. Further, students reported that they felt a sense of accomplishment in using technology (King, 2001), which may boost student self-efficacy and self-esteem. Students who participated in Web board discussions were more likely to engage in classroom discussions and chat with each other before and after class, as Web boards fostered a sense of community (King, 2001).

A few Web board barriers to successful conferencing include unequal access to technology, prior technical skill and comfort level, and lack of accompanying nonverbal communication and immediate delivery. In addition, instructors must be cautious in using this method and avoid using it as the only one, as the literature provides us with somewhat mixed messages regarding technology. For example, Van Wynen (2001) reported that older adults preferred to work in a formal learning environment, in the presence of an instructor, and with their peers during the late morning. Further, in regard to the use of technology, Knowles (1990) informed us that "adults have a deep need to know why they need to know something before they are willing to invest the time and energy in learning it. You need to explain to us why we need to know how to format, to move the cursor this way and that, etc., before asking us to memorize the commands" (p. 164).

Much of the identity of being "adult" is associated with the idea that adults are self-sufficient and responsible. Knowles (1990) reported older adults saying, "We resent being talked down to, having decisions imposed on us, controlled, directed, and otherwise treated like children" (p. 164). To avoid resentment and misunderstanding, and considering that adults are such a heterogeneous group when it comes to the use of technology (some have experience through work, whereas others may still utilize typewriters), it is imperative that instructors effectively assess the technological levels of their students and avoid stereotyping. The avoidance of stereotyping should not be limited to inferences about knowledge, skills, and abilities, but also the interpersonal communication that takes place between instructor and student. Instructors should be mindful of the positive and negative stereotypes associated with older adults and how they may influence communication and the self-identity of students, and the consequences they may have for power and control in the classroom (for in-depth discussion of stereotypes and their consequences, see chapters by Hummert, Shaner, and Garstka, and Harwood, Giles, and Ryan, this volume).

Power and Compliance Gaining

Teachers must be aware of the power- and compliance-gaining aspects that occur within the classroom and how these elements may be perceived differently by younger and older students. Barraclough and Stewart (1992) reviewed the definitions and bases of classroom power and defined it as "the potential or capacity to influence the behavior of some other person or persons. Compliance gaining, or behavior alteration, is the realization of that potential" (p. 4). French's and Raven's (1960) five power bases—reward, coercive, legitimate, referent, and expert—were adapted to the classroom context by McCroskey and Richmond (1983). McCroskey and Richmond (1983) explained coercive power as being exerted when students perceive punishment from the teacher for noncompliance. Reward power is opposite in that students perceive rewards for compliance with the teacher. Legitimate power is given to the teacher when a student perceives that the teacher has the right of compliance because of his or her assigned role. Referent power refers to student perceptions of identification with the teacher, whereas expert power is given when a student perceives that the instructor is competent and knowledgeable (McCroskey & Richmond, 1983).

McCroskey and Richmond (1983) reported that coercive and legitimate powers as demonstrated through teachers' messages and techniques are ineffective in today's college classrooms. This finding is particularly relevant to older adults who are motivated in the classroom for different reasons, which are usually internal. The power bases that may be more useful for instructors to employ for both younger and older students are referent and expert.

As mentioned, a student grants referent power to an instructor when he or she feels identification with the instructor (McCroskey & Richmond, 1983). Thus, for an older adult who may be very organized, he or she may more positively perceive and hence learn from an instructor who is also very organized. An instructor's organization may be demonstrated by assigning projects with ample time to complete them and then returning papers when promised. Expert power is granted when a student perceives the instructor as competent and knowledgeable in a given area (McCroskey & Richmond, 1983). The student believes that the instructor knows more than he or she does and as a result respects and wants to achieve that which is outlined by the instructor. To gain expert power may be extremely difficult for younger instructors who might need to "prove themselves" to be taken seriously. Although proving oneself poses a challenge, communication can assist younger instructors to overcome this barrier not just by knowing content well, but by managing the classroom in an overall effective manner. Managing the classroom more effectively can be done through the use of personal stories and examples. If an instructor is teaching an organizational communication class and has had the benefit of working in the "real world," he or she can draw on those experiences to provide examples and explain the class content. Instructors can also manage more effectively through a question-and-answer processes. Asking the older student to explain an experience or provide an example may contribute to referent power. McCroskey and Richmond (1983) stated that "Communication is central to the teaching process . . . [and that] some even argue that communication is the teaching process" (p. 176). We can think about power in the classroom as a negotiated dialogue between teacher and students. Referent and expert power are sources that lead to compliance and can be negotiated through the communication process in the classroom.

Although it is important for teachers to be granted power in the classroom, compliance is one of the ultimate goals. Demonstrating to students that they are appreciated as individuals can increase compliance. If older adults are provided a climate in which they are treated with respect, do not feel threatened, and perceive they are cared about their learning may be increased (Knowles, 1984; Wlodkowski, 1999). Instructors can provide such a climate by asking for the students' thoughts on matters and then by noting excellent contributions. As we next examine, immediacy also adds to this supportive climate.

A peculiar phenomenon occurs when older adults reenter the classroom. As we mentioned earlier, the more we mature, the more independent we become. However, when older adults return to the formal educational setting, they fall back on their earlier experiences in the classroom context, and in this situation some older learners will revert to dependence (Knowles, 1984). To encourage these adults who may wish to remain independent, instructors can follow our earlier suggestions and allow older adults to set their own goals and evaluation. By using their power in the classroom, educators can give older students a voice in their education, which ultimately leads to compliance gaining and student learning.

Immediacy

Immediacy behaviors are one way that teachers can gain compliance and increase student learning (Richmond & McCroskey, 1992). Immediacy behaviors are defined by Mehrabian (1969) as the communication behaviors that "enhance closeness to and nonverbal interaction with another" (p. 302). Approachable behaviors such as eye contact, head nods, forward leans, facial expressions, and vocal animation all contribute to immediacy (Andersen & Nussbaum, 1990). These behaviors allow older learners to feel supported, as instructors not only show interest and enthusiasm while lecturing, but they also demonstrate that they are approachable. Immediacy increases student liking of the instructor (Mehrabian, 1971) and student motivation and learning (McCroskey & Richmond, 1992).

Benefits of immediacy within the classroom include being able to hold students' attention for longer periods of time. With students being more cognitively aware, the likelihood for learning increases. Immediacy also increases classroom interaction and increases positive affect toward the content and the teacher. Drawbacks that may occur from immediacy are perceived loss of classroom control and perceptions that the instructor is too easygoing (Richmond & McCroskey, 1995). To older adults, this may indicate that the instructor is not serious about his or her job, and thus may decrease the instructor's credibility and power. As a result, teachers must be careful when they incorporate immediacy behaviors. Being attentive to the nonverbal behaviors that students exhibit can assist teachers in using immediacy behaviors appropriately. For example, if a student yawns or is looking out the window, the teacher is advised to think quickly and determine what may not be going so well. Perhaps he or she may need to add more vocal inflection to the lecture. Maybe the problem is abstract content that needs to be explained in a more concrete way. Indicators of boredom may simply be a sign to stop a lecture and move on to a classroom activity. Incorporating classroom activities where students can work together in groups facilitates a sense of inclusion that we have explained assists the older adult to feel connected with the instructor and other students, and may foster respect among students (Wlodkowski, 1999).

Wlodkowski (1999) has found that teaching older adults begins with respectful relationships. He stated that "for most adults, the first sense of the quality of the teacher–student relationship will be a feeling, sometimes quite vague, of inclusion or exclusion" (p. 90). If the teaching relationship begins with perceptions of exclusion, students will lose their motivation and enthusiasm. Wlodkowski (1999) advocated creating a sense of community in the classroom to foster inclusion. Immediacy is one such tool that instructors can use to facilitate this process and bridge the gap for older students who may feel as though they are part of an out-group. Wlodkowski (1999) recommended several strategies for enhancing connection among adults, such as allowing for introductions and indicating your cooperative intentions to help adults learn.

An instructor may spark connection by simply introducing himself or herself the first day of class. Although this may seem to be an obvious statement to some instructors, as students introduce themselves, many instructors may forget to extend this courtesy by skimming over their own introduction. Instructors can also ask students questions pertaining to their introductions. Wlodkowski (1999) stated that introductions emphasize the students' "importance and your interest in them as people" (p.100). Instructors can also disclose information about themselves, such as their research interests, which can show that they want their students to get to know them beyond just being "Professor X." In addition, such introduction of research interests can enhance instructor credibility and power.

Another way to build student–instructor relations is to support adult learners indirectly. Besides giving comments about how to improve, instructors can give older adults more control of their learning by informing them of additional resources they may use to master the material. This way, if the instructor is not able to succeed in offering assistance or tutoring, the older adult has options from which to choose. Adults often experience fear associated with failure even if they are putting forth their best efforts. For some students, this fear may be attributed to stepping out of their comfort zones and back into the classroom; others may have fears associated with evaluation and the feeling that their mind is not as "swift" as it used to be. Wlodkowski (1999) encouraged instructors to announce where students can seek additional assistance or tutoring beyond your availability to them, and remind them that seeking help is okay and that "their vulnerability will be safeguarded" (p. 103). Because of the need for dependence and the fear associated with evaluation, older adults may prefer to find other resources to aid them in understanding material even though a teacher is appropriately immediate with his or her students. It has been our personal experience that older as well as younger learners respond well to immediacy. However, there may be instances when students are uncomfortable when teachers appear too close. This may especially be the case if students are communicatively apprehensive and, in the case of older learners, the teacher is using a method they have not observed in the past.

Communication Apprehension

McCroskey (1978) defined communication apprehension as "an individual's level of fear or anxiety associated with either real or anticipated communication with another person or persons" (p. 192). Communication apprehension can be thought of as existing on a continuum ranging from traitlike (personality-type orientation) to statelike (situational) communication apprehension. When people experience the fear of communicating, they generally will withdraw, avoid, or overcommunicate. If they do communicate, they may experience disruptions in their verbal or nonverbal communication (i.e., stuttering or a nervous twitch). Instructors need to be aware of the potential for communication apprehension that may hinder learning in the classroom.

A summary by Richmond (1997) reviewed the findings that quietness has in contemporary society; among the contexts she reviewed are the classroom setting. Quiet students choose to sit along the sides or in the back of the classroom, where they perceive less focus from the instructor, as opposed to sitting in the front or center of the room. When placed in small groups, quiet students are less likely to have the ability to share brainstorming ideas even when they are able to generate as many ideas as verbal students do. If a quiet student is placed in a small group, he or she may be more worried about communicating in the setting, as opposed to worrying about the problem being discussed. Further, quiet students are less likely to disagree with other group members and more likely to go along with the consensus. The speech of quiet students will exhibit more pauses and interrogatives. Quiet students are less likely to enroll in classes that require public speaking or participation and, if enrolled, more likely to drop out within the first few weeks. We now look at implications these findings may have for quiet older learners.

Older students who are re-experiencing classroom learning may be afraid or feel uncomfortable in such a setting. This may cause them to avoid classroom discussions. There are several suggestions for instructors from McCroskey and McCroskey (2002) that will make the classroom environment bearable for younger and older students who experience apprehension. Teachers can reduce oral communication requirements (i.e., grading on participation and calling on students), praise students who participate and reward efforts, make assignments and the grading of them as clear as possible, and encourage students to ask questions. If possible, students should be provided options in regard to assignments and the presentation of them. Some students may prefer turning in written work as opposed to an oral presentation (McCroskey & McCroskey, 2002). Because quiet and verbal students are able to brainstorm effectively, before placing students in groups, instructors can provide them with time alone to jot down ideas. This may make quiet students more comfortable and increase participation in group discussions. The main idea is to make students comfortable and unafraid to communicate in the classroom setting. We speculate that instructors with communicatively apprehensive older learners would be few. If communication and higher education are determinants of quality of life, it is quite possible that older learners who suffer from communication apprehension do not reenter the classroom. Research is needed to determine whether individuals who have communication apprehension live as long as those who are less apprehensive.

SUMMARY

In summary, we have discussed older learners and the characteristics associated with them in the classroom context. We have highlighted the importance of realizing that older adults are a heterogeneous group consisting of various contextual ages and should be treated as such. Life experiences contribute to, and are ever present, in the learning environment, and instructors can drawn on them as a resource. In addition to life experience, there are age-related psychological and physical changes that influence the potential and amount of learning that can occur. Motivations and goals for learning change across the life span, are often different for older and younger students, and will affect learning and communication in the educational setting. The use of language, media, and technology in the classroom has the potential to pose both rewards and challenges for older students. This synthesis of knowledge provides us with a framework that enables us to empathize with our older students and instruct them appropriately. Further, we have explicated research findings and ideas associated with instructing older students that provide us with a direction to enhance classroom communication and learning.

Suggestions we have provided include being immediate with our students, realizing our power and control, and acknowledging that some students may be apprehensive about communicating. Strategies have been provided to assist us with instruction, taking into account the age-related psychological and physical changes and barriers that our older students may experience. How we teach is, on all counts, as important as what we teach and can have substantial implications for student motivation and learning, especially in regard to older learners.

REFERENCES

Andersen, J., & Nussbaum, J. (1990). Interaction skills in the classroom. In J. A. Daly, G. W. Freidrich, & A. L. Vangelisti (Eds.), *Teaching communication: Theory, research, and methods.* Hillsdale, NJ: Lawrence Erlbaum Associates.

Anderson, R. H., Bikson, T. K. Law, S. A., & Mitchell, B. M. (1995). *Universal access to e-mail: Feasibility and societal implications.* Santa Monica, CA: Rand Corporation.

Barraclough, R. A., & Stewart, R. A. (1992). Power and control: Social science perspectives. In V. P. Richmond & J. C. McCroskey (Eds.), *Power in the classroom: Communication, control, and concern* (pp. 1–18). Hillsdale, NJ: Lawrence Erlbaum Associates.

Belsky, J. K. (1990). *The psychology of aging* (2nd ed.). Pacific Grove, CA: Brooks/Cole.

Birren, J. E. (Ed.). (1996). *Encyclopedia of gerontology: Age, aging, and the aged* (Vol. 2). San Diego, CA: Academic Press.

Charles Schwab & Co. (1998). *Research on senior's computer and Internet usage: Report of a national survey* [Online]. Available: www.seniornet.org/php

Chesebro, J. L. (2002). Student listening behavior. In J. L. Chesebro & J. C. McCroskey (Eds.), *Communication for teachers* (pp. 8–18). Boston: Allyn & Bacon.

Cody, M. J., Dunn, D., Hoppin, S., & Wendt, P. (1999). Silver surfers: Training and evaluating Internet use among older adult learners. *Communication Education, 48,* 269–287.

Cross, K. P. (1981). *Adults as learners: Increasing participation and facilitating learning.* San Francisco: Jossey-Bass.

Daly, J. A., & Kreiser, P. O. (1992). Affinity in the classroom. In V. P. Richmond & J. C. McCroskey (Eds.), *Power in the classroom: Communication, control, and concern* (pp. 121–143). Hillsdale, NJ: Lawrence Erlbaum Associates.

Edwards, R., & Usher, R. (2001). Lifelong learning: A postmodern condition of education? *Adult Education Quarterly, 51,* 273–287.

Flowers, P. J., & Murphy, J. W. (2001). Talking about music: Interviews with older adults about their music education, preferences, activities, and reflections. *Applications of Research in Music Education, 20,* 26–32.

French, J. R. P., Jr. & Raven, B. (1960). The bases of social power. In D. Cartwright & A. Zander (Eds.), *Group Dynamics* (pp. 259–269). New York: Harper & Row.

Gibbons, A. C. (1977). Popular music preferences of elderly people. *Journal of Music Therapy, 14,* 180–189.

Graham, S. W. (1989). Assessing the learning outcomes for adults participating in formal credit programs. *Continuing Higher Education, 3*(2–3), 73–85.

Glendenning, F. (1995). Education for older adults: Lifelong learning, empowerment, and social change. In J. F. Nussbaum & J. Coupland (Eds.), *Handbook of Communication and Aging Research* (pp. 467–490). Mahwah, NJ: Lawrence Erlbaum.

Hately, B. J. (Winter, 1987–1988). Reaching in, reaching up. In H. R. Moody (Ed.), *Generations: Quarterly Journal of the American Society on Aging: Late-Life Learning, 12,* San Francisco: American Society on Aging.

Houle, C. O. (1961). *The inquiring mind: A study of the adult who continues to learn.* Madison: University of Wisconsin Press.

John, M. T. (1988). *Geragogy: A theory for teaching the elderly.* New York: Haworth.

Justice, E. M., & Dornan, T. (2001). Metacognitive differences between traditional-age and nontraditional-age college students. *Adult Education Quarterly, 51,* 236–249.

Kearney, P., Plax, T. G., Richmond, V. P., & McCroskey. (1984). Power in the classroom IV: Alternatives to discipline. In R. Bostrom (Ed.), *Communication yearbook 8* (pp. 724–746). Beverly Hills, CA: Sage.

Kearney, P., Plax, T. G., Richmond, & McCroskey, V. P. (1985). Power in the classroom III: Teacher communication techniques and messages. *Communication Education, 34,* 19–28.

Kidd, J. R. (1973) *How adults learn.* New York: Association Press.

King, K. P. (2001). Educators revitalize the classroom "bulletin board": A case study of the influence of online dialogue on face-to-face classes from an adult perspective. *Journal of Research on Computing in Education, 33,* 337–354.

Knowles, M. S. (1980). *The modern practice of adult education.* New York: Cambridge University Press.

Knowles, M. S. (1984). *Andragony in action.* San Francisco: Jossey-Bass.

Knowles, M. S. (1990). *The adult learner: A neglected species* (4th ed.). Houston, TX: Gulf.

McCroskey, J. C. (1978). Validity of the PRCA as an index of oral communication apprehension. *Communication Monographs, 45,* 192–203.

McCroskey, J. C., & Richmond, V. P. (1983). Power in the classroom I: Teacher and student perceptions. *Communication Education, 32,* 175–184.

McCroskey, J. C., & Richmond, V. P. (1992). Increasing teacher influence through immediacy. In V. P. Richmond & J. C. McCroskey (Eds.), *Communication in the classroom: Communication, control, and concern* (pp. 101–119). Hillsdale, NJ: Lawrence Erlbaum Associates.

McCroskey, J. C., Richmond, V. P., Plax, T. G., & Kearney, P. (1985). Power in the classroom V: Behavior alteration techniques, communication training, and learning. *Communication Education, 34,* 214–226.

McCroskey, L. L., & McCroskey, J. C. (2002). Willingness to communicate and communication apprehension in the classroom. In J. L. Chesebro & J. C. McCroskey (Eds.), *Communication for teachers* (pp. 3–7). Boston: Allyn & Bacon.

Mehrabian, A. (1969). Methods & designs: Some referents and measures of nonverbal behavior. *Behavioral Research Methods and Instrumentation, 1,* 203–207.

Mehrabian, A. (1971). *Silent messages* (pp.1–23, & 76–79). Belmont, CA: Wadsworth.

Millette, D. M. & Gorham, J. (2001). Teacher behavior and student motivation. In Joseph L. Chesebro & J. C. McCroskey (Eds.), *Communication for Teachers* (pp. 141–153). Boston, MA: Allyn and Bacon.

Nussbaum, J. F. (1992). Communicator style and teacher influence. In V. P. Richmond & J. C. McCroskey (Eds.), *Communication in the classroom: Communication, control, and concern* (pp. 145–158). Hillsdale, NJ: Lawrence Erlbaum Associates.

Nussbaum, J. F., Hummert, M. L., Williams, A., & Harwood, J. (1996). Communication and older adults. In B. R. Burleson (Ed.), *Communication Yearbook, 19,* 1–47.

Nussbaum, J. F., Pecchioni, L., Robinson, D., & Thompson, T. (2000). *Communication and aging* (2nd ed.). Mahwah, NJ: Lawrence Erlbaum Associates.

Patterson, I., & Pegg, S. (1999). Adult learning on the increase: The need for leisure studies programs to respond accordingly. *Journal of Physical Education and Dance, 70*(5), 45–49.

Peterson, D. A. (1983). *Facilitating education for older learners.* San Francisco: Jossey-Bass.

Plax, T. G., Kearney, P., McCroskey, J. C., & Richmond, V. P. (1986). Power in the classroom VI: Verbal control strategies, nonverbal immediacy and affective learning. *Communication Education, 35,* 43–55.

Richmond, V. P. (1997). Quietness in contemporary society: Conclusions and generalizations of the research. In J. A. Daly, J. C. McCroskey, J. Ayres, T. Hopf, & D. M. Ayres (Eds.), *Avoiding communication: Shyness, reticence, and communication apprehension* (2nd ed., pp. 257–284). Cresskill, NJ: Hampton.

Richmond, V. P., & McCroskey, J. C. (1984). Power in the classroom II: Power and learning. *Communication Education, 33,* 125–136.

Richmond, V. P., & McCroskey, J. C. (1995). *Nonverbal behavior in interpersonal relations* (3rd ed.). Needham Heights, MA: Allyn & Bacon.

Richmond, V. P., McCroskey, J. C., Kearney, P., & Plax, T. G. (1987). Power in the classroom VII: Linking behavior alteration techniques to cognitive learning. *Communication Education, 36,* 1–12.

Rubin, A. M., & Rubin, R. B. (1986). Contextual age as a life-position index. *International Journal of Aging and Human Development, 23,* 27–35.

Salthouse, T. A. (1988). Effects of aging on verbal abilities: Examination of the psychometric literature. In L. L. Light & D. Burke (Eds.), *Language, memory, and aging* (pp. 17–35). New York: Cambridge University Press.

Schaie, K. W., & Willis, S. L. (1996). *Adult development and aging* (4th ed.). New York: HarperCollins.

SeniorNet. (2000). *Survey about the Internet* [Online]. Available: www.seniornet.org/php/

Smith, R. M. (1990). *Learning to learn across the life span.* San Francisco: Jossey-Bass.

Thoits, P. (1982). Conceptual, methodological, and theoretical problems in studying social support as a buffer against life stress. *Journal of Health and Social Behavior, 23,* 145–159.

Van Wynen, E. A. (2001, September). A key to successful aging: Learning-style patterns of older adults. *Journal of Gerontological Nursing, 27*(9), 6–15.

Walkin, L. (1990). *Teaching and learning in further and adult education.* Avon, UK: Bath.

Wlodkowski, R. J. (1999). *Enhancing adult motivation to learn* (2nd ed.). San Francisco: Jossey-Bass.

Wright, K. B. (2000). Computer-mediated social support, older adults, and coping. *Journal of Communication, 50,* 100–118.

Author Index

Q

T

Subject Index

Printed in the United States
by Baker & Taylor Publisher Services